ANNUAL REPORTS ON
NMR SPECTROSCOPY

Volume 5B

ANNUAL REPORTS ON
NMR SPECTROSCOPY

Edited by

E. F. MOONEY

Anacon (Instruments) Limited, Bourne End, Buckinghamshire, England

Volume 5B

ACADEMIC PRESS
A Subsidiary of Harcourt Brace Jovanovich, Publishers
London and New York
1973

ACADEMIC PRESS INC. (LONDON) LTD.
24–28 Oval Road,
London, NW1 7DX

U.S. Edition published by

ACADEMIC PRESS INC.
111 Fifth Avenue,
New York, New York 10003

Copyright © 1973 by ACADEMIC PRESS INC. (LONDON) LTD.

All Rights Reserved

No part of this book may be reproduced in any form by photostat, microfilm, or any other means, without written permission from the publishers

Library of Congress Catalog Card Number: 68–17678
ISBN: 0–12–505345–2

PRINTED IN GREAT BRITAIN BY
WILLIAM CLOWES & SONS, LIMITED,
LONDON, BECCLES AND COLCHESTER

Volume 5B
NMR STUDIES OF PHOSPHORUS COMPOUNDS
(1965-1969)

by

G. MAVEL

IRCHA, 12 Quai Henri IV
Paris, France

ACKNOWLEDGMENTS

For permission to reproduce, in whole or in part, certain figures and diagrams we are grateful to the following publishers: American Chemical Society, North-Holland Publishing Company and JEOL Ltd.

Detailed acknowledgments are given in the legends to the figures.

PREFACE

The division of Volume 5 into two parts is a new venture in this series. It has become evident that these volumes are continually growing. Perhaps correctly, criticism has been made that certain topics are out-weighing the balance of the volumes. This is especially true with such topics as ^{19}F resonance.

Certainly had this chapter, which constitutes Volume 5B, been published with Volume 5A as one volume the number of pages would have been in excess of one thousand.

Therefore this Volume 5B is in some way an experiment which is reasonable as many would be interested in the ^{31}P data without necessarily wishing to have Volume 5A. This will become especially evident on inspection of the Compound Indices which are exhaustive and basically forming a catalogue of some hundred and ninety pages giving data on various phosphorus compounds. Comments on this new move would be most welcome and it may be more convenient in future years to run the second part of each volume on some special topic such as ^{19}F or ^{31}P resonance studies which, by virtue of the nature of the investigations, tends to become a vast catalogue.

Volume 6, which is now in preparation, is again a single volume work.

I must finally express my gratitude to Professor Mavel for his patience with my editing and for his efforts in preparing the manuscript and producing a very valuable contribution to this series.

ERIC F. MOONEY

Anacon (Instruments) Ltd,
Buckinghamshire,
October, 1973

NMR Studies of Phosphorus Compounds (1965–1969)

CONTENTS

Acknowledgments vi
Preface vii
Chemical Shift Scales xi

I. Introduction 1

II. Chemical Shifts in Phosphorus Compounds 2
    A. Phosphorus Chemical Shifts 2
        1. Experimental Determination of Phosphorus Chemical Shifts . . 2
        2. Theoretical Investigation of Phosphorus Chemical Shifts . . 4
        3. Empirical Correlations of Phosphorus Chemical Shifts . . . 5
        4. Relation of Observed Chemical Shifts to the Nature of Phosphorus bonds 6
    B. Proton and Fluorine Chemical Shifts 7
    C. Carbon-13 Chemical Shifts 7
    D. Boron-11 Chemical Shifts 10
    E. Other Nuclei 11
        1. Direct Observation of Other Nuclei 11
        2. Indirect Observation of Other Nuclei 11

III. Spin-Spin Couplings in Phosphorus Compounds 12
    A. General Considerations and Theoretical Approach 12
        1. Main Features and Phenomenological Description . . . 12
        2. Theoretical Descriptions 14
    B. Phosphorus Coupling to Protons and Fluorines 17
        1. One-bond P–H and P–F Couplings 17
        2. P–H and P–F Couplings through Carbon Atoms 25
        3. Couplings through Oxygen Atoms 37
        4. Couplings through Sulphur Atoms 42
        5. Couplings through Nitrogen Atoms 44
        6. Couplings through Other Nuclei 48
        7. Couplings through Unsaturated Systems 49
        8. Long-range Couplings 59
    C. Phosphorus–Phosphorus Couplings 60
        1. Direct ^{31}P–^{31}P Couplings 60
        2. Two-bond ^{31}P–^{31}P Couplings 64
        3. Higher ^{31}P–^{31}P Couplings 68
    D. Carbon–Phosphorus Couplings 68
        1. Direct ^{13}C–^{31}P Couplings 68
        2. Two-bond and Higher ^{13}C–^{31}P Couplings 71

CONTENTS

    E. Phosphorus Coupling to Other Nuclei. 71
    F. Couplings through Phosphorus 74
        1. Proton-Proton Coupling through Phosphorus 74
        2. Proton-Fluorine Coupling through Phosphorus 75
        3. Fluorine-Fluorine Coupling through Phosphorus 77

IV. Applications of NMR in Phosphorus Chemistry 78
    A. Applications to Structural Problems 78
        1. Carbon–Phosphorus Heterocyclic 78
        2. Heterocyclic containing phosphorus atom in ring 78
        3. Phosphonitrilics 78
    B. Applications to Stereochemical Problems 79
        1. Stereochemistry of Saturated Aliphatic Compounds 80
        2. Stereochemistry of Unsaturated Aliphatic Compounds . . . 81
        3. Stereochemistry of Acyclic Penta-coordinated Compounds . . 81
        4. Stereochemistry of One-ring Compounds 82
        5. Stereochemistry of Spiro-Compounds 83
        6. Stereochemistry of Other Cyclic Compounds 83
    C. Applications to Intermolecular Studies 83
        1. Solvent and Temperature Effects 83
        2. Hydrogen Bonding 84
        3. Adduct Formation 84
    D. Applications to Reaction Studies 85
    E. Miscellaneous Applications 87
        1. Relaxation Studies 87
        2. Investigations on Solid Samples 88
        3. Metal Complexes 88
        4. Biological Problems 89

V. Formula Indices 95

Formulaes Index I 97

Formulaes Index II 103

Structural formulae 289

References 307

Author Index 353

Subject Index 435

THE CHEMICAL SHIFT SCALES

Readers are reminded of the convention for the presentation of chemical shift data introduced into Volume Three of this series. This self-consistent convention has now been used for some two years in various places and is gradually being universally accepted.

Convention adopted for Chemical Shift Scales

1. All shifts will be denoted by the delta scale, low-field shifts being shown as positive and high-field shifts as negative values. In all cases the standard will take the reference shift of δ 0·0.
2. No other symbols to denote shifts at infinite dilution will be used.

NMR Studies of Phosphorus Compounds (1965–1969)

G. MAVEL

Ircha–12, Quai Henri IV, Paris, France

I. INTRODUCTION

THE IMPORTANCE of NMR in phosphorus chemistry is today not a matter for dispute—this was already clearly evident at the time of writing my earlier review on this topic.[(1966, 1)] This earlier review was concerned with nearly four hundred and fifty publications from the early days of NMR (about 1950) until the end of 1965. Papers specifically restricted to ^{31}P data were not included but these were, however, reviewed by Van Wazer *et al.*[(1967, 1)] who at approximately the same time presented results for about three thousand compounds.

The years 1966 to 1969† have seen a continuous increase in the interest of NMR studies of phosphorus compounds both from the practical, for the solution of structural and conformational problems, and the theoretical aspects, for a better understanding of the nature of the bonding in phosphorus compounds. About one thousand five hundred papers appeared during those four years and it was clear that the moment was ripe for an up-dated account of the NMR achievements in this field. In this review it will be found that this formidable volume of literature has answered many of the questions raised in the earlier review,[(1966, 1)] especially those dealing with phosphorus coupling constants and the relationship of these J values to the hybridisation of phosphorus and molecular geometry. Meanwhile, a number of new puzzling points have been revealed, challenging for theoreticians and intriguing for chemists attracted to a better understanding of structure and reactivity in phosphorus compounds. This essentially arises from the richness of phosphorus as a central atom, with its flexible hybridisation, and as a major element in inorganic, organic and biological chemistry.

† The present literature survey was concluded at the end of December 1969.

This review is intended to cover most of the aspects of NMR in phosphorus chemistry, especially stressing those aspects directly related to electronic and structural factors. Some subjects will only be briefly reviewed, e.g. resonances in solid compounds, in metal complexes [this subject has been treated in a previous article in this series[1969,1]] and biological applications. All the compounds reviewed are tabulated in either Formula Index I (inorganics) or Formula Index II (organics), the whole comprising of more than four thousand five hundred entries. These indices† list all available chemical shifts for phosphorus and fluorine; those for other nuclei (except protons) are reported in the text of this review.

II. CHEMICAL SHIFTS IN PHOSPHORUS COMPOUNDS‡

A. Phosphorus chemical shifts

1. *Experimental determination of phosphorus chemical shifts*

As the phosphorus-31 nucleus represents 100% of natural phosphorus, has a spin of $\frac{1}{2}$ and not too low a magnetic moment the direct observation of ^{31}P resonance is straightforward, as adequately reviewed by Van Wazer *et al.*[1967,1] The use of newer techniques is nevertheless interesting and, as initiated by Pudovik and other Russian authors,[1967, 2 to 4] the INDOR technique of Baker may be especially fruitful. In addition to its increased sensitivity it gives a convenient method for the determination of the signs of coupling constants [for a review see McFarlane[1968,1]]. Advantage of this technique has been taken over the past two years for investigating various typical compounds[1968, 2, 3; 1969, 2 to 4] as discussed later in Section III. Another method of obtaining spectra is the use of FOURIER Transform Spectroscopy as described by Klein and Phelps;[1967, 5] the sample under observation being submitted to pulse sequences, the FOURIER transform of its response-function is the absorption function (see vol. 5A, p. 557). Fig. 1 provides a typical FT spectrum of A.T.P.

Obviously, all this refers to observations on liquid samples; some work is nevertheless done on solid samples. With great care, one may obtain additional data using solid samples, especially *chemical shift anisotropy*. Such work has very recently been performed on anhydrous sodium and potassium pyrophosphates, for which $\sigma_\parallel - \sigma_\perp$ was measured as 200 ± 10

---

† The signs given in the indices strictly follow the conventions adopted in these volumes (see the General Foreword). In the text itself, however, the original conventions have been used for the sake of comparison with the reviewed literature.

‡ Throughout this section all chemical shifts will be discussed as given in the literature (i.e. with positive sign for *upfield* shifts), except where noted.

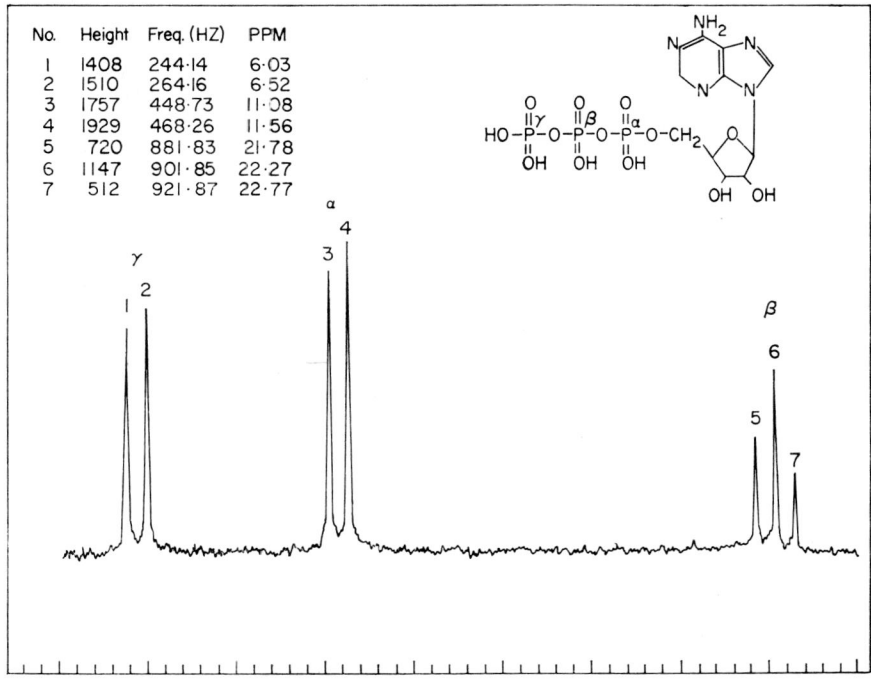

FIG. 1 ^{31}P FT spectrum of A.T.P. (Courtesy of Jeol. Co.)

ppm for phosphorus [1969, 5] and for polycrystalline $BaFPO_3$, for which a value of -145 ± 20 ppm for phosphorus and 182 ± 22 ppm for fluorine [1969, 6, 7] were obtained. A full discussion has been presented for $P(CN)_3$, P_4O_{10} and P_4S_{10}; for these Lucken and Williams [1969, 8] analysed the anisotropies of the different phosphorus types in terms of orbital populations.

Another interesting point is to be presented in relation to chemical shift evaluations. From the directly-measured *spin-rotation* constant of POF_3, diamagnetic and paramagnetic contributions to phosphorus chemical shifts were estimated (*ca.* $+1000$ and -640 p.p.m.). Extended to H_3PO_4, which is the commonly accepted reference compound, this gives *about 320 p.p.m. for the absolute ^{31}P chemical shift.*[1969, 9]

2. Theoretical investigation of phosphorus chemical shifts

The outstanding variation of the hybridisation of phosphorus, which may involve mixed s, p and d orbital participations, makes the interpretation of ^{31}P chemical shifts a very difficult task. As mentioned above (§ 1), the separation of dia- and para- magnetic terms is obtained from spin-rotation constants, in those compounds for which such observations are possible but this is far from being generally applicable. Another type of experimental data which may be found relevant, when discussing diamagnetic contributions, is to be expected from "chemical shifts" observed in ESCA spectroscopy;[1967, 6] these chemical shifts being of electrostatic origin. In any case, the paramagnetic term which is the dominating one (except, possibly, for tetrahedrally surrounded atoms) integrates the influence of too many factors to be described by a few gross parameters even for the simplest cases, such as attempted for PX_3 or YPX_3 compounds for which a relation of X–P–X bond angles and "magnetic configurations" was derived on the basis of an empirical modification of the Lamb formula.[1965, 1]† The same applies to attempted correlation of ^{31}P chemical shifts with theoretically calculated changes.[1969, 10]

Letcher and Van Wazer [1966, 2, 3; 1967, 1] have proposed to deal with s and p contributions, and especially their π-terms, as being by far the most important in relation to sizeable parameters, e.g. substituent electronegativities and bond angles, of phosphorus. As the latter are known in but a few compounds the use of derived formulae can only be qualitative as for phosphonate anions.[1967, 7] Of course, the consideration of only s and p orbitals is better when dealing with di- or tri- valent

† A clarification of this notation of "magnetic configurations" has been presented more recently by this author [1968, 4] as the first step for elucidating the ^{31}P chemical shifts on a firmer theoretical basis.

phosphorus compounds;† in these cases, however, the role of lone-pair electrons is so important that their contribution cannot be really properly appreciated.[1965, 2]

For phosphorus involving tetra-, penta- or hexa- coordination the importance of d electrons (even if quantitatively small) must not be underestimated;[1968, 5] it is especially easy to detect as it strongly decreases the paramagnetic shift from which a net high-field shift arises. To consider these d-contributions on a theoretical basis is rather complex[1964, 1] and is at present limited to qualitative conclusions.[1968, 5] The case of the assumed $PBr_4^{\ominus}$ ion, for which a shift of -150 p.p.m. is observed (compared with -225 for PBr_3) is rather interesting in this respect.[1969, 11]

3. Empirical correlations of phosphorus chemical shifts

The above facts explain the difficulties encountered when trying to set up *empirical correlations* for observed chemical shifts. First attempts were unsuccessful,[1956, 1] but Grim and co-workers have succeeded in establishing such a relation for P^{III} and $P^{IV\oplus}$ compounds. In defining additive constants σ^P for various groups these authors[1965, 3; 1967, 8; 1968, 6] established fair correlation for tertiary phosphines:

$$\delta = 62 - \sum_1^3 \sigma^P$$

and for secondary phosphines:

$$\delta = 99 - 1 \cdot 5 \sum_1^2 \sigma^P$$

For primary phosphines, Maier[1966, 4] similarly obtained the relation: $\delta = 163 \cdot 5 - 2 \cdot 5 \sigma^P$. Typical values of σ^P are shown in Table I.

TABLE I

Some typical additive constants σ^P for phosphorus-31 chemical shifts

R	σ^P	R	σ^P
CH_3	0	C_6H_5	+18
C_2H_5	+14	$C_6H_5CH_2$	+17
$iso C_3H_7$	+27	CN	−24·5
cyclo–C_6H_{11}	+23		

† Also applies to phosphonium compounds, although diamagnetic term is now more important and chemical shifts are essentially controlled by small changes in hybridisation arising from substituent electro-negativities.

As stated in the footnote on p. 5, phosphonium compounds afford a better example of chemical shifts which are essentially governed by inductive factors thus giving better correlations.

The following relations, having the same σ^P values, have been given:[1968, 7]

$$R_3PH^{\oplus}: \quad \delta = 3 - 0.5 \sum_1^3 \sigma^P$$

$$R_2(C_6H_5)PH^{\oplus}: \quad \delta = -1 - 0.57 \sum_1^2 \sigma^P$$

$$R(C_6H_5)_2PH^{\oplus}: \quad \delta = -1.4 - 0.57\, \sigma^P$$

More generally, in phosphonium ions the relationship:

$$\delta = -22.0 - 0.8\, n - 0.26 \sum \sigma^P$$

where n = the number of attached alkyl groups, is found to be valid.[1966, 5] It is seen that the contribution from phenyl groups is entirely different when dealing with P^{III} and $P^{IV\oplus}$ compounds. In the former case the phenyl groups accept electrons from the phosphorus lone-pair while this is not so in the latter case in which there is π-electron donation from the phenyl groups to the phosphorus d orbitals.[1968, 7]

Lastly, Olah[1969, 12] has proposed the following relationship for tertiary phosphonium ions:

$$\delta = 3.2 - 0.56 \sum_1^3 \sigma^P$$

It must be noted that all these σ^P values have no significant correlation with the corresponding Taft or Hammett constants;† the precise interpretation of these σ^P values remains open to doubt.

4. *Relation of observed chemical shifts to the nature of phosphorus bonds*

As previously stated, the analysis of ^{31}P chemical shifts for the elucidation of the nature of bonds is not straightforward, but qualitative relationships are possible. For example, in the recently described azaphosphatriptycene (**1**)[1969, 13] for which the δ_P of $+80$ is especially high for a P^{III} nucleus linked to three carbon atoms, the authors suggested that this indicated a high percentage s character for the lone pair. For the phosphonium homolog (**2**) the δ_P of $+4.8$ indicated a P–CH$_3$ bond with a high σ character, i.e. an electron density greater than sp^3 on phosphorus and less than s on methyl group. Similarly, in phosphiran (**3**)

† Correlations have, however, been described for phosphorus chemical shifts in $(C_6H_5)_2P(O)X$, $(C_6H_5)_2P^{\oplus}X_2$ etc. with the X substituent parameters.[1967, 9; 1968, 8]

the ^{31}P shift appears at an astonishingly high position of $+341$;$^{(1967, 10)}$ it should be noted that the PH group is neither mobile nor exchangeable.

These unusual shifts may be an indication of a significant d orbital participation, especially in phosphiran which probably has a nearly planar geometry and may possess some "aromatic" character as in the parent cyclopropanes [the P–H coupling in **3** of 155 Hz is between that found in $PH_2^{\ominus}$ (138 Hz) and PH_3 (182 Hz)]. The same conclusions would to some extent apply to organogermyl-, organostannyl- and organosilyl-phosphines recently discussed by Schumann.$^{(1969, 14)}$

Lastly, relevant routine discussions for the elucidation of electronegativity or conjugation effects from ^{31}P chemical data are fairly numerous;$^{(1965, 4; 1966, 6\ to\ 9; 1967, 3, 11\ to\ 14; 1968, 9\ to\ 12; 1969, 15)}$ cyano-, phenyl and perfluorophenyl groups have all been considered from this point of view as have halogeno groups [for which ^{31}P data compares with direct observation as for instance obtained from ^{35}Cl NQR data$^{(1967, 15)}$].

B. Proton and fluorine chemical shifts

Most of the applications of NMR to phosphorus compounds imply a discussion of the relevant proton or fluorine spectra. These deserve no special comment as they generally are not typical; mention will be made of some discussions of the *electronic structure* of molecules (stereochemical aspects will be considered in Section IV). Consideration of organic derivatives† is often based upon a Taft-type interpretation of resonance data, thus permitting a separation of σ- and π- electron distributions in substituent-phosphorus bonds (often through p_π–d_π conjugation; see 1967, 16 for a general discussion). Such investigations often make use of ^{19}F data involving either monofluoro- or perfluoro- substituents.$^{(1965, 5; 1966, 10; 1967, 17; 1968, 13, 14; 1969, 16\ to\ 19)}$ PMR may give gratifying results, e.g. in p-tolyl compounds$^{(1968, 15, 16)}$ and appears to be useful for discussing group electronegativities$^{(1967, 18; 1968, 17, 18)}$ or for estimating pK values.$^{(1967, 19)}$

C. Carbon-13 chemical shifts

Because of the low natural abundance, the low sensitivity of the resonance signal and the very long relaxation times associated with the nucleus, carbon-13 resonance has not been really investigated prior to 1965. The first spectra were observed using fast passage conditions;$^{(1957, 1; 1965, 6)}$

---

† Inorganic compounds present similar problems, e.g. for mixed halides of the type SPFXX′$^{(1968, 19)}$ in which the fluorine resonance gives an indication of π-bonding present in systems when considered in conjunction with phosphorus-resonances.

TABLE II

Carbon-13 chemical shifts for typical organo-phosphorus compounds

Type	Compound	Chemical shift[a]	Reference
Tricoordinated phosphorus	$(CH_3)_3P$	−116·8	1968, 3
		−111·4	1968, 22
	$(CH_3)_2PH$	−121·6	1968, 22
	CH_3PH_2	−133·1	1968, 22
	CH_3PCl_2	−113·5	1968, 20
	$(\overset{*}{C}H_3CH_2)_3P$	−104·0	1968, 3
	$(CH_3\overset{*}{C}H_2)_3P$	−113·6	1968, 3
	$(CH_3O)_3P$	−75	1968, 3
		−79·8	1968, 22
	$(\overset{*}{C}H_3CH_2O)_3P$	−65·6	1968, 3
		−70·7	1968, 22
	$(CH_3\overset{*}{C}H_2O)_3P$	−106·3	1968, 3
		−111·3	1968, 22
Phosphonium ions	$(CH_3)_4P^{\oplus}$	−111·9	1968, 3
	$(\overset{*}{C}H_3CH_2)_4P^{\oplus}$	−111·3	1968, 3
	$(CH_3\overset{*}{C}H_2)_4P^{\oplus}$	−117·3	1968, 3
Tetracoordinated phosphorus	$(CH_3)_3PS$	−100·4	1968, 3
	$(CH_3)_3PSe$	−100	1968, 3
	$CH_3P(O)Cl_2$	−97·0	1968, 20
	$CH_3P(O)F_2$	−107·0	1968, 20
	$CH_2P(S)Cl_2$	−87·5	1968, 20
	$HP(O)(OCH_3)_2$	−71·3	1968, 3
	$(CH_3O)_3PO$	−69·0	1968, 3
			1965, 7
	$(\overset{*}{C}H_3CH_2O)_3PO$	−59·8	1968, 3
	$(CH_3\overset{*}{C}H_2O)_3PO$	−107·3	1968, 3
	$(CH_3O)_3PS$	−68·5	1968, 3
	$(\overset{*}{C}H_3CH_2O)_3PS$	−62·3	1968, 3
	$(CH_3\overset{*}{C}H_2O)_3PS$	−107·4	1968, 3
	$(CH_3S)_3PS$	−110·4	1969, 20
	$[(CH_3)_2N]_3P$	−90·7	1969, 20
	$[(CH_3)_2N]_3PO$	−91·5	1969, 20
	$[(CH_3)_2N]_3PS$	−90	1969, 20
	$[(CH_2)_4N]_3P$	−84·3; −104·5	1969, 20

[a] In ppm downfield respect to C_6H_6.

the results (especially regarding resolution) were latter improved either by spectral accumulation [1968, 20, 21; 1969, 20] or by noise decoupling which washes out all structure arising by coupling to neighbouring protons.[1968, 22; 1969, 21] Another method is using the INDOR technique, as mentioned for ^{31}P resonance, which involves selective irradiation of the carbon-13 peaks while observing the proton spectrum.[1968, 3; 1969, 2 to 4]

Some typical carbon-13 shifts of phosphorus compounds are shown in Table II. As is well known (see 1969, 22), one of the interests in carbon-13 data is the sensitivity to charge distributions in molecules since these shifts are less susceptible to spurious effects than found for proton shifts (e.g. solvent effects, diamagnetic anisotropy of neighbours, etc.). For instance, comparison of carbon-13 shifts of methyl phosphines and methyl substituted methanes (Table III) at first sight exclude any hyperconjugation as a major factor in the P–CH$_3$ bond. A systematic study of phosphines and phosphonium compounds by Bucci,[1968, 17] which took into account the lone-pair contribution, revealed only a slight change in phosphorus electronegativity, from 2·4 to 2·2, in going from $(CH_3)_3P$ to $(CH_3)_4P^{\oplus}$ despite the presence of the positive charge on the phosphonium ion; a charge of similar magnitude but in the reverse direction was expected.† Thus some minor interaction in the P–CH$_3$ bond seems to be very likely.

This effect is better illustrated when considering conjugated and hetero-atomic systems. For instance, a carbon-13 shift of 85·5 p.p.m. is observed[1968, 20] in $(EtO)_2P(O)CH_2COCH_3$ compared to 104 in $(CH_3)_2CO$ and 57·0 and 28·0 ppm in $(EtO)_2P(O)C{\equiv}C\cdot CH_3$ compared to 46 in Et·C$\equiv$C·Et. In both these cases the observed changes support the presence of conjugation of the phosphonyl group with either the carbonyl or acetylenic group[1969, 23] as was suggested from consideration of proton and phosphorus resonance.[1967, 20 to 22] Similar trends are found in hetero-atomic molecules as shown in Table IV.[1968, 3; 1969, 20] These relations are especially neat for nitrogen in which, as discussed later (see III.B.5), P–N conjugation occurs. Conversely, it is noted that the observed changes are relatively insensitive to the nature of the element attached to phosphorus in the $\rangle$P$=$X group. This was attributed to the anisotropic effect of the phosphoryl or thiophosphoryl groups as these are highly polarisable.[1968, 3] This interpretation is questionable as it ignores the possible rehybridisation of phosphorus and, in consequence, of the neighbouring atoms.

† A similar effect is found in the comparison of tetramethylammonium ion and trimethylamine.[1961, 1]

TABLE III

Comparison of carbon-13 shifts of methyl phosphines and methyl substituted methanes

CH_3PH_2	134·3	$(CH_3)_2PH$	122·9	$(CH_3)_3P$	112·5
CH_3CH_3	122·8	$(CH_3)_2CH_2$	113·1	$(CH_3)_3CH$	104·3

TABLE IV

Comparison of carbon-13 shifts of some phosphorus compounds and some closely related ethers, sulphides and amines

$(CH_3O)_3P$	75 to 80	CH_3OH	81
$(CH_3O)_3P=O$	69 to 75	$(CH_3)_2O$	69·3
$(CH_3O)_3P=S$	68 to 73		
$[(CH_3)_2N]_3P$	90·7	$(CH_3)_3N$	81·2
$[(CH_3)_2N]_3P=O$	91·5		
$(CH_3S)_3P=S$	110·4	$(CH_3)_2S$	109

D. Boron-11 chemical shifts

Boron-11 resonance has been rather frequently investigated in boron containing phosphorus compounds which, in most instances, were borane- or boron halide-adducts (see Section IV). Some data are nevertheless available for typical boron-phosphorus compounds which may be discussed in relation to available ^{11}B literature (for a general review see 1969, 24).

Cage-like structures containing phosphorus have been assigned on the basis of NMR spectra. For instance, in $B_{10}H_{10}CHP$, identified as a 1,2- or 1,7- isomer, the shifts of the main resonance peaks for the isomers differ +14·7 and −9·6 in the former and +13·6 and −3·9 in the latter isomer,[1969, 25] or in $RR'PB_5H_8$, in which the phosphorus links B-2 and B-3 when $R = CH_3, R' = CH_3$ or CF_3 but resides on B-1 if $R = R' = CF_3$.[1968, 23]

Other interesting problems solved by ^{11}B resonance include monomer-dimer equilibria in molecules of the type $Cl_2BNRP(CF_3)_2$,[1968, 24] the structure of $(Me_3PBH_2NMe_3)^{\oplus}$ ($\delta^{11}B = +8·7$, ref. BF_3-etherate) which is written $Me_3\overset{\oplus}{P}B\overset{\ominus}{H}_2NMe_3^{\oplus}$ [1969, 26] and lastly the nature of the bonding in the compounds shown in Table V compared to corresponding simple boron esters.[1969, 27]

TABLE V

Comparison of boron-11 shifts of phosphorus containing boron esters with simple boron esters

	$\delta\,^{11}B$		$\delta\,^{11}B$
$B[OP(CF_3)_2]_3$	−17·2	$B(OCH_3)_3$	−18·3
$CH_3B[OP(CF_3)_2]_2$	−33·6	$CH_3B(OCH_3)_2$	−29·5
$(CH_3)_2BOP(CF_3)_2$	−60·3	$(CH_3)_2BOCH_3$	−53·0

E. Other nuclei

Obtaining resonance data on elements other than those mentioned above is not very easy and available data are therefore rather scarce. The other nuclei will be reviewed according to the method used for observing the resonance.

1. Direct observation of other nuclei

(a) *Nitrogen-14*. The ^{14}N resonance has been investigated in $[(CH_3)_2N]_3PO$ [1964, 2] and in $Et_2P(O)NHEt$ [1968, 25] but only the chemical shift of the former compound was reported (351 ± 8 p.p.m., ref. $NO_3^{\ominus}$). For comparison, one may note that the shifts in $(CH_3)_2NH$ and Et_2NCHO are 353 ± 1 and 259 respectively which, at first sight, might imply a negligible delocalisation of the nitrogen lone-pair to phosphorus in structures such as $>\overset{\oplus}{N}=P(O^{\ominus})$ (for a general review see 1969, 26). This implication is in contradiction with indirect conclusions drawn from estimation of Lewis basicity (see Section IV) and from dipole moment analysis.[1970, 1] A more profound analysis is certainly required in order to elucidate this problem.

(b) *Oxygen-17*. Christ and Diehl[1963, 1] collected data on a rather large number of compounds (Table VI) some years ago and, on this account, no detailed discussion is possible at present.

(c) *Aluminium-27*. This resonance has only been observed in adducts such as $(CH_3)_3P \cdot AlCl_3$ and $(CH_3)_3P \cdot AlBr_3$ (−108·2 and −100·8 respectively vs $AlCl_3$) in which ^{27}Al–^{31}P coupling appears.[1967, 23] This resonance has also been useful for following halogen exchange in Al_2X_6–$POCl_3$ mixtures [1969, 28] (see Vol. 5A, p. 465).

2. Indirect observations of other nuclei

The technical difficulties of observing the resonance of other nuclei are often deterring and an elegant solution may be found in the INDOR

TABLE VI

Oxygen-17 data on typical phosphorus compounds
[After Christ and Diehl [1963, 1]]

Compounds	Chemical Shifts[a]	Compounds	Chemical Shifts[a]
H_3PO_2	84	$(CH_3O)_2PO$	68, 19
H_3PO_3	102	$(C_2H_5O)_3PO$	66
H_3PO_4	79	$(CH_2O)_2P(O)H$	101, 38
$(CH_3O)_3P$	45	$(C_4H_9O)_2P(O)H$	84
$(C_2H_5O)_3P$	80	Cl_3PO	216

[a] In p.p.m. downfield vs. H_2O.

technique, previously mentioned for observing ^{13}C and ^{31}P shifts. The first, perhaps typical applications, have recently been reported for *silicon* and *tin* in $(CH_3)_3SiP(C_6H_5)_2$ and $(CH_3)_3SnP(C_6H_5)_2$[1969, 2] and for *selenium* in $CH_3SeP(CH_3)_2$ and $CH_3SeP(S)(CH_3)_2$.[1969, 4] As a common reference standard has not yet been defined (reference frequency being ^1H resonance of TMS or dioxane according to the authors), the measured shifts are not reported here; the technique certainly appears to be promising for future investigations.

III. SPIN-SPIN COUPLINGS IN PHOSPHORUS COMPOUNDS

A. General considerations and theoretical approach

1. *Main features and phenomenological description*

The spin-spin couplings observed in phosphorus compounds offer a number of interesting features many of which being particularly challenging for theoreticians. These will be detailed hereafter but to pinpoint the most important we may list:

(i) relation to the hybridisation of phosphorus
(ii) marked stereospecificity in many two- and more bond situations.
(iii) sign reversal appearing especially in typical one-bond couplings for which a positive (or at least, a stable) sign was previously taken as a rule.

The first two points permit the invoking of lone-pair contributions to explain the unusual behaviour in P^{III} compounds as previously proposed for ^{15}N couplings [1969, 28] and was satisfactorily accounted for in this case on the basis of the theoretical treatment of Yonezawa *et al.* [1967, 24; 1969, 29]

It is noticeable that lone-pair contributions may intervene in coupling not directly involving but passing *through* the heteroatom; this has been reported in pyrazolines.[1967, 25] Similarly, the stereospecificity of $H\cdots H$ couplings in **4** for which $J(trans)$ is *ca.* 12 Hz and J(gauche) *ca.* 2 Hz.[1968, 27]† We shall find a similar behaviour later when discussing $H\cdots H$ couplings in trioxaphosphabicyclooctane (see Section III.B.8); a clearer example is afforded by $F\cdots F$ couplings in P_2F_4—the fluorine atoms are inequivalent and the $FP-PF$ and $FP-PF'$ couplings are opposite in sign.[1969, 31]

Obviously, phosphorus lone-pair contribution cannot be invoked for explaining the stereospecificity observed in couplings involving P^V atoms, but some evidence has been presented supporting "through-space" contributions ["through-space" meaning via another route other than formal bonds and not arising from direct dipole-dipole coupling which involves a "direct coupling" contribution as discussed by Barfield and Karplus[1969, 32]]. This is especially clear when fluorine nuclei are involved; for instance, $P\cdots CF_3$ couplings in $(CF_3C_6H_4)_3P$ only appear when the trifluoromethyl group is *ortho*, i.e. sterically very close to the phosphorus atom.[1969, 33] Many examples of stereospecific proton-phosphorus couplings will be presented to support such "through-space" long-range mechanisms. Just to quote some striking cases, methyl groups in $(CH_3)_2NP(O)F(t-C_4H_9)$ are inequivalent, one with a $P\cdots CH_3$ coupling of 8·2 Hz, the other with a zero coupling (simultaneously, the $F\cdots CH_3$ couplings are 2·0 and 0·0 Hz);[1965, 8] methylene protons in $HP(C_2H_5)C_6H_5$ present two different couplings (+5·5 and 1·0 Hz) to phosphorus[1968, 28; 1969, 34] and R' groups have different couplings in $(RO)_2P(O)C(O)NR'_2$.[1967, 26; 1968, 29] Similarly, "through-space" couplings have been suggested in metal complexes.[1967, 27] In all cases, observed trends cannot agree with the simple description of Fermi contact terms by a "Hund" coupling mechanism.[1964, 3]

The same appears to be the case when couplings do not decrease with increasing bond separation as in $\diagdown P(O)C(O)CH_2CH_3$, in which $^3J(P-CH_2)$ is less than 0·5 Hz while $^4J(P-CH_3)$ is about 1 Hz. Similarly, the absence of sign reversal when comparing the couplings in a compound and the methyl homologue is an indication of anomalous behaviour. This behaviour has been observed in oximes (**5** and **6**) in which the absolute signs of coupling are known[1969, 35] for which the coupling involving the *cis* (H) is normal but not so for *trans* (H'); this is in agreement with the possible influence of the nitrogen lone-pair. A similar behaviour will be found in unsaturated phosphorus systems (see Section III.B.7).

† See also [1969, 30] for a Karplus-like relationship in secondary cyclic alcohols.

2. Theoretical descriptions

The preceeding approach is purely qualitative and the proposed mechanism is only phenomenological. Despite the difficulty of the description of phosphorus orbitals the problem is nevertheless an appealing one for theoreticians and to date two different approaches have been made to a non-empirical description.

(a) *Theoretical approach of Cowley and White.* This method starts from the Pople-Santry formalism [1964, 4; 1965, 9] of the general Ramsey formula for spin-spin coupling, thus avoiding the "mean excitation energy" approximation which is especially misleading as it necessarily predicts a stable sign for a given type of coupling. As a first step, Cowley, White and Manatt performed calculations for one bond couplings, using extended Huckel and CNDO-SCF molecular orbitals, thus obtaining some qualitative agreement with experimental data as shown in Table VII. [1967, 28]

TABLE VII

Comparison of some calculated and experimental phosphorus coupling constants

	EHMO	CNDO–SCF	Experimental
J(P–H) in PH_3	+189·4	+66·6	+182
$PH_4^{\oplus}$	+213·7	+192·6	+547
J(P–F) in PF_3	−736·2	−424·4	−1441
J(P–P) in "P_2"	−129·0	−30·5	−180 (in P_2Me_4)

Evidently, this approach did not take into account orbital overlap. These rather gratifying results and the puzzling question of the stereospecificity of coupling and sign reversal prompted the authors to refine their approach by including overlap and bond-geometry in one-bond and higher couplings; again use was made of semi-empirical LCAO–SCF orbitals after the Pople-Santry formalism. [1969, 36 to 38] They succeeded in reproducing changes in sign of ^{31}P–^{31}P and ^{31}P–^{13}C one-bond coupling and of ^{31}P–C–^{1}H two-bond couplings; the general predicted trends qualitatively agreed with experimental finds as shown in Table VIII. One shortcoming has nevertheless been found for P_2F_4 in which $J(^{31}P-^{31}P)$ is definitely negative; [1969, 31] further improvements therefore are necessary.

(b) *Theoretical approach of Jameson.* Jameson and Gutowsky [1969, 39] have been able to account for the reversal of sign of various one-bond

TABLE VIII
Typical couplings calculated by Cowley and White[1969, 37]

Molecules	Couplings	Calculated value	Experm. value
One-bond couplings			
$PH_2^\ominus$	P–H	+130.6	139
PH_3	,,	+146.3	+182
$PH_4^\oplus$	,,	+297.0	+547
P_2H_4 *gauche*	,,	+169.0	+186$_5$
trans	,,	+160.0	—
cis	,,	+167.6	—
CH_3PH_2	,,	+179.4	+186$_3$
$CH_3PH_3^\oplus$	,,	+246.3	—
CF_3PH_2	,,	+122.1	+199$_9$
PF_3	P–F	−1407.3	−1441
P_2F_4 *gauche*	,,	−1189.5	−1125
trans	,,	−1160.8	—
cis	,,	−1250.7	—
P_2H_4 *gauche*	P–P	−53.5	−108$_2$
trans	,,	+169.0	—
cis	,,	−84.6	—
P_2F_4 *gauche*	,,	+219.8	−227.4[a]
trans	,,	+707.4	—
cis	,,	+212.2	—
CH_3PH_2	P–C	−20.9	9$_3$[b]
CF_3PH_2	,,	+42.4	—
$CH_3PH_3^\oplus$	,,	+7.5	[c]
Two-bond couplings			
PH_3	H–P–H	−18.8	−13$_4$
CH_3PH_2	,,	−14.1	−12$_5$
CF_3PH_2	,,	−4.0	−13$_4$
HPF_2	H–P–F	+34.4	+41$_7$
CF_3PH_2	P–C–F	+47.9	+48$_3$
CH_3PH_2	P–C–H	+2.9$_5$	+3.9$_9$
$CH_3PH_3^\oplus$	,,	−7.7	

[a] (1969, 31).
[b] (1968, 22) probably negative.
[c] +55$_5$, +56$_5$ in $(CH_3)_4P^\oplus$ (1968, 3; 1969, 2).

(reduced) couplings $K_{N-N'}$ by separating the contact terms of N and N', a_N and $a_{N'}$, each of which may be written:

$$a_N = a_N(s) + a_N \text{ (core pol)}$$

This introduces, besides $a_N(s)$ (the direct Fermi contact term from the

s electrons in an open shell or in accessible configurations with open s shells), an indirect Fermi contact interaction from the polarisation of s electrons in the core. The latter could be the most important if the ground figuration of N is not an open s shell or if no low-lying configurations with open s shells are available. This is the case for Group VI and VII nuclei, such as ^{19}F, but not for Groups I to IV (e.g. ^1H, ^{11}B, ^{13}C); Group V appears to be on the borderline and, according to the state of hybridisation, a_N could be positive or negative.

This model at least provides a rationalisation of the observed trends throughout the table of elements. It has been extended by Jameson to account for two-bond couplings, especially those involving phosphorus. Using localized VB orbitals (s and p orbitals only with perfect pairing), she considered [1969, 40] reduced geminal couplings $^2K_{XYZ}$ as arising from a_X, a_Y and γ_{XYZ}, a transfer factor for "electron spin information" from the X bonding to the Z bonding orbitals.† The latter term is a function of exchange integrals so it is sensitive to the hybridisation of the Y atom, the bond angles, the substituents etc. The factor, γ, may be positive or negative—its average over all dihedral angles being probably negative (in agreement with "Hund coupling").

This phenomenological model has been used by Jameson [1969, 40] to explain some typical results which will be discussed in detail later. The first one is the great charge from P^{III} compounds (with a nearly p_3 hybridisation) to $P^{IV\oplus}$ and P^V compounds (nearly sp_3, with more p contribution in P=X bonds from X = Se, S to O). For the former, $a_P(s)$ is small and in bonds involving carbon a_P (core pol) becomes predominant, thus giving small positive P–C couplings; P–H couplings, governed by $a_P(s)$, are also small and positive. If substituents attached to phosphorus have an increasing electronegativity, the s character of the P–C bond is increased and the P–C coupling becomes positive and *P–C–H* coupling negative; similarly P–H coupling becomes increasingly positive. This whole picture agrees with the rationalisation of experimental data on the basis of the Walsh rule (see ref. 1966, 1, p. 265 *seq.*). For $P^{IV\oplus}$ and P^V compounds, $a_P(s)$ is greater towards C or H (increasing for $P^\oplus$, PSe, PS, PO); P–H coupling becomes more and more positive, P–C coupling positive and *P–C–H* coupling negative. As both P–C and P–C–H couplings have a_P as the dominant term, the close relationship between these two types (see Section III.D.1) is clear enough.

The Jameson model would similarly apply to *H–P–H* couplings which were investigated by Manatt *et al.*[1969, 41] In P^{III} compounds (PH_3,

† The importance of separating the effects of intervening atoms in the transmission of spin-spin couplings had similarly been stressed for couplings through carbon atoms in saturated chains.[1962, 1]

CH_3PH_2, etc.) $^2J(H\text{–}P\text{–}H)$ are negative and decrease when substituents on phosphorus have an increased electronegativity (e.g. $C_6H_5PH_2$, CF_3PH_2). In $P^{IV\oplus}$ compounds [$PH_4^\oplus$, $(CH_3)_2PH_2^\oplus$], $^2J(H\text{–}P\text{–}H)$ are small and positive; in $H_2PO_2^\ominus$, which has a tetrahedral hybridisation with strongly electronegative substituents on phosphorus, the H–P–H coupling is large and positive (despite the rather small H–P–H angle, ca. 92°). This latter example caused Manatt et al. some puzzlement, but now fits in fairly well with the Jameson approach.

Somewhat similar arguments may be used to discuss P–F (or other) couplings in penta-coordinated compounds which were discussed at some length in ref. 1966, 1, p. 281 seq. For the direct P–F couplings, $a_P(s)$ would be the dominating term and $J(P\text{–}F_{ax})/J(P\text{–}F_{eq})$ may be directly related to the s character of these two orbitals. For couplings through phosphorus (proton-proton, proton-fluorine or fluorine-fluorine), one would have $J(ax\text{–}ax)$, $J(eq\text{–}ax)$ and $J(eq\text{–}eq)$ in a ratio determined by $\gamma(ax\text{–}ax)$, $\gamma(eq\text{–}ax)$ and $\gamma(eq\text{–}eq)$. A closer examination of these transfer factors may be of interest for discussing this type of experimental data.

The various types of couplings encountered involving phosphorus in the order of practical importance, i.e. P–H, P–F, P–P, P–C and others, will now be considered.

B. Phosphorus coupling to proton and fluorine

1. *One-bond P–H and P–F couplings*

(a) *P–H couplings.* P–H couplings have been investigated in a large variety of compounds and cover a range from ca. 120 to 1180 Hz [in H_2PSiH_3 [1963, 2]]. Typical data are given in Table IX for the various types of phosphorus hybridisation. In all cases, where the signs of coupling were determined, these were found to be positive, [1966, 13; 1967, 31 and 40; 1968, 28 and 30; 1969, 47 and 48] in clear contradiction with the reasoning of Paolillo and Feretti.[1966, 54] It should be emphasised that such determinations were generally made on P^{III} compounds but no obvious reason exists for reversal of sign as no value around zero has yet been reported. Some theoretical considerations regarding these couplings have already been mentioned which agree with the stability of the sign of the coupling, variation in magnitude with the phosphorus hybridisation and substituent electronegativity. Vinogradov and Nicolaiev [1969, 49] have recently calculated the P–H coupling in PH_3 and $PH_4^\oplus$ using the Pople-Santry formalism and the function of the angle ϕ between the P–H bond and the symmetry axis of the molecule. These authors found that the effect of ϕ was marked for the tetrahedral conformation of $(P^{IV})^\oplus$ but was relatively of less importance for P^{III} conformations.

TABLE IX

Typical one-bond phosphorus-proton coupling constants

Compound	Coupling	References
$(P^{II})^{\ominus}$		
$PH_2^{\ominus}$	137 to 140	1965, 10; 1966, 11 and 12
$(BH_3)_2 \cdot PH_2^{\ominus}$	ca. 320	1969, 42
P^{III}		
PH_3	182 to 189	1966, 11; 1969, 42
H_2PPH_2	$+186 \cdot 5$	a
H_2PPF_2	191	1968, 31
H_2PCH_3	$186 \cdot 4$	a
H_2PGeH_3; H_2PSiH_3	180 to 181	1967, 29; 1968, 32
H_2PCF_3	$\pm 187 \cdot 8$	1966, 13
H_2PR ($R = C_2H_5$ to $C_{12}H_{25}$)	185 to 192	1965, 11; 1966, 14 and 15
$H_2PCH_2PH_2$	190 to 202	1966, 15 and 16
$H_2P(CH_2)_nPH_2$ ($n = 2$ to 41)	190 to 193	1966, 16
$H_2PC_6H_5$	195 to 201	1965, 11; 1966, 17 and 18; 1967, 30
$H_2PC_6F_5$	$205 \cdot 5$	1966, 19
HPF_2	$+181 \cdot 7$; $182 \cdot 4$	1966, 20; 1967, 31 and 32
$HP(CH_3)_2$	$+191 \cdot 6$	a
$HP(GeH_3)_2$; $HP(SiH_3)_2$	171 to 187	1967, 29; 1968, 32
$HP(CF_3)_2$	217 to 218	1969, 43
HP(cyclopropane ring)	155	1967, 10
HP(6-membered ring with O)	185	1969, 44
$HP(C_6H_5)_2$	214 to 218 (239)	1966, 21; 1967, 30; 1968, 33
$HP(C_6F_5)_2$	219	1966, 19
$HP(CH_3)C_6H_5$	$+204 \cdot 2$	1968, 33; 1969, 34
$HP(C_2H_5)C_6H_5$	$+205$	1965, 11; 1968, 28; 1969, 34
$(P^{IV})^{\oplus}$		
$PH_4^{\oplus}$	$547 \cdot 5$ to $548 \cdot 3$	a; 1969, 42
$(CH_3)_3PH^{\oplus}$	$+505 \cdot 5$	1966, 22
$(CH_3)_2PH_2^{\oplus}$	$516 \cdot 6$	1966, 22
$CH_3PH_3^{\oplus}$	$527 \cdot 1$	1966, 22
$(i\text{-}C_3H_7)_3PH^{\oplus}$	455	1968, 7
P^V		
$HP(O)F_2$	844 to 878	1967, 33; 1968, 34
$HP(O)F(OH)$	1030	1968, 34
$HP(O)(CH_3)C_6H_5$	$+467 \cdot 8$	1968, 33; 1969, 34

[a] Quoted in 1968, 30.

TABLE IX (continued)

Compound	Coupling	References
$HP(O)(C_6H_5)CH_2C_6H_5$	+474	1968, 35; 1969, 34
$HP(O)(C_6H_5)_2$	481 to 486	1968, 35 and 36
$HP(O)(OC_2H_5)_2$	688	1967, 34
$HP(O)[N(CH_3)_2]_2$	570	1966, 18
$[HP(O)(O^\ominus)]_2O$	+667·5	1968, 37
$HPO_3^{2\ominus}$	577	1967, 35
$HPO_2S^{2\ominus}$	548	1967, 35
$HP(S)F_2$	725·4	1967, 33
$HP(Se)(i\text{-}C_4H_9)_2$	420	1966, 23
$HP(Se)(C_6H_5)_2$	450	1966, 23
$HPF_4 \, (-16°)$	1075	1966, 24; 1967, 36
$H_2PF_3 \, (0°)$	841	1966, 24; 1967, 36
$H(CH_3)PF_3$	850	1968, 38 and 39
$HPF_2(CH_3)(OC_2H_5)$	850	1967, 37 and 38
$H[(CH_3)_2N]_2P=NR$	560 to 590	1969, 15

$$\begin{array}{c} H \quad R \\ \diagdown P \diagup \\ N \diagup \diagdown N \\ | \quad \quad | \\ (C_6H_5)_2P \diagdown \diagup P(C_6H_5)_2 \\ N \end{array}$$

$R=CH_3$	509	1968, 40
$R=OC_6H_5$	676	1968, 40

$$\begin{array}{c} \text{O} \quad \text{O} \\ \diagdown P \diagup \\ N \diagup | \diagdown N \\ \quad H \end{array}$$ 730 to 810 1967, 39; 1969, 45

$$\begin{array}{c} \text{O} \quad \text{O} \\ \diagdown P \diagup \\ N \diagup | \diagdown O \\ \quad H \end{array}$$ 780 1967, 39

$$\begin{array}{c} \text{O} \quad \text{O} \\ \diagdown P \diagup \\ O \diagup | \diagdown O \\ \quad H \end{array}$$ 830 1967, 39

(tribenzophosphole structure) 480 to 500 1966, 25; 1969, 46

$(P^{VI})^\ominus$
| $[HPF_4(CF_3)]^\ominus$ | 943 | 1967, 39 |
| $[HPF_3(CF_3)_2]^\ominus$ | 622 | 1968, 41 |

Previously Issleib et al.$^{(1966, 27)}$ had found a clear relationship of J(P–H) with the s character, calculated from LCAO–MO orbitals, for $PH_2^{\ominus}$ (ca. 140 Hz; 17% s), PH_3 (185 Hz; 18% s) and $PH_4^{\oplus}$ (550 Hz; 36% s).† Other factors, however, may contribute to these couplings, some arguments have been recently proposed $^{(1969, 50)}$ to prove the importance of the phosphorus *effective* charge, similar to that found for the effect of the carbon charge on ^{13}C–H coupling. $^{(1965, 11)}$

Related to these "electronic" factors, some examples of stereospecific coupling are known. In **7** the geminal protons on phosphorus are non-equivalent and exhibit slightly different couplings to phosphorus, both of which are temperature and solvent dependent. $^{(1969, 51)}$ A clearer example is given by the recent work of Mikołajczyk $^{(1969, 52)}$ on the two isomers **8** and **9**.

Isotope effects have also been investigated for the case of hydrogen-deuterium substitution in $(C_6H_5)_2PH$, $C_6H_5PH_2$, $^{(1967, 30)}$ $HPO_3^{2\ominus}$, $HP(O)(OCH_3)_2$ and other examples. $^{(1967, 42)}$ Lastly, temperature and solvent effects have also been considered for a variety of cases. $^{(1967, 43\ to\ 47; 1969, 51\ and\ 53)}$

(b) *P–F couplings.* P–F couplings vary from ca. 750 to 1500 Hz which is a much smaller range than for P–H couplings by nearly $\frac{1}{10}$ (see Table X). Relative signs have been established in $P^{III\ (1967, 28, 31, 40, 55\ and\ 57)}$ and in P^V compounds; $^{(1967, 57; 1969, 31)}$ all are *negative*, the same as for P–F coupling in phosphonitriles (see ref. 1967, 55 and 1969, 68). This has been definitely confirmed by analysing the line-shape of the ^{19}F resonance in solid $BaFPO_3$. $^{(1969, 6\ and\ 7)}$ It was seen in Section III.A, that both the Cowley-White and Jameson theoretical approaches properly account for this observation.

Tri- and tetra- coordinated phosphorus compounds will first be considered. On account of the extreme electronegativity of fluorine, substituent changes in P^{III} compounds do not alter the phosphorus hybridisation too markedly as far as inductive effects are concerned. This is reflected in the rather small range of variation (50%), and in the restricted influence on complexation, of the coupling constants; the latter effect is seen in the case of $(F_2P)_2O \cdot BH_3$ [1328 Hz compared to 1354 Hz in the free ligand $^{(1968, 34)}$]. This also accounts for the good correlation of the additivity of group contributions with J(P–F) in compounds of the form XYPF. $^{(1968, 12)}$ Similarly, the relation of J(P–F)

† Such an argument may be used for elucidating the nature of the phosphorus bonding in phosphirane $HP(CH_2)_2$ ($J=155$ Hz).$^{(1967, 10)}$ On the other hand, it explains changes between P^{III} compounds and their adducts, for instance PH_3: ca. 180 Hz and $PH_3 \cdot BH_2Br$: 405 Hz.$^{(1967, 41)}$

TABLE X
Some typical one-bond phosphorus-fluorine couplings

Compound	Coupling	References
P^{III}		
PF_3	1400 to 1440	1967, 32 and 48
P_2F_4	$-1194, -1199$	1967, 49; 1969, 31
$P(PF_2)_3$	1225	1968, 42
H_2PPF_2	1134	1968, 31
F_2PCl	1380 to 1390	1967, 31
F_2PBr	1388	1967, 50
F_2PCH_3	1131 to 1157	1967, 37, 38 and 51; 1968, 42
F_2PCCl_3	1285	1967, 48
F_2PCF_3	1245	1967, 48
$F_2PCF=CF_2$	-1202	1969, 54, 55
$F_2PC_6H_5$	1110 to 1170	1967, 51 and 52
$F_2PC_6F_5$	1222	1966, 19
F_2POCH_3	1280	1963, 3
$F_2PN(CH_3)_2$	1190 to 1197	1965, 12; 1968, 43
F_2PCN	1267 to 1273	1966, 28
F_2PNCO	1361	1969, 56
F_2PNCS	1336	1967, 53
F_2POPF_2	1354 to 1358	1966, 28, 1968, 34
$F_2\overset{*}{P}OP(O)F_2$	1032·5	1969, 57 and 58
$F_2\overset{*}{P}SP(S)F_2$	1321·5	1969, 59
F_2PSPF_2	-1306	1969, 31
$F_2P(NCH_3)PF_2$	$-1261; 1264$	1969, 31, 60 and 61
$F_2P(NC_6H_5)PF_2$	-1252	1969, 61
$FPCl_2$	1320	1967, 50
$FPBr_2$	1301	1967, 50
$FP(CH_3)_2$	820 to 830	1967, 54; 1968, 12 and 42
$FP(CF_3)_2$	-1013	1967, 48 and 55
$FP(OC_2H_5)_2$	1225	1969, 61
$FP[N(CH_3)_2]_2$	1023 to 1043	a
$FP(NCS)_2$	1252	1967, 53
P^V		
F_3PO	1055 to 1080	1966, 29; 1969, 62
$F_2P(O)H$	1114 to 1122	1967, 33; 1968, 34
$F_2P(O)Br$	1203	1967, 56
$F_2P(O)(t-C_4H_9)$	1169	1965, 8; 1966, 30
$F_2P(O)CF_3$	1215	1968, 44
$F_2P(O)C_6H_5$	1105	1968, 45

a Quoted in 1968, 30.

TABLE X (*continued*)

Compound	Coupling	References
P^V		
$F_2P(O)N(CF_3)_2$	1065	1966, 31
$F_2P(O)OC_2H_5$	1012	1967, 53
$F_2P^*(O)OPF_2$	1412	1969, 58
$F_2P(O)OP(O)F_2$	−1062·9	1969, 62
$FP(O)(OC_3H_7\text{-}i)$	−955	1967, 57
$FP(O)H(OH)$	1030	1968, 34
$FP(O)Br_2$	1263	1967, 56
$F_3P(S)$	1170 to 1184	1965, 13; 1966, 8
$F_2P(S)H$	1153	1966, 32; 1967, 33
$F_2P(S)Cl$	1220	1966, 8
$F_2P(S)OCH_3$	1126	1968, 45
$F_2P(S)SCH_3$	1207	1968, 45
$F_2P(S)N(CH_3)_2$	1079 to 1082	1966, 30; 1968, 46
$F_2P(S)OP(S)F_2$	−1168·2	1969, 17
$FP(S)(CF_3)_2$	1174·6	1968, 47
$FP(S)Cl_2$	1240	1967, 58
$FP(S)[N(CH_3)_2]_2$	1016	1968, 46
P^V pentacoordinate		
PF_5	938[b]	1969, 63
PF_4H	944 to 980	1966, 24; 1967, 36
$PF_4(CF_3)$	802 (F_e)	1968, 48
$PF_4(SCH_3)$	1032	1969, 62
$PF_4[N(CH_3)_2]$	836	1967, 59; 1969, 64
PF_4Cl	1085	1968, 49
PF_3H_2	860 to 877	1966, 24; 1967, 36
$PF_3H(CH_3)$	795 (F_a); 965 (F_e)	1968, 38, 39
	805 (F_a); 968 (F_e)	1968, 42
$PF_3(CF_3)_2$	968 (F_e)	1968, 48
$PF_3Cl[N(CF_3)_2]$	930 (F_a); 1030 (F_e)	1966, 31
$PF_3(CH_3)[N(CH_3)_2]$	804 (F_a); 957 (F_e)	1965, 12
$F_3P\underset{N(CH_3)}{\overset{N(CH_3)}{\diagup\!\!\diagdown}}PF_3$	894	1969, 65
$PF_2H(CH_3)_2$	535	1968, 42
$PF_2(CH_3)_3$	541	1967, 60
$PF_2(C_6H_5)_3$	659 to 664	1967, 61; 1968, 303
$PF_2(CF_3)_3$	881	1968, 48
$PF_2(OC_6H_5)_3$	721	1967, 61
$PF_2[N(CH_3)_2]_3$	707	1967, 61
$PF_2Cl_2(CF_3)$	1085	1969, 66

[b] Previously reported as 1080 from oxidised $PF_5(POF_3)$.

TABLE X (continued)

Compound	Coupling	Reference
P^V pentacoordinate		
$PFCl_3(CF_3)$	1000	1967, 62
$C_6H_5(F)P\underset{N}{\overset{PF_2}{\underset{N}{\diagup\diagdown}}}P(F)C_6H_5$	PF_2: 879 to 898 $P(F)C_6H_5$ 939 to 965	1968, 50
$PF_2(C_6H_5)=NCN$	1132 to 1140	1967, 63
$(P^{VI})^\ominus$		
$PF_6^\ominus$	706	1969, 67
$PF_5S^{2\ominus}(?)$	718 (F_a); 909 (F_e)	1968, 46
PF_5CF_3	810	1969, 67
$PF_4H(CF_3)$	858	1969, 67
	725 to 834 (F_e)	1968, 41
$PF_4(CF_3)_2$	884	1969, 67
$F_4P\underset{O-C\cdot CH_3}{\overset{O-C\cdot CH_3}{\diagup\diagdown}}CH$	714, 824	1966, 33

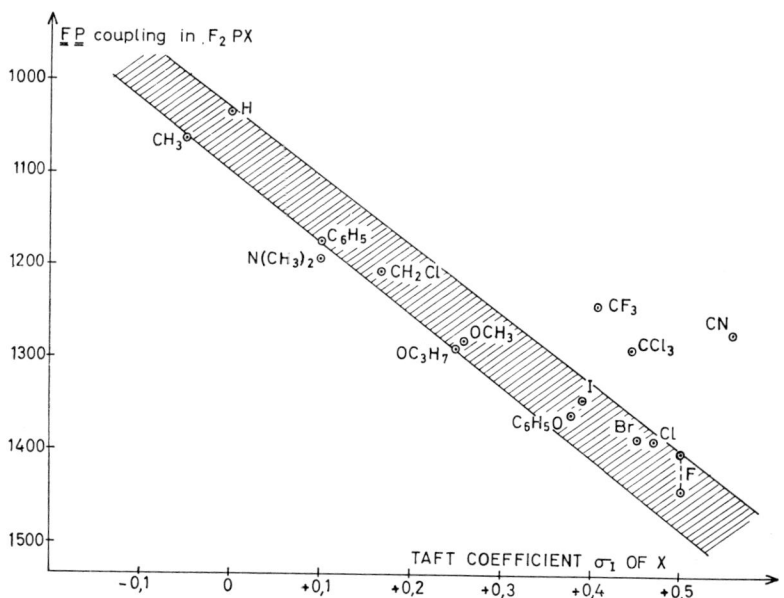

FIG. 2(a). P–F couplings as a function of the inductive effect of the X substituents for F_2PX.

to substituent constants is generally good (Figs. 2a and 2b); Walsh's rule was found to be generally applicable except for very bulky groups such as CF_3, CCl_3[1968, 30] and CN. Another cause of discrepancy may be the conjugation of substituents to phosphorus[1966, 34] which may be sufficient to stabilize conformers. In **10** two isomers were identified

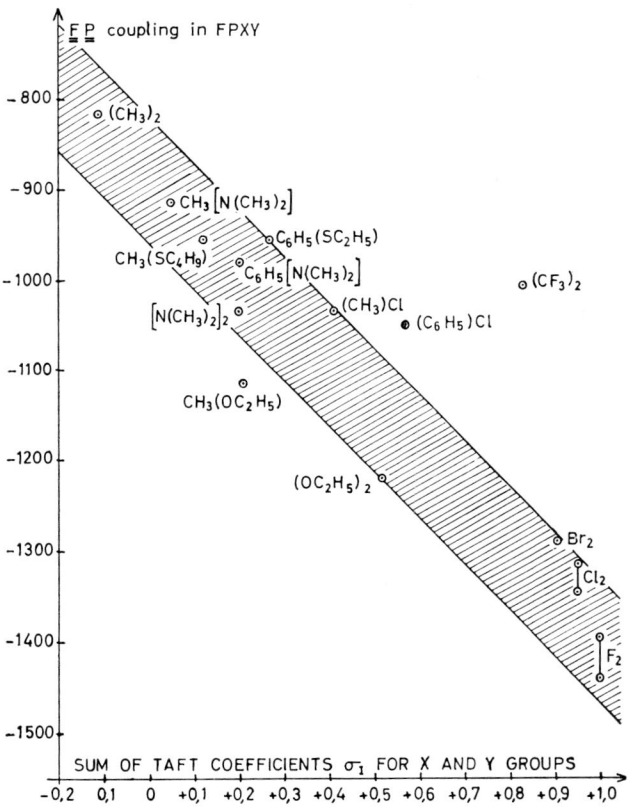

FIG. 2(b). P–F couplings as a function of the X and Y substituents for FPXY compounds.

which have different P–F couplings namely 1071 and 1011 Hz.[1965, 8] Related to this has been some solvent and variable temperature studies.[1968, 51; 1969, 31]

When considering P^V compounds, trends are not so clear as any change is accompanied by a complete rehybridisation of phosphorus, especially in phosphorylated series. Nevertheless, additivity rules are approximately obeyed in compounds of type XYP(Z)F,[1969, 69] for which Fluck determined the following increments (1200 Hz being taken as the reference value):

=O	0	–SR	–70
=S	+110	–OR	–90
–Br	+20	–NR$_2$ and –NCS	–130
–Cl	0	–NCO	–140
–F	–60	–NH$_2$	–160

Penta-coordinated fluorophosphoranes have been extensively considered by Schmutzler[1965, 12 and 14] and discussed in detail in ref. 1966, 1, p. 281 seq. The special interest was in the existence of two well-defined types of fluorine substituents, as well as for the hexa-coordinated $(P^{VI})^{\ominus}$ compounds. Fluorine exchange generally occurs by pseudorotation† but steric hindrance, substituent conjugation and low temperatures may stabilize fluorine atoms in their respective position. As a rule (see Section IV.B.3), the fluorine atoms first occupy the two axial positions—which may not be equivalent[1967, 64] as in $F_2P(C_6H_5)_2SCH_3$ in which the P–F couplings are 694 and 823 Hz[1969, 62]—then one or more of the three [four for $(P^{VI})^{\ominus}$] equatorial positions are occupied.

As was discussed in the earlier review,[1966, 1, p. 282 seq.] $P-F_a$ bonds are larger than $P-F_e$[1965, 15 and 16; 1966, 35; 1969, 70] and, having a lower s character, have smaller coupling constants which become even smaller when the electronegativity of the other substituents decreases. In addition to the data reported in 1966, 1, this point may be further illustrated by considering the $P-F_{ax}$ coupling in F_2PXYZ compounds (Fig. 3); the only major exceptions to the relationship of these couplings with substituent electronegativity arise from the presence of bulky substituents, such as OC_6H_5 or, to a lesser extent, OCH_3, or CF_3; the latter groups may occupy axial positions (see later Section IV.B.3) as in $(P^{VI})^{\ominus}$ compounds, for instance in **11**.[1968, 41]

2. *P–H and P–F couplings through carbon atoms*

When discussed in the earlier review[1966, 1, p. 266 seq.] *P–C–H, P–C–C–H* and higher couplings presented a number of interesting features in relation to the nature of the bonds in the relevant molecules; some gross trends were explained on the basis of hybridisation (and especially s character) changes induced by substituents. Similar arguments were used by Manatt and co-workers when considering newer data with signs of couplings.[1966, 13; see also 1968, 30] Ignoring the signs of couplings does not present any understandable correlations;[1965, 17; 1966, 36] as we shall see

---

† Line width measurements at variable temperature[1969, 64] agree with the mechanism proposed by Berry.[1960, 1]

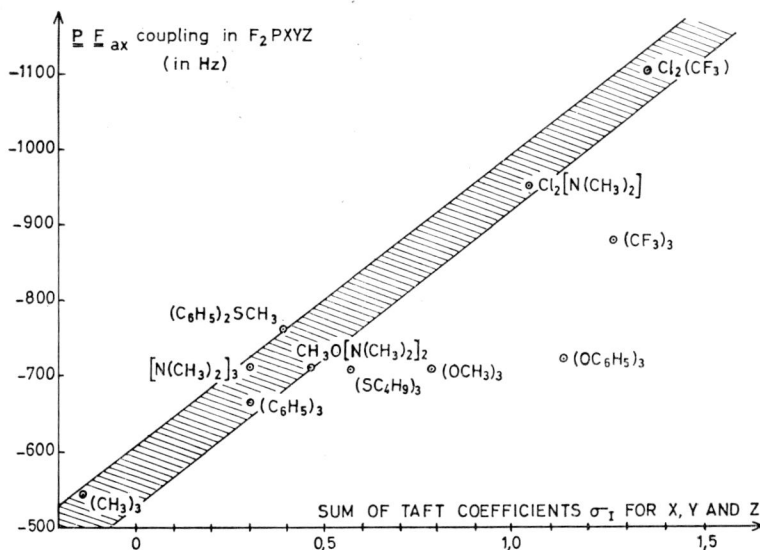

FIG. 3. P–F couplings in F_2PXYZ compounds as a function of the inductive effects of the X, Y and Z groups.

later this point is now better appreciated. Another point which has in the meantime been clarified is the strong stereospecific dependence of these couplings. In all cases, where some conformational preference exists, observed couplings reflect this aspect in addition to the dependence upon bond nature [and possibly on the net charge on phosphorus (see 1969, 50)]. This makes the discussion of P–H couplings through carbon atoms more interesting although naturally more difficult; the same applies to P–F couplings in similar situations. The various aspects of relevant couplings will each be considered in turn.

(a) *P–C–H couplings.* These couplings are known in a wide variety of compounds and some typical data are given in Table XI. Their dependence upon substituents has been discussed for various representative acyclic compounds by Gallagher [1968, 56] but, as earlier pointed out, the relevant hybridisation change cannot clearly be the only governing criteria. [1965, 19; 1966, 36]

This is especially true for trivalent compounds for which no clear correlation pattern could be found. When the signs are considered (these changes being more sensitive in P^{III} compounds as suggested in 1966, 1) it is then possible to highlight the importance of geometrical factors as will now be exemplified.

TABLE XI
Some typical values of P–C–H couplings

Compound	Coupling, Hz	References
P^{III}		
$P(CH_3)_3$	+2·7	1969, 2
$P(C_2H_5)_3$	−0·5	a
$P(CH_2OH)_3$	6·6	1968, 52
$P(CH_2O)_3P$	+9·3	1966, 37; 1969, 3
$P(CH_2O)_3CR$	7·8 to 8·5	1966, 37; 1968, 53; 1969, 71
$HP(CH_3)_2$	+3·6	a
$FP(CH_3)_2$	5·9 to 6·0	1967, 65 and 66
$ClP(CH_3)_2$	8·3 to 8·7	a
$C_6H_5P(CH_3)_2$	+3·0; 2·7 to 3·6	1968, 7 and 54
$C_6F_5P(CH_3)_2$	5·4	1966, 19
$(CH_3)_2NP(CH_3)_2$	5·5 to 5·7	1965, 18; 1966, 38; 1969, 72
$CH_3SeP(CH_3)_2$	+7·0	1969, 4
$(CH_3)_2PP(CH_3)_2$	+6·9	a
$(CH_3)_2P(S)P(CH_3)_2^*$	+4·1	a
$(CF_3)_2P \cdot P(CH_3)_2$	6·2	1968, 55
H_2PCH_3	+4·1	a
F_2PCH_3	10·2 to 10·4	a
Cl_2PCH_3	17·5 to 17·7	1966, 36
$(CF_3)_2PCH_3$	4·8	a
$H(H_3Si)PCH_3$	+3·7	1969, 73
$(C_6H_5)_2PCH_3$	4·0 to 5·0	1967, 65; 1968, 7 and 54
$Cl_2PCH_2CH_3$	15·0	a
$(P^{IV})^{\oplus}$		
$P^{\oplus}(CH_3)_4$	−14·6; 15·1	1966, 39; 1968, 54; 1969, 2
$P^{\oplus}(C_2H_5)_4$	−12·6	1966, 22
$HP^{\oplus}(CH_3)_3$	−15·7	1966, 22
$HP^{\oplus}(C_2H_5)_3$	−12·7	1966, 22
$(CH_3)_2NP^{\oplus}(CH_3)_3$	13·4	1965, 18
$H_2P^{\oplus}(CH_3)_2$	−17·0	1966, 22
$[(CH_3)_2N]_2P^{\oplus}(CH_3)_2$	13·5 to 13·8	1965, 18; 1967, 67
$H_3P^{\oplus}CH_3$	−17·6	1966, 22
$(t\text{-}C_4H_9)_3P^{\oplus}CH_3$	13	1967, 68
$(CH_3O)_3P^{\oplus}CH_3$	17·2	1967, 69
$[(CH_3)_2N]_3P^{\oplus}CH_3$	14·1 to 14·5	1965, 18; 1967, 67 and 70
P^V tetracoordinate		
$OP(CH_3)_3$	−12·8 and −13·4	1967, 71
$OP(CH_2O)_3PO$	7·8	1966, 37
$OP(CH_2O)_3CR$	5·6 to 7·4	1966, 37; 1968, 53
$OP(CH_3)F_2$	19 to 19·9	1969, 20 and 74
$OP(CH_3)Cl_2$	16·2 to 17	1969, 20 and 74

TABLE XI (continued)

Compounds	Coupling, Hz	References
P^V		
$OP(CH_3)(OCH_3)_2$	17.4	1969, 20
$OP(CH_3)(SCH_3)_2$	13.9	1969, 20
$OP(CH_3)(C_6H_5)H$	−13.8	1969, 34
$SP(CH_3)_3$	−13.0	a
$SP(CH_3)_2H$	14.4	1966, 40
$SP(CH_3)_2SeCH_3$	−12.7	1969, 4
$SP(CH_2O)_3CR$	5.3 to 7.3	1968, 53
$SP(CH_3)(t\text{-}C_4H_9)H$	13.4	1968, 2
$SP(CH_3)(t\text{-}C_4H_9)OH$	12.3	1968, 2
$SP(CH_3)(OH)(OR)$	15.8 to 16.0	1967, 72
P^V pentacoordinate		
$(CH_3)_3PF_2$	17.2	1967, 60
$(CH_3)_3PCl[N(CH_3)_2]$	14.0	1968, 54
$(CH_3)_3P{=}NSi\cdots$	12.4 to 12.5	1967, 73
$(C_2H_5)_3P{=}NSi\cdots$	10.2 to 10.5	1967, 73
$(CH_3)_2PF_2H$	18.3	1968, 42
CH_3PF_3H	17.6 to 18	1968, 38, 39 and 42
$C_2H_5PF_3H$	18.5	1968, 38 and 39

[a] Quoted in reference 1968, 30.

(i) *Signs* have been determined for various P^{III} compounds. $^2J(P\text{–}C\text{–}H)$ are positive in all cases involving methyl groups such as in $(CH_3)_3P$, $(CH_3)_2PH$, CH_3PH_2, $(CH_3)_2P\cdot P(CH_3)_2$, etc.; [1966, 13; 1968, 28, 33 and 57; 1969, 2, 4, 34 and 73] this is also true for ethyl groups. The two methylene protons of $C_2H_5PH(C_6H_5)$ or of $ClCH_2PH_2$ are non-equivalent and both have positive coupling constants. [1968, 28 and 58; 1969, 34] In both the cases of $ClCH_2PCl(NR_2)$ and $C_6H_5CH_2PH(C_6H_5)$, the two non-equivalent methylene protons have, besides one largely positive, one slightly negative coupling; [1968, 59; 1969, 34] both couplings, however, remain positive in $P(CH_2O)_3P$. [1969, 3]

For $(P^{IV})^{\oplus}$ ions, the only determination, carried out on $P(CH_3)_4^{\oplus}$, gave a negative sign for the coupling [1969, 2] in agreement with the suggestions of Allen *et al.*[1968, 60] Likewise, all known signs in P^V compounds, either with phosphoryl $[CH_3P(O)H(C_6H_5)$, etc.] or with thiophosphoryl groups $[(CH_3)_2P(S)SeCH_3]$, are negative. [1968, 33; 1969, 4 and 34]

Thus, it clearly appears that sign reversal occurs in this coupling system. As will be seen later, this is partly dependent upon molecular geometry and upon nature of substituents, but the main factor is the hybridisation of the phosphorus. This is illustrated by the change of sign observed in $P(CH_3)_3$ when complexed with some Lewis acid, e.g. $AlEt_3$; the phosphorus hybridisation changes from nearly p_3 to sp_3 and the ^2J(P–C–H) coupling changes from positive to negative when the $AlEt_3$ concentration is increased.[1966, 41; see also 1968, 53] This was taken as an indication of the existence of two competing mechanisms involving opposite signs, one at least being highly dependent upon the molecular geometry.[1968, 30] The absence of the reversal of sign from *P–C–H* to the relevant *P–C(CH₃)* coupling (both being positive for P^{III} compounds) and the rather general anomaly $-|J(P-C-C-H)| > |J(P-C-H)|-$ point to the same conclusion as discussed earlier (Section III.A.1), especially for trivalent phosphorus. This, however, would be valid in addition for other hybridisations as observed in such compounds as **12** in which ^3J(P–C–CH₃) is 13 Hz while in **13** the ^2J(P–CH) coupling is zero.[1967, 74]

Apart from this phenomenological description and that of Jameson, we have seen that the theoretical approach of Cowley and White was capable of reproducing this behaviour (Section III.A.2). It would therefore be interesting to similarly investigate the geometrical dependence which will now be considered.

(ii) The *stereospecificity* of ^2J(P–C–H) is clearly revealed by observations on various compounds having some degree of asymmetry such as in $ClCH_2PH_2$ and especially in $ClCH_2P(Cl)N(CH_3)_2$ in which hindered rotation affords different *P–C–H* couplings at room temperature (+19·4, +7·6 and +26, −3 Hz respectively).[1968, 58, 59] Similarly, the *P–C–H* couplings involving the axial and equatorial methyl groups 1-methyl in phosphorinanes also differ;[1967, 75] the same is found in $(P^{IV})^{\oplus}$, e.g. **14** and **15**, P^V, e.g. **16** and **17**[1968, 61] and pentacoordinated compounds (**18** and **19**)[1969, 75]; in **20** the two P–C–H couplings have been reported to be different.[1967, 76]

All this data has prompted various authors to present a discussion of *P–C–H* couplings as a function of the molecular geometry and more especially of the dihedral angle H–C–P: or H–C–P=X.[1967, 77; 1968, 30] A more complete analysis has been presented by Albrand et al.[1968, 62; 1969, 76] (Fig. 4) to improve the Karplus-like description of ^2J(P–C–H) values for trivalent phosphorus. As is seen zero coupling is possible with a dihedral angle of about 45°; the CH and CH_3 resonances appear as singlets in **21**[1966, 42] and **22**[1967, 78; 1969, 77] respectively.†

The same stereospecificity is probably the origin of the marked

† The extension of this curve for $(P^{IV})^{\oplus}$ and P^V compounds has not actually been considered. It would be necessary to take into account such unusual couplings as observed in **23** and **24**.[1968, 63]

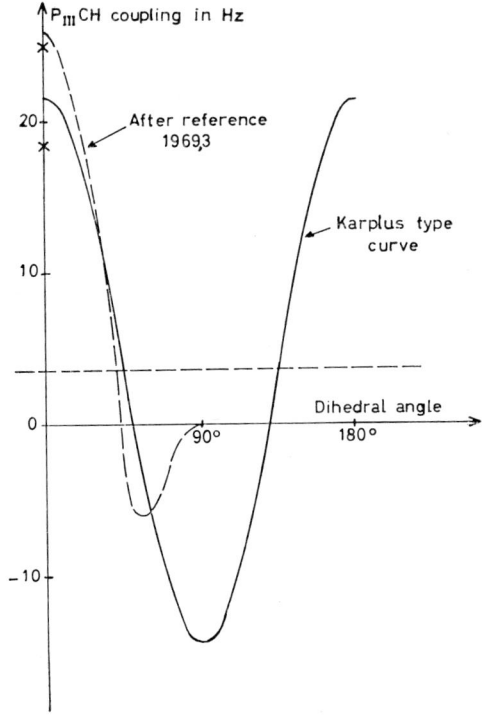

FIG. 4. Geometrical dependence of P–C–H couplings in derivatives of trivalent phosphorus.

changes observed when the conformation of the PCH group is perturbed, for instance by increasing chain length. This effect is especially sensitive in P^{III} compounds, for example in $CH_3P(C_6H_5)_2$, $^2J(P-CH)$ is 5·0 Hz while in $CH_3 \cdot CH_2 \cdot P(C_6H_5)_2$ the same coupling is only 1·5 Hz;[1967, 79] from Fig. 4 this would indicate the importance of a preferred conformation (**25** and **26**). (In agreement with this, an increase is expected for *P–C–C–H* couplings with increasing chain length; see below.) Similar effects are found in penta-coordinated phosphorus compounds, e.g. compare **27** and **28**, while in the homolog acyclics the difference in *J* value is by no means as great (**29** and **30**).

(b) *P–C–C–H couplings.* Typical values for this coupling are given in Table XII and have been found to be positive, in all cases where sign determination has been performed, in both P^{III} [e.g. $t\text{-}C_4H_9P(C_6H_5)_2$ and $C_2H_5P(C_6H_5)H$] and P^V compounds [e.g. $C_2H_5P(O)(OC_2H_5)C_6H_5$ and $ClCH_2CH_2P(S)Cl_2$].[1968, 28; 1969, 2, 34 and 79]

TABLE XII
Some typical values of $^3J(\text{P-C-CH})$

Compound	Coupling	References
P^{III}		
$P(C_2H_5)_3$	+13.7	a
$P(i\text{-}C_3H_7)_3$	11.5	1968, 7
$P(t\text{-}C_4H_9)_3$	9.8	1969, 78
$C_6H_5(C_2H_5)_2$	ca. 14	1968, 7
$C_6H_5P(i\text{-}C_3H_7)_2$	15.2	1968, 7
$C_6H_5P(t\text{-}C_4H_9)_2$	+12.3	1969, 2
$H_2N \cdot P(t\text{-}C_4H_9)_2$	11.0	1968, 64
$(C_6H_5)_2PC_2H_5$	16.5	1967, 79
$(C_6H_5)_2PC_4H_9\text{-}t$	12.3	1967, 79
$(H_2N)_2PC_4H_9\text{-}t$	12.5	1968, 65
$(P^{IV})^{\oplus}$		
$P^{\oplus}(C_2H_5)_4$	+18.1	1966, 22
$HP^{\oplus}(C_2H_5)_3$	+20.0	1966, 22
$HP^{\oplus}(i\text{-}C_3H_7)_3$	18.0	1968, 7
$CH_3P^{\oplus}(t\text{-}C_4H_9)_3$	15	1967, 68
$HP^{\oplus}(C_6H_5)(C_2H_5)_2$	21.5	1968, 7
$HP^{\oplus}(C_6H_5)(i\text{-}C_3H_7)_2$	20.6	1968, 7
P^V		
$P(O)(C_2H_5)_3$	+18	a
$CH_3OP(O)(i\text{-}C_3H_7)_2$	15.7	1968, 65
$ClP(O)(t\text{-}C_4H_9)_2$	16.2	1965, 8
$F_2P(O)C_4H_9\text{-}t$	19.0	1965, 8; 1966, 30
$Cl_2P(O)C_4H_9\text{-}t$	25.0	1965, 8
$P(S)(t\text{-}C_4H_9)_3$	14	1967, 68
$HP(S)(CH_3)C_4H_9\text{-}t$	18.5	1968, 2
P pentacoordinate		
$C_2H_5PF_3H$	26.4	1968, 39
$t\text{-}C_4H_9PCl(NHSi\cdots)=N\cdots$	20.4	1968, 65

[a] Quoted in reference 1966, 1.

Despite the smaller effects, compared with $P\text{-}C\text{-}H$ couplings, stereospecificity of vicinal $P\text{-}C\text{-}C\text{-}H$ couplings is well documented, in agreement with earlier proposals of Benezra when studying phosphonates.[1966, 43] Typical examples may be found in acyclic compounds with discernible rotamers as in $i\text{-}C_3H_7P(C_6H_5)N(CH_3)_2$ in which $P\text{-}C\text{-}CH_3$ couplings are 17.9 and 13.4 Hz for the two non-equivalent methyls of the isopropyl group[1967, 80] or in $(i\text{-}C_3H_7)_2PC_6H_5$ where the corresponding couplings are 14·7 and 11·0 Hz.[1968, 66] Likewise, significant differences have been observed in the isomers of t-$C_4H_9P(O)(F)N(CH_3)_2$ [16·9 and 15·6 Hz][1965, 8] or for the geminal methyl groups in $(C_6H_5)_2P(O)$-$C(CH_3)_2CH(C_6H_5)NHC_6H_5$ [16 and 14 Hz],[1967, 76] in $(CH_3)(C_6H_5)$-$P(O)C(CH_3)_2CH(CH_3)C(CH_3)C_6H_5$ [19·5 and 17·0 Hz][1968, 67] or in $CH_3P(O)(C_6H_5)C(CH_3)_2i$-$C_3H_7$ [16 and 15 Hz].[1968, 68] The effect involving the methylene groups is much larger in $(C_6H_5)_2P(S)CH$-$(CO_2H)CH_2 \cdot C_4H_9$-t [19·0 and 2·5 Hz].[1966, 44]

Phosphorus ring systems also present the same stereospecificity and, once more, facilitate the assignment of the relevant geometry. Such examples were first described by Churi[1967, 81] while studying epoxyalkylphosphonates (**31** and **32**) and by Cremer and Chorvat[1967, 82] in phosphetanes in which the non-equivalent ^3J(P–C–CH) couplings involving the methyl and methylene groups are shown in **33** and **34** respectively. This similar effect has been found in phosphonium ions (e.g. **35**), in the isomeric phosphorus(V) compounds (**36** and **37**)[1968, 61] and for the non-equivalent methyl groups in **38**[1968, 67] and in various dimethyl bicyclo[2.2.1]heptyl phosphonates.[1969, 80] These results permitted Benezra to propose a Karplus-like dependence for the relevant P–C–C–H couplings. Further data were obtained by Cox and Adelman[1969, 81] on various compounds of the form $(CH_3O)_2P(O)CH_2CH_2X$ and by Bothner-By and Cox[1969, 79] while analysing the rotamers in $Cl_2P(S)$-CH_2CH_2Cl during a detailed variable-temperature investigation. These authors concluded in proposing the dihedral dependence of P–C–C–H couplings shown in Fig. 5.

Such an angular dependence may be the origin of the trends observed when systematically comparing homologous couplings in n-alkyl groups; when these are larger such conformations as shown in **39** are probably stabilised, thus resulting in larger couplings following the upper curve. This is found, for instance, in $C_2H_5(t$-$C_4H_9)P(O)Cl$ in which the ^3J(P–H) is 23 Hz for the ethyl group and 26 Hz for the t-butyl group;[1966, 45] in agreement with this picture, the opposite change is expected (and observed) for the relevant P–C–H couplings as discussed earlier.

This behaviour is not restricted to triply- or quadruply- connected compounds as is seen in the pentacoordinated compound (**40**)[1967, 76] but experimental data on this latter type of compound are rather scarce.

(c) *Higher PC····CH couplings*. These are far less well documented but in general appear to present similar features.[1969, 82, 83] The stereospecificity of ^4J(P–C–C–C–H) has been clearly indicated from the observations of Churi[1967, 81] on **41**. Cremer and Chorvat have reported other examples in the phosphetane derivatives;[1967, 82] this, together

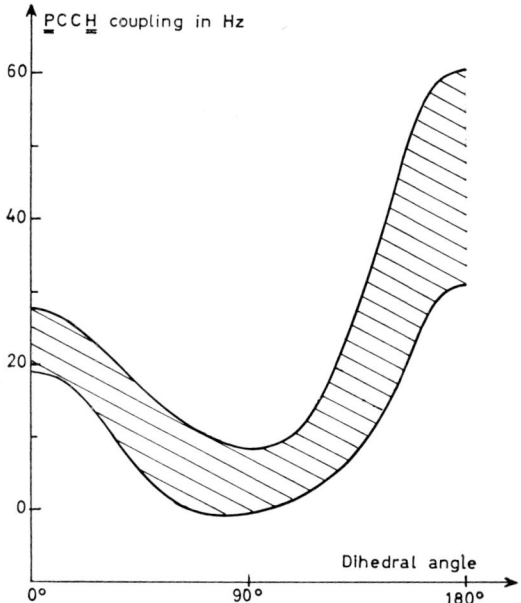

FIG. 5. Geometrical dependence of P–C–C–H couplings [From Bothner-By and Cox.[1969, 79]]

with the rigid systems such as bicyclo[2.2.1]heptyl derivatives investigated by Benezra,[1969, 80] are characteristic W-type couplings. Use of this stereospecificity has been made by Ross and Martz for the assignment of the conformation of cholestanyl phosphonate and related compounds.[1969, 33]

(d) *P–C–F couplings*. Typical values for these couplings are shown in Table XIII; it will be noted that no value is less than 40 Hz. In all cases, where the sign has been determined, the couplings were found to be positive as, for example, in $(CF_3)_2PF$ of $+89.6$ and $(CF_3)_2PSCF_3$ of $+83.8$ Hz[1967, 40, 55 and 57] based upon the negative sign for the P–F coupling which had been established earlier. Thus, in contrast to the corresponding P–C–H couplings, the P–C–F couplings undergo no reversal in sign. As discussed in Section III.A, the observed sign of coupling agrees with the theoretical conclusions of Cowley and White and of Jameson.

Relatively few examples of the stereospecificity of *P–C–F* couplings exist in the literature but the general features are similar to that found for *P–C–H* couplings. It is observed for instance, that in the neat sample of

TABLE XIII
Some typical values of P–C–F couplings

Compounds	Coupling	References
P^{III}		
$(CF_3)_3P$	85·5	a
$(CF_3)_2PH$	68·6 to 70·0	1969, 43
$(CF_3)_2PF$	+89·6	1967, 48 and 55
$(CF_3)_2PCl$	85·1	a
$(CF_3)_2PBr$	80·6	a
$(CF_3)_2PCN$	73·2	a
$(CF_3)_2PCH_3$	76·7	a
$(CF_3)_2PN(CH_3)_2$	85·6	a
$(CF_3)_2PN(CF_3)_2$	149·4	1968, 68
$(CF_3)_2P \cdot P(CH_3)_2$	64·1 to 64·2	1967, 83; 1968, 55
$(CF_3)_2POC_2H_5$	86·6	a
$(CF_3)_2PSCH_3$	77·8	a
$(CF_3)_2PSCF_3$	+83·8	1967, 48
$(CF_3)_2P \cdot S \cdot P(S)(CF_3)_2$	82·5	1969, 57
$(CF_3)_2PSeCF_3$	77·2	a
$(CF_3)_2P \cdot P(CH_3)_2$	64·1	1967, 83
CF_3PF_2	87·2	1967, 48
CF_3PCl_2	79·9	a
CF_3PBr_2	69·6	a
CF_3PI_2	52·1	a
CF_3PH_2	48·5	a
$CF_3P[N(CH_3)_2]_2$	87·4	a
$CF_3P[N(CF_3)_2]_2$	113·7	1968, 69
$CF_3P[P(CH_3)_2]_2$	40·2	1967, 83
$(CF_3CF_2CF_2)_2PF$	62·1	a
$CF_3CF_2CF_2PF_2$	90·5	a
P^V		
$(CF_3)_3PO$	113·4	a
$(CF_3)_2P(O)CH_3$	99	1969, 84
$(CF_3)_2P(O)OCH_3$	120	1969, 84
$CF_3P(O)F_2$	162	1968, 44
$CF_3P(O)Cl_2$	151	1968, 70
$(CF_3)_3PS$	108·7	a
$(CF_3)_2P(S)F$	128·6	1968, 47
$(CF_3)_2P(S)Cl$	123·1	1968, 47
$(CF_3)_2P(S)Br$	119·7	1968, 47
$(CF_3)_2P(S)I$	111·4	1968, 47
$(CF_3)_2P(S)N(CH_3)_2$	102·9	1968, 47
$(CF_3)_2P(S)SH$	110·0	1968, 47
$(CF_3)_2P(S) \cdot S \cdot P(CF_3)_2$	111·7	1969, 57

[a] Quoted in reference 1968, 30.

TABLE XIII (continued)

Compounds	Coupling	References
P^V pentacoordinate		
$(CF_3)_2PCl_2[N(CF_3)_2]$	187·3	1969, 85
$(CF_3)_2PCl_3$	193 (CF_3 axial)	1967, 62
$(CF_3CF_2CF_2)_2PF_3$	125	1969, 66
$CF_3PF_2Cl_2$	165	1969, 66
CF_3PFCl_3	149	1967, 62
CF_3PCl_4	154	1967, 62
$CF_3CF_2CF_2PF_4$	124	1969, 66
$CF_3CF_2CF_2PF_2Cl_2$	128	1969, 66
$(P^{VI})^\ominus$		
$(CF_3)_2PF_3H^\ominus$	132	1968, 41
$CF_3PF_4H^\ominus$	156	1969, 67
$CF_3PF_5^\ominus$	145	1969, 67
$(CF_3)_2PF_4^\ominus$	145	1969, 67

$CH_2ClCF_2P(Cl)N(CH_3)_2$ the geminal fluorine atoms of the CF_2 group are non-equivalent and have different couplings, but of the same sign, ±93 and ±53 Hz at room temperature; if the sample is, however, dissolved then both couplings merge and have a common value of ca. 70 Hz. Similarly, if the neat sample is heated to +150° the couplings become more similar (80 and 60 Hz); conversely these are more dissimilar at low temperature (116 and 40 Hz at −60°). The relation of these changes of coupling constant values to molecular geometry thus seems well established.[1968, 71] The same undoubtedly occurs in cyclic compounds, for which one example is known **(42)**[1962, 2] in which the two non-equivalent fluorine atoms have couplings of 128 and 54 Hz.

These data are not sufficient to build up a complete curve relating $P-C-F$ couplings and dihedral angles but such a relation has been sketched in Fig. 6 for P^{III} compounds by similarity with the corresponding behaviour of $P-C-H$ couplings. For this system one may postulate that preferred conformations such as **43** may explain the reduction in the value of the coupling constant generally observed between $P-CF_3$ to the analogous $P-CF_2-R$ groups. Similar perturbations of conformational equilibria presumably result in the observed solvent dependence of P–C–F couplings[1969, 43] which, in this case, involves hydrogen bonding with the solvent.

Considering the influence of substituents on the observed couplings it is found that for the best documented cases (Fig. 7) that increasing electronegativity results in increased J values; some discrepancies occur

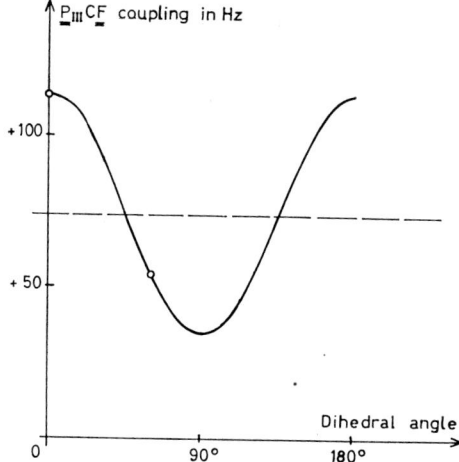

Fig. 6. Assumed geometrical dependence of P–C–F couplings in derivatives of trivalent phosphorus.

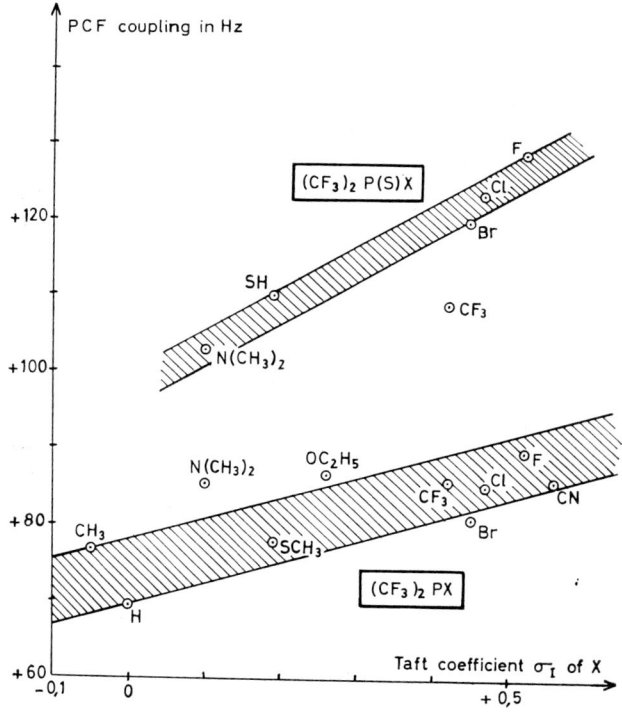

Fig. 7. P–C–F couplings in $(CF_3)_2PX$ and $(CF_3)_2P(S)X$ compounds as a function of the inductive effect of the substituent X.

for P^{III} compounds when mesomeric effects are operative, e.g. for $-OR$ or $-NR_2$ groups.

The interpretation of observations on penta- or hexa- coordinated compounds is not so clear as CF_3 groups may occupy *axial* positions, even when competing with fluorine substituents as in (**44**).[1968, 41]

(e) *P–C–C–F couplings.* These couplings have not been so extensively studied as for the corresponding proton couplings. The sign of $^3J(P-C-C-F)$ in $C_2F_5PCl_2$ and $C_3F_7PCl_2$ has been established by Manatt *et al.*[1967, 40] and was found to be positive, i.e. the same as the corresponding *P–C–F* coupling. Thus, there is no reversal of sign between ^2J and ^3J but, as previously noted,[1966, 1, p. 286] long-range anomalies are not so frequent, e.g. in most cases, but not all, $^3J(P-C-C-F) < {}^2J(P-C-F)$.

3. Couplings through oxygen atoms

A large number of POCH couplings† have been reported and typical examples are given in Table XIV. It is clearly seen that these couplings are not too sensitive to the nature of substitution. Four bond couplings are generally rather small (< 1 Hz) and the sign of the coupling constant affords the greatest interest. Exceptionally high values‡ have been found in $P(OC_2H_5)_4^\oplus$, of ca. 2 Hz,[1969, 86] in **45** to **49**[1966, 37, 46] and a similar value for the six bond coupling in **50**.[1969, 343] No similar data are available for the fluorine homologs and thus the following discussion is of necessity restricted to POCH couplings.

(i) Relative signs have first been established in P^V compounds. Using triple resonance, McFarlane assigned a *positive* value to $^3J(P-O-CH)$ in $HP(O)(OCH_3)_2$ ($+12.0$) in contrast to P–H and $^{13}C-H$ values. The same sign was proposed by Miyazima *et al.* for $^3J(P-O-CH)$ and $^4J(P-O-C-CH)$ in $Cl_2P(O)OC_2H_5$,[1967, 84] in agreement with previous determinations of relative signs by Duval and co-workers[1966, 47; 1967, 85] for the same couplings in:

	$OP(OCH_2CH_3)_3$	same sign	(8·40; 0·84)
	$SP(OCH_2CH_3)_3$	,, ,,	(9·85; 0·73)
	$P(OCH_2CH_2CH_3)_3$	,, ,,	(7·65; 0·80)
but	$P(OCH_2CH_3)_3$	opposite sign	($|7·9|$; $|0·55|$).

† P–O–H and P–C(O*H*) couplings have rarely been considered; one example of the latter has been reported in $(C_2H_5O)_2P(O)CH(OH)CH_2OC_2H_5$ of 10·0 Hz.[1969, 82]

‡ Values of 14 to 15 Hz have been reported for $P(O)OCH_2CH_3$ couplings;[1968, 72] this, however, requires confirmation.

TABLE XIV
Typical P–O–CH couplings

Compound	Couplings	References
P^{III}		
$(CH_3O)_3P$	10·8 to 11·8	1969, 86, 87
$(C_2H_5O)_3P$	7·3 to 8·0	1967, 86; 1969, 86
$(R \cdot CH_2O)_3P$ (R > CH_3)	6·1 to 7·5	[a]
$(C_6H_5CH_2O)_3P$	7·9	1968, 73
$(CH_3O)_2PCl$	11	1965, 20
$(CH_3O)_2PC_6H_5$	10·5 to 11·5	1969, 74, 87
$(C_2H_5O)_2PC_6H_5$	8·3	1969, 34
$CH_3OP(CF_3)_2$	13·0	1969, 84
$CH_3OP(C_6H_5)_2$	13·7	1969, 87
$CH_3OP\langle\text{(cyclic O,O)}\rangle$	13	[a]
$P(OCH_2)_3CR$	1·8 to 2·0	1966, 37, 48
$P(O-)_3$ (bicyclic)	6·3	1966, 37
(P^{IV})$^\oplus$		
$(CH_3O)_4P^\oplus$	11·2	1967, 69
$(CH_3O)_3P^\oplus CH_3$	11·5	1967, 69
$(C_2H_5O)_3P^\oplus C_2H_5$	7·5	[a]
P^V		
$(CH_3O)PO$	10·2 to 11·4	[a], 1969, 87
$(C_2H_5O)_3PO$	6·8 to 8·4	[a]
$(CH_3O)_2P(O)H$	12	1969, 74
$(CH_3O)_2P(O)CH_3$	11·0	[a]
$(CH_3O)_2P(O)C_6H_5$	11	1969, 87
$CH_3OP(O)H_2$	13	[a]
$CH_3OP(O)(C_6H_5)_2$	11·0	1969, 87
$CH_3OP(O)(CF_3)_2$	11·1	1969, 84
$OP(OCH_2)_3CR$	6·5	1966, 37, 48
$OP(O-)_3$ (bicyclic)	18·0	1966, 37
$(CH_3O)_3PS$	13·4	1969, 87
$(C_2H_5O)_3PS$	10·2	[a]
$(CH_3O)_2P(S)Cl$	11·0	[a]
$(CH_3O)_2P(S)CH_3$	10·2 to 10·7	[a]
$(CH_3O)_2P(S)C_6H_5$	13·8	1969, 87

TABLE XIV (continued)

Compound	Couplings	References
$CH_3OP(S)(C_6H_5)_2$	13·5 to 14	1967, 87; 1969, 87
$CH_3OP(S)F_2$	14·1	1968, 45
$CH_3OP(S)(OCH_2)CR_2$	13·8	1968, 74
$CH_3OP(O)\big<\!\!\begin{smallmatrix}O\\O\end{smallmatrix}\!\!\big>$= (catechol phosphate)	11·5	1967, 86
P^V pentacoordinate		
$(CH_3O)_5P$	12	a
$(CH_3O)_3PF_2$	15·2	1967, 61
cyclic $[O_2P(H)O_2]$ bis-dioxy with H	12·5	1967, 39
cyclic bis(N-CH₃) diamino-dioxy P-OCH₃	$POCH_3$: 14·5	1967, 163
$(CH_3O)_3P\big<\!\!\begin{smallmatrix}O\\O\end{smallmatrix}\!\!\big>$=	13·5	1965, 21
cyclic phosphazene with OCH₃	16·5	1966, 49
benzo-dioxaphosphole $(CH_3O)_2$	13·5	1968, 75
$(P^{VI})^\ominus$		
$F_4P(\text{ocacac})^\ominus$ (hexafluoro-acac complex)	$POCCH_3$: 2 $POCCH$: <2	1966, 33

[a] Quoted in reference 1967, 297.

A positive sign has also been found for $P-O-CH$ involving a pentacoordinated phosphorus (in phosphonitrilics).[1969, 68] As no $J(P-O-CH)$ values are in the vicinity of zero it is unlikely that this coupling experiences sign reversal; it may be taken as positive in all cases. As a consequence, this is not so for $J(P-O-C-CH)$, especially for the above mentioned phosphites.

Sign determinations have also been made on the stereospecific couplings to be discussed in section (ii) below. The O-methylene protons in $C_6H_5P(O)C_2H_5(OC_2H_5)$ are not equivalent but the two $J(P-O-CH)$ values are of the same sign.[1969, 34] Considering cyclic P^{III} compounds the $J(P-O-CH)$, $J(P-O-C-CH_A)$ and $J(P-O-C-CH_3)$ in the phosphorinanes (**51**), and $J(P-O-CH_A)$ and $J(P-O-CH_B)$ in phospholanes (**52**) are all of the same sign.[1969, 88] This is also true for P^V oxazaphospholanes (**53**) in which the $P-O-CH$ couplings are distinct and positive ($+11.15$ and $+8.05$).[1968, 76]

(ii) The stereospecificity of $P-O-CH$ couplings have been previously mentioned while discussing sign determination. Various other examples are known, for instance in acyclic compounds in which rotamers exist, such as $C_6H_5P(OC_2H_5)_2$ or $C_6H_5P(O)C_2H_5(OC_2H_5)$,[1969, 34] thus supporting the early suggestions of Verkade,[1964, 5; 1965, 22] Benezra[1966, 43] and Haake et al.[1968, 77] The relevant conformations are, however, not easily identified as the P–O–C bonds are rather flexible.[1969, 89] For the interpretation of the data it is perhaps better to consider the numerous examples of dioxa- phospholanes (**54**) and phosphorinanes (**55**) which have been investigated despite the possible controversy concerning the conformation at the phosphorus atom in **55**[1966, 50; 1968, 78, 79; 1969, 90] or in **54**.[1968, 77]

This topic has been discussed in detail by Gagnaire and co-workers[1967, 88; 1968, 80; 1969, 91] for P^{III} compounds; the stereo-specificity was attributed to the influence of the phosphorus lone-pair.† This parallels the explanation proposed to account for similar effects involving ^{15}N in oximes.[1967, 89] Additional data have been obtained in ethylene phosphites[1966, 50, 51; 1968, 77; 1969, 92 to 94] but, for the purposes of discussion, a clearer example have recently been investigated in great detail by Kainosho and Nakamura,[1969, 90] namely **56**. Complete assignment, based upon spin-decoupling gave the following values for the P–H couplings:

$P-H_x$	A	B	C	D	E	F	G
J	4.4	1.7	9.6	3.8	1.7	2.5	9.3

† The influence of the heteroatom lone-pairs on couplings through phosphorus, as proposed by Kainosho et al.[1969, 90] for explaining the $H_A \cdots H_F$ long-range couplings in the trioxaphosphabicyclo-octane (**56**), cannot be excluded.

and, by estimating dihedral angles using Dreiding models, these authors succeeded in establishing a Karplus-like curve for the angular dependence of $J(POCH)$ (Fig. 8).

In relation to the consideration of the stereospecificity of coupling in P^{III} compounds, an observation of Kainosho[1969,95] seems especially important. Complexing the phosphorus atom by addition of a paramagnetic ion removes any stereospecific phosphorus couplings. In the phosphorinane **55a**, $^3J(POCH_A) = 2·2$ and $^3J(POCH_B) = 10·5$ Hz but on increasing the concentration of nickel acetylacetonate in solution first

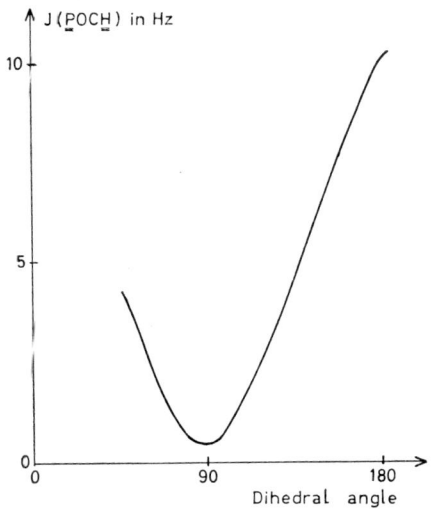

FIG. 8. Geometrical dependence of P–O–C–H couplings in derivatives of trivalent phosphorus [From Kainosho and Nakamura.[1969,90]]

results in decoupling of P–H_A then P–H_B. All these observations clearly indicate that *factors other than conventional "through-bond" mechanisms are important when considering such couplings.*[1969,94]

Considering now P^V compounds, various examples of stereospecific coupling have been reported. Following a complete study of conformational equilibria in phosphoric acid esters Tsuboi and co-workers [1967,90; 1968,81; 1969,95,96] were able to assign changes of coupling in dioxaphosphorinanes (**57**) as shown. Similar values were assigned (19 to 26 and 2 to 6 Hz) in various substituted trioxaphosphorinanes (**58**) [1967,91; 1968,82; 1969,92,97]. This also appears to be applicable to more complicated systems such as cytidine phosphate (**59**)[1969,98] and other "biological" structures.[1969,96]

The same features exist at a reduced level for *POCCH* couplings in phosphorinanes and these are about 2·6 Hz for an equatorial proton at C–5, while it is less than 1 Hz for an axial proton. Similarly, coupling of phosphorus to an equatorial methyl at C–4 is 2 to 3 Hz while that to axial methyl is less than 1 Hz;[1968, 83] similar trends have been found in the corresponding phosphonates.[1969, 52]

When comparing such values with the relevant data on ethyl (or higher) phosphates, it is clear that, in all acyclic compounds, the observed values for *J*(*POCCH*) are averaged over a range of 4 or 5 Hz (going down to negative values in P^{III} compounds) as rotation is completely free. This is not necessarily so when comparing *POCH* couplings in phosphites and phosphates of increasing chain length or, moreover, phosphites and phosphates with cage-like structures.[1966, 37] In the case of $P(OR)_3$ or $OP(OR)_3$ the coupling is *ca*. 11 Hz when R = CH_3, 6 to 8 Hz for higher alkyl group but in $OP(OR)_3$ with R = $CH_2C(CH_3)_3$ is 4·8 Hz. This strongly suggests that for higher alkyl groups the methylene groups are in a *gauche* conformation (**60**).[1969, 94] Such phenomena are certainly solvent and temperature dependent.[1967, 44]

Lastly, the pentacoordinated phosphorus compounds, in which Ramirez succeeded in observing isomers by freezing the pseudorotation at $-100°$, will be considered. Once again, the same stereospecific trends appear for the *POCH₃* and *POCH* coupling in **61a** and **b**; the $POC_{(5)}H$ couplings are 2·0 and 3·5 Hz respectively while the $PC_{(3)}H$ couplings are 17·0 and 6·0 Hz. The three methoxyl groups are also clearly identified and the couplings to phosphorus different, namely 10·8 Hz for the apical and 12·2 and 12·9 Hz for the non-equivalent equatorial methoxyl groups. The differences are greater in **62** in which the apical methoxyl has a coupling of 10·6 Hz compared to 12·7 and 14·0 for the equatorial methoxyl groups.[1968, 84]

4. *Couplings through sulphur atoms*

Coupling in *P–SH* groups has not yet been reported but may be present in the low temperature spectrum of $F_2P(S)SH$.[1969, 99] The *P–S–CH* couplings are better known (Table XV) but the range of information is by no means as complete as for the oxygen homologs; no sign determinations have been carried out,† but it is likely that all *P–S–CH* couplings have the same sign and probably the same as for *P–O–CH* (see 1966, 1, p. 270). The stereospecificity of these couplings has only been recognised

---

† The sign of *P–S–CF* in $CF_3SP(CF_3)_2$ has been determined and found to be positive.[1967, 55] As no further data on *PSCF* couplings are available this subject will not be further discussed.

TABLE XV

Some typical P–S–CH couplings

Compound	Coupling	References
P^{III}		
$(CH_3S)_3P$	9·8	1966, 1
$RC(CH_2S)_3P$	2·1	1968, 85; 1969, 71
$CH_3SP(OCH_3)Cl$	19	1965, 20
$(P^{IV})^\oplus$		
$CH_3SP^\oplus(NH_2)_3$	16·2	1968, 86
P^V		
$(CH_3S)_3PO$	15·1	1966, 1
$RC(CH_2S)_3PO$	11·0	1968, 85
$C_2H_5SP(O)(OC_2H_5)_2$	15	1966, 52
$CH_3SP(O)(NH_2)_2$	14·8	1968, 86
$(CH_3S)_3PS$	17·5	1966, 1
$RC(CH_2S)_3PS$	10·0	1968, 85
$CH_3SP(S)F_2$	20·3	1968, 45
$CH_3SP(S)(OCH_3)_2$	16	1965, 20
P^V pentacoordinate		
CH_3SPF_4	25·2	1969, 62
$CH_3S(C_6H_5)_2P=NP(O)-(C_6H_5)_2$	13·2	1967, 92
[ring structure with C_6H_5 and SCH_3]	18·0	1969, 100

in the S-methylene group of $CH_3P(O)(OR)SCH_2SR'$ [1969, 101] but it may be suspected in other cases as discussed earlier for P–O–CH couplings (Section 3.B.3(ii)). For instance this effect is seen when comparing free rotating groups and those of cage compounds. [1968, 85]

Comparison of P–S–CH couplings (Hz) in freely rotating groups and in related fixed conformations

$(CH_3S)_3P$	9·8	$(CH_3S)_3PO$	15·1	$(CH_3S)_3PS$	17·5
$CH_3C(CH_2S)_3P$	2·1	$CH_3C(CH_2S)_3PO$	11·0	$CH_3C(CH_2S)_3PS$	10·0

It should be noted that the range of P–S–CH couplings is much larger than for P–O–CH or P–N–CH; likewise these are found to be more sensitive to complex formation, e.g. the P–S–CH coupling in $CH_3SP(O)(OCH_3)_2$ and the corresponding BF_3 adduct differ by 1 Hz while the corresponding P–O–CH coupling only differs by 0·2 Hz.[1969, 102] This appears to arise from the decreased polarity in the P–X–CH chain on going from X = O to S (or Se) which facilitates the coupling transmission;† additionally this feature favours the transmission of long-range coupling effects such as found in **63**.[1966, 53]

5. *Couplings through nitrogen atoms*

(i) Two-bond P–NH couplings have only been observed in a few P^V compounds, such as $F_2P(O)NH_2$,[1967, 93] $(C_2H_5O)_2P(O)NHCH_3$,[1969, 20] $(C_6H_5)_2P(O)NHC(CH_2C_6H_5)=C(CN)C_6H_5$,[1968, 88] $FClP(S)NHSi(CH_3)_3$,[1969, 69] and $(P^{IV})^\oplus$ compounds, such as $C_6H_5P^\oplus(NH_2)$-$[NCH_3N(CH_3)_2]_2$.[1967, 94] No examples involving P^{III} have been reported; the doublet nature of the NH resonance in $C_6H_5P(NHC_3H_7$-$i)_2$ has been proved to arise from coupling with the neighbouring CH group.[1967, 95] Longer-range couplings between phosphorus and NH groups are not observed, as for instance might be possible in $RP'P(O)C(O)NH\cdots$ compounds.[1968, 29]

(ii) Three-bond PNCH couplings will now be considered and some typical values are shown in Table XVI.‡ A distinctive feature of the *PNCH* system (for which a positive sign is likely§) is the possibility of d_π–p_π conjugation between phosphorus and nitrogen; this shortens the P–N bond and facilitates the stabilisation of rotamers. These are readily observed in P^{III} compounds when steric hindrance is present or at sufficiently low temperatures. The interpretation of experimental data may, however, be obscured by various factors: repulsion of phosphorus and nitrogen lone-pairs seems a secondary factor in these P^{III} compounds,[1969, 105, 107] but the possible inversions at phosphorus and nitrogen have to be discussed further. Inversion at the phosphorus atom requires a much higher energy than that at nitrogen; moreover, when P–N conjugation occurs, the configuration about nitrogen which is normally pyramidal becomes planar[1969, 108] thus facilitating the *N*-inversion. Phosphorus, however, retains the pyramidal configuration as the *d*-

---

† Similarly, much larger couplings through sulphur than through oxygen are observed in 1,3-thioxanes.[1968, 87]

‡ *PNCF* couplings are known in too few compounds to warrant discussion here.[1966, 31]

§ Such a sign was suggested by Manatt *et al.*[1966, 13] as for *POCH* coupling; this has been established by Haigh[1969, 68] in phosphonitrilics.

TABLE XVI

Some typical values of P–N–CH couplings

Compound	Coupling	References
P^{III}		
$[(CH_3)_2N]_3P$	8·8 to 9·0	1965, 18; 1967, 65; 1969, 87
$[(CH_3)_2N(CH_3)N]_3P$	4·5	1967, 94
$[(C_2H_5)_2N]_3P$	8·3	1965, 18
$[(CH_3)_2N]_2PCl$	12·3	1965, 18; 1967, 65
$[(CH_3)_2N(CH_3)N]_2PCl$	7·3	1969, 103 and 104
$[(C_2H_5)_2N]_2PCl$	11·2	1965, 18
$[(CH_3)_2N]_2PCH_3$	8·7	1965, 18
$[(CH_3)_2N]_2PC_6H_5$	9·5	1969, 87
$(CH_3)_2NPCl_2$	12·9 to 13·0	1965, 18; 1966, 34; 1967, 65
$(CH_3)_2N(CH_3)NPCl_2$	7·2	1969, 103
$(C_2H_5)_2NPCl_2$	13·3	1965, 18
$(CH_3)_2NP(CH_3)_2$	9·8	1965, 18; 1969, 72
$(CH_3)_2NP(C_6H_5)_2$	9·5	1969, 87
$(CH_3)_2NP(OC_2H_5)_2$	9·4	1969, 20
$(CH_3)_2NP(C_6F_5)_2$	10·9	1966, 19
$CH_3N(PCl_2)_2$	3·0	1967, 96; 1968, 89
$[(CH_2)_2N]_3PO$	14	1964, 6
$(P^{IV})^{\oplus}$		
$[(CH_3)_2N]_3P^{\oplus}Cl$	13·0	1967, 65
$[(CH_3)_2N]_3P^{\oplus}CH_3$	10·0 to 10·1	1965, 18; 1967, 65
$[(CH_3)_2N]_3P^{\oplus}SCH_3$	11·2	1969, 105
$[(CH_3)_2N]_2P^{\oplus}(CH_3)_2$	10·3 to 10·5	1965, 54; 1967, 65
$(CH_3)_2NP^{\oplus}(CH_3)_3$	11·3 to 11·5	1965, 18; 1968, 54
$[(C_2H_5)_2N]_3P^{\oplus}Cl$	14	1968, 90
P^V		
$[(CH_3)_2N]_3PO$	9·4 to 9·5	1965, 18, 23; 1967, 65; 1968, 91; 1969, 87
$[(C_2H_5)_2N]_3PO$	9·9 to 14	1965, 18; 1968, 90
$[(CH_3)_2N]_2P(O)Cl$	12·9 to 13·0	1965, 18, 23; 1967, 65; 1968, 91
$[(C_2H_5)_2N]_2P(O)Cl$	13·4	1965, 18
$[(CH_3)_2N]_2P(O)C_6H_5$	9·8 to 9·9	1965, 23; 1969, 87
$(CH_3)_2NP(O)Cl_2$	15·8 to 15·9	1965, 18, 23; 1967, 65; 1968, 91
$(C_2H_5)_2NP(O)Cl_2$	17·2	1965, 18
$(CH_3)_2NP(O)(C_6H_5)_2$	10·0 to 10·2	1965, 23; 1968, 91; 1969, 87
$[(CH_3)_2N]_3PS$	11·0 to 11·3	1965, 18, 23; 1967, 65; 1968, 91; 1969, 87, 105

TABLE XVI (continued)

Compound	Coupling	References
P^V		
$[(C_2H_5)_2N]_3PS$	11·4	1965, 18
$[(CH_3)_2N]_2P(S)Cl$	15·3 to 15·4	1965, 18, 23; 1967, 65; 1968, 91
$[(CH_3)_2N]_2P(S)C_6H_5$	12·0 to 12·1	1968, 91; 1969, 87
$[(C_2H_5)_2N]_2P(S)Cl$	14·8	1965, 18
$(CH_3)_2NP(S)Cl_2$	17·2 to 17·7	1965, 18, 23; 1967, 65; 1968, 91
$(C_2H_5)_2NP(S)Cl_2$	18·9	1965, 18
$(CH_3)_2NP(S)(C_6H_5)_2$	14·3 to 14·6	1968, 91; 1969, 87
$[(CH_3)_2N]_3PSe$	11·6	1969, 87
$[(CH_3)_2N]_2P(Se)C_6H_5$	12·5	1969, 87
$(CH_3)_2NP(Se)(C_6H_5)_2$	15·0	1969, 87
P^V pentacoordinate		
$[(CH_3)_2N]_3PF_2$	10·6	1967, 61
$(CH_3)_2NPF_4$	2·1	1967, 59
$(CH_3)_2NPF_2Cl_2$	13·2	1968, 43
$[CH_3NPF_3]_2$	14·5	1967, 59
$\begin{array}{c}CH_3N\text{———}PCl_3\\ \;\;\;\;|\;\;\;\;\;\;\;\;\;\;\;\;\;\;| \\ Cl_3P\text{———}NCH_3\end{array}$	20·0	1966, 53
$[(CH_3)_2N]_2P\underset{N}{\overset{N\diagup\overset{|}{C}\diagdown N}{\bigcirc}}P[N(CH_3)_2]_2$	11·1	1969, 106

orbitals are involved when it participates in the P–N conjugation. An extensive variable temperature PMR study by Imbery and Friebolin [1968, 92] confirmed these views. In compounds which are asymmetric at the nitrogen but not at the phosphorus atom, e.g. $(R^1CH_2)R^2NPX_2$, no non-equivalence of the methylene group was observed, even at the lowest temperatures attained, on account of the rapid inversion at nitrogen. For the corresponding compounds $(R^1CH_2)R^2NPXX^1$ $(X \neq X^1)$ which are asymmetric at phosphorus, non-equivalence appears at some temperature; thus inversion at phosphorus is clearly slower than that at nitrogen. It is especially noticeable that this is accompanied by different couplings of the methylene protons to the phosphorus, e.g. 13·8 and 4·8 Hz in $C_6H_5CH_2(CH_3)NPCl(C_6H_5)$. Similar conclusions hold for a compound symmetrically substituted at nitrogen but asymmetric at phosphorus, e.g. in $(CH_3)_2NPCl(C_6H_5)$ in which the two $^3J(PNCH)$ values involving the methyl groups are 19·0 and 7·0 Hz, [1968, 92] or even if the phosphorus is symmetrically substituted but the temperature sufficiently low as, for example in $(CH_3)_2NPCl_2$ at $-80°$, the couplings are 19·3 and 6·6 Hz. [1967, 80] This clearly rules out the *anti*-isomer as predominating at these temperatures but favours the *gauche* isomer (64). Evidently, this remains true for cases in which the substituents on nitrogen are different, when lowering the temperature separates the conformers, e.g. in $CH_3(C_6H_5)NPF_2$; in one conformer the methyl is coupled to phosphorus by 4·2 Hz, while in the second conformer the coupling is 11·0 Hz. [1966, 34]

Such observations are naturally simpler in cyclic compounds, such as in ephedrine derivatives (65) when $^3J(PNCH)$ are 2·3 and 1·0 Hz depending upon the isomer. [1969, 109] This effect is even more pronounced in cyclic P^V compounds [1968, 76] such as 66 in which the two $^3J(PNCH)$ values are clearly distinguished and have opposite signs ($+11·5$ and $-0·31$ Hz).

Similar observations are possible with acyclic P^V compounds if the substituents of phosphorus are different and also rather bulky (and more especially if chelation can occur to stabilise the conformers). [1969, 110] In $(CH_3)_2NP(O)F(C_4H_9\text{-}t)$, the methyl groups are non-equivalent with $^3J(P-N-CH_3)$ of 8·2 and 0·0 Hz. [1965, 8] The phosphorus inversion remains the dominating phenomenon here. [1969, 111]

Nevertheless, although the stereospecificity is well documented in these *PNCH* systems, the underlying phenomena are probably more complicated than previously discussed for *PCH*, *PCCH* and *POCH* couplings, because of the role played by the π-electrons. [1968, 29]

(iii) On account of the stabilisation of the P–N bonds by the hybridisation features discussed above, the changes in J values accompanying modification of substituent groups, or changes in phosphorus coordination are rather regular; it is observed that, in a given series, the coupling modulus (as well as proton chemical shift) increases for the sequence P, PO, PS and $P^{\oplus}$. Anomalies are certainly linked to some perturbation of the conjugation scheme. In this respect, multiple bonding is especially important and enhances couplings as, for example, in $(RO)_2P(O)N=CHOR^1$ (15 to 17·5 Hz) [1966, 55] or in $(RO)_2P(S)N=CHNR^1_2$ (22 to 25 Hz). [1966, 43]

(iv) Longer-range couplings are known in non-conjugated systems such as in $(CH_3)_3CNHPCl_2$ (2·2 Hz) [1969, 112] or in $[(C_2H_5O)_2P(O)]_2\text{-}CHN(CH_3)_2$ (1·5 Hz); [1968, 93] this is also true when other heteroatoms are present, for instance $(t\text{-}C_4H_9)_2PNHX(CH_3)_3$ (X = Si, $J = 0·6$; X = Ge, $J = 0·3$ Hz). [1968, 64]

This long-range coupling becomes more predominant in compounds in which multiple bonding is present, e.g. as in **67** [1966, 56] or in **68** and **69** [1968, 88] and even more so when conjugation is present as in phosphonitrilics, where ^{14}N decoupling proves that $PNPNCH$ coupling exists [1965, 24] or in **70**, which when compared to **71** clearly shows the importance of conjugation; [1967, 97; 1968, 94] going further one has an example of five-bond coupling **72**. [1968, 94]

6. *Couplings through other nuclei*

Besides couplings through metal atoms observed in such compounds as **73** in which $^3J(PSnCH_3) = 4\cdot 4$ Hz, [1968, 95] the most interesting data come from the germyl and silyl homologs of the P^{III} compounds discussed earlier.

Relative signs have only been determined for *P–Si–H* coupling; this was found to be positive in $CH_3PH(SiH_3)$. [1969, 73] It is likely, contrary to expectation for these two bond couplings, that all couplings are similarly positive in derivatives of methylphosphines; [1967, 98; 1968, 32, 96; 1969, 113] these couplings are shown in Table XVII.

TABLE XVII

Some examples of two-bond P–X–H couplings (Hz), X = C, Ge and Si, in methyl phosphine derivatives

$P(CH_3)_3$	+2·8	$P(GeH_3)_3$	15·2 to 15·9	$P(SiH_3)_3$	17·0
$HP(CH_3)_2$	+3·6	$HP(GeH_3)_2$	15·7	$HP(SiH_3)_2$	15·7
H_2PCH_3	+4·1	H_2PGeH_3	15·3	H_2PSiH_3	15·3

In addition to the identical signs it is also noticed that, in the carbon derivatives, the $P \cdots H$ coupling increases with decreasing number of methyl groups while for the germanium and silicon derivatives the coupling basically decreases. This arises from the relative electronegativity of Si and Ge (1·8) compared to that of C (2·5) and P and H (2·1); replacing a methyl group by a proton decreases the *s* character of the remaining methyl groups while replacing a SiH_3 or GeH_3 groups by a proton increases the *s* character of the remaining groups. The same general rules apply for mixed groups, for instance it is observed [1969, 73] that *H–Si–P* coupling decreases in the series $H_3SiP(CH_3)(SiH_3)$; $H_3SiP(H)CH_3$, $H_3SiP(CH_3)_2$ with increasing *s* character.

The only recorded data for homologs of ethylphosphines [1969, 113] are shown in Table XVIII in which the data are compared to those of the triethylphosphine.

TABLE XVIII

Coupling constant data (Hz) for silicon derivatives of triethylphosphines

	$^2J(P-XH_2)$	$^3J(P-X-XH_3)$
$P(SiH_2SiH_3)_3$	17·5	?
$H_2PSiH_2SiH_3$	18·5	2·0
$P(CH_2CH_3)_3$	−0·5	+13·7

Lastly, comparison is possible between $P[C(CH_3)_3]_3$ for which $^3J(PCCH)=9·8$, $P[Si(CH_3)_3]_3$ with $^3J(PSiCH)=4·5$ and $P[Ge(CH_3)]_3$ with $^3J(PGeCH)=3·7$ Hz. [1966, 57; 1967, 29]

Having taken into account the change in electronegativity and size of the central atoms, it appears that the couplings are intrinsically larger for the carbon compounds. This helps clarify the anomalous behaviour which existed for coupling to α-CH groups which was discussed earlier (Section III.B.2.a).

7. Couplings through unsaturated systems

It has been realised for sometime that the presence of π-electrons in hydrocarbon chains enhances all the J values and this is also true for phosphorus compounds. In addition, the existence of isomers provides data on geometrically well-defined conformations and finally the participation of phosphorus in conjugated systems will give some unusual behaviours.

(a) Vinylic systems will be considered first; the P–H couplings somewhat parallel those of the corresponding H–H couplings. The double quantum analysis of Lancaster [1967, 99] has established that all the couplings have the same sign; these couplings constant values are shown in Table XIX. The data for **74** has been confirmed by Williamson et al. [1968, 97]

TABLE XIX

P–H couplings in vinyl phosphorus derivatives

	cis	trans	gem
$P(CH=CH_2)_3$	±13·6	±30·2	±11·8
$C_6H_5(CH_3)_2\overset{\oplus}{P}-CH=CH_2$	±25·3	±48·5	±24·0
$(C_2H_5O)_2P(O)CH=CH_2$ (**74**)	±25·1	±50·5	±11·8

using double resonance; they have carried out a detailed analysis of $(C_2H_5O)_2P(O)CX\!=\!CYZ$, where X, Y, Z = H or Cl. The observed trends are similar for all types of phosphorus hybridisation but once again it should be noted that there is a big variation in changing from P^{III} to $(P^{IV})^{\oplus}$ and P^V. It is also noticeable that in all cases the proton closer to phosphorus (i.e. *gem*) exhibits the smaller coupling and of the two β-protons that closer to phosphorus (i.e. *cis*) again has the smaller value.

Geometrical factors are also important as shown by Benezra et al. for the *cis* couplings in hindered compounds.[1966, 43, 58; 1967, 100] They proposed a Karplus-like behaviour for these couplings in compounds of the type **75**.

Considering now the corresponding methylated compounds it is found that for phosphines and phosphonates the *gem* $PC(CH_3)$ couplings are of similar importance to the corresponding *gem* P–C–H couplings (the relative sign is not known) as in *cis* and *trans* positions replacing CH by C·CH_3 gives a lower coupling (in modulus) with opposite sign as predicted by "Hund coupling"[1967, 101]. Thus, in methylated vinyl phosphines, the *gem*, *cis* and *trans* $J(P\cdots CH_3)$ values are respectively 12, 3 and 2·5 Hz while the corresponding values for $J(P\cdots CH)$ in the parent vinyl phosphines are 10, 12 and 30 Hz.

Similarly the variation in P^V compounds with α- and β- methyl substituents are shown in **76** and **77**[1966, 59] and **78** to **80**[1963, 4; 1966, 60].

The behaviour for substitution in the α-position is clearly contrary to predicted trends and supports the idea of two superimposed mechanisms acting in different directions when the expected coupling should be negative (*gem*) and in the same direction when expected to be positive.

Similar behaviour is found in the variously substituted vinylic systems. In P^{III} fluorinated vinyl compounds, for which Cowley and Taylor have obtained the relative signs,[1969, 54, 55] it is found that sign reversal may occur in *gem* couplings and that the *trans* couplings generally have the lowest values (Table XX).

TABLE XX

P–F couplings in fluorovinyl phosphines

	gem	cis	trans
$Cl_2PCF\!=\!CF_2$	−17·2	+85·7	+7·3
$F_2PCF\!=\!CF_2$ [a]	+20·9	+69·6	+4·8
$ClP(CF\!=\!CF_2)_2$	0	70	?
$P(CF\!=\!CF_2)_3$	24	53	8

[a] Stereospecificity is also found in the $^4J(F\text{–}P\text{–}C\text{–}CF)$ values: 2·8 Hz for *cis* and 13·2 Hz for *trans*.[1959, 55]

In P^V compounds results are only available for the *gem* positions [1968, 98] and these values cover a wide range:

$(CH_3O)_2P(O)CF{=}CClCF_3$	*cis* $^2J(P-CF)$	91·9
	trans $^2J(P-CF)$	90·0
$(C_6H_5)_2P(O)CF{=}CClCF_3$	*cis* $^2J(P-CF)$	14·2
	trans $^2J(P-CF)$	24·0

the same situation exists for the perfluoromethyl analogs:[1967, 102, 103; 1968, 98]

$(C_2H_5)_2PC(CF_3){=}CHCF_3$	$^3J(PCCF_3)$	*ca.*	0 Hz
	$^4J(PCCCF_3)$		53 Hz
$(C_6H_5)_2PC(CF_3){=}C(CF_3)P(C_6H_5)_2$	$^3J(PCCF_3)$		313 Hz
$(CH_3O)_2P(O)CF{=}CClCF_3$	*cis* $^4J(PCCCF_3)$		3·2 Hz
	trans $^4J(PCCCF_3)$		1·0 Hz
$(C_6H_5)_2P(O)CF{=}CClCF_3$	*cis* $^4J(PCCCF_3)$		63·7 Hz
	trans $^4J(PCCCF_3)$		5·5 Hz

and finally in phosphonium compounds:

$(C_6H_5)_3\overset{\oplus}{P}CH{=}C(CF_3)_2$	$^4J(PCCCF_3)$	1·5 to 5 Hz

This variability may be related to the sensitivity of fluorine to "through-space" coupling as mentioned in Section III.A.1.

Lastly, couplings have been reported for penta-coordinated phosphorus compounds having perfluorovinyl substituents **81**;[1969, 114] it will be observed that the stereospecificity of the couplings is retained.

(b) Allenic systems have received considerable attention from Simonnin and co-workers who established the following relative signs [1966, 61; 1967, 104; 1968, 99; 1969, 115]†

For allenic phosphines:

	$RR^1PCH{=}C{=}CH_2$	$^2J(P-CH)$	+
or	$RR^1PC(CH_3){=}C{=}CH_2$	$^3J(P-C-CH_3)$	+
		$^4J(P-CC-CH)$	±

the last sign varies according to the nature of the substituents on phosphorus and their ability to conjugate.

In allenic phosphine oxides:

$RR^1P(O)CR^2{=}C{=}CR_2^3$	$^2J(P-CH)$	+
	$^3J(P-C-CH_3)$	+
R^2 and $R^3 = H$ or CH_3	$^4J(PCCCH)$	−
	$^5J(PCCCCH_3)$	+

† Apart from cumulenes, non-conjugated polyenes have received far less attention but seem to have somewhat similar behaviour as regards long-range coupling.[1965, 25]

The change from P^{III} to P^V has rather smaller effects when compared to the effects of conjugation of phosphorus; this is especially true with amino substituents with which p_π–d_π interaction is likely to occur.

The consideration of allenic molecules and of the methylated homologs show anomalies similar to those discussed in the proceeding paragraph. Substituting a methyl for a proton it would be expected that there would be a reduction in the modulus, and a change in sign, of the relevant P····H coupling based on the simple "Hund coupling mechanism" of the Fermi contact term.[1964, 7] Actually, for the α-position to phosphorus, P–C–H and P–C–CH₃ couplings have the same sign and comparable magnitude, for instance in:

Cl₂PCH=C=CH₂	^2J(P–C–H)	±9·5 Hz
Cl₂PC(CH₃)=C=CH₂	^3J(P–C–CH₃)	±3·4 Hz
Cl₂P(O)CH=C=CH₂	^2J(P–C–H)	±16·4 Hz
Cl₂P(O)C(CH₃)=C=CH₂	^3J(P–C–CH₃)	±20·2 Hz

while, for the γ-position, the expected changes are observed, for instance in:

Cl₂P(O)CH=C=CH₂	^4J(PCCCH)	∓18·2
Cl₂P(O)CH=C=C(CH₃)₂	^5J(PCCCCH₃)	±12·1

The possibility of a "through-space" mechanism, which would be especially sensitive for the α-position for blocked systems such as these, cannot be ruled out.

(c) Acetylenic systems will be considered next for these may exist as a tautomeric form of the allenic compounds; allenic-acetylenic isomerisation readily occurs in various compounds as has been discussed by Russian authors (see ref. 1967, 21 for a brief review). On account of the non-existence of substituents α to phosphorus, sign determinations are not so easy and the only known result has been provided by Ionin et al.[1967, 105] in F₂P(O)C≡CCH₃, in which the P–F and P–H couplings were of opposite sign, which means the latter is positive.

It should be pointed out that sign reversals are likely to occur when the hybridisation of phosphorus, or the nature of the substituents, change in systems such as:[1966, 62]

(C₆H₅)₂PC≡CH	J < 0·5 Hz	(C₆H₅)₂P(C≡C)₂H	1·45 Hz
(C₆H₅)₂P(O)C≡CH	9·7		
(C₆H₅)₂PC≡CCH₃	1·7	(C₆H₅)₂P(C≡C)₂CH₃	1·3
(C₆H₅)₂P(O)C≡CCH₃	3·8	(C₆H₅)₂P(C≡C)₃CH₃	0·7
(C₆H₅)₂P(S)C≡CCH₃	4·1		
(C₆H₅)₃P⁺C≡CCH₃	4·6		

or, moreover when amino substituents are attached to the phosphorus atom as in: [1967, 106]

$(CH_3)_2NP(C\equiv CH)_2$	1·2 Hz	$(CH_3)_2NP(O)(C\equiv CH)_2$	11·8
$[(CH_3)_2N]_2PC\equiv CH$	1·5	$[(CH_3)_2N]_2P(O)C\equiv CH$	11·3
$(CH_3)_2NP(C\equiv CCH_3)_2$	2·7	$(CH_3)_2NP(O)(C\equiv CCH_3)_2$	4·2
$[(CH_3)_2N]_2PC\equiv CCH_3$	3·0	$[(CH_3)_2N]_2P(O)C\equiv CCH_3$	4·2

In the latter cases it might be anticipated that rehybridisation of the phosphorus occurs as there is possibility of conjugation between the phosphorus and neighbouring triple bond as discussed earlier (Section III.B.3).

(d) Allylic systems afford some interest on account of the well defined stereochemistry of the *cisoid-transoid* type.[1967, 107, 108] Assuming a negative sign for the phosphorus-methylene coupling, it is possible to establish the following signs:[1967, 101]

$-P(O)CH_2CH_\beta=CH_\gamma-$	$P-H_\beta$	+
	$P-H_\gamma$	−
$-P(O)CH_2C(CH_3)_\beta=C(CH_3)_\gamma-$	$P-(CH_3)_\beta$	−
	$P-(CH_3)_\gamma$	+

The regular behaviour previously sought for in vinylic and allenic systems is observed; couplings are reduced in magnitude and change sign in going from the proton to the methyl homolog. A similar comparison would be of interest in propargylic compounds but only the magnitude of the couplings are known in these compounds.[1966, 63]

(e) Phospholenic compounds are interesting homologs of the preceeding vinylic and allenic compounds, namely 2-phospholenes (**82**) and 3-phospholenes (**83**), and have received considerable attention.

(i) 2-Phospholenes have coupling constants rather similar to those observed in vinyl compounds. For P^V compounds typical data are shown in **84** and **85**.[1964, 8; 1966, 64, 65; 1967, 108; 1968, 100; 1969, 20] For the case of P^{III} and $(P^{IV})^\oplus$ compounds some $^2J(P-H)$ values are shown in **86** and **87**[1968, 100] and **88**.[1967, 108]

(ii) 3-Phospholenes are far better known. Relative signs of coupling in P^V compounds agree with results on the parent allyl compounds; being opposite in sign for $^2J(PCH)$ and $^3J(PCCH)$.[1967, 101] Typical values for P^V compounds are shown in **89**;[1964, 8; 1966, 64 to 66; 1967, 108 to 110; 1968, 61, 100 and 101; 1969, 20] couplings in related oxaphospholenes appear to be rather similar.[1969, 116] The corresponding data for $(P^{IV})^+$ and P^{III} compounds are shown in **90** and **91**.

There is an interesting example for comparison purposes in the pair of conformers, described by Katz *et al.*,[1966, 67] for which the data are shown in **92** and **93**; the data for the corresponding phosphoryl compounds are given in parenthesis. These data clearly show the marked stereospecificity, not only for the α-protons but also for those in the β-positions, and the influence of the phosphorus lone-pair.

For the α-protons an additional point is found when comparing the P····CH$_3$ coupling in methyl phospholenes with the relevant P····H coupling. For example, the data shown in **94** to **97** have been observed and demonstrates this point for the α-couplings, while for the β-position the PCCCH$_3$ couplings are between 1 and 2 Hz while the PCCH couplings are in a similar range to that shown in **94** to **97**.

Going one stage further phospholes present distinctive features on account of the aromatic character;[1967,78; 1969,77] this property is clearly reflected in the values of the coupling constants. The relevant data for 1-methyl phospholes have been published[1969,77] and are shown in **98**; these are obviously contrary to the trends previously reported for 2-phospholenes in which $|^2J(\text{P–CH})| < |^3J(\text{PCCH})|$. However, these are in agreement with observations on 1-phenyl-2,5-dimethylphosphole (**99**)[1967,70] and on 1-phenyl-3,4-dimethylphosphole (**100**);[1969,20] in the latter compound both couplings mentioned have been found to be of the same sign.

The corresponding $(\text{P}^{IV})^\oplus$ and P^V compounds exhibit the same general features and same representative data are shown in **101**, **102**[1968,101] and **103**.[1969,20] In **103** the signs of coupling are now opposite and, on account of the relative magnitude of both the couplings, it is likely that the long-range coupling has undergone a reversal in sign.

Analogues to these five-membered rings the six-membered ring compounds are also known. The family of phosphabenzenes are now well known and some typical values of coupling constants are shown in **104** to **107**;[1966,68; 1967,111; 1968,101] the data for a corresponding penta-coordinated homolog is shown in **108**.[1968,75]† The transmission of coupling interactions appears to be particularly facile and long-range values are especially high (unfortunately no sign determinations have been reported). The former point is especially clarified when comparing the conjugated species with the dihydrophosphazenes (**109**).[1967,112] The value of $^2J(\text{P–CH})$ is 13 Hz for $R = C_6H_5$ and 8 Hz for $R = C_4H_9$ but no three-bond P–C–CH coupling is observed; the couplings in the corresponding P^V compound are shown in **110**. The importance of conjugation for the transmission of couplings is also apparent when comparing the two values of $^3J(\text{PCCH})$ in **111**, the larger value is that involving the double bond, 36·5 as against 22 Hz.[1968,102]

† The conjugation in such compounds is confirmed by bond lengths determined by X-ray structural analysis.[1969,117]

(f) A number of interesting coupling constants are found when aromatic groups are attached to phosphorus atoms. In the case of unsubstituted rings double resonance and solvent effects have been used for clarifying spectra and obtaining values of $J(\text{P–H}_{ortho})$ when *ortho* and *meta-para* patterns are complex; [1968, 103] these signals are more widely separated when substituents on P^V are more electronegative thus permitting subspectral analysis. [1967, 113] From these and other investigations [1966, 21; 1967, 113, 114] values for the P–H$_{ortho}$ couplings obtained are shown in Table XXI. In all cases, except triphenylphosphine, free rotation of the

TABLE XXI

Same values of P–H (*ortho*) coupling constants

Compound	Hz	Compound	Hz
$\text{H}_2\text{PC}_6\text{H}_5$	6·0	$\text{OP}(\text{C}_6\text{H}_5)_3$	11·8
$\text{P}(\text{C}_6\text{H}_5)_3$	7·1	$\text{OP}(\text{C}_6\text{H}_5)_2\text{OCH}_3$	12·0
$\text{ClP}(\text{C}_6\text{H}_5)_2$	7·8	$\text{OP}(\text{C}_6\text{H}_5)_2\text{SCH}_3$	12·7
$\text{HP}(\text{C}_6\text{H}_5)_2$	7·9	$\text{OP}(\text{C}_6\text{H}_5)(\text{OCH}_3)_2$	13·3
$\text{Cl}_2\text{PC}_6\text{H}_5$	8·5 to 8·7	$\text{OP}(\text{C}_6\text{H}_5)\text{Cl}_2$	17·0

aromatic ring occurs under normal conditions and this has been taken as evidence for σ coupling. [1966, 69] In triphenylphosphine p_π–p_π bonding seems predominant owing to steric hindrance and this makes the spectrum very complex as the signals from *ortho*, *meta* and *para* protons all overlap.

The spectra are clearer in substituted compounds, especially in *para-* and poly-substituted aromatic systems. In the former case sign determination is feasible and McFarlane has obtained the values for some typical compounds as shown in Table XXII; the data (without signs) for the tris(4-fluorophenyl)phosphine oxide are also included.

Changes in values of coupling constants appear to be essentially dependent upon changes in hybridisation of phosphorus, especially for the capability of phosphorus to participate in p_π–p_π bonding and this effect is lessened from P^III to $(\text{P}^\text{IV})^\oplus$ or P^V compounds as the hybridisation changes from p_3 to sp_3. Other factors, such as the nature of the substituents on the ring, seem to be rather unimportant, for instance P^V compounds with tolyl, nitrophenyl or chlorophenyl groups give essentially the same couplings as given above. [1966, 70; 1967, 116; 1969, 119]

In the case of the tolyl derivatives it should be noted that coupling of the phosphorus occurs only with the *ortho*-methyl group, for instance in $o\text{-CH}_3\text{C}_6\text{H}_4\text{P(O)(OCH}_3)_2$ $^4J(\text{P–}o\text{CH}_3) = 1\cdot7$ Hz. [1966, 70; 1967, 117] As

TABLE XXII

Some values of P–H (*ortho*) and P–H (*meta*) in *para*-substituted phenylphosphine compounds

Compound	P–H (*ortho*)	P–H (*meta*)	Reference
(p-CH$_3$OC$_6$H$_4$)$_3$P	+6·5	+1·1	1969, 118
(p-CH$_3$OC$_6$H$_4$)$_3$PS	+12·5	+2·3	1969, 118
(p-CH$_3$OC$_6$H$_4$)$_3$P$^\oplus$CH$_3$, I$^\ominus$	+12·7	+2·6	1969, 118
(p-FC$_6$H$_4$)$_3$P	+6·7	+1·3	1969, 118
(p-FC$_6$H$_4$)$_3$PS	+12·4	+2·2	1969, 118
(p-FC$_6$H$_4$)$_3$P$^\oplus$CH$_3$, I$^\ominus$	+12·1	+2·5	1969, 118
(p-FC$_6$H$_4$)$_3$PO	12·4	3·2	1966, 70; 1967, 115

noted in Section III.A, for the relevant CF$_3$ compounds, this may be indicative of a "through-space" mechanism; a number of related observations on p-tolyl derivatives points to the same conclusion [1966, 71; 1967, 117, 118] and ^7J(P–CH$_3$) values of 2·3 and 3·0 Hz have been observed in p-CH$_3$C$_6$H$_4$CH$_2$P(O)(OCH$_3$)$_2$ and p-CH$_3$C$_6$H$_4$CH$_2$P$^\oplus$-(C$_6$H$_5$)$_3$ respectively; likewise in p-tolyl phosphates a value for ^7J(P–CH$_3$) of 0·6 Hz is observed in p-CH$_3$C$_6$H$_4$OP(O)(OCH$_3$)$_2$. Overlap hyperconjugation has been invoked in this respect, which is another way of depicting "through-space" contribution.

The case of higher conjugated systems is clearly different, for which a π-electron explanation seems well-suited; the relevant couplings are observed in similar ranges, for instance in **112** [1966, 72] and also retain the stereospecificity as in **113**. [1968, 104]

Fluoro- or perfluoro- aromatic compounds offer examples of coupling constants with higher moduli but the signs of which are only known in the *para*-fluorophenyl compounds; the results as obtained by McFarlane [1969, 118] are shown below:

(p-FC$_6$H$_4$)$_3$P ^5J(P—F) = −4·5 Hz
(p-FC$_6$H$_4$)$_3$PS ^5J(P—F) = +2·4 Hz
(p-FC$_6$H$_4$)$_3$P$^\oplus$CH$_3$ ^5J(P—F) = +1·3 Hz

Similarly, the ^5J(P–F) values are also known in:

(p-FC$_6$H$_4$)$_3$P=O   1·5 Hz   (1967, 115)
(p-FC$_6$H$_4$)$_2$PCl   5·2 Hz   (1969, 120)
p-FC$_6$H$_4$PCl$_2$ 5·4 Hz   (1969, 120)

On the basis of McFarlane's results the sign of the phosphine oxide would be positive while those of the chlorides would be negative. This change in the sign of coupling is especially noticeable—it would be interesting to investigate the P–F couplings in relevant perfluoromethyl derivatives but, as previously stated (Section III.A), no coupling is observed between the phosphorus and *meta-* and *para-* CF_3 groups; only the *ortho* compounds show this coupling as in $(o\text{-}CF_3C_6H_4)_3P$ $^4J(P-CF_3) = 55 \cdot 0$ Hz,[1969, 33] the marked positive temperature dependence of this coupling supports the proposed "through-space" mechanism.

Regarding other fluorine positions on the phenyl ring it is observed that *ortho*-fluorine atoms give the largest and most sensitive values; some reported values[1966, 19] are shown in Table XXIII. For the purposes of comparison the data of Nichols[1969, 121] shown in Table XXIV is of interest.

TABLE XXIII

Some typical values of *ortho*-fluorine coupling constants

Compound	$^3J(P-F)$, Hz	Compound	$^3J(P-F)$, Hz
$C_6F_5PF_2$	63·2	$(C_6F_5)_2PN(CH_3)_2$	40·1
$C_6F_5PCl_2$	43·6	$(C_6F_5)_2PH$	29
$C_6F_5P(CH_3)_2$	30·1	$(C_6F_5)_3P$	12·3[a]
$C_6F_5PH_2$	3·9		

[a] See footnote below.

TABLE XXIV

Comparison of P–F couplings in pentafluoroaromatic phosphine derivatives†

Compound	$^3J(P-F_o)$	$^4J(P-F_m)$	$^5J(P-F_p)$
$(C_6F_5)_3P$	36·5	<0·1	<0·1
$(C_6F_5)_3PO$	37·4	<1·0	<1·0
$(C_6F_5)_2PC_6H_5$	38·0	<0·5	<0·5
$(C_6F_5)_2P(O)C_6H_5$	6·5	3·6	2·3

Parent heteroaromatic structures have recently been described and typical couplings in the pyridine series, as reported by Thomas *et al.*[1969, 122, 123] are shown in **114**. Data for five-membered ring heterocyclics

† The value for $(C_6F_5)_3P$ does not agree with the figure given above by Barlow *et al.* in Table XXIII, but is supported by other measurements (39·4 Hz).[1966, 73]

have also been reported by Thomas et al. and also by Jakobsen et al.[(1969, 124 to 126)] and are shown in **115** to **122**. For both **121** and **122** Jakobsen and Nielson have established signs from the analysis of double-quanta transitions and all couplings were found to be of the same sign.

Lastly, reference will be made to the investigation of Elguero and Wolf [(1967, 119)] concerning phosphorus attached heteroaromatic systems by the heteroatom, in which couplings such as P–H$_5$ in pyrazoles (**123**) and P–H$_3$ in indazoles (**124**) are found.

(g) Unsaturated systems when attached to phosphorus through an heteroatom retain some of the previously described behaviours, for instance stereospecificity of the coupling in O-vinyl derivatives† as shown in Table XXV.[(1967, 120; 1969, 127)]

TABLE XXV

Stereospecificity of P–H couplings in O-vinyl derivatives

Compound	"gem" ^3J(P–H)	cis ^4J(P–H)	trans ^4J(P–H)
(C$_2$H$_5$O)$_2$P(O)OCH=CH$_2$	6·8	1·2	2·7
(CH$_3$O)$_2$P(O)OCH=CH$_2$	6·8	4·1	2·8
(CH$_3$O)$_2$P(O)OCH=CHCH$_3$	5·6	(7·0)^a	7·0
(CH$_3$O)$_2$P(O)OC(CH$_3$)=CH$_2$	(0·0)^a	0·2	0·3

^{a} Couplings involving CH$_3$ group.

The long-range coupling of phosphorus to methyl substituents are especially noticeable in these and similar compounds [(1967, 121)]. Chelation of the P=O group and of the γ-cis substituent has been invoked to explain this fact which would imply some "through-bond" mechanism. A further interesting point is the change of sign likely to occur for cis and trans P–H couplings.

Observations on O-phenyl compounds have been mentioned previously (Section III.B.7(f)); in these compounds the phosphorus coupling to the *para* protons is as great as that to the *ortho* protons [(1966, 75; 1967, 117)] which leads to similar conclusions regarding the geometry of such derivatives and the coupling mechanism.

(h) The last unsaturated systems to be discussed, methylene phosphoranes, present some interesting features.‡ In these compounds the coupling transmission is easier as seen in **125** for which the couplings

† S-vinyl compounds are not well known but it appears that the general behaviour differs from that of O-vinyl compounds. [(1966, 74)]

‡ One fluorine homolog [(CH$_3$)$_2$N]$_3$P=CF$_2$ [(1964, 9)] has been identified as [(CH$_3$)$_2$N]PF$_2$ by Ramirez et al. [(1966, 76)]

are as shown. The relative signs in **126** have also been determined (see 1968, 105 to 107) and are opposite in sign for P–CH$_3$ and P–CH$_2$. The former coupling is probably negative as for other pentacoordinated phosphorus compounds (see Section III.B.2) while that of the P=CH$_2$ group would be positive. Similarly the couplings in **127** have been investigated and the P=CH and P····CHO couplings have the same sign. (1969, 20)

As is known, (1968, 103) there is some evidence for the overlap of the empty d orbital of phosphorus with the methylene p orbital; this stabilizes the molecule as a partly ionic species. It would be worthwhile investigating the magnitude and signs of couplings in various ylides having a range of stabilisation.

Another typical aspect in this type of compound is the long-range couplings which are readily observed; some typical examples are shown in Table XXVI.

TABLE XXVI

Some examples of long-range couplings in ylides

Compound	J, Hz	Reference
(C$_6$H$_5$)$_3$P=C(COR)CH_2CO$_2$R	^3J(P–H), 28	1967, 122
(C$_6$H$_5$)$_3$P=C(CO$_2$R)CH_2C(O)····	^3J(P–H), 15 to 17	1964, 10
(C$_6$H$_5$)$_3$P=CH–CH=C(CH_3)COCH$_3$	^5J(P–H), 5	1968, 109

(C$_6$H$_5$)$_3$P = C——C(O)
            | \O
           CH_2—C(O)
	^3J(P–H), 0·8	1964, 10

(C$_6$H$_5$)$_3$P = C——C = O
            | |
   O = C——C—P$^\oplus$(C$_6$H$_5$)$_3$
            |
           CH$_3$
	^5J(P–H), 1·0	1967, 102 and 123
		1968, 110

(C$_6$H$_5$)$_3$P=C=C(CF$_3$)$_2$	^4J(P–F), 3·5	1967, 102

8. *Long-range couplings*

Finally, some typical phosphorus-proton or fluorine couplings through phosphorus or heteroatoms will be discussed.

(a) Some proton-phosphorus couplings of this type are given in Table XXVII. The sign of coupling in the *P–N–P–CH* group as found in (CH$_3$)$_2$P(S)NP(S)(C$_6$H$_5$)$_2$ is opposite to that of P–CH, (1968, 111) i.e. is probably positive.

TABLE XXVII
P–H couplings through phosphorus or other heteroatoms

Compound	J, Hz		Reference
H_2PPF_2	^2J(H–P–P)	17	1968, 31
H_2PPH_2	^2J(H–P–P)	11·9	1968, 31
$(CH_3)_2P–P(CH_3)_2$	^3J(P–P–CH$_3$)	+11·25	1967, 124 and 125
$(CH_3)P–P(CF_3)_2$	^3J(P–P–CH$_3$)	12·7	1968, 55
$(CH_3)_3P{=}CH–P(CH_3)_2$	^4J(P–C–P–CH$_3$)	0·7[a]	1968, 107
H–P(O)–O–P(O)H with $O^\ominus$ $O^\ominus$	^3J(P–O–PH)	+1·5	1968, 37
$[(CH_3)_2N](C_2H_5)_2C–P(O)(OC_2H_5)_2$	^4J(CH$_3$N·C–P)	2·7	1969, 20
$CH_3SCH_2–P^\oplus(CH_3)_3$	^4J(CH$_3$SC–P)	1	1969, 128
$(CH_3)_3\overset{\oplus}{N}–CH_2–P^\oplus(CH_3)_3$	^4J(CH$_3$NC–P)	1	1969, 128
$(CH_3)_2N–CH_2–P^\oplus(CH_3)_3$	^4J(CH$_3$NC–P)	0	1969, 128

[a] The P=CH–P(CH$_3$)$_2$ coupling is zero.

(b) Couplings involving fluorine have caused some interest. A series of compounds have been investigated by Rudolph, Newmark et al.[(1967, 40, 49; 1968, 42; 1969, 31, 60 and 61)] and these are compared to data for other compounds in Table XXVIII, in which is included the data on P–N=P compounds reported by Fluck and Heckman.[(1969, 69)] From a study of Cl$_3$P=NP(O)F$_2$ and Cl$_3$P=NP(O)FCl, Glemser et al.[(1969, 129)] found that the relative sign of ^3J(P–F) was opposite to that of ^1J(P–F), i.e. positive.

C. Phosphorus–phosphorus couplings

1. *Direct $^{31}P–^{31}P$ couplings*

The interest in considering such couplings in relation to the nature of P–P bonds has been discussed by Cowley in an earlier review.[(1965, 26)] Since then a considerable amount of experimental data has been obtained and sign determination on typical compounds has revealed another interesting fact, namely that sign reversal occurs in this one-bond coupling depending upon the hybridisation of phosphorus and possibly upon the substituents. As mentioned earlier (Section III.A) this has prompted some theoreticians to consider this problem in detail.

(a) *Determination of $^{31}P–^{31}P$ couplings*. Symmetrical compounds, such as (CH$_3$)$_2$PP(CH$_3$)$_2$, will be considered first. Harris has shown that

TABLE XXVIII
P–F couplings through phosphorus and other heteroatoms

Compound	Coupling	Compound	Coupling
F_2P-PF_2	$^2J(F-P)$, $+65^i$	F_2P-PH_2	$^2J(F-P)$, 82^a
$P(PF_2)_3$	$^2J(F-P)$, 61^i	$F_2P(O)OP(O)O_2^{2\ominus}$	$^3J(FP-O-P)$, -8^b
$F_2P-S-PF_2$	$^3J(FP-S-P)$, $+32^i$	$F_2P(O)OP(O)F_2$	$^3J(FP-O-P)$, $-2\cdot 9^c$
$F_2P-(NCH_3)PF_2$	$^3J(FP-N-P)$, $+41$, -52^i	$F_2PSP(S)F_2$	$^3J(FP-S-P)$, $22\cdot 0^{d,e}$
$F_2P-(NC_6H_5)PF_2$	$^3J(FP-N-P)$, 40^i	$F_2P(S)-S-PF_2$	$^3J(FP-S-P)$, $15\cdot 0^{d,e}$
$CF_3P[P(CH_3)_2]_2$	$^3J(FCP-P)$, $6\cdot 4^f$	$(CF_3)_2P-P(CH_3)_2$	$^3J(FCP-P)$, $7\cdot 8^g$
$(CF_3)_2PSP(S)(CF_3)_2$	$^4J(FCPSP)$, 0^d	$(CF_3)_2P(S)SP(CF_3)_2$	$^4J(FCPSP)$, $5\cdot 0^d$
$SPCl_2N=PClF_2$	$^3J(PN=PF)$, 5^h	$SPClFN=PF_3$	$^3J(PN=PF)$, $2\cdot 8^h$
$SPClFN=PF_3$	$^3J(FP-N=P)$, 27^h	$SPF_2N=PF_3$	$^3J(PN=PF)$, 3^h
$SPF_2N=PF_3$	$^3J(FPN=P)$, 30^h	$SPClFN=PCl_2(C_6H_5)$	$^3J(FPN=P)$, $19\cdot 5^h$
$SPClFN=PCl(C_6H_5)_2$	$^3J(FPN=P)$, $13\cdot 7^h$		

[a] 1968, 31. [b] 1968, 112. [c] 1969, 62. [d] 1969, 57. [e] 1969, 59. [f] 1967, 83. [g] 1968, 55. [h] 1969, 69.
[i] 1967, 40, 49; 1968, 42; 1969, 31, 60 and 61.

analysis of the relevant $X_nAA'Z_n$-type spectra (where A and $A' = {}^{31}P$) gives $^1J(P–P)$, in most cases from the proton or fluorine spectra, e.g. Harris and Finer have obtained $^1J(P–P)$ of -179.7 for $(CH_3)_2PP(CH_3)_2$; additionally $^2J(P–CH_3) = +2.9$ and $^3J(PPCH_3) = +11.25$; [1967, 124, 125; 1968, 57] the negative sign of this one-bond P–P coupling should immediately be noticed. This has been recently confirmed by direct tickling experiments [1969, 31] and it would appear that all P^{III}–P^{III} coupling constants have the same sign, but positive values are found when P^V atoms are involved.

Data for unsymmetrical compounds are obtained by directly observing the phosphorus AB spectra but the signs of coupling are obtained by spin-tickling or by analysis of higher order spectra.

(b) *Factors effecting ^{31}P–^{31}P one-bond couplings.* Typical data shown in Table XXIX clearly shows the dependence of the sign of coupling on the valency of phosphorus and the nature of substituents on phosphorus (see 1968, 57 and 113; 1969, 47). The importance of molecular geometry has been considered in recent variable temperature investigations of Rudolph and Newmark [1969, 31] on F_2PPF_2 (in parallel to F_2PSPF_2 and $MeN(PF_2)_2$ which will be discussed later). Despite the proximity of phosphorus lone-pairs no significant change is observed from -230.3 Hz at $-101°$ to -228.6 Hz at $-1°$ for the pure compound and -226.5 Hz at the same temperature for a 5% solution in $CFCl_3$.

(c) *Theoretical approach to ^{31}P–^{31}P one bond couplings.* The previously mentioned points are sufficiently intriguing to arouse the interest of theoreticians. Harris and Finer [1968, 113] were the first to discuss this matter and, from the available signs found in P_2H_4, P_2Me_4, $Me_2PP(S)Me_2$ (negative) and $HP_2O_5^{3\ominus}$ (positive), rationalised the dependence of J upon the substituent electronegativity on the assumption that the sign was positive in P_2F_4.

To account for this behaviour these authors proposed an extension of the theory, given by Pople and Santry, which predicted that the coupling constant depends upon the energies of the s relative to those of the p, d, etc. electrons of the coupled nuclei and additionally upon the resonance integral between the outer-shell s electrons of the two atoms. It was considered that the first factor remained fairly constant for the P–P bonds and the observed variation was qualitatively explained by the dependence of the second factor upon the electronegativity and the bulkiness of the substituent.

As discussed earlier (III.A.2(a)), Cowley, White and Manatt [1967, 28] had previously undertaken a quantitative determination of P–P coupling

TABLE XXIX

One-bond phosphorus-phosphorus couplings[a]

Compound	Coupling	Reference
P^{III}–P^{III} compounds		
H_2PPH_2	−108	1961, 2
H_2PPF_2	211	1968, 31
F_2PPF_2	−227·4	1968, 31; 1969, 31
$(CH_3)_2PP(CH_3)_2$	−179·7	1967, 124 and 125
$(CH_3)_2PP(CF_3)_2$	−256, 252	1967, 40; 1968, 55
$(C_2H_5)_2PP(C_6H_{11})_2$	282	1965, 11
$(C_6H_5)_2PP(C_6H_{11})_2$	224	1965, 11
$(C_2H_5)_2PP(C_2H_5)Li$	396	1965, 27
$C_2H_5P[P(C_2H_5)K]_2$	306	1965, 27
$[C_6H_5PP(C_6H_5)Li]_2$	216	1965, 27
$P(PF_2)_3$	323	1968, 42
$[CH_3P-C(CF_3)_2]_2$	55	1964, 11
$(CH_3P)_3(CCF_3)_2$	220	1964, 11
P*–P* (cyclic structure with S, S, S, P)	157	1969, 130
P^{III}–P^{V} compounds		
$(CH_3)_2PP(S)(CH_3)_2$	−220	1964, 12
$(P^{IV})^{\oplus}$–P^{V} compounds		
$(C_4H_9)_3P^{\oplus}P(S)(S^{\ominus})C_6H_5$	118	1966, 77; 1967, 126
$(C_4H_9)_3P^{\oplus}P(S)(S^{\ominus})C_{10}H_9$	96	1966, 77; 1967, 126
P^{V}–P^{V} compounds		
$(CH_3)_2P(S)P(S)(CH_3)_2$	18·7	1965, 28
$(C_2H_5)_2P(S)P(S)(C_6H_{11})_2$	69	1965, 11 and 29
$(C_2H_5)_2P(O)P(O)(C_2H_5)_2$	583	1967, 127
$(C_2H_5)_2P(O)P(S)(C_2H_5)_2$	590	1967, 127
$HP(O)(O^{\ominus})PO_3^{2\ominus}$	+465·5	1968, 113

[a] Coupling in Hz for the starred atoms, sign given where directly established.

in the P_2 fragment on the basis of extended Hückel and CNDO-SCF molecular orbitals within the framework of the Pople-Santry approach. These authors obtained −128·95 and −30·54 respectively for the two modes of calculation; figures which were approximately in agreement with the data for P_2Me_4 and P_2H_4. Cowley and White later proposed [1969, 36, 37] an improvement by using semi-empirical LCAO–SCF but including overlap integrals for the whole molecule again within the

Pople-Santry formalism. In this manner the molecular geometry could be taken into more detailed account and the following parameters were obtained:

P_2H_4	gauche	-53.50	P_2F_4	gauche	$+219.75$
	trans	$+169.03$		trans	$+707.36$
	cis	-84.63		cis	$+212.16$

The predicted reversal of the sign for the different stereorelationships of P_2H_4 should be noted as should the positive sign obtained for the coupling in P_2F_4 as was previously assumed by Finer and Harris.

All this therefore appears gratifying—unfortunately heteronuclear tickling experiments definitely established a negative sign for $^1J(P-P)$ in P_2F_4.[1969, 31] This therefore re-opens the question of a sound theoretical approach to such couplings and this problem remains exceptionally attractive.

No application of the Jameson-Gutowsky approach (see III.A.2(b)) to the study of P–P coupling has been considered to date.

2. *Two-bond $^{31}P-^{31}P$ couplings*

Such couplings have been reported occurring across both carbon- and hetero- atoms; two-bond P–P–P couplings are also known as in $P(PF_2)_3$ [36 Hz][1968, 42] and may also be present in **128** [75 Hz].[1966, 78]

When considering symmetrical compounds these couplings are obtained by analysis of the relevant $X_nAA'X_n$ spectra as previously discussed for direct couplings (see III.C.1(a)), e.g. in aminophosphines[1966, 79] or in various metal complexes.[1968, 57; 1969, 131] In the latter case these couplings were found to be especially useful in assigning the geometry of the complex and in the discussion of the phosphorus-metal hybridisation;[1969, 60, 132 to 134] this aspect will not be further discussed here.

In asymmetrical compounds direct observation gives the modulus of the coupling—sign determination being obtained from heteronuclear tickling experiments or analysis of higher-order spectra. As no general theory is applicable to the existing data each of the systems will be discussed separately.

(a) $J(P-C-P)$ have only been reported in **129** with a value of 47 ± 7 Hz[1967, 123] but as a related system, however, we may also quote **130** with a value of 37.2 Hz.[1967, 128]

(b) $J(P-O-P)$ are much better known. The relative sign of the coupling in the dianion **131** has been determined by McFarlane[1968, 37] and found to be negative (-10.7 Hz) but, as a number of values around zero have been found, a reversal of sign cannot be dismissed. As no further signs are known only the moduli of the couplings will be discussed. Following the early study of the trianion **132**$^{(1957, 2)}$ the tripolyphosphate **133** with $^2J(POP)$ of 21·5 Hz$^{(1965, 28)}$ and the ATP homolog (**134**), for which $J(P_\alpha-O-P_\beta) = 20·6$ and $J(P_\beta-O-P_\gamma) = 21·3$ (see 1967, 1, pp. 12 to 13), have been studied. This small difference in the value of $^2J(P-P)$ may arise from differences in the hybridisation of the phosphorus atoms (two bearing one $O^\ominus$ group and the third having two $O^\ominus$ substituents). A similar situation arises in **135** for which $^2J(P_\alpha-O-P_\beta)$ [12·1 Hz] is lower than $^2J(P_\beta-O-P_\gamma)$ [14·5 Hz].$^{(1966, 79)}$

Molecular conformation may additionally be of considerable importance and the effect of pH is another factor as is the nature of the counterions as was found in an earlier study of tripolyphosphate.$^{(1965, 30)}$ A clear relationship of the mean $^2J(POP)$ in ATP with pH has recently been found (Fig. 9) and is similar to that observed in ADP. Other examples have also been considered, the values of which are shown in Table XXX. No coupling has been found in $(C_6H_5)_3\overset{\oplus}{P}-O-\overset{\ominus}{P}Cl_5$.$^{(1969, 135)}$

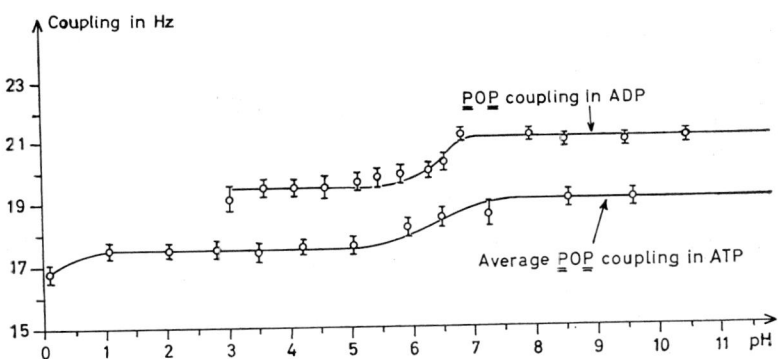

FIG. 9. pH Dependence of P–O–P couplings in ADP and ATP [From Ellenberger et al.$^{(1970, 2)}$]

The sensitivity of $^2J(POP)$ to changes in the hybridisation and electronegativity has been demonstrated for ATP. This effect is also noticeable when comparing ligands and related adducts, for example:$^{(1968, 34)}$

$$F_2(O)P-O-P(O)F_2 \qquad ^2J(POP) < 10 \text{ Hz}$$
$$F_2(O)P-O-P(O)F_2, BF_3 \qquad ^2J(POP) = 26·7 \text{ Hz}$$

(c) $J(P-S-P)$—although the number of known examples are fewer the general magnitude of the modulus of coupling appears to be greater. The sign of coupling (positive) is only known in F_2PSPF_2; this compound exhibits an astonishing temperature behaviour as $^2J(PSP)$ increases from $+274$ at $39°$ to $+392$ at $-120°$.$^{(1969, 31, 136)}$ This is a

TABLE XXX
Additional examples of ^2J(POP) couplings

	J (Hz)	Reference

HO(O)P–O–P(O)OH
    | |
    H   OH
	17·7	1967, 29

H(O)P–O–P(O)H
    | |
    O$^\ominus$ O$^\ominus$
	−10·7	1968, 37

F$_2$(O)P–O–P(O)F$_2$	4 0 <10 2·5	1966, 28 1969, 62 1969, 31 1968, 112

$^\ominus$O(O)P–O–P(O)F$_2$
    |
    O$^\ominus$		

(cyclic P–O–P with O and S)	ca. 28	1967, 91

F$_2$P–O–PF$_2$
  || ||
  S   S
	7·1	1969, 62

clear indication of a rotational equilibrium as sulphur affords a rather loose link between the two phosphorus atoms with $3p_\pi$–$3d_\pi$ bonding and additionally this conformational dependence may depend upon the overlap of phosphorus lone-pairs.

Other interesting examples are shown in **136**.[1969, 130] The increase of the mean value from P–O–P to P–S–P and P–Se–P coupling reflects the decreased electronegativity of the bridging atom; this reduces the polarity of the intervening bonds thus facilitating the "coupling transmission".[1966, 1, p. 217]

(d) J(P–N–P) couplings are known in many phosphonitrilic compounds and the relative sign has been found to be positive.[1967, 57; 1969, 68] Some typical values are shown in Table XXXI.

The available data and sensitivity to molecular environment prompted Finer and Harris to investigate these couplings further;[1967, 130; 1968, 57] they succeeded in accounting for the values, at least within a few percent, in compounds of the general type **137** by the following relationship:

$$J = (\lambda_A + \lambda_B)(\lambda_C + \lambda_D)$$

TABLE XXXI
Some typical values of $^2J(\text{P-N-P})$

Compound	J, Hz	Reference
$N_3P_3Cl_4(NH_2)[N{=}P(C_6H_5)_3]$	28	1968, 114
$N_3P_3Cl_{6-n}(NCH_3)_n$	34 to 48	1965, 24
$N_3P_3(OCH_2CF_3)_4(NHC_6H_5)_2$	72	1966, 80
$N_3P_3F_4(C_6H_5)_2$	cis 58	1968, 67
	trans 65	
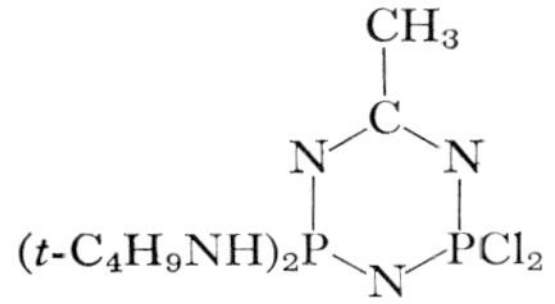 $(t\text{-}C_4H_9NH)_2P\underset{N}{\overset{N-C(CH_3)-N}{\diagup\diagdown}}PCl_2$	27	1969, 106

with typical λ contributions such as:

$$\lambda(F) = 7\cdot0 \qquad \lambda[N(CH_3)_2] = 3\cdot2$$
$$\lambda(OCH_3) = 5\cdot3 \qquad \lambda(SC_2H_5) = 0\cdot3$$
$$\lambda(Cl) = 3\cdot6$$

P–N–P couplings are, however, also known in systems other than phosphonitriles, for example:

$\begin{array}{c}CH_3S(O)P\text{—}N\text{—}CH_3\\ |\quad\quad|\\ CH_3N\text{—}P(O)SCH_3\end{array}$ $\qquad ^2J(\text{PNP}) = 31\cdot5$ Hz (1966, 53)

$P^{\oplus}(N{=}PCl_3)_4$ $\qquad ^2J(\text{PNP}) = 29\cdot9$ Hz (1969, 137)
$H[(CH_3)_2N]_2P{=}N\text{—}P(S)(OC_6H_5)_2$ $\qquad ^2J(\text{PNP}) = 51$ Hz (1969, 15)

and in the following series in which the electronegativity effects appear to be especially pronounced: (1969, 69, 129)

SPClFN = $PCl_2(C_6H_5)$	4·4 Hz		$SPCl_2N = PF_3$	75 Hz
SPClFN = $PCl(C_6H_5)_2$	12 Hz		SPClFN = PF_3	105 Hz
$SPCl_2$ = $NPClF_2$	60 Hz		$SPF_2N = PF_3$	137 Hz
$OPF_2N = PCl_3$	70 Hz		OPFClN = PCl_3	46·5 Hz

The most extensively investigated case is that of $F_2P\text{–}N(CH_3)\text{–}PF_2$ which, as for F_2PSPF_2 (see III.B.4), has been studied over a range of temperatures. In the latter compound the P(X)P coupling varied very considerably while for the case of $F_2P\text{–}N(CH_3)\text{–}PF_2$ the coupling is much more constant varying only from $+432$ to $+442$ Hz for temperature $+30$ to $-70°C$. (1969, 31) This was attributed to the rigidity of the

$2p_\pi$–$3p_\pi$ bonding which means there is restricted rotation around the P–N bonds and hence the absence of rotamers. For the phenyl homologue, $F_2P-N(C_6H_5)-PF_2$, the value of $^2J(PNP)$ is substantially lower, 371 Hz. (1969, 61) It is worth noting that the P–N–P couplings are positive for this P^{III}–P^{III} system as is the case for the P^V–P^V system of phosphonitrilics.

3. *Higher ^{31}P–^{31}P couplings*

These have only been observed for the P–O–C–P system in the following series of compounds: (1969, 3 and 138)

$P(OCH_2)_3P$	−37·2; −38·1
$SP(OCH_2)_3P$	+48·4
$OP(OCH_2)_3PO$	+139·2
$SP(OCH_2)_3PO$	+151·3

The observed trend may be interpreted on the basis of changes of s character but again sign reversal is present.

D. Carbon-phosphorus couplings

Such couplings may be determined from carbon-13 spectra using heteronuclear "tickling" (1967, 66; 1969, 2, 21 and 47) or by complete analysis of higher-order proton spectra. (1967, 125) Carbon-13 satellites in phosphorus resonance are used if sensitivity is sufficiently good or after signal enhancement, for instance using the nuclear Overhauser effect. (1967, 131) Of these methods "tickling" is the more useful as the relative signs are also obtained.

1. *Direct ^{13}C–^{31}P couplings*

The most significant results for ^{31}P–^{13}C couplings are presented in Table XXXII. It is evident that these couplings are very sensitive to: (i) the hybridisation of phosphorus (P^{III}, $P^{IV\oplus}$ or P^V), (ii) hybridisation of carbon (the largest $|J|$ being found in $(C_2H_5O)_2(O)P-C\equiv CH_3$ in which $^1J(^{13}C-^{31}P) = 304$ Hz), and (iii) the substituents attached to both atoms.

In particular, it is to be noted that in the case of P^{III} compounds a reversal of sign occurs for the direct coupling; a similar behaviour has also been found for direct phosphorus-phosphorus coupling (see III.C.1). Another feature which is also of significance is that, whenever it is possible to compare the corresponding $^{13}C-^{31}P$ and $^1H-C-^{31}P$ couplings, there is a fairly good correlation and that there is correspondence of sign reversal—from positive to negative for P–C when it is negative to positive for P–CH; (1968, 20; 1969, 47) this point is clearly indicated in

TABLE XXXII

Direct phosphorus-carbon couplings in typical organo-phosphorus compounds[a]

Compound	$J(^{31}P-^{13}C)$	Reference
Tricoordinated phosphorus		
$(CH_3)_3P$	−13·5; −14·0	1968, 3; 1969, 2
$(CH_3)_2PH$	11·6	1968, 22
$(\overset{*}{C}H_3)_2PC_6H_5$	−14	1967, 66
$(CH_3)_2PSeCH_3$	−25	1969, 4
CH_3PH_2	9·3	1968, 22
CH_3PCl_2	45	1968, 20
$(CH_3\overset{*}{C}H_2)_3P$	0 ± 4	1968, 3
$(C_3H_7\overset{*}{C}H_2)_3P$	−10·9	1969, 21
$(C_6H_5)_3P$	12·4	1969, 21
(2-thienyl)$_3$P	14·2	1969, 124
$\overset{*}{P}(CH_2O)_3P$	−24	1969, 3
Phosphonium compounds		
$(CH_3)_4P^\oplus$	+55·5; +56·5	1968, 3; 1969, 2
$(\overset{*}{C}H_3)_2PH(C_6H_5)$	+56	1967, 66
$(CH_3CH_2)_4P^\oplus$	+48·5	1968, 3
$(C_3H_7CH_2)_4P^\oplus$	+47·6	1969, 21
$(C_6H_5)_4P^\oplus$	88·4	1969, 21
Tetracoordinated phosphorus		
$(CH_3)_3PS$	+56·1	1968, 3
$(CH_3)_3PSe$	+48·5	1968, 3
$(CH_3)_2P(S)SeCH_3$	+50	1969, 4
$CH_3P(O)Cl_2$	104	1968, 20
$CH_3P(O)F_2$	147	1968, 20
$CH_3P(S)Cl_2$	81	1968, 20
$CH_3C\equiv\overset{*}{C}P(O)(OC_2H_5)_2$	304	1968, 20
$CH_3C(O)\overset{*}{C}H_2P(O)(OC_2H_5)_2$	127	1968, 20

[a] Coupling in Hz. Where there is possible ambiguity the carbon or phosphorus nucleus involved in the coupling is indicated by an asterisk.

Fig. 10. This feature undoubtedly reflects that there are similar factors effecting these couplings. Another observation which warrants mention is that geminal HPH couplings which have recently been investigated also demonstrate the same sign reversal, without any clear correlation with the H–P–H bond angle, but are very dependent upon the hybridisation of the phosphorus. [1969, 41]

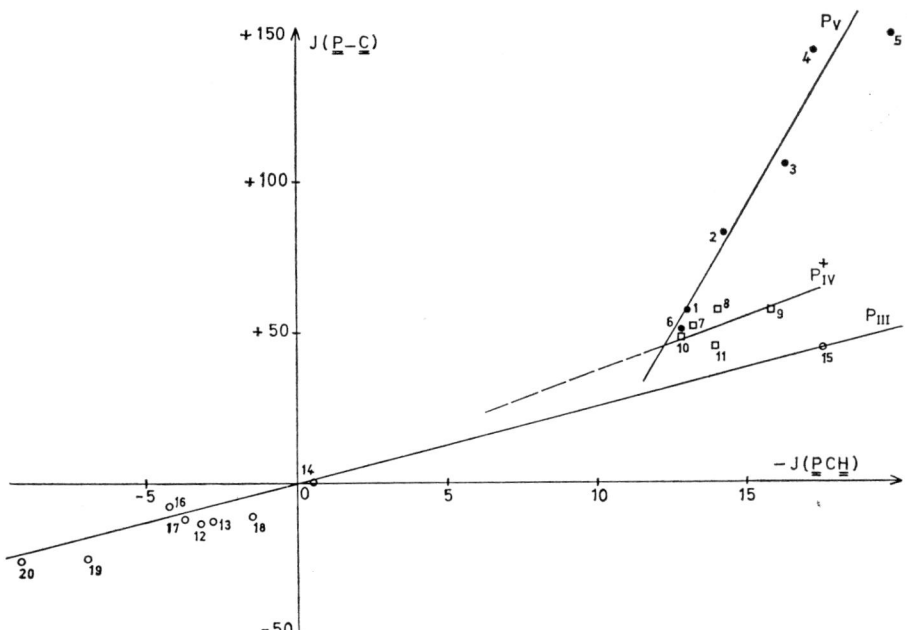

FIG. 10. Correlation of P–C and P–C–H couplings in typical derivatives of P^{III}, $P^{\oplus IV}$ and P^V derivatives.

If the general trend in all these couplings may be qualitatively explained on the basis of the Walsh rule (i.e., the s character of the phosphorus atom being concentrated in bonds with the most electropositive substituents) the explanation of the reversal of sign is more complex. As proposed by Finer and Harris,[1968, 113] for the ^{31}P–^{31}P direct couplings (see III.C.1(c)), the reversal may be reconciliated with the Pople-Santry theory, the dominating term in the coupling depending on the s level energies (approximately constant) and on the resonance integral between these s electrons which in turn depends upon the molecular hybridisation and geometry. Weigert and Roberts recently proposed[1969, 21] a similar explanation by noting that a negative one-bond coupling may occur when the energy difference between the valence s and p electrons become large, this being especially important for p^3 hybridised while $P^{IV\oplus}$ and P^{IV} are more nearly sp^3.

The most systematic approaches are these presented by Cowley and White and by Jameson (see above III.A.2). It was seen that predictions presented by the former authors (Table VIII) are in approximate agreement with observed values but far more data are required to establish the validity of the correlation. On the other hand, the Jameson model only qualitatively accounts for the observed trends.

2. *Two-bond and higher* $^{13}C-^{31}P$ *couplings*

Two-bond P–C couplings have been reported through both carbon and heteroatoms and in one case a P–P–C coupling has been identified. The relevant values are shown in Table XXXIII. Wherever comparisons are possible it is seen that reversal of sign again occurs; all couplings involving $P^{IV\oplus}$ and P^V are negative, as expected, while the P^{III} compounds without exception involve positive couplings either through carbon and phosphorus or through heteroatoms.

No theoretical approach has been proposed to explain these couplings. It is, however, interesting to note that the same conclusions do not apply to higher couplings, at least for the only well established series:

$$P(OC\overset{*}{H_2}CH_3)_3 \quad +4 \cdot 9 \quad (1968, 3 \text{ and } 22)$$
$$OP(OC\overset{*}{H_2}CH_3)_3 \quad +6 \cdot 8 \quad (1968, 3)$$
$$SP(OC\overset{*}{H_2}CH_3)_3 \quad +6 \cdot 4 \quad (1968, 3)$$

No clear relation between the related P⋯C and P⋯H couplings (the signs of which have been estimated to be negative—see III.B.3(a)) in these compounds:

	$^2J(P{-}O{-}C)$	$^3J(POCH)$	$^3J(POCC)$	$^4J(POCCH)$
$P(OCH_2CH_3)_3$	$+11 \cdot 3$	$+7 \cdot 9$	$+4 \cdot 9$	$-0 \cdot 5_5$
$OP(OCH_2CH_3)_3$	$-5 \cdot 8$	$+8 \cdot 4$	$+6 \cdot 8$	$+0 \cdot 8_4$
$SP(OCH_2CH_3)_3$	$-5 \cdot 2$	$+9 \cdot 85$	$+6 \cdot 4$	$+0 \cdot 7_3$

Other higher couplings, but of unknown signs, have been observed by Weigert and Roberts,[1969, 21] e.g.:

$^3J(PCCC)$	in $P(C_4H_9)_3$	12·5 Hz	$^4J(PCCCC)$	in $P(C_6H_5)_3$	0 Hz
	$P^\oplus(C_4H_9)_4$	15·4 Hz		$P^\oplus(C_6H_5)_4$	2·9 Hz
	$P(C_6H_5)_3$	6·7 Hz			
	$P^\oplus(C_6H_5)_4$	12·8 Hz			

E. Phosphorus coupling to other nuclei

Coupling of phosphorus to other nuclei have been reported in a variety of molecules but these examples have usually remained as curiosities rather than of practical use.

TABLE XXXIII
Two-bond phosphorus-carbon couplings in typical organo-phosphorus compounds[a]

Compound	$^2J(^{31}P-^{13}C)$	Reference
Coupling through carbon		
($\overset{*}{C}H_3CH_2)_3P$	+14·1	1968, 3
$(C_2H_5\overset{*}{C}H_2CH_2)_3P$	+11·7	1969, 21
$(\overset{*}{C}H_3)_3CP(C_6H_5)_2$	+18	1969, 2
$P(C_6H_5)_3$	19·6	1969, 21
$(\overset{*}{C}H_3CH_2)_4P^{\oplus}$	−4·3	1968, 3
$(C_2H_5\overset{*}{C}H_2CH_2)_4P^{\oplus}$	−4·3	1969, 21
$(C_6H_5)_4P^{\oplus}$	10·9	1969, 21
$\overset{*}{C}H_3C{\equiv}CP(O)(OC_2H_5)_2$	54	1968, 20
Coupling through phosphorus		
$(\overset{*}{C}H_3)_2P-\overset{*}{P}(CH_3)_2$	+7 (estimated)	1968, 57
Coupling through heteroatoms		
Oxygen		
$(CH_3O)_3P$	+10	1968, 3 and 22
$(\overset{*}{C}H_3CH_2O)_3P$	+11·3	1968, 3 and 22
$P(CH_2O)_3\overset{*}{P}$	⩽10	1969, 3
$(CH_3O)_3PO$	−5·8	1968, 3
$(CH_3O)_2P(O)H$	−6·0	1967, 132; 1968, 3
$(\overset{*}{C}H_3CH_2O)_3PO$	−5·8	1968, 3
$(CH_3O)_3PS$	−5·6	1968, 3
$(\overset{*}{C}H_3CH_2O)_3PS$	−5·2	1968, 3
Sulphur		
$(CH_3S)_3PS$	<10	1969, 20
Selenium		
$\overset{*}{C}H_3SeP(CH_3)_2$	+19·5	1969, 4
$\overset{*}{C}H_3SeP(S)(CH_3)_2$	−5	1969, 4
Nitrogen		
$[(CH_3)_2N]_3P$	17	1969, 20
$[(CH_3)_2N]_3PO$	<10	1969, 20
$[(CH_3)_2N]_3PS$	<10	1969, 20
$[(CH_2)_4N]_4P$	<10	1969, 20
Other atoms		
$(CH_3)_3SiP(C_6H_5)_2$	−13	1969, 2
$(CH_3)_3SnP(C_6H_5)_2$	+7·5	1969, 2

[a] Coupling in Hz. Where there is possible ambiguity the carbon or phosphorus nucleus involved in the coupling is indicated by an asterisk.

(a) *With N-14.* A value of 60 Hz has been reported for $^2J(^{14}\text{N}-^{31}\text{P})$ in F_2PCNS [1967, 53] and the value of $^1J(^{14}\text{N}-^{31}\text{P})$ in $(\text{EtO})_2\text{P(O)NHEt}$ has been reported to be "great" from a consideration of ^{14}N spectra but no further details are available.

(b) *With O-17.* Some typical values have been obtained by Christ and Diehl [1963, 1] but the data cannot be interpreted too deeply as the various oxygen sites were not always distinguished in this study. Some values are shown below:

H_3PO_2	115 Hz	$\text{P(OCH}_3)_3$	154 Hz
H_3PO_3	106 Hz	$\text{P(OCH}_2\text{CH}_3)_3$	150 Hz
$\text{OP(OCH}_3)_3$	{ 90 Hz 165 Hz	OPCl_3	203 Hz

In P–O–C and P–O–H groups the values of the P–O couplings appear in the ranges 150 to 160 and 100 to 120 Hz respectively. For phosphoryl compounds the value is more variable from 90 to 200 Hz. Clearly more data are required for a deeper understanding of the underlying effects.

(c) *With Si-29.* Data have been obtained for $(\text{C}_6\text{H}_5)_2\text{PSi(CH}_3)_3$ [21·5 Hz] [1969, 4] and for $\text{P(SiH}_3)_3$ [212 Hz?]. [1968, 96]

(d) *With Cl-35.* The coupling has been estimated by Winter [1959, 1] from relaxation studies in PCl_3 and in agreement with the recent estimation of Strange. [1969, 139]

(e) *With Se-77.* Reported in $(\text{CH}_3)_2\text{PSeCH}_3$ [−205 Hz] and $(\text{CH}_3)_2\text{P(S)SeCH}_3$ [−341 Hz]. [1969, 4]

(f) *With Br-79 and Br-81.* Couplings have been obtained for PBr_3 by analysis of relaxation mechanisms at variable temperature [1968, 115] and were 350 and 380 Hz for Br-79 and Br-81 respectively. These results do not agree with an earlier estimate by Winter [1959, 1] of 690 Hz.

(g) *With Sn-119.* Dreskamp et al. [1969, 2] have found a value of 598 ± 4 Hz for $^1J(\text{P-Sn})$ in $(\text{C}_6\text{H}_5)_2\text{PSn(CH}_3)_3$ using the INDOR method. The P–Sn couplings have also been observed in the P–O–Sn system [1969, 140] such as:

$(\text{CH}_3)_2\text{P(O)OSn(CH}_3)_3$	102 Hz
$\text{CH}_3\text{P(O)[OSn(C}_4\text{H}_9)_3]_2$	83 and 110 Hz
$\text{-[OP(O)(C}_8\text{H}_{17})[\text{OSn(C}_4\text{H}_9)_2]\text{]}_n$	144 and 172 Hz

For both of the last two compounds two different couplings were clearly identified corresponding to interaction between different sites in the first and to slow exchange of POSn bonds in the second as depicted by **138**.

(h) *With other metallic and non-metallic atoms.* Coupling of phosphorus to metal atoms has been observed in complexes, e.g. P–^{63}Cu, P–^{65}Cu, P–^{183}W and P–^{199}Hg, [1965, 31; 1967, 79; 1968, 116] in other systems to be discussed later (see IV.E.3), and boron and aluminium in appropriate adducts. [1966, 81; 1967, 23, 32, 133 and 134; 1968, 34; 1969, 26 and 141]

F. Couplings through phosphorus

The influence of a phosphorus atom on coupling constants is often apparent when the phosphorus atom is situated between the two coupling nuclei. Some typical cases where the effect is found will be discussed here.

1. *Proton-proton coupling through phosphorus*

(a) The case of *HPH* couplings has been recently well documented by Manatt *et al.*[1969, 41] This coupling has been mentioned earlier (III.A.2(b)) and especially for its sensitivity to the hybridisation of the phosphorus atom; the couplings being anomalously negative for P^{III} and positive for P$^{IV\oplus}$ and P^{V} compounds.

(b) Three bond *HPCH*† couplings do not appear to undergo change of sign and have been found to be positive, as expected, in all the cases in which sign determination has been made [1968, 28; 1969, 34]. Fortunately, however, the values for tri- and tetra- coordinated phosphorus are different and have the following values: +6·5 to 8 Hz for P^{III}, +5·5 Hz for P$^{IV\oplus}$ and +3 to 3·5 Hz for P^{V}. [1966, 14; 1967, 135; 1968, 28 and 35; 1969, 34 and 73] This behaviour‡ is contrary to the hybridisation changes predicted for the P–H and P–C bonds and cannot fit into the Jameson formalism (see III.A.2(b)). Additionally, some stereospecificity is observed in these couplings—the methylene protons of HP(CH$_2$CH$_3$)C$_6$H$_5$ are not equivalent and are coupled differently to the P–H proton [+6·56 and +7·10 Hz].[1968, 28] Likewise, it is possible to identify the conformation in cyclic compounds, e.g. in **139** the value of ^3J(HPCH) is 12 Hz for the *cis* proton but 2·5 Hz for the *trans*.[1969, 142] In a few instances couplings in pentacoordinated compounds have been reported, e.g. in CH$_3$PF$_3$H [1·2 Hz] and (CH$_3$)$_2$PF$_2$H [2·7 Hz].[1968, 42]

(c) Four bond couplings have only been reported for ^4J(HCPCH) in

† Some *HPSiH* couplings have similarly been described.[1969, 73]

‡ The same as for the relevant *CPCH* couplings, for which the value is +4·8 Hz in P(CH$_3$)$_3$ and +2·4 Hz in P$^{\oplus}$(CH$_3$)$_4$.[1969, 2]

P(CH$_3$)$_3$ [+0·15 Hz] and in P$^{\oplus}$(CH$_3$)$_4$ [+0·3 Hz]$^{(1969, 2)}$ and for ^4J(HCPSiH) in CH$_3$P(SiH$_3$)$_2$ and CH$_3$PHSiH$_3$ [0·4 and 0·3 Hz respectively];$^{(1969, 73)}$ as for the case of *HCPH* couplings the signs and magnitude were not as expected. In both cases "through-space" mechanism cannot be excluded which would involve some contribution from the lone-pair on P^{III}.

A longer-range coupling has been detected in **140**, between the ringed hydrogen atoms, and the stereospecificity used to assign the conformation.$^{(1968, 117)}$

2. *Proton-fluorine coupling through phosphorus*

(a) Two-bond *HPF* couplings. These were first recognised in simple tri-and tetra-coordinated compounds—HPF$_2$ +41·7 Hz, HP(O)F$_2$ 116 Hz and HP(S)F$_2$ 98·9 Hz.$^{(1966, 32; 1967, 31 \text{ and } 33; 1968, 118)}$ The signs are not known in the latter two compounds but, for HPF$_2$, the sign is coherent with that (negative) of the relevant PF coupling. Numerous data are available for penta-coordinated compounds; two examples in simple compounds are:

HPF$_4$	(*ca.* −20°C)	90, 92 Hz	1966, 24; 1967, 36
H$_2$PF$_3$	(*ca.* 0°C)	80 Hz	

In both cases it is not possible to distinguish between axial and equatorial fluorine atoms; this is only possible in those cases in which the phosphorus has heavier substituent groups$^{(1968, 38, 39 \text{ and } 42)}$ as for:

$$\text{HPF}_3(\text{CH}_3) \quad J(\text{H--F}_{\text{ax}}) = 117 \text{ Hz}$$
$$J(\text{H--F}_{\text{eq}}) = 29\cdot9 \text{ Hz}$$

When three substituents, are present then the two fluorine atoms, are both axial as in HPF$_2$(CH$_3$)$_2$ with $J(\text{H--F}_{\text{ax}}) = 98\cdot4$ Hz.$^{(1968, 42)}$ Despite the greater polarity of the P–F$_{\text{ax}}$ bonds it clearly appears that H–F$_{\text{ax}}$ couplings are of greater magnitude than those of the H–F$_{\text{eq}}$. This may arise from the fact of the closer proximity of the atoms in the H–F$_{\text{ax}}$ case; this has been interpreted fairly well on the basis of the Jameson-Gutowsky formalism (III.A.2(b)) by estimating the relevant exchange integral.

Lastly, for P$^{\mathrm{VI}\ominus}$ compounds it has been observed$^{(1968, 41)}$ that the *HPF* couplings for the two non-equivalent equatorial positions were quite distinct as shown in **141**. The *trans* H–F coupling in [PF$_5$H]$^{\ominus}$ is in fact a function of the H–P–F angle.$^{(1969, 143)}$

(b) The best documented three-bond coupling is that involving the FP–CH group.† Some typical values taken from the literature$^{(1966, 36; 1967, 60 \text{ and } 110; 1968, 38, 39 \text{ and } 42; 1969, 20, 62 \text{ and } 74)}$ are shown below:

† HP–CF coupling has only been reported for HP(CF$_3$)$_2$ and was 9·9 Hz.$^{(1969, 43)}$

(CH$_3$)$_2$PF	18·5 Hz	CH$_3$PF$_4$	7·2 Hz (average)
CH$_3$PF$_2$	20·0 Hz	CH$_3$PF$_3$H	F$_{ax}$ 12·6, F$_{eq}$ 4·2 Hz
CH$_3$P(O)F$_2$	6·0 to 6·2 Hz	CH$_3$PF$_2$(OC$_6$H$_5$)$_2$	F$_{ax}$ 12·0 Hz
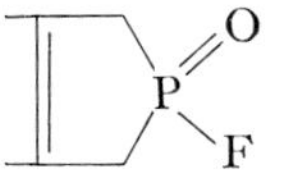 (cyclic P(O)F)	13 Hz	(CH$_3$)$_2$PF$_2$H	F$_{ax}$ 12·1 Hz
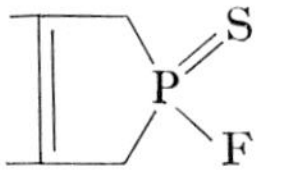 (cyclic P(S)F)	12 Hz	(CH$_3$)$_2$PF$_3$	F$_{ax}$ 12·9, F$_{eq}$ 3·3 Hz
		(CH$_3$)$_3$PF$_2$	F$_{ax}$ 12·0 Hz

A number of interacting factors precludes any generalisations to be made but it should be noted that the clear distinction of the proton couplings to axial and equatorial fluorine atoms in pentavalent compounds is again obtained. The same order is found as for ^2J(H–P–F) namely $|^3J(\text{H–F}_{ax})| \gg |^3J(\text{H–F}_{eq})|$.

Some unusual three-bond couplings have been found in the following compounds: F_2PNHCH$_3$ [10·2 Hz], [1968, 119] FClP(S)NHSi(CH$_3$)$_3$ [2 Hz] [1969, 69] and in F_2P(S)SH [1·2 Hz]. [1969, 99]

(c) The majority of interesting four-bond couplings are ^4J(H–F) found in FP–N–CH groups [1965, 8; 1966, 30; 1967, 59; 1968, 43 and 46; 1969, 31, 56, 61 and 103] for example:

(CH$_3$)$_2$NPF$_2$	3·6 Hz	(CH$_3$)$_2$NP(O)F$_2$	8·2 or 0·0† Hz
(CH$_3$)$_2$NP(CF=CF$_2$)$_2$	0·6 Hz	(CH$_3$)$_2$NP(S)F$_2$	2·1 Hz
CH$_3$N(PF$_2$)$_2$	1·5 Hz	[(CH$_3$)$_2$N]$_2$P(S)F	1·7 Hz
(CH$_3$)$_2$N(CH$_3$)NPF$_2$	3·1 Hz	(CH$_3$)$_2$NPF$_4$	2·1 Hz
CH$_3$O(CH$_3$)NPF$_2$	3·6 Hz	(CH$_3$)$_2$NPF$_2$Cl$_2$	2·7 Hz

The ^4J(H–F) values are usually lower when atoms other than nitrogen separate the two nuclei [1965, 8; 1966, 30; 1968, 45; 1969, 62 and 84] as for example in:

CH$_3$P(O)(CF$_3$)$_2$	0·78 Hz	(CH$_3$)$_3$CP(O)F$_2$	ca. 1 Hz
CH$_3$OP(S)F$_2$	0·7 Hz	CH$_3$SP(S)F$_2$	1·3 Hz
CH$_3$SP(C$_6$H$_5$)F$_3$	2·0 Hz	CH$_3$SPF$_4$ (−100°C)	F$_{ax}$ 3·0 Hz

Examples of longer-range H–F coupling have been found as follows:

^5J(H–F) in (CH$_3$)$_2$NP(CF$_3$)$_2$ [0·5 to 0·7 Hz], [1968, 47 and 55] in (CH$_3$)$_2$NP(CF=CF$_2$)$_2$ [0·6 Hz] and in [(CH$_3$)$_2$N]$_2$PCF=CF$_2$ [0·8 Hz], [1969, 55] in CH$_3$OP(CF$_3$)$_2$ [0·46 Hz] and CH$_3$OP(O)(CF$_3$)$_2$ [0·37 Hz] [1969, 84] and in (CH$_3$O)$_2$P(O)CF=CClCF$_3$ [0·45 Hz in cis and 0·40 in trans isomer]. [1968, 98]

^6J(H–F) in (CH$_3$)$_2$NP(C$_6$F$_5$)$_2$ and [(CH$_3$)$_2$N]$_2$PC$_6$F$_5$ [0·7 Hz]. [1966, 19]

† Depends upon isomer.

3. Fluorine–fluorine coupling through phosphorus

(a) The only observed two-bond couplings are found in penta-or hexa-coordinated fluorophosphoranes. The axial and equatorial positions, which differ appreciably when exchange is slowed down, are readily identified. For instance the following values have been obtained at low temperature: (1966, 31 and 82; 1968, 38, 39 and 42; 1969, 62 and 114)

$F_3PH(CH_3)$	19 to 20 Hz	$F_3P(CH_3)_2$	26·2 Hz
$F_3P(CF{=}CF_2)_2$	52 Hz	$F_3P(C_6H_5)SC_6H_5$	60·0 Hz
$F_3PCl[N(CF_3)_2]$	95 Hz		

The same appears to be true for $P^{VI\ominus}$ hexa-coordinated compounds such as $PF_5S^{\ominus}$ for which $^2J(F_{ax}-F_{eq})$ is 48 Hz (1968, 46) and in $(CF_3)_2PF_3H^{\ominus}$ in which all the fluorine atoms are equatorial but can be differentiated as shown in **142** (1968, 41) and finally **143**. (1966, 33)

In some instances the axial positions are not equivalent and it is possible to distinguish three different sites at $-100°$ in CH_3SPF_4 in which the couplings are $^2J(F_{ax}-F'_{ax}) = 19$, $^2J(F_{ax}-F_{eq}) = 104$ and $^2J(F'_{ax}-F_{eq}) = 91$ Hz, or two different sites at $-80°$ as in $CH_3SPF_2(C_6H_5)_2$ in which $^2J(F_{ax}-F'_{ax}) = 28$ Hz. (1969, 15)

(b) Higher F–F couplings through phosphorus have only been found in the following cases.

(i) Three-bond coupling. In P_2F_4 the fluorine atoms are not equivalent and it is found that the FPPF and FPPF′ couplings have opposite signs. (1969, 31) Other three-bond couplings for which the sign is known are in the compounds: $FP(CF_3)_2$ [$-3·5$ Hz], (1967, 55) $F_2P(O)CF_3$ [11 Hz], (1968, 44) $FPCl_3(CF_3)$ [8·4 Hz] (1967, 62) and $F_2PCF{=}CF_2$. (1967, 55)

(ii) Four-bond coupling. The values have been compared (1966, 31) in the following series of compounds: $F_2P(O)N(CF_3)_2$ [8·6 Hz], $F_2Cl_2PN(CF_3)_2$ [14·1 Hz] and $F_3ClPN(CF_3)_2$ [10·8 Hz]—this last value is an average of the axial and equatorial couplings and suggests that the value of $J(F-F_{ax})$ is longer than $J(F-F_{eq})$, as was previously observed for couplings involving protons (III.F.2(a)). The sign of $^4J(F-F)$ in the P^{III} compound, $(CF_3)_2PSCF_3$ has also been established [$+1·1$ Hz]. (1967, 55)

In fluorovinyl compounds a similar stereospecificity is observed for F–F coupling as for P–F coupling (III.B.7(a)) and in $F_2PCF{=}CF_2$ the cis and trans $^4J(F-F)$ are 2·8 and 13·2 Hz respectively. (1969, 55) Finally other example of four-bond couplings are found in $F_2P(O)OP(O)F_2$ and $F_2P(S)OP(S)F_2$ [2·4 and 3·7 Hz respectively] (1969, 62) and in $F_2P(O)N{=}SF_2$ [4·5 Hz]. (1967, 136)

(iii) Five-bond couplings have only been reported in $CF_3P[N(CF_3)_2]_2$ and $(CF_3)_2PN(CF_3)_2$ [5·5 and 4·7 Hz respectively]. (1968, 69)

IV. APPLICATIONS OF NMR IN PHOSPHORUS CHEMISTRY

A. Applications to structural problems

NMR is so widely used for structural assignments that it would be inappropriate to give a general review of these applications and the discussion will be confined to considering some aspects associated with cyclic compounds which are by far the most interesting. Apart from the cases of rings composed entirely of phosphorus atoms and organometallic structures such as **144** [1967, 76] the following systems will be considered.

1. *Carbon-phosphorus heterocyclics*

These have recently been reviewed, including NMR data, up to 1967 by Berlin and Hellwege [1969, 144] and some new examples, not included in the above review, are:

Phosphabenzene P^{III} **(145)** and P^V **(146)** derivatives, [1968, 75 and 101; 1969, 145] or the higher homologs such as **147** [1968, 120] which may be compared with the non-conjugated counterparts such as **148** and **149**. [1968, 75 and 102]

Phosphazolidines **(150)** have been investigated by Issleib and co-workers. [1967, 137]

Phosphetanes have previously been discussed with respect to the stereospecific coupling found in these compounds (III.B.2). [1967, 82; 1968, 67, 121 and 122]

Bicyclic compounds, such as **151**, have recently been synthetised. [1969, 146]

2. *Heterocyclics containing phosphorus atom in ring*

This type of compound **(152 to 155)** containing one P^{III}, or tetra- and penta-coordinated P^V atom have been described. [1967, 138 and 139; 1968, 123; 1969, 116] Heterocycles with two P-heteroatom bonds in the ring have been more commonly reported. In addition to the numerous P^V oxyphosphoranes and homologs, synthetised by Ramirez and co-workers (see 1968, 124 for a review of these compounds) and by others, [1967, 121, 140 and 141; 1969, 147] parent tri-, tetra- and penta- coordinated phosphorus heterocyclics with various structures **(156 to 162)** are known. [1965, 32; 1966, 83; 1967, 142 and 143; 1968, 76, 125 and 126; 1969, 65, 109 and 148] A somewhat related compound **163** has also been described. [1969, 149]

3. *Phosphonitrilics*

The greatest attention has been devoted to these conjugated systems (certainly as far as the pentamers). The understanding of the nature of the bonding and the molecular geometry is a very challenging problem as it presents an atypical case of aromaticity;† π-character of phosphorus has been considered in relation to chain mobility, [1966, 6] to electron withdrawal by substituents [1968, 127] and to chain planarity and lastly to substituent electronegativity. [1966, 49]

Numerous data have been reported by Shaw and co-workers, [1965, 24 and 33; 1966, 84 to 88; 1968, 91, 128 and 129; 1969, 150] by Moeller and co-workers [1966, 89; 1967, 144 and 145; 1968, 50, 130 to 133] and numerous other workers. [1965, 34 and 35; 1966, 6, 49, 80, 90 to 97; 1967, 130, 146 to 148; 1968, 114, 134 and 135; 1969, 151 to 153]

In all these cases the identification of the isomers could be readily established by using:

i) ^{31}P chemical shifts, e.g. in $N_3P_3F_5X$ it is possible to distinguish PF_2 and PFX groups:

PF_2	70·9	PF_2	71·2	PF_2	70·8
$PF[N(CH_3)_2]$	63·0	PFCl	32·0	PFBr	20·5

It was also possible to distinguish between the *cis* and *trans* isomers, e.g. in $N_3P_3F_4(C_6H_5)_2$:

PFC_6H_5 {	*cis*	^{31}P	−38·4	^{19}F	49·3
	trans		−38·3		51·5
PF_2 {	*cis*		−12·4		65·2 to 69·3
	trans		−10·8		67·7

ii) Coupling constants as in:

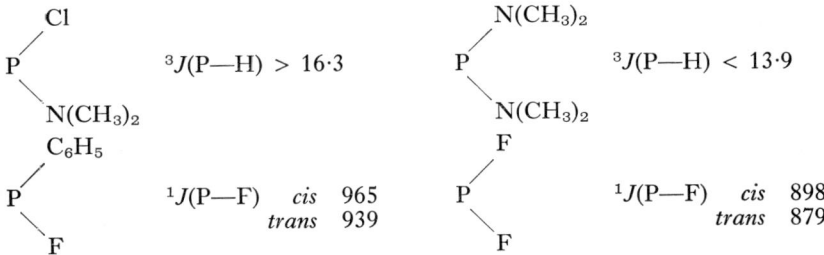

It should also be remembered, as discussed earlier (III.C.2(d)), that the P–N–P two bond couplings are also sensitive to nature of substituents.

Finally, it should be noted that various parent structures such as **164** and **165** have been reported. [1966, 98; 1967, 149 and 150; 1968, 136 and 137]

B. Applications to stereochemical problems

The importance of NMR in stereochemical studies of phosphorus compounds is now firmly established; this aspect was discussed at length

† For instance see ref. 1958, 1; 1960, 2; 1962, 3.

for organic phosphorus compounds in a review by Gallagher and Jenkins; [1968, 138] this is also true for the elucidation of various biological mechanisms such as pseudorotation in the hydrolysis of phosphate esters. [1968, 139] Once again, the topic is far too extensive to be considered in great detail—only typical examples will be presented neglecting studies of optically active compounds such as menthyl phosphinates, e.g. [1967, 151; 1968, 140; 1969, 154]

1. Stereochemistry of saturated aliphatic compounds

NMR is an especially sensitive tool for detecting non-equivalence and for studying the underlying phenomena (steric hindrance, solvent effects, etc.) and a number of examples were cited in the previous section. Thus, it was noted from work of Rudolph and Newmark [1969, 31] that conformational equilibrium exists in F_2PSPF_2 (in which sulphur acts as a rather loose link) but not in F_2PPF_2 or $F_2P(NCH_3)PF_2$ (see Sections III.C.1.(b), III.C.2(c) and III.C.2(e)). Similarly, Lambert et al. have investigated rotational isomerism in diphosphines and obtain the relevant activation parameters. [1966, 99; 1968, 141]

Mono-phosphorus compounds will now be considered and discussed in order of the bridging atom (non-equivalent group are underlined).

(a) *Carbon*. Steric hindrance effects the protons as in $ClC\underline{H}_2PH_2$ [1968, 58] and $ClC\underline{H}_2PCl[N(CH_3)_2]$. [1968, 59] This feature is more pronounced in compounds as **166**. [1969, 155] This effect is also found in $P^{IV\oplus}$ and P^V derivatives. [1968, 68; 1969, 156 to 158]

(b) *Oxygen*. Non-equivalence from steric factors exists in various P^{III} and P^V compounds such as $C_6H_5P(OC\underline{H}_2CH_3)_2$, $C_6H_5P(OC\underline{H}_2CH_3)(C\underline{H}_2CH_3)$. [1969, 34] The effect has been investigated as a function of temperature and solvent in **167**. [1969, 159] The effect can also arise from blocking by complexing in adducts; [1969, 160] other examples are to be found. [1965, 35 and 36; 1967, 152; 1968, 25, 142 and 143; 1969, 101, 161 and 162]

(c) *Sulphur*. The only detailed study, as a function of temperature and solvent, has been carried out on $CH_3P(O)(OR)SC\underline{H}_2R'$. [1969, 101]

(d) *Nitrogen*. As discussed earlier (III.B.5(b)) the phosphorus-nitrogen bond is very sensitive to d_π–p_π conjugation and this factor dominates the effects of inversion at nitrogen (always rapid) and at phosphorus (generally too slow). Trivalent phosphorus is very favourable for obtaining isomers at some accessible temperature such as in $RNHP(CF_3)_2$, [1968, 144] $R^1R^2PCl(C_6H_5)$, $CH_3(R^1)PR^2R^4$ [1968, 92] and other compounds. [1965, 8;

[1967, 80; 1968, 145] With P^V compounds, in which conjugation with nitrogen is balanced by the formation of a P=X bond, it appears that steric effects are predominant and initiate conformational stability which may arise from asymmetry of phosphorus [1969, 111] as observed in $C_6H_5P(X)(NHPr^i)_2$ [X = S or Se] and $(C_6H_5)_2P(S)NHPr^i$; the non-equivalence of the methyl groups of these compounds has been investigated at various temperatures in different solvents. [1968, 145] Internal hydrogen bonding may also add a stabilising effect when appropriate groups are present as in $Pr^iNHP(O)(OR)(OAr)$, in which H→O chelation occurring in the NHP(O) group is likely to occur. [1969, 110]

2. *Stereochemistry of unsaturated aliphatic compounds*

Stereochemical problems in these compounds are not specific to phosphorus compounds; the only additional feature is the existence of phosphorus couplings, the stereospecificity of which has been previously mentioned (III.B.7). It should be noted that phosphorus-bearing groups, and most especially phosphoryl groups, present on vinylic bonds create very important neighbouring effects on account of the magnetic anisotropy. These effects have recently been calculated, using the method of ApSimon, by Timofeeva *et al.* [1969, 163] for the β-proton or methyl in **168** and **169**.

Likewise, a systematic use of such anisotropy effects has been used to establish the conformation of small rings from the proton shielding in the corresponding allenic phosphine oxides, [1967, 153 to 156; 1968, 46; 1969, 164 to 166] for example in **170**.

3. *Stereochemistry of acyclic penta-coordinated compounds*

This question has been previously discussed in some detail (see ref. 1966, 1, p. 281) and is especially well-documented for fluorophosphoranes which have been extensively investigated by Schmutzler *et al.* [1965, 12] (see III.B.1). Actually, in many cases, the first problem to overcome is the observation of distinct isomers as fluorine exchange occurs rapidly when bulky substituents are absent.

The nature of this exchange is now better understood following the extensive variable temperature investigations of both proton and fluorine resonances. [1967, 157] For PF_5 it was concluded, from line width analysis, that intermolecular exchange occurred in solution; [1969, 167] for intramolecular exchange pseudorotation has been invoked [1967, 158] and the behaviour found to agree with the model of Berry. [1960, 1] It may be added that this pseudorotation has been systematically analysed, from a topological point of view, in trigonal bipyramids. [1968, 147; 1969, 168]

A last point is that the systematic assignment of fluorine atoms to the axial positions may not always be true despite the theoretical predictions of the extended Hückel calculation.[1966,100] It has recently been concluded, from study of microwave spectra, that the molecular symmetry of CF_3PF_4 was C_{3v}; this being only compatible with the presence of an axial CF_3 group.[1968,148]

4. Stereochemistry of one-ring compounds

This topic has been reviewed, for the period up to 1967, in some detail by Berlin and Hellwege.[1969,144] Various interesting examples will be considered later but many aspects have been discussed in Section III together with stereochemical aspects of spin-spin couplings. Additional instances will now be considered.

(a) Three membered rings have recently been described by Goldwhite et al.[1969,169] and in these, **171** to **174**, the isomers are distinguishable on account of the slow inversion at phosphorus.

(b) Four membered rings have been investigated by Cremer et al.,[1968,121; 1969,170] e.g. **175** and **176**; both inversion at phosphorus and cis-trans equilibrium have been characterised for various substituents.

(c) Five membered rings with incorporated heteroatoms are generally much better known.† With a P^{III} atom these compounds show a slow inversion at phosphorus as evidenced by the non-equivalence of methyl groups in **177**[1966,51] and the non-equivalence of protons in **178**.[1968,77] These and other data[1966,50] suggests that the different positions arise from a planar ring in which the lone-pair on phosphorus is cis to two substituents and trans to the other two. In fact a more detailed discussion by Cox et al. using the parent ethylene, propylene and butylene phosphites[1969,93] proves that the NMR data are only consistent with a twist-envelope conformation **179**, as proposed earlier by Haake et al.[1968,77] According to this work, the twist angle would be ca. 30° for the ethylene phosphite (from vicinal H–H couplings); in the isobutylene homologue this angle would be somewhat larger.

Other substituted phosphites have been considered[1967,115; 1969,85] as well as other heterocycles, for instance the oxaza analogues.[1969,171] Some P^V homologues have been investigated[1967,140; 1969,171] as well as the corresponding penta-coordinated compounds, e.g. **180** and **181**,[1967, 140 and 160; 1968, 76 and 123] in which the stereoisomerism may be washed out by pseudorotation.

(d) Six membered ring compounds—only dioxaphosphorinanes‡ have been investigated but present syn-anti isomerism with P^{III} [1966,46;

† The compound $(PCF_3)_5$ has been described in which scrambling occurs at high temperature by torsional vibration.[1967,159]

‡ The chair conformation has been definitely established by X-ray study.[1969,172]

1967, 161 and 162; 1968, 78 and 79; 1969, 92 and 172) or P^V. (1966, 46; 1967, 90; 1968, 79, 83, 149 and 150; 1969, 92, 95, 97, 173 to 177)

In all these compounds the dihedral dependence of both $POCH$ and $POCCH$ couplings are clearly seen; this is of some importance in applications, especially for biological compounds (1969, 96)—the only difficulty arises on account of the number of superimposed stereochemical factors as observed in the low temperature data of **182a** and **b**. (1969, 173) This means that the ring conformation is the major but not the only important factor.

5. *Stereochemistry of spiro-compounds*

(a) Spirophosphorus compounds, in which the phosphorus is attached to carbon atoms only, have been investigated by Hellwinkel (1965, 37) and the relevant stereochemistry was later discussed by the same author. (1966, 25 and 101) The underlying pseudorotation mechanism has been investigated by a variable-temperature study, a trigonal bipyramidal conformation being stable. (1968, 151)

(b) A large number of heteroatomic spirophosphoranes have been described and reviewed by Ramirez. (1967, 163; 1968, 152 and 153) Various other systems **183** to **185** have been reported by Burgada *et al.*; (1967, 39, 164 and 165; 1968, 154; 1969, 109 and 178) in **185** when $R=H$ a $P^V \leftrightarrow P^{III}$ tautomerism exists involving structure **186**. Many of these compounds exhibit a double diastereomerism (*cis-trans* and *quadrent* isomerism) which is detectable from the proton resonance of the R group; in this respect groups with $R=H$ and $CH(OH)R^1$ appear to very sensitive to molecular environment. Lastly, structural rearrangements have been elucidated for various H-spirophosphoranes **183**, $R=H$. (1969, 179)

6. *Stereochemistry of other cyclic compounds*

Various polycyclic or juxtacyclic compounds have been considered (1966, 102; 1967, 166; 1968, 117; 1969, 90 and 146) and **187** may be considered as a typical example in which the $^4J(POCCH)$ and $^5J(HPOCCH)$ values were used to assign the configuration shown. Some attention has been given to this aspect in steroid derivatives (1965, 38; 1966, 103 and 104; 1967, 100, 167 and 168) and the stereospecificity of long-range P–C–C–CH couplings have been used, in addition to Zürcher rules, for conformational assignment of **188** in which $^4J(P-H)$ was 0·8 Hz. (1969, 83)

C. Applications to intermolecular studies

1. Solvent and temperature effects have been especially studied in simple compounds—white phosphorus, (1969, 180) PH_3, (1967, 43) PF_3, (1968, 51), phosphorus acid, phosphines, phosphonium compounds etc. (1965, 39; 1967, 169; 1968, 155; 1969, 53) Phosphorus chemical shifts or line-widths,

proton or fluorine chemical shifts or coupling constants were in turn used for the measurements.

2. Hydrogen bonding has been considered in hydrofluoric or hydrochloric acid solutions of phosphates and other esters. [1967, 170; 1968, 156; 1969, 181 to 183] Less acidic media, such as alcohols, water and chloroform, have been used by Li et al. [1967, 171; 1968, 157 and 158] Comparison between the infrared and phosphorus resonance shifts have been found to be fruitful for understanding the interaction in the case of hydrogen phosphonates. [1967, 172] Thermodynamical parameters of the interaction were obtained using a variable temperature investigation. [1969, 184] To quote more "chemical" interactions, β-diketone-neutral ester systems have been considered, [1966, 105] as were mixtures of amines and acidic phosphates and pyrophosphates by Russian authors. [1969, 185]

3. Adduct formation results in well-defined species. Generally speaking, phosphorus compounds act as Lewis bases [exceptions being pentavalent phosphorus halides as reviewed by Webster [1966, 106]]; for other examples in which the relevant Lewis acids are metalloid derivatives see references 1966, 107 and 1969, 186. Adducts involving boron have recently been reviewed elsewhere [1969, 94 and 102] and are by far the most numerous and use has been made of phosphorus, boron, proton and fluorine resonances, in some cases at varying temperature.
[1965, 40; 1966, 20, 36, 81, 108 to 110; 1967, 32, 41, 133, 134, 173 and 174; 1968, 79, 158 to 160; 1969, 42, 72, 86, 102, 113, 141, 187 to 196] As previously mentioned (III.E), the most striking feature is the appearance in P^{III} ligands of boron-phosphorus coupling through the dative bond [ca. 30 Hz in $PH_3 \cdot BH_3$ to ca. 100 Hz in $(CH_3O)_3P \cdot BH_3$]. Simultaneously, phosphorus is coupled to some protons or fluorine atoms through the P→B bond as in $(CH_3)_3P \cdot BH_3$ in which ^2J(P–BH) is 26·8 Hz or in $(CH_3)_3P \cdot BF_3$ in which ^2J(P–BF) is 229 Hz. Additionally for both P^{III} and P^V derivatives, the specific chemical shifts and coupling constants of the ligand are modified. Boron resonances shift about 10 ppm upfield for BF_3 and about 40 ppm for BH_3, irrespective of the ligand and do not appear to be too sensitive to the "bond strength"; fluorine shifts give a better estimate of this, thus it is about 20 ppm upfield for the BF_3 complex which is far less than found in amine and ester complexes. For the phosphorus ligand nearly all shifts appear downfield on complex formation, in agreement with a σ-bonded type of adduct.† For protons the deshielding effect is nearly the same for all types of ligands (0·2 to 0·4 ppm), suggesting that this arises from the

† Assuming that no structural changes occur within the molecule. Halogen exchange is detected in some cases; [1968, 45; 1969, 28 and 102] similarly the influence of adduct formation on the thiono-thiolo isomerisation of thio- and seleno-phosphates has been noted. [1969, 189]

magnetic anisotropy of the boron ligand. Another interesting feature of the changes induced by adduct formation is afforded by the various phosphorus couplings. As noted earlier (III.B.1), phosphorus-fluorine couplings are rather insensitive on account of the high polarity of the bond while phosphorus-proton couplings are extremely sensitive. For example, P–H coupling is about 320 Hz in $PH_2^{\ominus}(BH_3)_2$ compared to about 140 Hz in the free ligand. As a rule the change is in the direction expected for an increase of the substituent electronegativity.

In molecules which can act as polydentates, one point of interest is the relative basic strength of the different basic sites. For $[(CH_3)_2N]_3P$ and other amino-phosphines, it was concluded that complexing to phosphorus was the major mechanism;[1969, 72 and 192] in $[(CH_3)_2N]_3PO$ there is competition between the phosphoryl group and nitrogen atoms for complex formation [1969, 94, 102 and 188] in agreement with previous hydrogen bonding studies (see 1966, 1, p. 295).

Adducts involving aluminium, gallium or indium derivatives as Lewis acids have also been described.[1965, 41; 1966, 111; 1967, 175 and 176; 1968, 161 and 162; 1969, 28, 197 to 199] Here again, in P^{III} ligands, phosphorus-aluminium coupling is found or phosphorus-proton coupling through aluminium or gallium. The formation of complexes actually affords a means of detecting changes, and especially in the signs, of coupling constants. This was used by Pidcock et al. to obtain evidence for reversal of sign in P····H couplings in aluminium complexes of $P(CH_3)_3$ by varying the concentration (adduct exchange is sufficiently fast at room temperature to give an average spectrum); no similar variation appears for the P····H coupling of $P(OCH_3)_3$.[1966, 41] This observation supported the early assumption that the *P–C–H* couplings depended upon two competing terms of opposite signs (see III.B.1).

Some attention has been given to the ligand exchange mechanism [1969, 198] and similarly bond-exchange has been detected in interhalogen mixtures, e.g. in Al_2F_6–$POCl_3$ from ^{27}Al resonance [1969, 28] or in BF_3–$POF_2(CH_3)$ from both fluorine and phosphorus resonances.[1969, 102]

In a few cases adducts other than these obtained with boron, aluminium gallium and indium have been described.[1962, 4; 1966, 112; 1968, 163 and 164; 1969, 200]

D. Applications to reaction studies

Brief consideration will be given to NMR investigation of intermediates, products and kinetics of reactions. Alkylation, arylation, metalation, reaction with organometallics, thermolysis and thermal rearrangements, isomerisation and many other reactions such as the Wittig, Michaelis and Arbuzov reactions, which are typical of organophosphorus chemistry, have been considered. (1963, 5; 1965, 42 to 45; 1966, 60, 62, 71, 109, 113 to 137; 1967, 121, 139, 141, 177 to 185; 1968, 126, 145, 165; 1969, 116, 146, 201 and 202) NMR has also been used for polymerisation studies and polymer characterisation; (1965, 42, 46, and 47; 1966, 138; 1967, 186 to 188; 1968, 166; 1969, 203) this has been systematically exploited by South African authors (1968, 167; 1969, 203 and 204) for investigating mixed polymers in various systems such as methyl polyphosphate + methyl orthosilicate, arsenite or phosphite. It now appears that NMR data for inorganic systems is in general rather sparse. (1967, 129, 189 and 190)

Typical examples of such researches on reaction studies are:

(i) Valence rearrangements of phosphorus: $P^{III} \leftrightarrow P^V$ interconversion as in the "phenylphosphinidene" (1965, 48) and in spirophosphoranes (1967, 165) or $P^{IV\oplus} \leftrightarrow P^V$ interconversions as in R_3PX_2 compounds (1967, 191; 1968, 168) to which may also be added metathetical reactions. (1966, 139)

(ii) pH Dependence of molecular species as observed at pH values above 2·0 with methyl and dimethyl phosphates; (1969, 205) the former exhibits abrupt changes in proton chemical shifts and coupling constants when going through the upper pK (6·4) and the second which has only one low pK (1·3) presents a pH-independent spectrum in the range investigated.

(iii) Reactions and rearrangements of methylene phosphorane compounds in which an acid-catalysed exchange has been detected (1968, 169 to 171) which creates an anomalous temperature-dependence of the P=CH coupling, first attributed to hindered rotation. (1966, 140; 1967, 192) The relevant stereochemistry has been further investigated by Snyder et al. (1969, 206 and 207)

(iv) Ligand exchange, a field pioneered by Van Wazer et al., has received some attention, (1965, 49 and 50; 1967, 193; 1969, 140) for instance in adducts with fast interconverting ligands as in **189**. (1966, 36) Most examples, however, deal with halogen exchange either through an intramolecular process (if any) in PF_5 (1967, 194; 1968, 172) or intermolecularly in adducts (1968, 45 and 173; 1969, 28 and 102) for which the lifetime may be calculated.

Regarding lifetime studies, the interest in exchange observations cannot be over estimated. Thus, Grim and McFarlane could not observe phosphorus chemical shifts or coupling constants in mixtures of HBr + $(C_6H_5)_3P$, or $(C_6H_5)_2PH$ or $C_6H_5PH_2$, which gave an upper limit for the phosphonium ion lifetime of $3 \cdot 10^{-3}$ sec. In contrast to this H–P–CH couplings are sometimes observed, depending upon the experimental conditions, for mixtures of HBr with $CH_3P(C_6H_5)_2$ or $(CH_3)_2PC_6H_5$; as these couplings were ca. 6 Hz the observation of these couplings implies a lifetime greater than 0·2 sec. and absence a lifetime lower than this value. (1968, 7) Studies on the protonation of phosphines have recently been extended by Olah and co-workers to similar systems. (1969, 12)

E. Miscellaneous applications

Various fields in which NMR appears to be of considerable value will briefly be considered.

1. Relaxation studies have been rather scarce despite the early investigations of Winter (1959, 1) on ^{31}P relaxation in PCl_3 and PBr_3.

Variable temperature measurement of T_1 in PCl_3, $POCl_3$ and POF_3 (including use of fluorine resonance in the latter compound) (1968, 174),† of T_1 and $T_{1\rho}$ (in the rotating frame) for PBr_3 (1968, 115) and PCl_3 (1969, 139) permitted a better understanding of relaxation mechanisms and a separation of contributions from scalar, spin-rotational and direct dipolar interactions ($T_{1\rho}$ depends only on the first term). From these, it may be concluded that the major mechanism (for diamagnetic liquids) arises from the spin-rotation interaction, especially at high temperature; at lower temperatures scalar interactions become of some importance. Similar studies have been performed on liquid and solid P_4O_6. (1968, 175; 1969, 208) Conversely, phosphorus relaxation rates have been used as a probe for following solvent effects on tributyl phosphate and complex formation with uranyl nitrate (1968, 109, and also see 1968, 176 and 1969, 209) and also for investigating ligand exchange in metal complexes, (1968, 177) ferroelectric changes in KH_2PO_4‡ (1968, 178) and electronic structures of phosphides. (1969, 211 and 212)

As a related topic, attempts to improve the signal to noise ratio in ^{31}P spectra using dynamic polarisation will be briefly described (for a review of the Overhauser effect see[1969, 213]). This was first used in high fields, using some radical probe such as tri-t-butylphenoxy, and the enhancement was found to be positive for P^{III} compounds but not so for P^V homologues. (1966, 141; 1967, 198; 1969, 130) Observations at low fields are more numerous but the results more complex (1968, 8; 1969, 214 and 215) being positive for phosphites and triphenyl phosphine but negative for diphenyl phosphine (all having trivalent phosphorus) and being negative for phosphates and thiophosphates but positive for dimethyl phosphonate, thiophosphoryl chloride and phenyl thiophosphoryl dichloride (all with

---

† Further analysis of T_1 for phosphorus (and fluorine) in POF_3 gives an estimate of *para*- and *dia*-magnetic terms in the relevant chemical shifts, thus permitting an appreciation of the *absolute* ^{31}P *chemical shift* (1969, 9) as discussed earlier (II.A.1).

‡ Proton or deuteron relaxation is also useful in this respect as shown for $NH_4PO_4H_2$ (1967, 195 and 196) and for the determination of rotational barriers in solid phosphonium halides. (1967, 197; 1968, 179 and 180; 1969, 210)

pentavalent phosphorus). It is suspected that, in addition to direct coupling by s electrons, there exists some indirect mechanism. When considering phosphonitrilics [1969, 216] a specificity of the radical probe has been found, bis-diphenylene phenyl allyl being especially satisfactory and DPPH being the least effective of the species investigated. The enhancement attained appeared to be directly dependent upon the ring delocalisation.

2. Investigations on solid samples have been reported for a rather large variety of compounds generally using phosphorus resonance but in some cases proton, fluorine, boron and even ^{23}Na resonances [1969, 217] and relaxation time measurements have been used. Phosphides, and especially simple or mixed paramagnetic phosphides, have often been considered, [1961, 3; 1966, 142 to 145; 1967, 199 to 205; 1968, 181 to 185; 1969, 212, 218 and 219] sometimes using the resonance of the metal atom (Y, La, Mn, Al, etc.). Phosphates also present a large number of interesting cases—mineral and meteoritic, paramagnetic and paraelectric, ferroelectric and antiferroelectric phosphates have been studied, special attention being paid to the phase changes. [1966, 146; 1967, 188, 195, 206 to 211; 1968, 178, 186 to 191; 1969, 27, 61, 220 to 224] Data are available for various other solids or glasses. [1966, 147 and 148; 1967, 212; 1969, 225 to 227]

To cite the most interesting general features:

(i) Fine structure determination, pioneered by Andrew, has been extended to polycrystalline samples. [1966, 149; 1967, 204] This has been found to be especially valuable when investigating solid-state adducts such as $PCl_5 \cdot TiCl_4$, which is formulated $PCl_4^{\oplus} TiCl_5^{\ominus}$. [1966, 150; 1967, 213 to 216; 1968, 192] Similarly, double resonance has been performed on solid samples of $CaHPO_4$ [1968, 193] and of $NH_4PO_4H_2$. [1969, 228]

(ii) Line shape analysis has been possible in some cases, thus affording chemical shift anisotropy for phosphates [1969, 5 and 221] and for $BaFPO_4$. [1969, 6 and 7] In the latter case the absolute sign of the P–F coupling could be unambiguously assigned on these grounds.

(iii) Variable temperature measurements were very useful for following phase transitions and molecular rearrangements as in hydroxyapatite [1969, 224] as well as the identification of various hydroxyl and water species in solid phosphates. [1969, 226]

3. Metal complexes

The importance of metal complexes of organo-phosphorus in homogeneous catalysis, or the role played by NMR in the structural assignment, needs no stressing. In ref. 1966, 1, a formulae index was presented for all these species studied using resonance other than that of phosphorus and more recently the ^{31}P resonances of these complexes has been fully discussed in an earlier review in this series by Nixon and Pidcock.[1969, 229] To keep abreast with this rapidly growing field of interest a summary is given in Table XXXIV according to the central atom and the resonance investigated and only a few outstanding points will be discussed.

(i) Phosphorus-phosphorus couplings—When more than one phosphorus ligand is present these couplings greatly assist in the stereochemical assignment, for instance being -40.5 and $+162$ Hz respectively in *cis-* and *trans-* $(CO)_4Mo\{P(OMe)_3\}_2$.[1969, 132] These important variations may be considered in terms of changes in the s-character.[1969, 133]

(ii) Ligand-exchange mechanisms are clarified using variable-temperature studies[1967, 34; 1968, 235 and 243] or relaxation rate measurements[1968, 177] and have been considered in relation to ion-pairing mechanisms.[1965, 54 to 56] Stereochemical conversions[1969, 279] have also been discussed in relation to intramolecular[1967, 262] and σ–π rearrangements.[1969, 309]

(iii) Chemical exchange spin decoupling in metal complexes, which removes couplings of phosphorus (i.e. to neighbouring protons) owing to its preferential relaxation by unpaired electrons,[1965, 63 and 64; 1968, 217; 1969, 237] has been investigated using temperature and solvent dependence by Frankel[1969, 237, 238 and 310] and the relevant theory established by Fackler and co-workers[1969, 291] using the density-matrix technique of Kaplan and Alexander. It is especially noteworthy that this effect (which only exists in P^{III} compounds) is selective[1969, 93] as non-equivalent couplings are not removed simultaneously, as previously discussed see III.B.3(b); thus, for H_A and H_B in **190** which have different coupling constants to phosphorus (2·2 and 10·5 Hz) are decoupled at different concentrations of added nickel acetylacetonate.

4. Biological problems

A detailed discussion of this topic is once again beyond the scope of this review. The field is particularly broad and includes the investigation of biological structures and configurations using various resonances, generally proton and phosphorus, but also ^{15}N[1966, 196], of the reactivities through complexing [especially by metal cations which may be used as resonance probes, such as ^{43}Ca[1969, 311] but also using other ions such as ^{35}Cl] or the interaction with other substrates [e.g. procaine-thiamine pyrophosphate interactions[1969, 312]]. To briefly mention the relevant references according to the phosphorus compounds data may be listed for:

(i) AMP, ADP and ATP.[1965, 72 to 74; 1966, 196 to 198; 1967, 263 and 264; 1968, 270 and 271; 1969, 311, 313 and 314]

TABLE XXXIV

NMR studies of metal complexes of phosphorus compounds[a]

Metal atom	Proton resonance	Fluorine resonance	Phosphorus resonance	Other resonance
Ag	1966, 41; 1967, 102; 1968, 162, 194 and 195	1966, 41	1966, 110; 1967, 217 and 218	
Al	1969, 199, 230 to 232		1967, 23; 1969, 199 and 230	^{27}Al-1967, 23
As			1968, 196	
Au	1966, 151; 1967, 219; 1969, 233			
Ca			1965, 30	
Cd	1966, 171		1966, 171; 1967, 221	
Co	1965, 51 to 57; 1966, 152 to 157; 1967, 27, 220 to 224; 1968, 197 to 205; 1969, 14, 78, 132, 234 to 243	1965, 57 and 58; 1967, 225; 1968, 57 and 206; 1969, 235, 240, 241, 244 and 245	1965, 53; 1967, 217 and 218; 1968, 203; 1969, 237 and 238	^{59}Co—1969, 246
Cr	1966, 72 and 158; 1967, 79, 226 and 227; 1968, 202, 207 to 209; 1969, 14, 78, 247 and 248	1969, 60 and 133	1967, 79, 227 and 228; 1968, 207, 210 and 211; 1969, 60, 249 to 251	

[a] This index follows in a more condensed form the Formulae Index V of ref. 1966, 1.

NMR STUDIES OF PHOSPHORUS COMPOUNDS (1965–1969)

Cu	1967, 229; 1968, 43 and 194; 1969, 252 and 253	1968, 43	1965, 31; 1966, 110; 1967, 217 and 218
Fe	1962, 5; 1965, 59; 1966, 158 to 161; 1967, 227; 1968, 53, 98, 201 to 204, 209, 212 to 215; 1969, 78, 237, 239 to 241, 254 to 260	1968, 98; 1969, 133, 134, 240, 241, 254 and 258	1967, 227; 1968, 203 and 216; 1969, 237
Hg	1968, 202		1966, 110; 1967, 221
Ir	1965, 52, 59 to 64; 1966, 162 to 164; 1967, 230 and 231; 1968, 217 to 224; 1969, 14, 236, 261 to 268	1969, 167	1969, 263
Li			1965, 30
Mg			1965, 30
Mn	1962, 5; 1965, 65; 1966, 165 and 166; 1967, 232; 1968, 201, 212, 225 to 228; 1969, 14, 78, 237, 240, 241, 269 to 271	1969, 240 and 241	1969, 237 ^{11}B—1966, 1672 ^{55}Mn—1969, 27
Mo	1965, 51; 1966, 72 and 158; 1967, 79, 226, 227, 233 to 235; 1968, 202, 207 to 209, 212, 229 to 232; 1969, 78, 104, 131, 239, 247, 273 to 275	1964, 13; 1966, 168; 1967, 235; 1968, 119; 1969, 60, 131, 133, 134 and 276	1967, 79, 227, 228 and 235; 1968, 119, 207, 210, 211, 230; 1969, 60, 249 to 251 and 276

TABLE XXXIV (*continued*)

Metal atom	Proton resonance	Fluorine resonance	Phosphorus resonance	Other resonance
Ni	1965, 54 to 56 and 66; 1966, 152 and 169 to 173; 1967, 211, 220, 221, 224, 235 to 242; 1968, 53, 199, 204, 205 and 233; 1969, 34, 78, 141, 153, 159, 234, 277 to 282	1966, 168 and 174 and 175; 1967, 48 and 235; 1968, 234; 1969, 133, 283 and 284	1965, 66 and 67; 1966, 110, 169, 171, 176 to 178; 1967, 48, 174, 217, 218, 235, 236, 240, 243 to 245; 1968, 211, 234 and 235; 1969, 282 to 286	
Os	1965, 59; 1968, 217 and 236; 1969, 195, 263, 287 to 289		1969, 263	
Pb	1966, 171		1966, 171	
Pd	1965, 68; 1966, 179; 1967, 246; 1968, 237 to 243; 1969, 155, 233, 280, 290 to 293	1967, 247 and 248; 1968, 245 and 246	1968, 244; 1969, 291	
Pt	1965, 59, 68 and 69; 1966, 179 to 182; 1967, 220, 249 to 254; 1968, 218, 219, 247 to 254; 1969, 26, 233, 236, 263, 277, 280, 290, 292, 294 to 297	1965, 69; 1966, 181, 183 and 184; 1967, 247 and 248; 1968, 219, 246, 253 and 255; 1969, 121, 298 and 299	1966, 185; 1967, 228, 253 and 255; 1968, 145, 247, 249 and 256; 1969, 263	^{195}Pt—1968, 251 and 257
Re	1965, 63 and 64; 1966, 186; 1968, 217; 1969, 263, 289 and 300		1969, 263	

Rh	1965, 52, 63, 64 and 70; 1966, 155, 163, 182, 187 to 191; 1967, 254, 256 to 259; 1968, 224, 258 to 262; 1969, 263, 301 to 305	1967, 248; 1968, 263; 1969, 121	1966, 192; 1969, 263	^{103}Rh high-resolution resonance 1969, 306
Ru	1965, 59 and 64; 1966, 193; 1967, 260; 1968, 217, 236, 259, 264 and 265; 1969, 263 and 307	1969, 307		
Sb			1968, 196	
Sn	1967, 211; 1969, 53			
Ti			1966, 194	
U	1965, 71; 1969, 74	1967, 261	1968, 109 (relaxation)	
V	1968, 212			
W	1962, 5; 1966, 72, 158 and 160; 1967, 79, 226 and 227; 1968, 207 to 209, 212, 230, 231, 266 and 267; 1969, 14, 71, 78 and 247	1968, 268; 1969, 60 and 133	1966, 195; 1967, 79, 227 and 228; 1968, 207, 210, 211, 230 and 269; 1969, 60, 71, 138, 249 to 251 and 308	
Zn	1968, 269		1967, 221	

(ii) Nucleotide and nucleoside phosphates, [1965, 75 and 76; 1966, 199; 1967, 265 to 267; 1968, 172, 272 to 276; 1969, 96, 98, 315 to 318] for which an extensive survey has been recently presented by Hollis and co-workers. [1969, 319]

(iii) Phospholipids. [1966, 200 and 201; 1967, 268 and 269; 1968, 277 to 279]

(iv) Phosphoproteins. [1966, 202; 1969, 320 and 321]

(v) Phosphorylated steroids and hormones, [1965, 38; 1966, 103 and 104; 1967, 167, 168 and 270] sugars and carbohydrates [1966, 203; 1967, 271 and 272; 1968, 280; 1969, 320 and 321] and enzymes. [1967, 257; 1968, 281]

(vi) Others include radioprotectors, [1966, 204] thiamine phosphate, [1966, 205; 1967, 273; 1969, 312] vitamin B6 [1967, 274] etc.—see also 1967, 275.

(vii) Lastly, phosphorus pesticides have received considerable attention [1961, 4; 1968, 282 and 283] and monographs have been recently published on these, one using 100 MHz spectra; [1968, 284 and 285; 1969, 322] see also review on the NMR of pesticides in Volume 4 of this series. [1971, 1]

5. *Miscellanous applications*

Finally, the following miscellaneous applications will be briefly mentioned:

(i) Analytical uses. [1967, 189; 1968, 286 and 287]

(ii) Extraction studies [1969, 74, 323 and 324] on transition metal ions and in relation to flotation reagents. [1969, 325] It appears that the interest in such investigations has now decreased (see ref. 1966, 1, pp. 295–296), in clear contrast with all of the other fields reviewed herein.

V. FORMULA INDICES

The following indices list by formulae all available investigations on inorganic and organic phosphorus compounds† (except biological and metal complexes) with the corresponding references. Observed resonances are given with the relevant chemical shifts for phosphorus and fluorine; most of the published chemical shifts for other nuclei (except proton) are in the text of the review. Phosphorus chemical shifts are referred to 85% H_3PO_4 and for fluorine to CCl_3F;‡ all are *positive* with increasing *frequencies* (i.e. for decreasing fields) according to the rules detailed in the General Foreword. When more than one nucleus of a given species exists in the reported molecule, chemical shifts are reported when available from left to right; when more than one nucleus exists on a phenyl ring these are reported in the sequence *ortho*, *meta* and *para*.

---

† No distinction has been made between deuterated compounds and the corresponding hydrogenated homologues.

‡ ^{31}P Chemical shifts referred to P_4O_6 have been corrected by $+112.5$ ppm; ^{19}F chemical shifts referred to CF_3CO_2H, $CF_3C_6H_5$ or F_2 have been corrected by -79.9, -63.8 and 427 ppm respectively.

FORMULAE INDEX I
NMR DATA ON INORGANIC COMPOUNDS[a]

Formula	Compound	Resonance	Reference
$PAlCl_8$	$PCl_4^{\oplus} AlCl_4^{\ominus}$	P (+86·5)	1967, 276
$PAl_2H_2O_2$	$Al_2(PO_2H_2)$	H	1968, 162
$PBBrH_5$	$PH_3 \cdot BH_2Br$	H	1967, 41
$PBGeH_8$	$GeH_3 \cdot PH_2 \cdot BH_3$ (and deuterated derivative)	H	1968, 158
PBH_8Si	$SiH_3 \cdot PH_2 \cdot BH_3$	H	1967, 173; 1968, 158
$PBH_{10}Si_2$	$Si_2H_5 \cdot PH_2 \cdot BH_3$	H	1969, 113
PBO_4	BPO_4	^{11}B	1961, 3
PB_2H_8Na	$H_2P(BH_3)_2Na$	^{11}B, H	1969, 194
$PB_4F_2H_9$	$PF_2H \cdot B_4H_8$	^{11}B, H, P (ca. 118), F(−121·7)	1969, 141
$PB_4F_3H_8$	$B_4H_8 \cdot PF_3$	^{11}B	1966, 109
$PBaFO_3$	$BaFPO_3$	F, P	1969, 7
$PBrFH_2NS$	$FBrP(S)NH_2$	H, F (−2·6)	1969, 325
$PBrCl_2O$	$OPCl_2Br$	P (−29·6)	1966, 9
$PBrF_2$	PF_2Br	P (+218), F (−40·1)	1967, 50; 1968, 19
$PBrF_2O$	OPF_2Br	P (−28), F (−32·5)	1967, 56; 1968, 19
$PBrF_2S$	SPF_2Br	P (28·6), F (−2·3)	1966, 8; 1968, 19
$PBrF_4$	PF_4Br	P (−72·6) F −(9·6)	1968, 49
PBr_2ClO	$OPBr_2Cl$	P (−64·8)	1966, 9
PBr_2F	PBr_2F	P (225), F (−70·4)	1967, 50; 1968, 19
PBr_2FO	$OPBr_2F$	P (−48·3), F (16·1)	1967, 56;[b] 1968, 19
PBr_2FS	$SPBr_2F$	P (−17·2), F (31·3)	1967, 58; 1968, 19
PBr_2F_3	PBr_2F_3	F (ca. +55)	1965, 77
PBr_3	PBr_3	P (relaxation; 227 to 229)	1959, 1; 1965, 1 and 2; 1967, 277; 1968, 19 and 115
PBr_3O	$OPBr_3$	P (−102, −103·4)	1965, 1; 1966, 9 and 206; 1967, 277; 1968, 19
PBr_3S	$SPBr_3$	P (−111·8, −112)	1965, 1 and 4; 1967, 277; 1968, 19
$PCaHO_4$	$CaHPO_4$	P (solid)	1968, 193
$PClF_2$	PF_2Cl	P (176), F (−38·6)	1966, 105; 1968, 19
$PClF_2O$	OPF_2Cl	P (−14·8), F (−48·1)	1966, 278; 1967, 14; 1968, 19

[a] Formulae have symbols arranged alphabetically except for phosphorus which is always placed first.
[b] Misprint in ^{31}P chemical shift originally quoted.

Formula	Compound	Resonance	Reference
$PClF_2S$	SPF_2Cl	P (50·0), F (−15·9)	1966, 8; 1968, 19
$PClF_3NO_2S$	$F_3P=NSO_2Cl$	F (−84·6)	1969, 326
$PClF_4$	PF_4Cl	F (−24·3)	1968, 49
PCl_2F	$PFCl_2$	P (224), F (−55·9 −57·8)	1967, 50; 1968, 19
PCl_2FO	$OPFCl_2$	P (0·0), F (−9·5)	1966, 9; 1967, 14; 1968, 19
PCl_2FS	$SPFCl_2$	P (43·0), F (15·6)	1967, 58; 1968, 19
$PCl_2F_2NO_2S$	$ClF_2P=NSO_2Cl$	F (−50·4)	1969, 326
PCl_2F_3	PCl_2F_3	P (−27·8), F (ca. 32)	1965, 77; 1968, 288
PCl_3	PCl_3	P (relaxation, 219 to 220)	1959, 1; 1965, 1 and 2; 1967, 235 and 277; 1968, 19, 42 and 174
PCl_3FNO_2S	$Cl_2P=NSO_2F$	F (59·6)	1967, 279
PCl_3O	$OPCl_3$	^{17}O, P (relaxation; 2·2 to 3·0)	1963, 1; 1965, 1 and 4; 1966, 9; 1967, 277; 1968, 19, 136 and 174
PCl_3S	$SPCl_3$	P (28·8, 30·8)	1965, 1; 1966, 111; 1967, 277; 1968, 19
PCl_5	PCl_5	P (−80, −80·9)	1968, 124 and 288
PCl_5O_4	$PCl_4^{\oplus}ClO_4^{\ominus}$	P (87·1)	1967, 276
PCl_6	$PCl_6^{\ominus}$	P (−295)	1968, 288
$PCl_{10}Sb$	$PCl_4^{\oplus}SbCl_6^{\ominus}$	P (87·9)	1967, 276
$PCsH_2$	$CsPH_2$	H, P	1969, 327
PFH_2O_2	$FP(O)H(OH)$	H, P (2·74), F (−64·3)	1968, 34
PFH_2O_3	$FP(O)(OH)_2$	P (−8), F (−77)	1967, 14
$PFH_8N_2O_2S$	$(NH_4)_2PO_2SF$	P (43·9)	1967, 280
PFO_3	$PO_3F^{2\ominus}$	P (0·75)	1966, 79
PF_2HO	OPF_2H, OPF_2D	H, P (−1), F (−61·9, −66·4)	1967, 33; 1968, 34
PF_2HO_2	$F_2P(O)OH$	H, P (−20·1), F (−85·6, −87·5)	1967, 14, 281 and 282
PF_2HS	SPF_2H, SPF_2D	H, F (−46·1)	1966, 32; 1967, 33
PF_2HS_2	$F_2P(S)SH$	H, F (−15·8)	1969, 99
PF_2H_2NO	$F_2P(O)NH_2$	H, P (−1·2, −6), F (−73·5)	1967, 93, 282; 1969, 129
PF_2H_2NS	$F_2P(S)NH_2$	H, P (66·2), F (−42·7)	1969, 328
PF_2H_4NOS	NH_4POSF_2	P (47·1)	1967, 280
PF_2I	PF_2I	P (242·2), F (−48)	1966, 59
PF_2IS	SPF_2I	P (−71), F (−11·2)	1968, 45

Formula	Compound	Resonance	Reference
PF$_3$	PF$_3$	P [97, 105(?)], F ($-33\cdot7$, $-37\cdot6$)	1965, 1 and 2; 1966, 28; 1967, 28, 48, 51 and 235; 1968, 19, 42 and 51
PF$_3$H$_2$	PF$_3$H$_2$	H, P (ca. -39), F (-50)	1966, 24; 1967, 157
	PF$_3$H$_2$, PF$_3$D$_2$	H, P ($-24\cdot1$), F ($-52\cdot5$)	1967, 36
PF$_3$O	OPF$_3$	P (relaxation; $-92\cdot3$, $-94\cdot3$) F (relaxation; $-35\cdot5$)	1966, 9 and 29; 1967, 14; 1968, 19 and 174; 1969, 63
PF$_3$O$_4$S	F$_2$P(O)OSO$_2$F	F ($46\cdot1$; $-81\cdot7$)	1966, 29
PF$_3$O$_7$S$_2$	FP(O)(OSO$_2$F)$_2$	F ($47\cdot1$, $-72\cdot3$)	1966, 29
PF$_3$O$_{10}$S$_3$	OP(OSO$_2$F)$_3$	P (165), F ($50\cdot1$)	1966, 29, 1967, 283
PF$_3$S	SPF$_3$	P ($32\cdot4$), F ($-51\cdot3$)	1965, 13; 1966, 8; 1968, 19
PF$_4$H	PF$_4$H, PF$_4$D	H, P (-50, $-53\cdot6$) F [$-49\cdot6$, -52, $7\cdot5$(?)]	1966, 24; 1967, 36; 1968, 42
PF$_4$NOS	F$_2$P(O)N=SF$_2$	F ($-69\cdot9$, $57\cdot1$)	1967, 136
PF$_4$NO$_2$S	F$_3$P=NSO$_2$F	F ($-86\cdot7$, $-60\cdot9$)	1969, 328
PF$_5$	PF$_5$	P^a ($-80\cdot3$), F ($-71\cdot4$)	1966, 207; 1968, 42; 1969, 63 and 167
PF$_5$HK	PF$_5$H$^\ominus$K$^\oplus$	H, F	1969, 143
PF$_5$S	PF$_5$S$^{2\ominus}$	F ($-22\cdot0$, $-71\cdot0$)	1968, 46
PF$_6$	PF$_6^\ominus$	P (-145), F	1966, 207; 1968, 288
PGeH$_5$	GeH$_3$, PH$_2$; GeD$_3$, PH$_2$ GeH$_3$, PD$_2$	H, P	1967, 284; 1968, 32 and 158
PGe$_2$H$_7$	(GeH$_3$)$_2$PH	H	1968, 32
PGe$_3$H$_9$	(GeH$_3$)$_3$P	H, P ($-332\cdot5$)	1965, 120; 1967, 98; 1968, 32
PHO$_3$	HPO$_3^{2\ominus}$, DPO$_3^{2\ominus}$	P ($+7\cdot1$, $+6\cdot9$)	1967, 42
PH$_2$	PH$_2^\ominus$	H	1966, 11
PH$_2$K	PH$_2$K	H, P ($-255\cdot3$)	1965, 10; 1966, 12; 1969, 203
PH$_2$Li	PH$_2$Li		1969, 327
PH$_2$Na	PH$_2$Na	H	1967, 46; 1969, 327
PH$_2$O$_2$	PH$_2$O$_2^\ominus$	H	1968, 162; 1969, 41
PH$_2$O$_3$	PH$_2$O$_3^\ominus$	H, P	1966, 131
PH$_2$Rb	PH$_2$Rb		1969, 203
PH$_3$	PH$_3$	H, P (-240, -241)	1965, 1 and 2; 1966, 11 and 13; 1967, 28, 43 and 47; 1968, 96 and 158; 1969, 41

^{a} Incorrectly reported as δP -35 from observations on oxidized PF$_5$ (see 1969, 63).

Formula	Compound	Resonance	Reference
PH_3O_2	PH_3O_2	^{17}O, H, P (10·5)	1963, 1; 1967, 45, 47 and 189
PH_3O_3	PH_3O_3	^{17}O, H, P (3·5)	1963, 1; 1966, 208; 1967, 45, 47 and 189; 1968, 155
PH_3O_4	H_3PO_4	^{17}O, P (0)	1963, 1; 1965, 1
	H_3PO_4 (117%)	P (−1·7)	1966, 117
PH_4	$PH_4^{\oplus}$	H, P	1966, 22; 1967, 28; 1969, 41
PH_5Si	SiH_3PH_2	H	1968, 158; 1969, 113
PH_6N_3O	$OP(NH_2)_3$	H	1968, 86
PH_7Si_2	$Si_2H_5PH_2$, $(SiH_3)_2PH$	H	1969, 113
PH_8IN_4	$IP(NH_2)_4$	P (31·6)	1969, 137
$PH_9N_2O_2S$	$(NH_4)_2PHO_2S$	P (34)	1967, 35
PH_9Si_3	$(SiH_3)_3P$	P (−378)	1968, 96
$PH_{15}Si_6$	$(Si_2H_5)_3P$	H	1969, 113
PI_3	PI_3	P (178)	1965, 1 and 2; 1967, 190
PO_4	$PO_4^{3\ominus}$	P (6·0)	1965, 78
P_2	P_2	P	1967, 28
P_2ClF_4NS	$F_3P=NP(S)FCl$	P (−39·0, 44·3), F (−4·8)	1969, 69 and 325
$P_2ClH_{12}N_7$	$[P(NH_2)_3]_2^{\oplus}N, Cl^{\ominus}$	P (15·6)	1967, 285; 1969, 137
$P_2Cl_2F_3NS$	$F_3P=NP(S)Cl_2$	P (−40·8, 29·7), F (−3·7)	1969, 69
$P_2Cl_3F_2NO$	$Cl_3P=NP(O)F_2$	P (6·5, −2·5), F (−70·3)	1969, 129
$P_2Cl_3F_2NS$	$F_2ClP=NP(S)Cl_2$	P (−16·7, 29·4), F (33·5)	1969, 69
P_2Cl_4FNO	$Cl_3P=NP(O)FCl$	P (3·9, −14·5), F (−30·2)	1969, 129
P_2Cl_5NO	$Cl_3P=NP(O)Cl_2$	P (−0·1, −14·2)	1967, 286; 1969, 129
$P_2F_2H_2$	H_2PPF_2	H, P (293·7, −137·6), F (−91·2)	1968, 31
$P_2F_2K_2O_5$	$K_2(P_2O_5F_2)$	P (−17·8), F (−72·6)	1968, 112
P_2F_4	P_2F_4	P (226), F (−115·2, −117·5)	1966, 28 and 209; 1967, 49; 1968, 113; 1969, 31
P_2F_4O	$O(PF_2)_2$	P (111), F (39·9)	1966, 28; 1968, 34; 1969, 31
$P_2F_4OS_2$	$P_2S_2OF_4$	P, F (−40·2)	1969, 62
$P_2F_4O_2$	$F_2P(O)OPF_2$	P (−9·8, −149·7), F (−80, −38·3)	1969, 57 to 58

Formula	Compound	Resonance	Reference
$P_2F_4O_3$	$O[P(O)F_2]_2$	P (−39·3), F (−80)	1967, 287; 1969, 62
P_2F_4S	$S(PF_2)_2$	F (−65)	1969, 31 and 136
$P_2F_4S_2$	$F_2P(S)SPF_2$	P (−40·7, 75·5), F (13·7, 60·5)	1969, 57 and 59
P_2F_5NS	$F_3P{=}NP(S)F_2$	F (−85·2, −38·0)	1969, 325 and 328
$P_2Ge_5H_{14}$	$[(GeH_3)_2P]_2GeH_2$	H, P (−36·4, 42·0), F (−5·5, 41·5)	1967, 98; 1969, 69
P_2HO_5	$HP_2O_5^{3\ominus}$	H, P	1967, 28; 1968, 113
P_2HO_6	$HP_2O_6^{2\ominus}$	P	1967, 129
$P_2H_2Na_2O_7$	$P_2O_7H_2Na_2$	P (−7·9)	1969, 321
$P_2H_2O_5$	$H_2P_2O_5^{2\ominus}$	H, P	1966, 131; 1968, 37
$P_2H_2O_6$	$H_2P_2O_6^{\ominus}$	P	1967, 129
$P_2H_3O_6$	$H_3P_2O_6$	P	1967, 129
P_2H_4	P_2H_4	H, P	1967, 28; 1968, 113
$P_2H_4O_5$	$O[P(OH)_2]_2$	H, P	1966, 208; 1967, 189
$P_2H_4O_6$	$H_4P_2O_6$	P	1967, 129
P_2I_4	P_2I_4	P	1967, 190
$P_2K_4O_7$	$K_4P_2O_7$	P	1969, 5
$P_2N_2O_6$	$[NP(O)O_2^{2\ominus}]_2$	P (3·3)	1965, 78
$P_2Na_4O_7$	$Na_4P_2O_7$	P	1969, 5
P_2Zn_3	Zn_3P_2	P	1969, 329
$P_3BrCl_5N_3$	$P_3N_3BrCl_5$	P (17·7; −7·8)	1966, 94
$P_3BrF_5N_3$	$P_3N_3F_5Br$	F (−20; −71)	1969, 151 and 152
$P_3Br_2Cl_4N_3$	$P_3N_3Br_2Cl_4$	P (16·1; −8·7; −38·6)	1966, 94
$P_3Br_3Cl_3N_3$	$P_3N_3Br_3Cl_3$	P (14·0; −10·0; −39·8)	1966, 94
$P_3Br_4Cl_2N_3$	$P_3N_3Br_4Cl_2$	P (13·9; −12·1; −41·3)	1966, 94
$P_3Br_5ClN_3$	$P_3N_3Br_5Cl$	P (−14·0; −42·5)	1966, 94
$P_3Br_6N_3$	$P_3N_3Br_6$	P (−45·4)	1966, 93
$P_3ClF_5N_3$	$P_3N_3F_5Cl$	F (−32; −71)	1969, 151 and 152
$P_3Cl_4F_2N_3$	$P_3N_3Cl_4F_2$	P	1966, 79; 1967, 130
$P_3Cl_4H_4N_5$	$P_3N_3Cl_4(NH_2)_2$	P (18; 9)	1967, 145 and 147; 1968, 114
$P_3Cl_5FN_3$	$P_3N_3Cl_5F$	P (14·4; 23); F	1966, 49; 1967, 130
$P_3Cl_5H_2N_4$	$P_3N_3Cl_5(NH_2)$	P (20·4; 19·0)	1967, 145
$P_3Cl_6N_3$	$P_3N_3Cl_6$	P (19)	1966, 6 and 49; 1967, 147; 1968, 136 and 144; 1969, 216
P_3FO_7	$P_3O_7F^{4\ominus}$	P (−5; −16·8; −20·4)	1966, 79
$P_3F_6N_3$	$P_3N_3F_6$	P (13·9); F (−7·3)	1968, 130
$P_3H_{12}N_9$	$P_3N_3(NH_2)_6$	P (18·8)	1968, 114
P_3O_{10}	$P_3O_{10}^{5\ominus}$	P (−6·8; −22·4)	1965, 30

Formula	Compound	Resonance	Reference
P_4	P_4	$P(-450; -462)$	1965, 1; 1968, 96
$P_4AlCl_{14}N_3$	$Cl(Cl_2P{=}N)_3PCl_3^{\oplus}, AlCl_4^{\ominus}$	P	1968, 166
$P_4BCl_{14}N_3$	$Cl(Cl_2P{=}N)_3PCl_3^{\oplus}, BCl_4^{\ominus}$	B; P	1968, 166
$P_4BrF_7N_4$	$P_4N_4F_7Br$	$F(-30 \cdot 6)$	1969, 151
$P_4ClF_7N_4$	$P_4N_4F_7Cl$	$F(-38 \cdot 9)$	1969, 151
$P_4Cl_8N_4$	$P_3N_3Cl_5(N{=}PCl_3)$	$P(20 \cdot 5; -2 \cdot 2; -3 \cdot 3)$	1967, 145
	$(PNCl_2)_4$	P	1969, 216

FORMULAE INDEX II
NMR DATA ON ORGANIC COMPOUNDS[a]

Formula	Compound	Resonance	Reference
$PCBr_2NS_2$	$SPBr_2(NCS)$	P (−62·8)	1965, 4
$PCCl_2F_3O$	$CF_3P(O)Cl_2$	F (−77·5)	1968, 70
$PCCl_2F_5$	$CF_3PF_2Cl_2$	P (−22·7); F (−77·0; −0·5)	1969, 66
$PCCl_2NOS$	$OPCl_2(NCS)$	P (−21)	1965, 4
$PCCl_3F_2$	CCl_3PF_2	F (−88); P (130·9)	1967, 48 and 51
$PCCl_3F_4$	CF_3PCl_3F	F (−79·8; 126·1)	1967, 62
$PCCl_3O_3$	$CCl_3PO_3^{2\ominus}$	P (8·1)	1967, 7
$PCCl_4F_3$	CF_3PCl_4	F (−81)	1967, 62
$PCCsF_8$	CF_3PF_5, Cs	F (−71·3; −75·5)	1968, 48
PCF_2N	F_2PCN	P (140·8); F (−91·6)	1966, 28
PCF_2NO	F_2PNCO	F (−39·2)	1969, 56
PCF_2NOS	$OPF_2(NCS)$	P (−36·4); F (−71·6)	1967, 53 and 287
PCF_2NO_2	$OPF_2(NCO)$	P (−29·5; −30); F (−72)	1967, 287 and 288
PCF_2NS	$F_2P(NCS)$	P (132); F (−54·2)	1967, 53; 1968, 289
PCF_2NS_2	$SPF_2(NCS)$	P (−28·6); F (−36·6)	1967, 53 and 289
PCF_5	CF_3PF_2	P (153·3); F (−87·7; −104·1)	1967, 48 and 51
PCF_5O	$CF_3P(O)F_2$	F (−73·9; −80·7)	1968, 44
$PCHCl_2O_3$	$CHCl_2PO_3^{2\ominus}$	P (8·2)	1967, 7
PCH_2ClF_2	CH_2ClPF_2	H; P (201); F (−99)	1966, 36; 1967, 51
PCH_2ClF_2O	$CH_2ClP(O)F_2$	H; P (12)	1966, 36; 1967, 14
PCH_2ClF_2S	$CH_2ClP(S)F_2$	H	1966, 36
PCH_2ClF_4	CH_2ClPF_4	H	1966, 36

[a] Formulae have symbols arranged alphabetically except for phosphorus which is always placed first followed immediately by carbon and hydrogen.

Formula	Compound	Resonance	Reference
PCH$_2$ClO$_3$	CH$_2$ClPO$_3^{2\ominus}$	P (11·3)	1967, 7
PCH$_2$Cl$_2$FO	CH$_2$ClP(O)FCl	P (32·0); F (−63)	1967, 14
PCH$_2$Cl$_3$	CH$_2$ClPCl$_2$	H	1966, 36
PCH$_2$Cl$_3$O	CH$_2$ClP(O)Cl$_2$	H	1966, 36
PCH$_2$Cl$_3$S	CH$_2$ClP(S)Cl$_2$	H	1966, 36
PCH$_2$F$_3$	CF$_3$PH$_2$	H; P; F	1967, 40 and 83
PCH$_3$ClF	CH$_3$PFCl	P (240)	1968, 12
PCH$_3$ClFO$_2$	CH$_3$OP(O)FCl	P (−5·4); F (−50)	1967, 14
PCH$_3$Cl$_2$	CH$_3$PCl$_2$	^{13}C; H; P (192)	1966, 36; 1968, 20, 42 and 290
PCH$_3$Cl$_2$O	CH$_3$P(O)Cl$_2$	^{13}C; H; P	1966, 36; 1967, 186; 1969, 74
PCH$_3$Cl$_2$O	CH$_3$OPCl$_2$	H; P (180·8)	1966, 36; 1967, 236
PCH$_3$Cl$_2$O$_2$	CH$_3$OP(O)Cl$_2$	^{13}C; P	1968, 3
PCH$_3$Cl$_2$S	CH$_3$P(S)Cl$_2$	^{13}C; H; P (79·4)	1966, 36 and 114; 1968, 20
PCH$_3$F$_2$	CH$_3$PF$_2$	H; P (244·2, 245, 250·7); F (−92·7, −92·9, −98·8?)	1966, 36 and 210; 1967, 38, 51, 235 and 290; 1968, 42 and 81
PCH$_3$F$_2$O	CH$_3$P(O)F$_2$	^{13}C; H; P (26·8, 27·4); F (−59·5)	1965, 12; 1966, 36; 1967, 14; 1968, 20; 1969, 24
PCH$_3$F$_2$OS	CHF$_2$P(O)H$_2$	H; F	1967, 103
PCH$_3$F$_2$S	CH$_3$OP(S)F$_2$	H; F (−48·0)	1968, 45
PCH$_3$F$_2$S	CH$_3$P(S)F$_2$	H	1966, 36
PCH$_3$F$_2$S$_2$	CH$_3$SP(S)F$_2$	H; F (−26·7)	1968, 45
PCH$_3$F$_4$	CH$_3$PF$_4$	H; P (−29·9); F (−45·6)	1966, 36; 1968, 42
PCH$_3$F$_4$S	CH$_3$SPF$_4$	H; P (−34); F (−14·1, −19·8, −66·1)	1968, 291; 1969, 62
PCH$_3$F$_5$	CH$_3$PF$_5^{\ominus}$	P (−126·4); F (−45·8, −57·6)	1965, 12
PCH$_3$Na$_2$O$_4$	CH$_3$OPO$_3$Na$_2$	H	1967, 90
PCH$_3$O$_3$	CH$_3$PO$_3^{2\ominus}$	P (20·1)	1967, 7
PCH$_3$O$_4$	CH$_2$OHPO$_3^{2\ominus}$	P (15·5)	1967, 7

PCH₄Cl	CH₂ClPH₂	H; P (18·5)	1966, 211; 1968, 58
PCH₄ClO₃	CH₂ClPO₃H₂	H; P (140·5); F (−70·6)	1965, 79; 1966, 212
PCH₄FN	CH₃NHPF	P (32·2); F (−63)	1968, 119
PCH₄FO₂	CH₃P(O)(OH)F	H; F (−175·8, −94·3)	1967, 14
PCH₄F₃	CH₃PF₃H (?)	H; P (−7·7); F (−15·1, −96·5)	1967, 157; 1968, 38
PCH₄F₃	CH₃PF₃H (?)	¹³C; H; P	1968, 42
PCH₅	CH₃PH₂		1966, 4 and 13; 1967, 11; 1968, 22 and 33; 1969, 41 and 330
PCH₅O₂	CH₃PO₂H₂	H	1966, 211; 1968, 72
PCH₅O₃	CH₃PO₃H₂	H	1966, 36
	CH₃OPO₂H₂	H	1966, 213
PCH₅O₄	CH₃OHPO₃H₂	H; P (23·5)	1965, 79; 1966, 212; 1967, 291
PCH₆	CH₃PH₃⁺	H	1966, 22
PCH₆NO₂S	CH₃SP(O)(OH)NH₂	H	1968, 86
PCH₆NO₃	⁺NH₃CH₂P(O)(OH)O⁻	H	1966, 214
PCH₆N₂NaO₃S	(NH₂)₂CCH₂SPO₃·Na	H	1968, 292
PCH₇N₂OS	CH₃SP(O)(NH₂)₂	H	1968, 86
PCH₇Si	CH₃P(SiH₃)H	H	1969, 73
PCH₈B₁₀Br₃	B₁₀H₇Br₃CHP	H	1969, 25
PCH₉B₁₀Br₂	B₁₀H₈Br₂CHP	¹¹B	1969, 25
PCH₉IN₃S	CH₃SP(NH₂)₃I⁺	H	1968, 86
PCH₉Si₂	CH₃P(SiH₃)₂	H	1969, 73
PCHB₁₀Br	B₁₀H₉BrCHP	H	1969, 25
PCH₁₁B₉	B₉H₁₀CHP⁻	¹¹B; H	1969, 25
PCH₁₁B₁₀	B₁₀H₁₀CHP (isom.)	¹¹B; H; P	1967, 292; 1969, 25
PCH₁₂B₉	B₉H₁₀CHPH	¹¹B; H	1969, 25
PC₂BrF₆S	(CF₃)₂P(S)Br	F (−70·1)	1968, 47
PC₂BrN₂S₃	BrP(S)(NCS)₂	P (−28·4)	1965, 4

FORMULAE INDEX II

Formula	Compound	Resonance	Reference
PC$_2$ClF$_6$	(CF$_3$)$_2$PCl	F (-61.4)	1962, 6; 1963, 6 and 7; 1968, 293
PC$_2$ClF$_6$S	(CF$_3$)$_2$P(S)Cl	F (-70.5)	1968, 47
PC$_2$ClN$_2$OS$_2$	ClP(O)(NCS)$_2$	P (-41.5)	1965, 4
PC$_2$ClF$_9$N	(CF$_3$)$_2$NPF$_3$Cl	F (-55.3; -5.5)	1966, 31
PC$_2$Cl$_2$F$_3$	CF$_2$=CFPCl$_2$	F (-82.5, -103.6, -186.4)	1969, 54 and 55
PC$_2$Cl$_2$F$_5$	C$_2$F$_5$PCl$_2$	P; F	1967, 40
PC$_2$Cl$_2$F$_8$N	(CF$_3$)$_2$NPF$_2$Cl$_2$	F (-52.9, 64.0)	1966, 31
PC$_2$Cl$_3$F$_6$	(CF$_3$)$_2$PCl$_3$	F (-79.8)	1967, 62
PC$_2$Cl$_4$F$_3$	CF$_2$ClCFClPCl$_2$	F (-60.3, -130.8)	1969, 55
PC$_2$CsF$_{10}$	(CF$_3$)$_2$PF$_4$Cs	F (-71.9, -78.5)	1968, 48
PC$_2$FN$_2$OS$_2$	FP(O)(NCS)$_2$	P (-47.7); F (-55)	1967, 287
PC$_2$FN$_2$S$_2$	FP(NCS)$_2$	P (126.5); F (-80.3)	1967, 53; 1968, 289
PC$_2$FN$_2$S$_3$	FP(S)(NCS)$_2$	P (12.2); F (-27.4)	1967, 53 and 289
PC$_2$F$_5$	CF$_2$=CFPF$_2$	F (-79.9, -112.0, -199.1, -103.6)	1969, 54 and 55
PC$_2$F$_5$O$_2$	CF$_3$C(O)OPF$_2$	F (-76.5, -49.9)	1969, 56
PC$_2$F$_6$IS	IP(S)(CF$_3$)$_2$	F (-70.1)	1968, 47
PC$_2$F$_7$	(CF$_3$)$_2$PF	P; F (-66.5, -219)	1967, 40, 48 and 55
	CF$_2$=CFPF$_4$	F (-64.7, -79.4, -177.8, -54.9)	1969, 114
PC$_2$F$_7$S	(CF$_3$)$_2$P(S)F	F (-72.3, -94.5)	1968, 47
PC$_2$F$_8$NO	(CF$_3$)$_2$NP(O)F$_2$	F (-54.8, -70.5)	1966, 31
PC$_2$HCl$_3$F$_3$	CF$_2$ClCFHPCl$_2$	H; F (-59.2, -196.2)	1969, 55
PC$_2$HF$_6$	(CF$_3$)$_2$PH	H; F; P (49.8)	1967, 40; 1969, 43
PC$_2$HF$_6$S$_2$	(CF$_3$)$_2$P(S)SH	H; F (-71.0)	1968, 47 and 294
PC$_2$HF$_9$	(CF$_3$)$_2$PF$_3$H$^\ominus$	H; F (-72.7, -94.3, -58.5)	1968, 41
PC$_2$H$_2$Cl$_2$F$_3$	CHF$_2$CHFPCl$_2$		1966, 26

NMR DATA ON ORGANIC COMPOUNDS 107

PC₂H₂F₆NS	(CF₃)₂P(S)NH₂	F (−73·5)	1968, 47
PC₂H₂NO₃	N≡CCH₂PO₃²⁻	P (13·1)	1967, 7
PC₂H₂O₅	⁻O₂C·CH₂PO₃²⁻	P (13·3)	1967, 7
PC₂H₃ClF₃	CHClFCF₂PH₂	H; F; P (−140·3)	1966, 215
PC₂H₃Cl₂F₂	CHF₂CH₂PCl₂	F	1965, 80
PC₂H₃F₂O₂	CH₃C(O)OPF₂	F (−54·4)	1969, 56
PC₂H₃F₄	CHF₂CH₂PF₂	F	1965, 80
PC₂H₄ClO₂	(CH₂O)₂PCl	H; P (167)	1966, 50; 1968, 77; 1969, 45
PC₂H₄Cl₃	CH₂ClCH₂PCl₂	H; P (185·0)	1969, 331
	CH₃CHClPCl₂	H; P (161·6)	1969, 331
PC₂H₄Cl₃O	CH₂ClCH₂P(O)Cl₂	H; P (42·9)	1966, 220; 1969, 331
	CH₃CHClP(O)Cl₂	H; P (45·5)	1969, 331
	(CH₂Cl)₂P(O)Cl	H; P (−49·3)	1969, 332
	CH₂ClCH₂OPCl₂	H; P (179·0)	1969, 331
	CH₂ClCH₂OP(O)Cl₂	H; P (5·9)	1969, 331
PC₂H₄Cl₃O₂	Cl₂P(S)CH₂CH₂Cl	H; P (78·8)	1969, 79 and 331
PC₂H₄Cl₃S	Cl₂P(S)CHClCH₃	H; P (86·9)	1969, 331
PC₂H₄FO₂	(CH₂O)₂PF	H; F (−41·3)	1963, 3; 1966, 50
PC₂H₄F₃	CHF₂CHFPH₂	H; F (−128·2, −195·6)	1966, 86
	CH₂FCF₂PH₂	H; F (−228·7, −96·3)	1966, 86
P₂C₂H₅	(CH₂)₂PH	P (−341)	1969, 169
P₂C₂H₅ClFO₂	C₂H₅OP(O)FCl	F (−47); P (−5·2)	1967, 14
PC₂H₅Cl₂O	C₂H₅P(O)Cl₂	P (53·0)	1966, 111
	C₂H₅OPCl₂	H; P (178)	1966, 216; 1967, 84 and 236; 1968, 25
PC₂H₅Cl₂OS	C₂H₅OP(S)Cl₂	H	1966, 216; 1967, 84; 1968, 25
PC₂H₅Cl₂O₂	C₂H₅OP(O)Cl₂	¹³C; H; P	1966, 216; 1967, 84; 1968, 3
PC₂H₅Cl₂S	C₂H₅P(S)Cl₂	P (95·4)	1966, 114
PC₂H₅F₂	C₂H₅PF₂	F (−106·5); P (234, 245)	1967, 38, 51 and 290; 1968, 12
PC₂H₅F₂O	C₂H₅P(O)F₂	P (29·2)	1967, 14
PC₂H₅F₂O₂	C₂H₅OP(O)F₂	H; F (−85·6); P (−20·9)	1967, 14 and 282

Formula	Compound	Resonance	Reference
PC$_2$H$_5$F$_4$S	C$_2$H$_5$SPF$_4$	F (-13.4, -21.4, -70.5); P (-34)	1968, 291; 1969, 62
PC$_2$H$_5$Na$_2$O$_4$	C$_2$H$_5$OPO$_3$Na$_2$	H	1967, 90
PC$_2$H$_5$O$_2$	HO$_2$C·CH$_2$PH$_2$	H; P (-142.6)	1967, 135
PC$_2$H$_5$O$_3$	C$_2$H$_5$PO$_3^{2\ominus}$	P (24.1)	1967, 7
PC$_2$H$_6$BrO$_3$	(CH$_3$O)$_2$P(O)Br	H	1969, 95
PC$_2$H$_6$BrS	(CH$_3$)$_2$P(S)Br	H	1966, 36
PC$_2$H$_6$Cl	(CH$_3$)$_2$PCl	H; P (96.5)	1966, 36; 1968, 42 and 290
PC$_2$H$_6$ClFN	(CH$_3$)$_2$NPFCl (?)	H; F (-70.1)	1969, 55
PC$_2$H$_6$ClO	(CH$_3$)$_2$P(O)Cl	H	1966, 36
PC$_2$H$_6$ClO$_2$	(CH$_3$O)$_2$PCl	H; P (169)	1965, 20; 1968, 211
	CH$_2$Cl(CH$_3$O)P(O)H	H	1966, 124
	CH$_3$(CH$_3$O)P(O)Cl	H; P	1967, 186
PC$_2$H$_6$ClO$_2$S	(CH$_3$O)$_2$P(S)Cl	H; P (37.1)	1966, 36; 1969, 95
	(CH$_3$O)(CH$_3$S)P(O)Cl	H; P (8)	1965, 20
PC$_2$H$_6$ClO$_3$	(CH$_3$O)$_2$P(O)Cl	H	1967, 69; 1969, 95
PC$_2$H$_6$ClS	(CH$_3$)$_2$P(S)Cl	H	1966, 36
PC$_2$H$_6$Cl$_2$F$_2$N	(CH$_3$)$_2$NPF$_2$Cl$_2$	H; F (54.3)	1968, 43
PC$_2$H$_6$Cl$_2$N	(CH$_3$)$_2$NPCl$_2$	H	1965, 18; 1966, 36
PC$_2$H$_6$Cl$_2$NO	(CH$_3$)$_2$NP(O)Cl$_2$	H; P (16)	1962, 7; 1965, 18 and 23; 1966, 36; 1968, 91 and 136
PC$_2$H$_6$Cl$_2$NS	CH$_3$O(CH$_3$)NPCl$_2$	H	1969, 103
	(CH$_3$)$_2$NP(S)Cl$_2$	H	1965, 18 and 23; 1968, 91
PC$_2$H$_6$F	(CH$_3$)$_2$PF	H; F (-195.5); P (185, 187)	1967, 54; 1968, 12, 42 and 290
PC$_2$H$_6$FO	(CH$_3$)$_2$P(O)F	H; F (-68.5); P (66.7)	1966, 36; 1967, 14
PC$_2$H$_6$FO$_2$	CH$_3$(CH$_3$O)P(O)F	F (-63.4; 2.2)	1967, 13 and 14; 1968, 295
PC$_2$H$_6$FO$_2$S	(CH$_3$O)$_2$P(S)F	H	1966, 36
PC$_2$H$_6$FO$_3$	(CH$_3$O)$_2$P(O)F	F (-87; 26)	1967, 13 and 14; 1968, 295
	C$_2$H$_5$OP(O)(OH)F	F (-84); P (-11)	1967, 14

NMR DATA ON ORGANIC COMPOUNDS

Formula	Compound	Nuclei (shifts)	References
PC_2H_6FS	$(CH_3)_2P(S)F$	H	1966, 36
$PC_2H_6F_2N$	$(CH_3)_2NPF_2$	H; F ($-65\cdot3$); P (143)	1966, 30 and 36
$PC_2H_6F_2NO$	$(CH_3)_2NP(O)F_2$	H; F (-82)	1966, 36; 1967, 54
	$CH_3O(CH_3)NPF_2$	H	1969, 103
$PC_2H_6F_2NS$	$(CH_3)_2NP(S)F_2$	H; F (-52); P ($75\cdot7$)	1966, 30; 1968, 46
$PC_2H_6F_3$	$(CH_3)_2PF_3$	H; F ($-2\cdot6, -86\cdot4$); P ($8\cdot0$)	1966, 36; 1967, 54; 1968, 42
	$C_2H_5PF_3H$	H; F ($-178\cdot8, -105\cdot2$)	1968, 38 and 39
$PC_2H_6F_3N_2O_2S$	$F_3P(NCH_3)_2SO_2$	H; F ($-77\cdot8$); P ($-76\cdot8$)	1968, 125; 1969, 65
$PC_2H_6F_3S$	$CH_3(CH_3S)PF_3$	F ($-75\cdot2, 1\cdot2$); P ($2\cdot0$)	1969, 62
$PC_2H_6F_4N$	$(CH_3)_2NPF_4$	H; F (-69); P	1966, 217; 1967, 59; 1969, 64
$PC_2H_6F_5O$	$(CH_3)_2OPF_5$	H	1966, 139
$PC_2H_6F_5S$	$(CH_3)_2SPF_5$	H	1966, 139
$PC_2H_6NaO_3S$	$(CH_3O)_2POSNa$	H	1969, 325
$PC_2H_6O_2$	$(CH_3)_2P(O)O^{\ominus}$	P ($37\cdot5$)	1967, 9
PC_2H_7	$(CH_3)_2PH$	^{13}C; H; P	1965, 3; 1966, 13; 1967, 11; 1968, 22 and 33; 1969, 330
	$C_2H_5PH_2$	P (-128)	1966, 4
$PC_2H_7F_2$	$(CH_3)_2PHF_2$	H; F ($-26\cdot6$); P ($-31\cdot7$)	1968, 42
$PC_2H_7F_3N$	$CH_3(CH_3NH)PF_3$	F ($-50, -75$)	1967, 64
$PC_2H_7F_5N$	$(CH_3)_2NHPF_5$	F ($-70\cdot8$)	1966, 217
$PC_2H_7O_2$	$(CH_3)_2PO_2H$	H	1967, 19; 1968, 65
	$CH_3(CH_3O)P(O)H$	P (32)	1967, 172
$PC_2H_7O_2S$	$(CH_3O)_2P(S)H$	P ($74\cdot9$)	1966, 218
	$CH_3(CH_3O)P(S)OH$		1967, 72
$PC_2H_7O_3$	$(CH_3O)_2P(O)H$	^{13}C; ^{17}O; H; P ($9\cdot5, 11\cdot2$)	1963, 1; 1964, 14; 1965, 81; 1966, 36; 1967, 42, 44, 198, 277 and 293; 1968, 3; 1969, 74
	$(CH_3O)_2P(O)D$		
$PC_2H_7O_4$	$(CH_2OH)_2P(O)OH$	H; P ($48\cdot7$)	1965, 79; 1967, 291
	$(CH_3O)_2P(O)OH$	H; P	1966, 138
PC_2H_7S	$(CH_3)_2P(S)H$	P (5)	1966, 40

Formula	Compound	Resonance	Reference
$PC_2H_7S_2$	$(CH_3)_2PS_2H$	H	1967, 19
PC_2H_8	$(CH_3)_2PH_2^{\oplus}$	H	1966, 22; 1969, 41
$PC_2H_8B_5F_6$	$(CF_3)_2PB_5H_8$	^{11}B; F (-49.9); P (-25.1)	1968, 23
PC_2H_8Ga	$(CH_3)_2PGaH$	H	1966, 125
$PC_2H_8NO_3$	$CH_3CH(NH_2)PO_3H_2$	P (17.6)	1968, 296
PC_2H_8NS	$(CH_3)_2PSNH_2$	H	1968, 281
$PC_2H_8N_2NaO_3S$	$(CH_3NH)(NH_2)CH_2SPO_3Na$	H	1968, 292
PC_2H_9Si	$(CH_3)_2PSiH_3$	H	1969, 73
$PC_2H_{10}ClN_2$	$(CH_3)_2P(NH_2)_2Cl$	H; P (42.3)	1967, 65
$PC_2H_{11}B_5F_3$	$CH_3(CF_3)PB_5H_8$ (isom.)	^{11}B; H; F (-59.7); P (-33.5)	1968, 23
$PC_2H_{14}B_4F_2N$	$(CH_3)_2NPF_2B_4H_8$	^{11}B; H; F (-72)	1969, 141
$PC_2H_{14}B_5$	$(CH_3)_2PB_5H_8$	^{11}B; H; P (-85)	1968, 23
$PC_2H_{14}B_9$	$B_9H_{10}CHPCH_3$ (isom.)	^{11}B; H; P	1969, 25
PC_3BrF_8O	$(CF_3)_2CBrOPF_2$	F (-77.8, -45.9)	1967, 294
$PC_3Cl_2F_7$	$C_3F_7PCl_2$	H; F; P	1967, 40
$PC_3Cl_2F_9$	$C_3F_7PF_2Cl_2$	F (-82.3; -124.0, -122.4, 40.2); P (-14.3)	1969, 66
$PC_3Cl_3F_4O$	$CF_2=CClCF_2P(O)Cl_2$	F (-69.3, -72.1, -104.7)	1967, 295
PC_3CsF_{12}	$(CF_3)_3PF_3Cs$	F (-68.9, -63.3)	1968, 48
$PC_3F_7O_2$	$CF_3CF_2C(O)OPF_2$	F (-83.8, -122.1, -49.7)	1969, 56
PC_3F_8IO	$(CF_3)_2ClOPF_2$	F (-74.5, -46.9)	1967, 294
PC_3F_9S	$(CF_3)_2P(SCF_3)$	F; P	1967, 55
PC_3F_9Se	$(CF_3)_2P(SeCF_3)$	F	1968, 293
PC_3F_{11}	$C_3F_7PF_4$	F (-80.2, -123.0, -114.0, -56.7)	1969, 66
$PC_3N_3OS_3$	$OP(NCS)_3$	P (-61)	1967, 277
$PC_3N_3O_3$	$P(NCO)_3$	P (97.0)	1967, 277
$PC_3N_3O_3S$	$SP(NCO)_3$	P (12.4)	1967, 277

NMR DATA ON ORGANIC COMPOUNDS

Formula	Compound	Nuclei	Reference
PC$_3$N$_3$O$_4$	OP(NCO)$_3$	P (−40·9)	1967, 277
PC$_3$N$_3$S$_3$	P(NCS)$_3$	P (86·6)	1967, 277
PC$_3$N$_3$S$_4$	SP(NCS)$_3$	P (−9·3)	1967, 277
PC$_3$HF$_8$O	(CF$_3$)$_2$CHOPF$_2$	H; F (−75·5, −50·5); P	1967, 294
PC$_3$H$_3$BCl$_2$F$_6$N	(CF$_3$)$_2$PN(CH$_3$)BCl$_2$	^{11}B; H; P (50·5)	1968, 24
PC$_3$H$_3$Cl$_2$	CH$_2$=C=CHPCl$_2$	H	1968, 99; 1969, 115
PC$_3$H$_3$Cl$_2$O	CH$_2$=C=CHP(O)Cl$_2$	H	1966, 127; 1967, 104 and 296; 1969, 115
PC$_3$H$_3$F$_2$O	CH$_2$=C=CHP(O)F$_2$	H; F (−70·7 or −66·4)	1966, 127; 1967, 296
	CH$_3$C=CP(O)F$_2$	H; F; P	1967, 105
PC$_3$H$_3$F$_6$	CH$_3$P(CF$_3$)$_2$	H; F (−66·9); P (−5·8)	1969, 84
PC$_3$H$_3$F$_6$O	CH$_3$P(O)(CF$_3$)$_2$	H; F (−72·4)	1969, 84
	CH$_3$OP(CF$_3$)$_2$	H; F (−66·2); P (94·8)	1969, 84
	CH$_3$OP(O)(CF$_3$)$_2$	H; F (−73·3); P (2·5)	1969, 84
PC$_3$H$_3$F$_6$O$_2$	CH$_3$P(CN)$_2$	P (−81·4)	1966, 4
PC$_3$H$_3$N$_2$	Cl$_2$P(O)CH$_2$C(O)Cl	H	1966, 219 and 220
PC$_3$H$_4$Cl$_3$O$_2$	CF$_3$P(OCH$_3$)CHF$_2$	H	1965, 11
PC$_3$H$_4$F$_5$O	(CF$_3$)$_2$PNHCH$_3$	H	1968, 144
PC$_3$H$_4$F$_6$N	Cl$_2$PCH$_2$CO$_2$CH$_3$	H	1968, 297
PC$_3$H$_5$Cl$_2$O$_2$	CH$_2$=CHCH$_2$P(S)Cl$_2$	P (86·4)	1966, 114
PC$_3$H$_5$Cl$_2$S	CH$_2$=CHCH$_2$OPF$_2$	F (−48·8)	1963, 3
PC$_3$H$_5$F$_2$O	CH$_2$=CHCH$_2$PO$_3^{2\ominus}$	P (18·6)	1967, 7
PC$_3$H$_5$O$_3$	CH(CH$_2$O)$_2$PO	H	1966, 221
PC$_3$H$_5$O$_4$	CH(CH$_2$O)$_2$P(O)O$^{\ominus}$	H	1966, 221
	CH$_3$C(O)CH$_2$PO$_3^{2\ominus}$	P (9·6)	1967, 7
PC$_3$H$_6$Br$_3$	CH$_3$CHBrCH$_2$PBr$_2$	H; P (183)	1966, 222
	CH$_3$CH(PBr$_2$)CH$_2$Br	H; P (189)	1966, 222
PC$_3$H$_6$Br$_3$O	(CH$_2$Br)$_3$PO	H	1967, 7
PC$_3$H$_6$Cl$_3$N$_2$O	OC(NCH$_3$)$_2$PCl$_3$	H	1965, 6

Formula	Compound	Resonance	Reference
$PC_3H_6Cl_3O$	$(CH_2Cl)_3PO$	H; P (39·3)	1968, 298; 1969, 333
$PC_3H_6F_4NS$	$CF_3P(S)(F)N(CH_3)_2$	F (−72·3, −76·8)	1968, 47
PC_3H_6N	$H_2PCH_2CH_2CN$	P (−135)	1966, 4
PC_3H_7	$CH_2{=}CH{\cdot}CH_2PH_2$	H	1969, 169
	$(CH_2)_2PCH_3$	H; P (−251)	1969, 169
	$CH_2CH(CH_3)PH$	H	1969, 169
PC_3H_7ClFO	$i\text{-}C_3H_7P(O)ClF$	F (−53); P (49·5)	1967, 14
PC_3H_7ClFOS	$CH_3(ClCH_2CH_2S)P(O)F$	F (−43); P (63)	1967, 14
$PC_3H_7Cl_2$	$i\text{-}C_3H_7PCl_2$	P (−198·6)	1968, 6
$PC_3H_7Cl_2O$	$C_3H_7OPCl_2$	H; P (179·5)	1967, 236
	$i\text{-}C_3H_7OPCl_2$	H	1969, 94
	$CH_3(CH_2Cl)_2PO$	H; P (41·6)	1967, 297; 1968, 299; 1969, 334
$PC_3H_7Cl_2O_2$	$(CH_3O)(CH_2Cl)_2PO$	H; P (40·9)	1968, 300; 1969, 332
	$i\text{-}C_3H_7OP(O)Cl_2$	H	1969, 74
$PC_3H_7Cl_2S$	$C_3H_7P(S)Cl_2$	P (90·1)	1966, 114
	$i\text{-}C_3H_7P(S)Cl_2$	P (107·0)	1966, 114
$PC_3H_7Cl_3NO_2$	$(CH_3)_2\overset{\oplus}{N}CHOP(O)Cl_2Cl$	H	1966, 223
$PC_3H_7F_2O$	$C_3H_7OPF_2$	F (−50); P (111·5)	1963, 3; 1967, 235; 1969, 381
	$i\text{-}C_3H_7OP(O)F_2$	F (−79); P (29·5)	1967, 14
$PC_3H_7I_2O$	$CH_3(CH_2I)_2PO$	H	1967, 297; 1968, 60
$PC_3H_7O_2$	$H_2P(CH_2)_2CO_2H$	P (−136·9)	1967, 135
$PC_3H_7O_3$	$CH_3OP(OCH_2)_2$	H; P (131·6)	1966, 50 and 224; 1968, 77 and 301
	$CH_3P(O)(OCH_2)_2$	P (53)	1967, 69
PC_3H_8ClFNO	$CH_2ClP(O)[N(CH_3)_2]F$	H; P	1967, 77
PC_3H_8ClFNS	$CH_2ClP(S)[N(CH_3)_2]F$	H; P	1967, 77

NMR DATA ON ORGANIC COMPOUNDS

PC_3H_8ClOS	$CH_3P(S)(OC_2H_5)Cl$	H	1967, 72
$PC_3H_8ClO_2$	$CH_3P(O)(OC_2H_5)Cl$	H	1969, 74
$PC_3H_8ClO_3$	$CH_2ClP(O)(OCH_3)_2$	P (28)	1967, 69
PC_3H_8ClS	$CH_3P(SC_2H_5)Cl$	P (151)	1969, 335
$PC_3H_8Cl_2N$	$CH_2ClP[N(CH_3)_2]Cl$	H; P (123·0)	1965, 12; 1968, 59; 1969, 107
$PC_3H_8Cl_2NO$	$CH_2ClP(O)[N(CH_3)_2]Cl$	H(T){P}	1967, 77
$PC_3H_8Cl_2NS$	$CH_2ClP(S)[N(CH_3)_2]Cl$	H{P}	1967, 77
PC_3H_8F	$CH_3(C_2H_5)PF$	P (182)	1968, 12
PC_3H_8FO	$CH_3(C_2H_5O)PF$	F (-101); P (218)	1968, 12
PC_3H_8FOS	$C_2H_5(CH_3S)P(O)F$	F (-57); P (67)	1967, 13 and 14; 1968, 295
	$CH_3(C_2H_5S)P(O)F$	F (-45); P (61·5)	1967, 14
$PC_3H_8FO_2$	$C_2H_5(CH_3O)P(O)F$	F (-74); P (32·9)	1967, 13 and 14; 1968, 295
	$CH_3(C_2H_5O)P(O)F$	F (-63)	1967, 13 and 14; 1968, 295
$PC_3H_8F_3S$	$CH_3(C_2H_5)PF_3$	F ($-72·7$, 1·2); P (1·0)	1969, 62
$PC_3H_8NO_4$	$O(CH_2)_2NCH_2PO_3H_2$	P (5·9)	1967, 291
$PC_3H_8NO_4S$	$H_2NCH_2CHOHCH_2SPO_3^{2\ominus}$	H	1966, 204
PC_3H_9	$(CH_3)_3P$	^{13}C; H; P (-62 to -66)	1965, 1, 2 and 3; 1966, 13, 36 and 41; 1967, 244; 1968, 3, 7, 17, 22, 42, 54, 211, 290, 302 and 303; 1969, 2
$PC_3H_9AlBr_3$	$CH_3(C_2H_5)PH$	P (-77)	1965, 3
$PC_3H_9AlCl_3$	$(CH_3)_3P, AlBr_3$	^{27}Al; P	1967, 23
PC_3H_9ClN	$(CH_3)_3P, AlCl_3$	^{27}Al; P	1967, 23
PC_3H_9ClNO	$CH_3PN(CH_3)_2]Cl$	H; P (150·2)	1965, 12; 1966, 36
$PC_3H_9Cl_2N_2$	$CH_3P(O)[N(CH_3)_2]Cl$	H	1965, 82
PC_3H_9FN	$Cl_2PN(CH_3)_2N(CH_3)_2$	H	1969, 103
PC_3H_9FNO	$CH_3PN(CH_3)_2]F$	H; F (-117); P (170)	1965, 12; 1966, 36
	$CH_3P(O)[N(CH_3)_2]F$	F (-60); P (38·4)	1965, 12; 1966, 36; 1967, 13 and 14; 1968, 295
$PC_3H_9F_2$	$(CH_3)_3PF_2$	H; F ($-4·8$); P (-158)	1967, 60; 1968, 356
$PC_3H_9F_2N_2$	$F_2PN(CH_3)N(CH_3)_2$	H	1969, 103

Formula	Compound	Resonance	Reference
$PC_3H_9F_2O$	$CH_3(C_2H_5O)PF_2H$	P (−20·5)	1967, 38
$PC_3H_9F_2O_3$	$(CH_3O)_3PF_2$	H; F (−72·8)	1967, 61
$PC_3H_9F_2S$	$(CH_3)_2PF_2(SCH_3)$	F (0·2)	1969, 62
$PC_3H_9F_2SiO_2$	$(CH_3)_3SiOP(O)F_2$	H; F (−82·8); P (−28·7)	1967, 282
$PC_3H_9F_3N$	$CH_3[(CH_3)_2N]PF_3$	H; F (−68·5, −27·5); P (−37·5)	1965, 12; 1968, 304
PC_3H_9O	$(CH_3)_3PO$	H; P (48·3)	1965, 78 and 83; 1967, 175
$PC_3H_9O_2$	$CH_3(C_2H_5O)P(O)H$	H; P (32·6)	1967, 172; 1968, 72; 1969, 162
	$(CH_3)_2(CH_3O)PO$	H	1968, 65
	$CH_3(CH_3O)_2P$	H	1968, 290
$PC_3H_9O_2S$	$CH_3(C_2H_5O)P(S)OH$	H; P (99·6)	1967, 72
$PC_3H_9O_2S_2$	$(CH_3O)_2(CH_3S)PS$	H	1965, 20; 1966, 218
$PC_3H_9O_3$	$(CH_3O)_3P$	^{13}C; H; ^{17}O; P (140)	1963, 1; 1965, 7, 21 and 84; 1966, 36, 141, 169, 206, 224 and 225; 1967, 198, 218, 277, 293, 298 and 299; 1968, 3, 22 and 211; 1969, 336
	$CH_3(CH_3O)_2PO$	^{13}C; H; P (35)	1966, 126, 133 and 226; 1967, 19, 120, 186 and 300; 1968, 3
$PC_3H_9O_3S$	$(CH_3O)_3PS$	H; P (73·4)	1966, 36, 133, 206 and 208; 1967, 277
	$(CH_3O)_2(CH_3S)PO$	^{13}C; H; P	1966, 133 and 226; 1968, 3
$PC_3H_9O_3Se$	$(CH_3Se)(CH_3O)_2PO$	H	1966, 226
$PC_3H_9O_4$	$(CH_3O)_3PO$	^{13}C; H; ^{17}O; P (2·4)	1963, 1; 1965, 21, 84 and 85; 1966, 36, 133 and 226; 1967, 69, 86, 198 and 277; 1968, 3 and 73; 1969, 86, 127 and 336

NMR DATA ON ORGANIC COMPOUNDS

$PC_3H_9O_5$	$CH_3CHOHCH_2OPO_3H_2$	P (4·04)	1966, 227
	$CH_3CH(OPO_3H_2)CH_2OH$	P (5·05)	1966, 227
PC_3H_9S	$(CH_3)_3PS$	^{13}C; H; P	1966, 36; 1968, 3
PC_3H_9SSe	$(CH_3)_2P(S)SeCH_3$	H{C; P; Se}	1969, 4
$PC_3H_9S_3$	$(CH_3S)_3P$	P (124·5)	1966, 208
PC_3H_9Se	$(CH_3)_3PSe$	^{13}C; H{C; P; Se}	1968, 3; 1969, 4
PC_3H_{10}	$(CH_3)_3PH^{\oplus}$	H	1966, 22
$PC_3H_{10}Br$	$(CH_3)_3PHBr$	H; P ($-2·9$)	1968, 7
$PC_3H_{10}F_2NOSi$	$F_2P(O)NHSi(CH_3)_3$	H; F ($-69·7$); P (-3)	1969, 129
$PC_3H_{10}NO_2$	$CH_3(CH_3O)P(O)NHCH_3$	H	1966, 124
	$(CH_3)_2NCH_2PO_2H_2$	P (9·6)	1967, 301
$PC_3H_{10}NO_3$	$(CH_3)_2NCH_2PO_3H_2$	H; P (8·8)	1966, 212; 1967, 291
$PC_3H_{11}ClN$	$(CH_3)_3PNH_2·Cl$	H; P (49·6)	1967, 65; 1968, 54
$PC_3H_{11}F_5N$	$CH_3PF_5, (CH_3)_2NH_2$	H; F ($-57·4$, $-46·8$); P ($-124·9$)	1966, 228
$PC_3H_{11}N_2O_3S$	$(CH_3O)_2P(O)C(NH_2)_2SH$	H	1967, 302
$PC_3H_{11}O_7S$	$(CH_3O)_3P^{\oplus}H, HSO_4^{\ominus}$	P (26)	1969, 50
$PC_3H_{11}Si$	$(CH_3)_3SiPH_2$	H; P ($-239·0$)	1967, 29
$PC_3H_{15}B_2Be$	$(CH_3)_3PBeB_2H_6$ (?)	^{11}B	1966, 45
$PC_3H_{17}B_2Be$	$(CH_3)_3BeB_2H_8$	^{11}B	1966, 45
PC_3N_3	$P(CN)_3$	P ($-135·7$)	1966, 4; 1969, 8
$PC_3N_3OS_3$	$OP(NCS)_3$	P (97·0)	1965, 50
$PC_3N_3O_3$	$P(NCO)_3$	P (67·0)	1965, 4
$PC_3N_3S_3$	$P(NCS)_3$	P (85·2)	1965, 50
$PC_3N_3S_4$	$SP(NCS)_3$	P ($-9·5$)	1965, 4 and 50; 1967, 289
PC_4AsF_{12}	$(CF_3)_2PAs(CF_3)_2$	F ($-44·9$; -42)	1968, 55
PC_4ClF_6	$CF_2=CF)_2PCl$	F ($-81·9$; $-103·9$; $-190·0$)	1969, 55
PC_4F_8I	$(CF_2)_4PI$	F	1962, 2
PC_4F_9	$(CF_2=CF)_2PF_3$	F ($-66·8$; $-83·0$; $-180·7$; $-41·3$; $-76·8$)	1969, 55, 114

116 FORMULAE INDEX II

Formula	Compound	Resonance	Reference
$PC_4F_9O_2$	$C_3F_7C(O)OPF_2$	F (-81.4; -127.2; -119.7; -49.7)	1969, 56
$PC_4F_{12}N$	$(CF_3)_2PN(CF_3)_2$	F (-55.8; -51.7)	1968, 69
$PC_4F_{19}Cl_2N$	$(CF_3)_2NP(CF_3)_2Cl_2$	F (-48.2; -77.7)	1969, 85
$PC_4H_3Cl_2O$	$CH_2=CHC\equiv CP(O)Cl_2$	H	1965, 86
$PC_4H_3F_8O_3$	$CHF_2(CF_2)_3PO_3H_2$	F	1962, 2
$PC_4H_4ClF_4$	$CF_2=CClCF_2PHCH_3$	H; F ($ca.$ -80; -91.6)	1967, 295
PC_4H_5ClFO	$FP(O)CH_2CH=CClCH_2$	H	1967, 110
PC_4H_5ClFS	$FP(S)CH_2C(Cl)=CHCH_2$	H	1967, 110
$PC_4H_5ClO_2$	$ClP(O)OCH(=CH_2)CH=CH$	H; P (15.4)	1966, 229
$PC_4H_5Cl_2$	$Cl_2PC(CH_3)=C=CH_2$	H	1968, 99
$PC_4H_5Cl_2$	$Cl_2P(O)CH=CHCH=CH_2$	H	1965, 86
$PC_4H_5Cl_2O$	$Cl_2P(O)C(CH_3)=C=CH_2$	H	1969, 115
	$Cl_2P(O)C\equiv CC_2H_5$	H	1965, 86
	$ClP(O)CH_2CH=C(Cl)CH_2$	H	1967, 110
$PC_4H_6AsF_6$	$(CF_3)_2PAs(CH_3)_2$	H; P (10.2); F (-42.9)	1968, 55
	$(CH_3)_2PAs(CF_3)_2$	H; P (-44.0); F (-43.8)	1968, 55
$PC_4H_6BF_6O$	$(CH_3)_2BOP(CF_3)_2$	^{11}B; H; P (31.2); F (-67)	1969, 27
PC_4H_6BrO	$BrP(O)CH_2CH=CHCH_2$	H	1964, 8
PC_4H_6ClO	$ClP(O)CH_2CH=CHCH_2$	H	1967, 109
$PC_4H_6ClO_2$	$ClP(O)OC(CH_3)=CHCH_2$	H	1967, 110

$PC_4H_6Cl_3N_4O_3S$	$Cl_2P^{\oplus}NC(O)N(CH_3)C^{\oplus}(N^{\ominus}SO_2)NCH_3, Cl^{\ominus}$	H; P (44·5)	1969, 380
$PC_4H_6Cl_3O_2$	$Cl_2P(O)CH(CH_3)CH_2C(O)Cl$	H	1966, 219
	$Cl_2P(O)CH_2C(CH_3)C(O)Cl$	H	1966, 219
PC_4H_6FO	$FP(O)(CH_2CH)_2$	H	1967, 110, 303
$PC_4H_6F_4N$	$CF_2=CFPF[N(CH_3)_2]$	H; F (−90·1; −113·0; −190·0; −126·3)	1969, 55
$PC_4H_6F_6N$	$(CF_3)_2PNCH_2H_5$	H (T)	1968, 144
$PC_4H_6F_6NS$	$(CF_3)_2P(S)N(CH_3)_2$	F (−68·3)	1968, 47
$PC_4H_7Br_2Cl_2O_4$	$(CH_3O)_2P(O)OCHBrCBrCl_2$	H	1968, 284, 285; 1969, 322
$PC_4H_7ClF_3$	$(CH_3)_2PCF_2CHFCCl$	H; P (−26·8); F	1966, 215
$PC_4H_7ClF_4N$	$CHF_2CF_2PCl[N(CH_3)_2]$	H; F (−134·9; −121·6)	1968, 71
$PC_4H_7Cl_2$	$Cl_2PCH=C(CH_3)_2$	H	1966, 94
	$Cl_2PCH_2C(CH_3)=CH_2$	H	1966, 94
$PC_4H_7Cl_2F_3N$	$CHCl_2CF_2PF[N(CH_3)_2]$	H; F (−115·6; −135·4)	1968, 71
$PC_4H_7Cl_2O_2$	$Cl_2P(O)CH=CHOC_2H_5$	H	1969, 288
$PC_4H_7Cl_2O_3$	$Cl_2P(O)(CH_2)_2COOCH_3$	H	1967, 313
$PC_4H_7Cl_2O_4$	$(CH_3O)_2P(O)OCH=CCl_2$	H	1968, 284
$PC_4H_7Cl_3F_2N$	$CHCl_2CF_2PCl[N(CH_3)_2]$	H; F (−112·6)	1968, 71; 1969, 107
$PC_4H_7F_5N$	$CHF_2CF_2PF[N(CH_3)_2]$	H; F (−134·7; −126·1; −136·5)	1968, 284
$PC_4H_7F_6N_2$	$(CH_3)_2NNHP(CF_3)_2$	H; F (−63·5)	1968, 145
$PC_4H_7O_3$	$P(OCH_2)_3CH$	H	1966, 37
	$P(CH_2O)_3CH$	H	1966, 37; 1968, 53
	$CH_2OPOCH(CH_2)_2O$	H	1966, 121; 1969, 90
$PC_4H_7O_3S$	$SP(CH_2O)_3CH$	H	1968, 53
$PC_4H_7O_4$	$OP(CH_2O)_3CH$	H	1966, 37; 1968, 53
	$OP(OCH_2)_3CH$	H	1966, 37

Formula	Compound	Resonance	Reference
$PC_4H_8ClO_2$	$(CH_2O)_2POC(O)CH_3$	H	1966, 50
	$ClPO(CH_2)_2CH(CH_3)O$	H (T)	1966, 46; 1969, 92
$PC_4H_8Cl_2N$	$ClP[OCH(CH_3)]_2$	H	1966, 50
	$C_4H_8NPCl_2$	H; P (164·5)	1966, 230
$PC_4H_8Cl_3$	$Cl_2PCH_2CCl(CH_3)_2$	H	1966, 60
	$Cl_2PC(CH_2Cl)C(CH_3)_2$	H	1966, 60
$PC_4H_8Cl_3O_4$	$(CH_3O)_2P(O)CH(OH)CCl_3$	H	1968, 283, 284, 285; 1969, 322
$PC_4H_8Cl_5$	$(CH_2Cl)_4PCl$	H; P (44·7)	1969, 63
$PC_4H_8F_2N$	$C_4H_8NPF_2$	H; P (146·2); F (−67·9)	1966, 230
$PC_4H_8F_2NS$	$C_4H_8NP(S)F_2$	P (65·0); F (−48·4)	1966, 30
$PC_4H_8NaO_3S$	$O(CH_2)_2CH(CH_3)OP(S)ONa$ (sterois.)	H; P (112·9; 120)	1969, 177
PC_4H_9	$HPCH_2CHC_2H_5$ (isom.)	H; P (−271, −788)	1969, 169
$PC_4H_9Cl_2$	$i\text{-}C_4H_9PCl_2$	H	1967, 393
	$t\text{-}C_4H_9PCl_2$	P (197·5)	1968, 28
$PC_4H_9Cl_2O$	$t\text{-}C_4H_9P(O)Cl_2$	H; P (65·6)	1965, 8; 1966, 36
	$C_4H_9OPCl_2$	P (180·0)	1967, 236
$PC_4H_9Cl_2O_2$	$C_2H_5P(O)(CH_2Cl)_2$	H; P (45·7)	1968, 299; 1969, 334
$PC_4H_9Cl_2O_3$	$C_2H_5OP(O)(CH_2Cl)_2$	H; P (39·7)	1968, 300; 1969, 332
	$HP(O)(OCH_2CH_2Cl)_2$	P (8·8)	1967, 277
$PC_4H_9Cl_2S$	$C_4H_9P(S)Cl_2$	P (90·6)	1966, 114
	$i\text{-}C_4H_9P(S)Cl_2$	P (89·0)	1966, 114
$PC_4H_9Cl_3N$	$s\text{-}C_4H_9N=PCl_3$	P (−38·7)	1966, 288
$PC_4H_9F_2O$	$t\text{-}C_4H_9P(O)F_2$	H; P (31·3); F (−81·6)	1965, 8; 1966, 36
	$C_4H_9P(O)F_2$	P (29·3)	1967, 14
	$C_4H_9OPF_2$	F (−50·8)	1969, 61

NMR DATA ON ORGANIC COMPOUNDS

$PC_4H_9F_3N$	$CHF_2CH_2PF[N(CH_3)_2]$	F (-110.2; -119.5)	1965, 80
	$F_3P(CH_3)NH(CH_2CH=CH_2)$	P (-36.4); F' (-75; -48)	1967, 64
$PC_4H_9Li_2O_4$	$t\text{-}C_4H_9OP(O)(OLi)_2$	H	1965, 88
PC_4H_9O	$(CH_3)_2PC(O)CH_3$	H; P (-22.2)	1967, 346; 1968, 307
	$HP(CH_2CH_2)_2O$	H; P (-79)	1969, 44
$PC_4H_9O_2$	$H_2PCH(C_2H_5)COOH$	P (-120.8)	1967, 135
	$H_2P(CH_2)_3COOH$	P (-138.0)	1967, 135
	$(CH_3)_2P(O)C(O)CH_3$	H	1968, 307
	$(CH_2)_4P(O)OH$	^{13}C	1968, 22
$PC_4H_9O_2$	$C_2H_5P(O)(OCH_2)_2$	H{P}, P (53)	1964, 14; 1967, 69
$PC_4H_9O_3$	$(CH_3)_2P(O)CH_2COOH$	H	1965, 87
	$CH_3OPOCH_2CH(CH_3)O$ (isom.)	H; P (139; 142)	1966, 46; 1969, 94
	$(CH_3O)_2P(O)CH=CH_2$	H	1969, 82
	$HP(O)O(CH_2)_2CH(CH_3)O$	H	1969, 52
$PC_4H_9O_4$	$(CH_2O)_4PH$	H; P (-26)	1967, 39; 1968, 154, 304; 1969, 45
	$(CH_2O)_2P(O)OC_2H_5$	P (17)	1967, 69
	$(CH_3O)_2P(O)OCH=CH_2$	H	1967, 120; 1969, 127
	$CH_3OP(O)OCH_2CH(CH_3)O$	H; P (16)	1969, 94
$PC_4H_9O_5$	$(CH_3O)_2P(O)OCH_2C(O)H$	H; P (1.2)	1969, 337
$PC_4H_{10}Ba_{1/2}$	$CH_3CH(OH)CH_2OP(O)(O^\ominus)(OCH_3)Ba^\oplus_{1/2}$	H	1965, 88
$PC_4H_{10}BrO_3$	$(CH_3O)_2P(O)(CH_2)_2Br$	H; P (28.3)	1969, 145
$PC_4H_{10}Br_2N$	$Br_2PN(C_2H_5)_2$	H	1965, 18
$PC_4H_{10}ClN_2$	$ClP[N(CH_3)CH_2]_2$	H; P (167.3)	1966, 224; 1967, 305
$PC_4H_{10}ClO_2$	$(C_2H_5O)_2PCl$	H; P (165)	1965, 20; 1966, 216; 1967, 198; 1968, 25
	$(C_2H_5)(C_2H_5O)P(O)Cl$	P (45)	1966, 111

FORMULAE INDEX II

Formula	Compound	Resonance	Reference
$PC_4H_{10}ClO_2S$	$(C_2H_5O)(C_2H_5S)P(O)Cl$	H; P (34·1)	1965, 20
	$(C_2H_5O)_2P(S)Cl$	H	1966, 216; 1967, 84; 1968, 25
$PC_4H_{10}ClO_3$	$(C_2H_5O)_2P(O)Cl$	^{13}C; H; P (4)	1966, 216; 1967, 69, 84, 198; 1968, 3, 25; 1969, 74
$PC_4H_{10}ClO_3S$	$(CH_3O)_2P(O)CH_2CH_2Cl$	H; P (28·5)	1969, 81
$PC_4H_{10}Cl_2N$	$(CH_3O)_2P(O)SCH_2CH_2Cl$	H	1966, 74
	$Cl_2PN(C_2H_5)_2$	H	1965, 18
	$Cl_2PNHC(CH_3)_3$	H	1969, 112
$PC_4H_{10}Cl_2NO$	$Cl_2P(O)N(C_2H_5)_2$	H	1965, 30
$PC_4H_{10}Cl_2NS$	$Cl_2P(S)N(C_2H_5)_2$	H	1965, 30
$PC_4H_{10}FO$	$(C_2H_5)_2P(O)F$	P (71·8); F (−83·6)	1967, 13, 14; 1968, 295
	$CH_3PF(OC_3H_7)$	P (214); F (−101)	1968, 12
$PC_4H_{10}FO_2$	$C_2H_5(C_2H_5O)P(O)F$	P (33·1); F (−68·5)	1967, 13, 14; 1968, 295
	$CH_3PF(Oi-C_3H_7)$	H; P (27·5); F (−58·4)	1967, 13, 14; 1969, 74
	$CH_3PF(OC_3H_7)$	P (26·0); F (−63)	1967, 14
	$(C_2H_5O)_2PF$	F (−60·4)	1969, 61
$PC_4H_{10}FO_2S$	$(C_2H_5O)(C_2H_5S)P(O)F$	P (26·6)	1967, 14
$PC_4H_{10}FO_3$	$(C_2H_5O)_2P(O)F$	P (−9·2); F (−82·3)	1967, 13, 14; 1968, 295
$PC_4H_{10}F_2N$	$(C_2H_5)_2NPF_2$	H; P (147;144); F (−64·8)	1966, 230; 1967, 48, 235
$PC_4H_{10}F_2NO$	$(C_2H_5)_2NP(O)F_2$	P (−3·6)	1967, 13
$PC_4H_{10}F_2NS$	$(C_2H_5)_2NP(S)F_2$	H; P (71·3); F (−48·9)	1966, 30
$PC_4H_{10}F_4N$	$(C_2H_5)_2NPF_4$	F (−68)	1966, 217
$PC_4H_{10}KO_2S_2$	$(C_2H_5O)_2PS_2K$	H	1969, 325
$PC_4H_{10}KO_3S$	$(C_2H_5O)_2POSK$	H	1969, 325
$PC_4H_{10}Li$	$(CH_3)_2(LiCH_2)P=CH_2$	H	1968, 106, 107
$PC_4H_{10}NO_2$	$(CH_2O)_2PN(CH_3)_2$	P (140)	1968, 301; 1969, 45

PC$_4$H$_{10}$NO$_3$	CH$_3$OPO(CH$_2$)$_2$NCH	H	1969, 171
PC$_4$H$_{10}$NaO$_2$S$_2$	HP[O(CH$_2$)$_2$O][O(CH$_2$)$_2$NH]	H; P (-36.4)	1968, 304; 1969, 45
PC$_4$H$_{10}$NaO$_2$S$_2$	(C$_2$H$_5$O)$_2$PS$_2$Na	H	1969, 325
PC$_4$H$_{11}$	CH$_3$(C$_3$H$_7$)PH	P (-84)	1965, 3
	(C$_2$H$_5$)$_2$PH	P (-57)	1965, 3
	(CH$_3$)$_2$PC$_2$H$_5$	P (-48)	1965, 3
	(CH$_3$)$_3$P=CH$_2$	H (T)	1967, 306; 1968, 105, 107, 305
PC$_4$H$_{11}$Cl$_2$OSi	(CH$_3$)$_3$SiCH$_2$P(O)Cl$_2$	H	1966, 231
PC$_4$H$_{11}$F$_5$N	PF$_5 \cdot$NH(C$_2$H$_5$)$_2$	F (-80.5; -68)	1966, 217
PC$_4$H$_{11}$N$_2$O$_2$	HP[O(CH$_2$)$_2$NH]$_2$	P (-54)	1969, 45
PC$_4$H$_{11}$O$_2$	(CH$_3$)$_2$P(O)OC$_2$H$_5$	H; P (50.3)	1966, 232; 1968, 65
	(C$_2$H$_5$)$_2$PO$_2$H	H	1968, 65
PC$_4$H$_{11}$O$_2$S	(C$_2$H$_5$O)$_2$P(S)H	P (68.4)	1966, 218; 1968, 25
	CH$_3$(C$_3$H$_7$O)POSH	H	1967, 72
PC$_4$H$_{11}$O$_2$S$_2$	(C$_2$H$_5$O)$_2$PS$_2$H	H	1967, 84
PC$_4$H$_{11}$O$_3$	(C$_2$H$_5$O)$_2$P(O)H	^{13}C; H; P (7.5)	1965, 81; 1966, 126; 1967, 42, 198, 277, 254, 293; 1968, 3, 25, 165, 306
PC$_4$H$_{11}$O$_3$S	t-C$_4$H$_9$PO$_3$H$_2$	H; P (37.0)	1965, 8
	(CH$_3$O)$_2$P(O)SC$_2$H$_5$	H	1966, 133
	(C$_2$H$_5$O)$_2$P(O)SH	H	1967, 307
PC$_4$H$_{11}$O$_4$	C$_4$H$_9$OPO$_3$H$_2$	P	1966, 117
	(C$_2$H$_5$O)$_2$PO$_2$H	P (0.6)	1969, 321
PC$_4$H$_{11}$S	(C$_2$H$_5$)$_2$P(S)H	P (31)	1966, 40
	(CH$_3$)$_2$PCH$_2$SCH$_3$	H	1969, 128
PC$_4$H$_{12}$	P(CH$_3$)$_4^{\oplus}$	H	1966, 22

Formula	Compound	Resonance	Reference
$PC_4H_{12}BF_4O_4$	$(CH_3O)_4P^{\oplus}BF_4^{\ominus}$	H; P (1·9)	1969, 86
$PC_4H_{12}Br$	$(CH_3)_4PBr$	P (25·2)	1966, 5
$PC_4H_{12}Cl$	$(CH_3)_4PCl$	H; P (24·4)	1966, 39; 1968, 54, 107
$PC_4H_{12}ClN_2$	$[(CH_3)_2N]_2PCl$	H; P (158·7)	1965, 12, 18; 1966, 36; 1967, 80
$PC_4H_{12}ClN_2O$	$[(CH_3)_2N]_2P(O)Cl$	H; P (30)	1965, 18, 23; 1966, 36; 1968, 91, 136
$PC_4H_{12}ClN_2O_2$	$ClP[N(CH_3)OCH_3]_2$	H	1969, 103
$PC_4H_{12}ClN_2S$	$[(CH_3)_2N]_2P(S)Cl$	H	1965, 18, 23; 1966, 36; 1968, 91
$PC_4H_{12}Cl_6O_3SSb$	$(CH_3S)(CH_3O)_3P, SbCl_6$	H; P (53·2)	1969, 87
$PC_4H_{12}Cl_6O_3Sb$	$CH_3(CH_3O)_3P, SbCl_6$	H; P (56)	1967, 69
$PC_4H_{12}Cl_6O_4Sb$	$(CH_3O)_4P, SbCl_6$	H; P (51·5)	1967, 69, 308
$PC_4H_{12}FN_2$	$[(CH_3)_2N]_2PF$	H; P (150·8); F (−99·6)	1965, 12; 1966, 36
$PC_4H_{12}FN_2O$	$[(CH_3)_2N]_2P(O)F$	H; P (23·9); F (−83·6)	1966, 36; 1967, 14, 309
$PC_4H_{12}FN_2O_2$	$FP[N(CH_3)OCH_3]_2$	H	1969, 103
$PC_4H_{12}FN_2S$	$FP(S)[N(CH_3)_2]_2$	H; F (−62·3)	1968, 46
$PC_4H_{12}FN_4O_4S_2$	$FP[(NCH_3)_2SO_2]_2$	H; P (−85·0); F (−88·5)	1968, 125; 1969, 65
$PC_4H_{12}F_3N_2$	$F_3P[N(CH_3)_2]_2$	H; F (−55·3; −74·0)	1966, 217
$PC_4H_{12}I$	$(CH_3)_4PI$	^{13}C; P (25·3)	1967, 244; 1968, 3; 1969, 2
$PC_4H_{12}N$	$(CH_3)_2NP(CH_3)_2$	H; P (39)	1965, 14; 1966, 38; 1968, 290; 1969, 72
$PC_4H_{12}NO_2$	$CH_3(CH_3O)P(O)N(CH_3)_2$	H	1966, 124
$PC_4H_{12}NO_3$	i-$C_3H_7CH(NH_2)PO_3H_2$	P (16·4)	1968, 296
$PC_4H_{12}NS$	$(CH_3)_2NP(S)(CH_3)_2$	H	1966, 36
$PC_4H_{12}N_3O$	$(CH_3)_2NP(NCH_3)_2CO$	H; P (93·6)	1968, 126
$PC_4H_{13}N_2$	$(CH_3)_3CP(NH_2)_2$	H	1968, 65
$PC_4H_{13}N_2O$	$[(CH_3)_2N]_2P(O)H$	H; P (20·5)	1966, 82; 1967, 172

Formula	Structure	Nuclei	References
PC$_4$H$_{14}$ClN$_2$	(CH$_3$)$_2$P[N(CH$_3$)$_2$]NH$_2$Cl	H; P (54·0)	1967, 65
PC$_4$H$_{14}$N$_3$O	[(CH$_3$)$_2$N]P(O)NH$_2$	H	1965, 23; 1968, 91
PC$_5$F$_{15}$N$_2$	[(CF$_3$)$_2$N]$_2$PCF$_3$	F (−58·6; −51·5)	1968, 69
PC$_5$H$_6$ClF$_4$	CF$_2$=CClCF$_2$PH(C$_2$H$_5$)	H; F (−80·0; −89·4)	1967, 295
PC$_5$H$_6$ClF$_4$O$_3$	CF$_2$=CClCF$_2$P(O)(OCH$_3$)$_2$	H; F (−74·8; −77·0; −108·0)	1967, 295
	CF$_3$CCl=CFP(O)(OCH$_3$)$_2$	H; F (cis −62·0; −98·2 trans −64·0; −106·8)	
PC$_5$H$_6$Cl$_3$OS	CH$_3$SCH=CHCCl=CHP(O)Cl$_2$	H	1968, 98
PC$_5$H$_6$Cl$_3$O$_2$	CH$_3$OCH=CHCCl=CHP(O)Cl$_2$	H	1965, 25
PC$_5$H$_7$	CH$_3$P(CH=CH)$_2$	H; P (−8·6)	1965, 25
PC$_5$H$_7$Cl$_2$O	(CH$_3$)$_2$C=C=CHP(O)Cl$_2$	H	1967, 78
			1966, 127; 1967, 104; 1969, 135
	CH$_2$=CHC(CH$_3$)=CHP(O)Cl$_2$	H	1965, 86; 1967, 310
	CH$_3$CH=CHCH=CHP(O)Cl$_2$	H	1965, 86
PC$_5$H$_7$Cl$_2$S	CH$_2$=CHC(CH$_3$)=CHP(S)Cl$_2$	H	1965, 86
PC$_5$H$_7$F$_4$O$_2$	F$_4$POC(CH$_3$)CHC(CH$_3$)O	H; F (−53·9; −73·8)	1966, 33
PC$_5$H$_7$F$_6$O$_2$	(CH$_3$)$_2$POCH(CF$_3$)$_2$	H; P (62·0)	1968, 302
PC$_5$H$_8$BrO	BrP(O)CH=C(CH$_3$)(CH$_2$)$_2$	H	1964, 8
PC$_5$H$_8$ClO	ClP(O)CH=C(CH$_3$)(CH$_2$)$_2$	H	1964, 8
PC$_5$H$_8$ClO$_2$	ClP(O)OC(CH$_3$)=C(CH$_3$)CH$_2$	H	1967, 311
	CH$_3$P(O)OC(CH$_3$)=C(Cl)CH$_2$	H	1967, 312
PC$_5$H$_8$ClO$_6$	(CH$_3$O)$_2$P(O)OCH(COCl)C(O)H	H; P (−2·9)	1969, 337
PC$_5$H$_8$FO	FP(O)CH$_2$CHCH=CHCH$_3$	H	1967, 110

Formula	Compound	Resonance	Reference
PC_5H_8FO	$FP(O)CH_2CH=C(CH_3)CH_2$	H	1967, 110, 303
PC_5H_8FS	$FP(S)CH_2CH=C(CH_3)CH_2$	H	1967, 110
PC_5H_9	$CH_3P(CH=CH_2)_2$	H; P (−41·8)	1967, 78; 1968, 100; 1969, 77
$PC_5H_9Br_2$	$CH_3P(CH_2CHBr)_2$	H; P (−41·5)	1967, 78; 1969, 77
$PC_5H_9ClN_3OS$	$CH_3N[C(O)NCH_3]_2P(O)Cl$	P (36)	1966, 233
$PC_5H_9ClO_2S$	$Cl_2P(O)(CH_2)_2COSC_2H_5$	H	1967, 313
$PC_5H_9Cl_2$	$CH_3P^{\oplus}Cl(CH=CH_2)_2Cl^{\ominus}$	H; P (112)	1968, 100
$PC_5H_9Cl_2O_3$	$Cl_2P(O)(CH_2)_2COOC_2H_5$	H	1967, 313
	$ClP(O)(OCH_2)_2C(CH_3)CH_2Cl$	H	1968, 149
$PC_5H_9Cl_3N_3O_2$	$CH_3N[C(O)NCH_3]_2P^{\oplus}Cl_2, Cl^{\ominus}$	P (50)	1966, 233
$PC_5H_9N_2O_3S$	$(CH_2O)_2P(S)O^{\ominus}C_3H_5N_2^{\oplus}$	P (154·3)	1969, 177
PC_5H_9O	$CH_3P(O)(CH=CH_2)_2$	H	1968, 100
	$CH_3P(O)(CH_2)_2CH=CH$	H	1968, 100
$PC_5H_9OS_3$	$(CH_3S)_2P(O)SCH_2C\equiv CH$	H; P (62·5)	1967, 21
	$OP(SCH_2)_3CCH_3$	H	1968, 231
$PC_5H_9O_2$	$CH_3OP(O)(CH=CH_2)_2$	H	1964, 8, 15; 1966, 133; 1967, 109
	$(CH_2)_2C(CH_3)=CHPO_2H$	H	1964, 8
	$(CH_3O)_2P(O)C=CCH_3$	H	1966, 63
	$(CH_3O)_2P(O)CH_2C\equiv CH$	H	1966, 63
	$(CH_3O)_2P(O)CH=C=CH_2$	H	1967, 296
	$CH_3OP(O)OC(CH_3)CH_2$	H	1967, 311

NMR DATA ON ORGANIC COMPOUNDS 125

$PC_5H_9O_3$	$P(OCH_2)_3CCH_3$	H	1966, 37; 1968, 85, 209
	$P(CH_2O)_3C(CH_3)$	H	1968, 53
$PC_5H_9O_3S$	$(CH_3O)_2P(O)SCH_2C\equiv CH$	H; P (27·6)	1967, 21
	$SP(CH_2O)_3CCH_3$	H	1968, 53
	$SP(OCH_2)_3CCH_3$	H	1968, 85
$PC_5H_9O_4$	$OP(OCH_2)_3CCH_3$	H	1966, 37
	$OP(CH_2O)_3CCH_3$	H	1968, 53
	$CH_3OP(O)O(CCH_3)_2O$	H; P (11·5)	1967, 86; 1968, 73, 85
PC_5H_9S	$(CH_3)_2P(S)C\!=\!CCH_3$	H	1965, 89
$PC_5H_9S_3$	$P(SCH_2)_3CCH_3$	H	1968, 85
$PC_5H_9S_4$	$SP(SCH_2)_3CCH_3$	H	1968, 85
	$(CH_3S)_2P(S)SCH_2C\equiv CH$	H; P (95)	1967, 21
$PC_5H_{10}Ba_{1/2}O_4$	$CH_3OP(O)(OBa_{1/2})OCH(CH_3)COCH_3$	H	1967, 181
$PC_5H_{10}BrO_2S$	$(CH_3)_2C(CH_2O)_2P(O)Br$	H	1969, 95
$PC_5H_{10}BrO4$	$(CH_3)_2O_2P(O)OC(CH_2Br)\!=\!CH_2$	H	1969, 127
	$(CH_3)_2O_2P(O)OC(CH_3)\!=\!CHBr$	H	1969, 127
$PC_5H_{10}ClO_2$	$(CH_3)_2C(CH_2O)_2PCl$	H; P (146·5)	1968, 80
	$CH_2(CH_3CHO)_2PCl$	H	1969, 88
$PC_5H_{10}ClO_2S$	$(CH_3)_2C(CH_2O)_2P(S)Cl$	H	1967, 91; 1968, 74; 1969, 95
$PC_5H_{10}ClO_3$	$(CH_3)_2C(CH_2O)_2P(O)Cl$	H	1967, 91; 1968, 74; 1969, 95
$PC_5H_{10}ClO_3S$	$(C_2H_5O)_2P(O)CH_2CH\!=\!CHCl$	H; P (24)	1967, 108
	$(CH_3O)_2P(O)SC(CH_3)\!=\!CHCl$	H	1966, 74
	$(CH_3O)_2P(O)SC(CH_2Cl)\!=\!CH_2$	H	1966, 74
$PC_5H_{10}ClO_4$	$Cl(CH_3O)P(O)(CH_2)_2COOCH_3$	H	1967, 313
$PC_5H_{10}Cl_3N_2O$	$CCl_3P(O)[N(CH_3)CH_2]_2$	H	1967, 314
$PC_5H_{10}Cl_3O_3$	$CCl_3P(O)(OC_2H_5)_2$	P (6·5)	1966, 234
$PC_5H_{10}FO_2$	$(CH_3)_2C(CH_2O)_2PF$	H; P (132·9)	1968, 80
$PC_5H_{10}F_2N$	$C_5H_{10}NPF_2$	H; P (140·5); F (−66·4)	1966, 230; 1967, 48

Formula	Compound	Resonance	Reference
$PC_5H_{10}NO_3$	$(CH_3O)_2P(O)(CH_2)_2CN$	H; P (29·1)	1969, 81
$PC_5H_{10}N_3O_4$	$CH_3N[C(O)NCH_3]_2P(O)OH$	P(0)	1966, 233
PC_5H_{11}	$C_5H_{10}PH$	H	1969, 142
$PC_5H_{11}Cl_2O$	$t\text{-}C_5H_{11}P(O)Cl_2$	H	1966, 231
$PC_5H_{11}Cl_2O_2$	$i\text{-}C_3H_7OP(O)(CH_2Cl)_2$	H; P (37·3)	1969, 332
$PC_5H_{11}F_2O$	$C_5H_{11}OPF_2$	F ($-50\cdot 4$)	1969, 61
$PC_5H_{11}NNaO_4$	$(C_2H_5O)_2P(O)N\ominus CHO,\ Na^\oplus$	H	1966, 235
$PC_5H_{11}O$	$(CH_3)_2P(O)C(CH_3)=CH_2$	H	1966, 119
$PC_5H_{11}O_2$	$CH_3(CH_3O)P(O)CH_2CH=CH_2$	P (31·5)	1967, 21
	$CH_3OPOCH_2C(CH_3)_2O$	H	1969, 93
	$C_5H_{10}PO_2H$	H	1969, 142
$PC_5H_{11}O_2S$	$(CH_3)_2C(CH_2O)_2P(S)H$	H	1967, 91
$PC_5H_{11}O_2S_2$	$(CH_3)_2C(CH_2O)_2P(S)SH$	H	1967, 91
$PC_5H_{11}O_3$	$CH_3OPO(CH_2)_2CH(CH_3)O$	H, {P}; P (301·4; 292·8) (stereois.)	1964, 14; 1969, 176
	$CH_3OP(OCH_2)_2CHCH_3$	H	1966, 46
	$(CH_3)_2C(CH_2O)_2P(O)H$	H	1967, 91
	$(CH_3O)_2P(O)CH_2CH=CH_2$	H; P (27·5)	1967, 21, 108
	$CH_3(CH_3O)P(O)OCH_2CH=CH_2$	H	1967, 21
	$CH_3OP[OCH(CH_3)]_2$	H; P (meso: 135; 150; dl: 140)	1969, 94
$PC_5H_{11}O_3S$	$CH_3OP(S)O(CH_2)_2(CH_3)O$	H; P	1969, 176
$PC_5H_{11}O_4$	$CH_3OP(O)(OCH_2)_2CHCH_3$	H	1966, 46
	$C_2H_5O(H)P(O)CH_2COOCH_3$	H	1968, 297
	$(CH_3)_2C(CH_2O)_2PO_2H$	H	1969, 95
	$(CH_3O)_2P(O)OCH=CHCH_3$	H	1969, 127
	$(CH_3O)_2P(O)OC(CH_3)=CH_2$	H	1969, 127
	$CH_3OP(O)[OCH(CH_3)]_2$	H; P (meso: 15; 17; dl: 141)	1969, 94

PC$_5$H$_{11}$O$_5$	(CH$_3$O)$_2$P(O)CH$_2$COOCH$_3$	H {P}	1966, 236
	(CH$_3$O)$_3$P(OCH)$_2$	H; P (-44.2)	1969, 337
PC$_5$H$_{12}$Ba$_{½}$O$_5$	CH$_3$CH(OH)CH$_2$OP(O)(O$^\ominus$)(OC$_2$H$_5$), Ba$_{½}^{\oplus}$	H	1965, 88
PC$_5$H$_{12}$Ba$_{½}$O$_2$	CH$_3$CH(OH)CH$_2$OP(O)(O$^\ominus$)(O(CH$_2$)$_2$OH), Ba$_{½}^{\oplus}$	H	1965, 88
PC$_5$H$_{12}$BrS	CH$_3$(t-C$_4$H$_9$)P(S)Br	H; P (103·1)	1968, 2
PC$_5$H$_{12}$ClN$_2$O	CH$_2$ClP(O)N(CH$_3$)(CH$_2$)$_2$NCH$_3$	H	1967, 314
PC$_5$H$_{12}$ClO$_3$	(C$_2$H$_5$O)$_2$P(O)CH$_2$Cl	H; P (21)	1966, 234; 1967, 69; 1968, 56
PC$_5$H$_{12}$ClO$_3$S	(CH$_3$O)$_2$P(O)SCH(CH$_3$)CH$_2$Cl	H	1966, 74
	(CH$_3$O)$_2$P(O)SCH$_2$CH(Cl)CH$_3$	H	1966, 74
PC$_5$H$_{12}$ClS	CH$_3$P (SC$_4$H$_9$)Cl	P (153)	1969, 335
PC$_5$H$_{12}$Cl$_2$N	CH$_2$ClPCl[N(C$_2$H$_5$)$_2$]	H	1968, 59
PC$_5$H$_{12}$Cl$_2$NO	CH$_2$ClP(O)Cl[N(C$_2$H$_5$)$_2$]	H {P}	1967, 77
PC$_5$H$_{12}$FN$_4$O$_3$S	FP[(NCH$_3$)$_2$SO$_2$][(NCH$_3$)$_2$CO]	H; P (-67); F ($-102·8$)	1968, 125; 1969, 65
PC$_5$H$_{12}$FO	CH$_3$(C$_4$H$_9$)P(O)F	P (65·2); F (-75)	1967, 14
	CH$_3$(i-C$_4$H$_9$)P(O)F	P (216); F (-103)	1968, 12
PC$_5$H$_{12}$FOS	C$_2$H$_5$(i-C$_3$H$_7$)SP(O)F	P (65·3); F ($-49·5$)	1967, 13, 14; 1968, 295
	C$_2$H$_5$(C$_3$H$_7$S)P(O)F	P (67·4); F (-54)	1967, 14
	i-C$_3$H$_7$(C$_2$H$_5$S)P(O)F	P (74·0)	1967, 14
PC$_5$H$_{12}$FO$_2$	t-C$_4$H$_9$P(O)F(OCH$_3$)	H; P (36·6); F ($-82·0$)	1965, 8; 1966, 36
	C$_2$H$_5$(C$_3$H$_7$O)P(O)F	F ($-68·6$)	1967, 13
	C$_2$H$_5$(i-C$_3$H$_7$O)P(O)F	P (32·2); F ($-66·4$)	1967, 13, 14; 1968, 295
PC$_5$H$_{12}$FS	CH$_3$(C$_4$H$_9$S)PF	P (214); F (-151)	1968, 12
PC$_5$H$_{12}$IO$_3$	CH$_2$IP(O)(OC$_2$H$_5$)$_2$	H	1968, 56
PC$_5$H$_{12}$Li	(C$_2$H$_5$)$_2$(LiCH$_2$)P=CH$_2$	H	1968, 106
PC$_5$H$_{12}$NO$_2$	(CH$_3$)$_2$NP(OCH$_2$)$_2$CH$_2$	P (143·4)	1968, 301
PC$_5$H$_{12}$NO$_3$	(CH$_2$O)$_2$P[NHCH$_2$CH(CH$_3$)O]H	H; P ($-39·5$)	1967, 164; 1969, 45
	O(CH$_2$)$_4$NCH$_2$PO$_2$H$_2$	P (8·7)	1967, 301

Formula	Compound	Resonance	Reference
PC$_5$H$_{12}$NO$_3$	(CH$_2$O)$_2$PO(CH$_2$)$_2$NHCH$_3$	P (135)	1967, 165
	(CH$_2$O)$_2$[N(CH$_3$)(CH$_2$)$_2$O]H	H; P (−44)	1967, 165
PC$_5$H$_{12}$NO$_3$S$_2$	(CH$_3$O)$_2$P(S)SCH$_2$C(O)NHCH$_3$	H	1968, 283 to 285
PC$_5$H$_{12}$NO$_4$	(C$_2$H$_5$O)$_2$P(O)NHCHO	H	1966, 235
	(CH$_3$O)$_2$P(O)C(O)NHC$_2$H$_5$	H	1967, 26
PC$_5$H$_{12}$NO$_4$S	(CH$_3$O)$_2$P(O)SCH$_2$C(O)NHCH$_3$	H	1968, 285
PC$_5$H$_{12}$N$_3$O	(CH$_3$)$_2$NP(NCH$_3$)$_2$CO	H; P (93·6)	1969, 109
PC$_5$H$_{12}$O	(CH$_3$)$_3$P$^\oplus$C(O)CH$_3$	H	1968, 307
PC$_5$H$_{13}$	CH$_3$P(C$_2$H$_5$)$_2$	P (−34)	1965, 3
	CH$_3$P(C$_4$H$_9$)H	P (−86)	1965, 3
	(CH$_3$)$_2$(C$_2$H$_5$)P=CH$_2$	H(T)	1968, 105 to 107, 305
	(CH$_3$)$_2$(C$_2$H$_5$)P=CH$_2$, LiCl	H	1968, 106, 107
PC$_5$H$_{13}$ClLi	i-C$_3$H$_7$PCl[N(CH$_3$)$_2$]	H	1969, 107
PC$_5$H$_{13}$ClN	t-C$_4$H$_9$NHP(CH$_3$)Cl	H	1969, 112
PC$_5$H$_{13}$ClNO	i-C$_3$H$_7$P(O)[N(CH$_3$)$_2$]Cl	H{P}, (T)	1967, 77
PC$_5$H$_{13}$ClNS	i-C$_3$H$_7$P(S)[N(CH$_3$)$_2$]Cl	H{P}	1967, 77
PC$_5$H$_{13}$FN	CH$_3$P[N(C$_2$H$_5$)$_2$]F	P (165); F (−101 or −112)	1967, 37; 1968, 12
PC$_5$H$_{13}$FNO	CH$_3$P(O)[N(C$_2$H$_5$)$_2$]F	P (37·2); F (−55)	1967, 13, 14; 1968, 295
	i-C$_3$H$_7$P(O)[N(CH$_3$)$_2$]F	H{P}	1967, 77
PC$_5$H$_{13}$F$_2$O	CH$_3$P(Oi-C$_4$H$_9$)HF$_2$	P (−22·5)	1967, 38
PC$_5$H$_{13}$F$_3$N	(C$_2$H$_5$)$_2$N(CH$_3$)PF$_3$	F (−72; −39)	1967, 64
	i-C$_4$H$_9$NH(CH$_3$)PF$_3$	F (−75; −42; −30)	1967, 64
	C$_4$H$_9$NH(CH$_3$)PF$_3$	F (−73; −42; −30)	1967, 64
PC$_5$H$_{13}$N$_2$O	CH$_3$OP[N(CH$_3$)CH$_2$]$_2$	P (123·2)	1966, 224; 1968, 153
PC$_5$H$_{13}$OS	(CH$_3$)$_2$P(O)SC$_3$H$_7$	P (52·7)	1967, 9
	CH$_3$(t-C$_4$H$_9$)P(S)OH	H	1968, 2
PC$_5$H$_{13}$O$_2$	(C$_2$H$_5$)$_2$P(O)OCH$_3$	H	1968, 65
	CH$_3$P(O)(Os-C$_4$H$_9$)H	P	1969, 162

NMR DATA ON ORGANIC COMPOUNDS

PC$_5$H$_{13}$O$_2$S	CH$_3$P(S)(OC$_4$H$_9$)OH	H	1967, 72
PC$_5$H$_{13}$O$_3$	(CH$_3$O)(t-C$_4$H$_9$O)P(O)H	P (3·2)	1967, 293
	(C$_2$H$_5$O)$_2$PCH$_3$	H	1968, 72
PC$_5$H$_{13}$O$_3$S	(C$_2$H$_5$O)$_2$P(O)SCH$_3$	H	1966, 133
PC$_5$H$_{13}$O$_4$	(CH$_3$O)$_2$P(O)(CH$_2$)$_2$OCH$_3$	H; P (32·9)	1969, 81
PC$_5$H$_{13}$S	CH$_3$(t-C$_4$H$_9$)P(S)H	H; P (36·2)	1968, 2
PC$_5$H$_{13}$S$_2$	CH$_3$(t-C$_4$H$_9$)P(S)SH	H; P (79·0)	1968, 2
PC$_5$H$_{14}$Br	(CH$_3$)$_3$(C$_2$H$_5$)PBr	P (28·0)	1966, 5; 1968, 6
PC$_5$H$_{14}$Cl	(CH$_3$)$_3$(C$_2$H$_5$)PCl	H	1968, 107
PC$_5$H$_{14}$Cl$_6$OSb	(CH$_3$)$_2$(CH$_3$O)(C$_2$H$_5$O)P, SbCl$_6$	H; P	1967, 308
PC$_5$H$_{14}$IS	(CH$_3$)$_3$P$^{\oplus}$CH$_2$SCH$_3$I$^{\ominus}$	H	1969, 128
PC$_5$H$_{14}$Li$_2$NO$_4$	CO(CHCH)$_2$NPO$_3$Li$_2$	H	1969, 338
PC$_5$H$_{14}$NO$_2$	(C$_2$H$_5$)$_2$NCH$_2$PO$_2$H$_2$	P (16·6)	1967, 301
PC$_5$H$_{14}$NO$_3$	(C$_2$H$_5$)$_2$NCH$_2$PO$_3$H$_2$	H; P (6·1)	1967, 291, 301, 315
	s-C$_4$H$_9$CH(NH$_2$)PO$_3$H$_2$	P (18·1)	1968, 296
	CH$_3$NHP[N(CH$_3$)CH$_2$]$_2$	H	1967, 305
PC$_5$H$_{14}$N$_3$	(CH$_3$)$_3$P$^{\oplus}$N(CH$_3$)$_2$, Cl$^{\ominus}$	H	1968, 54
PC$_5$H$_{15}$ClN	CH$_3$P$^{\oplus}$(N(CH$_3$)$_2$)$_2$F	P (71·4); F (−82·3)	1965, 12
PC$_5$H$_{15}$FN$_2$	[(CH$_3$)$_2$N]$_2$P(OCH$_3$)F$_2$	F (−55·5)	1967, 61
PC$_5$H$_{15}$F$_2$N$_2$O	(CH$_3$)$_3$NP$^{\oplus}$(CH$_3$)$_3$, I$^{\ominus}$	H	1965, 18
PC$_5$H$_{15}$IN	[(CH$_3$)$_2$N]$_2$PCH$_3$	H	1965, 18; 1968, 290
PC$_5$H$_{15}$N$_2$	(CH$_3$)$_2$P(O)OSn(CH$_3$)$_3$	P{H}	1969, 140
PC$_5$H$_{15}$O$_2$Sn	(CH$_3$)$_3$PNSi(CH$_3$)$_2$N$_3$	H	1967, 316
PC$_5$H$_{15}$N$_4$Si	(CH$_3$)P[N(CH$_3$)$_2$]$_2$NH$_2$Cl	H; P (53·0)	1967, 65
PC$_5$H$_{17}$ClN$_3$			
PC$_6$Br$_2$F$_5$	C$_6$F$_5$PBr$_2$	P (113·5)	1966, 7
PC$_6$Cl$_2$F$_5$	C$_6$F$_5$PCl$_2$	P (−137); F (−130; −160; −145)	1969, 17
PC$_6$F$_7$	C$_6$F$_5$PF$_2$	F (−140·1; −163·5; −148·8)	1966, 19
PC$_6$F$_7$O	C$_6$F$_5$P(O)F$_2$	F (−128·7; −157·6; −140·2)	1969, 17
PC$_6$F$_9$	(CF$_2$=CF)$_3$P	F (−82·6; −106·3; −175·9)	1969, 55

Formula	Compound	Resonance	Reference
PC_6F_{17}	$(C_3F_7)_2PF_3$	F (-80.7; -122.6; -122.8; -51.0)	1969, 66
$PC_6F_{18}N_3$	$[(CF_3)_2N]_3P$	F (-49.6)	1968, 69
$PC_6H_2Cl_5OS$	$C_6Cl_3H_2OP(S)Cl_2$	H	1967, 116
$PC_6H_2F_5$	$C_6F_5PH_2$	H; P (-183.1); F (-130.1; -163.0; -155.1)	1966, 7, 19; 1967, 11
$PC_6H_2F_5O_2$	$C_6F_5PO_2H_2$	H	1966, 237
$PC_6H_4BrCl_2O_2$	$OC_6H_7OPBrCl_2$	P (-75.8)	1966, 238
$PC_6H_4BrO_2$	OC_6H_4OPBr	P (195.6)	1966, 238; 1967, 277
$PC_6H_4BrO_2S$	$OC_6H_4OP(S)Br$	P (55.1)	1966, 238; 1967, 277
$PC_6H_4BrO_3$	$OC_6H_4OP(O)Br$	P (3.6)	1966, 238; 1967, 277
$PC_6H_4Br_2ClO_2$	$C_6H_4BrPO_2^{2\ominus}$	P (8.8)	1967, 7
	$OC_6H_4OPBrCl$	P (-131)	1966, 238; 1969, 45
$PC_6H_4Br_3$	$C_6H_4BrPBr_2$	P (147)	1968, 100
$PC_6H_4Br_3O_2$	$OC_6H_4OPBr_3$	P (-189.0)	1966, 238; 1969, 45
$PC_6H_4ClF_2$	$C_6H_4ClPF_2$	P (196.8)	1967, 51, 52, 290
$PC_6H_4ClO_2$	OC_6H_4OPCl	P (173)	1967, 235, 277, 290; 1969, 45
$PC_6H_4ClO_2S$	$OC_6H_4OP(S)Cl$	P (76.5)	1966, 238; 1967, 277
$PC_6H_4ClO_3$	$OC_6H_4OP(O)Cl$	P (18.0)	1966, 238; 1967, 277
$PC_6H_4ClO_3$	$C_6H_4ClPO_3^{2\ominus}$	P (9.9)	1967, 7

Formula	Compound	NMR	References
$PC_6H_4ClO_3S$	$(CH_3O)_2P(O)SCH_2C(CH_3)_2Cl$	H	1966, 74
	$(CH_3O)_2P(O)SC(CH_3)_2CH_2Cl$	H	1966, 74
$PC_6H_4Cl_2F$	$p\text{-}FC_6H_4PCl_2$	H; P (159); F (−106·3)	1967, 17; 1968, 14; 1969, 18, 120
			1969, 19, 120
$PC_6H_4Cl_2FO$	$m\text{-}FC_6H_4PCl_2$	P (152); F (−120·3)	1967, 17; 1969, 18, 19
$PC_6H_4Cl_2FS$	$m\text{-}$ and $p\text{-}FC_6H_4P(O)Cl_2$	F	1967, 17; 1969, 18, 19
$PC_6H_4Cl_3$	$m\text{-}$ and $p\text{-}FC_6H_4P(S)Cl_2$	F	1967, 114
	$m\text{-}$ and $p\text{-}ClC_6H_4PCl$	P (158·4)	1967, 114
$PC_6H_4Cl_3O_2$	$o\text{-}ClC_6H_4PCl_2$	P (151·9)	1965, 45; 1966, 238, 239
	$OC_6H_4OPCl_3$	P (−26)	
$PC_6H_4Cl_3S$	$p\text{-}ClC_6H_4P(S)Cl_2$	P (71·4)	1966, 114
$PC_6H_4FO_2$	OC_6H_4OPF	P (123·1)	1967, 235
$PC_6H_4F_3$	$FC_6H_4PF_2$	F	1968, 14
$PC_6H_4F_5$	$FC_6H_4PF_4$	F	1968, 14
$PC_6H_4IO_3$	$IC_6H_4PO_3^{2\ominus}$	P (10·1)	1967, 7
$PC_6H_5Br_2$	$C_6H_5PBr_2$	P (151·8)	1968, 100
PC_6H_5ClF	C_6H_5PClF	F (−74)	1967, 37
PC_6H_5ClFO	$C_6H_5P(O)ClF$	P; F	1968, 45
PC_6H_5ClFOS	$C_6H_5SP(O)ClF$	H; P (84·8); F (−2·2)	1968, 308
$PC_6H_5Cl_2$	$C_6H_5PCl_2$	^{13}C; H; P (166 or 161)	1965, 7; 1966, 21, 240; 1967, 198, 235, 277; 1968, 103; 1969, 50, 120
$PC_6H_5Cl_2O$	$C_6H_5P(O)Cl_2$	^{13}C; H; P (34)	1965, 7; 1966, 240; 1967, 186, 198, 277; 1968, 45, 103
$PC_6H_5Cl_2O_2$	$C_6H_5OP(O)Cl_2$	P (1·5)	1965, 45; 1966, 239
$PC_6H_5Cl_2S$	$C_6H_5P(S)Cl_2$	^{13}C; P (74·6)	1965, 7; 1966, 111, 114, 240; 1967, 256
$PC_6H_5Cl_4$	$C_6H_5PCl_4$	P (−39·3)	1968, 168, 288

Formula	Compound	Resonance	Reference
PC$_6$H$_5$Cl$_4$O$_4$	C$_6$H$_5$P$^{\oplus}$Cl$_3$, ClO$_4^{\ominus}$	P (103·0)	1967, 276
PC$_6$H$_5$Cl$_9$Sb	C$_6$H$_5$P$^{\oplus}$Cl$_3$, SbCl$_6^{\ominus}$	P (102·9)	1967, 276
PC$_6$H$_5$F$_2$	C$_6$H$_5$PF$_2$	P (205–208); F (−92–97)	1967, 51, 52, 235, 290
PC$_6$H$_5$F$_2$O	C$_6$H$_5$P(O)F$_2$	P (11·4); F (−65·3)	1965, 12; 1967, 14; 1968, 45
PC$_6$H$_5$F$_2$OS	C$_6$H$_5$OPF$_2$	F (−44·5)	1963, 3
PC$_6$H$_5$F$_2$O$_2$	C$_6$H$_5$SP(O)F$_2$	H; P (11·9); F (−49·2)	1968, 308
PC$_6$H$_5$F$_4$S	C$_6$H$_5$OP(O)F$_2$	P (−27·1)	1967, 14
	C$_6$H$_5$SPF$_4$	P (−38·3); F (−13·9; −16·9; −65·0)	1968, 291; 1969, 62
PC$_6$H$_5$F$_5$	C$_6$H$_5$PF$_5^{\ominus}$	P (−136·0)	1965, 12
PC$_6$H$_5$O$_3$	C$_6$H$_5$PO$_3^{2\ominus}$	P (10·9)	1967, 7
PC$_6$H$_6$Cl$_2$NO$_3$S	C$_6$H$_5$SO$_2$NHP(O)Cl$_2$	P	1967, 286
PC$_6$H$_6$F	m- and p-FC$_6$H$_4$PH$_2$	F	1967, 17; 1968, 14; 1969, 18, 87
PC$_6$H$_6$F$_6$N	(CF$_2$=CF)$_2$PN(CH$_3$)$_2$	H; F (−89·0; −110·0; −179·3)	1969, 55
PC$_6$H$_6$NO$_3$	p-NH$_2$C$_6$H$_4$PO$_3^{2\ominus}$	P (11·8)	1967, 7; 1969, 119
PC$_6$H$_7$	C$_6$H$_5$PH$_2$	H; P (−119 or −124)	1966, 4, 18, 21; 1967, 11, 30; 1969, 41
PC$_6$H$_7$Cl$_2$O$_4$S	C$_6$H$_5$P$^{\oplus}$HCl$_2$, HSO$_4^{\ominus}$	P (41)	1969, 50
PC$_6$H$_7$F$_2$N$_2$S	C$_6$H$_5$(NH)$_2$P(S)F$_2$	P (65·3); F (−54·3)	1967, 317
PC$_6$H$_7$OS$_2$	C$_6$H$_5$P(O)(SH)$_2$	P (40·2)	1967, 277
PC$_6$H$_7$O$_2$	C$_6$H$_5$PO$_2$H$_2$	H	1966, 18
PC$_6$H$_7$O$_2$S	C$_6$H$_5$P(S)(OH)$_2$	P (14·6)	1967, 277
PC$_6$H$_7$O$_3$	C$_6$H$_5$P(O)(OH)$_2$	H; P	1965, 79; 1967, 186
PC$_6$H$_7$S$_3$	C$_6$H$_5$P(S)(SH)$_2$	P (60·2)	1967, 277
PC$_6$H$_8$FS$_2$	CH$_3$(C$_2$H$_5$S)P(S)F	H	1966, 36

NMR DATA ON ORGANIC COMPOUNDS 133

Formula	Structure	Nucleus	References
$PC_6H_8F_3O_6$	$(CH_3O)_2P(O)OCH(COCF_3)C(O)H$	H; P (−3·2)	1969, 337
PC_6H_8N	$(CH_3)_2NP(C\equiv CH)_2$	H	1967, 106
PC_6H_8NO	$(CH_3)_2NP(O)(C\equiv CH)_2$	H; P (−20·7)	1967, 106
PC_6H_9	$P(CH=CH_2)_3$		1966, 241; 1967, 99; 1969, 172
$PC_6H_9BCl_2F_6N$	$(CF_3)_2PN(t\text{-}C_4H_9)BCl_2$	^{11}B; H; P	1968, 24
$PC_6H_9N_2$	$HP[(CH_2)_2CN]_2$	P (−75)	1966, 4
$PC_6H_9N_2O_7$	$(CH_3O)_2P(O)OCH[C(O)NH]_2CO$	H; P (1·3)	1966, 242
$PC_6H_9N_2S$	$C_6H_5P(S)(NH_2)_2$	H	1967, 318
$PC_6H_9O_2S$	[structure] $C_2H_5P(O)(H)$	H	1969, 74
$PC_6H_9O_3$	$(CH_3O)_2P(O)C\equiv CCH=CH_2$	H	1965, 86
	$P(OCH)_3(CH_2)_3$	H; P (137)	1966, 37; 1967, 218
	$C_6H_9PO_2^{\ominus}$	P (17·6)	1967, 7
	$(CH_2O)_2POC_2H_5$	P (135)	1969, 109
$PC_6H_9O_3S$	[structure] $(CH_3O)_2P(O)$-	H	1966, 130; 1969, 123
$PC_6H_9O_4$	$OP(OCH)_3(CH_2)_3$	H	1966, 37
	[structure] $(CH_3O)_2P(O)$-	H	1969, 123
PC_6H_9S	$(CH_3)_2P(S)C\equiv CCH=CH_2$	H	1965, 89
$PC_6H_{10}BrO$	$BrP(O)[CH_2C(CH_3)]_2$	H	1964, 16
$PC_6H_{10}BrO_3$	$(C_2H_5O)_2P(O)C\equiv CBr$	H	1967, 319
$PC_6H_{10}Br_3$	$C_6H_{10}BrPBr_2$ (isom.)	P (185; 202)	1966, 222
$PC_6H_{10}ClN_4O_3$	$ClP(O)[N(CH_2)_2NHCO]_2$	H	1961, 32

134 FORMULAE INDEX II

Formula	Compound	Resonance	Reference
$PC_6H_{10}ClO$	$ClP(O)CH=C(CH_3)CH(CH_3)CH_2$	H	1964, 8, 16; 1969, 339
$PC_6H_{10}ClO_2$	$ClP(O)OC(CH_3)=CHC(CH_3)_2$	H	1967, 311
$PC_6H_{10}ClO_3$	$(C_2H_5O)_2P(O)C=CCl$	H	1965, 90
	$ClP(O)[OC(CH_3)=CH_2]_2$	H	1968, 297
$PC_6H_{10}Cl_3$	$C_6H_{10}ClPCl_2$ (isom.)	P (187; 199)	1966, 222
$PC_6H_{10}Cl_3O_3$	$(CH_3)_2C(CH_2O)_2P(O)CCl_3$	H	1967, 91; 1968, 74
$PC_6H_{10}Cl_3O_5$	$(CH_3O)_2P(O)CH(CCl_3)OC(O)CH_3$	H; P (11·4)	1967, 3
	$(CH_2O)_4PCH(OH)CCl_3$	H; P (−18)	1968, 154
$PC_6H_{10}FO$	$FP(O)[CH_2C(CH_3)]_2$	H	1967, 110
$PC_6H_{10}F_6N$	$(CF_3)_2PNH$-t-C_4H_9	H(T)	1968, 144
$PC_6H_{10}NaO_2$	$NaOP(O)[CH_2C(CH_3)]_2$	H	1966, 66
PC_6H_{11}	$(C_2H_5)_2PC=CH$	H	1966, 243; 1967, 20
	$CH_3PCH_2CH=C(CH_3)CH_2$	H	1968, 100
	$C_2H_5P(CH=CH_2)_2$	P (−20·8)	1969, 77
$PC_6H_{11}ClN_3$	$C_6H_5P(NH_2)_3Cl$	H; P (29·4)	1967, 65
$PC_6H_{11}Cl_2S$	$C_6H_{11}P(S)Cl_2$	P (101·8)	1966, 114
$PC_6H_{11}FNO_3$	$CH_2FP(O)(OH)O^\ominus C_6H_5NH_5^\oplus$	H; F	1967, 320
$PC_6H_{11}N_2O_3S$	$OCH_2CH(CH_3)OP(S)O^\ominus, C_3H_5N_2^\oplus$	H; P (160)	1969, 177
$PC_6H_{11}O$	$(C_2H_5)_2P(O)C=CH$	H	1967, 20
	$CH_3P[(CH_2)_2]_2CO$	H	1967, 75
	$CH_3P(O)CH_2CH=C(CH_3)CH_2$	H	1968, 100
	$CH_3P(O)(CH_2)_2C(CH_3)=CH$	H	1968, 100

$PC_6H_{11}O_2$	$CH_3OP(O)CH_2C(CH_3)=CHCH_2$	H	1964, 8, 15; 1968, 100
	$C_2H_5OP(O)(CH_2CH_2)$	H; P (68)	1964, 8, 15; 1966, 133, 241; 1969, 146
	$HOP(O)CH_2CH(CH_3)C(CH_3)=CH$	H	1964, 8, 16
	$HOP(O)[CH_2C(CH_3)]_2$	H	1964, 16; 1966, 66; 1969, 339
	$CH_3OP(O)(CH_2)_2C(CH_3)=CH$	H	1968, 100
	$C_2H_5OP(O)(CH_2)_2CH=CH$	P (69)	1969, 146
$PC_6H_{11}O_3$	$(CH_2)_4CH-CPO_3H_2$	H; P (12·3)	1966, 241
	$P(OCH_2)_3CC_2H_5$	P (92)	1967, 218; 1968, 209
	$(C_2H_5O)_2P(O)C\equiv CH$	H; P (−10)	1967, 21, 319
	$CH_3OP(O)O[C(CH_3)]_2CH_2$	H	1967, 311
$PC_6H_{11}O_4$	$C_6H_{11}PO_3^{2\ominus}$	P (24·8)	1967, 7
	$HP(O)OCHCH_2CH(OH)CH_2CH(O)CH_2$	H	1966, 102; 1968, 117
$PC_6H_{11}O_6$	$(CH_3O)_2P(O)OCH(COCH_3)C(O)H$	H; P (−2·6)	1969, 337
$PC_6H_{12}BrO_5$	$(CH_3O)_2PCBr(CH_3)COCH_3$	H; P (−5·5)	1968, 309
$PC_6H_{12}Cl$	$(CH_3)_2P(CH_2CH_2)_2, Cl^{\ominus}$	H	1968, 61
$PC_6H_{12}ClN_2O_2$	$[NCH(CH_3)]_2P(O)O(CH_2)_2Cl$	H	1966, 220
$PC_6H_{12}ClO_2$	$[(CH_3)_2CO]_2PCl$	H(T)	1966, 51
	$(C_2H_5O)_2P(O)CCl=CH_2$	H	1966, 244; 1968, 25, 97
	$(C_2H_5O)_2P(O)CH=CHCl$	H	1966, 244; 1968, 25, 97
$PC_6H_{12}ClO_3$	$CH_3P(O)(OCH_3)CH_2CHClC(O)CH_3$	H	1967, 312

Formula	Compound	Resonance	Reference
$PC_6H_{12}ClO_3$	$CH_3POCH_2C(CH_3)(CH_2Cl)CH_2O$	H	1968, 85
$PC_6H_{12}ClO_3S$	$(CH_3O)_2P(O)SCH_2CHClCH=CH_2$	H	1966, 74
	$(CH_3O)_2P(O)SCH(CH_2Cl)CH=CH_2$	H	1966, 74
$PC_6H_{12}Cl_3O_3$	$P(OCH_2CH_2Cl)_3$	P (128·7)	1967, 277
$PC_6H_{12}Cl_3O_3S$	$SP(OCH_2CH_2Cl)_3$	P (53·0)	1967, 277
$PC_6H_{12}Cl_3O_4$	$OP(OCH_2CH_2Cl)_3$	P (2·0)	1967, 277
$PC_6H_{12}FO_2$	$CH_2=CH(i\text{-}C_4H_9)OP(O)F$	P (15·5); F (−70)	1967, 14
$PC_6H_{12}F_3N_2$	$CF_2=CFP[N(CH_3)_2]_2$	H; F (−89·4; −107·8; −181·3)	1969, 55
$PC_6H_{12}NO_2S$	$CH_3P(O)(OC_2H_5)(SCH_2CH_2CN)$	H	1969, 74
$PC_6H_{12}NO_3$	$(C_2H_5O)_2P(O)CH_2CN$	H	1965, 90
	$C_2H_5OP(O)(H)OC(CH_3)_2(CN)$	P (3·9)	1967, 293
	$(CH_2O)_2PN(CH_2CH_2)_2O$	P (140)	1969, 340
$PC_6H_{12}N_3O_9$	$(CH_3O)_2PO(CH_2)_3C(NO_2)_3$	H{P}	1964, 14
$PC_6H_{13}Cl_2O_2$	$C_4H_9OP(O)(CH_2Cl)_2$	H; P (37·8)	1968, 300; 1969, 332
$PC_6H_{13}Cl_2O_3$	$(C_2H_5O)_2P(O)CH_2CH_2Cl$	H	1966, 244
$PC_6H_{13}FN$	$CH_3PF(NC_5H_{10})$	P (165); F (−115)	1968, 12
$PC_6H_{13}F_2NS$	$C_6H_{11}(NH)_2P(S)F_2$	P (65·3); F (−54·3)	1966, 30
$PC_6H_{13}F_3N$	$F_3P(CH_3)[N(CH_2)_5]$	F (−69; −47)	1967, 64
$PC_6H_{13}N_2$	$[(CH_3)_2N]_2PC\equiv CH$	H	1967, 106
$PC_6H_{13}N_2O$	$[(CH_3)_2N]_2P(O)C\equiv CH$	H	1967, 106
$PC_6H_{13}N_2O_2$	$HP[OCH(CH_3)CH_2NH]_2$	H	1967, 165
$PC_6H_{13}N_2O_4$	$[O(CH_2)_2NCH_2]_2PO_2H$	P (26·8)	1967, 291
$PC_6H_{13}O$	$CH_3P[(CH_2)_2]_2CH(OH)$	H	1967, 75
	$CH_3C(O)P(C_2H_5)_2$	H	1969, 341
$PC_6H_{13}O_2$	$C_2H_5(CH_2=CH)P(O)OC_2H_5$	H	1966, 120
	$CH_3OPOCH(CH_3)C(CH_3)_2O$	H; P (137; 150)	1969, 94

$PC_6H_{13}O_3$	$CH_3OP(OCH_2)_2C(CH_3)_2$	H; P (122·7)	1966, 46; 1967, 88; 1968, 78, 80
	$CH_3OP[OCH(CH_3)]_2CH_2$	H	1966, 46; 1969, 88
	$CH_3P(O)(OCH_2)_2C(CH_3)_2$	H	1967, 91; 1968, 74
	$(C_2H_5O)_2P(O)CH=CH_2$	H	1967, 99; 1968, 25, 97
	$C_2H_5OPOCH(CH_3)(CH_2)_2O$	H(T)	1969, 92
	$i\text{-}C_3H_7OPOCH_2CH(CH_3)O$	H; P (139; 142)	1969, 94
$PC_6H_{13}O_3S$	$(CH_3O)_2P(O)SCH_2C(CH_3)CH_2$	H	1966, 74
	$CH_3OP(S)(OCH_2)_2C(CH_3)_2$	H; P (63)	1967, 91; 1968, 74
	$C_2H_5OP(S)OCH(CH_3)(CH_2)_2O$	H(T)	1969, 92
$PC_6H_{13}O_4$	$(CH_3O_2P(O)CHC(CH_3)_2O$	H	1966, 245; 1967, 81
	$(CH_3O)_2P(O)C(CH_3)_2C(O)H$	H	1966, 245; 1967, 81
	$CH_3OP(O)(OCH_2)_2C(CH_3)_2$	H; P (−6·8)	1966, 46; 1967, 91; 1968, 74
	$CH_3OP(O)[OCH(CH_3)]_2CH_2$	H	1966, 46
	$(CH_3O)_2P(O)C_4H_6(OH)$	H	1967, 100
	$(C_2H_5O)_2P(O)OCH=CH_2$	H	1967, 120
	$CH_3(CH_3O)P(O)OCH(CH_3)COCH_3$	H	1967, 181
	$(C_2H_5O)_2P(O)C(O)CH_3$	H	1967, 166; 1969, 82
	$(C_2H_5O)_2P(O)(CH_2)_2COCH_3$	H; P (34·3)	1967, 311; 1969, 81
	$C_2H_5OP(O)(H)CH_2COOC_2H_5$	H	1968, 297
	$CH_3(C_2H_5O)P(O)CH_2COOCH_3$	H	1968, 297
	$(C_2H_5O)_2P(O)CH_2C(O)H$	H	1969, 82
	$(C_2H_5O)_2P(O)CHCH_2O$	H{P}	1969, 82
	$(CH_3O)_2P(O)CH=CHOC_2H_5$	H	1969, 342

Formula	Compound	Resonance	Reference
$PC_6H_{13}O_4$	$i\text{-}C_3H_7OP(O)OCH_2CH(CH_3)O$	H; P (14)	1969, 94
$PC_6H_{13}O_5$	$(C_2H_5O)_2P(O)CH_2COOH$	H	1965, 87
	$(CH_3O)_2P(O)(CH_2)_2COOCH_3$	H; P (32·6)	1966, 220; 1967, 313; 1969, 81
	$(CH_3O)_2P(O)OCH(CH_3)C(O)CH_3$	P (−0·3)	1967, 86; 1968, 73
	$(CH_3O)_2P(O)OC(CH_3)=C(OH)CH_3$	P (−2·5)	1968, 73
	1-glucose phosphate	P (−2·4)	1969, 321
$PC_6H_{13}O_9$	$CH_3CH(OH)CH_2OP(O)(O^\ominus)[O(CH_2)_2OCH_3], Ba^{\oplus}_{1/2}$	H	1965, 88
$PC_6H_{14}Ba_{1/2}O_6$	$C_2H_5P(SC_4H_9)Br$	P (163)	1969, 335
$PC_6H_{14}BrS$	$C_2H_5P(SC_4H_9)Cl$	P (163)	1969, 335
$PC_6H_{14}ClS$	$[(CH_3)_2N^\oplus CHO]_2PCl, 2Cl^\ominus$	H	1966, 223
$PC_6H_{14}Cl_3N_2O_2$	$(i\text{-}C_3H_7)_2P(O)F$	P (78·2); F (−102)	1967, 14
$PC_6H_{14}FO$	$C_3H_7(C_3H_7S)P(O)F$	P (65·5); F (−51)	1967, 14
$PC_6H_{14}FOS$	$s\text{-}C_4H_9(C_2H_5)SP(O)F$	P (71·0)	1967, 14
$PC_6H_{14}FO_2$	$C_3H_7(C_3H_7O)P(O)F$	P (30); F (−65·4)	1967, 13, 14
	$C_3H_7(i\text{-}C_3H_7O)P(O)F$	P (26); F (−65·4)	1967, 14; 1968, 295
	$i\text{-}C_3H_7(i\text{-}C_3H_7O)P(O)F$	P (31·7); F (−75)	1967, 14
	$C_2H_5(C_4H_9O)P(O)F$	P (32·3)	1967, 14
	$C_2H_5(i\text{-}C_4H_9O)P(O)F$	P (32·4); F (−72)	1967, 14
	$(C_3H_7O)_2PF$	F (−60·3)	1969, 61
$PC_6H_{14}FO_3$	$(i\text{-}C_3H_7O)_2P(O)F$	H {P, F}; P (−4·5);	1967, 13, 14, 51, 185; 1968, 295; 1969, 74
	$(C_3H_7O)_2P(O)F$	P (−8·5); F (−86)	1967, 14
$PC_6H_{14}F_2NS$	$(C_3H_7)_2NP(S)F_2$	H; P (70·0); F (−48·9)	1966, 30
$PC_6H_{14}F_4N$	$F_4PN(C_3H_7)_2$	F (−67·5)	1966, 217
$PC_6H_{14}LiO_5$	$CH_3CH(OH)CH_2OP(O)(O^\ominus)(Oi\text{-}C_3H_7)Li^\oplus$	H	1965, 88
$PC_6H_{14}NO$	$(C_2H_5)_2NC(O)CH_2PH_2$	P (−150·2)	1967, 135
$PC_6H_{14}NO_2$	$C_5H_{10}NCH_2PO_2H_2$	P (9·2)	1967, 301

$PC_6H_{14}NO_3$	$(CH_3)_2NP[OCH(CH_3)]_2$ (isom.)	P (144·4; 147·8; 140·0)	1967, 321; 1968, 304
	$(CH_3)_2NP(O)[OCH(CH_3)_2](?)$	P (14·8)	1969, 343
	$(CH_2O)_2P(H)OCH_2C(CH_3)_2NH$	P (37 −41)	1969, 45, 178
$PC_6H_{14}NO_4$	$(C_2H_5O)_2P(O)CH_2C(O)NH_2$	H	1967, 22
	$(C_2H_5O)_2P(O)CH(OH)CH_2NH_2$	H	1969, 82
	$(CH_3O)_2P(O)C(O)NHi$-C_3H_7	H	1967, 26; 1968, 29
$PC_6H_{14}N_3$	$[(CH_3)_2N]_2PN(CH_2)_2$	H	1967, 305
PC_6H_{15}	$P(C_2H_5)_3$	^{13}C; H; P (−20; −19)	1965, 3, 39; 1967, 20, 244; 1968, 3, 7, 17, 302, 303
	$CH_3(C_2H_5)_2P{=}CH_2$	H	1968, 106, 107
$PC_6H_{15}AlBr_3$	$(C_2H_5)_3PAlBr_3$	^{27}Al; P	1967, 23
$PC_6H_{15}AlCl_3$	$(C_2H_5)_3PAlCl_3$	^{27}Al; P	1967, 23
$PC_6H_{15}ClLi$	$CH_3(C_2H_5)_2P{=}CH_2$, LiCl	H	1968, 106, 107
$PC_6H_{15}Cl_2$	$(C_2H_5)_3PCl_2$	H; P	1969, 344
$PC_6H_{15}FNO$	t-$C_4H_9P(O)F[N(CH_3)_2]$ (isom.)	H; P (42·5; 29·6); F (−88·4; −77·8)	1965, 8; 1968, 80
$PC_6H_{15}F_2N_2$	$CHF_2CH_2P[N(CH_3)_2]_2$	F	1965, 80
$PC_6H_{15}F_3N$	$F_3P(C_2H_5)[N(C_2H_5)_2]$	P (−35·5); F (−74; −45)	1967, 64
	$F_3P(CH_3)NHi$-C_5H_{11}	F (−73; 42; 30)	1967, 64
$PC_6H_{15}N_2O$	$C_2H_5P(O)[N(CH_3)CH_2]_2$	H	1967, 314
$PC_6H_{15}N_2O_2$	$HP[OCH_2CH(CH_3)NH]_2$	H; P (−56)	1967, 164
	$C_2H_5OP(O)[N(CH_3)CH_2]_2$	H	1967, 314
	$HP(OCH_2CH_2NCH_3)_2$	H; P (−63)	1968, 304
	$CH_3NH(CH_2)_2OPN(CH_3)CH_2OH_2O$	H; P (137·2)	1968, 304
	$HP[OCH(CH_3)CH_2NH]_2$	P (−54·6)	1969, 45

Formula	Compound	Resonance	Reference
$PC_6H_{15}O$	$(C_2H_5)_3PO$	H	1965, 83; 1967, 20
	$(C_2H_5)_2P(O)OC_2H_5$	H	1966, 120; 1968, 65
	$(i-C_3H_7)_2PO_2H$	H	1968, 65
$PC_6H_{15}O_2S_2$	$C_2H_5SP(S)(OC_2H_5)_2$	H; P (94·1)	1965, 20; 1966, 218
	$CH_3P(O)(OCH_3)SCH_2Si\text{-}C_3H_7$	H	1969, 101
$PC_6H_{15}O_3$	$P(OC_2H_5)_3$	^{13}C; ^{17}O; H; (137−139)	1963, 1; 1964, 9; 1966, 47, 141, 216, 225; 1967, 34, 85, 198, 277, 293; 1968, 3, 22, 25, 303, 306; 1969, 50, 336
	$C_2H_5P(O)(OC_2H_5)_2$	H; P (32·8)	1966, 51, 105, 133
	$HP(O)(Oi\text{-}C_3H_7)_2$	P (3)	1967, 277, 293; 1969, 101, 162
$PC_6H_{15}O_3S$	$HP(O)(OC_2H_5)(Ot\text{-}C_4H_9)$	P (1·4)	1967, 293
	$SP(OC_2H_5)_3$	^{13}C; H; P (68)	1966, 47, 133, 216; 1967, 84, 85, 198, 277, 293; 1968, 3, 25
	$C_2H_5SP(O)(OC_2H_5)_2$	H	1966, 52, 133; 1967, 307
$PC_6H_{15}O_3S_2$	$HSP(O)(Oi\text{-}C_3H_7)_2$	H	1967, 307
	$(CH_3O)_2P(O)SC_2H_4SC_2H_5$	H	1968, 284, 285; 1969, 322
$PC_6H_{15}O_3Se$	$(C_2H_5O)_3PSe$	H; P (72·1)	1967, 84, 293; 1968, 25
$PC_6H_{15}O_4$	$(C_2H_5O)_3PO$	^{13}C; ^{17}O; H; P (−1·5)	1963, 1; 1966, 47, 133, 138, 216; 1967, 69, 84, 85, 198, 277, 293; 1968, 3, 25, 306; 1969, 74, 86, 336

Formula	Structure	NMR	Reference
PC_6H_{16}	$(C_2H_5O)_2P(O)CH(OH)CH_3$	H	1969, 82
	$HP(C_2H_5)_3^{\oplus}$	H	1966, 22
$PC_6H_{16}As$	$(CH_3)_3P=CHAs(CH_3)_2$	H	1968, 287
$PC_6H_{16}B_{11}$	$B_{11}H_{11}PC_6H_5$	^{11}B; H	1967, 292
$PC_6H_{16}Br$	$HP(C_2H_5)_3Br$	H; P (19·7)	1968, 7
$PC_6H_{16}Cl$	$(CH_3)_2P(C_2H_5)_2Cl$	H	1968, 287
$PC_6H_{16}Cl_6O_2Sb$	$(CH_3)_2(C_2H_5O)_2P^{\oplus}$, $SbCl_6^{\ominus}$	H	1967, 308
$PC_6H_{16}I$	$(CH_3)_2P(C_2H_5)_2I$	H	1967, 169
$PC_6H_{16}N$	$(C_2H_5)_2PN(CH_3)_2$	P (43·5)	1967, 277
$PC_6H_{16}NO$	$(C_2H_5)_2P(O)N(CH_3)_2$	P (85·1)	1967, 277
$PC_6H_{16}NO_2$	$(C_2H_5O)_2PN(CH_3)_2$	P (143·4)	1964, 9
$PC_6H_{16}NO_3$	$(C_2H_5O)_2P(O)NHC_2H_5$	H	1968, 25; 1969, 110
	$CH_3(i-C_3H_7O)P(O)O(CH_2)_2NH_2$	H; P (33·1)	1967, 185
	$CH_3(i-C_3H_7O)P(O)NH(CH_2)_2OH$	H; P (32·8)	1967, 276
$PC_6H_{16}NO_4$	$(C_2H_5O)_2P(O)NH(CH_2)_2OH$	H; P (11·5)	1967, 185
$PC_6H_{16}NS$	$(C_2H_5)_2P(S)N(CH_3)_2$	P (61·3)	1967, 277
$PC_6H_{16}N_3$	$(CH_3)_2NP[N(CH_3)CH_2]_2$	P (114·2)	1966, 224; 1968, 303
	$[(CH_3)_2N]_2P(O)N(CH_3)_2$	H	1966, 246
$PC_6H_{16}N_3O$	$(CH_3)_2NP(O)[N(CH_3)CH_2]_2$	H; P (26·4)	1967, 314; 1968, 153
	$(CH_3)_2NP(NC_2H_5)_2CO$	P (87)	1968, 126
$PC_6H_{16}Sb$	$(CH_3)_3P=CHSb(CH_3)_2$	H	1968, 107
$PC_6H_{17}ClN$	$(CH_3)_3P^{\oplus}CH_2N(CH_3)_2$, $Cl^{\ominus}$	H	1969, 128
$PC_6H_{17}ClN_3O$	$[(CH_3)_3-N]_2P(O)NH(CH_2)_2Cl$	H	1966, 246
$PC_6H_{17}IN_3$	$CH_3(CH_3NH)P^{\oplus}[N(CH_3)CH_2]_2$, $I^{\ominus}$	H	1967, 305
$PC_6H_{17}NO_2$	$C_2H_5OP[N(CH_3)_2]_2$	P (135·2)	1964, 9
	$[(CH_3)_2NCH_2]_2PO_2H$	P (22·7)	1967, 291
$PC_6H_{17}N_2O_2$	$C_2H_5[(CH_3)_2N]P(O)(CH_2)_2NH_2$	H; P (20·1)	1967, 185
$PC_6H_{17}N_2O_3$	$C_2H_5O[(CH_3)_2NP(O)NH(CH_2)_2OH$	H; P (12·1)	1967, 185
$PC_6H_{17}N_4$	$(CH_3)_2NNHP[N(CH_3)CH_2]_2$	H	1967, 305
$PC_6H_{17}O_7S$	$(C_2H_5O)_3P^{\oplus}H$, $HSO_4^{\ominus}$	P (18)	1969, 50
$PC_6H_{18}AlBr_3NSi$	$(CH_3)_3SiN(AlBr_3)P(CH_3)_3$	H	1967, 176

Formula	Compound	Resonance	Reference
PC$_6$H$_{18}$AlCl$_3$NSi	(CH$_3$)$_3$SiN(AlCl$_3$)P(CH$_3$)$_3$	H	1967, 176
PC$_6$H$_{18}$AlI$_3$NSi	(CH$_3$)$_3$SiN(AlI$_3$)P(CH$_3$)$_3$	H	1967, 176
PC$_6$H$_{18}$AlO	(CH$_3$)$_3$POAl(CH$_3$)	H	1965, 83
PC$_6$H$_{18}$BF$_6$N$_2$	[(CH$_3$)$_2$NCH$_2$]$_2$BH$_2$, PF$_6$	H	1969, 26
PC$_6$H$_{18}$Br$_2$N$_3$	[(CH$_3$)$_2$N]$_3$P$^\oplus$Br, Br$^\ominus$	H	1965, 91; 1967, 67
PC$_6$H$_{18}$Br$_3$InSi	(CH$_3$)$_3$SiN(InBr$_3$)P(CH$_3$)$_3$	H	1967, 176
PC$_6$H$_{18}$ClN$_2$	(CH$_3$)$_2$P[N(CH$_3$)$_2$]Cl	H; P (68·4)	1967, 67
PC$_6$H$_{18}$ClN$_4$	[(CH$_3$)$_2$N(CH$_3$)N]$_2$PCl	H	1969, 103
PC$_6$H$_{18}$Cl$_2$N$_3$	[(CH$_3$)$_2$N]$_3$P$^\oplus$Cl, Cl$^\ominus$	H; P (50−53)	1967, 67; 1969, 59, 345
PC$_6$H$_{18}$Cl$_3$GaNSi	(CH$_3$)$_3$SiN(GaCl$_3$)P(CH$_3$)$_3$	H	1967, 176
PC$_6$H$_{18}$Cl$_3$InNSi	(CH$_3$)$_3$SiN(InCl$_3$)P(CH$_3$)$_3$	H	1967, 176
PC$_6$H$_{18}$FN$_4$	FP[N(CH$_3$)$_2$]N(CH$_3$)$_2$]$_2$	H	1969, 103
PC$_6$H$_{18}$F$_2$N$_3$	F$_2$PN[N(CH$_3$)$_2$]$_3$	H; P (−65·5); F (−54)	1966, 76; 1967, 61; 1968, 124
PC$_6$H$_{18}$Ga	Ga(CH$_3$)$_3$P(CH$_3$)$_3$	H (T)	1969, 198
PC$_6$H$_{18}$GeN	(CH$_3$)$_3$GeN=P(CH$_3$)$_3$	H	1967, 73
PC$_6$H$_{18}$IN$_2$	(CH$_3$)$_2$P$^\oplus$[N(CH$_3$)$_2$]$_2$I$^\ominus$	H	1965, 18
PC$_6$H$_{18}$NSi	(CH$_3$)$_3$SiN=P(CH$_3$)$_3$	H	1967, 73, 176, 322
PC$_6$H$_{18}$NSn	(CH$_3$)$_3$SnN=P(CH$_3$)$_3$	H	1967, 73
PC$_6$H$_{18}$N$_3$	P[N(CH$_3$)$_2$]$_3$	H; P (121·5 −123·0)	1964, 9; 1965, 12, 18, 84; 1966, 18, 36, 206; 1967, 277, 299; 1968, 153, 310; 1969, 72
PC$_6$H$_{18}$N$_3$O	OP[N(CH$_3$)$_2$]$_3$	^{14}N; H; P (23)	1964, 2; 1965, 18, 23, 78, 84; 1966, 206; 1967, 198, 277; 1968, 91, 153, 287, 310
PC$_6$H$_{18}$N$_3$S	SP[N(CH$_3$)$_2$]$_3$	H; P	1965, 18, 23; 1966, 36; 1967, 277; 1968, 91; 1969, 105

NMR DATA ON ORGANIC COMPOUNDS

Formula	Structure	NMR	Reference
$PC_6H_{19}ClN_3O$	$[(CH_3)_2N]_3PO, HCl$	H	1969, 181
$PC_6H_{19}Si$	$HP[Si(CH_3)_3]_2$	H; P (-237.4)	1967, 29
$PC_6H_{20}ClN_4$	$[(CH_3)_2N]PNH_2, Cl$	H; P (39.0)	1967, 65
$PC_7H_4ClO_3$	$ClPOC_6H_4C(O)O$	P (151.2)	1965, 45
$PC_7H_4ClO_4$	$ClP(O)OC_6H_4C(O)O$	P (-9.5)	1965, 45
$PC_7H_4Cl_3O_3$	$Cl_2P(O)OC_6H_4-m-C(O)Cl$	P (2.6)	1966, 239
	$Cl_2P(O)OC_6H_4-p-C(O)Cl$	P (1.4)	1966, 239
$PC_7H_5F_2N_2$	$C_6H_5PF_2=NCN$	P (15.3); F (-66.5)	1967, 63
$PC_7H_6ClO_2$	$ClPOC_6H_3(CH_3)O$	P (177)	1969, 45
$PC_7H_7Cl_2$	$p\text{-}CH_3C_6H_4PCl_2$	H	1968, 15
$PC_7H_7Cl_2S$	$C_6H_5CH_2P(S)Cl_2$	P (85.0)	1966, 114
	$p\text{-}CH_3C_6H_4P(S)Cl_2$	P (75.3)	1966, 114
$PC_7H_7F_2$	$CH_3C_6H_4PF_2$	P (205.3); F (-94)	1967, 52, 290
	$C_6H_5CH_2PF_2$	P (223.8); F (-99.5)	1967, 52, 290
$PC_7H_9Cl_2O_3$	$Cl_2P(O)CH=C-C(CH_3)COOC_2H_5$	H	1968, 311
$PC_7H_9F_3N$	$F_3P(C_6H_5)NHCH_3$	F $(-75; -43)$	1967, 64
$PC_7H_9N_2O_2$	$(CH_3)_2NPO(C_5H_3N)O$	P (148)	1969, 45
PC_7H_9O	$CH_3(C_6H_5)P(O)H$	H	1968, 33; 1969, 34
$PC_7H_9O_2$	$(CHCH_2)_2P(O)OCH_2C\equiv CH$	H	1969, 190
	$CH_3(C_6H_5)PO_2H$	H	1966, 36; 1968, 33, 56
	$CH_3O(C_6H_5)P(O)H$	H; P (25.2)	1967, 113, 172
	$p\text{-}CH_3C_6H_4PO_2H_2$	H	1968, 15
PC_7H_9S	$CH_3P(S)(C\equiv CCH_3)_2$	H	1965, 89
$PC_7H_{10}Cl_3O_3$	$Cl_2P(O)CH_2CCl=C(CH_3)COOC_2H_5$	H	1968, 287
$PC_7H_{10}NO_3$	$C_6H_5CH(NH_2)PO_3H_2$	H	1968, 142

Formula	Compound	Resonance	Reference
PC$_7$H$_{10}$NO$_5$	CH$_3$OP(O)(O$^\ominus$)CH(COOCH$_3$)N$^\oplus$(CH$_3$)$_3$	H; P	1969, 346
PC$_7$H$_{11}$F$_6$O$_2$	(C$_2$H$_5$)$_2$POCH(CF$_3$)$_2$	H; P (68·8)	1968, 302
PC$_7$H$_{11}$N$_2$O$_7$	(CH$_3$O)$_2$P(O)OC(C(O)NH)$_2$COCH$_3$	H; P (−1·2)	1966, 242
PC$_7$H$_{11}$O$_5$	HP(O)OCH(CH=CH$_2$)C(CH$_3$)C(O)CH$_3$)O	H; P (14·0)	1968, 312
PC$_7$H$_{12}$ClO$_3$	(C$_2$H$_5$O)$_2$P(O)C=CCH$_2$Cl	H; P (−9·8)	1967, 21
PC$_7$H$_{12}$ClO$_6$	(CH$_3$O)$_2$P(O)OC(CH$_3$)[C(O)CH$_3$]C(O)Cl	P (−3·2)	1968, 313
PC$_7$H$_{12}$Cl$_3$O$_5$	(CH$_3$O)$_2$P(O)OC(CCl$_3$)[C(O)CH$_3$]CH$_3$	H	1965, 35
PC$_7$H$_{12}$NO$_3$	(C$_2$H$_5$O)$_2$P(O)CH=CHCN	H	1967, 99
PC$_7$H$_{13}$	(C$_2$H$_5$)$_2$PC=CCH$_3$	H	1966, 243, 247
	(C$_2$H$_5$)$_2$PCH=C=CH$_2$	H	1967, 104; 1968, 99; 1969, 115
	CH$_3$P[CH$_2$C(CH$_3$)]$_2$	H	1968, 100
PC$_7$H$_{13}$Br$_2$O$_3$	(C$_2$H$_5$O)$_2$P(O)CH$_2$CPBr=CHBr	H; P (20)	1967, 101, 108, 323, 324
	(CH$_3$O)$_2$P(O)CBrCHBrCH(CH$_2$)$_3$	H	1966, 43
PC$_7$H$_{13}$Cl$_2$	CH$_3$P$^\oplus$(CH(CH$_3$)CH)$_2$Cl, Cl$^\ominus$	H	1968, 61
PC$_7$H$_{13}$O	(C$_2$H$_5$)$_2$P(O)C=CCH$_3$	H	1966, 243, 247
	(C$_2$H$_5$)$_2$P(O)CH=C=CH$_2$	H	1967, 104, 296; 1969, 115
	CH$_3$P(O)[CH$_2$C(CH$_3$)]$_2$	H	1968, 100
	CH$_3$P(O)CH$_2$[CH(CH$_3$)]$_2$CH	H	1968, 100
PC$_7$H$_{13}$O$_2$	CH$_3$OP(O)CH=C(CH$_3$)CH(CH$_3$)CH$_2$	H	1964, 8, 16

NMR DATA ON ORGANIC COMPOUNDS 145

	$CH_3OP(O)CH_2C(CH_3)=C(CH_3)CH_2$	H	1964, 16
	$C_2H_5OP(O)CH_2CH=C(CH_3)CH_2$	H	1964, 15
$PC_7H_{13}O_3$	$CH_3OP(O)[CH_2C(CH_3)]_2$	H; P (−45·8)	1966, 66
	$(C_2H_5O)_2PCH=C=CH_2$	H	1968, 99; 1969, 115
	$(C_2H_5O)_2P(O)C\equiv CCH_3$	^{13}C; H; P (−92)	1966, 63; 1967, 21; 1968, 20
	$(CH_3O)_2P(O)C_5H_7$	H	1967, 100
	$(C_2H_5O)_2P(O)CH-C=CH_2$	H; P (11·1)	1967, 21, 296, 324; 1969, 115
	$(CH_3O)_2P(O)CH=C(CH_3)CH=CH_2$	H	1967, 310
	$(C_2H_5O)_2P(O)CH_2C\equiv CH$	H; P (18·6)	1967, 21, 324
	$(C_2H_5O)_2P(O)CH_2CH=CH_2$	P (23·5)	1967, 108
	$CH_3OP(O)OC(CH_3)CHC(CH_3)_2$	H	1967, 311
	$CH_3P(O)[OC(CH_3)=CH_2]_2$	H	1968, 297
	$P(OCH_2)_2CC_3H_7$	H; P (93)	1967, 218; 1968, 209
	$CH_3OP[OC(CH_3)_2]_2$	H; P (147)	1969, 94
$PC_7H_{13}O_4$	$(C_2H_5O)_2P(O)C\equiv CCH_2OH$	H	1967, 21
	$(C_2H_5O)_2P(O)OCH_2C\equiv CH$	H; P (−2)	1967, 21
	$CH_3OP(O)[OC(CH_3)_2]_2$	H	1969, 94
$PC_7H_{13}O_5$	$CH_3P(O)(CH_2COOCH_3)_2$	H	1968, 297
	$CH_2=CHCH(OH)C(CH_3)PO_3H_2$	H; P (−4·1)	1968, 312
	$CH_3(CH_2O)_2P[OC(CH_3)]_2$	H; P(−27·0)	1968, 301
$PC_7H_{13}O_5$	$(CH_2O)_2POCH(CH_3)COOC_2H_5$	P (137)	1969, 109
$PC_7H_{13}O_6$	$(CH_3O)_2P(O)OC(CH_3)=CHCOOCH_3$	H; P (−6)	1961, 4; 1967, 325; 1968, 284, 285; 1969, 322
$PC_7H_{14}BrO_2$	$(C_2H_5O)_2P(O)CH_2CH=CHBr$	H; P (24)	1967, 108, 326
	$(C_2H_5O)_2P(O)CH=CHCH_2Br$	P (17)	1967, 326

146 FORMULAE INDEX II

Formula	Compound	Resonance	Reference
$PC_7H_{14}Br_3$	$(C_2H_5O)_2P(O)CH_2CBr=CH_2$	H; P (22)	1967, 101, 108, 323, 324
	$CH_3(CH_2)_4CH(PBr_2)CH_2Br$	P (191)	1966, 222
	$CH_3(CH_2)_4CHBrCH_2PBr_2$	P (178)	1966, 222
$PC_7H_{14}ClN_2O_2$	$[NCH(CH_3)]_2P(O)OCH(CH_3)CH_2Cl$	H	1966, 220
$PC_7H_{14}ClO$	$C_4H_9P(Cl)CH_2COCH_3$	H	1968, 297
$PC_7H_{14}ClO_2$	$t\text{-}C_4H_9CH(CH_2O)_2PCl$	H	1968, 78
	$C_4H_9OP(Cl)[OC(CH_3)=CH_2]$	H	1968, 297
$PC_7H_{14}ClO_3$	$OCH_2C(CH_3)[CH(CH_3)(OCH_3)]CH_2OPCl$	H; P (140·8)	1967, 161, 162
	$(C_2H_5O)_2P(O)CH_2CH=CHCl$	H; P (24·5)	1967, 323, 326
	$(C_2H_5O)_2P(O)CH=CHCH_2Cl$	P (17·4)	1967, 326
	$(C_2H_5O)_2P(O)CH_2C(Cl)=CH_2$	H; P (22·6)	1967, 323
$PC_7H_{14}ClO_3S$	$(CH_3O)_2P(O)SCH_2CHClC(CH_3)=CH_2$	H	1966, 74
	$(CH_3O)_2P(O)SCH(CH_2Cl)C(CH_3)=CH_2$	H	1966, 74
$PC_7H_{14}ClO_4$	$C_2H_5OP(O)(Cl)(CH_2)_2COOC_2H_5$	P (189)	1967, 313
$PC_7H_{14}Cl_3$	$CH_3(CH_2)_4CH(PCl_2)CH_2Cl$	P (191)	1966, 222
	$CH_3(CH_2)_4CHClCH_2PCl_2$		1966, 222
$PC_7H_{14}NO$	$C_2H_5OP(CH=C=CH_2)[N(CH_3)_2]$	H	1968, 99; 1969, 115
$PC_7H_{14}NO_5$	$(CH_3O)_2P(O)OC(CH_3)CHC(O)NHCH_3$	H	1968, 284
$PC_7H_{14}N_3O_3$	$OP(NHCOOC_2H_5)[N(CH_2)_2]_2$	H	1965, 42
$PC_7H_{14}O_3$	$(CH_3O)_2P(O)C(C_2H_5)=CHCH_3$	H	1966, 43
$PC_7H_{15}F_3N$	$F_3P(CH_3)NHC_6H_4$	$P(-36·4); F_A(-73; -43; -32)$	1967, 64
$PC_7H_{15}N_2$	$[(CH_3)_2N]_2PC\equiv CCH_3$	H	1967, 106
	$[(CH_3)_2N]_2PCH=C=CH_2$	H	1967, 104; 1968, 99; 1969, 115
$PC_7H_{15}N_2O$	$[(CH_3)_2N]_2P(O)C\equiv CCH_3$	H	1967, 106
	$[(CH_3)_2N]_2P(O)CH=C=CH_2$	H	1967, 104, 296; 1969, 115

NMR DATA ON ORGANIC COMPOUNDS

PC$_7$H$_{15}$N$_4$	[(CH$_3$)$_2$N]$_2$PN(CH$_3$)N	H	1967, 119
PC$_7$H$_{15}$OS	(C$_2$H$_5$)$_2$P(S)CH$_2$COCH$_3$	H	1968, 297
PC$_7$H$_{15}$O$_2$	C$_4$H$_9$P(O)O(CH$_2$)$_3$	H {P}	1964, 14
	C$_2$H$_5$P(O)(OC$_2$H$_5$)CH$_2$CH=CH$_2$	H	1966, 120
	(C$_2$H$_5$)$_2$P(O)CH$_2$COCH$_3$	H	1968, 297
	(C$_2$H$_5$)$_2$PCH$_2$COOCH$_3$	H	1968, 297
	i-C$_3$H$_7$OPOCH$_2$C(CH$_3$)$_2$O	H	1969, 93
	t-C$_4$H$_9$POCH$_2$CH(CH$_3$)O	H	1969, 93
PC$_7$H$_{15}$O$_3$	C$_3$H$_7$P(O)O(CH$_2$)$_4$O	H {P}	1964, 14
	CH$_3$OP[OC(CH$_3$)$_2$]$_2$	H(T)	1966, 51
	(CH$_3$O)$_2$P(O)C(C$_2$H$_5$)CHCH$_3$	H	1967, 100
	C$_2$H$_5$OP(O)[OC(CH$_3$)$_2$](CH=CH$_2$)H	P (1·7)	1967, 293
	(CH$_3$O)$_2$P(O)CH$_2$CH=C(CH$_3$)$_2$	H	1967, 327
	CH$_3$OPOCH$_2$CH(t-C$_4$H$_9$)O (isom.)	H; P (134; 142)	1969, 94
		H	1967, 21; 1968, 314; 1969, 163
	(C$_2$H$_5$O)$_2$P(O)CH=CHCH$_3$	H; P (2·5)	1967, 21
	(C$_2$H$_5$O)$_2$P(O)CH$_2$CH=CH$_2$	H	1968, 297
	(C$_2$H$_5$O)$_2$POC(CH$_3$)=CH$_2$	H; P (−81·0)	1969, 71
	P(CH$_2$O)$_3$CC$_4$H$_9$	H	1969, 93
	t-C$_4$H$_9$OPO(CH$_2$)$_3$O		
PC$_7$H$_{15}$O$_3$S	(C$_2$H$_5$O)$_2$P(O)SCH$_2$CH=CH$_2$	H	1967, 307
PC$_7$H$_{15}$O$_7$	(C$_2$H$_5$O)$_2$P(O)CH$_2$COCH$_3$	^{13}C; H; P (19)	1966, 234; 1967, 22, 278; 1968, 20

Formula	Compound	Resonance	Reference
	$C_2H_5OP(O)(OCH_2)_2C(CH_3)_2$	H	1967, 91; 1969, 86
	$(C_2H_5O)_2P(O)C(O)C_2H_5$	H	1967, 166
	$(CH_3O)_2P(O)CH_2CH(CH_3)COCH_3$	H	1967, 311
	$(C_2H_5O)_2PCH_2COOCH_3$	H	1968, 297
	$C_2H_5(C_2H_5O)P(O)CH_2COOCH_3$	H	1968, 297
	$CH_3OP(O)OCH_2CH(t\text{-}C_4H_9)O$	H; P (16)	1969, 94
	$(C_2H_5O)_2P(O)CH_2CHCH_2O$	H	1969, 249
	$t\text{-}C_4H_9OP(O)O(CH_2)_3O$	H	1969, 93
$PC_7H_{15}O_4S$	$(C_2H_5O)_2P(S)CH_2COOCH_3$	H	1968, 213
$PC_7H_{15}O_5$	$(CH_3O)_3P[OC(CH_3)]_2$ (stereois.)	H; P (-49; -55)	1965, 21, 45; 1966, 225, 239, 248; 1967, 86, 160; 1968, 73, 301; 1969, 147
	$(C_2H_5O)_2P(O)CH_2COOCH_3$	H	1968, 297
	$(C_2H_5O)_2P(O)CH(OH)CH_2OCH_3$	H	1969, 82
	$(C_2H_5O)_2P(O)CH_2CH(OH)CH_2OH$	H	1969, 82
$PC_7H_{16}ClO_3$	$(C_2H_7O)_2P(O)CH_2Cl$	H	1966, 36
$PC_7H_{16}ClO_3S$	$(CH_3O)_2P(O)SCH(CH_2Cl)C_3H_7$	H	1966, 74
	$(CH_3O)_2P(O)SCH_2CHClC_3H_7$	H	1966, 74
$PC_7H_{16}FO_2$	$CH_3P(O)(F)[OCH(CH_3)t\text{-}C_4H_9]$	H	1969, 74, 160
$PC_7H_{16}NO_2$	$(CH_3)_2NP(OCH_2)_2C(CH_3)_2$	H	1968, 80
$PC_7H_{16}NO_2S$	$(CH_3)_2NP(S)(OCH_2)_2C(CH_3)_2$	H	1968, 74
	$(C_2H_5O)_2P(O)CH_2CH_2CONH_2$	H	1966, 249
	$(C_2H_5O)_2P(O)C(O)NHC_2H_5$	H	1967, 26
	$(CH_3O)_2P(O)C(O)NHC_4H_9$	H	1967, 26
$PC_7H_{16}NO_4$	$(C_2H_5O)_2P(O)CH_2CH(OH)CH_2NH_2$	H	1969, 82

NMR DATA ON ORGANIC COMPOUNDS

$PC_7H_{16}N_2NaO_3S$	$NaPO_3SCH_2C(NH_2)NHC_6H_{11}$	H	1968, 292
$PC_7H_{16}N_3O$	$(CH_3)_2NP(NC_2H_5)_2CO$	H; P (81)	1969, 109
PC_7H_{17}	$(C_2H_5)_3P=CH_2$	H	1968, 105, 107, 305
$PC_7H_{17}N_2O$	$C_2H_5P(O)(NCH_3)_2(CH_2)_3$	H	1967, 314
$PC_7H_{17}N_2O_2$	$[(CH_3)_2N]_2PCH_2COOCH_3$	H	1968, 297
	$[(CH_3)_2N]_2P(O)CH=CHOCH_3$	H	1969, 342
$PC_7H_{17}N_2O_2S$	$(C_2H_5O)_2P(S)NCHN(CH_3)_2$	H	1966, 56
$PC_7H_{17}O_2$	$(C_2H_5)_2P(O)O\textit{i}\text{-}C_3H_7$	H	1968, 65
	$(\textit{i}\text{-}C_3H_7)_2P(O)OCH_3$	H	1968, 65
	$(\textit{i}\text{-}C_3H_7O)_2PCH_3$	H; P (−52)	1969, 192
$PC_7H_{17}O_2S_3$	$(C_2H_5O)_2P(S)SCH_2SC_2H_5$	H	1968, 284, 285; 1969, 322
$PC_7H_{17}O_3$	$(C_2H_5O)_2P(O\textit{i}\text{-}C_3H_7)$	P (137·2)	1967, 293
	$(CH_3O)_2P(O)CH_2CH(CH_3)C_2H_5$	H	1967, 310
	$(C_2H_5O)_2P(O)C_3H_7$	P (30)	1967, 21
	$CH_3P(O)(OC_6H_{13})(OH)$	H	1969, 74
$PC_7H_{17}O_4$	$(CH_3O)_2P(O)C(C_2H_5)_2OH$	H	1967, 100
	$(C_2H_5O)_2P(O)CH_2CH(OH)CH_3$	H; P (28·7)	1967, 22
PC_7H_{18}	$CH_3P^\oplus(C_2H_5)_3$	P (37·8)	1967, 7
$PC_7H_{18}Cl_6O_2Sb$	$CH_3O(C_2H_5)_2(C_2H_5O)P^\oplus, SbCl_6^\ominus$	H	1967, 308
$PC_7H_{18}Cl_6O_2Sb$	$CH_3OP^\oplus(OC_2H_5)_3, SbCl_6^\ominus$	H; P (50)	1967, 69
$PC_7H_{18}F_2N_3$	$[(CH_3)_2N]_3P=CF_2$	H; P (−65·5); F (−54·7)	1964, 9
$PC_7H_{18}I$	$(CH_3)_3(C_2H_5)_3PI$	P (36·5)	1967, 169, 244
$PC_7H_{18}NO_3$	$(C_2H_5O)_2P(O)NH\textit{i}\text{-}C_3H_7$	H	1969, 110
$PC_7H_{18}N_2O_2$	$\textit{i}\text{-}C_3H_7(C_2H_5O)P(O)N(CH_3)_2$	H	1965, 92
$PC_7H_{18}N_3O$	$[(CH_3)_2N]_2P(O)NCH_2CHCH_3$	H	1966, 246
$PC_7H_{19}Ge$	$(CH_3)_2NP(O)[N(CH_3)]_2(CH_2)_3$	H	1967, 314
	$(CH_3)_3P=CHGe(CH_3)_3$	H	1968, 107
$PC_7H_{19}Si$	$(CH_3)_3P=CHSi(CH_3)_3$	H	1967, 328; 1968, 107
$PC_7H_{19}Sn$	$(CH_3)_3P=CHSn(CH_3)_3$	H	1968, 107

Formula	Compound	Resonance	Reference
$PC_7H_{20}Al$	$(CH_3)_3P^{\oplus}CH_2Al^{\ominus}(CH_3)_3$	H	1968, 105, 107
$PC_7H_{20}ClGe$	$(CH_3)_3PCH_2Ge(CH_3)_3Cl$	H	1968, 107
$PC_7H_{20}ClSi$	$(CH_3)_3PCH_2Si(CH_3)_3Cl$	H	1968, 107
$PC_7H_{20}IN_2$	$CH_3(C_6H_5)P^{\oplus}[N(CH_3)_2]_2, I^{\ominus}$	H; P (59·4)	1969, 87
$PC_7H_{20}N_3O_2$	$[(CH_3)_2N]_2P(O)NH(CH_2)_2OCH_3$	H	1966, 246
$PC_7H_{21}ClN_3$	$CHP^{\oplus}[N(CH_3)_2]_3, Cl^{\ominus}$	H; P (54·4)	1967, 67
$PC_7H_{21}Cl_6N_3OSb$	$CH_3OP^{\oplus}[N(CH_3)_2]_3, SbCl_6^{\ominus}$	H; P (38·0)	1969, 87
$PC_7H_{21}Cl_6N_3SSb$	$CH_3SP^{\oplus}[N(CH_3)_2]_3, SbCl_6^{\ominus}$	H; P (66·7)	1969, 87
$PC_7H_{21}Cl_6N_3SbSe$	$CH_3SeP^{\oplus}[N(CH_3)_2]_3, SbCl_6^{\ominus}$	H; P (63·5)	1969, 87
$PC_7H_{21}IN_3$	$CH_3P^{\oplus}[N(CH_3)_2]_3, I^{\ominus}$	H	1965, 18, 91
$PC_7H_{21}IN_3S$	$CH_3SP^{\oplus}[N(CH_3)_2]_3, I^{\ominus}$	H	1969, 105, 347
$PC_8F_5N_2$	$C_6F_5P(CN)_2$	P (−121·3)	1968, 9
$PC_8F_5N_2O_2$	$C_6F_5P(NCO)_2$	P (70·5)	1968, 9
$PC_8F_5N_2S_2$	$C_6F_5P(NCS)_2$	P (62·2)	1968, 9
$PC_8F_5N_2S_3$	$C_6F_5P(S)(NCS)_2$	P (6·8)	1968, 9
$PC_8H_4FN_2$	$FC_6H_4P(CN)_2$	F	1968, 14
$PC_8H_4F_7$	$FC_6H_4P(CF_3)_2$	F	1968, 14
$PC_8H_5ClNO_2S$	$(C_2H_5O)_2P(O)NS(O)CH_2CH{=}CClCH_2$	H	1965, 93
$PC_8H_5Cl_2O$	$Cl_2P(O)C{\equiv}CC_6H_5$	H	1965, 86
PC_8H_6Cl	$(CH_3)_2P^{\oplus}[CH(CH_3)CH_2]_2, Cl^{\ominus}$	H	1968, 61
$PC_8H_6ClF_5N$	$C_6F_5P[N(CH_3)_2]Cl$	H; F (−129·0; −162·5; −151·0)	1966, 19
$PC_8H_6F_5$	$C_6F_5P(CH_3)_2$	H; P (−478); F (−133; −164; −154)	1966, 10, 19; 1967, 11; 1969, 17
$PC_8H_6F_5O$	$C_6F_5P(O)(CH_3)_2$	P (29·7)	1967, 11
$PC_8H_6F_6N$	$(CF_3)_2PNHC_6H_5$	H(T)	1968, 144
$PC_8H_7Cl_7O_5$	$(CH_3O)_2P(O)C_6Cl_4OH$	H; P(−2·7)	1968, 315

$PC_8H_7O_2S_2$	![thiophene with PO2H]	H	1966, 130
$PC_8H_7O_3$	$C_6H_5CH=CHPO_3^{2\ominus}$	P (10·4)	1967, 7
$PC_8H_8BrO_3$	$CHBr=C(C_6H_5)PO_3H_2$	H; P (11·5)	1966, 241
PC_8H_8ClO	$C_6H_5(CH_2=CH)P(O)Cl$	H	1966, 250
$PC_8H_8ClO_3$	$CHCl=C(C_6H_5)PO_3H_2$	P (12·3)	1966, 241
$PC_8H_8Cl_3O_3S$	$(CH_3O)_2P(S)OC_6H_2Cl_3$	H	1968, 284, 285; 1969, 322
$PC_8H_8NO_2$	$(CH_2O)_2PN(CH-CH)_2$	P (124)	1969, 109
PC_8H_9	$C_6H_5P(CH_2)_2$	H; P (−234)	1969, 169
$PC_8H_9ClNO_5S$	$(CH_3O)_2P(S)OC_6H_3Cl(NO_2)$	H	1968, 284, 285; 1969, 322
$PC_8H_9Cl_2O$	$C_6H_5P(O)(CH_2Cl)_2$	H; P (−33·9)	1968, 299; 1969, 334
$PC_8H_9Cl_3NO_2S$	$C_6H_2Cl_3OP(S)(OCH_3)NHCH_3$	H	1967, 116
$PC_8H_9F_6O_5$	$(CH_3O)_2P(O)OC(CH_3)[C(O)CF_3]_2$	H; P (−3·2)	1968, 313
$PC_8H_9Li_2$	$C_6H_5P(CH_2Li)_2$	P (−21)	1967, 329
$PC_8H_9N_2O_4$	$(CH_3O)_2P(O)C_6H_3,2,4-(NO_2)_2$	H	1969, 119
$PC_8H_9O_2S_4$	$CH_3OP(S)SC(S)CSC(CH_3)C(CH_3)CO$	H	1969, 148
$PC_8H_9O_3$	$C_6H_5OP(OCH_2)_2$	H; P (128)	1966, 50; 1968, 77, 301; 1969, 50
	$CH_2=C(C_6H_5)PO_3H_2$	H; P (14·4)	1966, 241
	$CH(C_6H_5)=CHPO_3H_2$	P (4·7)	1966, 241
	$CH_2C_6H_4CHPO_3H_2$	P (8·5)	1966, 241
$PC_8H_9O_4$	$C_6H_5PO(OCH_2)_2$	P (11·1)	1969, 343
	$C_6H_5C(O)OP(OCH_2)_2$	H	1966, 77
	$C_2H_5OP(O)OC_6H_4O$	P (10·0)	1966, 238

Formula	Compound	Resonance	Reference
PC$_8$H$_9$S$_2$	CH$_3$PSC$_6$H$_3$(CH$_3$)S	H	1967, 330
PC$_8$H$_9$S$_2$Se	CH$_3$P(Se)SC$_6$H$_3$(CH$_3$)S	H	1967, 330
PC$_8$H$_9$S$_3$	CH$_3$P(S)SC$_6$H$_3$(CH$_3$)S	H	1967, 330
PC$_8$H$_{10}$ClF$_4$	(C$_2$H$_5$)$_2$PC=CCl(CF$_2$)$_2$	H	1967, 331
PC$_8$H$_{10}$ClOS	CH$_3$P(SC$_6$H$_4$OCH$_3$)Cl	P (150)	1969, 335
PC$_8$H$_{10}$ClS	C$_6$H$_5$P(SC$_2$H$_5$)Cl	P (139)	1969, 335
PC$_8$H$_{10}$Cl$_3$N$_2$OS	C$_6$H$_5$Cl$_3$OP(S)(NHCH$_3$)$_2$	H	1967, 116
PC$_8$H$_{10}$F	m- and p-FC$_6$H$_4$P(CH$_3$)$_2$	F	1967, 17; 1969, 18, 19
PC$_8$H$_{10}$FO	m- and p-FC$_6$H$_4$P(O)(CH$_3$)$_2$	F	1967, 359; 1969, 18, 19
	C$_2$H$_5$(C$_6$H$_5$)P(O)F	P (53·4); F (−80)	1967, 14
PC$_8$H$_{10}$FOS$_2$	C$_2$H$_5$O(C$_6$H$_5$)SP(S)	H; P (−89·1); F (−28·7)	1968, 308
PC$_8$H$_{10}$FO$_2$	C$_2$H$_5$O(C$_6$H$_5$)P(O)F	P (14·9); F (−65·4)	1967, 13, 14; 1968, 295
	m- and p-FC$_6$H$_4$P(OCH$_3$)$_2$	F	1968, 14
PC$_8$H$_{10}$FO$_3$	m- and p-FC$_6$H$_4$P(O)(OCH$_3$)$_2$	H; F	1967, 115; 1969, 18, 19
PC$_8$H$_{10}$FS	C$_6$H$_5$PF(SC$_2$H$_5$)	P (196)	1968, 12
PC$_8$H$_{10}$F$_3$S	F$_3$P(SC$_2$H$_5$)C$_6$H$_5$	H; P (−18·3); F (−14; −21; −71)	1968, 277, 316; 1969, 62
PC$_8$H$_{10}$F$_5$	(C$_2$H$_5$)$_2$PC=CF(CF$_2$)$_2$	H	1967, 231
PC$_8$H$_{10}$Li	CH$_3$(C$_6$H$_5$)PCH$_2$Li	P (−22·6)	1966, 132; 1967, 329
PC$_8$H$_{10}$NO	H$_2$PCH$_2$C(O)NHC$_6$H$_5$	P (−144·5)	1967, 135
PC$_8$H$_{10}$NO$_2$	(CH$_3$)$_2$NPOC$_6$H$_4$O	P (150)	1969, 45

NMR DATA ON ORGANIC COMPOUNDS 153

Formula	Structure	Nuclei	References
$PC_8H_{10}NO_3$	$HP(OC_6H_4NH)(OCH_2CH_2O)$	H; P (−37·7)	1967, 165
$PC_8H_{10}NO_5$	$HP(OC_6H_4O)(OCH_2CH_2NH)$	P (−35)	1969, 45
$PC_8H_{10}NO_5$	$(CH_3O)_2P(O)C_6H_4NO_2$	H	1969, 119
$PC_8H_{10}NO_5S$	$CH_3O(CH_3SP(O)OC_6H_4NO_2$	H; P (25·2)	1965, 20
	$(CH_3O)_2P(S)OC_6H_4NO_2$	H	1968, 284, 285; 1969, 322
PC_8H_{11}	$(CH_3)_2PC_6H_5$	^{13}C; H; P (−46)	1965, 3; 1966, 41, 132; 1967, 66, 329; 1968, 6, 7, 33, 54, 303; 1969, 50
	$C_2H_5(C_6H_5)PH$	H	1968, 28; 1969, 34
	$CH_3(C_6H_5)CHPH_2$	H(T)	1969, 51
$PC_8H_{11}ClN$	$C_6H_5PCl[N(CH_3)_2]$	H; P (141)	1965, 12; 1966, 36; 1967, 80; 1968, 80, 92
$PC_8H_{11}ClNO$	$C_6H_5P(O)Cl[N(CH_3)_2]$	H	1966, 36; 1967, 80
$PC_8H_{11}Cl_2N_2$	$C_6H_5N=P[N(CH_3)_2]Cl_2$	H; P (−18·6)	1969, 348
$PC_8H_{11}Cl_3NO_3S$	$(C_2H_5O)_2P(S)OCClCH(CCl)_2N$	H	1968, 284
$PC_8H_{11}FN$	$C_6H_5PF[N(CH_3)_2]$	H; P (159·8); F (−128·5)	1965, 12; 1966, 36
$PC_8H_{11}FNO$	$C_6H_5P(O)F[N(CH_3)_2]$	H	1966, 36
$PC_8H_{11}FNS_2$	$C_6H_5SP(S)F[N(CH_3)_2]$	H; P (94·0); F (−40·3)	1968, 308
$PC_8H_{11}F_2O$	$F_2P(CH_3)_2(OC_6H_5)$	P (−11·7); F (−42)	1969, 62
$PC_8H_{11}F_3N$	$F_3P(C_6H_5)[N(CH_3)_2]$	H; P (−53); F (−40·1; −68·1)	1965, 12; 1966, 36; 1967, 50, 64
$PC_8H_{11}F_6$	$F_3P(CH_3)(NHC_6H_4CH_3)$	F (−75; −42; −30)	1967, 64
	$(C_2H_5)_2PC(CF_3)=CH(CF_3)$	H; F (−57; −61·8)	1967, 103
$PC_8H_{11}O$	$(CH_3)_2P(O)C_6H_5$	H	1967, 19
$PC_8H_{11}OS$	$C_2H_5O(C_6H_5)P(S)H$	P (65·3)	1966, 40; 1969, 162

Formula	Compound	Resonance	Reference
$PC_8H_{11}O_2$	$CH_3(CH_3O)P(O)C_6H_5$	H	1966, 133
	$C_2H_5OP(O)H(C_6H_5)$	H	1969, 74, 162
	$(CH_3O)_2PC_6H_5$	H; P (15·8)	1969, 74, 162
$PC_8H_{11}O_2S$	$CH_3O(C_6H_5SCH_2)P(O)H$	H	1966, 124
	$(CH_3O)_2P(S)C_6H_5$	H; P (9·0)	1966, 36; 1967, 113; 1969, 87
$PC_8H_{11}O_2Se$	$C_6H_5P(Se)(OCH_3)_2$	H; P (98)	1967, 113; 1969, 87
	$C_6H_5P(O)(OCH_3)(SeCH_3)$	H; P (38·7)	1967, 113
$PC_8H_{11}O_3$	$C_6H_5P(O)(OCH_3)_2$	H; P (19·3)	1966, 36, 133; 1967, 113
$PC_8H_{11}O_3S$	$C_6H_5SP(O)(OCH_3)_2$	H	1967, 87
$PC_8H_{11}O_4$	$C_6H_5OP(O)(OCH_3)_2$	^{13}C	1965, 7; 1966, 240
	$(CH_3O)_2P(O)C_6H_4OH$	H	1968, 317
$PC_8H_{11}O_5$	$(CH_3O)_2P(O)OC_6H_4OH$	P (−4·1)	1968, 318
$PC_8H_{11}O_7S$	$(CH_2O)_2P^{\oplus}(OC_6H_5)H, HSO_4^{\ominus}$	P (16)	1969, 50
$PC_8H_{11}S$	$C_6H_5P(S)(CH_3)_2$	H	1967, 19
$PC_8H_{11}S_3$	$C_6H_5P(S)(SCH_3)_2$	H; P (83·5)	1967, 113
$PC_8H_{12}Br$	$(CH_3)_2P^{\oplus}(C_6H_5)H, Br^{\ominus}$	^{13}C; H(P, C); P (−2·1)	1967, 66; 1968, 3, 7, 33
$PC_8H_{12}N$	$(CH_3)_2NP(C{\equiv}CCH_3)_2$	H	1967, 106
	$(CH_3)_2NP(O)(C{\equiv}CCH_3)_2$	H	1967, 106
$PC_8H_{12}NO$	$(CH_2O)_2PN(CH_2)_4$	H	1969, 109
$PC_8H_{12}NO_2$	$(i\text{-}C_3H_7)CHC(CN)P(O)(OC_2H_5)OH$	P (138)	1968, 319
$PC_8H_{12}NO_3$	$(CH_3O)_2P(O)C_6H_4NH_2$	H	1968, 317
$PC_8H_{12}NO_4$	$p\text{-}CH_3OC_6H_4CH(NH_2)PO_3H_2$	H; P (13·3)	1968, 142, 296
$PC_8H_{12}N_2NaO_3S$	$NaPO_3SCH_2C(NH_2)(NHCH_2C_6H_5)$	H	1968, 292
$PC_8H_{13}ClN$	$C_6H_5(CH_3)_2PNH_2, Cl$	H; P (42·8)	1967, 332; 1968, 54
$PC_8H_{13}N_2O_2$	$(CH_3NH)_2P(O)OC_6H_5$	H	1968, 284
$PC_8H_{13}N_2O_3$	$HP(O)[OC(CH_3)_2CN]_2$	P (−0·1)	1967, 293

$PC_8H_{13}N_2S$	$C_6H_5P(S)(NHCH_3)_2$	H	1967, 318
$PC_8H_{13}O_2$	$CH_2=CHCH_2OP(O)CH_2CHC(CH_3)CH_2$	H	1964, 15
$PC_8H_{13}O_2S$	$(C_2H_5O)_2P\begin{smallmatrix}\\S\end{smallmatrix}$	H	1969, 74
$PC_8H_{13}O_4S$	$(CH_3)_2(C_6H_5)P^{\oplus}H, HSO_4^{\ominus}$	H; P (−1)	1969, 50
$PC_8H_{13}O_5$	$(CH_3O)_2P(O)OCH(CH=CH_2)C(CH_3)(OH)C(O)CH_3$	H; P (0·5)	1968, 312
	$CH_3OP(O)OCH(CH=CH_2)C(CH_3)[C(O)CH_3]O$ (isom.)	P (14·7; 15·7)	1968, 312
$PC_8H_{13}O_6S$	$(CH_3O)_2(C_6H_5)P^{\oplus}H, HSO_4^{\ominus}$	P (54)	1969, 50
$PC_8H_{14}ClO_3$	$(C_2H_5O)_2P(O)C=CCHClCH_3$	H	1967, 21
	$[(CH_2)_2CHCH_2O]_2P(O)Cl$	H	1966, 213
$PC_8H_{14}Cl_3O_5$	$(C_2H_5O)_2P(O)CH(CCl_3)OCOCH_3$	H; P (9·6)	1966, 234
$PC_8H_{14}O_3$	$i\text{-}C_4H_9OP(O)OCH(=CH_2)CH=CH$	H; P (5·2)	1966, 229
$PC_8H_{15}Cl_2O_5$	$(C_2H_5O)_2P(O)CH(CHCl_2)OC(O)CH_3$	H; P (13·2)	1967, 3
$PC_8H_{15}N_2S_2$	$C_6H_5P(S)(S^{\ominus})NHCH_3CH_3NH_3^{\oplus}$	P (81·8)	1967, 126
$PC_8H_{15}O_3$	$(C_2H_5O)_2P(O)C=CC_2H_5$	H; P (−8·5)	1965, 90; 1967, 21
	$(CH_3O)_2P(O)P(O)C_6H_9$	H	1966, 43; 1967, 100
	$(C_2H_5O)_2P(O)CH=CHCH_3$	H; P (13·5)	1967, 21
	$(C_2H_5O)_2P(O)CH_2C=CCH_3$	H; P (21·4)	1967, 100, 324
$PC_8H_{15}O_4$	$C_3H_5C(O)P(O)(OC_2H_5)_2$	H	1967, 166
	$(C_2H_5O)_2P(O)C=CCH(OH)CH_3$	H; P (−3·5)	1967, 166
	$C_2H_5P(O)(COCH_3)CH_2COOC_2H_5$	H	1968, 297
	$[(CH_3)_2CO]_2POC(O)CH_3$	P (134·3)	1968, 304
$PC_8H_{15}O_5$	$(CH_3O)_2P(O)C(CH_3)CHC(O)OC_2H_5$	H	1967, 325

Formula	Compound	Resonance	Reference
	CH₃OP(O)OCH(C₂H₅)C(CH₃)(COCH₃)O⌐	P (13·9; 15·2)	1968, 312
	(stereois.)		
	C₂H₅P(O)(CH₂COOCH₃)₂	H	1968, 297
	(C₂H₅O)₂P(O)CH=CHCOOCH₃	H; P	1969, 349
PC₈H₁₅O₅S	C₂H₅OP(S)(CH₂COOCH₃)₂	H	1968, 297
PC₈H₁₅O₆	C₂H₅OP(O)(CH₂COOCH₃)₂	H	1968, 297
	(CH₃O)₂P(O)OC(CH₃)(COCH₃)₂	H; P (−3·0)	1968, 313
PC₈H₁₅O₇	(CH₃O)₂P(O)CH(COOCH₃)CH₂COOCH₃	H	1965, 94
	(CH₃O)₂P(O)OC(OC₂H₅)CHCOOCH₃	H; P (−6·6)	1965, 95
PC₈H₁₅O₈	(CH₃O)₂P(O)OC(CH₃)(COOCH₃)₂	H; P (4·6)	1967, 4
PC₈H₁₆BrO₃	(C₂H₅O)₂P(O)CH₂C(CH₃)=CHBr	H; P (22·9)	1967, 108, 326
	(C₂H₅O)₂P(O)CH=C(CH₃)CH₂Br	P(16·5)	1967, 326
	(C₂H₅O)₂P(O)CH₂CBr=CHCH₃	H; P (23)	1967, 73, 101, 108
PC₈H₁₆BrO₄	(C₂H₅O)₂P(O)CBr=CHOC₂H₅	H	1969, 127
PC₈H₁₆Br₂Cl	(CH₃)₂P⊕[CH(CH₃)CHBr]₂, Cl⊖	H	1968, 61
PC₈H₁₆Cl	ClP[C(CH₃)₂]₂CHCH₃	H(T)	1969, 350
PC₈H₁₆ClN₂O₂	[NCH(CH₃)₂]₂P(O)OCH(CH₃)(CH₂)₂Cl	H	1966, 220
	[NC(CH₃)₂]₂P(O)O(CH₂)₂Cl	H	1966, 220
PC₈H₁₆ClN₄	C₆H₅NHPN(CH₃)₂(NH₂)₂Cl	H; P (28·7)	1969, 348
PC₈H₁₆ClO₃	(C₂H₅O)₂P(O)CH₂CH=C(Cl)CH₃	H; P (25)	1967, 101, 108, 323
PC₈H₁₆ClO₅	(C₂H₅O)₂P(O)CH(CH₂Cl)OC(O)CH₃	H; P (17·5)	1967, 3
PC₈H₁₆Cl₃	CH₃(CH₂)₅CH(PCl₂)CH₂Cl	P (185)	1966, 222
	CH₃(CH₂)₅CHClCH₂PCl₂	P (191)	1966, 222
PC₈H₁₆NO₂	[(C₂H₅)₂N]P(O)OC(CH₃)CHCH₂	H; P (311)	1967, 311
PC₈H₁₆NO₂S	(C₂H₅O)₂P(O)NS(O)CH₂CH=CHCH₂⌐	H	1965, 93

$PC_8H_{16}NO_3$	$[(CH_2)_2CHCH_2O]_2P(O)NH_2$	H	1966, 213
	$(C_2H_5O)_2POC(CH_3)_2CN$	P (136·4)	1967, 293
$PC_8H_{16}NO_4$	$(C_2H_5O)_2P(O)OC(CH_3)_2CN$	P (−6·3)	1967, 293
	$[(CH_3)_2N](CH_2O)_2P[OC(CH_3)]_2$	P (−26·1)	1968, 301
$PC_8H_{16}NO_5$	$(CH_3O)_2P(O)OC(CH_3)=CHC(O)N(CH_3)_2$	H	1968, 283 to 285; 1969, 322
$PC_8H_{17}F_3N$	$F_3P(C_2H_5)NHC_6H_{11}$	F (−71; −47; −34)	1967, 64
$PC_8H_{17}NO_5$	$C_2H_5OP(O)(O^{\ominus})CH(COOCH_3)N^{\oplus}(CH_3)_3$	H; P(0)	1969, 346
$PC_8H_{17}N_2$	$[(CH_3)_2N]_2PC(CH_3)=C=CH_2$	H	1968, 99
$PC_8H_{17}N_2O$	$[(CH_3)_2N]_2P(O)C(CH_3)=C=CH_2$	H	1967, 104; 1969, 115
$PC_8H_{17}N_4$	$(CH_3)_2NPNCHC(CH_3)N$	H	1967, 119
	$(CH_3)_2NPNC(CH_3)CHCHN$	H	1967, 119
$PC_8H_{17}O$	$CH_3P[(CH_2)_2]_2C(C_2H_5)OH$	H	1967, 75
$PC_8H_{17}O_2$	$CH_3P[(CH_2)_2]_2C(CH_2CH_2OH)OH$	H	1967, 75
	$C_6H_{11}(C_2H_5O)P(O)H$	P (40·3)	1967, 172
	$t-C_4H_9OPOCH_2C(CH_3)_2O$	H	1969, 93
	$(C_2H_5)_2P(O)CH=CHOC_2H_5$	H	1969, 342
$PC_8H_{17}O_2S_2$	$(C_2H_5S)_2P(O)CH=CHOC_2H_5$	H	1969, 342
$PC_8H_{17}O_3$	$(C_2H_5O)_2P(O)CH=C(CH_3)_2$	H	1967, 327
	$(C_2H_5O)_2P(O)CH=CHC_2H_5$	H	1968, 314; 1969, 163
	$CH_3OP(OCH_2)_2CHt-C_4H_9$	H	1968, 78; 1969, 172
	$CH_3P(O)(OCH_2)_2CHt-C_4H_9$	H	1969, 173
	$P(OCH_2)_3CC_5H_{11}$	P (92·8)	1969, 71
$PC_8H_{17}O_3S$	$(C_2H_5O)_2P(O)C(SC_2H_5)=CH_2$	H	1967, 332
	$(C_2H_5O)_2P(O)SCH_2CH=CHCH_3$	H	1967, 307
	$(C_2H_5O)_2P(O)CH=CHSC_2H_5$	H	1967, 332
	$CH_3OP(S)(OCH_2)_2CHt-C_4H_9$	H	1969, 173
$PC_8H_{17}O_4$	$(CH_3O)_2P(O)CH_2C(O)C(CH_3)_3$	H	1966, 119

Formula	Compound	Resonance	Reference
PC$_8$H$_{17}$O$_4$	(CH$_3$O)$_2$P(O)OC(C$_4$H$_9$)=CH$_2$	H	1966, 119
	(C$_2$H$_5$O)$_2$P(O)(CH$_2$)$_2$COCH$_3$	H; P (30·8)	1966, 249; 1967, 278
	i-C$_3$H$_7$OP(O)(OCH$_2$)$_2$C(CH$_3$)$_2$	H	1967, 91
	CH$_3$OPOCH$_2$CH(CH$_3$)(OCH$_3$)CH$_2$O	H; P (137·9)	1967, 161
	CH$_3$P(O)OCH$_2$CH(CH$_3$)(OCH$_3$)C(CH$_3$)CH$_2$O	H; P (29·7)	1967, 161
	(C$_2$H$_5$O)$_2$P(O)C(O)i-C$_3$H$_7$	H	1967, 166
	(C$_2$H$_5$O)$_2$P(O)CH=CHOC$_2$H$_5$	H; P (21·4)	1967, 278; 1969, 127
	(C$_2$H$_5$O)$_2$P(O)CH$_2$COC$_2$H$_5$	H; P (20·1)	1967, 278
	(CH$_3$O)$_2$P(O)C(CH$_3$)$_2$CH$_2$COCH$_3$	H	1967, 311
	HP(OCH$_2$)$_2$[OC(CH$_3$)$_2$]$_2$	P ($-$30·4)	1968, 274; 1969, 179
	CH$_3$OP(O)(OCH$_2$)$_2$CHt-C$_4$H$_9$	H	1969, 173
	C$_2$H$_5$OP(O)[OC(CH$_3$)$_2$]$_2$	H	1969, 86
PC$_8$H$_{17}$O$_4$S$_2$	(C$_2$H$_5$O)$_2$P(S)SCH(CH$_3$)OCOCH$_3$	H	1967, 333
PC$_8$H$_{17}$O$_5$	(C$_2$H$_5$O)$_2$P(O)CH$_2$COOC$_2$H$_5$	H; P (19)	1967, 22
PC$_8$H$_{18}$	(C$_2$H$_5$)$_3$P$^\oplus$CH=CH$_2$	H	1967, 99
PC$_8$H$_{18}$Cl	(t-C$_4$H$_9$)$_2$PCl	P (144)	1968, 6
PC$_8$H$_{18}$ClO	(t-C$_4$H$_9$)$_2$P(O)Cl	H	1965, 8
PC$_8$H$_{18}$ClO$_3$	CH$_2$Cl(CH$_3$O)P(O)C(CH$_3$)(OC$_2$H$_5$)$_2$	H	1966, 124
PC$_8$H$_{18}$CsO	(C$_4$H$_9$)$_2$POCs	P (91·7)	1968, 320
PC$_8$H$_{18}$FO	[(CH$_3$)$_2$CHCH$_2$]$_2$P(O)F	P (71·3); F ($-$75·4)	1967, 13, 14; 1968, 295
PC$_8$H$_{18}$FO$_2$	(C$_4$H$_9$O)$_2$PF	F ($-$60)	1969, 61
PC$_8$H$_{18}$FO$_3$	(i-C$_4$H$_9$O)$_2$P(O)F	P ($-$9); F ($-$86)	1967, 14
PC$_8$H$_{18}$IN$_2$	[(CH$_3$)$_2$N]$_2$P$^\oplus$(CH$_3$)CCH$_3$I$^\ominus$	H	1967, 106
PC$_8$H$_{18}$KO	(C$_4$H$_9$)$_2$POK	P (94·2)	1968, 320
PC$_8$H$_{18}$NO$_2$	(CH$_3$)$_2$NP[OC(CH$_3$)$_2$]$_2$	H	1966, 51
PC$_8$H$_{18}$NO$_2$S	(CH$_3$)$_2$C(CH$_2$O)$_2$P(S)NHi-C$_3$H$_7$	H	1967, 91
PC$_8$H$_{18}$NO$_3$	(CH$_3$)$_2$C(CH$_2$O)$_2$P(O)NHC$_3$H$_7$	H	1967, 91; 1969, 97

NMR DATA ON ORGANIC COMPOUNDS

	Compound		Reference
	i-$C_4H_9(CH_3O)P(O)C(O)N(CH_3)_2$	H	1967, 26
	$HP[OC(CH_3)_2]_2(OCH_2CH_2NH)$	H; P (-47.3)	1968, 304
$PC_8H_{18}NO_4$	$(CH_3)_2NP(O)OC(CH_3)_2$	P (12·5)	1969, 343
	$(C_2H_5O)_2P(O)N=C(CH_3)C_2H_5$	P (3)	1969, 351
	$(C_2H_5O)_2PONC(CH_3)C_2H_5$	P (143)	1966, 55
	$(C_2H_5O)_2P(O)NCH(Oi\text{-}C_3H_7)$	H	1967, 34
	$(i\text{-}C_3H_7O)_2P(O)C(O)NHCH_3$	H	1969, 351
	$(C_2H_5O)_2P(O)N=C(CH_3)OC_2H_5$	P (0)	1969, 351
	$(C_2H_5O)_2PON=C(CH_3)OC_2H_5$	P (134)	1966, 224; 1968, 153,
$PC_8H_{18}N_3$	$C_4H_8NP[N(CH_3)CH_2]_2$	P (104·8)	303, 310
	$P[N(CH_3)CH_2]_3CCH_3$	H	1967, 334
$PC_8H_{18}N_3O$	$OP[N(CH_3)CH_2]_3CCH_3$	H	1967, 334
	$C_4H_8NP(O)[N(CH_3)CH_2]_2$	P (23·1)	1968, 153, 310
$PC_8H_{18}N_3S$	$SP[N(CH_3)CH_2]_3CCH_3$	H	1967, 334
$PC_8H_{18}ORb$	$(C_4H_9)_2PORb$	P (94·8)	1968, 320
PC_8H_{19}	$(C_4H_9)_2PH$	P (-69.5)	1965, 3
	$(i\text{-}C_4H_9)_2PH$	P (-82)	1965, 3
	$(t\text{-}C_4H_9)_2PH$	P (-20.1)	1968, 6
	$C_8H_{17}PH_2$	P (-138)	1966, 4
	$(C_2H_5)_3P=CHCH_3$	H(T)	1968, 105, 107, 305
$PC_8H_{19}BrNO_3$	$(i\text{-}C_3H_7O)_2P(O)NHCH_2CH_2Br$	H	1969, 352
$PC_8H_{19}ClN$	$t\text{-}C_4H_9(Cl)PNHt\text{-}C_4H_9$	H	1969, 112
$PC_8H_{19}N_2OS$	$[(CH_3)_2N]_2P(O)CH=CHSC_2H_5$	H	1969, 342
$PC_8H_{19}N_2O_2$	$HP[OCH_2C((CH_3)NH]_2$	P (-58)	1968, 304
$PC_8H_{19}O$	$(C_4H_9)_2P(O)H$	H	1968, 65
$PC_8H_{19}O_2S_2$	$CH_3P(O)(OCH_3)SCH_2SC_5H_{11}$	H	1969, 101
	$CH_3P(O)(OC_3H_7)SCH_2Si\text{-}C_3H_7$	H	1969, 101
	$CH_3P(O)(Oi\text{-}C_3H_7)SCH_2Si\text{-}C_3H_7$	H	1969, 101

160 FORMULAE INDEX II

Formula	Compound	Resonance	Reference
$PC_8H_{19}O_2S_3$	$(C_2H_5O)_2P(S)C(CH_2)_2SC_2H_5$	H	1968, 284, 285; 1969, 322
$PC_8H_{19}O_3$	$(C_4H_9O)_2P(O)H$	^{17}O; P (6·9)	1963, 1; 1967, 277
	$(s\text{-}C_4H_9O)_2P(O)H$	P (5·5)	1967, 277
	$(C_2H_5O)_2POC(CH_3)_3$	P (134·0)	1967, 293
$PC_8H_{19}O_3$	$C_2H_5OP(Oi\text{-}C_3H_7)_2$	P (137·3)	1967, 293
$PC_8H_{19}O_3S$	$(C_2H_5O)_2P(O)St\text{-}C_4H_9$	H	1966, 52
	$(C_2H_5O)_2P(S)Oi\text{-}C_3H_7$	P (59·0)	1967, 293
$PC_8H_{19}O_3S_2$	$(C_2H_5O)_2P(O)S(CH_2)_2SC_2H_5$	H	1968, 282, 284
$PC_8H_{19}O_4$	$(CH_3O)_2P(O)(CH_2)_2OC_4H_9$	H	1966, 251; 1967, 293
	$(C_4H_9O)_2PO_2H$	^{13}C	1968, 22
$PC_8H_{19}O_5$	$(C_2H_5O)_2P(O)CH(OH)CH_2OC_2H_5$	H	1969, 82
	$(C_2H_5O)_2P(O)CH_2CH(OH)CH_2OCH_3$	H	1969, 82
$PC_8H_{19}S$	$(i\text{-}C_4H_9)_2P(S)H$	P (13·4)	1966, 40
$PC_8H_{19}Se$	$(i\text{-}C_4H_9)_2P(Se)H$	P (−5·3)	1966, 23
PC_8H_{20}	$P^{\oplus}(C_2H_5)_4$	H; P (40·5)	1966, 22; 1967, 7
$PC_8H_{20}BF_4O_4$	$(C_2H_5O)_4P^{\oplus}BF_4^{\ominus}$	H; P (−2·4)	1969, 86
$PC_8H_{20}Br$	$(C_2H_5)_4PBr$	^{13}C; P (40·2)	1968, 3, 6
$PC_8H_{20}ClN_2$	$[(C_2H_5)_2N]_2PCl$	H	1965, 18
$PC_8H_{20}ClN_2O$	$[(C_2H_5)_2N]_2P(O)Cl$	H	1965, 18
$PC_8H_{20}ClN_2S$	$[(C_2H_5)_2N]_2P(S)Cl$	H	1965, 18
$PC_8H_{20}Cl_6O_2Sb$	$(C_2H_5)_2(C_2H_5O)_2P^{\oplus}SbCl_6^{\ominus}$	H	1967, 308
$PC_8H_{20}Cl_6O_3Sb$	$C_2H_5(C_2H_5O)_3P^{\oplus}SbCl_6^{\ominus}$	H	1967, 308
$PC_8H_{20}Cl_6O_4Sb$	$(C_2H_5O)_4P^{\oplus}SbCl_6^{\ominus}$	H; P (4)	1967, 308
$PC_8H_{20}F_3N_2$	$F_3P[N(C_2H_5)_2]_2$	F (−60·8; −69)	1966, 217
$PC_8H_{20}I$	$(C_2H_5)_4PI$	H; P (40·1)	1965, 39; 1967, 169, 244
$PC_8H_{20}N$	$(t\text{-}C_4H_9)_2PNH_2$	H	1968, 64
$PC_8H_{20}NO$	$CH_3P(Oi\text{-}C_3H_7)[N(C_2H_5)_2]$	H; P (−92·5)	1969, 192

PC$_8$H$_{20}$NO$_2$	(t-C$_4$H$_9$O)[(C$_2$H$_5$)$_2$N]P(O)H	P (6·5)	1967, 172
PC$_8$H$_{20}$NO$_3$	(CH$_3$O)$_2$P(O)CH(i-C$_3$H$_7$)N(CH$_3$)$_2$	H	1965, 36
	(CH$_3$O)P(O)(O$^\ominus$)CH(i-C$_3$H$_7$)N$^\oplus$(CH$_3$)$_3$	H	1965, 36
	(CH$_2$O)$_2$P(O)C(C$_2$H$_5$)$_2$N(CH$_3$)$_2$	P (50)	1969, 340
	(C$_2$H$_5$O)$_2$P(O)NH-C$_4$H$_9$	H	1969, 110
PC$_8$H$_{20}$NO$_3$S	O$^\ominus$P(S)O(CH$_2$)$_2$CH(CH$_3$)O, N$^\oplus$(CH$_3$)$_4$	H; P (112·8)	1969, 176
PC$_8$H$_{20}$NO$_4$	(i-C$_3$H$_7$O)$_2$P(O)O(CH$_2$)$_2$	H; P (−10·8)	1967, 185
	(i-C$_3$H$_7$O)$_2$P(O)NH(CH$_2$)$_2$OH	H; P (8·2)	1967, 185
PC$_8$H$_{20}$N$_3$O	[(CH$_3$)$_2$N]$_2$P(O)NCH$_2$C(CH$_3$)$_2$	H	1966, 246
	[(CH$_3$)$_2$N](C$_2$H$_5$O)PN(CH$_3$)CN(CH$_3$)$_2$	H; P (128·8)	1968, 321
	[(CH$_3$)$_2$N]$_2$P(O)CH=CHN(CH$_3$)$_2$	H	1969, 342
PC$_8$H$_{20}$N$_3$OS	[(CH$_3$)$_2$N](C$_2$H$_5$O)P(S)N(CH$_3$)CN(CH$_3$)$_2$	P (71·3)	1968, 321
PC$_8$H$_{21}$BN$_3$	CH$_3$C[CH$_2$(CH$_3$)N]$_3$PBH$_3$	H	1967, 334
PC$_8$H$_{21}$BN$_3$O	CH$_3$C[CH$_2$(CH$_3$)(BH$_3$)N][CH$_2$(CH$_3$)N]$_2$PO	H	1967, 334
PC$_8$H$_{21}$BN$_3$S	CH$_3$C[CH$_2$(CH$_3$)(BH$_3$)N][CH$_2$(CH$_3$)N]$_2$PS	H	1967, 334
PC$_8$H$_{21}$ClNO$_3$	(C$_2$H$_5$O)$_2$P(O)CH(i-C$_3$H$_7$)NH$_3$Cl	P (21·3)	1968, 296
PC$_8$H$_{21}$O$_3$Si	(C$_2$H$_5$O)[(C$_2$H$_5$)$_3$SiO]P(O)H	H	1967, 34
PC$_8$H$_{21}$Si	(CH$_3$)$_3$SiCH=P(CH$_3$)$_2$C$_2$H$_5$	H	1968, 105, 107, 322
PC$_8$H$_{22}$AlO	(CH$_3$)$_3$POAl(C$_2$H$_5$)$_3$	H	1965, 83
PC$_8$H$_{22}$ClSi	(CH$_3$)$_2$(C$_2$H$_5$)P$^\oplus$CH$_2$Si(CH$_3$)$_3$, Cl$^\ominus$	H	1968, 322
	(CH$_3$)$_3$P$^\oplus$CH(CH$_3$)Si(CH$_3$)$_3$, Cl$^\ominus$	H	1968, 107, 322
PC$_8$H$_{22}$N$_3$O	[(CH$_3$)$_2$N]$_2$P(O)NHt-C$_4$H$_9$	H	1968, 91
PC$_8$H$_{22}$N$_3$O$_2$	[(CH$_3$)$_2$N]$_2$P(O)NHCH(CH$_3$)CH$_2$OCH$_3$	H	1966, 246
PC$_8$H$_{24}$B$_2$N$_3$	BH$_3$P[N(CH$_3$)CH$_2$]$_2$[N(BH$_3$)CH$_2$]CCH$_3$	H	1967, 334
PC$_9$H$_6$ClF$_4$	CF$_2$CCl=CF$_2$PHC$_6$H$_5$	H; F (−80·8; −79·4; −90·6)	1967, 295
PC$_9$H$_9$Cl$_4$O$_5$	(CH$_3$O)$_3$POC$_6$Cl$_4$O	H; P (−46·2)	1968, 315
PC$_9$H$_9$F$_{12}$O$_2$	(CH$_3$)$_3$P[OC(CF$_3$)$_2$]$_2$	H; P (3·2); F (9·3)	1968, 302

Formula	Compound	Resonance	Reference
$PC_9H_9F_{12}O_2$	$(CH_3)_2[(CF_3)_2CHO]POC(CF_3)_2CH_2$	$H(T); P(-23.7);$ F	1968, 302
$PC_9H_9F_{12}O_5$	$(CH_3O)_3P[OC(CF_3)_2]_2$	$P(-50.1)$	1966, 225; 1967, 335; 1968, 124, 310
$PC_9H_9O_3$	$CH_2=C=CHP(O)(OCH_2C\equiv CH)_2$	H	1967, 104; 1969, 115
$PC_9H_9O_4$	$C_6H_5C(O)OP(OCH_2)_2$	H	1966, 50
PC_9H_9S	$(CH_3C\equiv C)_3PS$	H	1965, 96
$PC_9H_{10}ClO_2$	$C_2H_5P(CH_2COOCH_3)Cl$	H	1968, 297
	$ClPOC_6H_3(i\text{-}C_3H_7)O$	P (175)	1969, 45
$PC_9H_{10}ClO_2S$	$ClP(S)OCH(C_6H_5)(CH_2)_2O$	H	1969, 95
$PC_9H_{10}ClO_3$	$ClP(O)OCH(C_6H_5)(CH_2)_2O$	H	1969, 95
$PC_9H_{10}Cl_4NO_2S$	$C_6HCl_4OP(S)(OCH_3)NHC_2H_5$	H	1967, 116
$PC_9H_{10}KO_4$	$O(CH_2)_2CH(C_6H_5)OP(O)O^\ominus, K^\oplus$	H	1969, 95
$PC_9H_{11}Ba_{\frac{1}{2}}ClO_5$	$CH_3CHOHCH_2OP(O)(O^\ominus)(OC_6H_7Cl), Ba^\oplus_{\frac{1}{2}}$	H	1965, 88
$PC_9H_{11}Ba_{\frac{1}{2}}NO_7$	$CH_3CHOHCH_2OP(O)(O^\ominus)(OC_6H_4NO_2), Ba^\oplus_{\frac{1}{2}}$	H	1965, 88
$PC_9H_{11}Cl_2HgO_3$	$(CH_3O)_2P(O)CCl_2HgC_6H_5$	H	1966, 252
$PC_9H_{11}O$	$C_6H_5PO(CH_2)_3$	H; P (110.2)	1967, 138
$PC_9H_{11}OS$	$C_6H_5P(S)O(CH_2)_3$	H; P (102.4)	1967, 138
	$C_6H_5SPO(CH_2)_3$	H; P (79.0)	1967, 138
$PC_9H_{11}O_2$	$C_6H_5P(O)O(CH_2)_3$	H; P (58.4)	1967, 138

PC$_9$H$_{11}$O$_2$S$_4$	CH$_3$OP(S)SC(S)CSC(CH$_3$)C(C$_2$H$_5$)CO	H	1969, 148
PC$_9$H$_{11}$O$_3$	CH$_3$CH=C(C$_6$H$_5$)PO$_3$H$_2$ (*trans, cis*)	H; P (14·8; 8·9)	1966, 241
	C$_6$H$_4$(CH$_2$)$_2$CHPO$_3$H$_2$	H; P (11·0)	1966, 241
	C$_6$H$_5$P(O)OCH(CH$_3$)CH$_2$O	P (10·4; 11·0)	1969, 343
	C$_6$H$_5$OPO(CH$_2$)$_3$O	H	1969, 93
PC$_9$H$_{11}$O$_4$	C$_6$H$_5$OP(O)(OCH$_2$)$_2$CH$_2$	H	1968, 83; 1969, 93
	HCOC$_6$H$_4$P(O)(OCH$_3$)$_2$	H	1968, 317
	CH$_3$CHOHCH$_2$OP(O)(O$^\ominus$)(OC$_6$H$_5$), Ba$_{1/2}^{\oplus}$	H	1965, 88
	BrC$_6$H$_4$CH$_2$P(O)(OCH$_3$)$_2$	H	1967, 117
PC$_9$H$_{12}$Ba$_{1/2}$O$_5$	BrCH$_2$C$_6$H$_4$P(O)(OCH$_3$)$_2$	H	1966, 253
PC$_9$H$_{12}$BrO$_2$	BrCH$_2$C$_6$H$_4$OP(O)(OCH$_3$)$_2$	H	1966, 253
PC$_9$H$_{12}$BrO$_3$	(CH$_3$O)$_2$P(O)CHClHgC$_6$H$_5$	H	1966, 252
PC$_9$H$_{12}$ClHgO$_3$	(CH$_3$O)$_2$P(S)SCH$_2$SC$_6$H$_4$Cl	H	1968, 284, 285
PC$_9$H$_{12}$ClO$_2$S	CH$_3$O(C$_2$H$_5$NH)P(O)OC$_6$H$_3$Cl$_2$	H	1969, 110
PC$_9$H$_{12}$Cl$_2$NO$_3$	C$_6$H$_5$(i-C$_3$H$_7$O)P(O)F	F (−64·9)	1967, 13; 1968, 295
PC$_9$H$_{12}$FO$_2$	C$_6$H$_5$P(CH$_2$)$_2$NHCH$_2$	H	1967, 137
PC$_9$H$_{12}$N			
PC$_9$H$_{12}$NO$_2$	C$_6$H$_5$P(O)OCH$_2$CH$_2$NCH$_3$	H; P (35·7)	1969, 171
	(CH$_3$)$_2$NPOC$_6$H$_3$(CH$_3$)O	P (150)	1969, 45
PC$_9$H$_{12}$NO$_3$	HP(OC$_6$H$_4$O)[OCH(CH$_3$)CH$_2$NH]	P (−35)	1969, 45, 178
	HP(OC$_6$H$_4$O)[OCH$_2$CH$_2$NCH$_3$]	P (−99·5)	1969, 45
	OC$_6$H$_4$OPO(CH$_2$)$_2$NHCH$_3$	P (130)	1969, 45

Formula	Compound	Resonance	Reference
$PC_9H_{12}NO_3$	$C_6H_5OP(O)OCH_2CH_2NCH_3$	H; P (15·5)	1968, 76
$PC_9H_{12}NO_4$	$CH_3(C_2H_5O)P(O)C_6H_4NO_2$	H	1969, 119
$PC_9H_{12}N_3$	$P(CH_2CH_2CN)_3$	P (−23)	1966, 4; 1967, 244
$PC_9H_{12}N_3O_7$	$CH=CH(NH_2)CNC(O)NCHOCH(CH_2OPO_2H)CH(OH)CHO$	H	1969, 98
PC_9H_{13}	$CH_3C_6H_4P(CH_3)_2$	H	1968, 15
$PC_9H_{13}F$	$FC_6H_4P^{\oplus}(CH_3)_3$	F	1965, 5
$PC_9H_{13}FI$	$FC_6H_4P^{\oplus}(CH_3)_3I^{\ominus}$	F	1968, 13; 1969, 18, 19
$PC_9H_{13}N_2S$	$(CH_3)_2NPSC_6H_4NCH_3$	H	1968, 304
$PC_9H_{13}O_2$	$C_6H_5(i\text{-}C_3H_7O)P(O)H$	H	1969, 162
$PC_9H_{13}O_3$	$(CH_3O)_2P(O)C_6H_4CH_3$	H	1966, 70, 253; 1967, 117; 1968, 317
$PC_9H_{13}O_4$	$(CH_3O)_2P(O)OC_6H_4CH_3$	H	1967, 117; 1968, 317
	$(CH_3O)_2P(O)C_6H_4CH_2OH$	H	1966, 253
$PC_9H_{13}O_5$	$(CH_3O)_2P(O)OC_6H_4OCH_3$	P (−3·7)	1966, 242; 1968, 301, 318
	$(CH_3O)_3POC_6H_4O$	H	1968, 124
$PC_9H_{13}O_7$	$CH_3OP(O)OC(CH_3)(COCH_3)C(OCH_3)=CCOOH$ (isom.)	H; P (30·4; 29·8)	1969, 353
$PC_9H_{14}Br$	$(CH_3)_3P^{\oplus}C_6H_5, Br^{\ominus}$	P (23·3)	1968, 6
$PC_9H_{14}Cl_6O_2SSb$	$(CH_3O)_2(C_6H_5)P^{\oplus}SCH_3, SbCl_6^{\ominus}$	H; P (83·9)	1969, 87
$PC_9H_{14}Cl_6O_2SbSe$	$(CH_3O)_2(C_6H_5)P^{\oplus}SeCH_3, SbCl_6^{\ominus}$	H; P (84·6)	1969, 87
$PC_9H_{14}Cl_6O_3Sb$	$(CH_3O)_3(C_6H_5)P^{\oplus}SbCl_6^{\ominus}$	H; P (35·1)	1969, 87

NMR DATA ON ORGANIC COMPOUNDS

$PC_9H_{14}NO_4$	$OCH(C_6H_5)CH(CH_2)_2OPO_2^\ominus, NH_4^\oplus$	H	1967, 90
$PC_9H_{14}NO_5$	$(C_2H_5O)_2P(O)C_5H_4N$	H	1969, 119
$PC_9H_{14}N_3O_8$	$(CH_3O)_2C_6H_3CH(NH_2)PO_3H_2$	H	1968, 142
	$CH=CH(NH_2)CNC(O)NCHOCH(CH_2OH)CH(OH)CHOPO_3H_2$		
	 H		1969, 119
	$CH=CH(NH_2)CNC(O)NCHOCH(CH_2OPO_3H_2)CH(OH)CHOH$		
	 H		1969, 119
PC_9H_{15}	$P(CH_2CH=CH_2)_3$	P (−34·3)	1967, 244
	$P[CH(CH_2)_2]_3$	H; P (17·1)	1969, 354
$PC_9H_{15}N_2O$	$CH_3[(CH_3)_2N]P(O)NHC_6H_5$	H	1965, 82
$PC_9H_{15}N_2O_5S$	$NH_2NHSO_2C_6H_4CH_2P(O)(OCH_3)_2$	H	1966, 71
$PC_9H_{15}N_2O_7$	$(CH_3O)_2P(O)OCC(O)N(CH_3)C(OCH_3)NC(OCH_3)$	H; P (−1·3)	1966, 242
	$(CH_3O)_2P(O)OC[C(OCH_3)N]_2COCH_3$	H; P (−2·2)	1966, 242
$PC_9H_{15}O_3$	$CH_2=CHCH_2P(O)(OCH_2CH=CH_2)_2$	H	1966, 254
	$(C_2H_5O)_2P(O)C{\equiv}CC(CH_3)=CH_2$	H; P (−9·2)	1967, 21
$PC_9H_{15}O_3S$	$CH_3COCH_2P(S)[OC(CH_3)=CH_2]_2$	H	1968, 297
$PC_9H_{15}O_4$	$(CH_3O)_2P(O)C(OH)CH(CH_2)CH=CHCHCH_2$	H{P}	1969, 80
$PC_9H_{15}O_5S$	$CH_3COCH_2P(S)(CH_2COOCH_3)_2$	H	1968, 297
$PC_9H_{15}O_6$	$P(CH_2COOCH_3)_2$	H	1968, 297
$PC_9H_{16}BrO_4$	$(CH_3O)_2P(O)C(OH)CH(CHBr)(CH_2)_2CHCH_2$	H	1969, 80
$PC_9H_{16}ClN_3$	$C_6H_5P^\oplus(NH_2)[N(CH_3)N(CH_3)_2], Cl^\ominus$	H; P (39·9)	1967, 94
$PC_9H_{16}NO_3$	$(C_2H_5O)_2P(O)C(CN)=C(CH_3)_2$	H	1968, 323
$PC_9H_{17}Cl_3NO_7$	$(C_2H_5O)_2P(O)CH(OCONHCOCl_3)CH_2OCH_3$	H	1969, 82
$PC_9H_{17}N_2O_3S$	$[(CH_3)_2CO]_2P(S)O^\ominus, C_3H_5N_2^\oplus$	P (144·3)	1969, 177

Formula	Compound	Resonance	Reference
PC$_9$H$_{17}$O	C$_6$H$_{11}$PO(CH$_2$)$_3$	H; P (123·1)	1967, 138
PC$_9$H$_{17}$O$_2$	C$_6$H$_{11}$P(O)O(CH$_2$)$_3$	H; P (71·1)	1967, 138
PC$_9$H$_{17}$O$_3$	P(OCH$_2$)$_3$C(CH$_2$)$_4$CH$_3$	H	1966, 48
	CH$_2$=C=CHP(O)(Oi-C$_3$H$_7$)$_2$	H	1967, 104; 1969, 115
	C$_3$H$_7$C≡CP(O)(OC$_2$H$_5$)$_2$	H	1967, 21
	CH$_2$=CHC(CH$_3$)=CHP(O)(OC$_2$H$_5$)$_2$	H	1969, 163
PC$_9$H$_{17}$O$_3$S	SP(CH$_2$O)$_3$C(CH$_2$)$_4$CH$_3$	H	1968, 53
PC$_9$H$_{17}$O$_4$	(CH$_3$O)$_2$P(O)CHC(CH$_2$)$_5$O	H	1967, 81
	OP(OCH$_2$)$_3$C(CH$_2$)$_4$CH$_3$	H	1966, 48
	OP(CH$_2$O)$_3$C(CH$_2$)$_4$CH$_3$	H	1968, 53
	i-C$_3$H$_7$OP(O)(CH$_2$CHOCH$_2$)$_2$	H	1966, 254
	CH$_3$OP(O)[OCH$_2$CH(CH$_2$)$_2$]$_2$	H	1966, 213
	(CH$_3$O)$_2$P(O)C(CH$_2$)$_5$CHO	H	1967, 81
	(C$_2$H$_5$O)$_2$P(O)C(O)C$_4$H$_7$	H	1967, 166
	(CH$_3$O)P(O)C(O)C$_6$H$_{11}$	H	1967, 166
	(C$_2$H$_5$O)$_2$P(O)OC(CH$_3$)$_2$C≡CH	H; P (−7·3)	1967, 21
	(CH$_3$O)$_2$P(O)C(OH)CH(CH$_2$)(CH$_2$)$_2$CHCH$_2$	H	1969, 80
PC$_9$H$_{17}$O$_6$	(C$_2$H$_5$O)$_2$P(O)C(CH$_3$)CHCOCH$_3$	H	1969, 355
	(CH$_3$O)$_3$POC(=CH$_2$)C(CH$_3$)(COCH$_3$)O	P (−51·4)	1968, 313; 1969, 353
PC$_9$H$_{17}$S$_3$	P(SCH$_2$)$_3$CC$_5$H$_{11}$	H; P (32·8)	1969, 71
PC$_9$H$_{18}$	CH$_3$P[C(CH$_3$)$_2$]$_2$CCH$_3$	H(T)	1968, 121
PC$_9$H$_{18}$BrO$_3$	(C$_2$H$_5$O)$_2$P(O)CH$_2$CBr=C(CH$_3$)$_2$	H; P (24·2)	1967, 323

$PC_9H_{18}ClN_2O_2$	$[NC(CH_3)_2]_2P(O)OCH(CH_3)CH_2Cl$		H	1966, 220
$PC_9H_{18}Cl_3N_2O$	$Cl_3P(NC_4H_9)_2CO$		H	1965, 6
$PC_9H_{18}LiO_5$	$CH_3CHOHCH_2OP(O)(O)^{\ominus}(OC_6H_{11}), Li^{\oplus}$		H	1965, 88
$PC_9H_{18}NO_2$	$(C_2H_5)_2NP(O)O[C(CH_3)]_2CH_2$		H	1967, 311
$PC_9H_{18}NO_2S$	$(C_2H_5O)_2P(O)NS(O)CH(CH_3)CH=CHCH_2$		H	1965, 93
	$(C_2H_5O)_2P(O)NS(O)CH_2CH=CHCHCH_3$		H	1965, 93
	$(C_2H_5O)_2P(O)NS(O)CH_2CH=C(CH_3)CH_2$		H	1965, 93
$PC_9H_{18}NO_3$	$(CH_2O)_2P(O)C[N(CH_3)_2C_4H_8$		H; P	1968, 324
$PC_9H_{18}N_3O_3$	$CH_3CHCH_2N]_2P(O)NHCOOC_2H_5$		H	1965, 42
$PC_9H_{19}F_2O$	$(C_4H_9)_2P(O)CHF_2$		H	1965, 97
$PC_9H_{19}N_2O$	$[(CH_3)_2N]_2P(O)CH=C=C(CH_3)_2$		H	1969, 115
$PC_9H_{19}N_4$	$[(CH_3)N]_2PNCHC(CH_3)C(CH_3)N$		H	1967, 119
$PC_9H_{19}O$	$CH_3P(O)[C(CH_3)_2CHCH_3$		H	1967, 82
	$(C_3H_7)_2POC(CH_3)_{)}=CH_2$		H	1968, 297
	$C_2H_5P(CH_2CH_2)_2CH_2CH_2OH$		H	1969, 356
	$C_4H_9(C_2H_5O)PCH_2COCH_3$		H	1968, 297
	$C_5H_{10}P(O)OC_4H_9$		H	1969, 142
$PC_9H_{19}O_2S$	$C_4H_9(C_2H_5O)P(S)CH_2COCH_3$		H	1968, 297
	$C_4H_9(C_2H_5O)P(S)OC(CH_3)=CH_2$		H	1968, 297
$PC_9H_{19}O_2S_3$	$(C_2H_5S)_2P(O)(CH_2)_2C(O)SC_2H_5$		H	1967, 313
$PC_9H_{19}O_3$	$C_4H_9P(O)(OCH_2)_2C(CH_3)_2$		H	1967, 91
	$(C_2H_5O)_2POC(CH_3)_2CH=CH_2$		H	1967, 293
	$(C_2H_5O)_2P(O)CH=CH\text{-}i\text{-}C_3H_7$		H	1968, 314
$PC_9H_{19}O_3S$	$(C_2H_5O)_2P(O)SCH_2=CHC(CH_3)_2$		H	1966, 52
	$(C_2H_5O)_2P(O)SC(CH_3)_2CH=CH_2$		H	1966, 52

Formula	Compound	Resonance	Reference
$PC_9H_{19}O_3S$	$(C_2H_5O)_2P(O)SC_5H_9$	H	1966, 52
	$(C_2H_5O)_2P(O)CH_2CH=CHSC_2H_5$	H; P (24·7)	1967, 108
	$(i\text{-}C_3H_7O)_2P(O)SCH_2CH=CH_2$	H	1967, 307
$PC_9H_{19}O_4$	$t\text{-}C_4H_9OP(O)(OCH_2)_2C(CH_3)_2$	H	1967, 91
	$C_2H_5OPOCH_2[CH(CH_3)(OCH_3)]C(CH_3)CH_2O$	H; P (138·7)	1967, 161, 162
	$(C_2H_5O)_2P(O)C(O)t\text{-}C_4H_9$	H	1967, 166
	$(C_2H_5O)_2P(O)CH=C(CH_3)OC_2H_5$	H; P (21·6)	1967, 278
	$(C_2H_5O)_2P(O)CH_2CH=CHOC_2H_5$	H; P (27)	1967, 108, 278
$PC_9H_{19}O_5$	$(C_2H_5O)_2P(O)(CH_2)_2COOC_2H_5$	H	1966, 249; 1967, 313
	$(C_2H_5O)_2P(O)CH_2COCH_2OC_2H_5$	H; P (19)	1967, 22, 278
$PC_9H_{19}O_6$	$(C_2H_5O)_2P(O)CH(OC_2H_5)CH_2COOH$	H	1966, 249
$PC_9H_{20}BF_4O_4$	$(C_2H_5O)_2P^{\oplus}(OCH_2)_2C(CH_3)_2, BF_4^{\ominus}$	H; P (−4·3)	1969, 86
$PC_9H_{20}NO_2$	$(C_2H_5)_2P(O)CH_2C(O)CH_2N(CH_3)_2$	H	1968, 325
$PC_9H_{20}NO_2S$	$(CH_3)_2NP(S)(OCH_2)_2C(CH_3)_2$	H	1967, 91
$PC_9H_{20}NO_3$	$t\text{-}C_4H_9NHP(O)(OCH_2)_2C(CH_3)_2$	H	1967, 91; 1968, 74
	$HP[OC(CH_3)_2]_2OCH(CH_3)CH_2NH$	H; P (−49·5)	1967, 164
	$(C_2H_5O)_2P(O)CH=C(CH_3)N(CH_3)_2$	H; P (26·6)	1967, 278, 324
	$(CH_2O)_2P(O)C(C_2H_5)_2N(CH_3)_2$	H; P (50)	1968, 324
	$(C_2H_5O)_2P=NC(OCH_3)i\text{-}C_3H_7$	P (138)	1969, 357
	$(C_2H_5)_2NP(O)(OCH_2)_2C(CH_3)_2$	H	1969, 97
	$(C_2H_5O)_2P(O)NCH(O\text{-}i\text{-}C_4H_9)$	H	1966, 55
$PC_9H_{20}NO_4$	$(C_4H_9O)_2P(O)NHCHO$	H	1966, 235
	$(C_2H_5O)_2P(O)(CH_2)_2CONHC_2H_5$	H	1966, 249
	$(C_2H_5O)_2P(O)CH_2C(O)CH_2N(CH_3)_2$	H; P (19·7)	1967, 22
	$(CH_3O)_2P(O)C(O)N(i\text{-}C_3H_7)_2$	H	1967, 26; 1968, 29
	$(C_2H_5O)_2P(O)C(O)N(C_2H_5)_2$	H	1967, 26
	$(i\text{-}C_3H_7O)_2P(O)C(O)N(CH_3)_2$	H	1967, 26

$PC_9H_{20}NO_5$	$(i\text{-}C_3H_7O)_2P(O)C(O)NHC_2H_5$	H	1967, 26
PC_9H_{21}	$C_2H_5OP(O_2^{\ominus})C(CH_3)(COOCH_3)N^{\oplus}(CH_3)_3$	H	1969, 346
	$(C_3H_7)_3P$	$P(-33)$	1965, 3
	$(i\text{-}C_3H_7)_3P$	H; P (19·4)	1965, 3; 1967, 244, 277; 1968, 6, 7; 1969, 354
$PC_9H_{21}Cl_2$	$(C_3H_7)_3PCl_2$	H; P	1969, 282
$PC_9H_{21}FN$	$CH_3PF[N(C_4H_9)_2]$	P (167)	1967, 204; 1968, 12
$PC_9H_{21}F_3N$	$F_3P(CH_3)[N(C_4H_9)_2]$	P (−38·8); F (−71; −34)	1967, 64
$PC_9H_{21}N_2O_3$	$[(CH_3)_2N]_2POCH(CH_3)COOC_2H_5$	P (144)	1969, 109
$PC_9H_{21}O$	$(i\text{-}C_3H_7)_3PO$	P (55·0)	1967, 277
$PC_9H_{21}O_2$	$(i\text{-}C_3H_7)_2P(O)Oi\text{-}C_3H_7$	H	1968, 65
$PC_9H_{21}O_2S_2$	$CH_3P(O)(OC_4H_9)SCH_2Si\text{-}C_3H_7$	H	1969, 101
$PC_9H_{21}O_3$	$CH_3P(O)(Ot\text{-}C_4H_9)_2$	H	1967, 336
	$(i\text{-}C_3H_7O)_3P$	P (137)	1967, 277, 293; 1969, 50
	$(C_2H_5O)_2POCH_2C(CH_3)_3$	P (137·2)	1967, 277
	$CH_3OP(Ot\text{-}C_4H_9)_2$	P (130·5)	1967, 277
$PC_9H_{21}O_3S$	$(i\text{-}C_3H_7O)_3PS$	P (62·5)	1967, 277
	$(C_2H_5O)_2P(O)SC_5H_{11}$	H	1967, 307
$PC_9H_{21}O_3Se$	$(i\text{-}C_3H_7O)PSe$	P (67·9)	1967, 277
$PC_9H_{21}O_4$	$(C_3H_7O)_3PO$	H{P}	1967, 85
	$(i\text{-}C_3H_7O)_3PO$	$P(-3 \text{ to } -6)$	1967, 277, 293
$PC_9H_{21}O_5$	$(CH_3O)_3P[OC(CH_3)_2]_2$	$P(-48·9)$	1968, 318
	$(C_2H_5O)_2P(O)CH_2CH(OH)CH_2OC_2H_5$	H	1969, 82
$PC_9H_{21}S$	$(i\text{-}C_3H_7)_3PS$	P (74·0)	1967, 277
PC_9H_{22}	$i\text{-}C_3H_7P^{\oplus}(C_2H_5)_3$	P (42·8)	1967, 7
$PC_9H_{22}Br$	$HP^{\oplus}(i\text{-}C_3H_7)_3Br^{\ominus}$	H; P (39·6)	1968, 7
$PC_9H_{22}F_2N_3$	$[(C_2H_5)_2N]_3P=CF_2$	$P(-58·0)$	1964, 9
$PC_9H_{22}NO$	$(C_2H_5O)[(CH_3)_2N]PN(C_2H_5)CN(CH_3)_2$	H; P (128·1)	1968, 321

Formula	Compound	Resonance	Reference
$PC_9H_{22}NOS$	$(C_2H_5O)[(CH_3)_2NP(S)N(C_2H_5)CN(CH_3)_2$	H; P (69·4)	1968, 321
$PC_9H_{22}NO_2$	$(t\text{-}C_5H_{11}O)[(C_2H_5)_2N]P(O)H$	P (6·4)	1967, 172
$PC_9H_{22}NO_3$	$CH_3OP(O)(O^\ominus)CH(i\text{-}C_3H_7)$, $C_3H_7NH_2^\oplus$	H	1965, 36
	$(C_4H_9O)_2P(O)NHCH_3$	H	1966, 55, 235
$PC_9H_{22}N_2O_2$	$C_4H_9OP(O)N(i\text{-}C_3H_7)[N(CH_3)_2]$	P (100)	1965, 92
$PC_9H_{22}N_3O$	$[(CH_3)_2N]_2P=NC(OCH_3)C_3H_7$	H	1969, 357
$PC_9H_{22}ClN_3O$	$[(CH_3)_2N]_2P(O)NHCH(CH_3)CH_2Cl$	H	1966, 246
	$[(CH_3)_2N]_2P(O)NHCH_2CH(CH_3)Cl$	H	1966, 246
$PC_9H_{23}ClNO_3$	$(C_2H_5O)_2P(O)CH (s\text{-}C_4H_9)\overset{\oplus}{N}H_3\overset{\ominus}{Cl}$	P (21·3)	1968, 296
$PC_9H_{23}N_2$	$CH_3P[N(C_2H_5)_2]_2$	H; P (−145·5)	1969, 192
$PC_9H_{23}N_4Si$	$CH_3N(CH_2)_2N(CH_3)P[N(CH_2)_2NSi(CH_3)_3]$	H	1967, 305
$PC_9H_{23}OSi$	$(CH_3)_3SiO(CH_2)_4P(CH_3)_2$	H; P (−53·6)	1968, 326
$PC_9H_{23}O_2$	$CH_3P(O\text{-}s\text{-}C_4H_9)_2$(stereois.)	H; P (185·4; 182·8; 179·6)	1969, 162
$PC_9H_{23}O_3Si$	$C_3H_7O[(C_2H_5)_3SiO]P(O)H$	H	1967, 34
$PC_9H_{23}O_7S$	$HP^\oplus(OC_3H_7\text{-}i)$, $HSO_4^\ominus$	P (16)	1969, 50
$PC_9H_{24}AlO$	$(C_2H_5)_3POAl(CH_3)_3$	H	1965, 83
$PC_9H_{24}GeN$	$(CH_3)_3GeN=P(C_2H_5)_3$	H	1967, 73
$PC_9H_{24}NO_3Sn$	$(C_2H_5)_3SnN(CH_3)P(O)(OCH_3)_2$	H	1967, 299
$PC_9H_{24}NSi$	$(CH_3)_3SiN=P(C_2H_5)_3$	H	1967, 73, 322
$PC_9H_{24}NSn$	$(CH_3)_3SnN=P(C_2H_5)_3$	H	1967, 73
$PC_9H_{24}N_3O$	$[(CH_3)_2NCH_2]_3PO$	H; P (48·5)	1967, 291
$PC_9H_{24}N_3O_2$	$[(CH_3)_2N]_2P(O)NHC(CH_3)_2CH_2OCH_3$	H	1966, 246
$PC_9H_{27}AlNSi$	$[(CH_3)_3Si][(CH_3)_3Al][N=P(CH_3)_3]$	H	1967, 322
$PC_9H_{27}GaNSi$	$[(CH_3)_3Si][(CH_3)_3Ga]N=P(CH_3)_3$	H	1967, 322
$PC_9H_{27}Ge_3$	$P[Ge(CH_3)_3]_3$	H; P (−228)	1966, 57; 1967, 337; 1969, 14

PC$_9$H$_{27}$InNSi	[(CH$_3$)$_3$Si][(CH$_3$)$_3$In]N=P(CH$_3$)$_3$	H	1967, 322
PC$_9$H$_{27}$N$_6$	P[N(CH$_3$)N(CH$_3$)$_2$]$_3$	H; P (101·5)	1967, 94
PC$_9$H$_{27}$Si$_3$	P[Si(CH$_3$)$_3$]$_3$	H; P ($-251·2$)	1967, 29; 1968, 327
PC$_9$H$_{27}$Sn$_3$	P[Sn(CH$_3$)$_3$]$_3$	P (-330)	1967, 237; 1969, 14
PC$_{10}$H$_7$Cl$_2$S	β-C$_{10}$H$_7$P(S)Cl$_2$	P (74·4)	1966, 114
PC$_{10}$H$_9$	C$_6$H$_5$P(CH=CH)	H	1968, 101
PC$_{10}$H$_{10}$ClO$_5$	(CH$_3$O)$_2$P(O)CClC$_6$H$_4$C(O)O	H; P (9·7)	1967, 338
PC$_{10}$H$_{10}$F$_5$	C$_6$F$_5$P(C$_2$H$_5$)$_2$	H; P ($-23·4$) F ($-132·3$; $-164·3$; $-154·9$)	1966, 19; 1967, 11
PC$_{10}$H$_{11}$	C$_6$H$_5$P(CH$_2$CH)	H	1968, 101
PC$_{10}$H$_{11}$O	C$_6$H$_5$P(O)(CH$_2$CH)$_2$	H	1968, 101
PC$_{10}$H$_{11}$S	C$_6$H$_5$C=CP(S)(CH$_3$)$_2$	H	1965, 89
PC$_{10}$H$_{12}$F$_5$N$_2$	C$_6$F$_5$P[N(CH$_3$)$_2$]$_2$	H; F ($-140·2$; $-164·1$; $-157·1$)	1966, 19
PC$_{10}$H$_{12}$NO	C$_6$H$_5$P(O)(CH=CH$_2$)N(CH$_2$)$_2$	H	1966, 250
PC$_{10}$H$_{12}$N$_3$O$_3$S$_2$	(CH$_3$O)$_2$P(S)SCH$_2$NC(O)C$_6$H$_4$N=N	H	1968, 284, 285
PC$_{10}$H$_{12}$N$_3$O$_4$S	(CH$_3$O)$_2$P(O)SCH$_2$NC(O)C$_6$H$_4$N=N	H	1968, 285
PC$_{10}$H$_{13}$ClNO	C$_6$H$_5$(CH$_2$ClCH$_2$)P(O)N(CH$_2$)$_2$	H	1966, 250
PC$_{10}$H$_{13}$Cl$_3$NO$_2$S	C$_6$H$_2$Cl$_3$OP(S)(OCH$_3$)NHi-C$_3$H$_7$	H	1967, 116
PC$_{10}$H$_{13}$N$_2$O$_7$	(C$_2$H$_5$O)$_2$P(O)C$_6$H$_3$(NO$_2$)$_2$	H	1969, 119
PC$_{10}$H$_{13}$O	(CH$_3$)$_2$P(O)C(C$_6$H$_5$)=CH$_2$	H	1966, 119
	C$_6$H$_5$POCH$_2$C(CH$_3$)$_2$O	H	1969, 93
PC$_{10}$H$_{13}$O$_2$	C$_6$H$_5$OPOCH$_2$C(CH$_3$)$_2$O	H	1969, 93

172 FORMULAE INDEX II

Formula	Compound	Resonance	Reference
$PC_{10}H_{13}O_3$	$C_6H_5P(O)[OCH(CH_3)]_2$	P (8·4 – 7·1)	1969, 243
$PC_{10}H_{13}O_4$	$(CH_3O)_2P(O)CH_2C(O)C_6H_5$	H	1967, 121
	$(CH_3O)_2P(O)O(CH_2)CC_6H_5$	H	1967, 121
	$C_6H_5OP(O)O(CH_2)_2CH(CH_3)O$	H; P (−15)	1968, 83, 150; 1969, 175
	$(CH_3O)_2P(O)CHCH(C_6H_5)O$	H	1967, 81
$PC_{10}H_{14}$	$C_6H_5(CH_3)_2P^{\oplus}CH=CH_2$	H	1967, 99
$PC_{10}H_{14}Cl$	$t\text{-}C_4H_9(C_6H_5)PCl$	P (69·6)	1968, 6
$PC_{10}H_{14}ClO$	$t\text{-}C_4H_9(C_6H_5)P(O)Cl$	H	1965, 8
$PC_{10}H_{14}ClO_3$	chlorofenchene phosphonic acid	H	1967, 339
$PC_{10}H_{14}ClS_2$	$C_6H_5P(S)(SC_4H_9)Cl$	P (88·2)	1967, 126
$PC_{10}H_{14}Cl_2NO_2S$	$CH_3O(i\text{-}C_3H_7NH)P(S)OC_6H_3Cl_2$	H	1968, 284, 285; 1969, 110, 322
$PC_{10}H_{14}Cl_2NO_3$	$CH_3O(i\text{-}C_3H_7NH)P(O)OC_6H_3Cl_2$	H	1969, 110
$PC_{10}H_{14}FO_2$	$C_6H_5(C_4H_9O)P(O)F$	P (15·7); F (−68)	1967, 14
$PC_{10}H_{14}FO_3$	$FC_6H_4P(O)(OC_2H_5)_2$	F	1967, 17; 1969, 18, 19
$PC_{10}H_{14}F_{12}N_3O_2$	$(C_4H_8N)[CH_2N(CH_3)]_2P[OC(CF_3)]_2$	P (−31)	1968, 310
$PC_{10}H_{14}N$	$C_6H_5P(CH_2)_2NHCHCH_3$	H	1967, 137
$PC_{10}H_{14}NO_3$	$HP(OCH_2)_2[OCH(C_6H_5)CH_2NH]$	P (−40)	1969, 45, 178
	$HP(OC_6H_4O)[OCH_2C(CH_3)_2NH]$	P (−37)	1969, 45, 178
	$OC_6H_4OPO(CH_2)_2NHC_2H_5$	P (131)	1969, 45, 178
	$HP(OC_6H_4O)(OCH_2CH_2NC_2H_5)$	P (−100)	1969, 45, 178

	HP[OC$_6$H$_3$(CH$_3$)O]OCH(CH$_3$)CH$_2$NH	H; P (−37·6)	1969, 178
PC$_{10}$H$_{14}$NO$_5$	CH$_3$(i-C$_3$H$_7$O)P(O)OC$_6$H$_4$NO$_2$	H	1969, 160
	(C$_2$H$_5$O)$_2$P(O)OC$_6$H$_4$-o-NO$_2$	P (10·5)	1969, 119
PC$_{10}$H$_{14}$NO$_5$S	(C$_2$H$_5$O)(C$_2$H$_5$S)P(O)OC$_6$H$_4$NO$_2$	H; P (24·0)	1965, 20
	(C$_2$H$_5$O)$_2$P(S)OC$_6$H$_4$NO$_2$	H	1968, 284, 285; 1969, 270
PC$_{10}$H$_{14}$NO$_6$	(C$_2$H$_5$O)$_2$POC$_6$H$_4$NO$_2$	H	1968, 285; 1969, 270
PC$_{10}$H$_{14}$NO$_6$S	(CH$_3$O)$_2$P(O)C(O)NHSO$_2$C$_6$H$_4$CH$_3$	H	1968, 328; 1969, 79
PC$_{10}$H$_{15}$	C$_6$H$_5$P(C$_2$H$_5$)$_2$	H; P (−16, −17·8)	1965, 3, 11; 1966, 206; 1968, 7, 302, 303
	C$_6$H$_5$(t-C$_4$H$_9$)PH	P (−5·7)	1968, 6
PC$_{10}$H$_{15}$ClN	C$_6$H$_5$PCl[N(C$_2$H$_5$)$_2$]	H; P (140·4)	1965, 12; 1968, 92
	C$_6$H$_5$PCl[NH(t-C$_4$H$_9$)]	H	1967, 80
PC$_{10}$H$_{15}$Cl$_2$N$_2$	[(C$_2$H$_5$)$_2$N]Cl$_2$P=NC$_6$H$_5$	H; P (−21·6)	1965, 98; 1969, 348
PC$_{10}$H$_{15}$Cl$_6$N$_2$O$_9$	(C$_2$H$_5$O)$_2$P(O)CH(OCONHCOCl$_3$)CH$_2$(OCONHCOCl$_3$)	H	1969, 82
PC$_{10}$H$_{15}$FN	C$_6$H$_5$PF[N(C$_2$H$_5$)$_2$]	H; P (156·0); F (−125·7)	1965, 12
PC$_{10}$H$_{15}$FNS$_2$	C$_6$H$_5$SP(S)F[N(C$_2$H$_5$)$_2$]	H; P (94·0); F (−34·7)	1968, 308
PC$_{10}$H$_{15}$F$_3$N	C$_6$H$_5$PF$_3$[N(C$_2$H$_5$)$_2$]	H; P (−52·5); F (−43·5; −66·5)	1965, 12; 1967, 64
	C$_6$H$_5$PF$_3$(NH-i-C$_4$H$_9$)	P (−52·5); F (−73; −53; −41)	1967, 64
PC$_{10}$H$_{15}$F$_6$O$_6$	(CH$_3$O)$_3$POC(CF$_3$)$_2$C(CH$_3$)(COCH$_3$)O	H; P (−52·6)	1966, 248
PC$_{10}$H$_{15}$N$_2$O	C$_6$H$_5$P(O)[N(CH$_3$)CH$_2$]$_2$	H	1966, 225; 1967, 314
PC$_{10}$H$_{15}$O	C$_6$H$_5$P(O)(C$_2$H$_5$)$_2$	P (42·4)	1966, 206
	C$_6$H$_5$P(O)(C$_4$H$_9$)H	H	1968, 35

Formula	Compound	Resonance	Reference
$PC_{10}H_{15}OS_2$	$C_2H_5(C_2H_5O)P(S)SC_6H_5$	H	1968, 283, 284
$PC_{10}H_{15}O_2$	$C_6H_5P(OC_2H_5)_2$	^{13}C; H	1965, 7; 1966, 240; 1968, 25; 1969, 34
	$CH_3P(O)(OC_2H_5)CH_2C_6H_5$	H	1968, 304; 1969, 74
	$C_2H_5P(O)(OC_2H_5)C_6H_5$	H	1969, 34
$PC_{10}H_{15}O_3$	$C_6H_5P(O)(OC_2H_5)_2$	^{13}C; H	1965, 7; 1966, 240; 1968, 347; 1969, 34
	$(CH_3O)_2P(O)CH_2C_6H_4CH_3$	H	1967, 117
	$(CH_3O)_2P(O)C_6H_3(CH_3)_2$	H	1968, 317
$PC_{10}H_{15}O_3S_2$	$(CH_3O)_2P(S)OC_6H_3(CH_3)SCH_3$	H	1968, 284, 285; 1969, 322
$PC_{10}H_{15}O_7$	$(CH_3O)_2P(O)CH_2C_6H_4OCH_3$	H	1967, 117
$PC_{10}H_{15}O_5$	$(C_2H_5O)_2P(O)OC_6H_4OH$	H	1966, 42
	$(CH_3O)_2P(O)C_6H_3(OCH_3)_2$	H	1968, 317
$PC_{10}H_{15}O_7$	$(CH_3O)_2P(O)CC(O)OC(CH_3)(COCH_3)COCH_3$ (stereois.)	H; P (65·8)	1969, 353
$PC_{10}H_{15}S$	$C_6H_5P(S)(C_2H_5)_2$	P (52)	1966, 206
$PC_{10}H_{15}S_2$	$C_6H_5P(SC_2H_5)_2$	H	1966, 18
$PC_{10}H_{16}Br$	$(CH_3)_2P^{\oplus}(C_2H_5)C_6H_5, Br^{\ominus}$	P (28·4)	1968, 6
	$HP^{\oplus}(C_2H_5)_2C_6H_5, Br^{\ominus}$	H; P (18·3)	1968, 7
$PC_{10}H_{16}FN_2$	$C_6H_4FP[N(CH_3)_2]_2$	F	1968, 14
$PC_{10}H_{16}NO_3$	$C_6H_5P(O)(O^{\ominus})[O(CH_2)_2)\overset{\oplus}{N}H(CH_3)_2]$	H; P (−5·1)	1969, 343
$PC_{10}H_{17}ClN$	$(CH_3)_2P^{\oplus}(C_6H_5)N(CH_3)_2, Cl^{\ominus}$	H	1968, 54
$PC_{10}H_{17}FN_2$	$C_6H_5PF[N(CH_3)_2]_2$	H; P (56·0); F (−86·7)	1965, 12; 1966, 36
$PC_{10}H_{17}N_2$	$C_6H_5P(NHC_2H_5)_2$	H	1967, 95
	$C_6H_5P[N(CH_3)_2]_2$	H	1966, 36; 1967, 80
$PC_{10}H_{17}N_2O$	$C_6H_5P(O)[N(CH_3)_2]_2$	H	1965, 23; 1968, 91

NMR DATA ON ORGANIC COMPOUNDS

$PC_{10}H_{17}N_2O_2$	$C_6H_5OP(O)[N(CH_3)_2]_2$	H	1968, 91
$PC_{10}H_{17}N_2O_2S$	$CH_3[(CH_3)_2N]P(H)=NSO_2C_6H_4CH_3$	H; P (−1·4)	1969, 15
$PC_{10}H_{17}N_2O_3$	$C_2H_5OP[OC(CH_3)_2CN]_2$	P (135·6)	1967, 293
$PC_{10}H_{17}N_2O_4$	$C_2H_5OP(O)[OC(CH_3)_2CN]_2$	P (−11)	1967, 293
$PC_{10}H_{17}N_2S$	$C_6H_5P(S)[N(CH_3)_2]_2$	H	1965, 23; 1966, 36; 1968, 91
$PC_{10}H_{17}N_2Se$	$C_6H_5P(S)(NHC_2H_5)_2$	H	1967, 95, 318
	$C_6H_5P(Se)(NHC_2H_5)_2$	H	1967, 95
	$C_6H_5P(Se)[N(CH_3)_2]_2$	H; P (84·6)	1969, 87
$PC_{10}H_{17}O_8$	$(CH_3O)_2P(O)C(COOH)=C(OCH_3)C(CH_3)C(CH_3)-$ $-(COCH_3)OH$	H	1969, 353
$PC_{10}H_{18}ClO_3$	$(C_2H_5O)_2P(O)C_6H_8Cl$	H	1966, 256
$PC_{10}H_{18}Cl_3O_5$	$(C_3H_7O)_2P(O)CH(CCl_3)OC(O)CH_3$	H; P (10)	1967, 3
$PC_{10}H_{18}I$	$CH_3P^{\oplus}(CH_2CH=CH_2)_3I^{\ominus}$	P (25·2)	1967, 244
$PC_{10}H_{18}NO_3$	$(C_2H_5O)_2P(O)C(CN)=CH-i-C_3H_7$	H{P}	1968, 323
$PC_{10}H_{19}$	$(C_4H_9)_2PC=CH$	H	1966, 247; 1967, 20
	$(s-C_4H_9)_2PC=CH$	^{13}C; H	1968, 21
$PC_{10}H_{19}BrNO_3$	$C_5H_{10}NP(O)(OCH_2)_2C(CH_3)CH_2Br$	H	1968, 149
$PC_{10}H_{19}ClNO_3$	$C_5H_{10}NP(O)(OCH_2)_2C(CH_3)CH_2Cl$	H	1968, 149
$PC_{10}H_{19}ClNO_5$	$(CH_3O)_2P(O)OC(CH_3)=CClC(O)N(C_2H_5)_2$	H	1968, 285
$PC_{10}H_{19}ClN_3$	$C_6H_5P^{\oplus}[N(CH_3)_2]_2NH_2, Cl^{\ominus}$	H; P (41·0)	1967, 65
$PC_{10}H_{19}Cl_2O_5$	$(C_3H_7O)_2P(O)CH(CHCl_2)OC(O)CH_3$	H; P (13·8)	1967, 3
$PC_{10}H_{19}Cl_3NO_7$	$(C_2H_5O)_2P(O)CH(OCONHCOCl_3)CH_2OC_2H_5$	H	1969, 82
	$(C_2H_5O)_2P(O)CH_2CH(OCONHCOCl_3)CH_2OCH_3$	H	1969, 82
$PC_{10}H_{19}N_2O_2$	$HP[OCH(CH_3)CH(COOCH_3)NH]_2$	H	1969, 45
$PC_{10}H_{19}N_2S_2$	$C_6H_5P(S)(S^{\ominus})N(CH_3)_2, (CH_3)_2NH_2^{\oplus}$	P (92·8)	1967, 126
$PC_{10}H_{19}O$	$(C_4H_9)_2P(O)C=CH$	^{13}C; H	1967, 20; 1968, 21
$PC_{10}H_{19}O_3$	$(C_2H_5O)_2P(O)C_6H_9$	H	1966, 256
	$HP(O)[OC(CH_3)_2CH=CH_2]_2$	P (−3·2)	1967, 293
	$(C_2H_5O)_2P(O)C=CC_4H_9$	H	1967, 21

Formula	Compound	Resonance	Reference
$PC_{10}H_{19}O_3$	$CH_3P[(CH_2)_2]_2C(CH_2COOC_2H_5)OH$	H	1967, 75
	$C_4H_9OP[OC(CH_3)=CH_2]_2$	H	1968, 297
$PC_{10}H_{19}O_4$	$(C_2H_5O)_2P(O)C(O)C_5H_9$	H	1967, 166
	$C_2H_5P(CH_2COOC_2H_5)_2$	H	1968, 297
	$C_4H_9P(CH_2COOCH_3)_2$	H	1968, 297
$PC_{10}H_{19}O_5$	$(C_2H_5O)_2P(O)C(CH_3)CHC(O)C_2H_5$	H	1969, 355
$PC_{10}H_{19}O_6$	$(CH_3)_2C=C(COOC_2H_5)CH_2P(O)(OC_2H_5)OH$	H	1967, 327
	$(CH_3O)_3POCH(CH=CH_2)C(CH_3)O$	H; P (50·9)	1968, 312
$PC_{10}H_{19}O_6S_2$	$(CH_3O)_2P(S)SCH(COOC_2H_5)CH_2COOC_2H_5$	H	1968, 284, 285; 1969, 322
$PC_{10}H_{19}O_7S$	$(CH_3O)_2P(O)SCH(COOC_2H_5)(CH_2CO_2C_2H_5)H$	H	1968, 285; 1969, 322
$PC_{10}H_{20}ClN_2O_2$	$[NC(CH_3)_2]_2P(O)OCH(CH_3)(CH_2)_2Cl$	H	1966, 220
$PC_{10}H_{20}ClN_4$	$C_6H_5NHPN(C_2H_5)(NH_2)_2Cl$	H; P (26·0)	1969, 348
$PC_{10}H_{20}ClO_5$	$(C_3H_7O)_2P(O)CH(CH_2Cl)OC(O)CH_3$	H; P (17·8)	1967, 3
$PC_{10}H_{20}LiO_5$	$(CH_3)_2C(OH)CH_2OPO_3(C_6H_{11})^{\ominus}Li^{\oplus}$	H	1965, 99
	$(CH_3)_2C(OPO_3C_6H_{11})^{\ominus}CH_2OH, Li^{\oplus}$	H	1965, 99
$PC_{10}H_{20}NO_2$	$(C_2H_5)_2NP(O)OC(CH_3)CHC(CH_3)_2$	H	1967, 311
$PC_{10}H_{20}NO_2S$	$(C_2H_5O)_2P(O)NS(O)CH_2C(CH_3)=C(CH_3)CH_2$	H	1965, 93
$PC_{10}H_{20}NO_3$	$(C_2H_5O)_2P(O)C=CN(C_2H_5)_2$	H	1965, 100
	$(CH_2O)_2P(O)C[N(CH_3)_2]C_5H_{10}$	H; P (48)	1968, 324; 1969, 340
	$C_5H_{10}NP(O)(OCH_2)_2C(CH_3)_2$	H	1968, 149
$PC_{10}H_{20}NO_6$	$(C_2H_5O)_2P(O)CH(OCOCH_3)CH_2NH(OCH_3)$	H	1969, 82
$PC_{10}H_{21}ClNO_3$	$[CH_3(CH_2Cl)CHO][(C_2H_5)_2N]P(O)OC(CH_3)=CH_2$	H	1967, 340
$PC_{10}H_{21}Cl_2O$	$C_8H_{17}P(O)(CH_2Cl)_2$	H; P (45·4)	1968, 299; 1969, 334
$PC_{10}H_{21}O_2$	$(C_2H_5)_2P(O)C(CH_3)_2CH_2COCH_3$	H	1967, 341
$PC_{10}H_{21}O_3$	$(C_2H_5)_2P(O)C(CH_3)_2CH_2COOCH_3$	H	1967, 341

NMR DATA ON ORGANIC COMPOUNDS 177

$PC_{10}H_{21}O_3S$	$(C_2H_5O)_2P(O)C(CH_3)=CHCH_3$		H	1968, 314
	$(C_2H_5O)_2P(O)CH_2CH=C(CH_3)SC_2H_5$		H; P (25·9)	1967, 21, 108
	$(C_2H_5O)_2P(O)CH_2C(SC_2H_5)=ClICH_3$		H; P (24·2)	1967, 108
	$(C_2H_5O)_2P(O)CH_2C(CH_3)=CHSC_2H_5$		H; P (24·8)	1967, 108
	$(i\text{-}C_3H_7O)_2P(O)SCH_2CH=CHCH_3$		H	1967, 307
$PC_{10}H_{21}O_4$	$(C_2H_5)_2C(CH_2O)_2P(O)O\text{-}i\text{-}C_3H_7$		H	1967, 91
	$(C_2H_5O)_2P(O)CH_2C(OC_2H_5)=CHCH_3$		H; P (25·9)	1967, 278, 326
	$(C_2H_5O)_2P(O)CH=C(OC_2H_5)C_2H_5$		H; P (21·7)	1967, 278
	$(C_2H_5O)_2P(O)CH_2CH=C(CH_3)OC_2H_5$		H; P (27·7)	1967, 278
	$(C_2H_5O)_2P(O)CH_2C(CH_3)=CHOC_2H_5$		H; P (24·7)	1967, 108
	$(C_4H_9O)_2P(O)CH_2COOH$		H	1965, 87
$PC_{10}H_{21}O_5$	$(C_2H_5O)_3P[OC(CH_3)]_2$		P (−51·3)	1966, 225
	$(CH_3O)_2P(O)C(OH)CH_2C(CH_3)_2OC(CH_3)_2$		H	1966, 43
$PC_{10}H_{21}O_6$	$(CH_3O)_3POCH(C_2H_5)C(CH_3)(COCH_3)O$		P (−51·3)	1968, 312
	$(CH_3O)_3POC(CH_3)(COCH_3)C(CH_3)_2O$		H	1969, 147
	$(CH_3O)_3POC(CH_3)OC(CH_3)_2C(CH_3)O$		H	1969, 147
$PC_{10}H_{21}S$	$(C_4H_9)_2P(S)CH=CH_2$		P (44)	1966, 44
$PC_{10}H_{22}BF_4O_4$	$(C_2H_5O)_2P^{\oplus}[OC(CH_3)_2]_2, BF_4^{\ominus}$		H	1969, 86
$PC_{10}H_{22}Br$	$(CH_3)_2P^{\oplus}[C(CH_3)_2]_2CHCH_3, Br^{\ominus}$		H	1967, 82
$PC_{10}H_{22}NO_4$	$(C_2H_5O)_2P(O)CH_2C(O)N(C_2H_5)_2$		H	1965, 100
	$(i\text{-}C_3H_7O)_2P(O)C(O)NH\text{-}i\text{-}C_3H_7$		H	1967, 26
	$(C_3H_7O)_2P(O)C(O)NH\text{-}i\text{-}C_3H_7$		H	1968, 29
$PC_{10}H_{22}NO_5$	$(C_2H_5O)_2P(O)C^{\ominus}(COOCH_3)N^{\oplus}(CH_3)_3$		H; P (17)	1969, 346
$PC_{10}H_{22}N_3$	$P[N(CH_3)CH_2]_3CC_3H_7$		H	1969, 71
$PC_{10}H_{23}$	$C_{10}H_{21}PH_2$		H; P (−139)	1966, 14
	$C_2H_5P(C_4H_9)_2$		H	1967, 20

Formula	Compound	Resonance	Reference
$PC_{10}H_{23}BrNO_4$	$(C_2H_5O)_2P(O)CH(COCH_3)N^{\oplus}(CH_3)_3, Br^{\ominus}$	H	1969, 346
$PC_{10}H_{23}INO_4$	$(C_2H_5O)_2P(O)CH_2C(O)CH_2N^{\oplus}(CH_3)_3, I^{\ominus}$	H; P (17·9)	1967, 22
$PC_{10}H_{23}INO_5$	$(C_2H_5O)_2P(O)CH(COOCH_3)N^{\oplus}(CH_3)_3, I^{\ominus}$	H; P (10)	1969, 346
$PC_{10}H_{23}O$	$C_2H_5P(O)(C_4H_9)_2$	H	1967, 20
$PC_{10}H_{23}O_2S_2$	$CH_3P(O)(OC_5H_{11})SCH_2S(i\text{-}C_3H_7)$	H	1969, 101
	$CH_3P(O)(O\text{-}i\text{-}C_3H_7)SCH_2SC_5H_{11}$	H	1969, 101
$PC_{10}H_{23}O_3$	$HP(O)(OC_5H_{11})_2$	P (6·1)	1967, 277
	$HP(O)[OCH(CH_3)(C_3H_7)]_2$	P (3·5)	1967, 277
	$HP(O)[OCH(C_2H_5)_2]_2$	P (5·7)	1967, 277
	$HP(O)[OCH(CH_3)(i\text{-}C_3H_7)]_2$	P (4·8)	1967, 277
	$HP(O)[OC(CH_3)_2(C_2H_5)]_2$	P (−3·5)	1967, 293; 1969, 340
	$C_2H_5OP[OCH(CH_3)(C_2H_5)]_2$	P (138·4)	1967, 293
	$C_2H_5OP(O\text{-}t\text{-}C_4H_9)_2$	P (131·1)	1967, 293
$PC_{10}H_{23}O_3S$	$C_2H_5OP(S)(O\text{-}t\text{-}C_4H_9)_2$	P (50·2)	1967, 293
$PC_{10}H_{23}O_3Se$	$C_2H_5OP(Se)(O\text{-}t\text{-}C_4H_9)_2$	P (45·2)	1967, 293
$PC_{10}H_{23}O_4$	$(C_2H_5O)_2P(O)(CH_2)_2OC_4H_9$	H	1966, 251; 1969, 82
	$(i\text{-}C_3H_7O)_2P(O)O\text{-}t\text{-}C_4H_9$	H	1966, 257
	$C_2H_5OP(O)(O\text{-}t\text{-}C_4H_9)_2$	P (−10)	1967, 293
$PC_{10}H_{24}Br$	$CH_3P^{\oplus}(i\text{-}C_3H_7)_3, Br^{\ominus}$	P (45·5)	1966, 5; 1968, 6
$PC_{10}H_{24}Cl_6O_2Sb$	$(C_2H_5)_2(C_2H_5O)(i\text{-}C_4H_9O)P^{\oplus}SbCl_6^{\ominus}$	H	1967, 308
$PC_{10}H_{24}I$	$CH_3P^{\oplus}(i\text{-}C_3H_7)_3, I^{\ominus}$	P (44·2)	1967, 244
$PC_{10}H_{24}N$	$(C_6H_5)_2P(CH_2)_2N(C_2H_5)_2$	P (−20·4)	1965, 101
$PC_{10}H_{24}NO_3$	$(CH_3O)_2P(O)CH(i\text{-}C_3H_7)NHC_4H_9$	H	1965, 36
	$CH_3OP(O)(O^{\ominus})CH(i\text{-}C_3H_7)N^{\oplus}H(CH_3)C_4H_9$	H	1965, 36
	$C_2H_5OC(i\text{-}C_3H_7)NP[N(CH_3)_2]_2$	H; P (97·1)	1968, 321
	$(C_2H_5O)_2P(O)N(i\text{-}C_3H_7)C_3H_7$	H	1968, 29
$PC_{10}H_{24}N_3OS$	$C_2H_5OC(i\text{-}C_3H_7)NP(S)[N(CH_3)_2]_2$	P (70)	1968, 321
$PC_{10}H_{25}ClN_3O$	$[(CH_3)_2N]_2P(O)NHC(CH_3)_2CH_2Cl$	H	1966, 246
	$[(CH_3)_2N]_2P(O)NHCH_2C(CH_3)_2Cl$	H	1966, 246

$PC_{10}H_{25}N_2$	$[(C_2H_5)_2NCH_2]_2PH$	P (−101·9)	1966, 4
$PC_{10}H_{25}N_2O_2$	$[(C_2H_5)_2NCH_2]_2PO_2H$	P (30·3)	1967, 291
$PC_{10}H_{25}NO_4$	$[(CH_3)_2N]_2P(O)NH(CH_2)_2N(CH_2)_4$	H	1966, 246
$PC_{10}H_{25}O_3Si$	$(C_2H_5O)_2POSi(C_2H_5)_3$	H	1967, 34
$PC_{10}H_{25}Si$	$(C_2H_5)_3P=CHSi(CH_3)_3$	H	1968, 105, 107
$PC_{10}H_{26}N_3O$	$(CH_3)_2NP(O)(NHt\text{-}C_4H_9)_2$	H	1968, 91
$PC_{10}H_{27}GeSi$	$(CH_3)_3P=C[Si(CH_3)_3][Ge(CH_3)_3]$	H	1967, 316
$PC_{10}H_{27}SiSn$	$(CH_3)_3P=C[Si(CH_3)_3][Sn(CH_3)_3]$	H	1967, 176
$PC_{10}H_{27}Si_2$	$(CH_3)_3P=C[Si(CH_3)_3]_2$	H	1967, 316
$PC_{10}H_{27}Sn_2$	$(CH_3)_3P=C[Sn(CH_3)_3]_2$	H	1968, 107
$PC_{10}H_{28}ClN_2Si_2$	$t\text{-}C_4H_9P(Cl)[NHSi(CH_3)_3][=NSi(CH_3)_3]$	H	1968, 65
$PC_{10}H_{28}N_5$	$[(CH_3)_2N]_3P[N(CH_3)CH_2]_2$	P (122·0)	1966, 224
$PC_{10}H_{29}N_2Si_2$	$t\text{-}C_4H_9P[NHSi(CH_3)_3]_2$	H	1968, 65
$PC_{11}H_{10}ClNO_4$	$C_6H_4ClC(=NO_2)P(O)(OC_2H_5)_2$	H	1968, 142
$PC_{11}H_{10}N$	$CH_3(C_6H_5)NP(C\equiv CH)_2$	H	1967, 106
$PC_{11}H_{10}NS_2$	$C_6H_5P(S)(S^{\ominus})N^{\oplus}C_5H_5$	P (−82)	1966, 77; 1967, 126
$PC_{11}H_{11}F_6O_2$	$C_2H_5(C_6H_5)POCH(CF_3)_2$	H; P (53·5)	1968, 302
$PC_{11}H_{11}NO_4$	$C_6H_5C(=NOH)P(O)(OC_2H_5)_2$	H	1968, 142
$PC_{11}H_{12}NO_3$	$\beta\text{-}C_{10}H_7\text{CH}(NH_2)PO_3H_2$	H	1968, 142
	$C_6H_5CHC(CN)P(O)(OC_2H_5)OH$	H	1968, 319
$PC_{11}H_{12}NO_4S_2$	$(CH_3O)_2P(S)SCH_2N[C(O)]_2C_2H_4$	H	1968, 284, 285; 1969, 322
$PC_{11}H_{13}$	$C_6H_5PCH_2CH=C(CH_3)CH_2$	H	1967, 108; 1968, 100
	$C_6H_5PCH_2CH_2C(CH_3)=CH$	H	1968, 100
$PC_{11}H_{13}Br_2$	$C_6H_5P^{\oplus}(Br)CH_2CHC(CH_3)CH_2, Br^{\ominus}$	H; P (89)	1967, 108; 1968, 100
$PC_{11}H_{13}Cl_2$	$C_6H_5P^{\oplus}(Cl)CH_2CHC(CH_3)CH_2, Cl^{\ominus}$	H; P (99)	1967, 108; 1968, 100

Formula	Compound	Resonance	Reference
$PC_{11}H_{13}O$	$C_6H_5P(O)CH_2C(CH_3)=CHCH_2$	H	1967, 101, 108; 1968, 100
	$C_6H_5P(O)CH_2CH=CHCHCH_3$	H	1968, 100
	$C_6H_5P(O)CH_2CH_2CH=CCH_3$	H	1968, 100
	$C_6H_5P(O)CH_2CH_2C(CH_3)=CH$	H	1968, 100
$PC_{11}H_{13}O_6$	$C_6H_5P(CH_2CH_2)_2CO$	H	1969, 76
	$(CH_3O)_2P(O)OCH(COC_6H_5)CHO$	H; P (−2·6)	1969, 337
$PC_{11}H_{13}S$	$C_6H_5P(S)(CH_2)_2C(CH_3)=CH$	H; P (64·4)	1969, 20
$PC_{11}H_{14}ClO_3$	$C_6H_5CH_2P(O)O(CH_2)_2CH(CH_2Cl)O$	H	1966, 121
$PC_{11}H_{14}Cl_3N_2O$	$Cl_3PN(C_4H_9)C(O)NC_6H_5$	H	1965, 6
$PC_{11}H_{14}N$	$(CH_3)_2NP(C_6H_5)C\equiv CCH_3$	H	1967, 80, 106
	$(CH_3)_2NP(C_6H_5)CH=C=CH_2$	H	1968, 99; 1969, 115
$PC_{11}H_{14}NO$	$C_6H_5P(O)(CH=CH_2)NHCH_2CH=CH_2$	H	1966, 250
	$(CH_3)_2NP(O)(C_6H_5)CH=C=CH_2$	H	1969, 115
$PC_{11}H_{15}Cl_2O_2S_3$	$(C_2H_5O)_2P(S)SCH_2SC_6H_3Cl_2$	H	1968, 285
$PC_{11}H_{15}Cl_3NO_2S$	$C_6H_2Cl_3OP(S)(OCH_3)NH$-t-C_4H_9	H	1967, 116
$PC_{11}H_{15}F_3N$	$F_3P(C_6H_5)N(C_2H_5)_2$	F (−72, −47)	1967, 64
$PC_{11}H_{15}N_2O_3$	$CH_3OPO(CH_2)_2N(CH_3)C(O)NC_6H_5$	H	1969, 171
$PC_{11}H_{15}O_2$	$C_6H_5P(OCH_2)_2C(CH_3)_2$	H	1968, 80
$PC_{11}H_{15}O_3$	$C_6H_5OP(OCH_2)_2C(CH_3)_2$	H; P (115)	1967, 88; 1968, 80, 301

$PC_{11}H_{15}O_4$	$C_6H_5P(O)(OCH_2)_2C(CH_3)_2$	H	1968, 74
	$C_6H_5OP(O)(OCH_2)_2C(CH_3)_2$	H; P	1967, 91, 1968, 74, 83, 150
	$(C_2H_5O)_2P(O)C(O)C_6H_5$	P (-2)	1967, 338
	$C_6H_5OP(O)[OCH(CH_3)]_2CH_2$	H; P ($-13\cdot9$)	1968, 83
	$C_6H_5OP(O)OC(CH_3)_2(CH_2)_2O$	H; P ($-15\cdot9$)	1969, 175
	$(CH_3O)_2P(O)CHC(CH_3)/(C_6H_5)O$	H	1967, 81
$PC_{11}H_{15}O_5$	$(CH_3O)_3POC(C_6H_5)=CHO$	H; P ($-45\cdot4$)	1965, 21
	$(CH_2O)_4PCH(OH)C_6H_5$	H; P (-18)	1968, 154
	$(CH_3O)_2P(O)C_6H_4CH_2COOCH_3$	H	1966, 253
	$(C_2H_5O)_2P(O)CHCIHgC_6H_5$	H	1966, 252
$PC_{11}H_{16}ClHgO_3$	$HP(O)(C_4H_9)CH_2C_6H_4Cl$	H	1968, 35
$PC_{11}H_{16}ClO$	$(C_2H_5O)_2P(S)SCH_2SC_6H_4Cl$	H	1968, 284, 285; 1969, 322
$PC_{11}H_{16}ClO_2S_3$	$CH_3O(t-C_4H_9NH)P(O)OC_6H_3Cl_2$	H	1969, 110
$PC_{11}H_{16}Cl_2NO_3$	$C_6H_5P(CH_2)_2NHCHC_2H_5$	H	1967, 137
$PC_{11}H_{16}N$	$C_6H_5P(CH_2)_2NHC(CH_3)_2$	H	1967, 137
$PC_{11}H_{16}NO_2$	$(CH_3)_2NPOC_6H_3(i\text{-}C_3H_7)O$	P (148)	1969, 45
$PC_{11}H_{16}NO_3$	$HP[OC_6H_3(CH_3)O][OCH_2C(CH_3)_2NH]$	P ($-37\cdot7$)	1969, 45, 178
$PC_{11}H_{16}NO_5$	$(C_2H_5O)_2P(O)CH_2C_6H_4NO_2$	H	1965, 102; 1967, 116
$PC_{11}H_{17}F_3N$	$F_3P(CH_3C_6H_4)N(C_2H_5)_2$	F	1967, 64
$PC_{11}H_{17}N_2O$	$C_6H_5P(O)[N(CH_3)_2](CH_2)_3$	H	1967, 314
	$(CH_3)_2NPOCH(CH_3)CH_2NC_6H_5$ (stereois.)	P (121; 132)	1969, 109

Formula	Compound	Resonance	Reference
$PC_{11}H_{17}N_3$	$(CH_3)_2NPNCHC_6H_4N$	H	1967, 119
$PC_{11}H_{17}O$	$C_4H_9(C_6H_5CH_2)P(O)H$	H	1968, 35
	$CH_3(C_6H_5)P(O)t\text{-}C_4H_9$	H	1969, 154
$PC_{11}H_{17}O_3$	$(C_2H_5O)_2P(O)C_6H_4CH_3$	H	1968, 317
$PC_{11}H_{17}O_4$	$(C_2H_5O)_2P(O)CHOHC_6H_5$	H	1966, 42
	$(C_2H_5O)_2P(O)C_6H_4OCH_3$	H	1968, 317
$PC_{11}H_{18}ClO_3$	$(C_2H_5O)_2P(O)C_7H_8Cl$	H	1966, 256
$PC_{11}H_{18}Cl_2NO_3$	$(C_2H_5O)_2P(O)CH(NH_3Cl)C_6H_4Cl$	H	1968, 142
$PC_{11}H_{18}NO_3$	$t\text{-}C_4H_9C_6H_4CH(NH_2)PO_3H_2$	H	1968, 142
$PC_{11}H_{18}NO_4$	$C_6H_5OP(O)[N(CH_3)_2][OCH(CH_3)CH_2OH]$	H; P (27·3)	1969, 343
$PC_{11}H_{18}N_2O$	$C_6H_5P(O)N(i\text{-}C_3H_7)[N(CH_3)_2]$	H	1965, 92
$PC_{11}H_{19}ClNO_3$	$(C_2H_5O)_2P(O)CH(NH_3Cl)C_6H_5$	H; P (17·5)	1968, 142, 296
$PC_{11}H_{19}NO_2$	$C_6H_5P(O)(O^{\ominus})H, C_5H_{10}NH_3^{\oplus}$	H	1966, 18
$PC_{11}H_{19}N_2O_5S$	$(C_2H_5O)_2P(O)CH_2C_6H_4SO_2NHNH_2$	H	1966, 71
$PC_{11}H_{19}O_3$	$(C_2H_5O)_2P(O)C_7H_9$	H	1966, 256
$PC_{11}H_{19}O_7S_2$	$(CH_3O)_2P(S)SCH(COOC_2H_5)(CH_2C(O)COOC_2H_5)$	H	1968, 283
$PC_{11}H_{19}O_9$	$(C_2H_5O)_2P(O)OC(COCH_3)(COOCH_3)_2$	H; P (−2)	1967, 4
$PC_{11}H_{19}SSi$	$CH_3(C_6H_5)P(S)CH_2Si(CH_3)_3$	H; P (36·6)	1966, 132
$PC_{11}H_{20}Cl_6N_2OSb$	$C_6H_5(CH_3O)P^{\oplus}[N(CH_3)_2]_2, SbCl_6^{\ominus}$	H; P (57·5)	1969, 87
$PC_{11}H_{20}Cl_6N_2SSb$	$C_6H_5(CH_3S)P^{\oplus}[N(CH_3)_2]_2, SbCl_6^{\ominus}$	H; P (74·0)	1969, 87
$PC_{11}H_{20}Cl_6N_2SbSe$	$C_6H_5(CH_3Se)P^{\oplus}[N(CH_3)_2]_2, SbCl_6^{\ominus}$	H; P (68·9)	1969, 87
$PC_{11}H_{20}IN_2$	$CH_3(C_6H_5)P^{\oplus}(NHC_2H_5)_2, I^{\ominus}$	H	1967, 95
$PC_{11}H_{20}NS_2$	$C_6H_5(CH_3S)PS^{\ominus}, (CH_3)_4N^{\oplus}$	H	1967, 113
$PC_{11}H_{20}N_3O_2S$	$[(CH_3)_2N]_2P(H)=NSO_2C_6H_4CH_3$	H; P (21·5)	1969, 15
$PC_{11}H_{21}O_3$	$(C_2H_5O)_2P(O)C_6H_8CH_3$	H	1966, 256
	$(C_2H_5O)_2P(O)C\equiv CC_5H_{11}$	H	1967, 21
	$CH_3P(O)(OH)O\text{-}bornyl$	H	1967, 168

NMR DATA ON ORGANIC COMPOUNDS

PC$_{11}$H$_{21}$O$_3$S	(C$_2$H$_5$O)$_2$P(O)S(2-norbornyl)	H	1966, 52
PC$_{11}$H$_{21}$O$_4$	(C$_2$H$_5$O)$_2$P(O)CHC(CH$_2$)$_5$O	H	1966, 245; 1967, 81
	(C$_2$H$_5$O)$_2$P(O)C(CH$_2$)$_5$CHO	H	1966, 245; 1967, 81
	(C$_2$H$_5$O)$_2$P(O)C$_6$H$_8$OCH$_3$	H	1966, 256
	i-C$_3$H$_7$OP(O)[OCH$_2$CH(CH$_2$)$_2$]$_2$	H	1966, 213
	(C$_2$H$_5$O)$_2$P(O)C(O)C$_6$H$_{11}$	H	1967, 166
	(C$_2$H$_5$O)$_2$P(O)C≡CCH$_2$O-t-C$_4$H$_9$	H; P (−9·5)	1967, 21
	(C$_2$H$_5$O)$_2$P(O)C(CH$_3$)=CHC(O)i-C$_3$H$_7$	H	1969, 355
	(CH$_3$O)$_3$PO(C$_4$H$_6$)C(CH$_3$)(COCH$_3$)O	H; P (−53·1)	1966, 248
PC$_{11}$H$_{21}$O$_6$	(C$_2$H$_5$O)$_2$P(O)OC(OC$_2$H$_5$)CHCOOC$_2$H$_5$	H; P (−8·9)	1965, 95
PC$_{11}$H$_{21}$O$_7$	(C$_2$H$_5$O)$_2$P(O)OC(OC$_2$H$_5$)C(CH$_3$)COOCH$_3$	H	1965, 95
	(C$_2$H$_5$O)$_2$P[OC(CH$_3$)(COOC$_2$H$_5$)$_2$	P (12·8)	1965, 95
	(CH$_3$O)$_3$P[OC(CH$_3$)C(O)CH$_3$]$_2$ (stereois.)	H; P (−54·8; −52·6)	1966, 258
PC$_{11}$H$_{22}$N	(C$_2$H$_5$)$_2$PC(NC$_6$H$_{11}$)H	H	1968, 329
PC$_{11}$H$_{22}$NO$_3$	(C$_2$H$_5$O)$_2$P(O)CH$_2$C(O)CH$_2$(C$_4$H$_8$O)	H	1968, 325
PC$_{11}$H$_{22}$NO$_6$	(C$_2$H$_5$O)$_2$P(O)CH$_2$CH(OCOCH$_3$)CH$_2$NHCOCH$_3$	H	1969, 82
PC$_{11}$H$_{22}$NS	(C$_2$H$_5$)$_2$P(S)CH$_2$N(CH$_2$CH=CH$_2$)$_2$	P (51·4)	1966, 40
PC$_{11}$H$_{22}$N$_3$O$_3$	(CH$_3$CHCH$_2$N)$_2$P(O)NHCOOC$_2$H$_5$	H	1965, 42
PC$_{11}$H$_{22}$O$_4$	(C$_2$H$_5$O)$_2$P(O)OC$_6$H$_9$CH$_3$	H	1967, 121
PC$_{11}$H$_{22}$O$_5$	(CH$_3$O)$_2$P(O)C(OH)CH$_2$C(CH$_3$)$_2$OC(CH$_3$)$_2$	H	1969, 80
PC$_{11}$H$_{23}$O	(C$_4$H$_9$)$_2$PCH$_2$COCH$_3$	H	1968, 297
PC$_{11}$H$_{23}$O$_3$	(C$_2$H$_5$O)$_2$P(O)C$_6$H$_{10}$CH$_3$	H	1966, 256
	(C$_4$H$_9$O)$_2$PCH$_2$COCH$_3$	H	1968, 297
	(C$_4$H$_9$O)$_2$POC(CH$_3$)=CH$_2$	H	1968, 297
PC$_{11}$H$_{23}$O$_4$	(C$_4$H$_9$O)$_2$P(O)OC(CH$_3$)=CH$_2$	H	1968, 297

Formula	Compound	Resonance	Reference
$PC_{11}H_{23}O_5$	$(i\text{-}C_3H_7O)_3P[OC(CH_3)]_2$	P (−53·2)	1965, 21
	$(C_2H_5O)_2P(O)CH(OC_2H_5)CH_2COC_2H_5$	H	1966, 249
	$(C_2H_5O)_2P(O)CH=C(OC_2H_5)CH_2OC_2H_5$	H; P (25)	1967, 278
	$(C_2H_5O)_2P(O)CH_2C(O)CH_2Ot\text{-}C_4H_9$	H; P (18·7)	1967, 22
$PC_{11}H_{24}NO_2$	$(i\text{-}C_4H_9)_2P(O)C(O)N(CH_3)_2$	H	1967, 26
	$(C_2H_5)_2P(O)CH_2C(O)CH_2N(C_2H_5)_2$	H	1968, 325
$PC_{11}H_{24}NO_3$	$(C_2H_5)_2C(CH_2O)_2P(O)NHt\text{-}C_4H_9$	P (19·9)	1967, 91
	$(C_2H_5O)_2P(O)CH=CHCH_2N(C_2H_5)_2$	H; P (17·8)	1967, 108
$PC_{11}H_{24}NO_4$	$(C_2H_5O)_2P(O)CH_2C(O)CH_2N(C_2H_5)_2$	H	1967, 22
	$(CH_3O)_2P(O)C(O)N(i\text{-}C_4H_9)_2$	H	1967, 26
	$(CH_3O)_2P(O)C(O)N(s\text{-}C_4H_9)_2$	H	1967, 26
	$(C_2H_5O)_2P(O)C(O)N(i\text{-}C_3H_7)_2$	H	1967, 26; 1968, 29
	$(i\text{-}C_3H_7O)_2P(O)C(O)N(C_2H_5)_2$	H	1967, 26
	$(s\text{-}C_4H_9O)_2P(O)C(O)N(CH_3)_2$	H	1967, 26
	$(i\text{-}C_4H_9O)_2P(O)C(O)NHC_2H_5$	H	1967, 26
$PC_{11}H_{25}GeS_2$	$(C_2H_5)_3GeSCSP(C_2H_5)_2$	H	1967, 342
$PC_{11}H_{25}N_2O_3$	$(C_2H_5O)_2P(O)CH=C[N(CH_3)_2]CH_2N(CH_3)_2$	H; P (24·7)	1967, 278
$PC_{11}H_{25}O_3$	$CH_3P(O)(O\text{-}i\text{-}C_5H_{11})_2$	H; P	1968, 109, 156
$PC_{11}H_{25}O_3S$	$(i\text{-}C_3H_7O)_2P(O)SC_5H_{11}$	H	1967, 307
$PC_{11}H_{26}NO_3$	$(C_2H_5O)_2P(O)C[N(CH_3)_2](C_2H_5)_2$	H; P (32·5)	1968, 324; 1969, 340
$PC_{11}H_{27}ClNSi$	$(t\text{-}C_4H_9)_2P(Cl)=NSi(CH_3)_3$	H	1968, 64
$PC_{11}H_{27}OSi$	$(C_4H_9)_2POSi(CH_3)_3$	P (116·0)	1967, 343
$PC_{11}H_{27}N_4O$	$[(CH_3)_2N]_2P(O)NHCH(CH_3)CH_2N(CH_2)_4$	H	1966, 246
$PC_{11}H_{28}GeN$	$(t\text{-}C_4H_9)_2PNHGe(CH_3)_3$	H	1968, 64
$PC_{11}H_{28}NSi$	$(t\text{-}C_4H_9)_2PNHSi(CH_3)_3$	H	1968, 64
$PC_{11}H_{33}N_2Si_3$	$(CH_3)_2P[NSi(CH_3)_3]N[Si(CH_3)_3]_2$	H	1968, 65
$PC_{12}BrF_{10}$	$(C_6F_5)_2PBr$	P (13·0)	1966, 7

$PC_{12}ClF_{10}$	$(C_6F_5)_2PCl$	P (12 – 37); F (−129; −160; −148)	1966, 7, 19; 1969, 17
$PC_{12}F_{11}O$	$(C_6F_5)_2P(O)F$	F (−131·5; −158·2; −141·5)	1969, 17
$PC_{12}HF_{10}$	$(C_6F_5)_2PH$	H; P (−143); F (−130·0; −161·8; −151·9)	1966, 7, 19; 1967, 11
$PC_{12}HF_{10}O$	$(C_6F_5)_2P(O)H$	H	1966, 26
$PC_{12}H_5BrF_5$	$C_6H_5(C_6F_5)PBr$	H; P (39·3)	1966, 7
$PC_{12}H_5ClF_5$	$C_6H_5(C_6F_5)PCl$	H; P (57·1)	1966, 7
$PC_{12}H_6F_5$	$C_6H_5(C_6F_5)PH$	H; P (−92·2)	1966, 7; 1967, 11
$PC_{12}H_8ClF_2$	$(C_6H_4\text{-}p\text{-}F)_2PCl$	H; P (79); F	1969, 19, 120
	$(C_6H_4\text{-}m\text{-}F)_2PCl$	P (74·7); F	1969, 18, 120
$PC_{12}H_8ClF_2O$	$(C_6H_4F)_2P(O)Cl$	F	1969, 18, 19
$PC_{12}H_8ClO_4$	$(OC_6H_4O)_2PCl$	P (−9·7)	1966, 238
$PC_{12}H_9F_2$	$(C_6H_4F)_2PH$	F	1969, 18, 19
$PC_{12}H_9OS_3$	![structure](PO on bicyclic S ring)$_3$	H	1969, 123
$PC_{12}H_9S_3$	![structure](P on S ring)$_3$; ![structure](P on S ring)$_3$	H	1969, 124 to 126
$PC_{12}H_9S_4$	![structure](PS on S ring)$_3$; ![structure](PS on S ring)$_3$	H	1969, 124, 126
$PC_{12}H_{10}Br_3$	$(C_6H_5)_2P^{\oplus}Br_2, Br^{\ominus}$	P (55·6)	1968, 8

186 FORMULAE INDEX II

Formula	Compound	Resonance	Reference
$PC_{12}H_{10}Cl$	$(C_6H_5)_2PCl$	^{13}C; H; P (81·5)	1965, 7; 1966, 141, 240; 1967, 198, 277; 1968, 103; 1969, 50, 120
$PC_{12}H_{10}ClO$	$(C_6H_5)_2P(O)Cl$	P (42·7)	1967, 277
$PC_{12}H_{10}ClO_2$	$(C_6H_5O)_2PCl$	P (159)	1965, 45
$PC_{12}H_{10}ClO_3$	$(C_6H_5O)_2P(O)Cl$	P (−6·2)	1965, 45
	$C_6H_5(ClC_6H_3OH)PO_2H$	H	1965, 79
$PC_{12}H_{10}ClS$	$(C_6H_5)_2P(S)Cl$	P (79·5)	1966, 111; 1967, 277
$PC_{12}H_{10}Cl_3$	$(C_6H_5)_2PCl_3$	P (73)	1968, 168, 288
$PC_{12}H_{10}Cl_3O_4$	$(C_6H_5)_2P^{\oplus}Cl_2, ClO_4^{\ominus}$	P (93·2)	1967, 276
$PC_{12}H_{10}Cl_8Sb$	$(C_6H_5)_2P^{\oplus}Cl_2, SbCl_6^{\ominus}$	P (93·2)	1967, 276
$PC_{12}H_{10}Cs$	$(C_6H_5)_2PCs$	P (0·0)	1965, 103
$PC_{12}H_{10}CsO$	$(C_6H_5)_2POCs$	P (84·0)	1968, 320
$PC_{12}H_{10}FO$	$(C_6H_5)_2P(O)F$	P (40·5); F (4·1)	1967, 14, 103, 331
$PC_{12}H_{10}FO_3$	$(C_6H_5O)_2P(O)F$	P (−20·3); F (−82)	1967, 14
$PC_{12}H_{10}FS$	$(C_6H_5)_2P(S)F$	P (102·0)	1968, 8
$PC_{12}H_{10}FS_3$	$(C_6H_5S)_2P(S)F$	P (112); F (−26·4)	1968, 308
$PC_{12}H_{10}F_3$	$(C_6H_5)_2PF_3$	F	1967, 331
$PC_{12}H_{10}F_3S$	$C_6H_5(C_6H_5S)PF_3$	H; P (−23·6); F (−13·6; −71·3)	1968, 316; 1969, 62
$PC_{12}H_{10}F_4N$	$(C_6H_5)_2NPF_4$	F (−59·8)	1968, 268
$PC_{12}H_{10}K$	$(C_6H_5)_2PK$	P (−12·4)	1965, 103
$PC_{12}H_{10}KO$	$(C_6H_5)_2POK$	P (86·8)	1968, 320
$PC_{12}H_{10}Li$	$(C_6H_5)_2PLi$	P (−23·0)	1965, 103
$PC_{12}H_{10}LiO$	$(C_6H_5)_2POLi$	P (88·9)	1968, 320
$PC_{12}H_{10}Na$	$(C_6H_5)_2PNa$	P (−24·4)	1965, 103
$PC_{12}H_{10}NaO$	$(C_6H_5)_2PONa$	P (90·5)	1968, 320

Formula	Compound	Nucleus (δ)	Reference
$PC_{12}H_{10}NaS_2$	$C_6H_5P^{\oplus}(S^{\ominus})_2, Na^{\oplus}$	P (62·0)	1968, 8
$PC_{12}H_{10}ORb$	$(C_6H_5)_2PORb$	P (88·4)	1968, 320
$PC_{12}H_{10}O_2$	$(C_6H_5)_2PO_2^{\ominus}$	P (19·5)	1967, 9
$PC_{12}H_{10}Rb$	$(C_6H_5)_2PRb$	P (−7·8)	1965, 103
$PC_{12}H_{11}$	$(C_6H_5)_2PH$	H; P(−41)	1965, 3, 103; 1966, 204; 1967, 11, 30; 1968, 36
$PC_{12}H_{11}N_2O_2$	$HP(OC_6H_4NH)_2$	H; P (−47·5)	1967, 165
$PC_{12}H_{11}O$	$(C_6H_5)_2P(O)H$	H; P (25·9)	1967, 42; 1968, 35
		P (18·5)	1968, 36
$PC_{12}H_{11}OS$	$(C_6H_5)_2P(S)OH$	P (76·0)	1967, 277
$PC_{12}H_{11}O_2$	$(C_6H_5)_2PO_2H$	H; P (25·5)	1965, 79; 1967, 9, 277
$PC_{12}H_{11}O_3$	$C_6H_5(C_6H_4OH)PO_2H$	H	1965, 79
	$(C_6H_5O)_2POH$	P (0·0)	1967, 277
$PC_{12}H_{11}O_4$	$(C_6H_5O)_2PO_2H$	P (−11·5)	1967, 188
$PC_{12}H_{11}S$	$(C_6H_5)_2P(S)H$	P (19·6)	1966, 40
$PC_{12}H_{11}S_2$	$(C_6H_5)_2PS_2H$	P (56·5)	1967, 277
$PC_{12}H_{11}Se$	$(C_6H_5)_2P(Se)H$	P (5·8)	1966, 23
$PC_{12}H_{12}ClO_4S$	$(C_6H_5)_2P^{\oplus}(H)Cl, HSO_4^{\ominus}$	P (46)	1969, 50
$PC_{12}H_{12}Cl_2N$	$(C_6H_5)_2P^{\oplus}(NH_2)Cl, Cl^{\ominus}$	P (51)	1965, 34
$PC_{12}H_{12}NS$	$(C_6H_5)_2P(S)NH_2$	P (54)	1967, 344; 1968, 8
$PC_{12}H_{13}$	$C_6H_5P[C(CH_3)=CH]_2$	H	1967, 70
	$C_6H_5P[CH=C(CH_3)]_2$	H; P (−2·5)	1969, 20, 358
$PC_{12}H_{13}NO_5$	$CH_3OC_6H_4C(=NOH)P(O)(OC_2H_5)_2$	H	1968, 142
$PC_{12}H_{13}N_2$	$C_6H_5PCH_2C(CN)C(NH_2)(CH_2)_2$	H	1969, 76
$PC_{12}H_{13}N_6O$	$[(CH_3)_2N]_2P(O)NC(NH_2)(C_6H_5)N$	H	1968, 285
$PC_{12}H_{13}O_6S$	$(C_6H_5)_2P^{\oplus}(OH)_2, HSO_4^{\ominus}$	P (53·0)	1968, 8

Formula	Compound	Resonance	Reference
$PC_{12}H_{13}S$	$C_6H_5P(S)[C(CH_3)=CH]_2$	H; P (45·7)	1968, 101; 1969, 20, 358
$PC_{12}H_{14}ClN_2$	$(C_6H_5)_2P^{\oplus}(NH_2)_2, Cl^{\ominus}$	H; P (32)	1967, 65
$PC_{12}H_{14}ClO_3$	$C_6H_5CHCH=C(CH_3)OP(O)O(CH_2)_2Cl$	H	1966, 220
$PC_{12}H_{14}Cl_3O_4$	$(C_2H_5O)_2P(O)OC(C_6H_3Cl_2)CHCl$	H	1968, 284
$PC_{12}H_{14}NO$	$C_6H_5P(CH_2)_2NHCH(C_4H_3O)$	H	1967, 137
$PC_{12}H_{14}NO_4S$	$(C_2H_5O)_2P(S)NC(O)C_6H_4C(O)$	H	1966, 56
$PC_{12}H_{15}$	$C_6H_5P[CH_2C(CH_3)]_2$	H; P (−34·5)	1967, 291; 1969, 76, 358
	$CH_3P(CH_2)_2C(C_6H_5)=CHCH_2$	H	1967, 75
$PC_{12}H_{15}F_{12}O_2$	$(C_2H_5)_3P[OC(CF_3)_2]_2$	P (11·7)	1968, 302
	$(C_2H_5)_3[(CF_3)_2CHO]POC(CF_3)_2CHCH_3$ (stereios.)	P (−15·7); F (−79; −74·6; −75·9)	1968, 302
$PC_{12}H_{15}F_{12}O_5$	$(C_2H_5O)_3P[OC(CF_3)_2]_2$	P (−53·2); F (−70·8)	1966, 225; 1967, 335; 1968, 303; 1969, 20
$PC_{12}H_{15}O_2S$	$C_6H_5P(S)(CH_2COCH_3)_2$	H	1968, 297
$PC_{12}H_{15}O_3$	$(C_2H_5O)_2P(O)C=CC_6H_5$	H; P (−9)	1967, 21
$PC_{12}H_{15}O_4$	$C_6H_5P(CH_2COOCH_3)_2$	H	1968, 297
$PC_{12}H_{15}O_5$	$(C_2H_5O)_2P(O)CHC_6H_4C(O)O$	P (13·2)	1967, 338
	$C_6H_5OP(CH_2COOCH_3)_2$	H	1968, 297
	$CH_3O(CH_2O)_2POC(CH_3)=C(C_6H_5)O$	P (−28)	1968, 301

NMR DATA ON ORGANIC COMPOUNDS 189

PC$_{12}$H$_{15}$S	C$_6$H$_5$(CH$_2$O)$_2$P[OC(C$_6$H$_5$)]$_2$ (diastereois.)	P ($-30\cdot8$; $-36\cdot7$)	1968, 301
PC$_{12}$H$_{15}$S	C$_6$H$_5$P(S)[CH$_2$C(CH$_3$)]$_2$	H; P (45·3)	1969, 20
PC$_{12}$H$_{10}$BF$_6$N$_2$	(CH$_3$C$_5$H$_4$N)$_2$BH$_2$, PF$_6$	^{11}B; II	1968, 330
PC$_{12}$H$_{16}$Br	CH$_3$(C$_6$H$_5$CH$_2$)P$^{\oplus}$(CH$_2$CH)$_2$, Br$^{\ominus}$	H	1968, 100
PC$_{12}$H$_{16}$ClO$_2$	C$_6$H$_5$P(O)(OH)CH$_2$CCl=CHC$_3$H$_7$	H	1966, 259
PC$_{12}$H$_{16}$ClO$_2$S	C$_2$H$_5$OP(O)CH$_2$CH(SC$_6$H$_4$Cl)(CH$_2$)$_2$	H	1966, 134
PC$_{12}$H$_{16}$ClO$_3$	CH$_3$(CH$_2$Cl)C(CH$_2$O)$_2$P(O)CH$_2$C$_6$H$_5$	H	1967, 91
PC$_{12}$H$_{16}$ClO$_4$	(C$_2$H$_5$O)$_2$P(O)CHCH(C$_6$H$_4$Cl)O	H	1967, 81
PC$_{12}$H$_{16}$F$_{12}$N$_3$O$_2$	(CH$_3$)$_2$NP[N(CH$_3$)CH$_2$]$_2$O[C(CF$_3$)$_2$]$_2$O	P ($-28\cdot3$)	1968, 124, 186
PC$_{12}$H$_{16}$NO$_2$	C$_6$H$_5$(CH$_2$=CH)P(O)N(CH$_2$)$_4$O	H	1966, 250
PC$_{12}$H$_{16}$NO$_6$	(C$_2$H$_5$O)$_2$P(O)CH$_2$C(O)C$_6$H$_4$NO$_2$	H	1967, 121
PC$_{12}$H$_{16}$NO$_6$	(C$_2$H$_5$O)$_2$P(O)OC(=CH$_2$)C$_6$H$_4$NO$_2$	H	1967, 121
PC$_{12}$H$_{16}$N$_3$O$_3$S$_2$	(C$_2$H$_5$O)$_2$P(S)SCH$_2$NC(O)C$_6$H$_4$N=N	H	1968, 284
PC$_{12}$H$_{16}$N$_5$O$_3$	C$_6$H$_5$NHP(O)[N(CH$_2$)$_2$NHC(O)]$_2$	H	1961, 5
PC$_{12}$H$_{17}$	C$_6$H$_5$PCH$_2$CH(CH$_3$)C(CH$_3$)$_2$	H	1967, 82
PC$_{12}$H$_{17}$N$_2$O$_7$	(i-C$_3$H$_7$O)$_2$P(O)C$_6$H$_3$(NO$_2$)$_2$	H	1969, 119
PC$_{12}$H$_{17}$N$_2$S	C$_6$H$_5$P(S)(NCH$_2$CHCH$_3$)$_2$	H; P (101·5)	1967, 113
PC$_{12}$H$_{17}$O	C$_6$H$_5$P(O)CH$_2$CH(CH$_3$)C(CH$_3$)$_2$	H	1967, 82
PC$_{12}$H$_{17}$O$_2$	CH$_3$P[(CH$_2$)$_2$]$_2$C(C$_6$H$_5$)OH	H	1967, 75
PC$_{12}$H$_{17}$O$_2$	C$_6$H$_5$P[OC(CH$_3$)$_2$]$_2$	H(T)	1966, 51

190   FORMULAE INDEX II

Formula	Compound	Resonance	Reference
$PC_{12}H_{17}O_2S$	$C_2H_5OP(O)CH_2CH(SC_6H_5)CH_2CH_2$	H	1966, 134
$PC_{12}H_{17}O_3$	$C_6H_5OP[OC(CH_3)_2]_2$	H(T)	1966, 51
	$C_6H_5CH_2P(O)(OCH_2)_2C(CH_3)_2$	H	1967, 91; 1968, 74
	$C_6H_5P(O)[OC(CH_3)_2]_2$	P (6)	1969, 343
$PC_{12}H_{17}O_4$	$(C_2H_5O)_2P(O)CHCH(C_6H_5)O$	H	1966, 245; 1967, 81
	$(C_2H_5O)_2P(O)CH(C_6H_5)CHO$	H	1966, 245; 1967, 81
	$(C_2H_5O)_2P(O)CH_2C(O)C_6H_5$	H	1967, 121
	$(C_2H_5O)_2P(O)OC(=CH_2)C_6H_5$	H	1967, 121
	$(C_2H_5O)_2P(O)C(O)CH_2C_6H_5$	H	1968, 93
	$C_6H_5OP(O)OCH(CH_3)CH_2C(CH_3)_2O$	H; P (−14)	1968, 83, 150; 1969, 175
$PC_{12}H_{17}O_5$	$(CH_3O)_3POC(CH_3)C(C_6H_5)O$	H; P (−49.5)	1966, 258; 1968, 301
$PC_{12}H_{18}Br$	$CH_3(C_6H_5CH_2)P^{\oplus}(CH_2)_4, Br^{\ominus}$	H	1969, 356
$PC_{12}H_{18}NO$	$(C_2H_5)_2NP(O)(C_6H_5)CH=CH_2$	H	1966, 250
$PC_{12}H_{18}NO_3$	$C_6H_5(i-C_3H_7O)P(O)C(O)N(CH_3)_2$	H	1967, 26
	$(CH_2O)_2P(O)C(CH_3)(C_6H_5)N(CH_3)_2$	H; P (48)	1968, 324; 1969, 340
	$(C_2H_5O)_2P=NC(OCH_3)C_6H_5$	P (135)	1969, 357
	$HP[OCH(CH_3)CH_2NH][OC_6H_3(CH_3)O]$	H; P (−37.6)	1969, 178
$PC_{12}H_{18}NO_4$	$(CH_3O)_2P(O)C(O)N(C_2H_5)C_6H_4CH_3$	H	1968, 29
$PC_{12}H_{18}NO_5$	$(i-C_3H_7O)_2P(O)C_6H_4NO_2$	H; P (8.5)	1969, 119
$PC_{12}H_{18}NO_6S$	$(C_2H_5O)_2P(O)C(O)NHSO_2C_6H_4CH_3$	H	1968, 328; 1969, 161
$PC_{12}H_{18}N_3O$	$C_6H_5NHP(O)[C(CH_3)_2N]_2$	H	1966, 220
$PC_{12}H_{18}N_3O_3$	$P[OC(CH_3)_2(C\equiv N)]_3$	P (143.5)	1967, 293
$PC_{12}H_{18}N_3O$	$OP[OC(CH_3)_2(C\equiv N)]_3$	P (−15.4)	1967, 293

		H(T); P (10)	
$PC_{12}H_{19}$	$C_6H_5P(n-C_3H_7)_2$		1965, 3; 1967, 244; 1968, 6, 7, 66; 1969, 162
$PC_{12}H_{19}ClN$	$C_6H_5P(C_3H_7)_2$	$P(-27·7)$	1967, 8
	$(C_3H_7)_2NPCl(C_6H_5)$	H	1968, 92
$PC_{12}H_{19}ClNO_3$	$CH_3O(CH_3NH)P(O)OC_6H_3(Cl)t-C_4H_9$	H	1968, 284, 285; 1969, 322
$PC_{12}H_{19}N_2O$	$(CH_3)_2NPOCH(C_6H_5)CH(CH_3)NCH_3$	P (130)	1969, 109
$PC_{12}H_{19}O_2$	$(CH_3)_2C_6H_3P(O)(CH)O-s-C_4H_9$ (stereois.)	H; P (22·1; 20·9)	1969, 162
$PC_{12}H_{19}O_3$	$C_2H_5C_6H_4P(O)(OC_2H_5)_2$	H	1968, 317
	$C_6H_5P(O)(Oi-C_3H_7)_2$	H	1969, 101
	$Cl_3(C_6H_5)CHP(O)(OC_2H_5)_2$	H	1969, 51
$PC_{12}H_{19}O_3S$	$CH_3(C_6H_5)CHSP(O)(OC_2H_5)_2$	H	1966, 52, 260
$PC_{12}H_{19}O_5$	$(CH_3O)_2P(O)CH(OCH_3)C(CH_3)(C_6H_5)OH$	H	1967, 81
$PC_{12}H_{19}O_8$	$(C_2H_5O)_2P(O)CH[C(O)COOC_2H_5](CH_2)_2OCO$	H	1965, 104
$PC_{12}H_{20}Br$	$C_6H_5P^{\oplus}(C_2H_5)_3, Br^{\ominus}$	P (36·3)	1968, 6
	$C_6H_5(CH_3)_2P^{\oplus}C_4H_9, Br^{\ominus}$	P (26·2)	1968, 6
	$C_6H_5(i-C_3H_7)_2P^{\oplus}H, Br^{\ominus}$	H; P (32·0)	1968, 7
	$(C_3H_5)_4P^{\oplus}Br^{\ominus}$	H	1969, 354
$PC_{12}H_{20}N$	$(C_2H_5)_2NP(C_6H_5)_2$	H; P (60·8)	1969, 333
$PC_{12}H_{20}NO_3$	$(C_2H_5O)_2P(O)NHC_6H_4C_2H_5$	H	1965, 102
	$C_6H_5P(O)[N(CH_3)_2][O[CH(CH_3)]_2OH]$	P (24·2)	1969, 343
$PC_{12}H_{20}NO_4$	$(C_2H_5O)_2P(O)CH(OH)CH_2NHC_6H_5$	H	1969, 82
	$C_6H_5OP(O)(O^{\ominus})OCH_2CH(CH_3)\cdot(CH_3)_3NH^{\oplus}$	P (−6·9)	1969, 359
	$C_6H_5N^{\oplus}(CH_3)_3O^{\ominus}P(O)OCH_2CH(CH_3)O$	P (ca. 15)	1969, 359
$PC_{12}H_{20}N_3O$	$[(CH_3)_2N]_2P=N(OCH_3)C_6H_5$	P (98)	1969, 357
$PC_{12}H_{21}ClNO_4$	$CH_3OC_6H_4CH(NH_3Cl)P(O)(OC_2H_5)_2$	H; P (18·2)	1968, 142, 296
$PC_{12}H_{21}Cl_3NO_7$	$(C_2H_5O)_2P(O)CH_2CH(OCONHCOCCl_3)CH_2OC_2H_5$	H	1969, 82
$PC_{12}H_{21}N_2$	$C_6H_5P(NH-i-C_3H_7)_2$	H	1967, 95
$PC_{12}H_{21}N_2O$	$C_6H_5P(O)(NH-i-C_3H_7)_2$	H	1967, 95

192 FORMULAE INDEX II

Formula	Compound	Resonance	Reference
$PC_{12}H_{21}N_2O_3S_2$	$(C_2H_5O)_2P(S)OCNC(i\text{-}C_3H_7)NC(CH_3)CH$	H	1968, 284, 285; 1969, 322
$PC_{12}H_{21}N_2O_4$	$(C_2H_5O)_2P(O)OCNC(i\text{-}C_3H_7)NC(CH_3)CH$	H	1968, 296
$PC_{12}H_{21}N_2S$	$C_6H_5P(S)(NHC_3H_7)_2$	H	1967, 318
	$C_6H_5P(S)(NH\text{-}i\text{-}C_3H_7)_2$	H(T)	1967, 95; 1968, 145; 1969, 111
$PC_{12}H_{21}N_2Se$	$C_6H_5P(Se)(NH\text{-}i\text{-}C_3H_7)_2$	H	1968, 145
$PC_{12}H_{21}O_3Si$	$HP(O)[OSi(C_2H_5)_2C_6H_5](OC_2H_5)$	H	1967, 34
$PC_{12}H_{21}O_4$	$(CH_2CH_2CHO)_4PH$	P (−32)	1968, 154
$PC_{12}H_{21}O_5$	$(C_2H_5O)_2P(O)CH=C(CH_2)_4$	H	1967, 327
$PC_{12}H_{21}O_6$	$P(CH_2COOC_2H_5)_3$	H	1968, 297
$PC_{12}H_{22}Cl_3O_5$	$(i\text{-}C_4H_9O)_2P(O)CH(CCl_3)OC(O)CH_3$	H; P (11·4)	1967, 3
$PC_{12}H_{23}Cl_2O_5$	$(C_4H_9O)_2P(O)CH(CHCl_2)OC(O)CH_3$	H; P (13·5)	1967, 3
	$(i\text{-}C_4H_9O)_2P(O)CH(CHCl_2)OC(O)CH_3$	H; P (14·1)	1967, 3
$PC_{12}H_{23}N_4$	$C_6H_5P[N(CH_3)N(CH_3)_2]_2$	H; P (80·2)	1967, 94
$PC_{12}H_{23}O$	$(C_6H_{11})_2P(O)H$	P (46·4)	1967, 172
$PC_{12}H_{23}O_3$	$(C_2H_5O)_2P(O)C_6H_7(CH_3)_2$	H	1966, 256
	$C_2H_5OP[OC(CH=CH_2)(CH_3)]_2$	P (132·1)	1967, 293
	$(C_2H_5O)_2P(O)C=CC_6H_{13}$	H; P (−8·8)	1967, 21
	$CH_3(CH_3O)P(O)O\text{-bornyl}$	H	1967, 168
$PC_{12}H_{23}O_4$	$(C_2H_5O)_2P(O)C(O)C_7H_{13}$	H	1967, 166
	$[CH_2=CH(CH_3)_2CO]_2PO$	P (−13·1)	1967, 293
$PC_{12}H_{24}$	$t\text{-}C_4H_9P[C(CH_3)_2]_2CCH_3$	H(T)	1968, 121
$PC_{12}H_{24}ClO_5$	$(C_4H_9O)_2P(O)CH(CH_2Cl)OC(O)CH_3$	H; P (18·1)	1967, 3
	$(i\text{-}C_4H_9O)_2P(O)CH(CH_2Cl)OC(O)CH_3$	H; P (21·8)	1967, 3
$PC_{12}H_{24}Cl_2NO_3$	$[CH_3(CH_2Cl)CHO][(C_2H_5)_2N]P(O)OC(CH_2Cl)=CH_2$	H	1967, 340
$PC_{12}H_{24}NO_2$	$(C_2H_5)_2P(O)CH_2C(O)CH_2N(CH_2)_5$	H	1968, 325
$PC_{12}H_{24}N_3OS$	$[(CH_3)_2N]_2P(O)NH(CH_2)_2SC_6H_5$	H	1966, 246

$PC_{12}H_{25}ClNO_3$	$[CH_3(CH_2)_2CHO][(C_3H_7)_2N]P(O)OC(CH_3)=CH_2$	H	1967, 340
$PC_{12}H_{25}N_2O_2$	$(C_5H_{10}NCH_2)_2PO_2H$	P (16·0)	1967, 301
$PC_{12}H_{25}N_4S_2$	$(C_2H_5)_2NCNP(S)(C_2H_5)SC[N(C_2H_5)_2]N$	H; P (68·5)	1968, 137
$PC_{12}H_{25}O$	$C_8H_{17}P(CH_2CH_2)_2O$	P (−52)	1969, 44
$PC_{12}H_{25}O_3Sn$	$(C_2H_5O)_2P(O)C\equiv CSn(C_2H_5)_3$	H	1967, 319
$PC_{12}H_{25}O_4$	$HP[OC(CH_3)_2]_4$	P (−39·8)	1968, 304; 1969, 179
$PC_{12}H_{26}NO_3$	$(C_2H_5O)_2P(O)C[N(CH_3)_2]C_5H_{10}$	H; P (31)	1968, 324; 1969, 340
$PC_{12}H_{26}N_3$	$P[N(CH_3)CH_2]_3CC_5H_{11}$	H; P (86·6)	1969, 71
$PC_{12}H_{27}$	$C_{12}H_{25}PH_2$	H; P (−139)	1966, 4, 14
	$C_{10}H_{21}P(CH_3)_2$	H; P (−32)	1966, 14
	$(C_4H_9)_3P$	^{13}C; H; P (−325)	1965, 3; 1966, 206; 1967, 244, 277; 1968, 6, 7, 22, 164, 211, 303; 1969, 21, 50, 249, 354
	$(i\text{-}C_4H_9)_3P$	P (−45·3)	1965, 3; 1967, 8, 244
	$(s\text{-}C_4H_9)_3P$	P (7·9)	1967, 244; 1968, 6
	$t\text{-}C_4H_9P(C_4H_9)_2$	H; P (−4·6)	1968, 6
	$C_4H_9P(t\text{-}C_4H_9)_2$	H; P (26·6)	1968, 6
	$P(t\text{-}C_4H_9)_3$	H; P (63)	1969, 78
$PC_{12}H_{27}Br_2$	$(C_4H_9)_3PBr_2$	H; P (105)	1967, 345
$PC_{12}H_{27}Br_4Hg$	$(C_4H_9)_3P^{\oplus}Br, HgBr_3^{\ominus}$	H; P (102)	1967, 345
$PC_{12}H_{27}Cl_2$	$(C_4H_9)_3PCl_2$	H; P (105)	1967, 191, 345; 1968, 164
$PC_{12}H_{27}Cl_7Sb$	$(C_4H_9)_3PCl^{\oplus}, SbCl_6^{\ominus}$	H; P (102)	1967, 191, 345
$PC_{12}H_{27}F_2$	$(C_4H_9)_3PF_2$	F (−34·2)	1967, 61
$PC_{12}H_{27}F_2O_2Sn$	$(C_4H_9)_3SnOPOF_2$	P (−25·5); F (−82·3)	1967, 282
$PC_{12}H_{27}F_2S_3$	$(C_4H_9S)_3PF_2$	F (−72·3)	1967, 61
$PC_{12}H_{27}GeO$	$(C_2H_5)_3GeOC=(CH_2)P(C_2H_5)_2$	H	1969, 341

Formula	Compound	Resonance	Reference
$PC_{12}H_{27}O$	$(C_4H_9)_3PO$	P (43)	1965, 85; 1966, 106; 1967, 277, 345
$PC_{12}H_{27}OS_3$	$(C_4H_9S)_3PO$	H	1968, 285
$PC_{12}H_{27}O_3$	$(C_4H_9O)_3P$	P (140)	1967, 277
	$(t-C_4H_9O)_3P$	H; P (138·2)	1967, 293
	$[CH_3(C_2H_5)CHO]_3P$	P (139·1)	1967, 293
	$C_2H_5OP(O\ neo-C_5H_{11})_2$	P (137·2)	1967, 293
$PC_{12}H_{27}O_3S$	$(C_4H_9O)_3PS$	P (66·5)	1967, 277
	$(t-C_4H_9O)_3PS$	H; P (41·2)	1967, 293
$PC_{12}H_{27}O_3Se$	$(t-C_4H_9O)_3PSe$	H; P (31·1)	1967, 293
$PC_{12}H_{27}O_4$	$(C_3H_7O)_2P(O)(CH_2)_2OC_4H_9$	H	1966, 251
	$i-C_3H_7O)_2P(O)(CH_2)_2OC_4H_9$	H	1966, 251
	$(C_3H_7O)_2P(O)(CH_2)_2Oi-C_4H_9$	H	1966, 251
	$(C_4H_9O)_3PO$	^{13}C; H; P (1)	1966, 105; 1967, 277; 1968, 22, 109, 157; 1969, 74
	$(t-C_4H_9O)_3PO$	H; P (−13·3)	1967, 293
	$C_2H_5OP(O)(O\ neo-C_5H_{11})_2$	P (−1·0)	1967, 293
$PC_{12}H_{27}S$	$(C_2H_5)_2P(S)C_8H_{17}$	P (52·6)	1966, 40
	$(C_4H_9)_3PS$	P (53; 48)	1966, 106; 1967, 277
$PC_{12}H_{27}S_3$	$(t-C_4H_9S)_3PS$	H	1967, 68
$PC_{12}H_{28}$	$(C_4H_9S)_3P$	H	1968, 285
$PC_{12}H_{28}Br$	$(t-C_4H_9)_3P^{\oplus}H$	P (58·3)	1969, 12
	$(C_4H_9)_3P^{\oplus}H, Br^{\ominus}$	^{13}C; H; P (11·9)	1968, 7, 22
	$(C_3H_7)_4P^{\oplus}, Br^{\ominus}$	P (31·9)	1968, 6
$PC_{12}H_{28}Br_4N$	$(C_3H_7)_4NPBr_4$	P (ca. 150)	1969, 11
$PC_{12}H_{28}F_2NOS$	$(C_3H_7)_4N^{\oplus}, F_2P(O)S^{\ominus}$	P (46·5); F (−34)	1967, 281
$PC_{12}H_{28}NO_2$	$(t-C_5H_{11}O)_2PN(CH_3)_2$	P (135)	1969, 340
$PC_{12}H_{29}IN$	$C_2H_5(C_6H_5)_2P^{\oplus}(CH_2)_2N(C_2H_5)_2, I^{\ominus}$	P (30·2)	1965, 101

$PC_{12}H_{29}N_4O$	$[(CH_3)_2N]_2P(O)NHC(CH_3)_2CH_2N(CH_2)_4$	H	1966, 246
$PC_{12}H_{29}O_4S$	$(C_4H_9)_3\overset{\oplus}{P}H, HSO_4^{\ominus}$	H; P (12)	1969, 50
$PC_{12}H_{30}AlO$	$(C_2H_5)_3POAl(C_2H_5)_3$	H	1965, 83
$PC_{12}H_{30}Cl_2N_3$	$[(C_2H_5)_2N]_3PCl_2$	H	1968, 90
$PC_{12}H_{30}N_3$	$[(C_2H_5)_2N]_3P$	H	1965, 18
$PC_{12}H_{30}N_3O$	$[(C_2H_5)_2N]_3PO$	H	1965, 18; 1968, 90
$PC_{12}H_{30}N_3S$	$[(C_2H_5)_2N]_3PS$	H	1965, 18
$PC_{12}H_{31}O_3Si_2$	$HP(O)[OSi(C_2H_5)_3]_2$	H	1967, 34
$PC_{12}H_{33}AlNSi$	$(C_2H_5)_3PN[Al(CH_3)_3]Si(CH_3)_3$	H	1967, 322
	$(CH_3)_3PN[Al(C_2H_5)_3]Si(CH_3)_3$	H	1967, 322
$PC_{12}H_{33}GaNSi$	$(C_2H_5)_3PN[Ga(CH_3)_3]Si(CH_3)_3$	H	1967, 322
$PC_{12}H_{33}InNSi$	$(C_2H_5)_3PN[In(CH_3)_3]Si(CH_3)_3$	H	1967, 322
$PC_{12}H_{33}LiN_3S_2$	$t\text{-}C_4H_9[(CH_3)_2N]P[NLiSi(CH_3)_3]NSi(CH_3)_3$	H	1968, 65
$PC_{12}H_{33}N_4Sn$	$[(CH_3)_2N]_3P=NSn(C_2H_5)_3$	H	1967, 299
$PC_{12}H_{34}N_3Si_2$	$t\text{-}C_4H_9[(CH_3)_2N]P[NHSi(CH_3)_3]NSi(CH_3)_3$	H	1968, 65
$PC_{12}H_{36}ClSn_4$	$ClP[Sn(CH_3)_3]_4$	H	1969, 14
$PC_{13}F_{10}N$	$(C_6F_5)_2PC\equiv N$	P (−100·2)	1968, 9
$PC_{13}F_{10}NO$	$(C_6F_5)_2PN=C=O$	P (17·0)	1968, 9
$PC_{13}F_{10}NS$	$(C_6F_5)_2PN=C=S$	P (12·4)	1968, 9
$PC_{13}F_{10}NS_2$	$(C_6F_5)_2P(S)N=C=S$	P (−4·2)	1968, 9
$PC_{13}H_3F_{10}$	$(C_6F_5)_2PCH$	P (−52·2)	1967, 11
$PC_{13}H_3F_{10}O$	$(C_6F_5)_2P(O)CH_3$	P (18·7)	1967, 11
$PC_{13}H_8F_5$	$CF_3P(C_6H_4F)_2$	F	1968, 14
$PC_{13}H_9Cl_2O_4$	$Cl_2P(O)OC_6H_4COOC_6H_5$	P (3·7)	1965, 45
$PC_{13}H_9Cl_4O_2$	$Cl_2P(O)OC_6H_4CCl_2C_6H_5$	P (0·0)	1965, 45; 1966, 261
$PC_{13}H_9Cl_4O_3$	$Cl_2P(O)OC_6H_4CCl_2OC_6H_5$	P (0·0)	1965, 45; 1966, 261
$PC_{13}H_{10}Cl$	$ClPC_6H_4C_6H_4CH_2$	H	1968, 120
$PC_{13}H_{10}FN_2$	$(C_6H_5)_2PF=NCN$	P (50·3); F (−82·1)	1967, 63
$PC_{13}H_{10}N$	$(C_6H_5)_2PC\equiv N$	P (−33·8)	1966, 262

Formula	Compound	Resonance	Reference
$PC_{13}H_{10}NS$	$(C_6H_5)_2P(S)C\!\!=\!\!N$	P (23·9)	1966, 262
$PC_{13}H_{11}$	$HPC_6H_4C_6H_4CH_2$	H	1968, 120
$PC_{13}H_{11}Cl_2O_7S_2$	$CH_3P(O)(OC_6H_4SO_2Cl)_2$	H	1966, 75
$PC_{13}H_{11}F_2$	$CH_3P(C_6H_4F)_2$	F	1969, 18, 19
$PC_{13}H_{11}F_2O$	$CH_3P(O)(C_6H_4F)_2$	F	1969, 18, 19
	$CH_3OP(C_6H_4F)_2$	F	1969, 18, 19
$PC_{13}H_{11}F_2S$	$CH_3P(S)(C_6H_4F)_2$	F	1969, 18, 19
$PC_{13}H_{11}NNaO_2$	$(C_6H_5)_2P(O)N^{\ominus}CHO, Na^{\oplus}$	H	1966, 235
$PC_{13}H_{11}O_3$	$(C_6H_5)_2CHPO_3^{2\ominus}$	P (16·5)	1967, 7
$PC_{13}H_{12}BrO$	$C_6H_5(Br \cdot C_6H_4CH_2)P(O)H$	H	1968, 35
$PC_{13}H_{12}Cl$	$CH_2ClP(O)(C_6H_5)_2$	H	1968, 56
$PC_{13}H_{12}ClO_3$	$C_6H_5P(O)(OH)CH(C_6H_4Cl)OH$	H	1968, 56
$PC_{13}H_{12}Li$	$(C_6H_5)_2PCH_2Li$	P (2·6)	1967, 329
$PC_{13}H_{12}NO_2$	$(C_6H_5)_2P(O)NHCHO$	H	1966, 235
$PC_{13}H_{13}$	$CH_3P(C_6H_5)_2$	H; P (−28)	1965, 3; 1966, 41; 1967, 79, 244, 329; 1968, 7, 54, 259
	$HP(C_6H_5)CH_2C_6H_5$	H	1969, 360
$PC_{13}H_{13}ClN$	$CH_3(C_6H_5)NPCl(C_6H_5)$	H	1968, 92
$PC_{13}H_{13}Cl_2O_5$	$(CH_3O)_2P(O)O(C_{10}H_4Cl_2)OCH_3$	H; P (−3·8)	1968, 315
$PC_{13}H_{13}F_2O_2$	$CH_3PF_2(OC_6H_5)_2$	H; P; F (−18)	1968, 316; 1969, 62
$PC_{13}H_{13}F_2S$	$(C_6H_5)_2PF_2(SCH_3)$	H; P (−39); F (−25·0)	1968, 316; 1969, 62
$PC_{13}H_{13}O$	$CH_3P(O)(C_6H_5)_2$	H	1967, 19, 239; 1968, 56
	$HP(O)(C_6H_5)CH_2C_6H_5$	H	1968, 35; 1969, 34
	$CH_3OP(C_6H_5)_2$	P (115)	1969, 50

PC$_{13}$H$_{13}$OS	CH$_3$OP(S)(C$_6$H$_5$)$_2$	H; P (83·4)	1967, 87, 113; 1968, 8; 1969, 87
	CH$_3$SP(O)(C$_6$H$_5$)$_2$	H; P (42·5)	1967, 9, 113
PC$_{13}$H$_{13}$OSe	CH$_3$OP(Se)(C$_6$H$_5$)$_2$	H; P (88·5)	1967, 113; 1969, 87
PC$_{13}$H$_{13}$O$_2$	C$_6$H$_5$(C$_6$H$_5$CH$_2$)PO$_2$H	H	1965, 79; 1968, 56
	CH$_3$OP(O)(C$_6$H$_5$)$_2$	H; P (32·5)	1967, 9, 113; 1969, 87
	CH$_2$(OH)P(O)(C$_6$H$_5$)$_2$	H	1968, 56
PC$_{13}$H$_{13}$O$_5$	HOP(O)OC(CH$_3$)$_2$C(CH$_3$)(COCH$_3$)O	H	1967, 140
PC$_{13}$H$_{13}$S	CH$_3$P(S)(C$_6$H$_5$)$_2$	H	1967, 19
PC$_{13}$H$_{13}$S$_2$	CH$_3$SP(S)(C$_6$H$_5$)$_2$	H; P (65·2)	1967, 113; 1968, 8
PC$_{13}$H$_{14}$Br	CH$_3$(C$_6$H$_5$)$_2$PH, Br$^{\ominus}$	H; P (−2·2)	1968, 89
PC$_{13}$H$_{14}$N	CH$_3$(C$_6$H$_5$)NP(C≡CCH$_3$)$_2$	H	1967, 80, 106
PC$_{13}$H$_{14}$NO$_4$	C$_6$H$_5$P(O)(O$^{\ominus}$)O(CH$_2$)$_2$NH(C$_5$H$_5$)	P (−5·7)	1969, 359
PC$_{13}$H$_{14}$NS	CH$_3$NHP(S)(C$_6$H$_5$)$_2$	P (61·2)	1968, 8
PC$_{13}$H$_{15}$ClN	CH$_3$(C$_6$H$_5$)$_2$P$^{\oplus}$(NH$_2$), Cl$^{\ominus}$	H; P (38)	1967, 65; 1968, 54
PC$_{13}$H$_{15}$ClNO$_3$	(C$_2$H$_5$O)$_2$P(O)C(CN)=CH(C$_6$H$_4$Cl)	H {P}	1968, 323
PC$_{13}$H$_{15}$NO$_6$	(C$_2$H$_5$O)$_2$P(O)CH(NH$_3$Cl)C$_6$H$_5$(OCH$_3$)$_2$	H	1968, 142
PC$_{13}$H$_{15}$N$_2$O	CH$_3$(O)(NHC$_6$H$_5$)$_2$	H	1965, 82
PC$_{13}$H$_{15}$O$_5$	(CH$_3$O)$_2$P(O)C$_{10}$H$_6$(OCH$_3$)	H; P (−3·3)	1968, 315
	(CH$_3$O)$_3$POC$_{10}$H$_6$O	H; P (−45·5)	1968, 315
PC$_{13}$H$_{15}$O$_5$S	CH$_3$O(C$_6$H$_5$)$_2$P$^{\oplus}$H, HSO$_4^{\ominus}$	H; P (56)	1969, 50
PC$_{13}$H$_{16}$ClO$_3$	C$_6$H$_5$CHCH=C(CH$_3$)OP(O)OCH(CH$_3$)CH$_2$Cl	H	1966, 220
PC$_{13}$H$_{16}$FO$_6$	(CH$_3$O)$_2$P(O)OC(CH$_3$)(COCH$_3$)[C(O)C$_6$H$_4$F]	H; P (−2·9)	1968, 313
PC$_{13}$H$_{16}$NO$_3$	(C$_2$H$_5$O)$_2$P(O)C(CN)=CH(C$_6$H$_5$)	H {P}	1968, 323
PC$_{13}$H$_{16}$NO$_8$	(CH$_3$O)$_2$P(O)OC(CH$_3$)(COCH$_3$)C$_6$H$_4$NO$_2$	H; P (−2·7)	1968, 313
PC$_{13}$H$_{17}$N$_4$O$_7$S$_2$	CH$_3$P(O)(OC$_6$H$_4$SO$_2$NHNH$_2$)$_2$	H	1966, 71

Formula	Compound	Resonance	Reference
PC$_{13}$H$_{17}$O$_6$	(CH$_3$O)$_2$P(O)OC(CH$_3$)(COCH$_3$)C(O)C$_6$H$_5$	H; P (−3·0)	1968, 313
PC$_{13}$H$_{18}$Br	CH$_3$(C$_6$H$_5$CH$_2$)P$^\oplus$CH$_2$CH=C(CH$_3$)CH$_2$, Br$^\ominus$	H	1968, 100
PC$_{13}$H$_{18}$ClO$_3$	C$_6$H$_5$CH$_2$P(O)(OCH$_2$)$_2$C(CH$_2$Cl)C$_2$H$_5$	H	1967, 91
PC$_{13}$H$_{18}$N	C$_6$H$_5$P(CH$_2$)$_2$NHC(CH$_2$)$_4$	H	1967, 137
PC$_{13}$H$_{18}$NO	C$_6$H$_5$(CH$_2$=CH)P(O)N(CH$_2$)$_5$	H	1966, 250
PC$_{13}$H$_{18}$NO$_4$	(CH$_3$)$_2$NP[O(CH$_2$)$_2$O]OC(CH$_3$)=C(C$_6$H$_5$)O	P (−269)	1968, 301
PC$_{13}$H$_{18}$NO$_5$	(C$_2$H$_5$O)$_2$P(O)C(OH)C$_6$H$_4$N(CH$_3$)CO	H	1966, 263
PC$_{13}$H$_{18}$NO$_5$S	(C$_2$H$_5$O)$_2$P(S)NHCOCOOCH$_3$	H	1966, 264
PC$_{13}$H$_{19}$	C$_6$H$_5$PCH$_2$C(CH$_3$)$_2$C(CH$_3$)$_2$	H	1967, 82; 1969, 76
PC$_{13}$H$_{19}$ClNO$_5$	CH$_3$[t-C$_4$H$_9$(CH$_3$)CHO]P(O)OC$_6$H$_3$Cl(NO$_2$)	H	1969, 160
PC$_{13}$H$_{19}$O	C$_6$H$_5$P(O)CH$_2$C(CH$_3$)$_2$C(CH$_3$)$_2$	H	1967, 82
PC$_{13}$H$_{19}$O$_2$S	C$_2$H$_5$OP(O)CH$_2$CH(SC$_6$H$_4$CH$_3$)CH$_2$CH$_2$	H	1966, 134
PC$_{13}$H$_{19}$O$_3$	CH$_3$(C$_6$H$_5$)CHP(O)(OCH$_2$)$_2$C(CH$_3$)$_2$	H	1967, 91; 1968, 74
PC$_{13}$H$_{19}$O$_3$S	(C$_2$H$_5$O)$_2$P(O)S-indanyl	H	1966, 52, 260
PC$_{13}$H$_{19}$O$_4$	(C$_2$H$_5$O)$_2$P(O)CHC(CH$_3$)(C$_6$H$_5$)O	H	1966, 245; 1967, 81
	(C$_2$H$_5$O)$_2$P(O)C(CH$_3$)(C$_6$H$_5$)CHO	H	1966, 245; 1967, 81
	(C$_2$H$_5$O)$_2$P(O)OC(C$_6$H$_5$)=CHCH$_3$	H	1967, 121
	(C$_2$H$_5$O)$_2$P(O)CH(CH$_3$)CO(C$_6$H$_5$)	H	1967, 121
	(C$_2$H$_5$O)$_2$P(O)CH$_2$CH=CHOC$_6$H$_5$	H; P (25·5)	1967, 101, 108

	$C_6H_5OP(O)OCH(i-C_3H_7)C(CH_3)_2CH_2O$ (stereois.)	H; P (−10·3; −14·4)	1969, 175
$PC_{13}H_{19}O_5$	$(C_2H_5O)_2P(O)CH_2COOCH_2C_6H_5$	H	1965, 87
	$(C_2H_5O)_2P(O)CH_2C(O)C_6H_4OCH_3$	H	1967, 121
	$(C_2H_5O)_2P(O)OC(=CH_2)C_6H_4OCH_3$	H	1967, 121
	$(CH_3O)_2(C_6H_5CH_2O)P[OC(CH_3)]_2$	H; P (−49·8)	1968, 73
$PC_{13}H_{19}O_7$	$(CH_3O)_3POC(CH_3)(COCH_3)C(C_5H_4O)O$	H; P (−50·3)	1968, 318
$PC_{13}H_{20}Br$	$CH_3(C_6H_5)P\oplus CH_2CH(CH_3)C(CH_3)_2, Br^\ominus$	H	1967, 82
	$CH_3(C_6H_5CH_2)P\oplus(CH_2)_3CHCH_3, Br^\ominus$	H	1969, 356
$PC_{13}H_{20}N$	$C_6H_5P(CH_2)_2NHC(C_2H_5)_2$	H; P (0·0)	1967, 137
$PC_{13}H_{20}NO_3$	$HP[OC_6H_3(i-C_3H_7)O]OCH_2C(CH_3)_2NH$	H; P (−37·4)	1969, 178
$PC_{13}H_{20}NO_5$	$CH_3[CH_3CH(t-C_4H_9)O]P(O)C_6H_4NO_2$	H	1969, 160
$PC_{13}H_{21}N_2O_3$	$(C_2H_5O)_2PN(C_2H_5)CONHC_6H_5$	P	1969, 361
$PC_{13}H_{21}N_4O_5$	$[(CH_3)_2N]_2P(O)NH(CH_2)_2OC(O)C_6H_4NO_2$	H	1966, 246
$PC_{13}H_{21}O$	$CH_3(C_6H_5)P(O)C(CH_3)_2CH(CH_3)_2$	H	1968, 68
$PC_{13}H_{22}Br$	$CH_3(i-C_3H_7)_2P\oplus C_6H_5, Br^\ominus$	P (42·0)	1968, 6
$PC_{13}H_{22}I$	$CH_3(i-C_3H_7)_2P\oplus C_6H_5, I^\ominus$	P (40·9)	1967, 244
$PC_{13}H_{22}NO_4$	$(C_2H_5O)_2P(O)CH_2CH(OH)CH_2NHC_6H_5$	H	1969, 82
	$C_6H_5N\oplus(CH_3)_3O\ominus P(O)O[CH(CH_3)]_2O$	P (14)	1969, 359
$PC_{13}H_{22}N_3O$	$[(CH_3)_2N]_2PN=C(C_6H_5)OC_2H_5$	H; P (99·5; 104·4)	1968, 321
$PC_{13}H_{22}N_3OS$	$[(CH_3)_2N]_2P(S)N=C(C_6H_5)OC_2H_5$	P (69·8)	1968, 321
$PC_{13}H_{23}$	$(C_4H_9)_2PC_6H_5$	P (−26·2)	1968, 6
	$(i-C_4H_9)_2PC_6H_5$	P (−34·2)	1968, 6
	$(s-C_4H_9)_2PC_6H_5$	P (2)	1968, 6

Formula	Compound	Resonance	Reference
PC$_{13}$H$_{23}$ClNO$_5$	(C$_2$H$_5$O)$_2$P(O)CH(NH$_3^{\oplus}$Cl)C$_6$H$_3$(OCH$_3$)$_2$	H	1968, 142
PC$_{13}$H$_{23}$O$_5$	(C$_2$H$_5$O)$_2$P(O)CH=C(CH$_2$)$_5$	H	1967, 327
PC$_{13}$H$_{23}$O$_9$	(C$_2$H$_5$O)$_2$P(O)OC(COCH$_3$)(COOC$_2$H$_5$)$_2$	H; P (−3·9)	1967, 4
PC$_{13}$H$_{24}$IN$_2$	CH$_3$(C$_6$H$_5$)P$^{\oplus}$(NHi-C$_3$H$_7$)$_2$I$^{\ominus}$	H	1967, 95
	(C$_5$H$_9$)$_2$P(O)C(O)N(CH$_3$)$_2$	H	1967, 26
	(C$_5$H$_9$)$_2$P(O)C(O)NHC$_2$H$_5$	H	1967, 26
PC$_{13}$H$_{24}$N$_3$OS	[(CH$_3$)$_2$N]$_2$P(O)NHCH(CH$_3$)CH$_2$SC$_6$H$_5$	H	1966, 246
	[(CH$_3$)$_2$N]$_2$P(O)NHCH$_2$CH(CH$_3$)SC$_6$H$_5$	H	1966, 246
PC$_{13}$H$_{25}$O$_4$	C$_6$H$_{11}$C(O)P(O)(O-i-C$_3$H$_7$)$_2$	H	1967, 166
PC$_{13}$H$_{25}$O$_6$	(CH$_3$O)$_3$PO(C$_6$H$_{10}$)C(CH$_3$)(COCH$_3$)O	H; P (−54·3)	1966, 248
PC$_{13}$H$_{26}$NO$_3$	(C$_2$H$_5$O)$_2$P(O)CH=C(CH$_3$)CH$_2$NC$_5$H$_{10}$	H; P (16·5)	1967, 101, 108
PC$_{13}$H$_{26}$N$_3$O$_3$	OP(NHCOOC$_2$H$_5$)[NCH(CH$_3$)C(CH$_3$)$_2$]$_2$	H	1965, 42
PC$_{13}$H$_{27}$O$_7$	(CH$_3$O)$_3$P[OC(CH$_3$)COCH$_3$]$_2$ (diastereois.)	P (−54·8; −52·6)	1968, 318
PC$_{13}$H$_{28}$IN$_4$S$_2$	(C$_2$H$_5$)$_2$N$^{\oplus}$CNP(SCH$_3$)(C$_2$H$_5$)$_2$N,I$^{\ominus}$	P (63·0)	1968, 137
PC$_{13}$H$_{28}$NO$_2$	[(C$_2$H$_5$)$_2$CH]$_2$P(O)C(O)N(CH$_3$)$_2$	H	1967, 26
	[(C$_2$H$_5$)$_2$CH]$_2$P(O)C(O)NHC$_2$H$_5$	H	1967, 26
PC$_{13}$H$_{28}$NO$_4$	(C$_4$H$_9$O)$_2$P(O)NCH(O-s-C$_4$H$_9$)	H	1966, 55
	(C$_2$H$_5$O)$_2$P(O)C(O)N(i-C$_4$H$_9$)$_2$	H	1967, 26
	(i-C$_3$H$_7$O)$_2$P(O)C(O)N(i-C$_3$H$_7$)$_2$	H	1967, 26; 1968, 29
	(i-C$_4$H$_9$O)$_2$P(O)C(O)N(C$_2$H$_5$)$_2$	H	1967, 26
	(s-C$_4$H$_9$O)$_2$P(O)C(O)N(C$_2$H$_5$)$_2$	H	1967, 26
	[(CH$_3$)$_2$CH(CH$_3$)CHO]$_2$P(O)C(O)N(CH$_3$)$_2$	H	1967, 26
	[(CH$_3$)$_2$CH(CH$_3$)CHO]$_2$P(O)C(O)NHC$_2$H$_5$	H	1967, 26
PC$_{13}$H$_{28}$N$_3$O$_5$	[(CH$_3$)$_2$N]$_3$POC$_3$(COOC$_2$H$_5$)$_2$	H; P (38·2)	1965, 84

NMR DATA ON ORGANIC COMPOUNDS 201

Formula	Structure	Nucleus (shift)	Year, ref
$PC_{13}H_{29}$	$CH_3(C_{12}H_{25})PH$	P (-85.3)	1966, 14
$PC_{13}H_{29}O$	$CH_3(C_{12}H_{25})P(O)H$	P (33.9)	1966, 14
$PC_{13}H_{29}O_2$	$CH_3(C_{12}H_{25})PO_2H$	P (51)	1966, 14
$PC_{13}H_{30}$	$CH_3(C_4H_9)_3P^{\oplus}$	P (32.3)	1967, 7
$PC_{13}H_{30}Br$	$CH_3(s\text{-}C_4H_9)_3P^{\oplus}Br^{\ominus}$	P (44.2)	1968, 6
	$CH_3(t\text{-}C_4H_9)(C_4H_9)_2P^{\oplus}Br^{\ominus}$	P (41.5)	1968, 6
$PC_{13}H_{30}I$	$CH_3(s\text{-}C_4H_9)_3P^{\oplus}I^{\ominus}$	P (43.4)	1967, 244
	$CH_3(i\text{-}C_4H_9)_3P^{\oplus}I^{\ominus}$	P (28.0)	1967, 244
	$CH_3(C_4H_9)_3P^{\oplus}I^{\ominus}$	^{13}C; P (31.3)	1967, 16, 244
	$CH_3(t\text{-}C_4H_9)_3P^{\oplus}I^{\ominus}$	H	1967, 68
$PC_{13}H_{30}NS$	$(i\text{-}C_4H_9)_2P(S)CH_2N(CH_2)_5$	P (47.4)	1966, 40
$PC_{14}H_5F_{10}$	$C_2H_5P(C_6F_5)_2$	P (-44.0)	1967, 11
$PC_{14}H_5F_{10}O$	$C_2H_5P(O)(C_6F_5)_2$	P (19.3)	1967, 11
$PC_{14}H_6F_{10}N$	$(CH_3)_2NP(C_6F_5)_2$	H; F (-135.6; -162.7; -153.5)	1966, 19
$PC_{14}H_{11}$	$(C_6H_5)_2PC{=}CH$	H	1966, 243, 247, 265
$PC_{14}H_{11}F_2O_2$	$(CH_3)_2(C_6H_5)P[OC(CF_3)_2]_2$	H; P (-10.9); F (9.7)	1968, 303
$PC_{14}H_{11}O$	$(C_6H_5)_2P(O)C{=}CH$	H	1966, 243, 247
$PC_{14}H_{11}S$	$C_6H_5P(S)(C{=}CCH{=}CH_2)_2$	H	1965, 96
$PC_{14}H_{12}Cl_3S_3$	$C_6H_5P(S)(SCH_3)SCH_2C_6H_2Cl_3$	H	1967, 113
$PC_{14}H_{12}NaO_4$	$(C_6H_5CHO)_2PO_2^{\ominus}Na^{\oplus}$	H	1967, 90
$PC_{14}H_{13}$	$C_6H_5PCH(CH)_6CH$	H; P (-181)	1966, 67
	$C_6H_5PCH(CH)_4CH(CH)_2$ (stereois.)	H; P (-79; -14)	1966, 67
	$CH_2{=}CHP(C_6H_5)_2$	P (-11.7)	1967, 8
$PC_{14}H_{13}Cl_2$	$(C_6H_4)_2P^{\oplus}CH_3(CH_2Cl), Cl^{\ominus}$	H	1968, 60
$PC_{14}H_{13}I_2$	$(C_6H_4)_2P^{\oplus}CH_3(CH_2I), I^{\ominus}$	H	1968, 60

Formula	Compound	Resonance	Reference
$PC_{14}H_{13}N_2O_3S$	$(C_2H_5O)_2P(S)OCN(CH)_2NCH$	H	1968, 285
$PC_{14}H_{13}O$	$C_6H_5P(O)CH(CH)_4CH(CH)_2^r$(stereois.)	H; P (38; 26)	1966, 67
$PC_{14}H_{13}O_3$	$(C_6H_5)_2PCOCH_3$	H	1967, 346; 1968, 307
	$(C_6H_5)_2P(O)CH_2COOH$	H	1965, 87
	$C_6H_5CH=C(C_6H_5)PO_3H_2$	P (11·0)	1966, 241
	$(C_6H_5)_2P(O)OC(O)CH_3$	H	1967, 347
$PC_{14}H_{13}O_4$	$CH_3C_6H_4OP(O)OC_6H_4CH_2O$	H	1969, 74
	$C_6H_5OP(O)OC_6H_3(CH_3)CH_2O$	H	1969, 74
	$OCH(C_6H_5)CH(C_6H_5)OPO_2H$	P (16)	1969, 359
$PC_{14}H_{13}S$	$(C_6H_5)_2P(S)CH=CH_2$	P (37·5)	1966, 216
$PC_{14}H_{14}BrO$	$(C_6H_4)_2P^\oplus CH_3(CH_2OH), Br^\ominus$	H	1968, 60
$PC_{14}H_{14}I$	$(C_6H_4)_2P^\oplus (CH_3)_2, I^\ominus$	H	1968, 60
$PC_{14}H_{14}NO_3$	$HP(OC_6H_4O)OCH(C_6H_5)CH_2NH$	P (−36)	1969, 45, 178
$PC_{14}H_{14}NO_4S$	$C_6H_5(C_2H_5O)P(S)OC_6H_4NO_2$	H	1968, 284, 285
$PC_{14}H_{14}NS$	$(C_6H_5)_2P(S)CH_2CH_2CN$	P (34·7)	1966, 40
$PC_{14}H_{15}$	$C_2H_5P(C_6H_5)_2$	H; P (−12)	1965, 3; 1967, 79; 1968, 7, 302, 303
$PC_{14}H_{15}BrNO_3$	$(C_6H_5O)_2P(O)NHCH_2CH_2Br$	H	1969, 352
$PC_{14}H_{15}ClN$	$CH_3(C_6H_5CH_2)NPCl(C_6H_5)$	H	1968, 92
	$C_2H_5(C_6H_5)NPCl(C_6H_5)$	H	1968, 92
$PC_{14}H_{15}Cl_2N_2$	$C_6H_5N=P[N(CH_3)CH_2C_6H_5]Cl_2$	H; P (−19·8)	1969, 348
$PC_{14}H_{15}F_2S$	$C_2H_5S(C_6H_5)_2PF_2$	H; P (−39·2); F (−25·2)	1968, 316; 1969, 62

NMR DATA ON ORGANIC COMPOUNDS

PC$_{14}$H$_{15}$O	C$_2$H$_5$OP(C$_6$H$_5$)$_2$	^{13}C; P (109·8)	1965, 7; 1966, 106, 225, 240; 1968, 25
	(CH$_3$)$_2$P(O)C$_6$H$_4$OC$_6$H$_5$	H	1967, 348
	(C$_6$H$_5$CH$_2$)$_2$P(O)H	P (35)	1967, 172; 1968, 35
PC$_{14}$H$_{15}$OS	C$_2$H$_5$OP(S)(C$_6$H$_5$)$_2$	P (79·8)	1966, 106
	CH$_3$SCH$_2$P(O)(C$_6$H$_5$)$_2$	H	1967, 349
PC$_{14}$H$_{15}$O$_2$	C$_2$H$_5$OP(O)(C$_6$H$_5$)$_2$	P (31·1)	1965, 85; 1966, 106, 266; 1969, 119
	CH$_3$OCH$_2$P(O)(C$_6$H$_5$)$_2$	H	1967, 349
PC$_{14}$H$_{15}$O$_5$	CH$_3$OP(O)OC(CH$_3$)$_2$C(CH$_3$)(COCH$_3$)O	H	1967, 140
PC$_{14}$H$_{15}$O$_6$	C$_6$H$_5$P(O)OC(CH$_3$)(COCH$_3$)C(OCH$_3$)=CCOOH (stereois.)	H; P (43·5; 42·2)	1969, 353
PC$_{14}$H$_{15}$S	CH$_3$SCH$_2$P(C$_6$H$_5$)$_2$	H	1967, 349
PC$_{14}$H$_{16}$Br	(CH$_3$)$_2$P$^\oplus$(C$_6$H$_5$)$_2$, Br$^\ominus$	P (22·1)	1966, 5
	C$_2$H$_5$P$^\oplus$H(C$_6$H$_5$)$_2$, Br$^\ominus$	H; P (8·3)	1968, 7
PC$_{14}$H$_{16}$ClO$_5$S	(C$_2$H$_5$O)$_2$P(S)OC$_6$H$_3$OCOCClC(CH$_3$)	H	1968, 284, 285; 1969, 322
PC$_{14}$H$_{16}$Cl$_6$OSSb	CH$_3$O(CH$_3$S)P$^\oplus$(C$_6$H$_5$)$_2$, SbCl$_6^\ominus$	H; P (86·6)	1969, 87
PC$_{14}$H$_{16}$OSbSe	CH$_3$O(CH$_3$Se)P$^\oplus$(C$_6$H$_5$)$_2$, SbCl$_6^\ominus$	H; P (87·8)	1969, 87
PC$_{14}$H$_{16}$O$_2$Sb	(CH$_3$O)$_2$P$^\oplus$(C$_6$H$_5$)$_2$, SbCl$_6^\ominus$	P (61·3)	1968, 8; 1969, 87
PC$_{14}$H$_{16}$S$_2$Sb	(CH$_3$S)$_2$P$^\oplus$(C$_6$H$_5$)$_2$, SbCl$_6^\ominus$	P (77·3)	1968, 8
PC$_{14}$H$_{16}$Cl$_7$NSb	(CH$_3$)$_2$NP$^\oplus$Cl(C$_6$H$_5$)$_2$, SbCl$_6^\ominus$	P (72·5)	1968, 8
PC$_{14}$H$_{16}$N	(CH$_3$)$_2$NP(C$_6$H$_5$)$_2$	H; (T); P (66·2)	1966, 36; 1967, 277; 1969, 105
PC$_{14}$H$_{16}$NO	(CH$_3$)$_2$NP(O)(C$_6$H$_5$)$_2$	H; P (3·0 or 29·6)	1965, 23; 1967, 9, 92; 1968, 8, 91; 1969, 87
PC$_{14}$H$_{16}$NO$_3$	(CH$_3$)$_2$NP(O)(C$_6$H$_5$)$_2$	H	1968, 91
PC$_{14}$H$_{16}$NO$_4$	C$_6$H$_5$P(O)(O$^\ominus$)OCH$_2$CH(CH$_3$), C$_5$H$_5$N$^\oplus$H	P (−7·1)	1969, 359

Formula	Compound	Resonance	Reference
$PC_{14}H_{16}NO_4$	$C_6H_5NC_5H_5^{\oplus}O^{\ominus}P(O)OCH_2CH(CH_3)O$	P (ca. 15)	1969, 359
$PC_{14}H_{16}NS$	$(CH_3)_2NP(S)(C_6H_5)_2$	H; (T); P (41·0 or 70·9)	1967, 92, 277; 1968, 8, 91; 1969, 87, 105
$PC_{14}H_{16}NSe$	$(CH_3)_2NP(Se)(C_6H_5)_2$	H; P (72·0)	1969, 87
$PC_{14}H_{16}N_2S_2$	$(CH_3)_2NCNP(S)(C_6H_5)_2S^{\ominus}$	P (37; 40)	1967, 344
$PC_{14}H_{17}ClNO_3$	$(C_2H_5O)_2P(O)C(CN)=C(CH_3)C_6H_4Cl$	H	1968, 323
$PC_{14}H_{17}ClNO_5$	$(CH_3O)_2P(O)OC=C(Cl)C(CH_3)(C_6H_5)C(O)NCH_3$	H	1969, 345
$PC_{14}H_{17}N_2S_2$	$(CH_3)_2NC(S)NHP(S)(C_6H_5)_2$	P (50·9)	1967, 344
$PC_{14}H_{17}O$	$(C_6H_5)_2P(O)CH(CH_3)CH=CH_2$	H	1966, 267
$PC_{14}H_{18}ClN_2$	$(CH_3)_2NP^{\oplus}(NH_2)(C_6H_5)_2Cl^{\ominus}$	H; P (39·9)	1967, 65
$PC_{14}H_{18}ClO_3$	$C_6H_5CHCH=C(CH_3)OP(O)OCH(CH_3)(CH_2)_2Cl$	H	1966, 220
$PC_{14}H_{18}F_{12}N_3O_2$	$(CH_2)_4NP[N(CH_3)CH_2]_2[OC(CF_3)_2]_2$	P (−30·9)	1968, 303
$PC_{14}H_{18}NO_3$	$(C_2H_5O)_2P(O)C(CN)=C(CH_3)C_6H_5$	H	1968, 323
$PC_{14}H_{18}NO_4$	$(C_2H_5O)_2P(O)C(CN)=CH(C_6H_4OCH_3)$	H{P}	1968, 323
$PC_{14}H_{18}NO_5S$	$(CH_3)_2NP^{\oplus}(OH)(C_6H_5)_2, HSO_4^{\ominus}$	P (50·0)	1968, 8
$PC_{14}H_{18}NO_6S$	$(CH_2=CHCH_2O)_2P(O)C(O)NHSO_2C_6H_4CH_3$	H	1968, 328; 1969, 161
$PC_{14}H_{19}O_4$	$C_6H_5OP(O)[OCH_2CH(CH_3)_2]_2$	H	1966, 213
	$(C_2H_5O)_2P(O)C(O)CHCH_2CHC_6H_5$	H	1967, 166
$PC_{14}H_{19}O_5$	$C_2H_5OP(O)(OH)CH_2C(COOC_2H_5)=CHC_6H_5$	H	1967, 327
$PC_{14}H_{19}O_6$	$(CH_3O)_2P(O)OC(CH_3)=CHC(O)OCH(CH_3)C_6H_5$	H	1968, 284, 285; 1969, 123
$PC_{14}H_{19}O_7$	$(CH_3O)_2P(O)OC(CH_3)(COCH_3)C(O)C_6H_4OCH_3$	H; P (−3·0)	1968, 313
$PC_{14}H_{20}$	$C_6H_5P[C(CH_3)_2]_2CCH_3$	H(T)	1968, 121
$PC_{14}H_{20}BF_6N_2$	$[NC_5H_3(CH_3)_2]_2BH_2^{\oplus}PF_6^{\ominus}$	^{11}B; H	1968, 330

$PC_{14}H_{20}BI$	$CH_3(C_6H_5CH_2)P^{\oplus}[CH_2C(CH_3)]_2, Br^{\ominus}$	H	1968, 100
$PC_{14}H_{20}ClN_4$	$C_6H_5NHPN[(CH_3)CH_2C_6H_5](NH_2)_2Cl$	H; P (27·8)	1969, 348
$PC_{14}H_{20}F_5N_2$	$C_6F_5P(NH-t-C_4H_9)_2$	H; F (−137·7; −162·1; −155·7)	1966, 19
$PC_{14}H_{20}N$	$C_6F_5P[N(C_2H_5)_2]_2$	H; P (79·2)	1966, 7
	$C_6H_5P(CH_2)_2NHC(CH_2)_5$	H	1967, 137
$PC_{14}H_{20}NO_4$	$(CH_3)_2N[O(CH_2)_3O]POC(CH_3)=C(C_6H_5)$	H; P	1968, 301
$PC_{14}H_{20}NO_6$	$(CH_3O)_2P(O)OC(CH_3)(COCH_3)[C(O)NHC_6H_4CH_3]$	H; P (−2·5)	1968, 313
$PC_{14}H_{21}$	$C_6H_5PC(CH_3)_2CH(CH_3)C(CH_3)_2$	H	1967, 82
$PC_{14}H_{21}O$	$C_6H_5P(O)C(CH_3)_2CH(CH_3)C(CH_3)_2$	H	1967, 82
	$(CH_3)_2P(O)C_{12}H_{25}$	H	1967, 329
	$C_6H_5P(O)[C(CH_3)_2]_2CHCH_3$	H	1968, 122
	$CH_3(C_6H_5)P(O)OC_6H_{10}(CH_3)$		1967, 151; 1968, 140
$PC_{14}H_{21}O_2$	$(C_2H_5O)_2P(O)CH_2C(CH_3)=CHSC_6H_5$	H; P (24·2)	1967, 108
$PC_{14}H_{21}O_3S$	$(C_2H_5O)_2P(O)CH(C_6H_5)CH_2COCH_3$	H	1966, 249
$PC_{14}H_{21}O_4$	$(C_2H_5O)_2P(O)OC(C_6H_5)C(CH_3)_2$	H	1967, 121
	$C_6H_5OP(O)OCHC(CH_3)_2CH(C_3H_7)O$	P (−16·5; −10·9)	1968, 150
$PC_{14}H_{21}O_5$	$(C_2H_5O)_2P(O)CH(CH_3)COC_6H_4OCH_3$	H	1967, 81
	$(C_2H_5O)_2P(O)C(CH_3)(CHO)C_6H_4OCH_3$	H	1967, 81
$PC_{14}H_{21}O_6$	$(CH_3O)_3POCH(C_6H_5)C(CH_3)(COCH_3)O$	H; P (−51·5)	1967, 350, 351
$PC_{14}H_{22}Br$	$CH_3(C_6H_5)P^{\oplus}CCH_2C(CH_3)_2C(CH_3)_2, Br^{\ominus}$	H	1967, 82
$PC_{14}H_{22}NO_3$	$HP[OC(CH_3)_2]_2OCH(C_6H_5)CH_2NH$	H; P (−49·5)	1967, 164
	$(C_2H_5O)_2P(O)C[N(CH_3)_2]=CHC_6H_5$	H	1968, 93

Formula	Compound	Resonance	Reference
$PC_{14}H_{22}NO_6S$	$(C_3H_7O)_2P(O)C(O)NHSO_2C_6H_4CH_3$	H	1968, 328; 1969, 161
$PC_{14}H_{23}$	$C_6H_5P(C_4H_9)_2$	P (−26)	1965, 3; 1967, 244, 277; 1968, 7; 1969, 249
	$C_6H_5P(i-C_4H_9)_2$	P (−34·2)	1967, 8
	$C_6H_5P(s-C_4H_9)_2$	P (1·8)	1967, 8
	$P[C(t-C_4H_9)CH]_2C-t-C_4H_9$	H	1968, 101
$PC_{14}H_{23}ClN$	$(C_4H_9)_2NPCl(C_6H_5)$	H	1968, 92
$PC_{14}H_{23}Cl_2N_2$	$C_6H_5N=P[N(C_4H_9)_2]Cl_2$	H; P (−22·1)	1969, 348
$PC_{14}H_{23}N_4O_4S$	$PC_{14}H_{23}N_4O_4S$	H	1967, 352
$PC_{14}H_{23}N_4O_5$	$[(CH_3)_2N]_2P(O)NHCH(CH_3)CH_2OC(O)C_6H_4NO_2$	H	1966, 246
$PC_{14}H_{23}O$	$C_6H_5P(O)(C_4H_9)_2$	P (45·2)	1967, 277
$PC_{14}H_{23}O_2$	$C_6H_5P(OC_4H_9)_2$	H; P (155−7)	1966, 225; 1969, 333
$PC_{14}H_{23}S$	$C_6H_5P(S)(C_4H_9)_2$	P (47·6)	1967, 277
$PC_{14}H_{24}Br$	$C_2H_5(C_6H_5)P^{\oplus}(i-C_3H_7)_2, Br^{\ominus}$	P (42·2)	1966, 5; 1968, 6
	$H(C_6H_5)P^{\oplus}(C_4H_9)_2, Br^{\ominus}$	H; P (12·2)	1968, 7
	$C_2H_5(C_6H_5)P^{\oplus}(C_3H_7)_2, Br^{\ominus}$	P (32·1)	1968, 6
$PC_{14}H_{24}NO_3$	$(C_2H_5O)_3P=NC_6H_4C_2H_5$	H	1965, 102
	$(C_2H_5O)_2P(O)C[N(CH_3)_2](CH_3)C_6H_5$	P (26)	1969, 340
$PC_{14}H_{24}NO_4$	$C_6H_5P(O)[N(CH_3)_2][OC(CH_3)]_2OH$	P (21·1)	1969, 343
$PC_{14}H_{24}NO_4S_3$	$(i-C_3H_7O)_2P(S)S(CH_2)_2NHSO_2C_6H_5$	H	1968, 284
$PC_{14}H_{24}N_3O$	$C_2H_5OC(C_6H_5CH_2)NP[N(CH_3)_2]_2$	H	1968, 321
	$C_2H_5OC(CH_3C_6H_4)NP[N(CH_3)_2]_2$	H; P (99·7; 104·4)	1968, 321
$PC_{14}H_{25}BrNOS_3$	$C_6H_5(CH_3)SP(S)S(CH_2)_2N^{\oplus}(CH_3)_2C_4H_{10}O, Br^{\ominus}$	H	1967, 113
$PC_{14}H_{25}ClN_3$	$C_6H_5N=P[N(C_2H_5)_2]_2Cl$	H	1965, 98
$PC_{14}H_{25}N_2$	$C_8H_{17}P(CH_2CH_2C\equiv N)_2$	P (−25·7)	1966, 4; 1967, 244
	$C_6H_5P(NH-t-C_4H_9)_2$	H	1967, 95
$PC_{14}H_{25}N_2S$	$C_6H_5P(S)(NHC_4H_9)_2$	H	1967, 318
$PC_{14}H_{25}O_2$	$C_4H_5P(Os-C_4H_9)_2$ (stereois.)	H; P (160·7; 159·2; (158·1)	1969, 162

Formula	Structure	Nucleus	Reference
$PC_{14}H_{26}NO_2$	$(CH_3)_2NP[OC(C_5H_{10})]_2$	P (147)	1969, 362
$PC_{14}H_{27}O_3$	$(C_4H_9O)_2P(O)C\!=\!CC_4H_9$	H	1967, 21
	$(CH_3O)_2P(O)CCH(CH_2)_{10}$	H	1967, 100
$PC_{14}H_{27}SSi_2$	$C_6H_5P(S)[CH_2Si(CH_3)_3]_2$	H; P (38·8)	1966, 132
$PC_{14}H_{28}ClN_4$	$C_6H_5NHPN(C_4H_9)_2(NH_2)_2Cl$	H; P (26·6)	1969, 348
$PC_{14}H_{28}Cl_2NO_3$	$[CH_3(CH_2Cl)CHO][(C_4H_9)_2NP(O)OC(CH_2Cl)]\!=\!CH_2$	H	1967, 340
$PC_{14}H_{29}ClNO_3$	$[CH_3(CH_2Cl)CHO][(C_4H_9)_2NP(O)OC(CH_3)]\!=\!CH_2$	H	1967, 340
$PC_{14}H_{29}Cl_2O$	$C_{12}H_{25}P(O)(CH_2Cl)_2$	H	1968, 299; 1969, 334
$PC_{14}H_{29}O_4$	$(CH_3O)_2P(O)C_{12}H_{22}(OH)$	H	1967, 100
$PC_{14}H_{30}Li$	$CH_3(C_{12}H_{25})PCH_2Li$	P (−35·2)	1967, 329
$PC_{14}H_{30}NO_2$	$(C_2H_5)_2C_6H_{12}(i\text{-}C_3H_7)OP(O)C(O)N(CH_3)_2$	H	1967, 26
$PC_{14}H_{31}$	$C_{12}H_{25}P(CH_3)_2$	H; P (−52·6)	1966, 14; 1967, 329
$PC_{14}H_{31}GeO$	$(C_2H_5)_2PCH(CH_3)CH\!=\!CHOGe(C_2H_5)_3$	H	1968, 331
$PC_{14}H_{31}O$	$C_{12}H_{25}P(O)(CH_3)_2$	H; P (47)	1966, 14, 268
$PC_{14}H_{31}O_4$	$(C_4H_9O)_2P(O)(CH_2)_2OC_4H_9$	H	1966, 251
$PC_{14}H_{32}$	$C_2H_5P^{\oplus}(C_4H_9)_3$	P (35·5)	1967, 7
$PC_{14}H_{32}BF_4O$	$C_2H_5OP^{\oplus}(C_4H_9)_3, BF_4^{\ominus}$	H; P (98)	1968, 332
$PC_{14}H_{32}Br$	$C_2H_5P^{\oplus}(C_4H_9)_3Br^{\ominus}$	P (36·6)	1966, 5
$PC_{14}H_{32}N$	$C_{12}H_{25}(CH_3)_2P\!=\!NH$	P (18)	1966, 268
$PC_{14}H_{33}GeO$	$(C_2H_5)_2PCH(C_3H_7)OGe(C_2H_5)_3$	H	1968, 331
$PC_{14}H_{33}O$	$C_6H_5(C_4H_9)_2P(O)C(CH_3)\!=\!C\!=\!CC_6H_5(CH_3)_5$	H	1969, 166
$PC_{14}H_{35}O_3Si_2$	$C_2H_5OP[OSi(C_2H_5)_3]_2$	H	1967, 34
$PC_{14}H_{36}GeN_2Si$	$(t\text{-}C_4H_9)_2PN[Si(CH_3)_3]\!=\!NGe(CH_3)_3$	H	1968, 64
	$(t\text{-}C_4H_9)_2PN[Ge(CH_3)_3]\!=\!NSi(CH_3)_3$	H	1968, 64
$PC_{14}H_{44}O$	$(C_6H_{11})_2P(O)C(CH_3)\!=\!C\!=\!CC_6H_5(CH_3)_5$	H	1969, 166
$PC_{15}H_{10}ClF_4$	$(C_6H_5)_2PCF_2CCl\!=\!CF_2$	H; F (−94·7; −79·8; −78·0)	1967, 295
	$(C_6H_5)_2PCF\!=\!CClCF_3$	H; F (cis −58·8; −83·6 trans −60·8; −91·4)	1968, 98

Formula	Compound	Resonance	Reference
$PC_{15}H_{10}ClF_4O$	$(C_6H_5)_2P(O)CF_2CCl=CF_2$	H; F (−105·9; −76·4; −74·8)	1967, 295
$PC_{15}H_{11}F_6O_2$	$(C_6H_5)_2POCH(CF_3)_2$	H; P (39·2)	1968, 302
$PC_{15}H_{13}$	$(C_6H_5)_2PC≡CCH_3$	H	1966, 243, 247, 265
	$(C_6H_5)_2PCH=C=CH_2$	H	1968, 99; 1969, 115
$PC_{15}H_{13}ClNO_6$	$CH_3(NO_2C_6H_4O)P(O)OCH_2C(O)OC_6H_4Cl$	H (T)	1969, 159
$PC_{15}H_{13}N_2O_8$	$CH_3(NO_2C_6H_4O)P(O)OCH_2C(O)OC_6H_4NO_2$	H (T)	1969, 159
$PC_{15}H_{13}O$	$(C_6H_5)_2P(O)C≡CCH_3$	H	1966, 243, 247
	$(C_6H_5)_2P(O)CH=C=CH_2$	H	1966, 61; 1967, 104; 1968, 56; 1969, 115
$PC_{15}H_{13}S$	$(C_6H_5)_2P(S)C≡CCH_3$	H	1966, 243, 247
$PC_{15}H_{14}NOS_3$	$CH_3OP(S)SC(S)C(C_6H_4)CNC_6H_4CH_3$	H	1969, 363
$PC_{15}H_{14}NO_2S_2$	$CH_3OP(S)SC(O)C(C_6H_4)CNC_6H_4CH_3$	H	1969, 363
$PC_{15}H_{14}NO_2S_3$	$CH_3OP(S)SC(S)C(C_6H_4)CNC_6H_4OCH_3$	H	1969, 363
$PC_{15}H_{14}NO_3S_2$	$CH_3OP(S)SC(O)C(C_6H_4)CNC_6H_4OCH_3$	H	1969, 363
$PC_{15}H_{14}NO_6$	$CH_3(NO_2C_6H_4O)P(O)OCH_2C(O)C_6H_5$	H (T)	1969, 159
$PC_{15}H_{15}$	$(C_6H_5)_2PCH_2CH=CH_2$	P (−17·1)	1967, 8
	$(C_6H_5)_2PCH(CH_2)_2$	H; P (1·9)	1968, 333; 1969, 354
$PC_{15}H_{15}N_2O_6$	$NO_2C_6H_4OP(O)(CH_3)OCH_2C(C_6H_5)NOH$	H	1968, 334
$PC_{15}H_{15}N_3$	$(C_5H_5N)_3P$	H	1969, 122
$PC_{15}H_{15}N_3O$	$(C_5H_5N)_3PO$	H	1969, 122
$PC_{15}H_{15}O$	$(C_6H_5)_2P(O)CH=CHCH_3$	H	1965, 47
	$(C_6H_5)_2P(O)CH_2CH=CH_2$	H	1965, 87; 1968, 56
	$(C_6H_5)_2PCH_2COCH_3$	H	1968, 297
	$(C_6H_5)_2P(O)CH(CH_2)_2$	H	1968, 333

PC$_{15}$H$_{15}$O$_2$	(C$_6$H$_5$)$_2$P(O)CH$_2$COCH$_3$	H	1965, 105; 1968, 297
	(C$_6$H$_5$)$_2$P(O)OC(CH$_3$)=CH$_2$	H	1965, 105; 1968, 297
PC$_{15}$H$_{16}$ClO	(C$_6$H$_4$)$_2$P$^\oplus$CH$_3$(CH$_2$OCH$_3$), Cl$^\ominus$	H	1968, 60
PC$_{15}$H$_{16}$I	(C$_6$H$_4$)$_2$P$^\oplus$CH$_3$(C$_2$H$_5$), I$^\oplus$	H	1968, 60
PC$_{15}$H$_{16}$KN$_2$S$_2$	(C$_6$H$_5$)$_2$P(S)N=C(S$^\ominus$)N(CH$_3$)$_2$, K$^\oplus$	H; P (40·7)	1967, 221, 244, 353
PC$_{15}$H$_{16}$N	C$_6$H$_5$P(CH$_2$)$_2$NHCHC$_6$H$_5$	H; P ($-$2·9)	1967, 137
PC$_{15}$H$_{16}$NO$_3$	HP[OC$_6$H$_3$(CH$_3$)O]OCH(C$_6$H$_5$)CH$_2$NH	H; P ($-$35·2)	1969, 178
	OC$_6$H$_4$OPO(CH$_2$)$_2$NHCH$_2$C$_6$H$_5$	P (132)	1969, 45
	HP(OC$_6$H$_4$O)O(CH$_2$)NCH$_2$C$_6$H$_5$	P ($-$100)	1969, 45
PC$_{15}$H$_{17}$	(C$_6$H$_5$)$_2$P-i-C$_3$H$_7$	P (0·2)	1965, 3; 1968, 6, 7; 1969, 354
	(C$_6$H$_5$)$_2$PC$_3$H$_7$	P ($-$17·6)	1967, 8
	CH$_3$P(C$_6$H$_4$CH$_3$)$_2$	H	1968, 15
PC$_{15}$H$_{17}$Br	(C$_6$H$_5$)$_2$P$^\oplus$C$_3$H$_7$, Br$^\ominus$	H	1967, 354
PC$_{15}$H$_{17}$N$_2$S$_2$	(C$_6$H$_5$)$_2$P(S)NHC(S)N(CH$_3$)$_2$	H	1967, 353
PC$_{15}$H$_{17}$N$_4$	(C$_6$H$_5$)$_2$PNCHNN(CH$_3$)$_2$N	H; P (26·4)	1967, 355
PC$_{15}$H$_{17}$O	CH$_3$P(O)(C$_6$H$_7$CH$_3$)$_2$	H	1968, 15
PC$_{15}$H$_{17}$OS	(C$_6$H$_5$)$_2$P(S)(CH$_2$)$_3$OH	P (42·4)	1966, 40
PC$_{15}$H$_{17}$O$_2$	(C$_6$H$_5$)$_2$P(O)Oi-C$_3$H$_7$	H	1966, 266
	(C$_6$H$_5$)$_2$P(O)OC$_3$H$_7$	H	1967, 52
	C$_6$H$_5$C$_6$H$_4$P(O)CH$_3$(CH$_2$OCH$_3$)	H	1968, 60
	(C$_6$H$_5$)$_2$P(O)(CH$_2$)$_3$OH	H	1968, 118
PC$_{15}$H$_{17}$O$_6$	C$_6$H$_5$(CH$_3$O)P(O)CC(O)OC(CH$_3$)(COCH$_3$)COCH$_3$ (stereois.)	H; P (80·9; 84·0)	1969, 353
PC$_{15}$H$_{18}$Br	H(i-C$_3$H$_7$)P$^\oplus$(C$_6$H$_5$)$_2$, Br$^\ominus$	H; P (14·0)	1968, 7

Formula	Compound	Resonance	Reference
$PC_{15}H_{18}IS$	$CH_3(C_6H_5)_2P^{\oplus}CH_2SCH_3, I^{\ominus}$	H	1967, 349
$PC_{15}H_{18}N$	$(CH_3)_2NCH_2P(C_6H_5)_2$	H	1967, 349
$PC_{15}H_{18}NO$	$(CH_3)_2NCH_2P(O)(C_6H_5)$	H	1967, 349
$PC_{15}H_{18}NO_4$	$\beta\text{-}C_{10}H_7C(=NOH)P(O)(OC_2H_5)_2$	H; P (-7.4)	1968, 142
	$C_6H_5OP(O)(O^{\ominus})OCH_2CH(CH_3)N^{\oplus}H_2C_6H_5$	P (14)	1969, 359
	$C_6H_5NC_5H_5^{\oplus}O^{\ominus}P(O)O[CH(CH_3)]_2O$		1969, 359
	$(C_6H_5)_2NH_2^{\oplus}O^{\ominus}P(O)OCH_2CH(CH_3)O$	P ($ca.$ 14)	1969, 359
$PC_{15}H_{18}NS$	$(C_6H_5)_2P(S)(NH\text{-}i\text{-}C_3H_7)$	H	1968, 145
$PC_{15}H_{18}N_3O$	$[CH_3NCH=CHCH=C]_3PO$	H	1969, 123
$PC_{15}H_{18}O_3S$	$[OC(CH_3)=CHCH=C]_3PS$	H	1969, 124
$PC_{15}H_{19}ClN$	$CH_3(C_6H_5)_2P^{\oplus}N(CH_3)_2, Cl^{\ominus}$	H	1968, 54
$PC_{15}H_{19}Cl_6NOSb$	$CH_3O(C_6H_5)_2P^{\oplus}N(CH_3)_2, SbCl_6^{\ominus}$	H; P (57·5)	1968, 8; 1969, 87
$PC_{15}H_{19}Cl_6NSSb$	$CH_3S(C_6H_5)_2P^{\oplus}N(CH_3)_2, SbCl_6^{\ominus}$	H; P (67·5)	1968, 8; 1969, 87
$PC_{15}H_{19}Cl_6NSbSe$	$CH_3Se(C_6H_5)_2P^{\oplus}N(CH_3)_2, SbCl_6^{\ominus}$	H; P (65·7)	1969, 87
$PC_{15}H_{19}IN$	$CH_3P^{\oplus}(C_6H_5)_2N(CH_3)_2, I^{\ominus}$	H; P (50·0)	1969, 87
$PC_{15}H_{19}NO_4$	$(C_2H_5O)_2P(O)C(=NOH)C_6H_4\text{-}t\text{-}C_4H_9$	H	1968, 142
$PC_{15}H_{19}N_2O_5S_2$	$CH_3O[oC_6H_4(OH)]P(O)$ S-benzylthiuronium	H	1967, 143
$PC_{15}H_{19}N_2S_2$	$(C_6H_5)_2P(S)N=C(SCH_3)N(CH_3)_2$	P (41·6)	1967, 221, 344
	$(C_6H_5)_2P(SCH_3)=NC(S)N(CH_3)_2$	P (32·0)	1967, 221, 344
$PC_{15}H_{19}OSi$	$(C_6H_5)_2POSi(CH_3)_3$	P (94·1)	1967, 343
$PC_{15}H_{19}O_4Si_2$	$CH_3P(O)[OSi(CH_3)C_6H_5]_2O$	H	1965, 106
$PC_{15}H_{19}O_7$	$C_6H_5(CH_3O)P(O)C(COOH=C(OCH_3)C(CH_3)(COCH_3)OH$	H	1969, 353
$PC_{15}H_{19}Si$	$(C_6H_5)_2PSi(CH_3)_3$	H{Si, P}	1969, 2
$PC_{15}H_{19}Sn$	$(C_6H_5)_2PSn(CH_3)_3$	H{Sn, P}	1969, 2, 14

$PC_{15}H_{20}F_3O_6$	$(CH_3O)_3POC(CF_3)(C_6H_5)C(CH_3)(COCH_3)O$ (stereois.)	H; P (−52.5; −49.6)	1966, 248
$PC_{15}H_{20}NO_4$	$(C_2H_5O)_2P(O)C(CN)=C(CH_3)C_6H_4OCH_3$	H	1968, 323
$PC_{15}H_{20}O_4$	$C_6H_5OP(O)O[CH(CHCH_3)]_3O$	H; P	1968, 83
$PC_{15}H_{21}ClNO_3$	$(C_2H_5O)_2P(O)CH(NH_3Cl)\beta\text{-}C_{10}H_7$	H	1968, 142
$PC_{15}H_{21}O_2$	$(CH_3)_3P^\oplus CH(C_6H_5)=C(COCH_3)C(O^\ominus)CH_3$	P (109)	1968, 124, 152
$PC_{15}H_{21}O_5$	$(CH_3O)_3POC(CH_3)C(COCH_3)CH(C_6H_5)$	H; P (−27.9)	1966, 106; 1967, 160; 1968, 124, 152, 312
	$C_2H_5OP(O)(OH)CH_2C'(COOC_2H_5)=C(C_6H_5)CH_3$	H	1967, 327
	$C'_6H_5O[(CH_3)_2C(CH_2O)_2P[OC(CH_3)]_2$	P (−56.6)	1968, 301
	$(CH_3O)_3POCH(COCH_3)C(CH_3)(COCH_3)O$	H; P (−50.1)	1965, 21; 1966, 258
$PC_{15}H_{21}O_7$	$(CH_3O)_3POCH(C_6H_4CHO)C(CH_3)(COCH_3)O$ (stereois.)	H; P (−49.6; −51.1)	1967, 350, 351
$PC_{15}H_{22}N$	$C_6H_5P(CH_2)_2NHC(CH_2)_6$	H	1967, 137
$PC_{15}H_{22}NO_8$	$(CH_3O)_3POC(CH_3)(C_6H_4NO_2)C(CH_3)(COCH_3)O$ (stereois.)	H; P (−54.0; −52.4)	1966, 248
$PC_{15}H_{23}N_2O_4S$	$(C_2H_5O)_2P(S)NHCOC(O)NH\text{-}i\text{-}C_3H_7$	H	1966, 264
$PC_{15}H_{23}O$	$C_6H_5P(O)C(CH_3)_2CH(CH_3)_2CH$	H	1967, 356
	$CH_3P(O)C(CH_3)_2CH(CH_3)C(CH_3)_2C_6H_4$	H{P}	1968, 67
$PC_{15}H_{23}O_3$	$(CH_3)_2C(CH_2O)_2P(O)CH_2C_6H_2(CH_3)_3$	H	1968, 74
$PC_{15}H_{23}O_5$	$(C_2H_5O)_2P(O)CH(C_6H_5)CH_2COOC_2H_5$	H	1966, 249

212 FORMULAE INDEX II

Formula	Compound	Resonance	Reference
$PC_{15}H_{23}O_6$	$(CH_3O)_3POC(CH_3)(C_6H_5)C(CH_3)(COCH_3)O$	H; P (−54·2)	1966, 248
$PC_{15}H_{24}Br$	$CH_3(C_6H_5)P^{\oplus}C(CH_3)_2CH(CH_3)C(CH_3)_2$, $Br^{\ominus}$	H	1967, 82
$PC_{15}H_{24}NO$	$C_6H_5CH_2NHP(O)[C(CH_3)_2]_2CHCH_3$	H	1969, 350
$PC_{15}H_{24}NO_3$	$(CH_3)_2C(CH_2O)_2P(O)CH(NHC_3H_7)C_6H_5$	H	1968, 74
	$C_2H_5OC(C_6H_5)P(O)C(O)N(i\text{-}C_3H_7)_2$	H	1968, 29
$PC_{15}H_{24}NO_5S_3$	$C_6H_5P(S)(SCH_3)S(CH_2)_2N^{\oplus}H(C'_4H_{10}O)$, $COO^{\ominus}COOH$	H	1967, 113
$PC_{15}H_{25}N_6O$	$[(CH_3)_2N]_2P(O)NC(CH_3)NNCNHC_6H_5$	H	1966, 137
$PC_{15}H_{25}O$	$CH_3P(O)C(CH_3)_2CH(CH_3)C(CH_3)_2C_6H_6$	H;{P}	1967, 356; 1968, 67
$PC_{15}H_{26}Br$	$C_3H_7(C_6H_5)P^{\oplus}(i\text{-}C_3H_7)_2$, $Br^{\ominus}$	P (39·9)	1966, 5; 1968, 6
	$C_6H_5P^{\oplus}(C_3H_7)_2$, $Br^{\ominus}$	P (29·3)	1968, 6
	$CH_3(C_6H_5)P^{\oplus}(i\text{-}C_4H_9)_2$, $Br^{\ominus}$	P (26·5)	1968, 6
$PC_{15}H_{26}I$	$CH_3(C_6H_5)P^{\oplus}(C_4H_9)_2$, $Br^{\ominus}$	P (28·2)	1967, 244
$PC_{15}H_{27}$	$(C_5H_9)_3P$	P (4·5–1·0)	1967, 8, 244, 277; 1968, 6
$PC_{15}H_{27}ClNO_3$	$(C_2H_5O)_2P(O)CH(NH_3Cl)CHC_6H_4\text{-}t\text{-}C_4H_9$	H; P (17·4)	1968, 142, 296
$PC_{15}H_{27}Ge_3$	$P[C\!=\!\!=\!CGe(CH_3)_3]_3$	H	1969, 364
$PC_{15}H_{27}O$	$(C_5H_9)_3PO$	P (67·5)	1967, 277
$PC_{15}H_{27}O_3$	$[(CH_3)_2(CH_2\!=\!\!CH)CO]_3P$	P (140·2)	1967, 293
$PC_{15}H_{27}O_4$	$[(CH_3)_2(CH_2\!=\!\!CH)CO]_3PO$	P (−13·1)	1967, 293
$PC_{15}H_{27}S$	$(C_5H_9)_3PS$	P (70·1)	1967, 277
$PC_{15}H_{27}Si_3$	$P[C\!=\!\!=\!CSi(CH_3)_3]_3$	H	1969, 364
$PC_{15}H_{28}IN_2$	$CH_3(C_6H_5)P^{\oplus}[N(C_2H_5)_2]_2$, $I^{\ominus}$	H	1962, 4
	$CH_3(C_6H_5)P^{\oplus}(NH\text{-}t\text{-}C_4H_9)_2$, $I^{\ominus}$	H	1967, 95
$PC_{15}H_{28}NO_2$	$(C_5H_9)_2P(O)C(O)N(C_2H_5)_2$	H	1967, 26

Formula	Structure	Notes	Ref
$PC_{15}H_{29}N_2O_2$	$O(CH_2)_2N(C_6H_{11})POCH_2NHC_6H_{11}$	H	1968, 304
$PC_{15}H_{29}O_3$	$(CH_3O)_2P(O)CCH(CH_2)_{11}$	H	1967, 100
$PC_{15}H_{30}$	$(C_5H_{10})_3P$	P (-34)	1966, 4
$PC_{15}H_{30}BrO_2$	$(C_4H_9)_3P^{\oplus}CH=CHCOOH, Br^{\ominus}$	H	1969, 349
$PC_{15}H_{30}NO_7$	$C_2H_5O[(C_2H_5)_2NP(O)OC(C_2H_5)(COOC_2H_5)]_2$	H; P (-0.3)	1967, 4
$PC_{15}H_{30}N_3O$	$[(CH_2)_4NCH_2]_3PO$	H; P (48·4)	1967, 291
$PC_{15}H_{30}N_3O_3$	$OP(NHCOOC_2H_5)[NC(CH_3)_2C(CH_3)_2]_2$	H	1965, 42
$PC_{15}H_{30}N_3O_4$	$[(C_2H_5)_2C=NO]_2P(O)N=C(C_2H_5)_2$	P (0·0)	1968, 335
$PC_{15}H_{31}O_8$	$[O(CH_2)_2NCH_2]_3PO$	H; P (50·6)	1967, 291
$PC_{15}H_{31}O_4$	$CH_3(C_{12}H_{25})P(O)CH_2COOH$	H; P (48·7)	1967, 329
$PC_{15}H_{31}O_4$	$(CH_3O)_2P(O)C_{13}H_{24}(OH)$	H	1967, 100
$PC_{15}H_{32}$	$CH_2=CHCH_2P^{\oplus}(C_4H_9)_3$	P (32·4)	1967, 7
$PC_{15}H_{32}I$	$C_2H_5(C_4H_9)_3PI$	^{13}C	1968, 22
$PC_{15}H_{32}NO_2$	$(s\text{-}C_5H_{11})_2P(O)C(O)N(C_2H_5)_2$	H	1967, 26
$PC_{15}H_{32}NO_2$	$(i\text{-}C_3H_7O)_2P(O)C(O)N(i\text{-}C_4H_9)_2$	H	1967, 26
$PC_{15}H_{32}NO_4$	$(C_6H_{13}O)_2P(O)C(O)N(CH_3)_2$	H	1967, 26
	$[CH_3)_2C_4H_8O]_2P(O)C(O)N(CH_3)_2$	H	1967, 26
	$(C_6H_{13}O)_2P(O)C(O)NHC_2H_5$	H	1967, 26
$PC_{15}H_{33}O_3$	$(C_5H_{11}O)_3P$	P (134·5)	1967, 277
	$(s\text{-}C_5H_{11}O)_3P$	P (140·5)	1967, 277
	$(i\text{-}C_5H_{11}O)_3P$	P (143·5)	1967, 277
	$[CH_3(i\text{-}C_3H_7)CHO]_3P$	P (142·0)	1967, 277
	$[C_2H_5(CH_3)_2CO]_3P$	P (138·6)	1967, 293
$PC_{15}H_{33}O_3S$	$(C_5H_{11}O)_3PS$	P (64·2)	1967, 277
	$(s\text{-}C_5H_{11}O)_3PS$	P (62·8)	1967, 277
	$(i\text{-}C_5H_{11}O)_3PS$	P (64·3)	1967, 277
	$[CH_3(i\text{-}C_3H_7)CHO]_3PS$	P (66·8)	1967, 277
	$(t\text{-}C_4H_9CH_2O)_3PS$	P (68·5)	1967, 293
$PC_{15}H_{33}O_3Se$	$(t\text{-}C_4H_9CH_2O)_3PSe$	P (72·6)	1967, 293

Formula	Compound	Resonance	Reference
$PC_{15}H_{33}O_4$	$(C_5H_{11}O)_3PO$	P (−4·0)	1967, 277
	$(s\text{-}C_5H_{11}O)_3PO$	P (−4·4)	1967, 277
	$(i\text{-}C_5H_{11}O)_3PO$	P (−4·0)	1967, 277
	$[CH_3(i\text{-}C_3H_7)CHO]_3PO$	P (−2·2)	1967, 277
	$(t\text{-}C_4H_9CH_2O)_3PO$	P (−0·6)	1967, 293
	$[C_2H_5(CH_3)_2CO]_3PO$	P (−12·9)	1967, 293
$PC_{15}H_{34}$	$i\text{-}C_3H_7(C_4H_9)_3P^{\oplus}$	P (38·0)	1967, 7
$PC_{15}H_{34}I$	$(CH_3)_3P^{\oplus}C_{12}H_{25}, I^{\ominus}$	H; P (26·3)	1966, 14
$PC_{15}H_{36}F_2N_3$	$[(C_2H_5)_2NCH_2]_3PF_2$	F (−51·0)	1967, 61
$PC_{15}H_{36}NSi$	$(C_4H_9)_3P=NSi(CH_3)_3$	P (9)	1966, 268
$PC_{15}H_{36}N_3$	$[(C_2H_5)_2NCH_2]_3P$	P (−65·5)	1966, 4; 1967, 244, 291
$PC_{15}H_{36}N_3O$	$[(C_2H_5)_2NCH_2]_3PO$	H; P (43·2)	1967, 291
$PC_{15}H_{39}AlNSi$	$(C_2H_5)_3PN[Al(C_2H_5)_3][Si(CH_3)_3]$	H	1967, 322
$PC_{16}H_{10}FN$	$(C_6F_5)_2PNH\text{-}t\text{-}C_4H_9$	H; F (−136·7; −162·1; −152·5)	1966, 19
$PC_{16}H_{11}$	$(C_6H_5)_2P(C\!\equiv\!C)_2H$	H	1966, 234, 247
$PC_{16}H_{11}F_6$	$(C_6H_5)_2PC(CF_3)\!=\!CHCF_3$	H	1965, 107
$PC_{16}H_{13}O_3$	$C_6H_5[C_{10}H_6(OH)]PO_2H$	H	1965, 79
$PC_{16}H_{15}$	$(C_6H_5)_2PC(CH_3)\!=\!C\!=\!CH_2$	H	1966, 61
$PC_{16}H_{15}Cl_2O_3$	$(C_2H_5)_2(C_6H_5)P^{\oplus}CC(O)C(O)(CCl)_2CO^{\ominus}$	P (25·6)	1965, 85
$PC_{16}H_{15}F_5N$	$C_6H_5(C_6F_5)PN(C_2H_5)_2$	H; P (47·0)	1966, 7
$PC_{16}H_{15}F_{12}O_2$	$C_2H_5(C_6H_5)[(CF_3)_2CHO]POC(CF_3)_2CHCH_3$ (stereois.)	P (−21·7; −30·3); F	1968, 302
$PC_{16}H_{15}O$	$(C_2H_5)_2(C_6H_5)P[OC(CF_3)_2]_2$	P (−1·1); F (9·8)	1968, 14, 303
	$(C_6H_5)_2P(O)C(CH_3)\!=\!C\!=\!CH_2$	H	1966, 61; 1969, 115

PC$_{16}$H$_{16}$BrO$_5$	(CH$_3$O)$_2$P(O)CBr(C$_6$H$_5$)COC$_6$H$_5$	H; P (−4·1)	1968, 309
PC$_{16}$H$_{16}$ClO$_4$	(C$_2$H$_5$)$_2$(C$_6$H$_5$)P$^{\oplus}$CC(O)C(O)CClC(OH)CO$^{\ominus}$	P (23)	1965, 85
PC$_{16}$H$_{16}$I	CH$_3$(C$_6$H$_5$)$_2$P$^{\oplus}$C≡CCH$_3$, I$^{\ominus}$	H	1966, 73, 247
PC$_{16}$H$_{16}$NOS$_3$	CH$_3$C$_6$H$_4$NC(C$_6$H$_4$)CC(S)SP(S)OC$_2$H$_5$	H	1969, 363
PC$_{16}$H$_{16}$NO$_2$S$_2$	CH$_3$C$_6$H$_4$NC(C$_6$H$_4$)CC(O)SP(S)OC$_2$H$_5$	H	1969, 363
PC$_{16}$H$_{16}$NO$_2$S$_3$	CH$_3$OC$_6$H$_4$NC(C$_6$H$_4$)CC(S)SP(S)OC$_2$H$_5$	H	1969, 363
PC$_{16}$H$_{16}$NO$_3$S$_2$	CH$_3$OC$_6$H$_4$NC(C$_6$H$_4$)CC(O)SP(S)OC$_2$H$_5$	H	1969, 363
PC$_{16}$H$_{16}$NO$_6$	CH$_3$(NO$_2$C$_6$H$_4$O)P(O)OCH$_2$C(O)C$_6$H$_4$CH$_3$	H(T)	1969, 159
PC$_{16}$H$_{16}$NO$_7$	CH$_3$(NO$_2$C$_6$H$_4$O)P(O)OCH$_2$C(O)OCH$_2$C(O)C$_6$H$_4$OCH$_3$	H(T)	1969, 159
PC$_{16}$H$_{16}$N$_3$O$_8$	CH=CH(NHCOC$_6$H$_5$)CNC(O)N–	H	1969, 98
	–CHOCH(CH$_2$OPO$_3$H$_2$)[CH(OH)]$_2$	H	1969, 98
PC$_{16}$H$_{17}$N$_2$O$_3$	C$_6$H$_5$PO(CH$_2$)$_2$N(CH$_3$)C(O)NC$_6$H$_5$	H; P (116)	1969, 171
PC$_{16}$H$_{17}$N$_2$O$_9$	(CH$_3$O)$_2$P(O)OCH(C$_6$H$_4$NO$_2$)CH(OH)C$_6$H$_4$NO$_2$	H; P (−0·5)	1967, 163
PC$_{16}$H$_{17}$O$_4$	(CH$_3$O)$_2$P(O)CHC(C$_6$H$_5$)$_2$O	H	1966, 245
PC$_{16}$H$_{18}$F$_2$N	(CH$_3$O)$_2$P(O)C(C$_6$H$_5$)$_2$CHO	H	1966, 245
PC$_{16}$H$_{18}$N	(C$_2$H$_5$)$_2$NP(C$_6$H$_4$F)$_2$	F	1969, 18, 19
	C$_6$H$_5$P(CH$_2$)$_2$NHCCH$_3$(C$_6$H$_5$)	H	1967, 137
PC$_{16}$H$_{18}$NO$_2$	(C$_6$H$_5$)$_2$P(O)NCH(O-t-C$_4$H$_9$)	H	1966, 55
	(CH$_3$)$_2$NP[OCH(C$_6$H$_5$)]$_2$	P (145)	1969, 362
PC$_{16}$H$_{18}$NO$_3$	C$_6$H$_5$OP(O)CH(C$_6$H$_5$)CH(CH$_3$)NCH$_3$	H; P (13·4)	1968, 76

216   FORMULAE INDEX II

Formula	Compound	Resonance	Reference
$PC_{16}H_{19}$	$C_6H_5OP(O)CH_2CH(CH_3)NCH_2C_6H_5$ (stereois.)	H; P (15·2; 17)	1968, 76
	$C_4H_9P(C_6H_5)_2$	P (−26·2)	1965, 3
		P (−17·1)	1968, 7; 1969, 249
	i-$C_4H_9P(C_6H_5)_2$	P (−21)	1967, 8; 1968, 6
	s-$C_4H_9P(C_6H_5)_2$	P (−3·2)	1967, 8; 1968, 6
	t-$C_4H_9P(C_6H_5)_2$	H; P (17·1)	1967, 8, 79; 1968, 6
$PC_{16}H_{19}ClN$	i-$C_3H_7(C_6H_5CH_2)NPCl(C_6H_5)$	H{P, C}	1969, 190
$PC_{16}H_{19}N_2O$	$(CH_3)_2NPOCH(C_6H_5)CH_2NC_6H_5$ (stereois.)	H	1968, 92
		P (120; 130)	1969, 109
$PC_{16}H_{19}N_2O_2$	$HP[OCH(C_6H_5)CH_2NH]_2$	H; P (−54·6)	1967, 164; 1969, 45
$PC_{16}H_{19}N_2S_2$	$(C_6H_5)_2P(S)N=C(SCH_3)N(CH_3)_2$	H	1967, 353
	$(C_6H_5)_2P(SCH_3)=NC(S)N(CH_3)_2$	H	1967, 353
$PC_{16}H_{19}O$	$C_2H_5P(O)(C_6H_4CH_3)_2$	H	1968, 15
	$C_4H_9OP(C_6H_5)_2$	H; P (111·1)	1969, 333
$PC_{16}H_{19}O_2$	$(C_6H_5)_2P(O)O$-t-C_4H_9	H	1966, 257
	$(C_6H_5)_2P(O)O$-i-C_4H_9	H	1967, 152
$PC_{16}H_{20}Br$	$CH_3(i$-$C_3H_7)P^{\oplus}(C_6H_5)_2$, $Br^{\ominus}$	P (30·9)	1966, 5; 1968, 6
	$(C_2H_5)_2P^{\oplus}(C_6H_5)_2$, $Br^{\ominus}$	P (31·3)	1968, 6
	$H(C_4H_9)P^{\oplus}(C_6H_5)_2$, $Br^{\ominus}$	H; P (2·1)	1968, 7
	$(CH_3)_2P^{\oplus}(C_6H_5)CH(CH_3)C_6H_5$, $Br^{\ominus}$	H{T}{P}	1969, 157
$PC_{16}H_{20}I$	$(CH_3)_2P^{\oplus}(C_6H_4CH_3)_2$, $I^{\ominus}$	H	1968, 15
$PC_{16}H_{20}NO$	$(C_6H_5)_2PO(CH_2)_2N(CH_3)_2$	P (112·2)	1968, 336
$PC_{16}H_{20}NO_4$	$(C_6H_5)_2NH_2^{\oplus}$, $O^{\ominus}P(O)O[CH(CH_3)]_2O$	P (ca. 15)	1969, 359
$PC_{16}H_{21}O_6$	$(CH_3O)_3POC(CH_3)_2C(CH_3)(COCH_3)O$	H	1967, 140

NMR DATA ON ORGANIC COMPOUNDS 217

Formula	Compound	Nucleus (δ)	Reference
$PC_{16}H_{23}$	$(C_5H_9)_2PC_6H_5$	P (1.6)	1967, 8; 1968, 6
$PC_{16}H_{23}Ge_2$	$C_6H_5P[C\equiv CGe(CH_3)_3]_2$	H	1969, 364
$PC_{16}H_{23}O_7$	$(CH_3O)_3POC(C_6H_5)(COCH_3)C(CH_3)(COCH_3)O$ (stereois.)	H; P (-54.7; -52.4)	1966, 258
$PC_{16}H_{23}Pb_2$	$C_6H_5P[C\equiv CPb(CH_3)_3]_2$	H	1969, 364
$PC_{16}H_{23}Si_2$	$C_6H_5P[C\equiv CSi(CH_3)_3]_2$	H	1969, 364
$PC_{16}H_{23}Sn_2$	$C_6H_5P[C\equiv CSn(CH_3)_3]_2$	H	1969, 364
$PC_{16}H_{25}N_2$	$C_6H_5P(NC_5H_{10})_2$	H	1966, 18
$PC_{16}H_{25}N_2O$	$(CH_3)_2NPOCH(C_6H_5)CH_2NC_6H_4$ (stereois.)	P (132; 135)	1969, 109
$PC_{16}H_{25}N_4S_3$	$(C_2H_5)_2NCNP(S)(SC_6H_5)SC[N(C_2H_5)_2]N$	P (61.0)	1968, 137
$PC_{16}H_{25}O_2$	$CH_3(C_6H_5)P(O)OOC_6H_{10}(i-C_3H_7)$	H	1967, 151
$PC_{16}H_{26}Br$	$CH_3(C_6H_5CH_2)P(C_6H_5)P^{\oplus}C(CH_3)_2CH(CH_3)C(CH_3)_2$, $Br^{\ominus}$	H	1967, 82; 1969, 170
$PC_{16}H_{26}NO_6S$	$(C_4H_9O)_2P(O)C(O)NHSO_2C_6H_4CH_3$	H	1968, 328; 1969, 161
$PC_{16}H_{27}N_4O_4S$	Diethyl 2-{3-(2-methyl-4-aminopyrimidin-5-yl)methyl-3a-methylperhydrofuro[2,3-d]thiazole}phosphonate		1967, 352
$PC_{16}H_{27}O_2$	$(CH_3)_2C_6H_3PH(O)(Os-C_4H_9)_2$ (stereois.)	H; P (158.3; 156.6; 153.3)	1969, 162
$PC_{16}H_{28}Br$	$C_2H_5(C_6H_5)P^{\oplus}(i-C_4H_9)_2Br^{\ominus}$	P (29.9)	1968, 6
	$C_2H_5(C_6H_5)P^{\oplus}(s-C_4H_9)_2Br^{\ominus}$	P (41.2)	1968, 6
$PC_{16}H_{29}N_2$	$C_6H_5P[CH_2N(C_2H_5)_2]_2$	P (-51.3)	1966, 4; 1967, 244
$PC_{16}H_{29}O$	$(C_2H_5)_2P(O)C(t-C_4H_9)=C=C(CH_2)_3CHCH_3$	H	1969, 365
$PC_{16}H_{30}Br$	$CH_3P^{\oplus}(C_5H_9)_3$, $Br^{\ominus}$	P (40.0)	1966, 5; 1968, 6
$PC_{16}H_{30}GeNO$	$(C_2H_5)_2PC(O)N(C_6H_5)Ge(C_2H_5)_3$	H	1967, 342
$PC_{16}H_{30}GeNS$	$(C_2H_5)_2PC(S)N(C_6H_5)Ge(C_2H_5)_2$	H	1967, 342
$PC_{16}H_{30}I$	$CH_3P^{\oplus}(C_5H_9)_3$, $I^{\ominus}$	P (40.2)	1967, 244
$PC_{16}H_{30}NO_4$	$(C_6H_{11}O)_2P(O)C(O)N(CH_2)_4$	H	1968, 29
$PC_{16}H_{32}BrO_2$	$(C_4H_9)_3P^{\oplus}CH=CHCOOCH_3$, $Br^{\ominus}$	H	1969, 349

Formula	Compound	Resonance	Reference
$PC_{16}H_{32}ClO_2$	$(C_4H_9)_2P^{\oplus}CH=CHCOOCH_3$, $Cl^{\ominus}$	H	1969, 349
$PC_{16}H_{32}NO_2$	$(CH_3)_2NP[OCHC(CH_3)_2]_2$ (isom.)	P (144;150)	1969, 362
$PC_{16}H_{33}O$	$C_{12}H_{25}P(CH_2CH_2)_2O$	P (-52)	1969, 44
$PC_{16}H_{34}ClO_2$	$(C_4H_9)_3P^{\oplus}CH_2COOC_2H_5$, $Cl^{\ominus}$	H	1965, 87
$PC_{16}H_{35}$	$C_{12}H_{25}P(C_2H_5)_2$	H; P (-25)	1966, 14
	$HP(C_8H_{17})_2$	P (-71.5)	1966, 51
$PC_{16}H_{35}O$	$C_{12}H_{25}P(O)(C_2H_5)_2$	H; P (-56.1)	1966, 14, 40
	$C_{14}H_{29}P(O)(CH_3)_2$	P (37.2)	1966, 40
	$HP(O)(OC_8H_{17})_2$	P (7.4)	1967, 277
$PC_{16}H_{35}O_3$	$(C_3H_7O)_2P(O)(CH_2)_2OC_8H_{17}$	H	1966, 251
$PC_{16}H_{35}O_4$	$C_{12}H_{25}P(S)(C_2H_5)_2$	P (51.6)	1966, 40
$PC_{16}H_{35}S$	$C_{14}H_{29}P(S)(CH_3)_2$	P (34.1)	1966, 40
$PC_{16}H_{36}$	$(C_4H_9)_4P^{\oplus}$	P (33.9)	1967, 7
$PC_{16}H_{36}Br$	$(C_4H_9)_4P^{\oplus}Br^{\ominus}$	^{13}C; P (33.9)	1966, 5; 1968, 22; 1969, 21
$PC_{16}H_{36}Br_4N$	$(C_4H_9)_4N^{\oplus}PBr_4^{\ominus}$	P (150)	1969, 11
$PC_{16}H_{37}NOS$	$C_{12}H_{25}(CH_3)_2P=N=S(O)(CH_3)_2$	H	1967, 357
$PC_{16}H_{37}O_2$	$(C_4H_9)_3P(OC_2H_5)_2$	H; P (-38)	1968, 332
$PC_{17}H_{11}F_{10}O_5$	$(CH_3O)_3POCH(C_6F_5)OCHC_6F_5$ (isom.)	H; P (-34.1; -38.4)	1969, 75
$PC_{17}H_{13}$	$(C_6H_5)_2P(C=C)_2CH_3$	H	1966, 243, 247
$PC_{17}H_{13}F_6$	$C(C_6H_5)[CH=C(CH_3)]_2PC(CF_3)=CCF_3$	H	1968, 168
$PC_{17}H_{13}O$	$(C_6H_5)_2P(O)(C=C)_2CH_3$	H	1966, 243, 247
$PC_{17}H_{13}S$	$(C_6H_5)_2P(S)(C=C)_2CH_3$	H	1966, 243, 247
	$CH_3P(S)(C=CC_6H_5)_2$	H	1965, 89
$PC_{17}H_{15}O_5$	$(CH_3O)(CH_2O)_2P[OC(C_6H_4)]_2$	H; P	1968, 301
$PC_{17}H_{16}NO_3$	$(C_6H_5)_2C=C(CN)P(O)(OC_2H_5)OH$	H	1968, 319

NMR DATA ON ORGANIC COMPOUNDS

Formula	Structure	Nucleus; shift	References
$PC_{17}H_{17}O$	$(C_6H_5)_2P(O)CH=C=C(CH_3)_2$	H	1966, 61
$PC_{17}H_{17}O_2$	$(CH_3)_3POC_{14}H_8O$ (phenanthr.)	H; P (-3.0)	1968, 303
$PC_{17}H_{17}O_3$	$CH_3P(O)CH(CH_2COOCH_3)C_6H_4C_6H_4$	H; P (33)	1967, 358
$PC_{17}H_{17}O_5$	$(CH_3O)_3P[OC(C_6H_4)]_2$	H; P (-45)	1965, 21; 1966, 225, 242; 1968, 301, 309; 1969, 366
$PC_{17}H_{17}O_6$	$(CH_3O)_2P(O)O(C_{14}H_8)OCH_3$	H; P (-2.8)	1968, 315
$PC_{17}H_{18}Br$	$(CH_3O)(CH_2O)_2P[OC(C_6H_5)]_2$	H; P (-28.1)	1968, 301
$PC_{17}H_{19}$	$(CH_3O)_3P(O)C_{14}H_8O_2$	P (-44.7)	1965, 84
	$C_6H_5(C_6H_5CH_2)P^{\oplus}(CH_2CH)_2, Br^{\ominus}$	H	1968, 101
	$C_5H_9P(C_6H_5)_2$	P (-3.9)	1967, 8; 1968, 6
$PC_{17}H_{19}N_2O_9$	$(CH_3O)_3P[OCH(C_6H_4\text{-}p\text{-}NO_2)]_2$ (stereois.)	H; P $(-49.6; -50.2)$	1967, 163, 350; 1969, 75
$PC_{17}H_{19}O_2$	$(CH_3O)_3P(O)CH[C_6H_4\text{-}o\text{-}NO_2]_2$	H; P (-50.1)	1967, 28, 350
	$CH_3(C_6H_5)_2P[OC(CH_3)]_2$	P (-27.2)	1968, 152
$PC_{17}H_{19}O_5$	$(CH_3O)_3P[OC(C_6H_5)]_2$	P (-53)	1965, 45; 1966, 239
		H; P (-49.5)	1965, 21; 1966, 225; 1968, 124, 301
$PC_{17}H_{20}$	$(C_6H_5)_2P^{\oplus}(CH_2)_5$	P (-16.5)	1966, 269
$PC_{17}H_{20}N$	$C_6H_5P(CH_2)_2NHC(C_2H_5)C_6H_5$	H	1967, 137
$PC_{17}H_{20}NOS$	$(C_6H_5)_2P(S)CH_2N(C_2)H_2O(CH_2)_2$	P (35.3)	1966, 40
$PC_{17}H_{20}NO_2$	$(C_6H_5)_2P(O)CH_2N(CH_2)_4O$	H	1968, 56
$PC_{17}H_{20}NS$	$(C_6H_5)_2P(S)CH_2N(CH_2)_4$	P (36.1)	1966, 40
$PC_{17}H_{21}$	$(t\text{-}C_5H_{11})P(C_6H_5)_2$	P (15.4)	1967, 8; 1968, 6
	$(neo\text{-}C_5H_{11})P(C_6H_5)_2$	P (-23.9)	1967, 8; 1968, 6
$PC_{17}H_{21}O$	$(C_6H_5)_2P(O)CH_2C(CH_3)_3$	H	1965, 108

Formula	Compound	Resonance	Reference
PC$_{17}$H$_{21}$O$_2$	(C$_6$H$_5$)$_2$P(O)O-neo-C$_5$H$_{11}$	H	1967, 152
PC$_{17}$H$_{21}$O$_4$	(C$_6$H$_5$)$_2$CHOP(O)(OC$_2$H$_5$)$_2$	H	1969, 74
PC$_{17}$H$_{21}$O$_5$	(CH$_3$O)$_3$P[OCH(C$_6$H$_5$)]$_2$	P	1968, 124
PC$_{17}$H$_{21}$O$_7$	(CH$_3$O)$_3$POC(CH$_3$)(COCH$_3$)C(C$_9$H$_6$O)O	H; P (−50)	1968, 318
PC$_{17}$H$_{22}$Br	C$_2$H$_5$(i-C$_3$H$_7$)P$^\oplus$(C$_6$H$_5$)$_2$Br$^\ominus$	P (36·2)	1966, 5; 1968, 6
	CH$_3$(i-C$_4$H$_9$)P$^\oplus$(C$_6$H$_5$)$_2$Br$^\ominus$	P (22·5)	1968, 6
	CH$_3$(s-C$_4$H$_9$)P$^\oplus$(C$_6$H$_5$)$_2$Br$^\ominus$	P (30·2)	1968, 6
	CH$_3$(t-C$_4$H$_9$)P$^\oplus$(C$_6$H$_5$)$_2$Br$^\ominus$	H; P (33·0)	1968, 6
	CH$_3$(C$_2$H$_5$)P$^\oplus$(CH$_2$C$_6$H$_5$)$_2$Br$^\ominus$	H	1969, 156
	CH$_3$(i-C$_3$H$_7$)P$^\oplus$(CH$_2$C$_6$H$_5$)$_2$C$_6$H$_5$Br$^\ominus$	H	1969, 156
PC$_{17}$H$_{22}$Cl$_6$N$_2$S$_2$Sb	CH$_3$S(C$_6$H$_5$)$_2$P$^\oplus$N═C(SCH$_3$)[N(CH$_3$)$_2$], SbCl$_6^\ominus$	P (36·8)	1967, 221, 344
PC$_{17}$H$_{22}$N	(C$_6$H$_5$)$_2$PCH$_2$N(C$_2$H$_5$)$_2$	P (−27·3)	1966, 4; 1967, 244; 1968, 36
PC$_{17}$H$_{22}$NO	(C$_6$H$_5$)$_2$P(O)CH$_2$N(C$_2$H$_5$)$_2$	H; P (24·7)	1968, 36
PC$_{17}$H$_{22}$NO$_3$	(C$_2$H$_5$O)$_2$P(O)CH(NHC$_6$H$_5$)C$_6$H$_5$	P (23·5)	1969, 340
PC$_{17}$H$_{22}$NS	(C$_6$H$_5$)$_2$P(S)CH$_2$N(C$_2$H$_5$)$_2$	P (34·4)	1966, 40
PC$_{17}$H$_{24}$N$_3$S$_2$	(C$_6$H$_5$)$_2$P(S)NC(S$^\ominus$)N(CH$_3$)$_2$, (CH$_3$)$_2$NH$^\oplus$	H	1967, 353
PC$_{17}$H$_{25}$N$_2$O$_2$	[(CH$_3$)$_2$N]$_2$P(O)CH(C$_6$H$_5$)CH$_2$COC$_6$H$_5$	H	1966, 42
PC$_{17}$H$_{26}$Br	CH$_3$(C$_5$H$_9$)$_2$P$^\oplus$C$_6$H$_5$, Br$^\ominus$	P (36·4)	1966, 5; 1968, 6
PC$_{17}$H$_{27}$N$_4$OS$_2$	(C$_2$H$_5$)$_2$NCNP(S)(C$_6$H$_4$OCH$_3$)SC[N(C$_2$H$_5$)$_2$]N	P (56·5)	1968, 137
PC$_{17}$H$_{27}$O$_2$	CH$_3$(C$_6$H$_5$)P(O)O-menthyl	H	1967, 151; 1968, 140
PC$_{17}$H$_{27}$O$_5$	(C$_4$H$_9$)$_2$P(O)CH$_2$COOCH$_2$C$_6$H$_5$	H	1965, 87
PC$_{17}$H$_{28}$IN$_4$S$_2$	(C$_2$H$_5$)$_2$N$^\oplus$CNP(SCH$_3$)(SC$_6$H$_5$)SC[N(C$_2$H$_5$)$_2$]N, I$^\ominus$		1968, 137
PC$_{17}$H$_{29}$O	(C$_2$H$_5$)$_2$P(O)C(CH$_3$)═C═CC$_{10}$H$_{16}$	H	1969, 272
PC$_{17}$H$_{30}$O	(C$_2$H$_5$)$_2$P(O)C(CH$_3$)═C═C$_{10}$H$_{17}$	H	1967, 156
PC$_{17}$H$_{31}$GeO	(C$_2$H$_5$)$_2$PCH(C$_6$H$_5$)OGe(C$_2$H$_5$)$_3$	H	1968, 331

Formula	Structure	Shift	References
$PC_{17}H_{32}NO_2$	$(C_6H_{11})_2P(O)C(O)N(C_2H_5)_2$	H	1967, 26
$PC_{17}H_{32}N_3O_4$	$(C_5H_4N)OPO_3(NH_3C_6H_{11})_2$	H	1969, 338
$PC_{17}H_{33}O_2$	$CH_3(C_6H_{11})P(O)O$-menthyl	H	1967, 151; 1968, 140
$PC_{17}H_{33}O_3$	$(CH_3O)_2P(O)CCH(CH_2)_{13}$	H	1967, 100
$PC_{17}H_{35}N_2O_6$	$(C_2H_5)_2P(O)OC(C_2H_5)(COOC_2H_5)_2$	H; P (−2·6)	1967, 4
$PC_{17}H_{35}O_4$	$(CH_3O)_2P(O)C_{15}H_{28}OH$	H	1967, 100
$PC_{17}H_{35}Si_3S$	$[(CH_3)_3Si]_2CHP(S)(C_6H_5)CH_2Si(CH_3)_3$	H	1966, 132
$PC_{17}H_{36}ClO_3$	$CH_2ClP(O)[OCH_2CH(C_2H_5)C_4H_9]_2$	H	1969, 162
$PC_{17}H_{36}NO_2$	$(C_5H_{11})_2P(O)C(O)N(i\text{-}C_3H_7)_2$	H	1968, 29
$PC_{17}H_{36}NO_3$	$C_8H_{17}(C_4H_9O)P(O)C(O)(C_2H_5)_2$	H	1967, 26
$PC_{17}H_{36}NO_4$	$(C_6H_{13}O)_2P(O)C(O)NHC_4H_9$	H	1967, 26
	$(i\text{-}C_4H_9O)_2P(O)C(O)N(s\text{-}C_4H_9)_2$	H	1967, 26
	$(C_6H_{13}O)_2P(O)C(O)N(C_2H_5)_2$	H	1967, 26
$PC_{17}H_{38}O$	$(C_4H_9)_3P^{\oplus}OCH_2C(CH_3)_3$	P (−3·8)	1967, 345
$PC_{17}H_{40}NSi$	$(CH_3)_2(C_{12}H_{25})P=NSi(CH_3)_3$	H; P (6)	1966, 268
$PC_{18}F_{15}$	$(C_6F_5)_3P$	P (−77·9; −75·5); F (−131; −161; −149)	1966, 7, 19, 73; 1967, 11; 1968, 9, 263; 1969, 17, 121
$PC_{18}F_{15}Cl_2$	$(C_6F_5)_3PCl_2$	P (−104·7); F (−128·9, −158·9, −146·3)	1966, 73
$PC_{18}F_{15}O$	$(C_6F_5)_3PO$	P (−8); F (−132·5; −158·4; −142·6)	1966, 63, 73, 237; 1969, 17, 121
$PC_{18}F_{15}S$	$(C_6F_5)_3PS$	P (−8·6); F (−132·2; −158·9; −144·5)	1966, 73; 1969, 17
$PC_{18}F_{17}$	$(C_6F_5)_3PF_2$	F (−132·4; −159·4; −146·0)	1969, 17
$PC_{18}H_5F_{10}$	$C_6H_5P(C_6F_5)_2$	P (−48·7); F (−129·2; −160·8; −150·5)	1966, 10; 1967, 11; 1968, 263; 1969, 121, 276

Formula	Compound	Resonance	Reference
$PC_{18}H_5F_{10}O$	$C_6H_5P(O)(C_6F_5)_2$	$P(0·7); F$	1967, 11; 1968, 314
$PC_{18}H_5F_{12}O_2$	$C_6H_5PF_2(OC_6F_5)_2$	$P(-63·6); F(-41·0)$	1969, 62
$PC_{18}H_6F_9$	$(C_6H_2F_3)_3P$	$P(-78·5)$	1969, 121
$PC_{18}H_9F_6$	$(2,5-F_2C_6H_3)_3P$	$P(-34·5)$	1969, 121
	$(2,6-F_2C_6H_3)_3P$	$P(-78·5)$	1969, 121
$PC_{18}H_{10}BCl_3F_5$	$C_6F_5P(C_6H_5)_2BCl_3$	$F(-120·4; -158·7; -143·8)$	1969, 17
$PC_{18}H_{10}F_5$	$C_6F_5P(C_6H_5)_2$	$P(-26·3);$ $F(-127·7; -161·0; -150·6)$	1966, 10; 1967, 11; 1968, 263; 1969, 17, 121
$PC_{18}H_{10}F_5O$	$C_6F_5P(O)(C_6H_5)_2$	$P(18·7); F(-128·8; -160·1; -147·3)$	1967, 11; 1969, 17, 121
$PC_{18}H_{10}F_5S$	$C_6F_5P(S)(C_6H_5)_2$	$F(-128·1; -160·1; -148·6)$	1969, 17
$PC_{18}H_{10}F_7O$	$C_6F_5OPF_2(C_6H_5)_2$	$P(-46·6)$	1969, 62
$PC_{18}H_{11}ClF_5$	$C_6F_5P^{\oplus}(C_6H_5)_2H, Cl^{\ominus}$	$F(-125·6; -156·6; -140·4)$	1966, 10; 1969, 17
$PC_{18}H_{12}Cl_3O_{10}S_3$	$OP(OC_6H_4SO_2Cl)_3$	H	1966, 75
$PC_{18}H_{12}F_3$	$(p-FC_6H_4)_3P$	$H(P,F); P(-8·8); F$	1967, 17; 1968, 13; 1969, 18, 118, 120
	$(m-FC_6H_4)_3P$	$P(-6·5); F$	1969, 19, 120
$PC_{18}H_{12}F_3O$	$(p-FC_6H_4)_3PO$	$H; F$	1967, 17, 115; 1968, 13; 1969, 18, 19
$PC_{18}H_{12}F_3S$	$(p-FC_6H_4)_3PS$	$H(P); F$	1967, 17; 1969, 118
$PC_{18}H_{12}N$	$P(C_6H_4)_3N$	$P(-80)$	1969, 13
$PC_{18}H_{13}F_2$	$C_6H_5P(C_6H_4F)_2$	F	1969, 19
$PC_{18}H_{13}F_2S$	$C_6H_5SP(C_6H_4F)_2$	F	1969, 18, 19
$PC_{18}H_{13}O_5$	$C_6H_5OP(OC_6H_4O)_2$	$P(-29·8)$	1968, 124

NMR DATA ON ORGANIC COMPOUNDS 223

$PC_{18}H_{14}F$	$C_6H_4FP(C_6H_5)_2$	F	1967, 17; 1968, 14; 1969, 18
$PC_{18}H_{14}FO$	$C_6H_4FP(O)(C_6H_5)_2$	F	1968, 13, 14; 1969, 18, 19
$PC_{18}H_{14}FS$	$C_6H_4FP(S)(C_6H_5)_2$	F	1967, 17; 1969, 18, 19
$PC_{18}H_{14}NO_3$	$(C_6H_5)_2P(O)C_6H_4NO_2$	H	1969, 119
$PC_{18}H_{15}$	$(C_6H_5)_3P$	^{13}C; H; P (-6 to -8)	1965, 3, 7, 85, 103; 1966, 69, 78, 141, 240, 262; 1967, 198, 244, 277, 359; 1968, 22, 103, 303, 332; 1969, 50, 120, 354
$PC_{18}H_{15}AlBr_3$	$(C_6H_5)_3PAlBr_3$	^{27}Al	1967, 23
$PC_{18}H_{15}AlCl_5$	$(C_6H_5)_3P^{\oplus}Cl, AlCl_4^{\ominus}$	P (65)	1967, 191, 345
$PC_{18}H_{15}Cl_2$	$(C_6H_5)_3PCl_2$	P (63)	1967, 345; 1968, 168, 288
$PC_{18}H_{15}Cl_2O_4$	$(C_6H_5)_3P^{\oplus}Cl, ClO_4^{\ominus}$	P (65·7)	1967, 276
$PC_{18}H_{15}Cl_7Sb$	$(C_6H_5)_3P^{\oplus}Cl, SbCl_6^{\ominus}$	P (65·0)	1967, 191, 276, 345
$PC_{18}H_{15}FI$	$H(C_6H_4F)P^{\oplus}(C_6H_5)_2, I^{\ominus}$	F	1968, 14
$PC_{18}H_{15}FIO$	$C_6H_4FP^{\oplus}(OH)(C_6H_5)_2, I^{\ominus}$	F	1968, 14
$PC_{18}H_{15}F_2$	$(C_6H_5)_3PF_2$	P ($-58·1$); F ($-39·8$)	1967, 61; 1968, 303
$PC_{18}H_{15}F_2O$	$(C_6H_5)_2PF_2(OC_6H_5)$	H; P ($-34·4$); F ($-33·3$)	1968, 316; 1969, 62
$PC_{18}H_{15}F_2O_2$	$C_6H_5PF_2(OC_6H_5)_2$	H; P (-69); F (-35)	1968, 316; 1969, 62
$PC_{18}H_{15}F_2O_3$	$(C_6H_5O)_3PF_2$	F ($-73·0$)	1967, 61
$PC_{18}H_{15}F_2S$	$C_6H_5SPF_2(C_6H_5)_2$	H; P; F ($-23·5$)	1968, 316; 1969, 62
$PC_{18}H_{15}O$	$(C_6H_5)_3PO$	^{13}C; H; P ($-27·0$)	1965, 7, 78, 85; 1966, 106, 240, 270; 1967, 92, 198, 277; 1968, 103, 110, 303, 332, 337
$PC_{18}H_{15}O_2$	$C_6H_5OP(C_6H_5)_2$	P (111)	1967, 329
	$C_6H_5(C_6H_5C_6H_4)P(O)H$	H	1968, 35
	$C_6H_5P(OC_6H_5)_2$	P (157·9)	1966, 225

Formula	Compound	Resonance	Reference
$PC_{18}H_{15}O_3$	$P(OC_6H_5)_3$	^{13}C; P (127·8)	1966, 141, 225; 1967, 198, 277; 1968, 22, 301; 1969, 50
$PC_{18}H_{15}O_3S$	$SP(OC_6H_5)_3$	P (53·4)	1967, 277
$PC_{18}H_{15}O_4$	$OP(OC_6H_5)_3$	P (−17 to −18)	1965, 45; 1966, 239; 1967, 198, 277
$PC_{18}H_{15}O_5$	$(CH_3O)_2P(O)OC_{16}H_8(OH)$	H; P (0·3)	1968, 315
$PC_{18}H_{15}S$	$SP(C_6H_5)_3$	P (42·6)	1966, 106, 111; 1967, 277; 1968, 338
$PC_{18}H_{16}NO$	$(C_6H_5)_2P(O)CH_2(C_5H_4N)$	H	1966, 271
	$(C_6H_5)_2P(O)C_6H_4\text{-}o\text{-}NH_2$	H	1969, 119
$PC_{18}H_{17}ClN$	$(C_6H_5)_3P^{\oplus}NH_2, Cl^{\ominus}$	H; P (30)	1967, 65
$PC_{18}H_{17}N_2S$	$C_6H_5P(S)(NHC_6H_5)_2$	H	1967, 318
$PC_{18}H_{17}O_4S$	$HP^{\oplus}(C_6H_5)_3, HSO_4^{\ominus}$	H; P (5)	1969, 50
$PC_{18}H_{17}O_7S$	$HP^{\oplus}(OC_6H_5)_3, HSO_4^{\ominus}$	P (10)	1969, 50
$PC_{18}H_{18}BF_4N_2S_2$	$P(CSC_6H_4NCH_3)_2^{\oplus}, BF_4^{\ominus}$	H	1966, 272
$PC_{18}H_{18}NO_4$	$[(CH_3)_2N(CH_2O)_2P[OC(C_6H_4)]_2$	H; P (−21·3)	1968, 301
$PC_{18}H_{19}O_5$	$(CH_3O)_2P(O)CH(COC_6H_5)(CH_2COC_6H_5)$	P (24·8)	1966, 106
$PC_{18}H_{19}O_6$	$(CH_3O)_3POC(C_6H_5)(COC_6H_5)O$	H; P (−49·3)	1965, 21, 84; 1966, 242
$PC_{18}H_{20}Br$	$C_6H_5(C_6H_5CH_2)P^{\oplus}CH_2CH=C(CH_3)CH_2, Br^{\ominus}$	H	1968, 100
	$C_6H_5(C_6H_5CH_2)P^{\oplus}CH_2CH_2C(CH_3)=CH, Br^{\ominus}$	H	1968, 100
$PC_{18}H_{20}NO_3$	$(C_2H_5O)_2P(O)N=C=C(C_6H_5)_2$	H	1965, 109
	$(C_2H_5)_2P(O)\text{-}2\text{-phenyl-3-indolyl}$	H	1965, 102
$PC_{18}H_{20}NO_4$	$(CH_3)_2N(CH_2O)_2P[OC(C_6H_5)]_2$	P (−26·4)	1968, 301
$PC_{18}H_{21}$	$C_6H_{11}P(C_6H_5)_2$	P (−4·4)	1965, 3; 1967, 8; 1968, 6, 7

NMR DATA ON ORGANIC COMPOUNDS

$PC_{18}H_{21}N_2O_4S$	$(C_2H_5O)_2P(S)NHCOC(O)NHC_6H_5$	H	1966, 264
$PC_{18}H_{21}N_2O_9$	$(C_2H_5O)_2P(O)OCH(C_6H_4NO_2)CH(OH)C_6H_4NO_2$	H	1967, 163
$PC_{18}H_{21}N_4O_6S$	$(C_2H_5O)_2P(S)NHCOC(O)(NH)_2C_6H_4NO_2$	H	1966, 264
$PC_{18}H_{21}N_6O_{10}S_3$	$OP(OC_6H_4SO_2NHNH_2)_3$	H	1966, 71
$PC_{18}H_{21}O$	$(C_2H_5)_2POCCH(C_6H_5)_2$	H	1969, 341
$PC_{18}H_{21}O_2$	$(C_2H_5)_2P(O)CH=C(CH_3)CH(CH_3)OCH_3$	H	1967, 154
	$(C_2H_5)_2POC(O)CH(C_6H_5)_2$	H	1969, 341
$PC_{18}H_{21}O_3$	$C_2H_5O(C_6H_5)_2P[OC(CH_3)]_2$	P ($-27\cdot7$)	1966, 225
$PC_{18}H_{21}O_4$	$(C_2H_5O)_2P(O)CH-C(C_6H_5)_2O$	H	1966, 245; 1967, 81
	$(C_2H_5O)_2P(O)C(C_6H_5)_2CHO$	H	1966, 245; 1967, 81
	$(C_2H_5O)_2P(O)OC(C_6H_5)=CHC_6H_5$	H	1967, 121
$PC_{18}H_{21}S$	$C_6H_{11}P(S)(C_6H_5)_2$	P ($48\cdot6$)	1966, 40
$PC_{18}H_{22}$	$C_6H_5P^{\oplus}(CH_2C_6H_5)[CH=C(CH_3)](CH_2)_2$	H	1967, 108
	$C_6H_5P^{\oplus}(CH_2C_6H_5)CH_2C(CH_3)CH=CH_2$	H	1967, 108
$PC_{18}H_{22}Br$	$CH_3(C_5H_9)P^{\oplus}(C_6H_5)_2, Br^{\ominus}$	P ($29\cdot8$)	1966, 5; 1968, 6
	$H(C_6H_{11})P^{\oplus}(C_6H_5)_2, Br^{\ominus}$	H; P ($10\cdot5$)	1968, 7
	$C_6H_5(C_6H_5CH_2)P^{\oplus}(CH_2)_3CHCH_3, Br^{\ominus}$	H	1969, 356
$PC_{18}H_{22}F_2N$	$(C_3H_7)_2NP(C_6H_4F)_2$	F	1969, 18, 19
$PC_{18}H_{22}NS$	$(C_6H_5)_2P(S)CH_2N(CH_2)_5$	P ($34\cdot6$)	1966, 40
$PC_{18}H_{23}F_6$	$CH_3C[CH=C(t-C_4H_9)]_2PC(CF_3)=CCF_3$	H	1968, 339
$PC_{18}H_{23}S$	$C_6H_{13}P(S)(C_6H_5)_2$	P ($41\cdot5$)	1966, 40, 44
$PC_{18}H_{24}Br$	$CH_3(t-C_5H_{11})P^{\oplus}(C_6H_5)_2, Br^{\ominus}$	P ($32\cdot9$)	1968, 6
	$CH_3(neo-C_5H_{11})P^{\oplus}(C_6H_5)_2, Br^{\ominus}$	P ($20\cdot0$)	1968, 6
	$C_2H_5(i-C_4H_9)P^{\oplus}(C_6H_5)_2, Br^{\ominus}$	P ($28\cdot0$)	1968, 6
	$C_2H_5(s-C_4H_9)_2P^{\oplus}(C_6H_5)_2, Br^{\ominus}$	P ($35\cdot8$)	1968, 6
	$C_2H_5(t-C_4H_9)P^{\oplus}(C_6H_5)_2, Br^{\ominus}$	P ($39\cdot9$)	1968, 6

226 FORMULAE INDEX II

Formula	Compound	Resonance	Reference
$PC_{18}H_{27}$	$(C_3H_7)_2P^{\oplus}(C_6H_5)_2, Br^{\ominus}$	P (26·2)	1968, 6
$PC_{18}H_{27}F_{12}O_2$	$CH_3(C_4H_9)P^{\oplus}(CH_2C_6H_5)C_6H_5, Br^{\ominus}$	H	1969, 156
	$(C_5H_{11})_2PC_6H_5$	P (2·5)	1965, 3; 1967, 8; 1968, 7
	$(C_4H_9)_3P[OC(CF_3)_2]_2$	P (7·3); F (−70)	1968, 303
$PC_{18}H_{28}Br$	$C_2H_5(C_5H_9)_2P^{\oplus}C_6H_5, Br^{\ominus}$	P (37·3)	1966, 5; 1968, 6
	$H(C_6H_{11})_2P^{\oplus}C_6H_5, Br^{\ominus}$	H; P (20·8)	1968, 7
$PC_{18}H_{29}O_4$	$(C_4H_9O)_2(C_6H_5)P[OC(CH_3)]_2$	P (−36·1)	1966, 225
$PC_{18}H_{30}IN_4OS_2$	$(C_2H_5)_2-$		1968, 137
	$N^{\oplus}CNP(SCH_3)(C_6H_4OCH_3)SC[N(C_2H_5)_2]N, I^{\ominus}$		
$PC_{18}H_{31}Ge$	$(C_2H_5)_2PC(C_6H_5)=CHGe(C_2H_5)_3$	H	1967, 342, 360
	$(C_2H_5)_2PCH=C(C_6H_5)Ge(C_2H_5)_3$	H	1967, 342, 360
$PC_{18}H_{31}O_2$	$(t-C_4H_9)CCH_2CHCH(t-C_4H_9)PO_2H$	H	1968, 102
$PC_{18}H_{32}Br$	$C_6H_5P^{\oplus}(C_4H_9)_3, Br^{\ominus}$	P (30·5)	1968, 6
$PC_{18}H_{32}NO_3$	$(i-C_4H_9O)_2P(O)N(i-C_3H_7)CH_2C_6H_5$	H	1968, 29
$PC_{18}H_{32}O_6$	$(C_2H_5O)_2P(O)O$ — [structure]	H	1969, 367
$PC_{18}H_{33}$	$(C_6H_{11})_3P$	P (7)	1965, 3; 1967, 8, 244; 1968, 6, 7
$PC_{18}H_{34}Br$	$HP^{\oplus}(C_6H_{11})_3, Br^{\ominus}$	H; P (27·9)	1968, 7
$PC_{18}H_{35}N_4$	$C_6H_5N=P[N(C_2H_5)_2]_3$	H	1965, 98

$PC_{18}H_{35}O_4S$	$HP^{\oplus}(C_6H_{11})_3, HSO_4^{\ominus}$	H; P (2)	1969, 50
$PC_{18}H_{36}N_3O$	$[C_5H_{10}NCH_2]_3PO$	H; P (40–50)	1967, 291; 1968, 36
$PC_{18}H_{38}NO_3$	$(t\text{-}C_5H_{11}O)_2P(O)C[N(CH_3)_2]C_5H_{10}$	P (23·6)	1969, 340
$PC_{18}H_{38}O$	$(C_4H_9)_3P^{\oplus}CH_2C(O)C(CH_3)_3$	H	1966, 119
$PC_{18}H_{39}O_4$	$C_8H_{17}O(CH_2)_2P(O)(OC_4H_9)_2$	H	1966, 251
$PC_{18}H_{39}O_7$	$(C_4H_9OCH_2CH_2O)_3PO$	P	1967, 198
$PC_{18}H_{40}Cl$	$C_{12}H_{25}P^{\oplus}(C_2H_5)_3Cl^{\ominus}$	H; P (37)	1966, 14
$PC_{18}H_{45}O_3Si_3$	$P[OSi(C_2H_5)_3]_3$	H	1967, 34
$PC_{19}H_8F_{10}N$	$(C_6F_5)_2PNHCH_2C_6H_5$	H	1966, 19
$PC_{19}H_{13}$	$(C_6H_5)_2P(C\equiv C)_3CH_3$	H	1966, 243, 247
	$P(C_6H_4)_2CC_6H_5$	H	1968, 340
$PC_{19}H_{14}Br_2N$	$C_6H_5P(C_6H_3Br)_2NCH_3$	H	1967, 361
$PC_{19}H_{14}F_3$	$(C_6H_4F)_3P=CH_2$	F	1968, 13
	$(C_6H_5)_2PC_6H_4CF_3$	P (−10·9)	1969, 33
$PC_{19}H_{14}NO$	$C_6H_5(C_6H_5OC_6H_4)PCN$	P (−35·5)	1966, 262
$PC_{19}H_{14}NOS$	$C_6H_5(C_6H_5OC_6H_4)P(S)CN$	P (23·0)	1966, 262
$PC_{19}H_{15}ClIO$	$CH_3(C_6H_5)P^{\oplus}C_6H_4OC_6H_3Cl, I^{\ominus}$	H	1968, 341
$PC_{19}H_{15}F_3I$	$CH_3P^{\oplus}(C_6H_4F)_3N, I^{\ominus}$	H; P (−4·8)	1969, 13
$PC_{19}H_{15}O$	$(C_6H_5)_2PC_6H_4(HCO)$	H	1965, 110
	$C_6H_5P(O)C_6H_4C_6H_4CH_2$	H	1967, 362
$PC_{19}H_{15}O_2$	$(C_6H_5)_2PC_6H_4COOH$	H	1968, 342
$PC_{19}H_{16}BrO$	$(C_6H_5)_2P(O)C_6H_4CH_2Br$	H	1966, 62
$PC_{19}H_{16}F$	$C_6H_4F(C_6H_5)_2P=CH_2$	F	1968, 13, 14
$PC_{19}H_{16}IO$	$CH_3(C_6H_5)P^{\oplus}(C_6H_4)_2O, I^{\ominus}$	H	1968, 341
$PC_{19}H_{16}N$	$C_6H_5P(C_6H_4)_2NCH_3$	H	1967, 361
$PC_{19}H_{16}NaO_2$	$(C_6H_5)_2CHP(O)(C_6H_5)ONa$	H	1966, 123
$PC_{19}H_{17}$	$(C_6H_5)_3P=CH_2$	H; P (20·3)	1967, 192, 363; 1968, 108, 169, 337

228 FORMULAE INDEX II

Formula	Compound	Resonance	Reference
	$C_6H_5CH_2P(C_6H_5)_2$	$P\ (-10.4)$	1967, 8; 1968, 6
	$CH_3C_6H_4P(C_6H_5)_2$	H	1968, 15, 16
$PC_{19}H_{17}Br_2$	$CH_2BrP^{\oplus}(C_6H_5)_3, Br^{\ominus}$	H; P (22·6)	1966, 5; 1968, 56, 60
$PC_{19}H_{17}Cl_2$	$CH_2ClP^{\oplus}(C_6H_5)_3, Cl^{\ominus}$	H; P (23·8)	1966, 5; 1968, 56, 60
$PC_{19}H_{17}FI$	$CH_3(C_6H_4F)P^{\oplus}(C_6H_5)_2, I^{\ominus}$	F	1968, 13, 14
$PC_{19}H_{17}I_2$	$CH_2IP^{\oplus}(C_6H_5)_3, I^{\ominus}$	H	1968, 56, 60
$PC_{19}H_{17}O$	$(C_6H_5)_2P(O)C_6H_4CH_3$	H	1966, 70, 253; 1967, 117; 1968, 15, 16
	$CH_3(C_6H_5)P(O)C_6H_4OC_6H_5$	H	1967, 348
	$(C_6H_5)_2P(O)OCH_2C_6H_5$	H	1968, 56
$PC_{19}H_{17}O_2$	$C_6H_5CH_2C_6H_4P(O)(C_6H_5)OH$	H	1966, 123
	$(C_6H_5)_2P(O)OCH_2C_6H_5$	H	1966, 253
	$(C_6H_5)_2P(O)CH(OH)C_6H_5$	H	1968, 56
	$(C_6H_5)_2CHP(O)(C_6H_5)OH$	H	1968, 56
$PC_{19}H_{17}O_5$	$(CH_3O)_3POC_{16}H_8O$	H; P (−44·5)	1968, 315
$PC_{19}H_{17}S$	$CH_3C_6H_4P(S)(C_6H_5)_2$	H	1968, 15, 16
	$CH_3SC_6H_4P(C_6H_5)_2$	H	1968, 232
$PC_{19}H_{18}$	$CH_3P^{\oplus}(C_6H_5)_3$	P (22·6)	1967, 7
$PC_{19}H_{18}Br$	$CH_3P^{\oplus}(C_6H_5)_3, Br^{\ominus}$	P (22·7)	1966, 5; 1967, 363; 1968, 170
$PC_{19}H_{18}ClO$	$(C_6H_5)_3P^{\oplus}CH_2OH, Cl^{\ominus}$	H; P (17·7)	1966, 5; 1968, 56, 60
$PC_{19}H_{18}Cl_6OSb$	$CH_3OP^{\oplus}(C_6H_5)_3, SbCl_6^{\ominus}$	H; P (65·0)	1967, 92; 1969, 87
$PC_{19}H_{18}Cl_6O_4Sb$	$CH_3OP^{\oplus}(OC_6H_5)_3, SbCl_6^{\ominus}$	H	1967, 69
$PC_{19}H_{18}Cl_6SSb$	$CH_3SP^{\oplus}(C_6H_5)_3, SbCl_6^{\ominus}$	H; P (46·6)	1969, 87
$PC_{19}H_{18}Cl_6SbSe$	$CH_3SeP^{\oplus}(C_6H_5)_3, SbCl_6^{\ominus}$	H; P (35·8)	1969, 87
$PC_{19}H_{18}I$	$CH_3P^{\oplus}(C_6H_5)_3, I^{\ominus}$	^{13}C; H; P (20)	1967, 244; 1968, 22, 169; 1969, 13
$PC_{19}H_{18}IO_2$	$CH_3(C_6H_5)P^{\oplus}(OC_6H_5)_2, I^{\ominus}$	P (73·0)	1967, 244

PC$_{19}$H$_{18}$IO$_3$	CH$_3$P$^{\oplus}$(OC$_6$H$_5$)$_3$, I$^{\ominus}$	P (39·2)	1967, 244
PC$_{19}$H$_{19}$ClNO$_5$	(CH$_3$O)$_2$P(O)OC=CClC(C$_6$H$_5$)$_2$C(O)NCH$_3$	H	1969, 345
PC$_{19}$H$_{19}$O$_6$	CH$_3$OP(O)OC(C$_6$H$_5$)(COOC$_6$H$_5$)C(CH$_3$)(COCH$_3$)O	H	1966, 258
PC$_{19}$H$_{20}$Br	C$_6$H$_5$(C$_6$H$_5$CH$_2$)P$^{\oplus}$[C(CH$_3$)=CH]$_2$, Br$^{\ominus}$	H	1968, 101
PC$_{19}$H$_{20}$NO$_4$	[(CH$_3$)$_2$N][O(CH$_2$)$_3$O]P[OC(C$_6$H$_4$)]$_2$	P (−40·7)	1968, 301
PC$_{19}$H$_{20}$N$_2$O$_3$	[CH$_2$N(CH$_3$)]$_2$POC(COC$_6$H$_5$)=C(C$_6$H$_5$)O	P (−31·9)	1966, 242
PC$_{19}$H$_{21}$NO	C$_6$H$_5$[(CH$_3$)$_2$N]P(O)C(C$_6$H$_5$)=C=C((CH$_3$)$_2$	H	1969, 165
PC$_{19}$H$_{21}$N$_2$O$_3$	CH$_3$O[CH$_2$N(CH$_3$)]$_2$P[OC(C$_6$H$_4$)]$_2$	P (−33·5)	1968, 301
PC$_{19}$H$_{21}$O$_2$	(C$_6$H$_5$)$_2$P(O)CH=C=C(CH$_3$)CH(OCH$_3$)CH$_3$	H	1967, 155
PC$_{19}$H$_{21}$O$_5$	(CH$_3$O)$_3$P=C(COC$_6$H$_5$)(CH$_2$COC$_6$H$_5$)	H; P (56·2)	1965, 1; 1966, 106
PC$_{19}$H$_{21}$O$_7$	(CH$_3$O)$_3$P[OCH(COC$_6$H$_5$)]$_2$	H; P (−46·9)	1965, 21
PC$_{19}$H$_{22}$	CH$_3$(C$_6$H$_9$)P$^{\oplus}$(C$_6$H$_5$)$_2$	H	1966, 273
PC$_{19}$H$_{23}$N$_2$O$_3$	CH$_3$O[CH$_2$N(CH$_3$)]$_2$P[OC(C$_6$H$_5$)]$_2$	H; P (−38·6)	1968, 153, 301
PC$_{19}$H$_{23}$O$_2$	(C$_6$H$_5$)$_2$P(O)C(CH$_3$)=C(CH$_3$)CH(CH$_3$)OCH$_3$	H	1967, 154
PC$_{19}$H$_{23}$O$_2$S	(C$_6$H$_5$)$_2$P(S)CH(COOH)C$_5$H$_{11}$	H; P (46)	1966, 44
	(C$_6$H$_5$)$_2$P(S)CH(COOH)CH$_2$-t-C$_4$H$_9$	H; P (48·3)	1966, 44
PC$_{19}$H$_{23}$O$_5$	CH$_3$O(C$_6$H$_5$CH$_2$O)$_2$P[OC(CH$_3$)]$_2$	H; P (−50·6)	1968, 73
PC$_{19}$H$_{24}$Br	C$_2$H$_5$(C$_5$H$_9$)P$^{\oplus}$(C$_6$H$_5$)$_2$, Br$^{\ominus}$	P (34·8)	1966, 5; 1968, 6
	CH$_3$(C$_6$H$_{11}$)P$^{\oplus}$(C$_6$H$_5$)$_2$, Br$^{\ominus}$	P (26·7)	1968, 6
PC$_{19}$H$_{26}$Br	C$_2$H$_5$(neo-C$_6$H$_{11}$)P$^{\oplus}$(C$_6$H$_5$)$_2$, Br$^{\ominus}$	P (25·8)	1968, 6
	C$_2$H$_5$(C$_3$H$_7$)P$^{\oplus}$(CH$_2$C$_6$H$_5$)$_2$, Br$^{\ominus}$	H	1969, 156
	C$_2$H$_5$(C$_4$H$_9$)P$^{\oplus}$(CH$_2$C$_6$H$_5$)C$_6$H$_5$, Br$^{\ominus}$	H	1969, 156
PC$_{19}$H$_{26}$NO$_3$	(CH$_3$)$_3$C$_6$H$_2$C$_6$H$_4$NHP(O)(OC$_2$H$_5$)$_2$	H	1966, 274
PC$_{19}$H$_{27}$N$_4$OS$_2$	(C$_2$H$_5$)$_2$NCNP(S)(C$_6$H$_4$OCH$_3$)SC[N(CH$_2$)$_5$]N	P (57·5)	1968, 137
PC$_{19}$H$_{27}$O$_5$	(CH$_2$CH$_2$CHO)$_4$PCH(OH)C$_4$H$_5$	H; P (−20)	1968, 154
PC$_{19}$H$_{30}$Br	C$_3$H$_7$(C$_5$H$_9$)$_2$P$^{\oplus}$C$_6$H$_5$, Br$^{\ominus}$	P (35·6)	1966, 5; 1968, 6

Formula	Compound	Resonance	Reference
$PC_{19}H_{31}O_2$	$CH_3(C_6H_{11})_2P^{\oplus}C_6H_5$, $Br^{\ominus}$	P (34·0)	1968, 6
$PC_{19}H_{33}GeO$	$C_6H_5(C_3H_7)P(O)O$-menthyl	H	1967, 151; 1968, 140
$PC_{19}H_{33}O_5$	$(C_2H_5)_2PCH(C_6H_5)CH=CHOGe(C_6H_5)_3$	H	1968, 331
	$(i\text{-}C_3H_7O)_3POC_6(CH_3)_4O$	P (−49·7)	1968, 303
$PC_{19}H_{36}I$	$CH_3P^{\oplus}(C_6H_{11})_3$, $I^{\ominus}$	P (33·9)	1967, 244
$PC_{19}H_{37}O_3$	$(CH_3O)_2P(O)CCH(CH_2)_{15}$	H	1967, 100
$PC_{19}H_{39}O$	$(C_4H_9)_2P(O)(CH_2)_9CH=CH_2$	P (51·0)	1968, 336
	$(C_4H_9)_2P(O)(CH_2)_9CH=CH_2$	P (131·2)	1968, 336
$PC_{19}H_{39}O_4$	$(CH_3O)_2P(O)C_{17}H_{32}(OH)$	H	1967, 100
$PC_{19}H_{40}NO_4$	$(C_6H_{13}O)_2P(O)C(O)N(i\text{-}C_3H_7)_2$	H	1968, 29
$PC_{20}H_{13}F_6$	$C_6H_5P(C_6H_4CF_3)_2$	P (−14·5)	1969, 33
$PC_{20}H_{15}ClNO$	$(C_6H_5)_2P(O)CH(CN)C_6H_4Cl$	H	1968, 88
$PC_{20}H_{15}F_{12}O_2$	$C_2H_5(C_6H_5)_2P[OC(CF_3)_2]_2$	P (−6·1); F (−69)	1968, 302, 303
	$(CF_3)_2CHO(C_6H_5)_2POC(CF_3)_2CHCH_3$	P (−32·1); F (−77·9; −72·5; −74·9; −75·7)	1968, 302
$PC_{20}H_{15}F_{12}O_3$	$C_2H_5O(C_6H_5)_2P[O(CF_3)_2]_2$	P (−21·6)	1966, 225
$PC_{20}H_{15}O$	$(C_6H_5)_3P=C=C=O$	P (2·6)	1966, 270; 1968, 110
$PC_{20}H_{15}S$	$(C_6H_5)_3P=C=C=S$	P (−7·7)	1966, 270; 1968, 110
$PC_{20}H_{16}ClN_4O$	$(C_6H_5)_2P(O)NNHNNCCHC_6H_4Cl$	H	1968, 88
$PC_{20}H_{16}N$	$(C_6H_5)_3P=CH(CN)$	H	1967, 192
$PC_{20}H_{17}$	$(C_6H_5)_2PCH=CHC_6H_5$	H; P (13)	1965, 111; 1966, 275; 1967, 364
	$(C_6H_5)_2PC_6H_4CH=CH_2$	$H\{P\}$	1968, 104
$PC_{20}H_{17}F_{10}O_5$	$(C_2H_5O)_3P[OCH(C_6F_5)]_2$	P (−50·5; −54·0)	1968, 124, 303

$PC_{20}H_{17}N_4O$	$(C_2H_5O)_3POCH(C_6F_5)OCHC_6F_5$	$P(-37.7; -41.6)$	1968, 124, 303
$PC_{20}H_{17}O$	$(C_6F_5)_2P(O)NNHNNCCHC_6H_5$	H	1968, 88
	$(C_6H_5)_2P(O)CH=CHC_6H_5$	H	1965, 112; 1967, 99
	$(C_6H_5)_3P=CHC(O)H$	H	1967, 192
$PC_{20}H_{18}Br$	$(C_6H_5)_3P^{\oplus}CH=CH_2, Br^{\ominus}$	H	1966, 135
$PC_{20}H_{18}BrO$	$(C_6H_5)_2P(O)CH_2CHBrC_6H_5$	H	1965, 112
$PC_{20}H_{18}ClO_2$	$(C_6H_5)_3P^{\oplus}CH_2COOH, Cl^{\ominus}$	H	1965, 87
	$CH_3O(C_6H_5)P(O)(C_6H_5)_2Cl$	H	1966, 82
$PC_{20}H_{18}IO$	$C_2H_5(C_6H_5)P^{\oplus}(C_6H_4)_2O, I^{\ominus}$	H	1968, 341
$PC_{20}H_{18}NO_4S$	$(C_6H_5)_2P(O)C(O)NHSO_2C_6H_4CH_3$	H	1968, 328; 1969, 161
$PC_{20}H_{18}NO_6S$	$(C_6H_5O)_2P(O)C(O)NHSO_2C_6H_4CH_3$	H	1968, 328; 1969, 161
$PC_{20}H_{18}S$	$(C_6H_5)_2P(S)CH=CHC_6H_5$	H	1965, 112
$PC_{20}H_{19}$	$C_6H_5P(CH_2C_6H_5)_2$	$P(-12.1)$	1967, 8; 1968, 6
	$(C_6H_5)_3P=C_2H_4$	$H; P(14.6)$	1967, 363; 1968, 169, 337
	$C_6H_5P(C_6H_4CH_3)_2$	H	1968, 15, 16
$PC_{20}H_{19}BrNO$	$(C_6H_5)_3P^{\oplus}C(CH_3)=NOH, Br^{\ominus}$	H	1965, 113
$PC_{20}H_{19}Br_2$	$(C_6H_5)_3P^{\oplus}(CH_2)_2Br, Br^{\ominus}$	H	1966, 135
$PC_{20}H_{19}O$	$(C_6H_5)_2P(O)(CH_2)_2C_6H_4$	H	1965, 112
	$C_6H_5P(O)(C_6H_4CH_3)_2$	H	1968, 15, 16
$PC_{20}H_{19}O_5$	$(C_6H_5)_2P(O)CC(O)OC(CH_3)(COCH_3)COCH_3$	$H; P(81.7)$	1969, 353
$PC_{20}H_{19}S$	$C_6H_5P(S)(C_6H_4CH_3)_2$	H	1968, 15, 16
$PC_{20}H_{19}S_2$	$C_6H_5P(C_6H_4SCH_3)_2$	H	1968, 232
$PC_{20}H_{20}$	$C_2H_5P^{\oplus}(C_6H_5)_2$	$P(26.2)$	1967, 7
$PC_{20}H_{20}BF_4O$	$C_2H_5OP^{\oplus}(C_6H_5)_3, BF_4^{\ominus}$	$H; P(62)$	1968, 332
$PC_{20}H_{20}Br$	$C_2H_5P^{\oplus}(C_6H_5)_3, Br^{\ominus}$	$H; P(26.2)$	1966, 5; 1967, 363; 1968, 56
	$CH_3(C_6H_5CH_2)P^{\oplus}(C_6H_5)_2, Br^{\ominus}$	$P(22.8)$	1968, 6
$PC_{20}H_{20}BrO$	$(C_6H_5)_3P^{\oplus}CH_2CH_2OH, Br^{\ominus}$	H	1966, 135
$PC_{20}H_{20}ClO$	$(C_6H_5)_3P^{\oplus}CH_2OCH_3, Cl^{\ominus}$	H	1968, 60

Formula	Compound	Resonance	Reference
$PC_{20}H_{20}I$	$CH_3(C_6H_5)_2P^{\oplus}(C_6H_4CH_3)$, $I^{\ominus}$	H	1968, 15, 16
$PC_{20}H_{20}O$	$(C_6H_5)_2P(O)(CH_3)=C=C(CH_2)_2CCH_3$	H	1967, 156
$PC_{20}H_{21}O$	$(C_6H_5)_2P(O)CH=C=C_5H_7CH_3$	H	1969, 165, 365
$PC_{20}H_{21}O_2$	$(CH_3O)_2P[C(C_6H_5)CH_2CCH_3$	P (178·2)	1968, 75
$PC_{20}H_{21}O_5$	$(C_2H_5O)_2P(O)CH=C(C_6H_4)_2$	H	1967, 327
$PC_{20}H_{21}O_6$	$(C_6H_5)_2P(O)CC(O)O[C(CH_3)OH]_2COCH_3$	H	1969, 353
	$(C_6H_5)_2P(O)C(COOH)=C(OCH_3)C(CH_3)(OH)-COCH_3$	H	1969, 353
$PC_{20}H_{22}BF_4N_2S_2$	$P(CSC_6H_4NC_2H_5)_2^{\oplus}$, $BF_4^{\ominus}$	P (24·9)	1966, 272
$PC_{20}H_{23}F_{12}O_4$	$C_6H_5(C_4H_9O)_2P[OC(CF_3)_2]_2$	P (−33·4)	1966, 225
$PC_{20}H_{23}O_2$	$(C_6H_5)_2P(O)C(CH_3)=C=C(CH_3)CH(OCH_3)CH_3$	H	1967, 155
	$C_6H_5(C_2H_5)_2POC_{10}H_8O$	P (0·9)	1968, 124
$PC_{20}H_{23}O_4$	$C_6H_5(CH_3O)_2PCH(C_6H_5)C(COCH_3)=C(CH_3)O$ (stereois.)	H; P (−16·7; −13·3)	1968, 124, 152, 303
$PC_{20}H_{23}O_5$	$(C_2H_5O)_3POC_{14}H_8O$	P (−47·1)	1966, 225
$PC_{20}H_{23}O_7$	$(C_2H_5O)_2P(O)CH=C(C_6H_5)_2$	H	1967, 327
	$(CH_3O)_3POC(C_6H_5)(COCH_3)CH(COC_6H_5)O$	H; P (−49·4)	1966, 258
$PC_{20}H_{24}Br$	$CH_3(C_6H_5)P^{\oplus}(CH_2C_6H_5)C_6H_9$, $Br^{\ominus}$	H	1969, 156
$PC_{20}H_{24}N_3O_2$	$(CH_3)_2NP[N(CH_3)CH_2]_2OC_{14}H_8O$	P (−29·8)	1968, 124, 301
$PC_{20}H_{25}$	$(C_6H_5)_2PCH=CH(CH_2)_5CH_3$	H	1966, 275
$PC_{20}H_{25}N_2O_4$	$C_6H_5OP(O)(O^{\ominus})O(CH_2)_2NHC_6H_5$, $C_6H_5NH_3^{\oplus}$	H; P (−6·9)	1969, 359

$PC_{20}H_{25}O_5$	$(C_2H_5O)_3P[OC(C_6H_5)]_2$	$P\ (-52\cdot2)$	1966, 225
$PC_{20}H_{26}Br$	$C_3H_7(C_5H_9)P^{\oplus}(C_6H_5)_2,\ Br^{\ominus}$	$P\ (32\cdot5)$	1966, 5; 1968, 6
	$C_2H_5(C_6H_{11})P^{\oplus}(C_6H_5)_2,\ Br^{\ominus}$	$P\ (32\cdot8)$	1968, 6
$PC_{20}H_{26}N_3O_2$	$[(CH_3)_2N]_3POC_{14}H_8O$	$H;\ P\ (-38\cdot5)$	1965, 84; 1968, 124, 301
$PC_{20}H_{27}N_2O_2$	$(CH_3)_2N[(CH_3)NCH_2]_2P[OC(C_6H_5)]_2$	$P\ (-36\cdot9)$	1968, 124, 153, 301
	$HP[OCH(C_6H_5)CH(CH_3)NCH_3]_2$	$H;\ P\ (-71)$	1969, 45
$PC_{20}H_{27}O_4$	$CH_3(CH_3O)P(O)O(C_{18}H_{21}O)$	H	1967, 168
$PC_{20}H_{28}Br$	$(C_4H_9)_2P^{\oplus}(C_6H_5)_2,\ Br^{\ominus}$	$P\ (26\cdot6)$	1968, 6
$PC_{20}H_{28}N_3O_2$	$[(CH_3)_2N]_3P^{\oplus}OC(C_6H_5)=C(C_6H_5)_2]$	$H;\ P\ (-30\cdot2)$	1966, 224; 1968, 73, 124, 153
	$[(CH_3)_2N]_3P^{\oplus}OC(COC_6H_5)C(C_6H_5)O^{\ominus}$	$P\ (-13\cdot1)$	1968, 153, 301
$PC_{20}H_{28}N_3O_3$	$[(CH_3)_2N]_3P^{\oplus}OC(COC_6H_5)C(C_6H_5)O^{\ominus}$	$P\ (35\cdot9)$	1968, 153
$PC_{20}H_{29}O_4$	$CH_3(CH_3O)P(O)O(C_{18}H_{23}O)$	H	1967, 168
$PC_{20}H_{30}IN_4OS_2$	$(CH_2)_5N^{\oplus}CNP(SCH_3)(C_6H_4OCH_3)SC[N(CH_2)_5]N,\ I^{\ominus}$	$P\ (51\cdot7)$	1968, 137
$PC_{20}H_{32}Br$	$C_2H_5(C_6H_{11})_2P^{\oplus}C_6H_5,\ Br^{\ominus}$	$P\ (34\cdot6)$	1968, 6
$PC_{20}H_{32}O$	$(C_6H_{11})_2P(O)C(CH_3)=C=C(CH_2)_2CCH_3$	H	1967, 156
$PC_{20}H_{33}BrNO$	$(C_4H_9)_3PCH_2C(C_6H_4Br)=NO$	H	1968, 343
$PC_{20}H_{34}Br_2NO$	$(C_4H_9)_3P^{\oplus}CH_2C(=NOH)C_6H_4Br,\ Br^{\ominus}$	H	1968, 343
$PC_{20}H_{34}NO$	$(C_4H_9)_3PCH_2C(C_6H_5)=NO$	H	1968, 343
$PC_{20}H_{35}BrNO$	$(C_4H_9)_3P^{\oplus}CH_2C(=NOH)C_6H_5,\ Br^{\ominus}$	H	1968, 343
$PC_{20}H_{36}NO_3$	$(C_5H_{11})_2P(O)N(i\text{-}C_3H_7)CH_2C_6H_5$	H	1968, 29
$PC_{20}H_{36}O$	$(C_4H_9)_3P^{\oplus}CH_2C_6H_4CH_3$	H	1967, 117
$PC_{20}H_{43}$	$C_{12}H_{25}P(C_4H_9)_2$	$H;\ P\ (-32\cdot4)$	1966, 14
	$C_4H_9P(C_8H_{17})_2$	$P\ (-33\cdot3)$	1967, 244
$PC_{20}H_{43}S$	$C_{12}H_{25}P(S)(i\text{-}C_4H_9)_2$	$P\ (44\cdot2)$	1966, 40
$PC_{20}H_{45}GeO$	$(C_4H_9)_2POGe(C_4H_9)_3$	$P\ (116\cdot8)$	1967, 343
$PC_{20}H_{45}OSn$	$(C_4H_9)_2POSn(C_4H_9)_3$	$P\ (65\cdot6)$	1967, 343

Formula	Compound	Resonance	Reference
$PC_{21}H_{12}F_9$	$P(C_6H_4CF_3)_3$	P (o: -18.5; m: -5.0; p: -6.0)	1969, 33
$PC_{21}H_{17}BrN$	$(C_6H_5)_3P^{\oplus}CH=CHCN, Br^{\ominus}$	H	1969, 349
$PC_{21}H_{17}O$	$(C_6H_5)_2P(O)CH=C=CHC_6H_5$	H	1966, 61
$PC_{21}H_{18}BrO_2$	$(C_6H_5)_3P^{\oplus}CH=CHCOOH, Br^{\ominus}$	H	1969, 349
$PC_{21}H_{19}$	$(C_6H_5)_2PC_6H_4CH_2CH=CH_2$	H	1966, 72; 1968, 194
$PC_{21}H_{19}BrClO$	$(C_6H_5)_3P^{\oplus}CH(CH_3)C(O)Cl, Br^{\ominus}$	H	1966, 115
$PC_{21}H_{19}O$	$(C_6H_5)_3P=CHCOCH_3$	H	1967, 192, 365
$PC_{21}H_{19}O_2$	$(C_6H_5)_3P=CHCOOCH_3$	H; P (17.6)	1966, 270; 1967, 192; 1968, 170
$PC_{21}H_{19}O_3$	$(C_6H_5)_2P(O)C_6H_4CH_2COOCH_3$	H	1966, 253
$PC_{21}H_{19}O_4$	$C_6H_5OP(O)OCH_2C(C_6H_5)_2CH_2O$	H; P (-15.4)	1969, 175
$PC_{21}H_{20}$	$CH_2=CHCH_2P^{\oplus}(C_6H_5)_3$	P (21.4)	1967, 7
$PC_{21}H_{20}Br$	$CH_2=CHCH_2P^{\oplus}(C_6H_5)_3Br^{\ominus}$	P (21.1)	1966, 5
	$CH_3CH=CHP^{\oplus}(C_6H_5)_3, Br^{\ominus}$	H	1966, 135
	$CH_2=C(CH_3)P^{\oplus}(C_6H_5)_3, Br^{\ominus}$	H	1966, 135
	$CH_2CH_2CHP^{\oplus}(C_6H_5)_3, Br^{\ominus}$	H	1968, 333
$PC_{21}H_{20}BrO$	$CH_3C(O)CH_2P^{\oplus}(C_6H_5)_3, Br^{\ominus}$	P (19.4)	1966, 5; 1969, 19
$PC_{21}H_{20}BrO_2$	$CH_3OC(O)CH_2P^{\oplus}(C_6H_5)_3, Br^{\ominus}$	P (20.3)	1966, 5
$PC_{21}H_{20}Cl$	$CH_2=CHCH_2P^{\oplus}(C_6H_5)_3, Cl^{\ominus}$	H	1968, 56
$PC_{21}H_{20}ClO$	$CH_3C(O)CH_2P^{\oplus}(C_6H_5)_3, Cl^{\ominus}$	H	1967, 118
$PC_{21}H_{20}ClO_2$	$CH_3C(O)OCH_2P^{\oplus}(C_6H_5)_3, Cl^{\ominus}$	H	1968, 56
$PC_{21}H_{20}FO$	$CH_3C(O)CH_2P^{\oplus}(C_6H_5)_3, F^{\ominus}$	H	1969, 19
$PC_{21}H_{20}I$	$CH_3(CH_2=CHC_6H_4)P^{\oplus}(C_6H_5)_2, I^{\ominus}$	H	1968, 104
$PC_{21}H_{20}N$	$C_6H_5P[C_6H_3(CH_3)]_2NCH_3$	H	1967, 361

$PC_{21}H_{20}NO$	$(C_6H_5)_3PCH_2C(CH_3)=NO$	H	1968, 343
$PC_{21}H_{21}$	$P(CH_2C_6H_5)_3$	$P(-13; -10)$	1967, 8, 244; 1968, 6
	$C_2H_5CH=P(C_6H_5)_3$	$P(12\text{ to }10)$	1967, 359, 363; 1968, 337
	$i\text{-}C_3H_6=P(C_6H_5)_3$	$P(11\cdot3)$	1967, 363
	$P(C_6H_4CH_3)_3$	H	1968, 15, 16, 103, 260
	$(C_6H_5)_3P=C(CH_3)_2$	$P(10\text{ to }11)$	1968, 337
$PC_{21}H_{21}Br_2$	$(C_6H_5)_3P^\oplus(CH_2)_3Br, Br^\ominus$	$P(24\cdot0)$	1966, 5
$PC_{21}H_{21}ClNO$	$(C_6H_5)_3P^\oplus CH_2C(=NOH)CH_3, Cl^\ominus$	H	1968, 343
$PC_{21}H_{21}O$	$(C_6H_5)_2P(O)CH=C=C_7H_{10}$	H	1967, 153
	$OP(C_6H_4CH_3)_3$	H	1967, 117; 1968, 15, 16
	$(C_6H_5)_3P(CH_2)_3O$	H	1967, 366
$PC_{21}H_{21}O_3$	$P(OCH_2C_6H_5)_3$	$H; P(138\cdot6)$	1968, 73
	$P(C_6H_4OCH_3)_3$	$H\{P\}$	1969, 118
$PC_{21}H_{21}O_3S$	$SP(C_6H_4OCH_3)_3$	$H\{P\}$	1969, 118
$PC_{21}H_{21}O_4$	$OP(OC_6H_4CH_3)_3$	H	1967, 117; 1969, 74
$PC_{21}H_{21}S$	$SP(C_6H_4CH_3)_3$	H	1968, 15, 16
$PC_{21}H_{21}S_3$	$P(C_6H_4SCH_3)_3$	H	1968, 232
$PC_{21}H_{22}$	$i\text{-}C_3H_7P^\oplus(C_6H_5)_3$	$P(30\cdot9)$	1967, 7
$PC_{21}H_{22}BF_4S$	$(C_6H_5)_3P^\oplus CH^\ominus S^\oplus(CH_3)_2, BF_4^\ominus$	H	1967, 367
$PC_{21}H_{22}Br$	$C_3H_7P^\oplus(C_6H_5)_3, Br^\ominus$	$P(24\cdot1)$	1966, 5; 1967, 363
	$i\text{-}C_3H_7P^\oplus(C_6H_5)_3, Br^\ominus$	$P(30\cdot9)$	1966, 5; 1967, 363
	$C_2H_5(C_6H_5CH_2)P^\oplus(C_6H_5)_2, Br^\ominus$	$P(28\cdot3)$	1968, 6
	$CH_3(C_6H_5)P^\oplus(CH_2C_6H_5)_2, Br^\ominus$	$P(27\cdot4)$	1968, 6
	$CH_3(C_6H_5)P^\oplus(CH_2C_6H_5)C_6H_4CH_3, Br^\ominus$	H	1969, 156
$PC_{21}H_{22}BrO$	$(C_6H_5)_3P^\oplus CH(CH_3)CH_2OH, Br^\ominus$	H	1966, 135
	$C_6H_5CH_2(C_6H_5)_2P^\oplus CH_2OCH_3, Br^\ominus$	H	1967, 349
$PC_{21}H_{22}Cl$	$C_3H_7P^\oplus(C_6H_5)_3, Cl^\ominus$	H	1967, 118

Formula	Compound	Resonance	Reference
$PC_{21}H_{22}I$	$i\text{-}C_3H_7P^{\oplus}(C_6H_5)_3, I^{\ominus}$	H	1966, 122
	$CH_3(C_6H_5)P^{\oplus}(C_6H_4CH_3)_2, I^{\ominus}$	H	1968, 15, 16
$PC_{21}H_{22}NO$	$(C_6H_5)_2P(O)CH_2C_6H_4N(CH_3)_2$	H	1967, 368
$PC_{21}H_{23}B_2F_8S$	$(C_6H_5)_3P^{\oplus}CH_2S^{\oplus}(CH_3)_2, 2BF_4^{\ominus}$	H	1967, 367
$PC_{21}H_{23}O$	$(C_6H_5)_2P(O)C(CH_3)=C=C_5H_7(CH_3)$	H	1969, 164, 165
	$(C_6H_5)_2P(O)CH=C=C_6H_9(CH_3)$	H	1969, 165
$PC_{21}H_{23}O_2$	$(C_6H_5)_2P(O)CH=C=CCH(OCH_3)(CH_2)_4$	H	1969, 164
$PC_{21}H_{24}ClOSi$	$(C_6H_5)_3P^{\oplus}OSi(CH_3)_3, Cl^{\ominus}$	P (42·2)	1969, 368
$PC_{21}H_{24}GeN$	$(C_6H_5)_3PNGe(CH_3)_3$	H	1967, 73
$PC_{21}H_{24}NSi$	$(C_6H_5)_3PNSi(CH_3)_3$	H	1967, 73, 322
$PC_{21}H_{24}NSn$	$(C_6H_5)_3PNSn(CH_3)_3$	H	1967, 73
$PC_{21}H_{25}N_2O_4$	$C_6H_5OP(O)(O^{\ominus})OCH_2CH(CH_3)NHC_6H_5, C_6H_5NH_3^{\oplus}$	H; P ($-7\cdot7$)	1969, 359
$PC_{21}H_{25}N_2O_9$	$(C_2H_5O)_3P[OCH(C_6H_4NO_2)]_2$	H; P ($ca.\ -50$)	1967, 163
$PC_{21}H_{25}O_2$	$(C_6H_5)_2P(O)CH=C=C(CH_3)CH(OCH_3)\ i\text{-}C_3H_7$	H	1967, 155
$PC_{21}H_{25}O_5$	$i\text{-}C_3H_7O)_3P[OC(C_6H_4)]_2$	P ($-48\cdot9$)	1965, 21
$PC_{21}H_{25}O_7$	$(CH_3O)_3POC(C_6H_5)(COC_6H_5)C(CH_3)(COCH_3)O$ (stereois.)	H; P ($-54\cdot7;\ -52\cdot7$)	1966, 258
$PC_{21}H_{26}BIN$	$[(C_6H_5)_3P(CH_3)_3NBH_2]^{\oplus}, I^{\ominus}$	H	1969, 26
$PC_{21}H_{27}O_5$	$(i\text{-}C_3H_7O)_3P[OC(C_6H_5)]_2$	P ($-53\cdot9$)	1965, 21
$PC_{21}H_{27}O_9$	$P[OC(CH_3)(COOC_2H_5)C=CH]_3$	H	1968, 311
	$OP[OC(CH_3)(COOC_2H_5)C=CH]_2-$		
	$[CH=C=C(CH_3)COOC_2H_5]$		
$PC_{21}H_{28}Br$	$C_6H_5(C_6H_5CH_2)P^{\oplus}[C(CH_3)_2]_2CHCH_3, Br^{\ominus}$	H	1968, 311
$PC_{21}H_{28}N_3O_3$	$[(CH_3)_2N]_3P^{\oplus}OC^{\ominus}(COC_6H_5)_2$	H; P (35·9)	1969, 170
			1965, 84; 1966, 242; 1968, 309
$PC_{21}H_{29}O$	$CH_3(C_6H_5)P(O)C(CH_3)_2CH(CH_3)C(CH_3)_2C_6H_5$	H{P}	1968, 67
$PC_{21}H_{30}Br$	$C_4H_9(t\text{-}C_5H_{11})P^{\oplus}(C_6H_5)_2, Br^{\ominus}$	P (38·5)	1968, 6

$PC_{21}H_{30}I$	$C_6H_5(CH_3)_2P^{\oplus}CH_2C(CH_3)_2C(CH_3)_2C_6H_5$	H{P}	1968, 67
$PC_{21}H_{30}NO$	$C_6H_5[(CH_3)_2N]P(O)C(CH_3)=C=C_{10}H_{16}$	H	1969, 165
$PC_{21}H_{30}N_5O_7$	$[(CH_2)_2CHCH_2O]_2P(O)$ (2′,3′-isopropylidene-adenosine-5)	H	1966, 213
$PC_{21}H_{31}O_4$	$CH_3(CH_3OP(O)O(C_{19}H_{25}O)$	H	1967, 168
$PC_{21}H_{33}O_4$	$CH_3(CH_3O)P(O)O(C_{19}H_{27}O)$	H	1967, 168
$PC_{21}H_{35}O_5$	$(CH_3O)_2P(O)$ (hydroxyandrostanone)	H	1967, 167
$PC_{21}H_{42}N_3O$	$[CH_3CH(CH_2)_4NCH_2]_3PO$	P (50·9)	1968, 36
	$[CH(CH_3)(CH_2)_4NCH_2]_3PO$	P (49·4)	1968, 36
	$[CH_2CH(CH_3)(CH_2)_3NCH_2]_3PO$	P (50·4)	1968, 36
$PC_{21}H_{44}NO_4$	$(C_8H_{17}O)_2P(O)C(O)N(C_2H_5)_2$	H	1967, 26
	$(C_7H_{15}O)_2P(O)C(O)N(i-C_3H_7)_2$	H	1968, 29
$PC_{21}H_{46}I$	$CH_3(C_4H_9)P^{\oplus}(C_8H_{17})_2, I^{\ominus}$	P (30·2)	1967, 244
$PC_{22}H_{15}F_2O_2$	$(C_6H_5)_3P=CC(O)CF_2CO$	P (3·0)	1968, 110
$PC_{22}H_{15}F_6$	$(C_6H_5)_3P=C=C(CF_3)_2$	P (4·1); F (−61·4)	1967, 102
$PC_{22}H_{15}S$	$C_6H_5P(S)(C=CC_6H_5)_2$	H	1965, 96
$PC_{22}H_{16}BF_{10}$	$(C_6H_5)_3P^{\oplus}CH=C(CF_3)_2, BF_4^{\ominus}$	H; P (17·3); F (−58·4; −64·3; −151·7)	1967, 102
$PC_{22}H_{16}ClF_6$	$(C_6H_5)_3P^{\oplus}CH=C(CF_3)_2, Cl^{\ominus}$	H; P (17·3); F (−58·1; −63·2)	1967, 102
$PC_{22}H_{17}$	$C_6H_5P[C(C_6H_5)=CH]_2$	H	1969, 77
$PC_{22}H_{17}O_2$	$(C_6H_5)_3P=CC(O)CH_2CO$	H; P (−4·4)	1968, 110
$PC_{22}H_{17}O_3$	$(C_6H_5)_3P=CC(O)OC(O)CH_2$	H; P (13)	1964, 10; 1968, 344
	$(C_6H_5)_3P=CC(O)CH_2OCO$	H	1967, 369

238 FORMULAE INDEX II

Formula	Compound	Resonance	Reference
$PC_{22}H_{17}O_5$	$C_6H_5O(CH_2O)_2P[OC(C_6H_4)]_2$	$P(-27.0)$	1968, 301
$PC_{22}H_{18}NO_2$	$(C_6H_5)_3P^{\oplus}C[C(O)NHC(O)]^{\ominus}CH_2$	H	1968, 344
$PC_{22}H_{19}N_2O_2$	$(C_6H_5)_2P(O)NNC(CH_3)OC=CHC_6H_5$	H	1968, 88
$PC_{22}H_{19}O_2$	$C_6H_5(CH_3)_2POC_{14}H_8O$	$P(-9.6)$	1968, 303
$PC_{22}H_{19}O_3$	$(C_6H_5)_3P^{\oplus}[CH_2C(O)]_2^{\ominus}O$	H	1963, 8
$PC_{22}H_{19}O_5$	$C_6H_5(CH_2O)_2P[OC(C_6H_5)]_2$	$P(-32.2)$	1968, 301
$PC_{22}H_{20}BrO_2$	$(C_6H_5)_3P^{\oplus}CH=CHCOOCH_3, Br^{\ominus}$	H	1969, 349
$PC_{22}H_{20}ClO_2$	$(C_6H_5)_3P^{\oplus}CH=CHCOOCH_3, Cl^{\ominus}$	H	1969, 349
$PC_{22}H_{21}$	$(C_6H_5)_2PC_6H_4(CH_2)_2CCH_3$	H	1968, 194
$PC_{22}H_{21}Cl_2O$	$Cl_2P(O)C(C_6H_4CH_3)_3$	H	1967, 117
$PC_{22}H_{21}N_2O_7S$	$(C_6H_5)_3P^{\oplus}N(COOCH_3)N(SO_3^{\ominus})COOCH_3$	$P(53)$	1969, 324
$PC_{22}H_{21}O_2$	$(C_6H_5)_3P=C(CH_3)COOCH_3$	H (T)	1966, 140
	$(C_6H_5)_3P=CHCOOC_2H_5$	H; $P(16.8)$	1966, 270; 1967, 192; 1968, 171
$PC_{22}H_{21}O_3$	$(C_6H_5)_2P(O)CH(C_6H_5)CH_2COOCH_3$	H	1969, 369
$PC_{22}H_{21}O_4$	$C_6H_5(C_6H_5O)_2P[OC(CH_3)]_2$	$P(-42.4)$	1966, 225
$PC_{22}H_{21}O_5$	$(C_6H_5O)_3P[OC(CH_3)]_2$	$P(-64.7)$	1966, 225; 1968, 301
$PC_{22}H_{22}Br$	$CH_3CH=C(CH_3)P^{\oplus}(C_6H_5)_3, Br^{\ominus}$	H	1966, 135
$PC_{22}H_{22}BrO$	$CH_3C(O)CH(CH_3)P^{\oplus}(C_6H_5)_3, Br^{\ominus}$	H	1966, 135
$PC_{22}H_{22}BrO_2$	$(C_6H_5)_3P^{\oplus}CH_2COOC_2H_5, Br^{\ominus}$	H	1968, 170
$PC_{22}H_{22}Cl$	$CH_2=C(CH_3)CH_2P^{\oplus}(C_6H_5)_3, Cl^{\ominus}$	$P(21.0)$	1966, 5
	$CH_3CH=CHCH_2P^{\oplus}(C_6H_5)_3, Cl^{\ominus}$	H	1966, 276
$PC_{22}H_{22}ClN_2O_4$	$P[C(CH)_2C_6H_4NC_2H_5]_2, ClO_4^{\ominus}$	$P(48.8)$	1966, 272
$PC_{22}H_{23}$	$(C_6H_5)_3P=CHC_3H_7$	$P(13)$	1967, 165, 363; 1968, 337

NMR DATA ON ORGANIC COMPOUNDS

PC$_{22}$H$_{23}$Br$_2$	(C$_6$H$_5$)$_3$P=C(CH$_3$)C$_2$H$_5$		P (10·5)	1967, 363
PC$_{22}$H$_{23}$Br$_2$	(C$_6$H$_5$)$_3$P$^{\oplus}$(CH$_2$)$_4$Br, Br$^{\ominus}$		P (24·2)	1966, 5
PC$_{22}$H$_{23}$N$_2$O	(C$_6$H$_5$)$_2$PN(C$_6$H$_5$)C(O)NHC$_3$H$_7$		H	1967, 370
PC$_{22}$H$_{23}$O	(C$_6$H$_5$)$_2$P(O)C(CH$_3$)=C−C$_7$H$_{10}$		H	1967, 153
PC$_{22}$H$_{24}$	C$_4$H$_9$P$^{\oplus}$(C$_6$H$_5$)$_3$		P (24·0)	1967, 7
PC$_{22}$H$_{24}$BF$_4$S	(C$_6$H$_5$)$_3$P$^{\oplus}$CH$^{\ominus}$S$^{\oplus}$(CH$_3$)C$_2$H$_5$, BF$_4^{\ominus}$		H	1967, 367
PC$_{22}$H$_{24}$Br	C$_4$H$_9$P$^{\oplus}$(C$_6$H$_5$)$_3$, Br$^{\ominus}$		P (24·0)	1966, 5; 1967, 363
	s-C$_4$H$_9$P$^{\oplus}$(C$_6$H$_5$)$_3$, Br$^{\ominus}$		H; P (30·2)	1966, 5, 135; 1967, 363
	i-C$_3$H$_7$(C$_6$H$_5$CH$_2$)P$^{\oplus}$(C$_6$H$_5$)$_2$, Br$^{\ominus}$		P (35·9)	1968, 6
	C$_2$H$_5$(C$_6$H$_5$CH$_2$)$_2$P$^{\oplus}$C$_6$H$_5$, Br$^{\ominus}$		H; P (30·1)	1968, 6; 1969, 156
	CH$_3$(C$_6$H$_5$)P$^{\oplus}$(CH$_2$C$_6$H$_5$)CH(CH$_3$)C$_6$H$_5$, Br$^{\ominus}$		H	1969, 156
PC$_{22}$H$_{24}$BrO	(C$_6$H$_5$)$_3$P$^{\oplus}$CH(CH$_3$)CH(OH)CH$_3$		H	1966, 135
PC$_{22}$H$_{24}$I	t-C$_4$H$_9$P$^{\oplus}$(C$_6$H$_5$)$_3$, I$^{\ominus}$		P (34·7)	1966, 5
	CH$_3$P$^{\oplus}$(CH$_2$C$_6$H$_5$)$_3$, I$^{\ominus}$		P (26·5)	1967, 244
	CH$_3$P$^{\oplus}$(C$_6$H$_4$CH$_3$)$_3$, I$^{\ominus}$		H	1968, 15, 16
PC$_{22}$H$_{24}$IO$_3$	CH$_3$P$^{\oplus}$(C$_6$H$_4$OCH$_3$)$_3$, I$^{\ominus}$		H{P}	1969, 118
PC$_{22}$H$_{24}$N	(C$_6$H$_5$)$_3$P=N-t-C$_4$H$_9$		H	1965, 109
PC$_{22}$H$_{25}$Br$_2$F$_8$S	(C$_6$H$_5$)$_3$P$^{\oplus}$CH$_2$S$^{\oplus}$(CH$_3$)C$_2$H$_5$, 2 Br$^{\ominus}$		H	1967, 367
PC$_{22}$H$_{25}$ClN	(C$_6$H$_5$)$_3$P$^{\oplus}$NH-t-C$_4$H$_9$, Cl$^{\ominus}$		H	1965, 109
PC$_{22}$H$_{25}$IN	(C$_6$H$_5$)$_2$P(CH$_3$)(CH$_2$)$_2$N(CH$_3$)C$_6$H$_5$, I$^{\ominus}$		H	1967, 371
PC$_{22}$H$_{25}$N$_2$O$_7$	(C$_2$H$_5$O)$_3$P[OC(C$_6$H$_4$)NHC(O)]$_2$		H	1966, 263
PC$_{22}$H$_{25}$O	(C$_6$H$_5$)$_2$P(O)C(CH$_3$)=C−C$_6$H$_9$CH$_3$		H	1969, 165
PC$_{22}$H$_{25}$O$_2$	(C$_6$H$_5$)$_3$P(OC$_2$H$_5$)$_2$		H; P (−54)	1968, 332
	(C$_6$H$_5$)$_2$P(O)C(CH$_3$)=C=CCH(OCH$_3$)(CH$_2$)$_4$		H	1969, 164
PC$_{22}$H$_{25}$Si	(C$_6$H$_5$)$_3$P=CHSi(CH$_3$)$_3$		H	1967, 328
PC$_{22}$H$_{26}$ISi	(C$_6$H$_5$)$_3$P$^{\oplus}$CH$_2$Si(CH$_3$)$_3$, I$^{\ominus}$		H	1965, 108
PC$_{22}$H$_{27}$O	(C$_6$H$_5$)$_2$P(O) geranyl		H	1966, 267
PC$_{22}$H$_{27}$O$_2$	(C$_6$H$_5$)$_2$P(O)C(CH$_3$)=C=C(CH$_3$)CH(OCH$_3$)i-C$_3$H$_7$		H	1967, 155
PC$_{22}$H$_{28}$NO$_4$	(CH$_3$)$_2$NP[OC(CH$_3$)]$_2$[OC(C$_6$H$_5$)]$_2$		P (−35·5)	1968, 304

Formula	Compound	Resonance	Reference
$PC_{22}H_{28}N_3O_2$	$(CH_2)_4NP[N(CH_3)CH_2]_2[OC(C_6H_5)]_2$	H; P (−41·8)	1968, 153
$PC_{22}H_{28}N_3O_6$	$(CH_2)_4NP[N(CH_3)CH_2]_2[OCH(C_6H_4)NO_2]_2$	H; P (o: −42·4; m: −41·4; p: −42·3)	1968, 310
$PC_{22}H_{29}O$	$(C_6H_5)_2POC_6H_{10}C(CH_3)_3$	H	1966, 62
$PC_{22}H_{29}O_2$	$(C_6H_5)_2P(O)OC_6H_{10}C(CH_3)_3$	H	1966, 62;
	$(C_6H_5)_2P(O)O$ menthyl	H	1967, 151; 1968, 140
$PC_{22}H_{30}N_3O_2$	$[(CH_3)_2N]_3P=C(COC_6H_5)(CH_2COC_6H_5)$	P (63·2)	1965, 84
$PC_{22}H_{31}ClN_3O_2$	$[(CH_3)_2N]_3P^{\oplus}CH(COC_6H_5)(CH_2COC_6H_5), Cl^{\ominus}$	P (54·9)	1966, 106
$PC_{22}H_{32}I$	$(CH_3)_2(C_6H_5)P^{\oplus}C(CH_3)_2CH(CH_3)C(CH_3)_2C_6H_5, I^{\ominus}$	H; P	1968, 67
$PC_{22}H_{45}O$	$(C_2H_5)_2PO(CH_2)_8CH=CH(CH_2)_7CH_3$	P (135·3)	1968, 336
	$(C_2H_5)_2P(O)(CH_2)_8CH=CH(CH_2)_7CH_3$	P (51·0)	1968, 336
$PC_{23}H_{17}$	$P[C(C_6H_5)=CH]_2CC_6H_5$	H; P (178·2)	1966, 68
$PC_{23}H_{17}O_5$	$(CH_3O)_3POC_{14}H_8O$	P (−44·7)	1968, 303
$PC_{23}H_{19}O_3$	$[(C_6H_5)_2CCH]_2CH(C_6H_5)PO_2H$	H	1968, 102
	$(C_6H_5)_3P^{\oplus}CC(O^{\ominus})OC(O)CH(CH_3)$	H	1968, 344
$PC_{23}H_{21}BrNO_2$	$(C_6H_5)_3P^{\oplus}CH=CHC(=NOH)COCH_3, Br^{\ominus}$	H	1968, 345
$PC_{23}H_{21}O$	$(C_6H_5)_2P(O)(CH=CH_2)CH_2CH_2C_6H_5$	H	1966, 267
	$(C_6H_5)_2P(O)(C_6H_5)=C=C(CH_3)_2$	H	1969, 165
$PC_{23}H_{21}O_3$	$(C_6H_5)_3P=C(COOCH_3)CH_2CHO$	H	1964, 10
$PC_{23}H_{22}BrO$	$(C_6H_5)_3P^{\oplus}CH_2CH=CHCOCH_3, Br^{\ominus}$	H	1968, 345
$PC_{23}H_{22}BrO_2$	$(C_6H_5)_3P^{\oplus}CH_2CH=CHCOOCH_3, Br^{\ominus}$	H	1968, 345
$PC_{23}H_{22}ClO_2$	$(C_6H_5)_3P^{\oplus}CH_2CH=CHCOOCH_3, Cl^{\ominus}$	H	1968, 346
$PC_{23}H_{22}O$	$(C_6H_5)_3P=CHCH=CHCOCH_3$	H	1968, 345

NMR DATA ON ORGANIC COMPOUNDS

Formula	Structure	Nucleus	Reference
$PC_{23}H_{23}$	$(C_6H_5)_3P=C_5H_8$	P (4·8)	1967, 363
$PC_{23}H_{23}O_2$	$(C_6H_5)_3P=C(CH_3)COOC_2H_5$	H	1968, 171
$PC_{23}H_{24}BF_4O_2$	$(C_6H_5)_3P^{\oplus}CH=C(OCH_3)OC_2H_5, BF_4^{\ominus}$	H	1969, 206
$PC_{23}H_{24}Br$	$C_5H_9P^{\oplus}(C_6H_5)_3, Br^{\ominus}$	H; P (30·7)	1966, 5, 277; 1967, 363
$PC_{23}H_{25}$	$(C_6H_5)_3P=C_5H_{10}$	P	1967, 363; 1968, 337
$PC_{23}H_{26}BF_4S$	$(C_6H_5)_3P^{\oplus}CH^{\ominus}S^{\oplus}(C_2H_5)_2, BF_4^{\ominus}$	H	1967, 367
$PC_{23}H_{26}Br$	$(C_6H_5)_3P^{\oplus}CH(C_2H_5)_2, Br^{\ominus}$	P (30·1)	1966, 5; 1967, 363
	$C_3H_7(C_6H_5)P^{\oplus}(CH_2C_6H_5)_2, Br^{\ominus}$	P (24·4)	1968, 6
	$C_2H_5P^{\oplus}(CH_2C_6H_5)_3, Br^{\ominus}$	P (28·7)	1968, 6
	$CH_3(C_6H_5)P^{\oplus}(CH_2C_6H_5)C_6H_2(CH_3)_2$	H	1969, 156
$PC_{23}H_{26}I$	$neo\text{-}C_5H_{11}P^{\oplus}(C_6H_5)_3, I^{\ominus}$	H	1965, 108
	$C_2H_5(CH_3)_2CP^{\oplus}(C_6H_5)_3, I^{\ominus}$	P (36·7)	1966, 5
	$C_2H_5P^{\oplus}(C_6H_4CH_3)_3, I^{\ominus}$	H	1968, 15
$PC_{23}H_{26}O$	$(C_6H_5)_2P(O)CH=C=C_7H_7(CH_3)_3$	H	1969, 166
	$(C_6H_5)_3P^{\oplus}OCH_2C(CH_3)_3$	P (−4)	1967, 345
$PC_{23}H_{27}B_2F_8S$	$(C_6H_5)_3P^{\oplus}CH_2S^{\oplus}(C_2H_5)_2, 2BF_4^{\ominus}$	H	1967, 367
$PC_{23}H_{28}ISi$	$(C_6H_5)_3P^{\oplus}CH(CH_3)Si(CH_3)_3, I^{\ominus}$	H	1965, 108
$PC_{23}H_{28}I_2N$	$(C_6H_5)_2P^{\oplus}(CH_3)(CH_2)_2N^{\oplus}(CH_3)_2C_6H_5, 2I^{\ominus}$	H	1967, 257
$PC_{23}H_{28}N_3O_3$	$(CH_3)_4N[CH_2(CH_3)N]_2POC(COC_6H_5)C(C_6H_5)O$	P (−37·5)	1968, 153
$PC_{23}H_{29}O_5$	$(i\text{-}C_3H_7O)_3POC_{14}H_8O$	H	1969, 366
$PC_{23}H_{31}N_4O_5S$	Diethyl 2-{3-(2-methyl-4-benzoylaminopyrimidin-5-yl)-methyl-3a-methylperhydrofuro[2,3-d]thiazole}phosphate	H	1967, 352
$PC_{23}H_{31}O_2$	$C_6H_5(C_6H_5CH_2)P(O)O$ menthyl	H	1968, 35
$PC_{23}H_{32}NO_2$	$C_6H_5[O(CH_2)_2N]P(O)C(CH_3)=C=C_{10}H_{16}$	H	1969, 165
$PC_{23}H_{33}O$	$(C_6H_{11})_2P(O)C(C_6H_5)=C=C(CH_3)_2$	H	1969, 165
$PC_{23}H_{35}O_4$	$CH_3(CH_3O)P(O)O(C_{21}H_{29}O)$	H	1967, 168
$PC_{23}H_{37}O_6$	$(CH_3O)_2P(O)$ hydroxypregnanedione	H	1967, 167
$PC_{23}H_{48}NO_4$	$[(CH_3)_3C_5H_8O]_2P(O)C(O)N(i\text{-}C_3H_7)_2$	H	1967, 26

Formula	Compound	Resonance	Reference
$PC_{24}H_{15}Br_2$	$C_6H_5P[C(C_6H_4Br)=CH]_2$	H	1967, 70
$PC_{24}H_{15}F_{12}O_2$	$(C_6H_5)_3P[OC(CF_3)_2]_2$	H; P ($-22\cdot 2$ $-18\cdot 6$); F ($-65\cdot 2$)	1966, 225, 278; 1967, 102; 1968, 303
$PC_{24}H_{15}F_{12}O_4$	$C_6H_5(C_6H_5O)_2P[OC(CF_3)_2]_2$	P ($-40\cdot 1$)	1966, 225
$PC_{24}H_{15}F_{12}O_5$	$(C_6H_5O)_3P[OC(CF_3)_2]_2$	P ($-64\cdot 1$)	1966, 225; 1967, 335
$PC_{24}H_{15}S$	$C_6H_5C{=}C]_3PS$	H	1965, 96
$PC_{24}H_{17}$	$(C_6H_4)_4PH$	H	1966, 25
	$C_6H_5P[C(C_6H_5)=CH_2]$	H	1967, 70
$PC_{24}H_{17}N_2O_6$	$(C_6H_4)_4P^\oplus H(NO_3)_2^\ominus$	H	1966, 279
$PC_{24}H_{19}FI$	$CH_3(C_6H_4F)P^\oplus(C_6H_5)_2, I^\ominus$	F	1968, 14
$PC_{24}H_{19}O_5$	$(C_6H_5O)_3POC_6H_4O$	P ($-60\cdot 8$)	1968, 124
$PC_{24}H_{20}Br$	$(C_6H_5)_4P^\oplus Br^\ominus$	^{13}C	1969, 21
$PC_{24}H_{20}I$	$(C_6H_5)_2P^\oplus I^\ominus$	P ($23\cdot 2$)	1966, 5; 1967, 359
$PC_{24}H_{21}ClN$	$(C_6H_5)_3P^\oplus NHC_6H_5, Cl^\ominus$	H	1966, 280
$PC_{24}H_{21}ClNO_5$	$(CH_3O)_2P(O)OC{=}CClC(C_6H_5)_2C(O)NC_6H_5$	H	1969, 345
$PC_{24}H_{21}Cl_2O_6$	$(C_6H_3Cl_2OCH_2CH_2O)_3P$	H	1968, 285
$PC_{24}H_{21}IN$	$(C_6H_5)_3P^\oplus CH_2(C_5H_4N), I^\ominus$	H	1966, 271
$PC_{24}H_{21}O_3$	$CH_3OP(O)[C(C_6H_5)CH]_2C(OH)C_6H_5$	H	1968, 75
$PC_{24}H_{23}BrNO_3$	$(C_6H_5)_3P^\oplus C(CH_3)=CHC(=NOCH_3)COOH, Br^\ominus$	H	1968, 345
$PC_{24}H_{23}O$	$(C_6H_5)_2P(O)CH=C=C(CH_3)CH(CH_3)C_6H_5$	H	1967, 155
	$(C_6H_5)_3P=CHCH-C(CH_3)COCH_3$	H	1968, 345
$PC_{24}H_{23}O_2$	$C_2H_5(C_6H_5)_2POC_{10}H_8O$	P ($-9\cdot 3$)	1968, 124
	$C_6H_5(C_2H_5)_2POC_{14}H_8O$	P ($0\cdot 9$)	1968, 303
$PC_{24}H_{23}O_3$	$(C_6H_5)_3P=C(COCH_3)CH_2COOCH_3$	H	1967, 122

PC$_{24}$H$_{23}$O$_4$	(C$_6$H$_5$)$_3$P=C(COCH$_3$)COOC$_2$H$_5$	H	1967, 372
PC$_{24}$H$_{24}$	(C$_6$H$_5$)$_3$P=C(COOCH$_3$)CH$_2$COOCH$_3$	H	1964, 10
PC$_{24}$H$_{24}$Br	C$_6$H$_9$P$^\oplus$(C$_6$H$_5$)$_3$	H	1966, 273
PC$_{24}$H$_{24}$BrO	CH$_3$(C$_6$H$_5$)P$^\oplus$(CH$_2$C$_6$H$_5$-α-C$_{10}$H$_9$)	H	1969, 156
	C$_6$H$_9$OP$^\oplus$(C$_6$H$_5$)$_3$, Br$^\ominus$	H	1966, 116
PC$_{24}$H$_{24}$ClO$_3$	(C$_6$H$_5$)$_3$P$^\oplus$CH$_2$CH=C(CH$_3$)COCH$_3$, Br$^\ominus$	H	1968, 345
PC$_{24}$H$_{24}$O$_3$	(C$_6$H$_5$)$_3$CP$^\oplus$(OCH$_2$)$_2$C(CH$_3$)CH$_2$Cl	H	1967, 91
PC$_{24}$H$_{25}$	(C$_6$H$_5$)$_3$P$^\oplus$CH(COCH$_3$)COOC$_2$H$_5$	H	1967, 372
	(C$_6$H$_5$)$_3$P=C(CH$_2$)$_5$	P (6·4)	1967, 363; 1968, 337
PC$_{24}$H$_{25}$O$_2$	(CH$_3$C$_6$H$_4$)$_3$P=CHCOOCH$_3$	H	1967, 192
	(C$_6$H$_5$)$_3$P=C=C(OC$_2$H$_5$)$_2$	H	1969, 206
PC$_{24}$H$_{25}$O$_3$	(C$_6$H$_5$)$_3$CP(O)(OCH$_2$)$_2$C(CH$_3$)$_2$	H	1967, 91; 1968, 74
PC$_{24}$H$_{26}$Br	C$_6$H$_{11}$P$^\oplus$(C$_6$H$_5$)$_3$, Br$^\ominus$	P (26·6)	1966, 5; 1967, 363
PC$_{24}$H$_{26}$BrO$_2$	C$_6$H$_5$P$^\oplus$(CH$_2$C$_6$H$_5$)$_2$CH$_2$COOC$_2$H$_5$	H	1969, 156
PC$_{24}$H$_{27}$	(C$_6$H$_5$)$_3$P=C$_6$H$_{12}$	P	1967, 363
PC$_{24}$H$_{27}$O	(C$_6$H$_5$)$_2$P(O)CH=C=C$_{10}$H$_{16}$	H	1969, 165
PC$_{24}$H$_{28}$Br	C$_6$H$_{13}$P$^\oplus$(C$_6$H$_5$)$_3$, Br$^\ominus$	P (24·4)	1966, 5; 1967, 363
	C$_2$H$_5$(C$_3$H$_7$)CHP$^\oplus$(C$_6$H$_5$)$_3$, Br$^\ominus$	P (30·3)	1967, 363
	C$_4$H$_9$(C$_6$H$_5$)P$^\oplus$(CH$_2$C$_6$H$_5$)$_2$, Br$^\ominus$	P (28·7)	1968, 6
	t-C$_4$H$_9$(C$_6$H$_5$)P$^\oplus$(CH$_2$C$_6$H$_5$)$_2$, Br$^\ominus$	H	1969, 156
PC$_{24}$H$_{28}$I	(C$_6$H$_5$)$_3$P$^\oplus$CH(CH$_3$)-t-C$_4$H$_9$, I$^\ominus$	H	1965, 108
PC$_{24}$H$_{28}$O	(C$_6$H$_5$)$_2$P(O)C(CH$_3$)=C=C$_7$H$_7$(CH$_3$)$_3$	H	1969, 299
PC$_{24}$H$_{29}$O	(C$_6$H$_5$)$_2$P(O)C(t-C$_4$H$_9$)=C=C$_5$H$_7$CH$_3$	H	1969, 164, 165
	(C$_6$H$_5$)$_2$P(O)CH=C=C$_6$H$_9$-t-C$_4$H$_9$	H	1969, 165
PC$_{24}$H$_{30}$NSn	(C$_6$H$_5$)$_3$P=NSn(C$_6$H$_5$)$_3$	H	1967, 299
PC$_{24}$H$_{31}$O$_6$	(i-C$_3$H$_7$O)$_3$POC(C$_6$H$_5$)C(COC$_6$H$_5$)O	H; P (−49·3)	1965, 21
PC$_{24}$H$_{33}$AlNSi	(C$_6$H$_5$)$_3$PN[Al(CH$_3$)$_3$][Si(CH$_3$)$_3$]	H	1967, 322
PC$_{24}$H$_{33}$GaNSi	(C$_6$H$_5$)$_3$PN[Ga(CH$_3$)$_3$][Si(CH$_3$)$_3$]	H	1967, 322
PC$_{24}$H$_{33}$N$_2$O$_2$	C$_4$H$_9$[(CH$_3$)$_2$N]$_2$P=C(COC$_6$H$_5$)CH$_2$COC$_6$H$_5$	H; P (63·2)	1966, 106
PC$_{24}$H$_{35}$GeO	(C$_2$H$_5$)$_2$P(O)[C(C$_6$H$_5$)]$_2$COGe(C$_2$H$_5$)$_3$	H	1969, 341

Formula	Compound	Resonance	Reference
PC$_{24}$H$_{35}$N$_2$O$_2$	(C$_2$H$_5$)$_2$NP[OC$_6$H$_3$(CH$_3$)CH$_2$]OC$_6$H$_3$(CH$_3$)CHN(C$_2$H$_5$)$_2$	H; P (-28)	1967, 373
PC$_{24}$H$_{35}$S	C$_{12}$H$_{25}$P(S)(C$_6$H$_5$)$_2$	P (41·5)	1966, 40
PC$_{24}$H$_{37}$O$_7$	(CH$_3$O)$_2$P(O)(C$_{22}$H$_{31}$O$_4$)	H	1967, 270
PC$_{24}$H$_{42}$NO$_6$S	(C$_8$H$_{17}$O)$_2$P(O)C(O)NHSO$_2$C$_6$H$_4$CH$_3$	H	1968, 328; 1969, 161
PC$_{24}$H$_{51}$	HP(C$_{12}$H$_{25}$)$_2$	H; P (-71)	1966, 14
	P(C$_8$H$_{17}$)$_3$	P ($-31·8$)	1966, 4; 1967, 244, 277
PC$_{24}$H$_{51}$Cl$_2$	P[CH$_2$C(C$_2$H$_5$)(CH$_2$)$_3$CH$_3$]$_3$	P ($-48·5$)	1967, 244
	(C$_8$H$_{17}$)$_3$PCl$_2$	H; P	1969, 344
PC$_{24}$H$_{51}$O	(C$_8$H$_{17}$)$_3$PO	H; P (47·4)	1966, 105; 1967, 277
PC$_{24}$H$_{51}$S	(C$_8$H$_{17}$)$_3$PS	P (54·8)	1967, 277
PC$_{25}$H$_{15}$F$_6$O$_2$	(C$_6$H$_5$)$_3$P=CC(O)[C(CF$_3$)$_2$]CO	P (1·9); F ($-57·1$)	1968, 110
PC$_{25}$H$_{17}$F$_6$N$_2$	(C$_6$H$_5$)$_3$P$^\oplus$CH$_2$C(CF$_3$)$_2$C$^\ominus$(CN)$_2$	H; P (24·8); F ($-68·3$)	1967, 102
PC$_{25}$H$_{18}$Cl$_3$O$_4$	(C$_6$H$_5$O)$_2$P(O)C$_6$H$_3$ClCCl$_2$C$_6$H$_5$	P ($-19·6$)	1966, 261
PC$_{25}$H$_{19}$	CH$_3$P(C$_6$H$_4$)$_4$	H	1966, 281
PC$_{25}$H$_{19}$ClO	C$_6$H$_5$(C$_6$H$_5$CH$_2$)P$^\oplus$(C$_6$H$_4$)$_2$O, Cl$^\ominus$	H	1968, 46
PC$_{25}$H$_{19}$Cl$_2$O$_5$	(C$_6$H$_5$O)$_2$P(O)OC$_6$H$_4$CCl$_2$OC$_6$H$_5$	P ($-18·9$)	1965, 45
PC$_{25}$H$_{19}$O$_6$	(C$_6$H$_5$O)$_2$P(O)OC$_6$H$_4$C(O)OC$_6$H$_5$	P ($-18·5$)	1965, 45; 1966, 239
PC$_{25}$H$_{20}$AsF$_8$	CF$_3$PF$_5$, As(C$_6$H$_5$)$_4$	F($-70·9$; $-75·7$)	1969, 67
PC$_{25}$H$_{21}$BrNO$_2$	(C$_6$H$_5$)$_3$P$^\oplus$CH$_2$C$_6$H$_4$NO$_2$, Br$^\ominus$	P ($+24·0$)	1966, 5
PC$_{25}$H$_{21}$NO$_2$	(C$_6$H$_5$)$_3$P$^\oplus$CH$_2$C$_6$H$_4$NO$_2$	H	1967, 117
PC$_{25}$H$_{21}$O	(C$_6$H$_5$)$_2$P(O)CH(C$_6$H$_5$)$_2$	H	1966, 82; 1968, 56
PC$_{25}$H$_{22}$Br	C$_6$H$_5$CH$_2$P$^\oplus$(C$_6$H$_5$)$_3$, Br$^\ominus$	P (23·5)	1966, 5
	C$_7$H$_7$P$^\oplus$(C$_6$H$_5$)$_3$, Br$^\ominus$	P (22·6)	1966, 5
PC$_{25}$H$_{22}$Cl	C$_6$H$_5$CH$_2$P$^\oplus$(C$_6$H$_5$)$_3$, Cl$^\ominus$	H	1967, 118; 1968, 170

PC$_{25}$H$_{23}$O$_2$	(CH$_3$O)$_2$P[C(C$_6$H$_5$)CH]$_2$CC$_6$H$_5$	H; P (62·2)	1968, 75
PC$_{25}$H$_{23}$O$_3$	CH$_3$OP(O)[C(C$_6$H$_5$)CH]$_2$C(OCH$_3$)C$_6$H$_5$	H	1968, 75
PC$_{25}$H$_{23}$O$_5$	C$_6$H$_5$O[(CH$_3$)$_2$C(CH$_2$O)$_2$]P[OC(C$_6$H$_4$)]$_2$	H; P (−48·7)	1968, 301
PC$_{25}$H$_{24}$O	(C$_6$H$_5$)$_2$P(O)C(CH$_3$)=C=C(CH$_3$)C$_6$H$_5$	H	1967, 155
PC$_{25}$H$_{25}$ClNO$_2$	(C$_6$H$_5$)$_3$P$^\oplus$CH=CHC(=NOH)CO-i-C$_3$H$_7$, Cl$^\ominus$	H	1968, 345
PC$_{25}$H$_{25}$O$_2$	(C$_6$H$_5$)$_2$P(O)C(C$_6$H$_5$)=C=C(CH$_3$)CH(OCH$_3$)CH$_3$	H	1967, 155
PC$_{25}$H$_{25}$O$_3$	CH$_3$O(C$_6$H$_5$)$_2$PCH(C$_6$H$_5$)C(COCH$_3$)C(CH$_3$)O	H; P (−25·7)	1968, 124, 152
PC$_{25}$H$_{25}$O$_5$	(CH$_3$O)$_2$P(O)C(CH$_2$COC$_6$H$_5$)C(OCH$_2$C$_6$H$_5$)C$_6$H$_5$, Br$^\ominus$	H; P (22·0)	1966, 106
	C$_6$H$_5$O[(CH$_3$)$_2$C(CH$_2$O)]$_2$P[OC(C$_6$H$_5$)]$_2$	P (−53·7)	1968, 301
PC$_{25}$H$_{26}$BrO	(C$_6$H$_5$)$_3$P$^\oplus$O norbornyl, Br$^\ominus$	H	1965, 114
PC$_{25}$H$_{26}$NO$_3$	(C$_6$H$_5$)$_2$P(O)CH(C$_6$H$_5$)CH$_2$C(=NOH)CO-i-C$_3$H$_7$	H	1968, 345
PC$_{25}$H$_{27}$O$_5$	C$_6$H$_5$CH$_2$O)$_3$P[OC(CH$_3$)]$_2$	H; P (−51·3)	1968, 73
PC$_{25}$H$_{28}$I	(C$_6$H$_5$)$_3$P$^\oplus$C(CH$_3$)$_2$C$_5$H$_{10}$, I$^\ominus$	P (35·1)	1966, 5
PC$_{25}$H$_{29}$O	(C$_6$H$_5$)$_2$P(O)C(CH$_3$)=C=C$_{10}$H$_{16}$	H	1968, 146; 1969, 165
PC$_{25}$H$_{30}$O	(C$_6$H$_5$)$_2$P(O)C(CH$_3$)=C=C$_{10}$H$_{17}$	H	1967, 156
PC$_{25}$H$_{33}$Ge$_2$	(C$_6$H$_5$)$_3$P=C[Ge(CH$_3$)$_3$]$_2$	H	1967, 328
PC$_{25}$H$_{33}$Sn$_2$	(C$_6$H$_5$)$_3$P=C[Sn(CH$_3$)$_3$]$_2$	H	1967, 328
PC$_{25}$H$_{35}$	(C$_4$H$_9$)$_3$P=C$_{13}$H$_8$ (fluorenylidene)	H	1967, 374
PC$_{25}$H$_{53}$	CH$_3$P(C$_{12}$H$_{25}$)$_2$	H; P (−34)	1966, 14
PC$_{25}$H$_{54}$I	CH$_3$P$^\oplus$(C$_8$H$_{17}$)$_3$, I$^\ominus$	P (30·5)	1967, 244
PC$_{25}$H$_{57}$O$_3$Sn$_2$	CH$_3$P(O)[OSn(C$_4$H$_9$)$_3$]$_2$	P{H}	1969, 140
PC$_{26}$H$_{17}$F$_{12}$O$_4$	(C$_6$H$_5$)$_3$P$^\oplus$CHC(CF$_3$)$_2$, (CF$_3$COO)$_2$H$^\ominus$	H; P (−7·5); F (−58·6; −64·4; −76·0)	1967, 102
PC$_{26}$H$_{18}$N$_3$O	(C$_6$H$_5$)$_3$P=C$_8$H$_3$N$_3$O	H	1965, 115
PC$_{26}$H$_{19}$N$_2$O$_2$	(C$_6$H$_5$)$_3$P=C=C=NC$_6$H$_4$NO$_2$	P (1·6)	1968, 110
PC$_{26}$H$_{20}$AsF$_{10}$	(CF$_3$)$_2$PF$_4$, As(C$_6$H$_5$)$_4$	P (−70·8; −78·9)	1969, 67
PC$_{26}$H$_{20}$N	(C$_6$H$_5$)$_3$P=C=C=NC$_6$H$_5$	P (2·4)	1968, 110, 338

Formula	Compound	Resonance	Reference
$PC_{26}H_{21}$	$(C_6H_5)_2PC(C_6H_5)=CHC_6H_5$	H	1966, 282
	$C_6H_5P[C(C_6H_4CH_3)=CH_2]_2$	H	1967, 70
$PC_{26}H_{21}BrNO$	$(C_6H_5)_3PCH_2C(C_6H_4Br)=NO$	H	1968, 343
$PC_{26}H_{21}O_2$	$(C_6H_5)_2P(O)C_6H_4CH_2COC_6H_5$	H	1966, 266
$PC_{26}H_{22}Br$	$(C_6H_5)_3P^{\oplus}CHC_6H_4CH_2, Br^{\ominus}$	H	1967, 300
$PC_{26}H_{22}BrO$	$(C_6H_5)_3P^{\oplus}CH_2COC_6H_5, Br^{\ominus}$	H	1967, 118
$PC_{26}H_{22}Br_2N_2O$	$(C_6H_5)_3P^{\oplus}CH_2C(=NOH)C_6H_4Br, Br^{\ominus}$	H	1968, 343
$PC_{26}H_{22}NO$	$(C_6H_5)_3PON=C(C_6H_5)CH_2$	H; P ($-37 \cdot 0$)	1967, 139; 1968, 343
$PC_{26}H_{22}O$	$(C_6H_5)_3P^{\oplus}CH_2C(O)C_6H_5$	H	1966, 119
$PC_{26}H_{23}BrNO$	$(C_6H_5)_3P^{\oplus}CH_2C(=NOH)C_6H_5, Br^{\ominus}$	H	1967, 375; 1968, 343, 347
$PC_{26}H_{23}O$	$(C_6H_5)_2P(O)CH(C_6H_5)CH_2C_6H_5$	H	1967, 376; 1969, 369
$PC_{26}H_{23}OS$	$(C_6H_5)_2P(S)CH_2C(OH)(C_6H_5)_2$	H; P (34·6)	1967, 329
$PC_{26}H_{23}O_2$	$(C_6H_5)_2P(O)CH_2C(OH)(C_6H_5)_2$	H; P (33·1)	1967, 329
$PC_{26}H_{23}O_5$	$(C_2H_5O)_3POC_{14}H_8O$	P ($-47 \cdot 1$)	1968, 303
$PC_{26}H_{24}$	$(C_6H_5)_3P^{\oplus}CH_2C_6H_4CH_3$	H	1967, 117
$PC_{26}H_{24}Br$	$(C_6H_5)_3P^{\oplus}CH_2C_6H_4CH_3, Br^{\ominus}$	H{P}	1967, 118; 1969, 118
	$(C_6H_5)_2P^{\oplus}(CH_2C_6H_5)_2, Br^{\ominus}$	P (27·0)	1968, 6
$PC_{26}H_{24}ClO_4$	$(C_6H_5)_3P^{\oplus}C(COOCH_3)C(COOCH_3)_2, Cl^{\ominus}$	H	1967, 372
$PC_{26}H_{24}N_3O_2$	$(CH_3)_2N[N(CH_3)CH_2]_2POC_{14}H_8O$	P ($-29 \cdot 8$)	1968, 303
$PC_{26}H_{24}O$	$(C_6H_5)_3P^{\oplus}CH_2C_6H_4OCH_3$	H	1967, 117
$PC_{26}H_{25}N_2O_3$	$(C_6H_5)_3P^{\oplus}CC(O^{\ominus})N(NC_4H_8O)C(O)CH_2$	H	1968, 344
$PC_{26}H_{25}O$	$(C_6H_5)_2P(O)C(C_6H_5)=C=C(CH_2)_3CHCH_3$	H	1969, 164, 165
$PC_{26}H_{27}O_2$	$C_6H_5(C_2H_5)_2P=C(COC_6H_5)CH_2COC_6H_5$	H; P (21·1)	1966, 106

PC$_{26}$H$_{28}$ClO$_2$	C$_6$H$_5$(C$_2$H$_5$)$_2$P$^\oplus$CH(COC$_6$H$_5$)CH$_2$COC$_6$H$_5$, Cl$^\ominus$	P (34·1)	1966, 106
PC$_{26}$H$_{31}$O$_2$	C$_6$H$_5$(β-C$_{10}$H$_7$)P(O)O menthyl	H	1968, 140
PC$_{26}$H$_{33}$OSi	(C$_4$H$_9$)$_2$POSi(C$_6$H$_5$)$_3$	P (122·8)	1967, 343
PC$_{26}$H$_{35}$O$_5$	(C$_4$H$_9$O)$_3$POC$_{14}$H$_8$O	P (−47·3)	1966, 225
PC$_{26}$H$_{56}$I	(CH$_3$)$_2$P$^\oplus$(C$_{12}$H$_{25}$)$_2$, I$^\ominus$	H; P (29·5)	1966, 14
PC$_{27}$H$_{17}$F$_6$	C$_6$H$_5$C[CH=C(C$_6$H$_5$)]$_2$PC(CF$_3$)=CCF$_3$	H; P (−65)	1968, 339
PC$_{27}$H$_{17}$F$_{12}$O$_2$	(C$_6$H$_5$)$_3$P$^\oplus$CHC(CF$_3$)$_2$, (CF$_3$CO)$_2$CH$^\ominus$	H; P (18·1); F (−58·6; −63·8; −76·7)	1967, 102
PC$_{27}$H$_{20}$I	C$_6$H$_5$P$^\oplus$(C$_6$H$_4$CH$_3$)$_3$, I$^\ominus$	H	1968, 15
PC$_{27}$H$_{21}$O	(C$_6$H$_5$)$_2$P(O)CH=C=C(C$_6$H$_5$)$_2$	H	1966, 61
	(C$_6$H$_5$)$_2$P(O)CHC(C$_6$H$_5$)CC$_6$H$_5$	H	1968, 63
PC$_{27}$H$_{22}$N	(C$_6$H$_5$)$_3$P=C=C=NC$_6$H$_4$CH$_3$	P (3·0)	1968, 110
PC$_{27}$H$_{23}$O$_5$	(C$_6$H$_5$O)$_3$POC(CH$_3$)=C(C$_6$H$_5$)O	P	1968, 301
PC$_{27}$H$_{24}$Br	(C$_6$H$_5$)$_3$P$^\oplus$CH$_2$CH=CHC$_6$H$_5$, Br$^\ominus$	P (22·2)	1966, 5
	CH$_3$(C$_6$H$_5$)$_2$P$^\oplus$(CH$_2$C$_6$H$_5$) fluorenyl	H	1969, 156
PC$_{27}$H$_{24}$BrO	(C$_6$H$_5$)$_3$P$^\oplus$CH$_2$COC$_6$H$_4$CH$_3$, Br$^\ominus$	H	1967, 118
PC$_{27}$H$_{24}$ClO$_2$	(C$_6$H$_5$)$_3$P$^\oplus$CH$_2$COOCH$_2$C$_6$H$_5$, Cl$^\ominus$	H	1965, 87
PC$_{27}$H$_{24}$NO	(C$_6$H$_5$)$_3$PCH(CH$_3$)C(C$_6$H$_5$)=NO	H	1968, 343, 347
PC$_{27}$H$_{24}$O$_2$	(C$_6$H$_5$)$_3$P$^\oplus$CH$_2$C$_6$H$_7$OC(O)CH$_3$	H	1967, 117
PC$_{27}$H$_{25}$BrNO	(C$_6$H$_5$)$_3$P$^\oplus$CH(CH$_3$)C(=NOH)C$_6$H$_5$, Br$^\ominus$	H	1967, 375; 1968, 347
PC$_{27}$H$_{25}$ClNO	(C$_6$H$_5$)$_3$P$^\oplus$CH(CH$_3$)C(=NOH)C$_6$H$_5$, Cl$^\ominus$	H	1968, 343
PC$_{27}$H$_{25}$O	(C$_6$H$_5$)$_2$P(O)C(C$_6$H$_5$)=C=C$_7$H$_{10}$	H	1967, 153
PC$_{27}$H$_{25}$O$_2$	(C$_6$H$_5$)$_2$P(O)CH(C$_6$H$_5$)CH(C$_6$H$_5$)OCH$_3$	H	1967, 376
PC$_{27}$H$_{26}$BrO	(C$_6$H$_5$)$_3$P$^\oplus$CH(CH$_3$)CH(C$_6$H$_5$)OH, Br$^\ominus$	H	1967, 359

Formula	Compound	Resonance	Reference
$PC_{27}H_{27}$	$C_4H_9P[C(C_6H_5)=CH]_2CHC_6H_5$	H	1967, 112
$PC_{27}H_{27}O_2$	$(C_2H_5O)_2P[C(C_6H_5)=CH]_2CC_6H_5$	P	1968, 75
	$(C_6H_5)_2P(O)C(C_6H_5)=C=CCH(OCH_3)/(CH_2)_4$	H	1969, 164
$PC_{27}H_{28}IO$	$CH_3(C_6H_5)_2P^{\oplus}CH_2C(OH)(C_6H_5)_2, I^{\ominus}$	H; P (20·7)	1967, 329
$PC_{27}H_{29}O_2$	$(C_6H_5)_2P(O)(C_6H_5)=C=C(CH_3)CH(OCH_3)-i-C_3H_7$	H	1967, 155
$PC_{27}H_{30}ClO_4$	$(C_6H_5)_3P^{\oplus}CHC(C_3H_7)=CC_3H_7, ClO_4^{\ominus}$	H	1968, 63
$PC_{27}H_{33}O$	$(C_6H_5)_2P(O)CH_2(CH)_4C_9H_{17}$	H	1966, 267
$PC_{27}H_{37}N_4O_7$	$(CH_3O)_2P(O)$ methyloestrenone	H	1967, 270
$PC_{27}H_{38}ClO_4$	$(C_4H_9)_3P^{\oplus}CHC(C_6H_5)=CC_6H_5, ClO_4^{\ominus}$	H	1968, 63
$PC_{27}H_{39}N_4O_7$	$(CH_3O)_2P(O)$ hydroxymethyloestrene	H	1967, 270
$PC_{27}H_{60}N_3O$	$[(C_4H_9)_2NCH_2]_3PO$	H	1967, 291
$PC_{28}H_{18}N_3$	$(C_6H_5)_3PC(CH)_3CC(CN)=C(CN)_2$	H	1967, 377
$PC_{28}H_{19}ClN_3O_4$	$(C_6H_5)_3P^{\oplus}C(CH)_3CC(CN)=C(CN)_2, ClO_4^{\ominus}$	H	1967, 377
$PC_{28}H_{21}F_2N_2O$	$(C_6H_5)_2P(O)NHC(CH_2C_6H_4F)C(CN)C_6H_4F$	H	1968, 88
$PC_{28}H_{22}NO$	$(C_6H_5)_3P^{\oplus}CC(O^{\ominus})N(C_6H_5)C(O)CH_2$	H	1968, 344
$PC_{28}H_{23}N_2O$	$(C_6H_5)_2P(O)NHC(CH_2C_6H_5)C(CN)C_6H_5$	H	1968, 88
	$(C_6H_5)_3POC_{10}H_8O$	P (−15·6)	1968, 124
$PC_{28}H_{23}O_2$	$C_2H_5(C_6H_5)_2POC_{14}H_8O$	P (−9·3)	1968, 303

PC$_{28}$H$_{23}$O$_3$	(C$_6$H$_5$)$_2$P(O)CH(COC$_6$H$_5$)CH$_2$COC$_6$H$_5$	P (27·5)	1966, 106
	C$_2$H$_5$O(C$_6$H$_5$)$_2$POC$_{14}$H$_8$O	P (−16·3)	1966, 225
PC$_{28}$H$_{24}$NO$_2$	(C$_6$H$_5$)$_3$P$^\oplus$C(COCH$_3$)=CO$^\ominus$NHC$_6$H$_5$	H	1967, 365
PC$_{28}$H$_{25}$	C$_4$H$_9$P(C$_6$H$_4$)$_4$	H	1966, 281
PC$_{28}$H$_{25}$O	(C$_6$H$_5$)$_3$P=CHCO(CH$_2$)$_2$C$_6$H$_5$	H	1967, 192
PC$_{28}$H$_{25}$O$_2$	(C$_6$H$_5$)$_3$P=C(C$_6$H$_5$)COOC$_2$H$_5$	H	1968, 171
PC$_{28}$H$_{25}$O$_3$	C$_2$H$_5$O(C$_6$H$_5$)$_2$P[OC(C$_6$H$_5$)]$_2$	P (−27·0)	1966, 225
PC$_{28}$H$_{25}$O$_4$S	(C$_6$H$_5$)$_3$P=CHSO$_2$CH(CH$_3$)COOC$_6$H$_5$	H	1967, 74
PC$_{28}$H$_{25}$O$_7$S$_2$	CH$_3$P(O)(C$_6$H$_4$SO$_2$OC$_6$H$_4$)$_2$C(CH$_3$)$_2$	H	1968, 348
PC$_{28}$H$_{26}$Cl	(C$_6$H$_5$)$_3$P$^\oplus$CH$_2$CH=CHC$_6$H$_4$CH$_3$, Cl$^\ominus$	H	1966, 276
PC$_{28}$H$_{26}$F$_3$O$_4$	(C$_6$H$_5$)$_3$P$^\oplus$CH(COOC$_2$H$_5$)$_2$, CF$_3$COO$^\ominus$	H	1967, 378
PC$_{28}$H$_{26}$N$_3$O$_2$	(CH$_2$)$_4$NP[N(CH$_3$)CH$_2$]$_2$OC$_{17}$H$_8$O	P (−35·1)	1968, 303
PC$_{28}$H$_{31}$O$_4$	C$_6$H$_5$(C$_4$H$_9$O)$_2$POC$_{14}$H$_8$O	P (−29·6)	1966, 225
PC$_{28}$H$_{32}$Cl	(C$_6$H$_5$)$_3$P$^\oplus$ geranyl, Cl$^\ominus$	H	1968, 346
PC$_{28}$H$_{33}$O$_4$	C$_6$H$_5$(C$_4$H$_9$O)$_2$P[OC(C$_6$H$_5$)]$_2$	P (−36·6)	1966, 225
PC$_{28}$H$_{34}$BrO	(C$_6$H$_5$)$_3$P$^\oplus$CH$_2$CH=C(CH$_3$)(CH$_2$)$_3$C(CH$_3$)$_2$OH, Br$^\ominus$	H	1968, 346
PC$_{28}$H$_{39}$O$_2$	(C$_4$H$_9$)$_3$P=C(COC$_6$H$_5$)CH$_2$COC$_6$H$_5$	H; P (21·3)	1966, 106
PC$_{28}$H$_{40}$ClO$_2$	(C$_4$H$_9$)$_3$P$^\oplus$CH(COC$_6$H$_5$)CH$_2$COC$_6$H$_5$, Cl$^\ominus$	P (35·7)	1966, 106
PC$_{28}$H$_{49}$N$_4$S$_2$	(C$_6$H$_{11}$)$_2$NCNP(S)(C$_2$H$_5$)SC[N(C$_6$H$_{11}$)$_2$]N	P (65·0)	1968, 137
PC$_{29}$H$_{20}$NO$_4$	(C$_6$H$_5$)$_3$P=CC(O)C(=CHC$_6$H$_6$NO$_2$)CO	H; P (1·2)	1968, 110
PC$_{29}$H$_{23}$	C$_6$H$_5$P[C(C$_6$H$_5$)=CH]$_2$CHC$_6$H$_5$	H	1967, 112
PC$_{29}$H$_{23}$O	C$_6$H$_5$P(O)[C(C$_6$H$_5$)=CH]$_2$CHC$_6$H$_5$	H	1967, 112

250 FORMULAE INDEX II

Formula	Compound	Resonance	Reference
$PC_{29}H_{24}NO_2$	$(C_6H_5)_3P^{\oplus}CC(O^{\ominus})N(C_6H_5)CHCH_3$	H	1968, 344
$PC_{29}H_{25}O_2$	$C_6H_5P(O)C(C_6H_5)(OH)CHC(C_6H_5)CH_2CHC_6H_5$	H; P (30)	1966, 283
	$C_6H_5P^{\ominus}(OH)_2C(C_6H_5)CHC(C_6H_5)CHCC_6H_5, H^{\oplus}$	H; P (-22)	1966, 283
$PC_{29}H_{25}O_5$	$(CH_3O)_3P(OC_{13}H_8)_2$	H; P (-46.7)	1967, 335
$PC_{29}H_{27}$	$CH_3P[C_6H_3(CH_3)]_4$	H(T)	1968, 151
$PC_{29}H_{27}O_4S$	$(C_6H_5)_3P=C(CH_3)SO_2CH(CH_3)COOC_6H_5$	H	1967, 74
$PC_{29}H_{28}NO$	$(C_6H_5)_3PON=C[C_6H_2(CH_3)_3]CH_2$	H; P (-39.4)	1967, 139
	$(C_6H_5)_3PON=C(C_6H_5CHC_3H_7)$	H	1968, 343
$PC_{29}H_{29}BrNO$	$(C_6H_5)_3P^{\oplus}CH(C_3H_7)C(=NOH)C_6H_5, Br^{\ominus}$	H	1968, 343
$PC_{29}H_{29}O$	$(CH_3C_6H_4)_2P(O)CH(C_6H_4CH_3)CH_2C_6H_5$	H	1967, 376
$PC_{29}H_{31}O_3$	$CH_3OP(O)OC_{27}H_{26}CH_2$	H	1967, 270
$PC_{29}H_{31}O_4$	$(C_2H_5O)_2P[C(C_6H_4OCH_3)=CH]_2CC_6H_5$	P	1968, 75
$PC_{29}H_{34}ClO_4$	$(C_6H_5)_3P^{\oplus}CHC(C_4H_9)CC_4H_9, ClO_4^{\ominus}$	H	1968, 63
$PC_{29}H_{37}BrN_3O_2$	$[(CH_3)_2N]_3P^{\oplus}C(CH_2COC_6H_5)C(C_6H_5)OCH_2C_6H_5, Br^{\ominus}$	H; P (50.4)	1966, 106
$PC_{29}H_{49}O_3$	$CH_3(CH_3O)P(O)OC_{27}H_{43}$	H	1967, 168
$PC_{29}H_{50}BrO_4$	$(CH_3O)_2P(O)$ bromoketocholestane	H	1967, 167
$PC_{29}H_{51}Br_2O_3$	$(CH_3O)_2PO$ dibromocholestane	H	1966, 43; 1967, 167
$PC_{29}H_{51}O_3$	$(CH_3O)_2P(O)$cholestene	H	1967, 167
$PC_{29}H_{51}O_4$	$(CH_3O)_2P(O)$ epoxycholestane	H	1967, 167
$PC_{29}H_{52}BrO_4$	$(CH_3O)_2P(O)$ bromohydroxycholestane	H	1966, 43; 1967, 167
$PC_{29}H_{52}IN_4S_2$	$(C_6H_{11})_2N^{\oplus}CNP(SCH_3)(C_2H_5)SC[N(C_6H_{11})_2]N, I^{\ominus}$	P (58.8)	1968, 137

$PC_{29}H_{53}O_3$	$(CH_3O)_2P(O)$ cholestane	H	1967, 167
$PC_{29}H_{53}O_4$	$(CH_3O)_2P(O)$ hydroxycholestane	H	1967, 167
$PC_{29}H_{62}NS$	$(C_2H_5)_2P(S)CH_2N(C_{12}H_{25})_2$	P (52·2)	1966, 40
$PC_{30}H_{19}ClF_3O_2$	$(C_6H_5)_3P=CC(O)C[C(CF_3)(C_6H_4Cl)]CO$	P (1·9); F (−57·8)	1968, 110
$PC_{30}H_{20}Cl_5$	$(C_6H_4Cl)_5P$	P (−86·8)	1968, 288
$PC_{30}H_{20}NO_2$	$(C_6H_5)_3P=CC(O)C(CHC_6H_4CN)CO$	H; P (0·0)	1968, 110
$PC_{30}H_{20}N_3$	$(C_6H_5)_3P=C(C_6H_5)C(CN)=C(CN)_2$	H	1965, 115
$PC_{30}H_{21}$	$C_6H_5P(C_6H_4)_4$	H	1966, 281
$PC_{30}H_{21}O_3$	$(C_6H_5)_3PCC(O)C(CHC_6H_4CHO)CO$	H	1968, 110
$PC_{30}H_{24}BrO$	$(C_6H_5)_3P^{\oplus}CH_2CO(\beta\text{-}C_{10}H_7), Br^{\ominus}$	H	1968, 343
$PC_{30}H_{24}NO$	$(C_6H_5)_3PCH_2C(\beta\text{-}C_{10}H_7)=NO$	H	1968, 343
$PC_{30}H_{25}$	$P(C_6H_5)_5$	P (−88·7)	1968, 288
$PC_{30}H_{25}As_2$	$C_6H_5P[As(C_6H_5)_2]_2$	P (36·1)	1969, 14
$PC_{30}H_{25}BrNO$	$(C_6H_5)_3P^{\oplus}CH_2C(=NOH)\beta\text{-}C_{10}H_7, Br^{\ominus}$	H	1968, 343
$PC_{30}H_{25}OSi$	$(C_6H_5)_2POSi(C_6H_5)_3$	P (98·0)	1967, 343
$PC_{30}H_{25}O_3$	$(C_6H_5)_3P=C(COC_6H_5)C(=CH_2)COOCH_3$	H	1965, 116
$PC_{30}H_{25}O_5$	$P(OC_6H_5)_5$	P (−85·6)	1968, 124
$PC_{30}H_{25}Sn$	$(C_6H_5)_2PSn(C_6H_5)_3$	P (−56)	1967, 337
$PC_{30}H_{26}NSi$	$(C_6H_5)_2PNHSi(C_6H_5)_3$	P	1966, 284
$PC_{30}H_{27}N_2O_3$	$(C_6H_5)_2P(O)NHC(CH_2C_6H_4OCH_3)C(CN)C_6H_4OCH_3$	H	1968, 88
$PC_{30}H_{27}N_4O_3$	$(C_6H_5)_3P=NCNN(COOCH_3)C[OC_6H_3(CH_3)_2]N$	P (18·5)	1969, 360
$PC_{30}H_{27}O$	$(C_6H_5)_2P(O)C(C_6H_5)=C=C(CH_3)CH(CH_3)C_6H_5$	H	1967, 109
$PC_{30}H_{27}O_3$	$C_2H_5O(C_6H_5)_2P=C(COC_6H_5)(CH_2COC_6H_5)$	H; P (54·2)	1966, 106
$PC_{30}H_{28}N_2O_8$	$(C_6H_5)_3P=C(CO_2CH_3)CHC(CO_2CH_3)CH[N(CO_2CH_3)]_2$	H; P (21·5)	1969, 360

Formula	Compound	Resonance	Reference
$PC_{30}H_{29}$	$C_2H_5[C_6H_3(CH_3)]_4$	H(T)	1968, 151
$PC_{30}H_{31}O$	$(C_6H_5)_2P(O)C(C_6H_5)=C=C_{10}H_{16}$	H	1968, 146; 1969, 165
$PC_{30}H_{32}Cl$	$CH_3(t\text{-}C_4H_9)(C_6H_5)P^{\oplus}C_6H_4CH(C_6H_5)_2, Cl^{\ominus}$	H	1966, 285
$PC_{30}H_{29}O_4$	$OP(OC_6H_4\text{-}t\text{-}C_4H_9)_3$	P	1967, 198
$PC_{30}H_{47}O_3$	$HP(O)(OC_6H_4C_9H_{19})_3$	P (8·8)	1967, 277
$PC_{30}H_{47}O_5$	$(i\text{-}C_3H_7O)_2P(O)C_{24}H_{32}O_2$	H	1966, 104
$PC_{30}H_{51}O_4$	$CH_3P(O)(OH)O(C_{29}H_{47}O)$	H	1967, 168
$PC_{30}H_{53}O_4$	$CH_3P(O)(OH)O(C_{29}H_{49}O)$	H	1967, 168
$PC_{31}H_{20}F_3$	$C_{13}H_8=P(C_6H_4F)_3$	F	1968, 13
$PC_{31}H_{21}F_3I$	$C_{13}H_9P^{\oplus}(C_6H_4F)_3, I^{\ominus}$	F	1968, 13
$PC_{31}H_{23}$	$CH_3C_6H_4P(C_6H_4)_4$	H	1966, 281
	$C_{13}H_8=P(C_6H_5)_3$	H	1967, 374
$PC_{31}H_{23}O$	$CH_3OC_6H_4P(C_6H_4)_4$	H	1966, 281
$PC_{31}H_{24}N_3O$	$(C_6H_5)_3P=C(C_6H_5)CC(CN)C(O)NCNHCH_3$	H	1965, 115
$PC_{31}H_{25}IN_3O$	$(C_6H_5)_3P=C(C_6H_5)CC(CN)C(O)NCNH_2^{\oplus}CH_3, I^{\ominus}$	H	1965, 115
$PC_{31}H_{25}O$	$(C_6H_5)_2P(O)C_6H_4CH(C_6H_5)_2$	H	1966, 285
$PC_{31}H_{26}Br$	$(C_6H_5)P^{\oplus}CH(C_6H_5)_2, Br^{\ominus}$	P (21·4)	1966, 5
$PC_{31}H_{29}O_7$	$(CH_3O)_3P[OC(C_6H_5)COC_6H_5]_2$	P (−48·0)	1967, 335; 1968, 318
$PC_{31}H_{53}O_4$	$CH_3(CH_3O)P(O)O(\alpha\text{-amyryl})$	H	1967, 168
$PC_{31}H_{54}BrO_5$	$(CH_3O)_2P(O)$ bromoacetoxycholestanyl	H	1966, 43; 1967, 167
$PC_{31}H_{55}O_4$	$CH_3(CH_3O)P(O)O$ dimethylcholestenyl	H	1967, 168
$PC_{32}H_{23}Br_8O_5$	$(C_2H_5O)_3P(OC_{13}H_4Br_4)_2$	H; P (−48)	1967, 141
$PC_{32}H_{23}N_2$	$(C_6H_5)_3P=C(CH)_3C(C_6H_5)=C(CN)_2$	H	1967, 170
$PC_{32}H_{23}O_2$	$(C_6H_5)_3POC_{14}H_8O$	P (−15·6)	1966, 225; 1968, 303

PC$_{32}$H$_{23}$O$_7$	C$_6$H$_5$(C$_6$H$_5$O)$_2$POC$_{14}$H$_8$O	P (−38·1)	1966, 225
PC$_{32}$H$_{23}$O$_5$	(C$_6$H$_5$O)$_3$POC$_{14}$H$_8$O	P (−58·7)	1965, 21; 1966, 225; 1968, 301, 309
PC$_{32}$H$_{24}$N$_3$O$_2$	(C$_6$H$_5$)$_3$P=C(CH)$_3$CH(C$_6$H$_4$NO$_2$)CH(CN)$_2$	H; P	1967, 377
PC$_{32}$H$_{24}$N$_3$O$_7$	(C$_6$H$_5$)$_3$P$^{\oplus}$CH=CHC$_6$H$_5$, picrate	H	1967, 376
PC$_{32}$H$_{25}$N$_2$	(C$_6$H$_5$)$_3$P=C(CH)$_3$CH(C$_6$H$_5$)CH(CN)$_2$	H; P	1967, 377
PC$_{32}$H$_{25}$O$_4$	C$_6$H$_5$(C$_6$H$_5$O)$_2$P[OC(C$_6$H$_5$)]$_2$	P (−43·2)	1966, 225
PC$_{32}$H$_{25}$O$_5$	(C$_6$H$_5$O)$_3$P[OC(C$_6$H$_5$)]$_2$	P (−62·5)	1965, 21; 1966, 225; 1968, 301, 309
PC$_{32}$H$_{26}$NO	(C$_6$H$_5$)$_3$P=NC(O)CH(C$_6$H$_5$)$_2$	H	1965, 109; 1966, 280
PC$_{32}$H$_{26}$NO$_2$	(C$_6$H$_5$)$_3$P$^{\oplus}$C(COCH$_3$)=CO$^{\ominus}$NHC$_{10}$H$_7$	H	1967, 365
PC$_{32}$H$_{27}$ClNO	(C$_6$H$_5$)$_3$P$^{\oplus}$=NHC(O)CH(C$_6$H$_5$)$_2$, Cl$^{\ominus}$	H	1965, 109; 1966, 280
PC$_{32}$H$_{27}$O$_5$	(C$_6$H$_5$)$_3$PCHC(C$_6$H$_5$)O[C(COOCH$_3$)]$_2$	H	1965, 116
PC$_{32}$H$_{28}$NO	(C$_6$H$_5$)$_3$PON(C$_6$H$_5$)CH(C$_6$H$_5$)CH$_2$	H; P (−58·6)	1967, 76
PC$_{32}$H$_{29}$O$_5$	(C$_6$H$_5$)$_3$P=C(COOCH$_3$)CH(COOCH$_3$)CH$_2$COC$_6$H$_5$	H	1965, 116
PC$_{32}$H$_{31}$O$_5$	(C$_2$H$_5$O)$_3$P(OC$_{13}$H$_8$)$_2$	H; P (−49·3)	1967, 335
PC$_{32}$H$_{44}$BrO	(C$_6$H$_5$)$_3$CCH$_2$OP$^{\oplus}$(C$_4$H$_9$)$_3$, Br$^{\ominus}$	H	1966, 286
PC$_{33}$H$_{25}$BrNO$_2$	(C$_6$H$_5$)$_3$P$^{\oplus}$C(COC$_6$H$_4$Br)=C(O$^{\ominus}$)NHC$_6$H$_5$	H	1967, 365
PC$_{33}$H$_{26}$ClO$_4$	(C$_6$H$_5$)$_3$P$^{\oplus}$CHC(C$_6$H$_5$)CC$_6$H$_5$, ClO$_4^{\ominus}$	H	1968, 63
PC$_{33}$H$_{26}$NO$_2$	(C$_6$H$_5$)$_3$P$^{\oplus}$C(COC$_6$H$_5$)=C(O$^{\ominus}$)NHC$_6$H$_5$	H	1967, 365
PC$_{33}$H$_{27}$	(C$_6$H$_5$)$_3$P=C$_{13}$H$_6$(CH$_3$)$_2$	H	1967, 374
PC$_{33}$H$_{27}$N$_2$O	(C$_6$H$_5$)$_3$PC(CH)$_3$CH(C$_6$H$_4$OCH$_3$)CH(CN)$_2$	H; P	1967, 377
PC$_{33}$H$_{27}$O	(C$_6$H$_5$)$_3$P=C(C$_6$H$_5$)C(O)CH$_2$C$_6$H$_5$	H	1967, 139
PC$_{33}$H$_{29}$ClNS	(C$_6$H$_5$)$_2$CHC(SCH$_3$)NP$^{\oplus}$(C$_6$H$_5$)$_3$, Cl$^{\ominus}$	H	1966, 377
PC$_{33}$H$_{30}$Br	CH$_3$(C$_6$H$_5$)P$^{\oplus}$(CH$_2$C$_6$H$_5$)C$_{13}$H$_{11}$, Br$^{\ominus}$	H	1969, 156
	C$_6$H$_5$P$^{\oplus}$(CH$_2$C$_6$H$_5$)$_2$C$_{13}$H$_{11}$, Br$^{\ominus}$	H	1969, 156

Formula	Compound	Resonance	Reference
$PC_{33}H_{30}ClSn$	$(C_6H_5)_2PCH(CH_2Cl)CH_2Sn(C_6H_5)_3$	H	1965, 117
$PC_{33}H_{33}AlNSi$	$(CH_3)_3SiNP(C_6H_5)_2C_6H_4Al(C_6H_5)_2$	H	1967, 379
$PC_{33}H_{33}GaNSi$	$(CH_3)_3SiNP(C_6H_5)_2C_6H_4Ga(C_6H_5)_2$	H	1967, 379
$PC_{33}H_{34}BrO_2$	$C_6H_5(C_2H_5)_2P^{\oplus}C(CH_2COC_6H_5)C(OCH_2C_6H_5)C_6H_5, Br^{\ominus}$	H; P (28·0)	1966, 106
$PC_{33}H_{63}O_4$	$(C_2H_5O)_2P(O)$O-methyl cholestenyl	H	1969, 367
$PC_{34}H_{23}F_2N_2O_3$	$(C_6H_5)_3PC[C(O)NC_6H_4F]_2CO$	H; P (17·6); F (−114·3)	1968, 110
$PC_{34}H_{23}N_7O_7$	$(C_6H_5)_3PC[C(O)NC_6H_4NO_2]_2CO$	H; P (17·6)	1968, 110
$PC_{34}H_{24}N_3$	$(C_6H_5)_3PC(CH)_3CC(CN)(C_6H_5)CH(CN)_2$	H	1967, 377
$PC_{34}H_{25}N_2O_3$	$(C_6H_5)_3PC[C(O)NC_6H_5]_2CO$	H; P (17·1)	1968, 110
$PC_{35}H_{46}BrO_2$	$(C_4H_9)_3P^{\oplus}C(CH_2COC_6H_5)C(OCH_2C_6H_5)C_6H_5, Br^{\ominus}$	H; P (28·8)	1966, 106
$PC_{36}H_{23}N_4O_3$	$(C_6H_5)_3PC[C(O)NC_6H_4CN]_2CO$	H; P (18·1)	1968, 110
$PC_{36}H_{29}N_2O_3$	$(C_6H_5)_3PC[C(O)NC_6H_4CH]_2CO$	H; P (17·4)	1968, 110
$PC_{36}H_{30}$	$CH_3(C_6H_5)P^{\oplus}[C(C(C_6H_5))=CH]_2C(C_6H_5)_2$	H	1966, 287
$PC_{36}H_{30}As_3$	$P[As(C_6H_5)_2]_3$	P (−59·1)	1969, 14
$PC_{36}H_{75}$	$(C_{12}H_{25})_3P$	H; P (36)	1966, 14
$PC_{36}H_{75}O$	$(C_{12}H_{25})_3PO$	H; P (53)	1966, 14
$PC_{37}H_{27}BrNO_2$	$(C_6H_5)_3P^{\oplus}C(COC_6H_4Br)=CO^{\ominus}NH\text{-}\alpha\text{-}C_{10}H_7$	H	1967, 365

$PC_{37}H_{28}NO_2$	$(C_6H_5)_3P^{\oplus}C(COC_6H_5)=CO^{\ominus}NH\text{-}\alpha\text{-}C_{10}H_7$	H	1967, 365
$PC_{37}H_{30}Cl$	$(C_6H_5)_3P^{\oplus}C_6H_4C(C_6H_5)_2$, $Cl^{\ominus}$	H	1966, 285
$PC_{37}H_{35}$	$i\text{-}C_3H_7C_6H_4P(C_6H_4CH_3)_4$	H(T)	1967, 158; 1968, 151
$PC_{37}H_{78}Br$	$CH_3P^{\oplus}(C_{12}H_{25})_3$, $Br^{\ominus}$	H; P (32)	1966, 14
$PC_{37}H_{79}NOS$	$C_{18}H_{37}(CH_3)_2P=N=S(O)CH_3(C_{16}H_{33})$	H	1967, 357
$PC_{38}H_{29}Cl_2N_2$	$(C_6H_5)_2CHC(NC_6H_3Cl_2)N=P(C_6H_5)_3$	H	1966, 247
$PC_{38}H_{30}ClO$	$(C_6H_5)_3P^{\oplus}OC(C_6H_5)C(C_6H_5)_2$, $Cl^{\ominus}$	H; P (63)	1965, 109
$PC_{38}H_{30}Cl_3N_2$	$(C_6H_5)_3P^{\oplus}NC(NHC_6H_3Cl_2)CH(C_6H_5)_2$, $Cl^{\ominus}$	H	1966, 247
$PC_{38}H_{30}N_3O_2$	$(C_6H_5)_3P=NC(NC_6H_4NO_2)CH(C_6H_5)_2$	H	1966, 247
$PC_{38}H_{31}$	$C_{10}H_7P[C_6H_3(CH_3)]_4$	H(T)	1966, 101; 1968, 151
$PC_{38}H_{31}ClN_3O_2$	$(C_6H_5)_3P^{\oplus}NC(NHC_6H_4NO_2)CH(C_6H_5)_2$, $Cl^{\ominus}$	H	1966, 247
$PC_{38}H_{31}N_2$	$(C_6H_5)_3P=NC(NC_6H_5)CH(C_6H_5)_2$	H	1966, 247
$PC_{38}H_{31}Sn$	$(C_6H_5)_2PC(CH_2Cl)=CHSn(C_6H_5)_3$	H	1965, 117
$PC_{38}H_{32}ClN_2$	$(C_6H_5)_3P^{\oplus}NC(NHC_6H_5)CH(C_6H_5)_2$, $Cl^{\ominus}$	H	1966, 247
$PC_{38}H_{32}NO$	$(C_6H_5)_3PON(C_6H_5)CH(C_6H_5)CHC_6H_5$	H; P (−57.9)	1967, 76
$PC_{38}H_{33}Sn$	$(C_6H_5)_2PCH(C_6H_5)CH_2Sn(C_6H_5)_3$	H	1965, 117
$PC_{39}H_{32}N$	$(C_6H_5)_3P=C(C_6H_5)C(CH_2C_6H_5)=NC_6H_5$	H; P (11)	1967, 139, 375
$PC_{39}H_{39}AlNSi$	$(C_6H_5)_3PN[Al(C_6H_5)_3]Si(CH_3)_3$	H	1967, 379
$PC_{39}H_{39}GaNSi$	$(C_6H_5)_3PN[Ga(C_6H_5)_3]Si(CH_3)_3$	H	1967, 379
$PC_{39}H_{72}N_3O$	$[(C_6H_{11})_2NCH_2]_3PO$	H	1967, 291
$PC_{40}H_{33}$	$C_4H_4P[C_6H_3(CH_3)]_4$	H(T)	1966, 101; 1968, 151
$PC_{40}H_{36}N_3$	$(C_6H_5)_3P=NC[NC_6H_4N(CH_3)_2]CH(C_6H_5)_2$	H	1966, 280
$PC_{40}H_{37}ClN_3$	$(C_6H_5)_3P^{\oplus}NC[NHC_6H_4N(CH_3)_2]CH(C_6H_5)_2$, $Cl^{\ominus}$	H	1966, 280
$PC_{41}H_{34}BrO_2$	$(C_6H_5)_3P^{\oplus}C(CH_2COC_6H_5)C(OCH_2C_6H_5)C_6H_5$, $Br^{\ominus}$	H; P (20.9)	1966, 106
$PC_{41}H_{59}O_3$	$C_6H_5CH_2O)_2P(O)$ cholestenyl	H	1967, 167
$PC_{42}H_{33}$	$C_{14}H_9P[C_6H_3(CH_3)]_4$	H(T)	1966, 101

Formula	Compound	Resonance	Reference
$PC_{42}H_{35}Sn_2$	$C_6H_5P[Sn(C_6H_5)_3]_2$	P (−163)	1967, 337; 1969, 324
$PC_{42}H_{38}N$	$(C_6H_5)_3P=C[C_6H_2(CH_3)_3]C(O)CH_2C_6H_5$	H; P (11·5)	1967, 139
$PC_{45}H_{69}O_3$	$P(OC_6H_4 \cdot C_9H_{19})_3$	P (141·0)	1967, 277
$PC_{45}H_{69}O_3S$	$SP(OC_6H_4 \cdot C_9H_{19})_3$	P (74·0)	1967, 277
$PC_{45}H_{69}O_4$	$OP(OC_6H_4 \cdot C_9H_{19})_3$	P (−4·3)	1967, 277
$PC_{47}H_{37}N_2$	$[(C_6H_5)_2N]_2P[C(C_6H_5)=CH]_2CC_6H_5$	H; P (29·5)	1969, 145
$PC_{48}H_{31}N_2Si_2$	$(C_6H_5)_3SiNHP(C_6H_5)_2=NSi(C_6H_5)_3$	P	1966, 284
$PC_{54}H_{45}Sn_3$	$P[Sn(C_6H_5)_3]_3$	P (−323)	1967, 337; 1969, 14
P_2			
$P_2CH_2O_6$	$CH_2(PO_3^{2\ominus})_2$	P (15·4)	1967, 123
$P_2CH_3Cl_4N$	$CH_3N(PCl_2)_2$	H; P	1967, 96; 1968, 89; 1969, 370
$P_2CH_3Cl_4N_3O_2S$	$CH_3C(PCl_2=N)_2SO_2$	P (15·1)	1966, 98
$P_2CH_3F_4N$	$CH_3N(PF_2)_2$	H; P (141·5); F (−74·6)	1967, 96; 1968, 89; 1969, 31, 61, 370
$P_2CH_4Cl_2O_6$	$CCl(PO_3H_2)_2$	P (7·9)	1967, 12, 380; 1968, 349
$P_2CH_4Na_2O_7$	$CH(PO_3NaH)_2OH$	P (15·0)	1967, 380
$P_2CH_4Na_2O_8$	$C(PO_3NaH)_2(OH)_2$	P (14·5)	1967, 380
$P_2CH_4O_7$	$O=C(PO_3H_2)_2$	P (13·0)	1967, 12
P_2CH_6	$CH_2(PH_2)_2$	H; P (−122; −126)	1965, 118; 1966, 15, 16

$P_2CH_6O_6$	$CH_2(PO_3H_2)_2$	H; P (17 to 18)	1967, 12, 371; 1968, 56, 349; 1969, 332
$P_2CH_6O_7$	$CH(PO_3H_2)_2OH$	P (16·1)	1967, 12
$P_2CNa_4O_7$	$OC(PO_3Na_2)_2$	P (0·0)	1967, 380
$P_2C_2H_4F_4O_2$	$(F_2POCH_2)_2$	F (−49·0)	1963, 3
$P_2C_2H_4O_6$	$(CH_2PO_3^{\ominus})_2$	P (23·2)	1967, 7
$P_2C_2H_5Cl_4N$	$C_2H_5N(PCl_2)_2$	H; P	1967, 96; 1968, 89
$P_2C_2H_5F_4N$	$C_2H_5N(PF_2)_2$	H; P (145·3); F (−72·5)	1968, 89; 1969, 60, 61
$P_2C_2H_6BCl_6N_3$	$N_3P_2Cl_6B(CH_3)_2$	P (28)	1968, 136
$P_2C_2H_6BaO_8$	$[(CH_3O)_2PO^{\ominus}]_2, Ba^{2\oplus}$	H	1967, 90
$P_2C_2H_6Cl_2N_2O_2$	$[CH_3NP(O)Cl]_2$ (isom.)	H; P (−3; 0)	1966, 53
$P_2C_2H_6Cl_2N_2S_2$	$[CH_3NP(S)Cl]_2$	P (51·5)	1966, 83
$P_2C_2H_6Cl_6N_2$	$(CH_3NPCl_3)_2$	H; P (−77·5)	1965, 98; 1966, 53, 83, 288; 1968, 136
$P_2C_2H_6F_6N_2$	$(CH_3NPF_3)_2$	H; P (−70); F (−79)	1966, 79; 1967, 59; 1968, 57; 1969, 65
$P_2C_2H_6O_7$	$(CH_3O)_2P_2O_5^{2\oplus}$	^{13}C; P (−2·7)	1968, 3
$P_2C_2H_8$	$(CH_2PH_2)_2$	H; P (−130·8)	1965, 118; 1966, 147
$P_2C_2H_8O_6$	$(CH_2PO_3H_2)_2$	H; P (27·4)	1967, 371; 1969, 332
$P_2C_2H_8O_7$	$CH_3CH(PO_3H_2)_2$	P (22·6)	1967, 12; 1968, 349
	$CH_3C(OH)(PO_3H_2)_2$	P (19·8)	1967, 12
	$(CH_3OPO_2H)_2O$	H	1968, 57
$P_2C_2H_{10}Ge$	$(CH_3)_2Ge(PH_2)_2$	H	1969, 136
$P_2C_2H_{10}Si$	$(CH_3)_2Si(PH_2)_2$	H	1969, 136
$P_2C_2H_{16}ClN_7$	$(NH_2)_3P=NP(NH_2)(NHCH_3)_2Cl$	P (16·5)	1967, 285
$P_2C_3H_6Cl_4O_2$	$(CH_3)_2C[P(O)Cl_2]_2$	H; P (45·1)	1966, 15
$P_2C_3H_6F_3$	$CF_3P=P^{\oplus}(CH_3)_2$	H{P, F}	1967, 40
$P_2C_3H_6O_3$	$P(OCH_2)_3P$	H{^{13}C, P}	1966, 37; 1969, 3

Formula	Compound	Resonance	Reference
$P_2C_3H_6O_5$	$OP(OCH_2)_3PO$	H	1966, 37
$P_2C_3H_6O_6$	$(CH_2)_3(PO_2^\ominus)_2$	P (22·4)	1967, 7
$P_2C_3H_8O_6$	$CH_2{=}CHCH_2(PO_3H_2)_2$	P (20·5)	1968, 349
$P_2C_3H_8O_8$	$CH(CH_2COOH)(PO_3H_2)_2$	P (20·0)	1968, 349
$P_2C_3H_9ClO_4$	$CH_3(CH_3O)P(O)(CH_3)P(O)Cl$	H; P	1967, 186
$P_2C_3H_9O_9$	$CH(CH_2PO_3H_2)_2$	P (24·19)	1968, 349
$P_2C_3H_{10}$	$(CH_3)_2C(PH_2)_2$	H; P (-74)	1966, 15
	$(CH_2)_3(PH_2)_2$	H; P ($-138·6$)	1965, 118; 1966, 16
$P_2C_3H_{10}ClFNSSi$	$(CH_3)_3SiNHP(S)ClF$	H; P (68·4); F (73·4)	1969, 69
$P_2C_3H_{10}F_2NSSi$	$(CH_3)_3SiNHP(S)F_2$	H; P (66·6); F (40·2)	1969, 69
$P_2C_3H_{10}O_6$	$(CH_3)_2C(PO_3H_2)_2$	P (27)	1967, 12; 1968, 349
$P_2C_3H_{10}O_7$	$(CH_3O)_2P(O)O(CH_3O)PO_2H$	H; P	1966, 138
$P_2C_3H_{11}NO_6$	$CH_3N(CH_2PO_3H_2)_2$	H; P (ca. 9)	1966, 212
$P_2C_4F_8I_2$	$[(CF_2)_2PI]_2$	F	1962, 2
$P_2C_4F_{12}S_2$	$(CF_3)_2P(S)SP(CF_3)_2$	F ($-58·9$; $-53·8$)	1969, 57
$P_2C_4H_6Cl_4N_4O_2$	$[Cl_2PNN{=}C(CH_3)O]_2$	H; P ($-52·7$)	1969, 149
$P_2C_4H_6F_6$	$(CH_3)_2PP(CF_3)_2$	H; P (-56; 12·1); F ($-47·4$)	1967, 28, 40; 1968, 55
$P_2C_4H_9F_3$	$(CH_3)_3PPCF_3$	H; P; F	1967, 28
$P_2C_4H_{10}Cl_2F_4N_2$	$(CH_3NPF_2CH_2Cl)_2$	H	1966, 36
$P_2C_4H_{10}Cl_2N_2O_2$	$[C_2H_5NP(O)Cl]_2$ (isom.)	H; P ($-5·7$; $-3·1$)	1966, 53
$P_2C_4H_{10}Cl_6N_2$	$(C_2H_5N{=}PCl_3)_2$	H; P ($-79·0$)	1965, 98; 1966, 53, 288
$P_2C_4H_{10}F_2S_4Zn$	$Zn[SP(S)F(C_2H_5)]_2$	H; F ($-49·7$)	1968, 350
$P_2C_4H_{10}K_2$	$[P(C_2H_5)K]_2$	P ($-79·6$)	1965, 27

$P_2C_4H_{12}$	$(CH_2)_4(PH_2)_2$	$P\,(-138)$	1965, 11; 1966, 16
	$[(CH_3)_2P]_2$	$^{13}C;\ H\{P,\ ^{13}C\}$	1967, 28, 54, 83, 124, 125; 1968, 22, 57, 113
$P_2C_4H_{12}BaO_8$	$[(CH_3O)_2P(O)O^\ominus]_2,\ Ba^{2\oplus}$	H	1969, 95
$P_2C_4H_{12}Cl_2O_6$	$[(CH_3O)_2PO]_2CCl_2$	H; P (10·0)	1968, 349
$P_2C_4H_{12}F_6O_3$	$CH_3(CH_3O)_3P^\oplus,\ PF_6^\ominus$	H; P; F	1965, 43
$P_2C_4H_{12}NNaS_2$	$[(CH_3)_2P(S)]_2^\ominus N,\ Na^\oplus$	H; P (43·8)	1968, 111
$P_2C_4H_{12}N_2O_2S_2$	$[CH_3SP(O)NCH_3]_2$	H	1966, 53
$P_2C_4H_{12}N_4O_2$	$[(CH_3)_2NP(O)N]_2$	P (18·7)	1965, 78
$P_2C_4H_{12}O$	$O(CH_2CH_2PH_2)_2$	H	1969, 44
$P_2C_4H_{12}O_4S_3$	$[(CH_3O)_2P(S)]_2S$	P (83·5)	1966, 218
$P_2C_4H_{12}O_4S_4$	$[(CH_3O)_2P(S)]_2S_2$	P (83·5)	1966, 218
$P_2C_4H_{12}O_5$	$[CH_3(CH_3O)P(O)]_2O$	H; P	1967, 186
$P_2C_4H_{12}O_6$	$[(CH_3O)_2P(O)]_2$	H	1967, 381; 1968, 57
$P_2C_4H_{12}O_7$	$[(CH_3O)_2P(O)]_2O$	H	1967, 381
$P_2C_4H_{12}S$	$(CH_3)_2P(S)P(CH_3)_2$	H; P	1967, 28; 1968, 113
$P_2C_4H_{12}S_2$	$[(CH_3)_2P(S)]_2$	H	1965, 28; 1968, 113
$P_2C_4H_{13}NO_6$	$C_2H_5N(CH_2PO_3H_2)_2$	H; P (ca. 8)	1967, 315
$P_2C_4H_{14}N_4S_2$	$[CH_3NHP(S)NCH_3]_2$	P (60)	1966, 83
$P_2C_4H_{20}ClN_7$	$[C_2H_5NHP(NH_2)_2]_2N^\oplus,\ Cl^\ominus$	P (ca. 15)	1966, 288
	$C_2H_5NHP(NH_2)_2NHP(NH_2)_2\!=\!NC_2H_5,\ Cl^\ominus$	P	1966, 288
	$C_2H_5NHP(NH_2)_2NHP(NH_2)\!=\!NH(NHC_2H_5)$	P	1966, 288
$P_2C_5H_{20}N_7$			
$P_2C_5H_3BF_{12}O_2$	$CH_3B[OP(CF_3)_2]_2$	^{11}B; H; P (81·3); F ($-67·5$)	1969, 27
$P_2C_5H_9Cl_5O_2$	$C_4H_9CCl[P(O)Cl_2]_2$	H; P (32·9)	1966, 15
$P_2C_5H_{10}Cl_4O_2$	$C_4H_9CH[P(O)Cl_2]_2$	H; P (34·6)	1966, 15
$P_2C_5H_{12}$	$C_5H_{10}PH_2$	P (-139)	1966, 15
$P_2C_5H_{13}NaO_6$	$[(CH_3O)_2P(O)]_2CH^\ominus,\ Na^\oplus$	H; P (45·5)	1967, 371; 1968, 349
$P_2C_5H_{13}N_3$	$N[(CH_3)_2PN]_2CH$	H; P (27·1)	1967, 355; 1968, 94

260   FORMULAE INDEX II

Formula	Compound	Resonance	Reference
$P_2C_5H_{14}$	$C_4H_9CH(PH_2)_2$	H; P (-109)	1966, 15
$P_2C_5H_{14}O_6$	$C_4H_9CH(PO_3H_2)_2$	P (23·3)	1966, 15; 1968, 349
	$[(CH_3O)_2P(O)]_2CH_2$	H; P (23·0)	1967, 371; 1968, 349
$P_2C_5H_{15}NS_2$	$(CH_3)_2P(S)N=P(SCH_3)(CH_3)_2$	H; P (45·3; 34·6)	1968, 111
$P_2C_5H_{19}N_5S_2$	$[(CH_3NH)_2P(S)]_2NCH_3$	P (73)	1966, 83
$P_2C_6H_4ClF_4N$	$C_6H_4Cl(PF_2)_2$	P (131·3); F ($-68·1$)	1969, 61
$P_2C_6H_4F_4O_2$	$C_6H_4(OPF_2)_2$	F ($-44·9$)	1963, 3
$P_2C_6H_5Cl_3FNS$	$C_6H_5PCl_2=NP(S)ClF$	P (20·6; 49·1); F (75·9)	1969, 69
$P_2C_6H_5F_4N$	$C_6H_5(PF_2)_2$	P (132·7); F ($-68·2$)	1969, 61
$P_2C_6H_6B_2Cl_4F_{12}N_2$	$[(CF_3)_2PN(CH_3)BCl_2]_2$	^{11}B; P	1968, 24
$P_2C_6H_{12}O_6$	$[(CH_2O)_2POCH_2]_2$	P (134)	1968, 154
$P_2C_6H_{14}Cl_2N_2O_2$	$[C_3H_7NP(O)Cl]_3$ (isom.)	H; P ($-2·4$; $-1·1$)	1966, 53
$P_2C_6H_{14}Cl_6N_2$	$(C_3H_7NPCl_3)_2$	H; P (-79)	1965, 98; 1966, 53, 288
$P_2C_6H_{14}O_6$	$[(CH_3O)_2P(O)CH]_2$	H	1967, 128
$P_2C_6H_{15}ClF_6$	$(C_2H_5)_3P^\oplus Cl, PF_6^\ominus$	F	1967, 382
$P_2C_6H_{15}ClO_5$	$(CH_3O)_2P(O)CH_2P(OC_2H_5)CH_2Cl$	P (21·7; 38·9)	1969, 332
$P_2C_6H_{15}Li$	$(C_2H_5)_2PP(C_2H_5)Li$	P ($-17·1$; $-112·8$)	1965, 27
$P_2C_6H_{15}N_3$	$N[P(CH_3)_2N]_2CCH_3$	H; P (29·9)	1968, 94
$P_2C_6H_{15}N_3O_4$	$CH_3C[NP(OCH_3)_2]_2N$	H; P (26·6)	1969, 106
$P_2C_6H_{16}$	$[CH_3(C_2H_5)P]_2$	P ($-46·2$)	1968, 11
	$(CH_3)_3P=CHP(CH_3)_2$	H	1968, 107
$P_2C_6H_{16}ClN_3$	$N[P(CH_3)_2N]_2^\oplus CCH_3, Cl^\ominus$	H; P (36·4)	1968, 94
$P_2C_6H_{16}F_4N_2$	$[CH_3NPF_2C_2H_5]_2$	H	1966, 36
$P_2C_6H_{16}F_6S_2$	$(CH_3)_2P^\oplus(CH_2SCH_3)_2, PF_6^\ominus$	H	1969, 128

Formula	Structure	Nuclei (shifts)	Ref
$P_2C_6H_{16}N_2O_4$	$[(CH_2)_2NCH_2PO_2H_2]_2$	P (9.5)	1967, 301
$P_2C_6H_{16}O_6$	$[(CH_3O)_2P(O)CH_2]_2$	H	1967, 371
$P_2C_6H_{16}O_7$	$(CH_3O)_2P(O)OCH(CH_3)P(O)(OCH_3)_2$	H; P (1·1; 22·3)	1967, 178
$P_2C_6H_{18}Al_2Br_4N_2$	$[(CH_3)_3PNAlBr_2]_2$	H	1967, 176
$P_2C_6H_{18}Br_3N_3OS$	$[(CH_3)_2N]_3P^{\oplus}OP(S)Br_2, Br^{\ominus}$	P (47·6; −89·0)	1969, 368
$P_2C_6H_{18}Cl_3N_3OS$	$[(CH_3)_2N]_3P^{\oplus}OP(S)Cl_2, Cl^{\ominus}$	P (53·7; 40·1)	1969, 368
$P_2C_6H_{18}Cl_3N_3O_2$	$[(CH_3)_2N]_3P^{\oplus}OP(O)Cl_2, Cl^{\ominus}$	P (55·8; −9·1)	1969, 368
$P_2C_6H_{18}F_6N_2$	$CH_3PF_5^{\ominus}, CH_3P^{\oplus}F[N(CH_3)_2]_2$	H; P(−126·8; 70·9); F (−82·2; −57·5; −46·4)	1966, 228
$P_2C_6H_{18}INS_2$	$[CH_3SP(CH_3)_2]_2 \overset{\oplus}{N}, I^{\ominus}$	H; P (43·5)	1968, 111
$P_2C_6H_{18}N_4O_2$	$[CH_3NP(O)N(CH_3)_2]_2$	H; P (7·0)	1966, 53
$P_2C_6H_{18}N_4S_2$	$[[(CH_3)_2N]_2P(S)NCH_3]_2$	P (62)	1966, 83
$P_2C_6H_{18}N_6$	$P[N(CH_3)N(CH_3)]_3P$	H	1967, 371; 1968, 57
$P_2C_6H_{18}N_6O_2$	$OP[N(CH_3)N(CH_3)]_3PO$	H	1967, 371
$P_2C_6H_{18}N_6S_2$	$SP[N(CH_3)N(CH_3)]_3PS$	H	1967, 371; 1968, 57
$P_2C_6H_{20}BF_6N$	$[(CH_3)_3P(CH_3)_3NBH_2]^{\oplus}, PF_6^{\ominus}$	^{11}B; H	1969, 26
$P_2C_6H_{24}ClN_7$	$[C_3H_7NHP(NH_2)_2]_2N^{\oplus}, Cl^{\ominus}$	P (ca. 15)	1966, 288
$P_2C_6H_{24}N_7$	$C_3H_7NHP(NH_2)_2NHP(NH_2)(=NH)NHC_3H_7$	P (ca. 15)	1966, 288
$P_2C_7H_5Cl_4N_3$	$C_6H_5C(NPCl_2)_2N$	P (41·6)	1969, 106
$P_2C_7H_{10}O_6$	$C_6H_5CH_2(PO_3H_2)_2$	P (21·0)	1968, 349
$P_2C_7H_{18}Cl_2$	$CH_3P(Cl)P(C_2H_5)_3, Cl$	H	1969, 344
$P_2C_7H_{18}I_2$	$(CH_3)_2C[P(CH_3)_2]_2^{2\oplus}, 2I^{\ominus}$	H	1968, 351
$P_2C_7H_{18}H_4$	$N[P(CH_3)_2N]_2CN(CH_3)_2$	H; P (31·3)	1968, 94
$P_2C_7H_{21}NO_4$	$(CH_3O)(C_2H_5)_2P=NP(O)(OCH_3)_2$	P (2; 53)	1969, 371
$P_2C_8H_2F_{18}$	$CF_3CH_2C[P(CF_3)_2]_2CF_3$	H; F	1967, 103
$P_2C_8H_{10}O_4$	$HOP(O)CHCH(CH=CHPO_2H)CHCHCH=CH$	H	1969, 146
$P_2C_8H_{12}F_4$	$[(CH_3)_2PC(CF_2)]_2$	H; F (−109·6)	1967, 331

Formula	Compound	Resonance	Reference
$P_2C_8H_{18}Cl_2N_2$	$[(CH_3)_3CNPCl]_2$	H	1969, 112
$P_2C_8H_{18}Cl_6N_2$	$[C_4H_9N=PCl_3]_2$	H; P ($-79 \cdot 3$)	1965, 98; 1966, 288
$P_2C_8H_{18}N_4O_2$	$[(CH_3)_2NPN(COCH_3)]_2$	H; P (154)	1968, 126
	$[OC(CH_3)=NNP(CH_3)_2]_2$	H; P (-425)	1969, 149
$P_2C_8H_{18}O_8$	$[(CH_3O)_2P(O)OC(CH_3)]_2$	H; P ($-2 \cdot 8$; $-4 \cdot 1$)	1968, 309
	$(CH_3O)_2P(O)OC(CH_3)(COCH_3)P(O)(OCH_3)_2$	P ($-6 \cdot 9$; $19 \cdot 7$)	1968, 309
$P_2C_8H_{20}$	$[(C_2H_5)_2P]_2$	P ($-34 \cdot 2$)	1965, 11; 1967, 126
$P_2C_8H_{20}BaO_8$	$[(C_2H_5)_2PO_2^{\ominus}]_2, Ba^{2\oplus}$	H	1967, 90
$P_2C_8H_{20}Cl_2S_4Sn$	$Cl_2Sn[S(S)P(C_2H_5)_2]_2$	H	1967, 177
$P_2C_8H_{20}F_6N_2$	$C_2H_5[CH_2N(CH_3)]_2P^{\oplus}F, C_2H_5PF_5^{\ominus}$	P ($76 \cdot 7$; $-126 \cdot 5$); F ($-67 \cdot 1$; $-57 \cdot 4$; $-56 \cdot 0$)	1968, 352
$P_2C_8H_{20}O_4S_2$	$[(C_2H_5O)_2P(S)]_2$	H; P ($-37 \cdot 2$)	1967, 127
$P_2C_8H_{20}O_4S_3$	$[(C_2H_5O)_2P(S)]_2S$	P ($78 \cdot 0$)	1966, 218
$P_2C_8H_{20}O_4S_4$	$[(C_2H_5O)_2P(S)S]_2$	P ($84 \cdot 0$)	1966, 218
$P_2C_8H_{20}O_5S$	$(C_2H_5O)_2P(S)OP(O)(OC_2H_5)_2$	H; P ($-56 \cdot 1$; $16 \cdot 0$)	1967, 127
	$(C_2H_5O)_2P(S)P(O)(OC_2H_5)_2$	H; P ($-39 \cdot 5$; $-106 \cdot 7$)	1967, 127; 1968, 113
$P_2C_8H_{20}O_5S_2$	$[(C_2H_5O)_2P(S)]_2O$	H; P ($-60 \cdot 1$)	1967, 127; 1968, 25, 284
	$(C_2H_5O)_2P(S)SP(O)(OC_2H_5)_2$	H; P ($-34 \cdot 0$; $-98 \cdot 4$)	1967, 127
$P_2C_8H_{20}O_6$	$[(C_2H_5O)_2P(O)]_2$	H; P ($-105 \cdot 8$)	1967, 127, 381
	$(C_2H_5O)_2P(O)OP(O)(OC_2H_5)_2$	H; P (-122; $15 \cdot 2$)	1967, 127
	$(C_2H_5O)_2P(S)OP(O)(OC_2H_5)_2$	H; P ($-58 \cdot 4$; $-126 \cdot 4$)	1967, 127
$P_2C_8H_{20}O_6S$			
$P_2C_8H_{20}O_7$	$[(C_2H_5O)_2P(O)]_2O$	H; P (-11 to -16)	1967, 69, 185, 381; 1968, 282

Formula	Structure	NMR	References
$P_2C_8H_{20}S_4Zn$	$Zn[SP(S)(C_2H_5)_2]_2$	H; P (-125)	1967, 127
$P_2C_8H_{21}Cl$	$(C_2H_5)_3PP(CH_3)_2Cl$	H	1967, 177
$P_2C_9H_{22}Si$	$[(C_2H_5)_2P]_2SiI_2$	H; P (-42.2; 37.7)	1969, 344
$P_2C_8H_{24}N_2O_3$	$[(CH_3)_2N_2P(O)]_2O$	H	1967, 182
$P_2C_8H_{24}N_2Si$	$[(CH)_3PN]_2Si(CH_3)_2$	H	1968, 284
$P_2C_8H_{28}ClN_7$	$C_4H_9NHP(NH_2)_2N^\oplus$, $Cl^\ominus$	H	1967, 316
$P_2C_8H_{28}N_7$	$C_4H_9NHP(NH_2)_2NHP(NH_2)(=NH)NHC_2H_5$	P (ca. 15)	1966, 288
$P_2C_9H_{20}Br_2O_6$	$[(C_2H_5O)_2P(O)]_2CBr_2$	P (ca. 15)	1966, 288
$P_2C_9H_{20}Cl_2O_6$	$[(C_2H_5O)_2P(O)]_2CCl_2$	H; P (8·5)	1968, 349
$P_2C_9H_{21}NaO_6$	$[(C_2H_5O)_2P(O)]_2CH^\oplus$, $Na^\ominus$	H; P (8·5)	1968, 349
$P_2C_9H_{21}N_3O_3$	$[CH_3N(CH_2)_2OP]_2(OCH_2CH_2)NCH_3$	H; P (41·5)	1967, 371; 1968, 349
		H; P	1968, 349
$P_2C_9H_{22}$	$(C_4H_9)_2C(PH_2)_2$	H; P (-91)	1966, 15
$P_2C_9H_{22}O_4S_4$	$[(C_2H_5O)_2P(S)S]_2CH_2$	H	1968, 284, 285; 1969, 322
$P_2C_9H_{22}O_6$	$[(C_2H_5O)_2P(O)]_2CH_2$	H; P (19·0)	1967, 371; 1968, 56, 349; 1969, 332
$P_2C_9H_{24}Si$	$(PO_3H_2)_2C(C_4H_9)_2$	P (26·0)	1968, 349
	$(CH_3)_3PC[P(CH_3)_2]Si(CH_3)_3$	H	1968, 107
$P_2C_9H_{26}N_4O_2$	$\{[(CH_3)_2N]_2P(O)\}_2CH_2$	H	1969, 372
$P_2C_{10}H_{10}N_2S_5$	$[C_5H_5N^\oplus P(S)S^\ominus]_2S$	P (105·5)	1967, 126
$P_2C_{10}H_{13}NaO_4$	$NaOP(O)CHCH(CH=CHP(O)OC_2H_5)CHCHCH=CH$	P (78)	1969, 146
$P_2C_{10}H_{14}O_4$	$HOP(O)CHCH[CH=CHP(O)OC_2H_5]CHCH=CH$	H	1969, 146
$P_2C_{10}H_{15}Li$	$C_2H_5(C_6H_5)PP(C_2H_5)Li$	H	1968, 113
$P_2C_{10}H_{16}O_6S_2$	$[(CH_3O)_2P(S)O]_2C_6H_4$	H	1968, 284, 285

264 FORMULAE INDEX II

Formula	Compound	Resonance	Reference
$P_2C_{10}H_{20}O_4S_3$	$[(CH_3)_2C(CH_2O)_2P(S)]_2S$	H	1967, 91
$P_2C_{10}H_{20}O_5S_2$	$[(CH_3)_2C(CH_2O)_2P(S)]_2O$	H	1967, 91
	$(CH_3)_2C(CH_2O)_2P(S)SP(O)(COCH_2)_2C(CH_3)_2$	H	1967, 91
$P_2C_{10}H_{20}O_6$	$[(C_2H_5O)_2P(O)Cl]_2$	H	1965, 90
$P_2C_{10}H_{20}O_6S$	$(CH_3)_2C(CH_2O)_2P(O)OP(S)(OCH_2)_2C(CH_3)_2$	H; P $(-24; 44)$	1967, 91
	$[(CH_3)_2C(CH_2O)_2P(O)]_2O$	H	1967, 91
$P_2C_{10}H_{20}O_7$	$[C_5H_{11}N=PCl_3]_2$	H; P (-78)	1966, 288
$P_2C_{10}H_{22}Cl_6N_2$	$[(C_2H_5O)_2P(O)CH]_2$	H	1967, 383
$P_2C_{10}H_{22}O_6$	$CH_3C=NPCl_2=NP(NH\text{-}t\text{-}C_4H_9)_2=N$	H; P $(39\cdot3; 10\cdot3)$	1969, 106
$P_2C_{10}H_{23}Cl_2N_5$			
$P_2C_{10}H_{24}$	$C_9H_{19}CH(PH_2)_2$	P (-105)	1966, 15
$P_2C_{10}H_{24}O_6$	$[(C_2H_5O)_2P(O)CH_2]_2$	H; P $(26\cdot8)$	1967, 371, 384; 1968, 143; 1969, 322
$P_2C_{10}H_{25}NO_3$	$(C_2H_5)_3P=NP(O)(OC_2H_5)_2$	P $(1\cdot3; 32)$	1969, 371
$P_2C_{10}H_{25}NO_4$	$(C_2H_5O)_3P=NP(O)(C_2H_5)_2$	P $(0; 29)$	1969, 371
	$C_2H_5O(C_2H_5)_2P=NP(O)(OC_2H_5)_2$	P $(0; 52)$	1969, 371
$P_2C_{10}H_{25}NO_6$	$(C_2H_5O)_3P=NP(O)(OC_2H_5)_2$	P $(3; 3)$	1969, 371
	$C_2H_5OP(O_2^\ominus)CHP(O)(OC_2H_5)_2]N^\oplus(CH_3)_3$	P $(11; -4)$	1969, 346
$P_2C_{10}H_{26}Si$	$(CH_3)_2SiP[P(C_2H_5)_2]_2$	H	1967, 182
	$(CH_3)_2Si[(CH_2)_2P(CH_3)_2]_2$	H	1969, 373
$P_2C_{10}H_{27}ISi$	$(CH_3)_3PC[Si(CH_3)_3]P^\oplus(CH_3)_3, I^\ominus$	H	1968, 107
$P_2C_{10}H_{30}Al_2N_2$	$[(CH_3)_3PNAl(CH_3)_2]_2$	H	1967, 176
$P_2C_{11}H_{24}O_7$	$[(C_2H_5O)_2P(O)CH_2]_2CO$	H; P $(19\cdot4)$	1967, 278
$P_2C_{11}H_{26}$	$C_{10}H_{21}CH(PH_2)_2$	P $(-10\cdot8)$	1966, 15
$P_2C_{11}H_{26}NNaO_6$	$[(C_2H_5O)_2P(O)]_2C^\ominus N(CH_3)_2, Na^\oplus$	P (33)	1968, 353; 1969, 346
$P_2C_{11}H_{26}N_2$	$CH_2[PCH_2N(C_2H_5)_2]_2$	H	1965, 118
$P_2C_{11}H_{26}O_6$	$(CH_2)_3[P(O)(OC_2H_5)_2]_2$	P $(29\cdot3)$	1969, 332

$P_2C_{11}H_{27}Cl$	$(C_3H_7)_3PP(CH_3)_2Cl$	H; P (−47.9; 22.5)	1969, 344
$P_2C_{11}H_{27}NO_6$	$[(C_2H_5O)_2P(O)]_2CHN(CH_3)_2$	H; P (18)	1968, 93, 353
$P_2C_{11}H_{33}AlN_2Si$	$(CH_3)_3PN[Al(CH_3)_3]Si(CH_3)_2NP(CH_3)_3$	H	1967, 316
$P_2C_{11}H_{33}GaN_2Si$	$(CH_3)_3PN[Ga(CH_3)_3]Si(CH_3)_2NP(CH_3)_3$	H	1967, 316
$P_2C_{12}H_{10}Cl_2FNS$	$(C_6H_5)_2PCl=NP(S)ClF$	P (30.5; 50.7); F (78.7)	1969, 69
$P_2C_{12}H_{10}Cl_3NO$	$(C_6H_5)_2PCl=NP(O)Cl_2$	P	1967, 286
$P_2C_{12}H_{10}Cl_5N$	$(C_6H_5)_2PCl_2)_2NCl$	P (417)	1967, 385
$P_2C_{12}H_{10}N_2O_2$	$[C_6H_5P(O)N]_2$	P (29.4)	1965, 78
$P_2C_{12}H_{10}S_4$	$[C_6H_5P(S)S]_2$	P (71.3)	1967, 277
$P_2C_{12}H_{12}K_2O_4$	$[C_6H_5P(O)OK]_2$	P (5.2)	1967, 386
$P_2C_{12}H_{12}K_2S_4$	$[C_6H_5P(S)SK]_2$	P (95.2)	1967, 386
$P_2C_{12}H_{12}O_3$	$[C_6H_5P(O)H]_2O$	H	1965, 48; 1966, 18
$P_2C_{12}H_{12}O_5$	$[C_6H_5PO_2H]_2O$	P (9.8)	1967, 277
$P_2C_{12}H_{15}N_8O_2$	$\{[(CH_3)_2N]_2PNN=C(CH_3)O\}_2$	H; P (−40.5)	1969, 149
$P_2C_{12}H_{18}O_4$	$C_2H_5OP(O)CHCH[CH=CHP(O)OC_2H_5]CHCHCH=CH$	H; P (70; 83)	1969, 146
$P_2C_{12}H_{19}N_3$	$N[P(CH_3)_2N]_2CCH_2C_6H_5$	H; P (31.1)	1968, 94
$P_2C_{12}H_{22}O_4$	$C_2H_5OP(O)CHCH[(CH_2)_2P(O)OC_2H_5]CHCH(CH_2)_2$	P (70; 77)	1969, 146
$P_2C_{12}H_{26}Cl_6N_2$	$[C_6H_{13}N=PCl_3]_2$	H; P (−79.1)	1966, 288
$P_2C_{12}H_{26}O_6S_4$	$[(C_2H_5O)_2P(S)SCHOCH_2]_2$	H	1968, 285
$P_2C_{12}H_{27}Cl_3O_2$	$(C_4H_9)_3P^{\oplus}OP(O)Cl_2, Cl^{\ominus}$	P (104.5; −10.1)	1969, 368
$P_2C_{12}H_{27}Cl_5O$	$(C_4H_9)_3POPCl_5$	P (103.5; −296.0)	1969, 135
$P_2C_{12}H_{28}$	$[C_2H_5(C_4H_9)P]_2$	P (−37.5)	1965, 29
	$[(C_2H_5)_2P]_2(CH_2)_4$	P (−46.0)	1965, 29
	$C_{11}H_{23}CH(PH_2)_2$	P (−107)	1966, 15
$P_2C_{12}H_{28}BaO_8$	$[(C_3H_7O)_2PO_2^{\ominus}]_2, Ba^{2\oplus}$	H	1967, 90

Formula	Compound	Resonance	Reference
$P_2C_{12}H_{28}N_2$	$[(C_2H_5)_2NCH_2PCH_2]_2$	H	1965, 118
$P_2C_{12}H_{28}O_2$	$[(C_2H_5)_2P(O)(CH_2)_2]_2$	P (55·9)	1968, 336
$P_2C_{12}H_{28}O_5S_2$	$[(C_3H_7O)_2P(S)]_2O$	H	1968, 285
$P_2C_{12}H_{28}O_6$	$[(C_2H_5O)_2P(O)(CH_2)_2]_2$	P (31·5)	1969, 332
$P_2C_{12}H_{28}S_2$	$[(C_2H_5)_2P(S)(CH_2)_2]_2$	P (39·4)	1965, 11
	$C_2H_5(C_4H_9)P(S)]_2$	P (47·6)	1965, 29
$P_2C_{12}H_{29}NO_6$	$[(C_2H_5O)_2P(O)]_2C^{\oplus}N^{\oplus}(CH_3)_3$	H; P (28)	1968, 353; 1969, 346
$P_2C_{12}H_{29}N_5O_2$	$CH_3C=NP(OCH_3)_2=NP(NHt\text{-}C_4H_9)_2=N$	H; P (19·6)	1969, 106
$P_2C_{12}H_{30}INO_6$	$[(C_2H_5O)_2P(O)]_2CHN^{\oplus}(CH_3)_3, I^{\ominus}$	H; P (11)	1968, 353; 1969, 346
$P_2C_{12}H_{30}N_6$	$[CH_3N(CH_2)_2NCH_3PN(CH_3)CH_2]_2$	H	1967, 305
$P_2C_{12}H_{36}Cl_2N_6-$ O_8S_2	$[(CH_3)_2N]_3P^{\oplus}(S)]_2, ClO_4^{\ominus}$	H	1969, 105, 247
$P_2C_{13}H_{28}Br_2O_6$	$[i\text{-}C_3H_7O)_2P(O)]_2CBr_2$	H; P (7·0)	1967, 380; 1968, 349
$P_2C_{13}H_{28}Cl_2O_6$	$[i\text{-}C_3H_7O)_2P(O)]_2CCl_2$	H; P (6·5)	1968, 349
$P_2C_{13}H_{28}I_2O_6$	$[i\text{-}C_3H_7O)_2P(O)]_2CI_2$	H; P (10·5)	1968, 349
$P_2C_{13}H_{28}O_7$	$(C_2H_5O)_2P(O)CH_2CH=C(CO_2H_5)P(O)(OC_2H_5)_2$	H; P (30·6; 27·7)	1967, 278
$P_2C_{13}H_{29}BrO_6$	$[(i\text{-}C_3H_7O)_2P(O)]_2CHBr$	H; P (12·0)	1968, 349
		P (11·8)	1967, 380
$P_2C_{13}H_{29}ClO_6$	$[(i\text{-}C_3H_7O)_2P(O)]_2CHCl$	H; P (11·5)	1968, 349
$P_2C_{13}H_{29}IO_6$	$[(i\text{-}C_3H_7O)_2P(O)]_2CHI$	H; P (13·5)	1968, 349
$P_2C_{13}H_{29}NaO_6$	$[(i\text{-}C_3H_7O)_2P(O)]_2CHNa$	H; P (40·5)	1968, 349
$P_2C_{13}H_{30}$	$[(i\text{-}C_3H_7)_2PJ_2CH_2$	P (17·8)	1966, 15
$P_2C_{13}H_{30}Cl_2$	$(C_4H_9)_3P^{\oplus}P(CH_3)Cl, Cl^{\ominus}$	H; P ($-14·7$; 16·5)	1969, 344
$P_2C_{13}H_{30}N_2$	$[(C_2H_5)_2NCH_2P]_2(CH_2)_3$	H	1965, 118
$P_2C_{13}H_{30}O_6$	$[(C_2H_5O)_2P(O)]_2CHC_4H_9$	P (23·4)	1966, 15
	$[(i\text{-}C_3H_7O)_2P(O)]_2CH_2$	H; P (17·5)	1968, 349
		P (17·8)	1967, 380
$P_2C_{13}H_{31}NO_4$	$C_3H_7O(C_2H_5)_2P=NP(O)(OC_3H_7)_2$	P (0; 51)	1969, 371

$P_2C_{13}H_{33}NO_{11}S$	$(C_3H_7O)_3P=NP(O)(C_2H_5)_2$	P (0; 32)	1969, 371
	$[(C_2H_5O)_2P(O)]_2CHN^{\oplus}(CH_3)_3, SO_3CH_3^{\ominus}$	H; P (11)	1969, 346
$P_2C_{14}H_{14}O_2$	$(CH_3OC_6H_4P)_2$	P (−22·4)	1967, 126
$P_2C_{14}H_{16}$	$[CH_3(C_6H_5)P]_2$	H(T); P (−41·7)	1965, 11; 1966, 99; 1968, 141
$P_2C_{14}H_{16}F_4N_2$	$[CH_3NP(C_6H_5)F_2]_2$	H	1966, 36, 53
$P_2C_{14}H_{16}NNaS_2$	$(CH_3)_2P(S)N=P(C_6H_5)_2S^{\ominus}, Na^{\oplus}$	H; P (43·3; 37·6)	1968, 111
$P_2C_{14}H_{16}N_2$	$[CH_3(C_6H_5)PN]_2$	H; P (62·5)	1969, 199
$P_2C_{14}H_{18}N_4S_2$	$[C_6H_5NHP(S)NCH_3]_2$	P (54)	1966, 83
$P_2C_{14}H_{30}Cl_6N_2$	$(C_7H_{15}N=PCl_3)_2$	H; P (−79·7)	1966, 288
$P_2C_{14}H_{32}N_2$	$[(C_2H_5)_2NCH_2P(CH_2)_2]_2$	H	1965, 118
$P_2C_{14}H_{33}Cl$	$(C_4H_9)_3PP(CH_3)_2Cl$	H; P (−51·0; 21·2)	1969, 344
$P_2C_{14}H_{33}ClN_6$	$CH_3C=NP(NHt-C_4H_9)Cl=NP(NHt-C_4H_9)=N$	H; P (12·4)	1969, 106
$P_2C_{14}H_{40}N_6$	$[(CH_3)_2N)_3P^{\oplus}CH_2]_2$	H	1967, 371
$P_2C_{15}H_{19}NS_2$	$CH_3S(CH_3)_2P=NP(S)(C_6H_5)_2$	H; P (37·2; 43·2)	1968, 111
	$(CH_3)_2P(S)N=P(SCH_3)(C_6H_5)_2$	H; P (44·7; 24·2)	1968, 111
$P_2C_{15}H_{25}Cl_2N_5$	$C_6H_5C=NPCl_2=NP(NHt-C_4H_9)_2=N$	H; P (13·2)	1969, 106
$P_2C_{15}H_{29}N_7$	$C_6H_5C\{NP[N(CH_3)_2]_2\}_2N$	H; P (32·3)	1969, 106
$P_2C_{15}H_{34}O_6$	$[(i-C_3H_7O)_2P(O)]_2C(CH_3)_2$	H; P (25·6)	1966, 15
$P_2C_{16}H_{14}F_{10}N_2$	$[CH_3NP(C_6H_4CF_3)F_2]_2$	H	1966, 36
$P_2C_{16}H_{19}N_3$	$(CH_3)_2PNC(CH_3)NP(C_6H_5)_2N$	H; P (31·3; 14·7)	1968, 94
$P_2C_{16}H_{20}$	$[C_2H_5(C_6H_5)P]_2$ (isom.)	P (−21·5; −28·3)	1965, 11
$P_2C_{16}H_{20}ClN_3$	$(CH_3)_2PNH^{\oplus}C(CH_3)NP(C_6H_5)_2N, Cl^{\ominus}$	H; P (16·8; 40·3); F (40·3)	1968, 94
$P_2C_{16}H_{20}F_6N_2$	$[CH_2N(CH_3)]_2P^{\oplus}(C_6H_5)F, C_6H_5PF_5^{\ominus}$	H; P (56·1; −138); F (−78·7; −61·0; −57·4)	1968, 352

268 FORMULAE INDEX II

Formula	Compound	Resonance	Reference
$P_2C_{16}H_{20}S_2$	$(C_6H_5)_2P(S)P(S)(C_2H_5)_2$	P (26·0; 57·7)	1965, 29
$P_2C_{16}H_{22}F_6N_2$	$[(CH_3)_2N]_2P^{\oplus}(C_6H_5)F$, $C_6H_5PF_5^{\ominus}$	P (56·0; −136·0); F (−57·3; −60·9; −86·7)	1966, 228
$P_2C_{16}H_{22}INS_2$	$CH_3S(CH_3)_2P=NP^{\oplus}(SCH_3)(C_6H_5)_2$, $I^{\ominus}$	H; P (46·7; 34·7)	1968, 111
$P_2C_{16}H_{22}N_4$	$C_6H_5P[N(CH_3)N(CH_3)]_2PC_6H_5$	H; P (29·9)	1969, 199
$P_2C_{16}H_{23}N_3O$	$[(CH_3)_2N]_2P(H)=NP(O)(C_6H_5)_2$	H; P (22·4; 12·0)	1969, 15
$P_2C_{16}H_{23}N_3O_2S$	$[(CH_3)_2N]_2P(H)=NP(S)(OC_6H_5)_2$	H; P (23·3; 52·7)	1969, 15
$P_2C_{16}H_{23}N_3S$	$[(CH_3)_2N]_2P(H)=NP(S)(C_6H_5)_2$	H; P (24·4; 42·5)	1969, 15
$P_2C_{16}H_{26}N_2O_6S_2$	$(C_2H_5O)_2P(S)NHCOC(O)NHP(S)(OC_2H_5)_2$	H	1966, 264
$P_2C_{16}H_{32}$	$(C_2H_5)_2PP(C_6H_{11})_2$	H; P (−42·2; −13·8)	1965, 11, 29; 1968, 113
$P_2C_{16}H_{32}S_2$	$(C_2H_5)_2P(S)P(S)(C_6H_{11})_2$	H; P (47·6; 57·1)	1965, 29; 1968, 113
$P_2C_{16}H_{34}Cl_6N_2$	$(C_8H_{17}N=PCl_3)_2$	P (−79·8)	1966, 288
$P_2C_{16}H_{36}$	$[t-C_4H_9)_2P]_2$	P (40)	1966, 34
$P_2C_{16}H_{36}BaO_8$	$[(C_4H_9O)_2PO_2^{\ominus}]_2$, $Ba^{2\oplus}$	H	1967, 90
$P_2C_{16}H_{36}S_3$	$[t-C_4H_9)_2P(S)]_2S$	P (118·0)	1966, 34
$P_2C_{17}H_{25}Cl_2N_5$	$C_6H_5C[NP[N(CH_2)_5]Cl]_2N$	H; P (38·3)	1969, 106
$P_2C_{17}H_{38}O_6$	$[i-C_3H_7O)_2P(O)]_2CHC_4H_9$	P (21·6)	1966, 15
$P_2C_{18}H_{15}Cl_3OS$	$(C_6H_5)_3P^{\oplus}OP(S)Cl_2$, $Cl^{\ominus}$	P (65·2; 42·8)	1969, 368
$P_2C_{18}H_{15}Cl_3O_2$	$(C_6H_5)_3P^{\oplus}OP(O)Cl_2$, $Cl^{\ominus}$	P (65·0; −7·3)	1969, 368
$P_2C_{18}H_{15}Cl_5O$	$(C_6H_5)_3POPCl_5$	P (67·2; −297)	1969, 135
$P_2C_{18}H_{15}NS_3$	$C_6H_5P(S)N(C_6H_5)P(S)(C_6H_5)S$	P (61·0)	1967, 126, 142
$P_2C_{18}H_{22}O_6$	$[(CH_3O)_2P(O)C_6H_4CH]_2$	H	1966, 253
$P_2C_{18}H_{22}O_8$	$[(CH_3O)_2P(O)OC_6H_5]_2$	H; P (−3·1)	1968, 309
$P_2C_{18}H_{24}O_2$	$[(CH_3O)_2P(O)C_6H_4CH_2]_2$	H	1966, 253
$P_2C_{18}H_{25}NO_2$	$(C_2H_5O)_3P-NP(O)(C_6H_5)_2$	P (0; 8)	1969, 371

$P_2C_{18}H_{32}S_2$	$C_2H_5O(C_6H_5)_2P=NP(O)(OC_2H_5)_2$	P (0; 19)	1969, 371
$P_2C_{18}H_{38}Cl_6N_2$	$C_6H_5P(S)(S^{\ominus})P^{\oplus}(C_4H_9)_3$	P (6·8; 60·0)	1966, 77; 1967, 126
	$[C_9H_{19}N=PCl_3]_2$	P (−79·8)	1966, 288
$P_2C_{18}H_{40}O_2$	$[(C_4H_9)_2P(O)CH_2]_2$	P (54·3)	1968, 336
$P_2C_{18}H_{43}N_7$	$CH_3[NP(NHt\text{-}C_4H_9)_2]_2N$	H; P (13·6)	1969, 106
$P_2C_{19}H_{18}F_2OS$	$CH_3(C_6H_5)_3P^{\oplus}P(O)F_2(S^{\ominus})$	F (−34·2)	1967, 281
$P_2C_{19}H_{34}OS_2$	$CH_3OC_6H_4P(S)(S^{\ominus})P^{\oplus}(C_4H_9)_3$	P (7·4; 60·0)	1966, 77; 1967, 126
$P_2C_{19}H_{37}N_7$	$C_6H_5C=NP[N(CH_3)_2]_2=NP(NHt\text{-}C_4H_9)_2=N$	H; P (18·8)	1969, 106
$P_2C_{20}H_{18}O_2$	$[C_6H_5P(O)(CH=CH_2)]_2$	H	1968, 101
$P_2C_{20}H_{21}O_9$	$[(CH_3O)_2P(O)O]_2C_{16}H_8OH$	H; P (−1·0)	1968, 315
$P_2C_{20}H_{28}$	$[C_2H_5)_2PC_6H_4]_2$	H	1969, 155
$P_2C_{20}H_{36}OS_2$	$C_2H_5OC_6H_4P(S)(S^{\ominus})P^{\oplus}(C_4H_9)_3$	P (6·8; 60·9)	1966, 77; 1967, 126
$P_2C_{20}H_{42}Cl_6N_2$	$(C_{10}H_{21}N=PCl_3)_2$	P (−77·9)	1966, 288
$P_2C_{20}H_{44}O_2$	$[(C_2H_5)_2P(O)(CH_2)_6]_2$	P (55·5)	1968, 336
$P_2C_{21}H_{26}BF_6N$	$[(C_6H_5)_3P(CH_3)_3NBH_2]^{\oplus}, PF_6^{\ominus}$	H	1969, 26
$P_2C_{21}H_{37}N_7$	$C_6H_5C\{[NP(CH_3)_2]N(CH_2)_5\}_2N$	H; P (30·1)	1969, 106
$P_2C_{21}H_{46}O_6$	$[i\text{-}C_3H_7O)_2P(O)]_2C(C_4H_9)_2$	P (24·5)	1966, 15
	$[i\text{-}C_3H_7O)_2P(O)]_2CHC_8H_{17}$	P (22·2)	1966, 15
$P_2C_{21}H_{50}N_4$	$\{[(C_2H_5)_2NCH_2]_2P\}_2CH_2$	H; P (−56·5)	1966, 16
$P_2C_{22}H_{16}F_{12}$	$(C_6H_5)_3P^{\oplus}CHC(CF_3)_2, PF_6^{\ominus}$	H; P (17·8; −150·0); F(−59·3; −65·3; −72·9)	1967, 102
$P_2C_{22}H_{24}O_{10}$	$[(CH_3O)_3P(O)O]_2C_{16}H_8$	H; P (−44·3)	1968, 315
$P_2C_{22}H_{34}BrN_3O$	$(CHO_3)_2P(O)O[C(C_6H_5)_2]OP^{\oplus}[N(CH_3)_2]_3, Br^{\ominus}$	P (−4·2; 34·2)	1968, 309
$P_2C_{22}H_{34}O_6$	$CH_3(CH_3O)P(O)O]_2C_{18}H_{22}$	H	1967, 168
$P_2C_{22}H_{34}S_2$	$\beta\text{-}C_{10}H_6P(S)(S^{\ominus})P^{\oplus}(C_4H_9)_3$	P (9·5; 61·6)	1967, 126
$P_2C_{22}H_{36}O_{10}$	$[(CH_3O)_3POC(CH_3)(COCH_3)CH_2]_2C_6H_4$ (isom.)	P (−49·3; −49·5)	1967, 351
$P_2C_{22}H_{38}O_{12}$	$[(CH_3O)_3POC(CH_3)(COCH_3)CHO]_2C_6H_4$	H; P (−51·1)	1967, 350, 351

270   FORMULAE INDEX II

Formula	Compound	Resonance	Reference
$P_2C_{22}H_{36}O_{12}$	$(CH_3O)_3POCH(C_6H_4OCHO)CH–[C_6H_4CHC(CH_3)(COCH_3)OP(OCH_3)_3O]O$	H; P (−50·1)	1967, 350
$P_2C_{22}H_{40}O_2$	m-$[(C_4H_9)_2P(O)]_2C_6H_4$	P (133·0)	1968, 336
	p-$[(C_4H_9)_2P(O)]_2C_6H_4$	P (134·6)	1968, 336
$P_2C_{22}H_{52}N_4$	$\{[(C_2H_5)_2NCH_2]_2P\}_2(CH_2)_2$	H; P (−48·7)	1966, 16
$P_2C_{23}H_{54}N_4$	$\{[(C_2H_5)_2NCH_2]_2P\}_2(CH_2)_3$	H; P (−52·9)	1966, 16
$P_2C_{24}H_{20}$	$(C_6H_5)_4P_2$	H; P	1969, 344
$P_2C_{24}H_{20}Br_2N$	$[(C_6H_5)_2PBr]_2N^{\oplus}$	P (33·2)	1967, 9
$P_2C_{24}H_{20}ClFOS$	$(C_6H_5)_4P^{\oplus}S^{\ominus}P(O)FCl$	F (−43·6)	1967, 58
$P_2C_{24}H_{20}ClNS$	$(C_6H_5)_2P(S)$=$NP(C_6H_5)_2Cl$	P (45·1; 30·2)	1967, 92
$P_2C_{24}H_{20}Cl_4N$	$[(C_6H_5)_2PCl]^{\oplus}_2N$, $Cl^{\ominus}$	P (43·6)	1967, 92
$P_2C_{24}H_{20}Cl_8NSb$	$[(C_6H_5)_2PCl]^{\oplus}_2N$, $SbCl_6^{\ominus}$	P (43·5)	1967, 92
$P_2C_{24}H_{20}KNOS$	$[(C_6H_5)_2P(S)NP(O)(C_6H_5)_2]^{\ominus}$, $K^{\oplus}$	P (34·7; 12·4)	1967, 92, 344
$P_2C_{24}H_{20}KNS_2$	$[C_6H_5)_2P(S)]_2^{\ominus}$, $K^{\oplus}$	P (35·6)	1967, 344; 1968, 111
$P_2C_{24}H_{20}NOS$	$(C_6H_5)_2P(S)NP(C_6H_5)_2O^{\ominus}$	P (35·, 38)	1967, 344
$P_2C_{24}H_{20}NO_2$	$[(C_6H_5)_2PO]_2N^{\oplus}$	P (12·7)	1967, 9
$P_2C_{24}H_{20}NS_2$	$(C_6H_5)_2P(S^{\ominus})$=$N(C_6H_5)_2P(S)$	P (35 to 38)	1966, 232; 1967, 344
$P_2C_{24}H_{20}O_5$	$[(C_6H_5)_2P(O)]_2O$	P (32·2)	1967, 277
$P_2C_{24}H_{21}NOS$	$(C_6H_5)_2P(S)NHP(O)(C_6H_5)_2$	P (54·2; 22·1)	1967, 92, 344
$P_2C_{24}H_{21}NS_2$	$[(C_6H_5)_2P(S)]_2NH$	P (55·1)	1966, 232; 1967, 344
$P_2C_{24}H_{22}NO_2$	$[C_6H_5)_2POH]_2N^{\oplus}$	P (25·1)	1967, 9
$P_2C_{24}H_{22}N_2S$	$(C_6H_5)_2P(S)N$=$P(C_6H_5)_2NH_2$	P (42·6; 20·6)	1967, 92
$P_2C_{24}H_{23}ClN_2S$	$(C_6H_5)_2P(S)NHP^{\oplus}(C_6H_5)_2NH_2$, Cl	P (55·9; 40·7)	1967, 92
$P_2C_{24}H_{32}$	$(C_6H_5)_2PP(C_6H_{11})_2$	H; P (−28·8; −8·2)	1965, 11, 29; 1968, 113
$P_2C_{24}H_{32}S_2$	$(C_6H_5)_2P(S)P(S)(C_6H_{11})_2$	P (20·9; 64·0)	1965, 29

NMR DATA ON ORGANIC COMPOUNDS

Formula	Structure	Nucleus	References
$P_2C_{24}H_{52}O_6$	$C_{11}H_{23}CH[P(O)(Oi\text{-}C_3H_7)_2]_2$	P (22·2)	1966, 15
$P_2C_{24}H_{56}N_4$	$\{[(C_2H_5)_2NCH_2]_2P(CH_2)_2\}_2$	P (−53·2)	1966, 16
$P_2C_{25}H_{18}Br_4O_2$	$[(C_6H_4Br)_2P(O)]_2CH_2$	H	1968, 18
$P_2C_{25}H_{18}Cl_4O_2$	$[(C_6H_4Cl)_2P(O)]_2$	H	1968, 18
$P_2C_{25}H_{20}O_2$	$[(C_6H_5)_2P(O)CH]_2$	H	1967, 371
$P_2C_{25}H_{20}S_2$	$[(C_6H_5)_2P(S)CH]_2$	H	1967, 371
$P_2C_{25}H_{21}N_3$	$N[(C_6H_5)_2PN]_2CH$	H; P (14·4)	1967, 196, 335
$P_2C_{25}H_{22}$	$[(C_6H_5)_2P]_2CH_2$	H; P (−22·2)	1967, 329, 371, 387; 1968, 297; 1969, 374
$P_2C_{25}H_{22}F_4$	$[(C_6H_5)_2PF_2]_2CH_2$	H; F (−27·6)	1969, 375
$P_2C_{25}H_{22}N_4$	$N[P(C_6H_5)_2N]_2CNH_2$	H; P (21·7)	1968, 94
$P_2C_{25}H_{22}O$	$(C_6H_5)_2P(O)(CH_2)_2P(C_6H_5)_2$	H	1967, 371
$P_2C_{25}H_{22}O_2$	$[(C_6H_5)_2P(O)]_2CH_2$	H	1967, 371, 387; 1968, 18, 297
$P_2C_{25}H_{22}S_2$	$[(C_6H_5)_2P(S)]_2CH_2$	H; P (34·5)	1967, 329, 387
$P_2C_{25}H_{22}Se_2$	$[(C_6H_5)_2P(Se)]_2CH_2$	H	1967, 387
$P_2C_{25}H_{23}NOS$	$(C_6H_5)_2P(S)N=P(C_6H_5)_2OCH_3$	H; P (41·8; 28·9)	1967, 92, 344
	$CH_3S(C_6H_5)_2P=NP(O)(C_6H_5)_2$	H; P (26·7; 13·3)	1967, 92, 344
$P_2C_{25}H_{23}NS_2$	$CH_3S(C_6H_5)_2P=NP(S)(C_6H_5)_2$	H; P (42·4; 29·2)	1966, 232; 1967, 344; 1968, 111
$P_2C_{25}H_{26}Cl_2N_2$	$[(C_6H_5)_2PNH_2]_2CH_2Cl_2$	H; P (22)	1969, 374
$P_2C_{25}H_{46}O_2$	$[(C_6H_{11})_2P(O)]_2CH_2$	H	1965, 71
$P_2C_{25}H_{54}O_2$	$[(C_6H_{13})_2P(O)]_2CH_2$	H	1965, 71
$P_2C_{26}H_{22}$	$[(C_6H_5)_2PCH]_2$	H	1967, 371; 1969, 347
$P_2C_{26}H_{23}N_3$	$N[(C_6H_5)_2PN]_2CCH_3$	H; P (15·6)	1967, 97; 1968, 94
$P_2C_{26}H_{24}$	$[(C_6H_5)_2PCH_2]_2$	H	1967, 371, 387
$P_2C_{26}H_{24}ClN$	$(C_6H_5)_2PCH=CHP(NH_2)(C_6H_5)_2$, Cl	H	1969, 374
$P_2C_{26}H_{24}ClN_3$	$CH_3C[NP(C_6H_5)_2]_2NH^\oplus$, $Cl^\ominus$	H; P (19·0)	1967, 97; 1968, 94
$P_2C_{26}H_{24}O$	$[(C_6H_5)_2PCH_2]_2O$	H	1967, 349

Formula	Compound	Resonance	Reference
$P_2C_{26}H_{24}O_2$	$[(C_6H_5)_2P(O)CH_2]_2$	H; P (35·8)	1967, 371, 387; 1968, 336
$P_2C_{26}H_{24}S_2$	$[(C_6H_5)_2POCH_2]_2$	P (115·2)	1968, 336
	$[(C_6H_5)_2P(S)CH_2]_2$	H; P (79·2)	1967, 371, 387; 1968, 336
$P_2C_{26}H_{24}Se_2$	$[(C_6H_5)_2P(Se)CH_2]_2$	H	1967, 387
$P_2C_{26}H_{26}Cl_6-$ $-NOSSb$	$CH_3S(C_6H_5)_2P^{\oplus}N=P(C_6H_5)_2OCH_3, SbCl_6^{\ominus}$	H; P (38·8; 35·3)	1967, 344
$P_2C_{26}H_{26}Cl_6-$ $-NS_2Sb$	$[CH_3S(C_6H_5)_2P]_2^{\oplus}N, SbCl_6^{\ominus}$	H; P (38·4)	1967, 344; 1968, 111
$P_2C_{26}H_{26}NO_2$	$[(C_6H_5)_2POCH_3]_2N^{\oplus}$	P (35·0)	1967, 9
$P_2C_{26}H_{26}NS_2$	$[CH_3S(C_6H_5)_2P]_2^{\oplus}N$	P (38·9)	1966, 232
$P_2C_{26}H_{26}N_2S$	$[CH_3NP(C_6H_5)_2]_2S$	H	1967, 388
$P_2C_{26}H_{26}O_2Si$	$[(C_6H_5)_2PO]_2Si(CH_3)_2$	P (98·1)	1967, 100
$P_2C_{26}H_{26}S_4Sn$	$[(C_6H_5)_2P(S)S]_2Sn(CH_3)_2$	H	1967, 177
$P_2C_{26}H_{28}N_2$	$[(C_6H_5)_2PNH_2NCH_2]_2$	H	1969, 374
$P_2C_{26}H_{28}N_2S$	$CH_3S[NP(C_6H_5)_2]_2CH$	H	1967, 149
$P_2C_{26}H_{28}N_3$	$[(C_6H_5)_2PNHCH_3]_2N^{\oplus}$	P (20·9)	1967, 9
$P_2C_{26}H_{29}BrN_2S$	$CH_3S[NP(C_6H_5)_2]_2^{\oplus}CH_2, Br^{\ominus}$	H	1967, 149
$P_2C_{26}H_{30}Br_2N_2S$	$(CH_3)_2S[NP^{\oplus}(C_6H_5)_2]_2, 2Br^{\ominus}$	H	1967, 149
$P_2C_{26}H_{30}O_4$	$[C_6H_5(C_2H_5)_2P^{\oplus}C{\equiv}C(O)C(O)]_2$	P (26)	1965, 85
$P_2C_{26}H_{30}Cl_2N_4$	$[(C_6H_5)_2PNH_2]_2NN(CH_3)_2, Cl_2$	H; P (1·5)	1967, 94
$P_2C_{26}H_{57}Cl$	$(C_8H_{17})_3PP(CH_3)_2Cl$	H; P (−25·9; 14·5)	1969, 344
$P_2C_{27}H_{25}O_4$	$C_6H_5P(O)(OCH_3)C_6H_4CH(C_6H_5)P(O)(OCH_3)C_6H_5$	H	1966, 82
$P_2C_{27}H_{26}Br_2O$	$(C_6H_5)_2P^{\oplus}CH_2P^{\oplus}(C_6H_5)_2CH_2OCH_2, 2Br^{\ominus}$	H	1967, 349

$P_2C_{27}H_{26}Cl_2O$	$(C_6H_5)_2P^\oplus CH_2P^\oplus(C_6H_5)_2CH_2OCH_2, 2Cl^\ominus$	H	1967, 349
$P_2C_{27}H_{26}N_4$	$N[P(C_6H_5)_2N]_2CN(CH_3)_2$	H; P (19·3)	1968, 94
$P_2C_{27}H_{26}O_2$	$[(C_6H_5)_2P(O)]_2(CH_2)_3$	P (34·4)	1968, 336
$P_2C_{27}H_{27}I$	$[CH_3(C_6H_5)_2P]_2^\oplus CH, I^\ominus$	H	1967, 371
$P_2C_{27}H_{27}IO$	$(C_6H_5)_2P^\oplus(CH_3)(CH_2)_2P(O)(C_6H_5)_2, I^\ominus$	H	1967, 371
$P_2C_{27}H_{28}I_2$	$[(C_6H_5)_2P^\oplus CH_3]_2CH_2, 2I^\ominus$	H	1967, 371, 387
$P_2C_{28}H_{20}F_4$	$[(C_6H_5)_2PC(CF_2)]_2$	H; F (−108·6)	1967, 331
$P_2C_{28}H_{20}F_6$	$[(C_6H_5)_2PC(CF_3)]_2$	H; F (−58)	1967, 103
$P_2C_{28}H_{26}Br_2$	$(C_6H_5)_2P^\oplus(CH)_2P^\oplus(C_6H_5)_2(CH_2)_2, 2Br^\ominus$	H; P (2·8)	1966, 204; 1967, 364, 371
$P_2C_{28}H_{27}BrO$	$(C_6H_5)_2P(O)(CH)_2P^\oplus(C_6H_5)_2C_2H_5, Br^\ominus$	H	1967, 389
$P_2C_{28}H_{28}$	$[(C_6H_5)_2P(CH_2)_2]_2$	P (14·4)	1966, 204
$P_2C_{28}H_{28}Br_2$	$[(C_6H_5)_2P^\oplus CH_2CH_2]_2, 2Br^\ominus$	H	1967, 371
$P_2C_{28}H_{28}Br_2O$	$[(C_6H_5)_2P^\oplus(CH_2)CH_2]_2O, 2Br^\ominus$	H	1967, 349
$P_2C_{28}H_{28}I_2$	$[CH_3P^\oplus(C_6H_5)_2CH]_2, 2I^\ominus$	H	1967, 371, 387
$P_2C_{28}H_{28}O_2$	$[(C_6H_5CH_2)_2P(O)]_2$	H	1966, 289
	$C_6H_5(C_6H_5CH_2)P(O)CH_2]_2$	H	1967, 371
$P_2C_{28}H_{30}I_2$	$[CH_3C_6H_5)_2P^\oplus CH_2]_2, 2I^\ominus$	H	1967, 371
$P_2C_{28}H_{32}N_3$	$[(CH_3)_2NP(C_6H_5)_2]_2N^\oplus$	P (25·4)	1967, 9
$P_2C_{28}H_{44}O_2$	$[(C_4H_9)_2POC_6H_4]_2$	P (133·7)	1968, 336
$P_2C_{29}H_{18}F_{12}O_2$	$[(CF_3C_6H_4)_2P(O)]_2CH_2$	H	1968, 18
$P_2C_{29}H_{24}F_6O_4$	$(CH_3OC_6H_4)_2P(O)CH_2P(O)(C_6H_4CF_3)_2$	H	1968, 18
$P_2C_{29}H_{30}O_2$	$[(CH_3C_6H_4)_2P(O)]_2CH_2$	H	1968, 18
$P_2C_{29}H_{30}O_6$	$[(CH_3OC_6H_4)_2P(O)]_2CH_2$	H	1968, 18
$P_2C_{29}H_{32}I_2$	$[CH_3(C_6H_5)_2P^\oplus]_2(CH_2)_3, 2I^\ominus$	H	1967, 371
$P_2C_{29}H_{62}O_6$	$[(i-C_3H_7O)_2P(O)]_2C(C_8H_{17})_2$	P (25·5)	1966, 15
$P_2C_{29}H_{65}NO$	$[C_{12}H_{25}(CH_3)_2P]_2N^\oplus CH_3O^\ominus$	P (32)	1966, 268
$P_2C_{30}H_{25}As$	$C_6H_5As[P(C_6H_5)_2]_2$	P (−15·3)	1969, 14

Formula	Compound	Resonance	Reference
$P_2C_{30}H_{26}ClN$	$(C_6H_5)_2PC_6H_4P(NH_2)(C_6H_5)_2$, Cl	H	1969, 374
$P_2C_{30}H_{28}Cl_2N_2$	$[(C_6H_5)_2PNH_2]_2C_6H_4$, Cl_2	H	1969, 374
$P_2C_{30}H_{34}Br_2S_2$	$[(C_6H_5)_2(CH_3SCH_2)P^{\oplus}CH_2]_2$, $2Br^{\ominus}$	H	1967, 349
$P_2C_{30}H_{34}Cl_2S_2$	$[(C_6H_5)_2(CH_3SCH_2)P^{\oplus}CH_2]_2$, $2Cl^{\ominus}$	H	1967, 349
$P_2C_{30}H_{34}I_2$	$[CH_3(C_6H_5)(C_6H_5CH_2)P^{\oplus}CH_2]_2$, $2I^{\ominus}$	H	1967, 371
$P_2C_{31}H_{25}N_3$	$N[P(C_6H_5)_2N]_2CC_6H_5$	H; P (17·3)	1967, 97; 1968, 94
$P_2C_{31}H_{26}ClN_3$	$HN[P(C_6H_5)_2N]^{\oplus}_2CC_6H_5$, $Cl^{\ominus}$	H; P (21·9)	1967, 97; 1968, 94
$P_2C_{31}H_{26}O_3$	$(C_6H_5)_2P(O)CH(C_6H_5)OP(O)(C_6H_5)_2$	H	1967, 368
$P_2C_{31}H_{50}O_2$	$[(C_4H_9)_2POC_6H_4]_2C(CH_3)_2$	P (130·7)	1968, 336
$P_2C_{32}H_{27}N_3$	$N[P(C_6H_5)_2N]_2CCH_2C_6H_5$	H; P (15·9)	1967, 97; 1968, 94
$P_2C_{32}H_{28}ClN_3$	$HN[P(C_6H_5)_2N]^{\oplus}_2CCH_2C_6H_5$, $Cl^{\ominus}$	H; P (19·9)	1967, 97; 1968, 94
$P_2C_{32}H_{28}O_4$	$(C_6H_5)_2P(O)CH(C_6H_4OCH_3)OP(O)(C_6H_5)_2$	H	1967, 368
$P_2C_{32}H_{30}I_2$	$[CH_3(C_6H_5)_2P^{\oplus}]_2C_6H_4$, $2I^{\ominus}$	H	1967, 371
$P_2C_{32}H_{32}O_4$	$[(C_6H_5)_2P(CH_3)=C(COOCH_3)]_2$	H	1967, 378
$P_2C_{32}H_{70}O_4Sn$	$[(C_6H_{13})_2P(O)O]_2Sn(C_4H_9)_2$	P{H}	1969, 140
$P_2C_{33}H_{29}Br$	$(C_6H_5)_2PCH=CHP^{\oplus}(CH_2C_6H_5)(C_6H_5)_2$, $Br^{\ominus}$	H; P (−3·2; 4·6)	1966, 204
$P_2C_{33}H_{32}I_2$	$(C_6H_5)_3P^{\oplus}(CH_2)_2P^{\oplus}(C_6H_5)_2CH_3$, $2I^{\ominus}$	H	1967, 371
$P_2C_{34}H_{31}BrO_5$	$(CH_3O)_2P(O)O[C(C_6H_5)]_2OP^{\oplus}(C_6H_5)_3$, $Br^{\ominus}$	P (−4·3; 65·9)	1968, 309
$P_2C_{35}H_{74}O_6$	$[(i\text{-}C_3H_7O)_2P(O)]_2C(C_{11}H_{23})_2$	P (25·5)	1966, 15
$P_2C_{36}H_{30}ClN$	$[(C_6H_5)_3P=NP(C_6H_5)_3Cl$	H; P (20·1)	1968, 254
$P_2C_{36}H_{30}Cl_4N_3Sb$	$[(C_6H_5)_3P]_2N_3^{\oplus}$, $SbCl_4^{\ominus}$	P	1967, 390
$P_2C_{36}H_{44}O_2$	$[(C_6H_5)_2P(O)(CH_2)_6]_2$	P (35·3)	1968, 336
	$[(C_6H_5)_2PO(CH_2)_6]_2$	P (110·1)	1968, 336
$P_2C_{37}H_{30}$	$[(C_6H_5)_3P]_2C$	P (−4·3)	1966, 78
$P_2C_{37}H_{31}Br$	$[(C_6H_5)_3P]^{\oplus}_2CH$, $Br^{\ominus}$	H	1967, 371

$P_2C_{37}H_{32}Br_2$	$[(C_6H_5)_3P^\oplus]_2CH_2, 2Br^\ominus$	H	1967, 371
$P_2C_{37}H_{32}N_2$	$[(C_6H_5)_2P(=NC_6H_5)]_2CH_2$	H	1966, 113
$P_2C_{38}H_{30}O_2$	$[(C_6H_5)_3P]_2C^\oplus CO_2^\ominus$	P (3·6)	1968, 110
$P_2C_{38}H_{37}$	$[(C_6H_5)_3P^\oplus CH_2]_2$	P (24·7)	1966, 269
$P_2C_{38}H_{34}Br_2$	$[(C_6H_5)_3P^\oplus CH_2]_2, 2Br^\ominus$	H	1967, 371
$P_2C_{39}H_{33}IO_2$	$[(C_6H_5)_3P]_2^\oplus CCOOCH_3, I^\ominus$	H; P (21·4)	1966, 290
$P_2C_{39}H_{36}Br_2$	$[(C_6H_5)_2P^\oplus CH_2C_6H_5]_2CH_2, 2Br^\ominus$	H	1966, 113
	$[(C_6H_5)_3P]_2(CH_2)_3, 2Br^\ominus$	P (23·2)	1966, 5
$P_2C_{40}H_{30}F_6O$	$(C_6H_5)_3POC(CF_3)_2C=P(C_6H_5)_3$	H; P (7·3; −54·0); F (−71·2)	1967, 123
$P_2C_{40}H_{31}BF_{10}O$	$[(C_6H_5)_3P]_2^\oplus CC(CF_3)_2OH, BF_4^\ominus$	H; P (22); F (−70·3; −150·3)	1967, 102
$P_2C_{40}H_{32}Br_2$	$(C_6H_5)_2P^\oplus CH=C(C_6H_5)P^\oplus(C_6H_5)_2CHCC_6H_5, 2Br^\ominus$	H; P (−3·5)	1967, 364
$P_2C_{40}H_{33}ClO_3$	$(C_6H_5)_3P^\oplus CH(COOH)COCHP(C_6H_5)_3, Cl^\ominus$	H	1966, 291
$P_2C_{40}H_{36}$	$[C_6H_5CH_2P^\oplus(C_6H_5)_2CH]_2$	H; P (3·8)	1966, 269
$P_2C_{40}H_{36}Br_2$	$[(C_6H_5)CHCH_2P^\oplus(C_6H_5)_2]_2, 2Br^\ominus$	H; P (20·4)	1967, 364
$P_2C_{40}H_{38}Br_2$	$[C_6H_5CH_2P^\oplus(C_6H_5)_2CH_2]_2, 2Br^\ominus$	H	1967, 371
$P_2C_{40}H_{38}Br_2O$	$[C_6H_5CH_2P^\oplus(C_6H_5)_2CH_2]_2O, 2Br^\ominus$	H	1967, 349
$P_2C_{40}H_{42}Al_2N_2$	$[(C_6H_5)_3P=NAl(CH_3)_3]_2$	H	1965, 119
$P_2C_{40}H_{42}Ga_2N_2$	$[(C_6H_5)_3P=NGa(CH_3)_3]_2$	H	1965, 119
$P_2C_{40}H_{42}In_2N_2$	$[(C_6H_5)_3P=NIn(CH_3)_3]_2$	H	1965, 119
$P_2C_{41}H_{33}BF_4O_2$	$(C_6H_5)_3P^\oplus C(CH_3)C(O)C[P(C_6H_5)_3]CO, BF_4^\ominus$	H	1968, 110
$P_2C_{41}H_{33}IO_2$	$(C_6H_5)_3P^\oplus C(CH_3)C(O)C[P(C_6H_5)_3]CO, I^\ominus$	H; P (0·0; 25·8)	1968, 110
$P_2C_{41}H_{35}ClO_3$	$(C_6H_5)_3PCH(COOCH_3)COCH_2P^\oplus(C_6H_5)_3, Cl^\ominus$	H	1966, 291
$P_2C_{42}H_{36}O_4$	$[(C_6H_5)_3P=C(COOCH_3)]_2$	H(T)	1967, 378

Formula	Compound	Resonance	Reference
$P_2C_{42}H_{42}Br_2$	$[C_6H_5(C_6H_5CH_2)_2P^{\oplus}CH_2]_2, 2Br^{\ominus}$	H	1967, 371
$P_2C_{44}H_{32}F_{12}O_5$	$[(C_6H_5)_3P^{\oplus}_2CC(CF_3)_2OH, (CF_3COO)_2H^{\ominus}$	H; P (21·2); F (−71·4; −76·2)	1967, 102
$P_2C_{44}H_{35}NS$	$[(C_6H_5)_3P^{\oplus}]_2CCS^{\ominus}NC_6H_5$	P (12·1)	1968, 338
$P_2C_{44}H_{36}Br_2$	$[(C_6H_5)_3P^{\oplus}CH]_2C_6H_4, 2Br^{\ominus}$	H	1967, 300
$P_2C_{44}H_{38}Br_2$	$[(C_6H_5)_3P^{\oplus}CH_2]_2C_6H_4, 2Br^{\ominus}$	P (22·8)	1966, 5
$P_2C_{45}H_{37}ClO_6$	$[(C_6H_5)_3PCC(O)OCH_2CO][(C_6H_5)_3PC(COCH_3)C(O)CH_2Cl]$	H	1967, 369
$P_2C_{46}H_{36}O_3$	$[(C_6H_5)_3CCH]_2CH(C_6H_5)P(O)_2O$	H	1968, 102
$P_2C_{46}H_{39}ClO_6$	$[(C_6H_5)_3PCC(O)OCH_2CO][(C_6H_5)_3PC(COC_2H_5)C(O)CH_2Cl]$	H	1967, 369
$P_2C_{48}H_{42}O_8$	$[(C_6H_5)_3P=C(COOCH_3)C(COOCH_3)]_2$	H; P (22·9)	1967, 372
$P_2C_{48}H_{44}Cl_2O_{16}$	$[(C_6H_5)_3P^{\oplus}CH(COOCH_3)C(COOCH_3)]_2, 2ClO_4^{\ominus}$	H; P (28·9)	1967, 372
$P_2C_{50}H_{44}O_7$	$CH_3OC_6H_4CCH(C_6H_5)CCHCH(C_6H_4OCH_3)P(O)]_2O$	H	1968, 102
$P_2C_{53}H_{41}IN_4O_4$	$(C_6H_5)_3P^{\oplus}C(CH_3)C(=NC_6H_4NO_2)C[=P(C_6H_5)_3]C=$ $=NC_6H_4NO_2, I^{\ominus}$	H; P (27·7; −3·7)	1968, 33
$P_2C_{53}H_{43}IN_2$	$(C_6H_5)_3P^{\oplus}C(CH_3)C(=NC_6H_5)C[=P(C_6H_5)_3]C=NC_6H_5, I^{\ominus}$	H; P (28·1; −4·4)	1968, 110, 338
$P_2C_{55}H_{47}IN_2$	$(C_6H_5)_3P^{\oplus}C(CH_3)C(=NC_6H_4CH_3)C[=P(C_6H_5)_3]C=$ $=NC_6H_4CH_3, I^{\ominus}$	H; P (28·2; −4·4)	1968, 110, 338

NMR DATA ON ORGANIC COMPOUNDS 277

P₃

$P_3CH_3Cl_5N_3O$	$N_3P_3Cl_5(OCH_3)$	H; P (16·7; 22·5)	1966, 49; 1967, 130
$P_3CH_4Cl_5N_4$	$N_3P_3Cl_5(NHCH_3)$	P (22·9)	1968, 114
$P_3CH_4F_5N_4$	$N_3P_3F_5(NHCH_3)$	H; F (−62·5; −70·7)	1969, 152
$P_3C_2H_2Cl_5F_3N_3O$	$N_3P_3Cl_5(OCH_2CF_3)$	H; P (16·5; 22·7)	1966, 49; 1967, 130
$P_3C_2H_5Cl_5N_3O$	$N_3P_3Cl_5(OC_2H_5)$	H; P (13·6; 21·3)	1966, 49; 1967, 130
$P_3C_2H_6Br_5N_4$	$N_3P_3Br_5[N(CH_3)_2]$	H; P (−39·3; 4·5)	1966, 93; 1968, 91
$P_3C_2H_6Cl_5N_4$	$N_3P_3Cl_5[N(CH_3)_2]$	H; P (22·4; 20·5)	1965, 24; 1966, 49, 93; 1967, 130; 1968, 91
$P_3C_2H_6F_5N_4$	$N_3P_3F_5[N(CH_3)_2]$	H; F (−63·0; −70·9)	1969, 152
$P_3C_2H_8Cl_4N_5$	$N_3P_3Cl_4(NHCH_3)_2$	H; P (22·2; 21·6; 20·1; 12·3)	1967, 148; 1968, 114
$P_3C_2H_9O_8$	$PO_2H(CH_2PO_3H_2)_2$	H; P (37·3; 17·5)	1968, 300; 1969, 303
$P_3C_2H_{12}Cl_2N_7$	$N_3P_3Cl_2(NH_2)_2(NHCH_3)_2$	P (23; 16·5)	1968, 114
$P_3C_3H_7Cl_5N_3O$	$N_3P_3Cl_5(Oi\text{-}C_3H_7)$	H; P (12·6; 21·7)	1966, 49; 1967, 130
$P_3C_3H_9F_9N_3$	$(F_3PNCH_3)_3$	H	1966, 36
$P_3C_3H_{11}O_7$	$CH_3P(O)(CH_2PO_3H_2)_2$	H	1969, 334
$P_3C_3H_{12}NO_9$	$N(CH_2PO_3H_2)_3$	H; P(ca. 8·5)	1966, 212; 1967, 315, 391
$P_3C_3H_{12}NO_{10}$	$ON(CH_2PO_3H_2)_3$	H; P (ca. 5·5)	1967, 391
$P_3C_4H_{10}Cl_4N_3S_2$	$N_3P_3Cl_4(SC_2H_5)_2$	P	1967, 130
$P_3C_4H_{12}Br_4N_5$	$N_3P_3Br_4[N(CH_3)_2]_2$	H; P (ca. −36·7; 10·0)	1966, 93; 1968, 91
$P_3C_4H_{12}Cl_2F_2N_5$	$N_3P_3Cl_2F_2[N(CH_3)_2]_2$	H	1969, 153

Formula	Compound	Resonance	Reference
$P_3C_4H_{12}Cl_4N_5$	$N_3P_3Cl_4[N(CH_3)_2]_2$	H; P	1965, 24; 1966, 93; 1967, 130, 146; 1968, 91, 129
$P_3C_4H_{12}F_4N_2$	$N_3P_3F_4[N(CH_3)_2]_2$	H	1969, 151
$P_3C_4H_{13}O_{10}$	$CH_3P(O_2H)OP(O)(OCH_3)OP(O)(OCH_3)_2$	H; P	1966, 138
$P_3C_5H_{10}Cl_5N_4$	$N_3P_3Cl_5NC_5H_{10}$	H	1966, 88
$P_3C_5H_{12}F_3$	$CF_3[P(CH_3)_2]_2$	F	1967, 83
$P_3C_5H_{15}O_5$	$CH_3P(O)[OP(O)(OCH_3)CH_3]_2$	H; P	1967, 186
$P_3C_6BF_{18}O_3$	$B[OP(CF_3)_2]_3$	^{11}B; H; P (36·3) F (−66·8)	1969, 27
$P_3C_6F_{10}N_3$	$N_3P_3F_5(C_6F_5)$	F	1968, 127
$P_3C_6H_5Cl_5N_3O$	$N_3P_3Cl_5(OC_6H_5)$	P	1967, 130
$P_3C_6H_6Cl_4N_5$	$N_3P_3Cl_4(NH)_2C_6H_4$	H; P (21·5)	1966, 90
$P_3C_6H_{15}K_2$	$C_2H_5P[P(C_2H_5)K]_2$	P (−78·5; −23·9)	1965, 27, 29
$P_3C_6H_{15}Na_2$	$C_2H_5P[P(C_2H_5)Na]_2$	P	1965, 27
$P_3C_6H_{15}O_4S_3$	$[C_2H_5P(S)O]_3O$	H	1968, 355
$P_3C_6H_{16}Cl_4N_5$	$N_3P_3Cl_3(NHC_2H_5)(NHt\text{-}C_4H_9)$	H	1968, 129
$P_3C_6H_{17}F_{12}N$	$(CH_3)_2P^{\oplus}CH_2N^{\oplus}(CH_3)_3, 2PF_6^{\ominus}$	H	1969, 128
$P_3C_6H_{18}Br_3N_6$	$N_3P_3Br_3[N(CH_3)_2]_3$	H; P (15·4)	1966, 93; 1968, 91
$P_3C_6H_{18}Cl_3N_6$	$N_3P_3Cl_3[N(CH_3)_2]_3$	H; P (27·6; 26·1; 21·6)	1965, 24; 1966, 93; 1968, 91, 136
$P_3C_6H_{18}N_3O_6$	$N_3P_3(OCH_3)_6$	P (23·2)	1966, 6
$P_3C_6N_9S_6$	$N_3P_3(NCS)_6$	P (30)	1968, 114
$P_3C_7H_{19}O_3$	$CH_3P(O)[CH_2P(O)(CH_3)_2]_2$	P (39·6; 42·4)	1969, 334
$P_3C_7H_{19}O_8$	$CH_3OP(O)[CH_2P(O)(OCH_3)_2]_2$	H	1969, 332
$P_3C_8H_{11}Cl_5N_5$	$N_3P_3Cl_5[NHC_6H_4N(CH_3)_2]$	H	1966, 92
$P_3C_8H_{20}Cl_2N_3S_4$	$N_3P_3Cl_2(SC_2H_5)_4$	H	1966, 84

NMR DATA ON ORGANIC COMPOUNDS 279

Formula	Compound	Nuclei	References
$P_3C_8H_{20}Cl_4N_5$	$N_3P_3Cl_4(NHt\text{-}C_4H_9)_2$	H	1968, 129
$P_3C_8H_{21}O_8$	$C_2H_5OP(O)[CH_2P(O)(OCH_3)_2]_2$	H	1969, 332
$P_3C_8H_{24}Cl_2N_7$	$N_3P_3Cl_2[N(CH_3)_2]_4$	H; P (23·9; 30·5)	1965, 24; 1968, 91
$P_3C_8H_{28}N_9$	$N_3P_3[N(CH_3)_2]_4(NH_2)_2$	H	1966, 88
$P_3C_{10}H_{17}Cl_3N_5O$	$N_3P_3Cl_3(OC_6H_5)[N(CH_3)_2]_2$	H	1967, 146
$P_3C_{10}H_{20}Cl_4N_5$	$N_3P_3Cl_4(NC_5H_{10})_2$	H	1966, 88
$P_3C_{10}H_{28}Cl_2N_7$	$N_3P_3Cl_2[N(CH_3)_2]_2(NHi\text{-}C_3H_7)_2$	H	1966, 85
$P_3C_{11}H_{27}O_7$	$CH_3P(O)[CH_2P(O)(OC_2H_5)_2]_2$	H; P (34·5; 21·3)	1968, 299; 1969, 332
$P_3C_{12}H_{10}Cl_4N_3S_2$	$N_3P_3Cl_4(SC_6H_5)_2$	P	1967, 130
$P_3C_{12}H_{10}F_4N_3$	$N_3P_3F_4(C_6H_5)_2$	P (38; 12; 11); F (−49; −50; −65; −70)	1967, 144; 1968, 50, 130
$P_3C_{12}H_{12}Cl_4N_5$	$N_3P_3Cl_4(NHC_6H_5)_2$	P (−20·4; −2·3)	1966, 80; 1967, 130
$P_3C_{12}H_{12}F_{18}N_3O_6$	$N_3P_3(OCH_2CF_3)_6$	P (19·2)	1966, 6
$P_3C_{12}H_{29}O_7$	$C_2H_5P(O)[CH_2P(O)(OC_2H_5)_2]_2$	H; P (38·6; 21·1)	1968, 299; 1969, 334
$P_3C_{12}H_{29}O_8$	$C_2H_5OP(O)[CH_2P(O)(OC_2H_5)_2]_2$	H; P (37·3; 20·0)	1967, 2; 1969, 332
$P_3C_{12}H_{30}LiSi$	$[(C_2H_5)_2P]_3SiLi$	H	1967, 182
$P_3C_{12}H_{30}N_3O_6$	$N_3P_3(OC_2H_5)_6$	H; P (16·8)	1966, 6, 87; 1968, 57
$P_3C_{12}H_{31}Si$	$[(C_2H_5)_2P]_3SiH$	H	1967, 182
$P_3C_{12}H_{36}N_9$	$N_3P_3[N(CH_3)_2]_6$	H; P (25)	1965, 24; 1968, 91, 136
$P_3C_{13}H_{33}Si$	$[(C_2H_5)_2P]_3SiCH_3$	H	1967, 182
$P_3C_{13}H_{38}N_9$	$N_3P_3[N(CH_3)_2]_5NHi\text{-}C_3H_7$	H	1966, 85
$P_3C_{14}H_{16}Cl_3N_4$	$N_3P_3Cl_3(C_6H_5)_2[N(CH_3)_2]$	H	1969, 150
$P_3C_{14}H_{17}O_6$	$PO_2H[CH_2P(O)(C_6H_5)OH]_2$	H	1969, 332
$P_3C_{14}H_{33}O_7$	$C_{12}H_{25}P(O)[CH_2PO_2H]_2$	H	1969, 334
$P_3C_{14}H_{33}O_8$	$C_4H_9OP(O)[CH_2P(O)(OC_2H_5)_2]_2$	H; P (45·8; 16·3)	1968, 300; 1969, 332
$P_3C_{14}H_{40}N_9$	$N_3P_3[N(CH_3)_2]_4(NHi\text{-}C_3H_7)_2$	H; P (37·4; 20·4)	1966, 85
$P_3C_{14}H_{42}BF_4N_9$	$N_3P_3[N(CH_3)_2]_4[N^\oplus(CH_3)_3]_2, 2BF_4^\ominus$	H	1968, 135

Formula	Compound	Resonance	Reference
$P_3C_{15}H_{30}Cl_3N_6$	$N_3P_3Cl_3(NC_5H_{10})_3$	H	1966, 88
$P_3C_{15}H_{33}O_8$	$C_2H_5OP(O)[CH_2P(O)(OC_2H_5)]_2-$	H	1967, 2
	$-[CH(CH_2CH=CH_2)P(O)(OC_2H_5)_2]$	H; P (76·4; 7·9; 17·9)	1969, 100
$P_3C_{15}H_{34}N_7OS_2$	$CH_3OC_6H_4P(S)(SH)\{N=P[N(CH_3)_2]_2\}_2NH_2$	H	1966, 88
$P_3C_{15}H_{40}N_9$	$N_3P_3[N(CH_3)_2]_5NC_5H_{10}$	H	1967, 146; 1969, 150
$P_3C_{16}H_{22}Cl_2N_5$	$N_3P_3(C_6H_5)_2[N(CH_3)_2]_2Cl_2$	H	1969, 150
$P_3C_{16}H_{26}N_7$	$N_3P_3(C_6H_5)_2[N(CH_3)_2]_2C(NH_2)_2$	H	1966, 86
$P_3C_{16}H_{35}N_8O$	$N_3P_3[N(CH_3)_2]_5(C_6H_5)$	H	1965, 33
$P_3C_{16}H_{44}N_9$	$N_3P_3[N(CH_3)_2]_4(NHt-C_4H_9)_2$	H	1966, 85
	$N_3P_3[N(CH_3)_2]_2(NHi-C_3H_7)_4$		
$P_3C_{18}H_{12}N_3O_6$	$N_3P_3(C_6H_4O_2)_3$	P (13·3)	1966, 90
$P_3C_{18}H_{15}Br_3N_3$	$N_3P_3Br_3(C_6H_5)_3$ (isom.)	P (16·3; 17·8; 196·6)	1968, 133
$P_3C_{18}H_{15}F_3N_3$	$N_3P_3F_3(C_6H_5)_3$ (isom.)	P (31·2; 27·3; 8·6); F (−49·1; −68)	1968, 130
$P_3C_{18}H_{15}K_2$	$(C_6H_5P)_3K_2$	P	1966, 292
$P_3C_{18}H_{18}N_9$	$N_3P_3[C_6H_4(NH)_2]_3$	H; P (24·5)	1966, 90
$P_3C_{18}H_{21}N_6O_3$	$N_3P_3(NH_2)_3(OC_6H_5)_3$	H	1966, 129
$P_3C_{18}H_{37}O_8$	$C_2H_5OP(O)[CH(CH_2CH=CH_2)P(O)(OC_2H_5)_2]_2$	H	1967, 2
$P_3C_{18}H_{41}O_7$	$C_8H_{17}P(O)[CH_2P(O)(OC_2H_5)_2]_2$	H; P (37·2; 20·8)	1969, 334
$P_3C_{18}H_{41}O_8$	$C_4H_9OP(O)[CH_2P(O)(Oi-C_3H_7)_2]_2$	H; P (37·3; 18·5)	1968, 300; 1969, 332
$P_3C_{18}H_{42}N_3O_9$	$(C_6H_{11}NHPO_3H_2)_3$	P (−21·9)	1966, 65
$P_3C_{18}H_{44}N_9$	$N_3P_3[N(CH_3)_2]_4(NC_5H_{10})_2$	H	1966, 88
$P_3C_{20}H_{18}Cl_2-$ $F_{12}N_5O_4$	$N_3P_3(OCH_2CF_3)_4(NHC_6H_4Cl)_2$	H; P (6·9; 16·2)	1966, 80

NMR DATA ON ORGANIC COMPOUNDS

$P_3C_{20}H_{20}F_{12}N_5O_4$	$N_3P_3(OCH_2CF_3)_4(NHC_6H_5)_2$	H; P (6·4; 16·4)	1966, 80; 1967, 130
$P_3C_{20}H_{22}Cl_2N_5$	$N_3P_3Cl_2(C_6H_5)_2[N(CH_3)_2](NHC_6H_5)$	H	1969, 150
$P_3C_{20}H_{30}N_3O_4$	$N_3P_3(C_6H_5)_2(OC_2H_5)_4$	H	1966, 87; 1968, 57
$P_3C_{20}H_{34}N_7$	$N_3P_3(C_6H_5)_2[N(CH_3)_2]_4$	H	1969, 150
$P_3C_{20}H_{34}N_7O_2$	$N_3P_3(OC_6H_5)_2[N(CH_3)_2]_4$	H	1966, 86
$P_3C_{20}H_{40}Cl_2N_7$	$N_3P_3Cl_2(NC_5H_{10})_4$	H	1966, 88
$P_3C_{20}H_{48}N_3O_7$	$C_6H_{11}NH_2P(O_2H)[CH_2PO_3H_2, C_6H_{11}NH_2]_2$	P (28·7; 15)	1969, 332
$P_3C_{21}H_{18}N_3S_6$	$N_3P_3[SC_6H_3(CH_3)]_3$	H	1966, 90
$P_3C_{21}H_{41}O_8$	$C_2H_5OP(O)[CH(CH_2CH=CH_2)P(O)(OC_2H_5)_2]-$	H	1967, 2
	$-[CH_2CH=CH_2)P(O)(OC_2H_5)_2]$		
$P_3C_{21}H_{48}N_9$	$N_3P_3[N(CH_3)_2]_3(NC_5H_{10})_3$	H	1966, 88
$P_3C_{22}H_{24}F_{12}N_5O_4$	$N_3P_3(OCH_2CF_3)_4(NHC_6H_4CH_3)_2$	H	1966, 80
$P_3C_{22}H_{27}N_6$	$N_3P_3(C_6H_5)_3[N(CH_2)_2]_3$	H	1966, 96
$P_3C_{22}H_{49}O_7$	$C_{12}H_{25}OP(O)[CH_2P(O)(OC_2H_5)_2]_2$	H; P (36·9; 21·1)	1968, 299; 1969, 334
$P_3C_{22}H_{49}O_8$	$C_4H_9OP(O)[CH_2P(O)(OC_4H_9)_2]_2$	P (37·4; 20·7)	1968, 300; 1969, 332
$P_3C_{24}H_{20}Cl_2N_3$	$N_3P_3Cl_2(C_6H_5)_4$	P (17·2)	1968, 114
$P_3C_{24}H_{20}Cl_2N_3O_2$	$N_3P_3Cl_2(C_6H_5)_2(OC_6H_5)_2$	P (16·6; 21·9)	1966, 95
$P_3C_{24}H_{20}Cl_8N$	$[(C_6H_5)_2PCl]_2^{\oplus}N, PCl_6^{\ominus}$	P (43·3)	1967, 92
$P_3C_{24}H_{20}F_2N_3$	$N_3P_3F_2(C_6H_5)_4$	P (27·3; 62)	1968, 130
$P_3C_{24}H_{24}Cl_2N_3O_2$	$N_3P_3Cl_2(C_6H_5)_2(OC_6H_5)_2$	H	1966, 97
$P_3C_{24}H_{24}N_5$	$N_3P_3(C_6H_5)_4(NH_2)_2$	H	1966, 97
$P_3C_{24}H_{24}N_5O_2$	$N_3P_3(C_6H_5)_2(OC_6H_5)_2(NH_2)_2$	H	1966, 97
$P_3C_{24}H_{24}N_5O_4$	$N_3P_3(C_6H_5)_4(NH_2)_2$	H	1966, 97
$P_3C_{24}H_{33}N_6O_3$	$N_3P_3(OC_6H_5)_3[N(CH_3)_2]_3$	H	1966, 86
$P_3C_{24}H_{33}Sn_3$	$[C_6H_5PSn(CH_3)_2]_3$	H	1968, 95
$P_3C_{24}H_{34}N_7$	$N_3P_3(C_6H_5)_2[N(CH_3)_2]_2NHC_6H_5$	H	1969, 150
$P_3C_{24}H_{37}O_6$	$C_2H_5OP(O)[CH_2P(O)(C_6H_5)OC_4H_9]_2$	H; P (38·8; 33·6)	1968, 300
$P_3C_{24}H_{45}O_8$	$C_2H_5OP(O)[C(CH_2CH=CH_2)P(O)(OC_2H_5)_2]_2$	H	1967, 2
$P_3C_{24}H_{52}N_9$	$N_3P_3[N(CH_3)_2]_2(NC_5H_{10})_4$	H	1966, 88

Formula	Compound	Resonance	Reference
$P_3C_{25}H_{24}N_3$	$N_3P_3(C_6H_5)_4(CH_3)H$	H; P (12·9; 6·9)	1968, 40
$P_3C_{25}H_{57}O_3Si$	$[(C_4H_9)_2PO]_3SiCH_3$	P (119·0)	1967, 343
$P_3C_{26}H_{25}O_4$	$[(C_6H_5)_2P(O)CH_2]_2PO_2H$	H	1969, 332
$P_3C_{26}H_{26}N_3$	$N_3P_3(C_6H_5)_4(C_2H_5)H$	H; P (15·5; 12·5)	1968, 40
$P_3C_{26}H_{27}ClN_3$	$(CH_3)_2P^{\oplus}NH[P(C_6H_5)_2N]_2, Cl^{\ominus}$	P (39·6; 18·8)	1969, 376
$P_3C_{26}H_{28}N_5$	$N_3P_3(C_6H_5)_4(NHCH_3)_2$	P (19·5; 18)	1968, 114
$P_3C_{26}H_{33}Sn_2$	$C_4H_9Sn(PC_6H_5)_3SnC_4H_9$	P (−5·2)	1969, 377
$P_3C_{26}H_{41}O_5$	$C_{12}H_{25}P(O)[CH_2P(O)(C_6H_5)OH]_2$	H	1969, 334
$P_3C_{28}H_{29}O_4$	$C_2H_5OP(O)[CH_2P(O)(C_6H_5)_2]_2$	H	1969, 332
$P_3C_{28}H_{30}N_3O_2$	$N_3P_3(C_6H_5)_4(OC_2H_5)_2$	H	1966, 87; 1968, 57
$P_3C_{28}H_{32}N_5$	$N_3P_3(C_6H_5)_4[N(CH_3)_2]_2$	H	1969, 150
$P_3C_{28}H_{32}N_5O_2$	$N_3P_3(C_6H_5)_2(OC_6H_5)_2[N(CH_3)_2]_2$	H	1966, 95
$P_3C_{28}H_{32}N_5O_4$	$N_3P_3(OC_6H_5)_4[N(CH_3)_2]_2$	H	1966, 86
$P_3C_{28}H_{34}N_7$	$N_3P_3(C_6H_5)_2[N(CH_3)_2]_2(NHC_6H_5)_2$	H	1969, 150
$P_3C_{30}H_{25}ClN_3O_5$	$N_3P_3(OC_6H_5)_5Cl$	P	1967, 130
$P_3C_{30}H_{25}N$	$(C_6H_5)_2PP(C_6H_5)=NP(C_6H_5)_2$	P	1967, 385
$P_3C_{30}H_{25}Sn_2$	$(C_6H_5P)_3(SnC_6H_5)_2$	P	1969, 14
$P_3C_{30}H_{26}N_3O$	$N_3P_3(C_6H_5)_4(OC_6H_5)H$	H; P (14·5; 5·8)	1968, 40
$P_3C_{30}H_{60}N_9$	$N_3P_3(NC_5H_{10})_6$	H	1966, 88
$P_3C_{31}H_{28}N_3OS$	$N[P(C_6H_5)_2N]_2P(SCH_3)(SC_6H_5)$	H	1969, 100
$P_3C_{31}H_{28}N_3S_2$	$N[P(C_6H_5)_2N]_2P(C_6H_4OCH_3)SH$	P (21·6; 45·8)	1969, 100
$P_3C_{31}H_{29}IN_3S_2$	$(C_6H_5S)(CH_3S)P^{\oplus}NH[(C_6H_5)_2N]_2, I^{\ominus}$	H	1969, 100
$P_3C_{32}H_{29}O_3$	$C_6H_5P(O)[CH_2P(O)(C_6H_5)_2]_2$	P (27·6)	1969, 334

P₃C₃₂H₃₀N₃OS	N₃P₃(C₆H₅)₄(C₆H₄OCH₃)(SCH₃)	H; P (14·1; 27·0)	1969, 100
P₃C₃₂H₃₁IN₃OS	(CH₃OC₆H₄)(CH₃S)P⊕NH[P(C₆H₅)₂N]₂I⊖	H; P(21·4; 38·1)	1969, 100
P₃C₃₂H₃₁N₄O₅	N₃P₃(OC₆H₅)₅[N(CH₃)₂]	H	1966, 86
P₃C₃₂H₃₄N₇	N₃P₃(C₆H₅)₂(NHC₆H₅)₃[N(CH₃)₂]	H	1969, 150
P₃C₃₃H₅₁O₁₇	(CH₃O)₃POCH[C₆H₄CHC(CH₃)(COCH₃)OP(OCH₃)₃O]₂O	H; P (−49·9)	1967, 350
P₃C₃₄H₅₇O₅	C₁₂H₂₅P(O)[CH₂P(O)(OC₄H₉)C₆H₅]₂	H; P (37·7; 34·9)	1968, 299; 1969, 334
P₃C₃₆H₃₀As	As[P(C₆H₅)₂]₃	P (−15·2)	1969, 14
P₃C₃₆H₃₀N₃	N₃P₃(C₆H₅)₆	P (14·2)	1965, 34; 1968, 114
P₃C₃₆H₃₀N₃O₆	N₃P₃(OC₆H₅)₆	P (10·5)	1966, 6, 90
P₃C₃₆H₃₄ClN₄	H₂NP(C₆H₅)₂N]₂P(C₆H₅)₂NH₂, Cl	P (31·6; 18·1)	1969, 378
P₃C₃₇H₃₃O₃Si	[(C₆H₅)₂PO]₃SiCH₃	P (97·7)	1967, 343
P₃C₃₇H₃₇Cl₂N₄	[(C₆H₅)₂PNH]₂NNCH₃[P(C₆H₅)₂], Cl₂	H; P (36·7)	1967, 94
P₃C₃₇H₃₈Cl₂N₅	[(C₆H₅)₂PNH][(C₆H₅)₂PNH₂]NNCH₃[H₂NP(C₆H₅)₂], Cl₂	H; P (126)	1967, 94
P₃C₃₈H₄₉O₃	C₁₂H₂₅P(O)[CH₂P(O)(C₆H₅)₂]₂	H; P (42·6; 28·6)	1968, 299; 1969, 334
P₃C₄₀H₃₁F₁₂O	[(C₆H₅)₃P]₂⊕CC(CF₃)₂OH, PF₆⊖	H; P (−144·8; 22); F (−71·6; −72·5)	1967, 102
P₃C₄₃H₃₅Cl₃	[(C₆H₅)₃P]₂⊕CP(C₆H₅)Cl, 2Cl⊖	P (67·9; 25·9)	1966, 78
P₃C₄₈H₆₆N₁₅	N₃P₃[NHC₆H₄N(CH₃)₂]₆	H	1966, 92
P₃C₄₉H₄₀Cl	[(C₆H₅)₃P]₂⊕CP(C₆H₅)₂, Cl⊖	H; P 26·0; −1·5)	1966, 78
P₃C₄₉H₄₀ClO	[(C₆H₅)₃P]⊕CP(O)(C₆H₅)₂, Cl⊖	P (23·2; 26·9)	1966, 78
P₃C₅₀H₄₃I₂	[(C₆H₅)₃P]₂C⊕P⊕(C₆H₅)₂CH₃, 2I⊖	H; P (25·6; 22·6)	1966, 78
P₃C₅₁H₄₆ClOS	[(C₆H₅)₃P]₂C⊕P(S)(C₆H₅)₂, Cl⊖	H; P (24·7; 44·0)	1966, 78

Formula	Compound	Resonance	Reference
$P_3C_{54}Cl_3N_3S$	$[(C_6H_5)_3PN]_3S, Cl_3$	H; P (16·0)	1968, 254
$P_3C_{55}H_{45}F_8N_2$	$[(C_6H_5)_3P]_2^{\oplus}C^{\oplus}P(C_6H_5)_2N_2C_6H_5, 2BF^{\ominus}$	P (38·9; 24·1)	1966, 78
$P_3C_{56}H_{46}I_2$	$[(C_6H_5)_3P]_2^{\oplus}C^{\oplus}P(C_6H_5)_2CH_2C_6H_5, 2I_4^{\ominus}$	H; P (27·0; 28·5)	1966, 78

P_4

Formula	Compound	Resonance	Reference
$P_4C_2Cl_{14}N_6$	$(Cl_3P=NN=CClNPCl_3)_2$	P (15·2; −78·8)	1965, 32
$P_4C_3H_{12}O_{10}$	$OP(CH_2PO_3H_2)_3$	H; P (37·8; 15·3)	1968, 298; 1969, 333
$P_4C_4F_{12}$	$(PCF_3)_4$	P (−74·8); F(T)	1967, 159
$P_4C_6F_{12}N_4$	$N_4P_4F_7(C_6F_5)$	F	1968, 127
$P_4C_6H_{18}Cl_4N_6O_2$	$[ClP(O)(NCH_3)_2P(NCH_3)Cl]_2$	P (4; −73)	1966, 83
$P_4C_6H_{18}Cl_4N_6S_2$	$[ClP(S)(NCH_3)_2P(NCH_3)Cl]_2$	P (55·3; −68)	1966, 83
$P_4C_6H_{18}F_5N_7$	$N_4P_4F_5[N(CH_3)_2]_3$	H	1969, 379
$P_4C_6H_{20}N_2O_{12}$	$(CH_2)_2[N(CH_2PO_3H_2)_2]_2$	P	1966, 212
$P_4C_7H_{22}N_2O_{12}$	$(CH_2)_3[N(CH_2PO_3H_2)_2]_2$	P	1966, 212
$P_4C_8H_{20}$	$(PC_2H_5)_4$	P (−15·7)	1965, 29
$P_4C_8H_{20}Na_2$	$[C_2H_5PP(C_2H_5)Na]_2$	P (−29)	1965, 27
$P_4C_8H_{20}O_{12}$	$[C_2H_5OP(O)O]_4$	P (−29)	1963, 5
	$(C_2H_5O)_2P(O)OP(O)[OP(O)OC_2H_5]_2O$	P (−16; −42; −29)	1963, 5
$P_4C_8H_{24}F_4N_8$	$N_4P_4F_4[N(CH_3)_2]_4$	H	1969, 379
$P_4C_8H_{24}N_2O_{12}$	$(CH_2)_4[N(CH_2PO_3H_2)_2]_2$	P	1966, 212
$P_4C_8H_{24}N_4$	$N_4P_4(CH_3)_8$	H{^{13}C; ^{31}P}	1968, 3
$P_4C_9H_{24}O_4$	$OP[CH_2P(O)(CH_3)_2]_3$	H; P (36·9; 43·1)	1969, 333
$P_4C_9H_{24}O_{10}$	$OP[CH_2P(O)(OCH_3)_2]_3$	H; P (31·0; 22·8)	1968, 298; 1969, 333

$P_4C_9H_{26}N_2O_{12}$	$(CH_2)_5[N(CH_2PO_3H_2)_2]_2$	P	1966, 212
$P_4C_{10}H_{28}N_2O_{12}$	$(CH_2)_6[N(CH_2PO_3H_2)_2]_2$	P	1966, 212
$P_4C_{11}H_{30}N_2O_{12}$	$(CH_2)_7[N(CH_2PO_3H_2)_2]_2$	P	1966, 212
$P_4C_{12}H_{30}N_2O_{12}$	$CH_2C_6H_{10}CH_2[N(CH_2PO_3H_2)_2]_2$	P	1966, 212
$P_4C_{12}H_{32}N_2O_{12}$	$(CH_2)_8[N(CH_2PO_3H_2)_2]_2$	P	1966, 212
$P_4C_{13}H_{34}N_2O_{12}$	$(CH_2)_9[N(CH_2PO_3H_2)_2]_2$	P	1966, 212
$P_4C_{14}H_{36}N_2O_{12}$	$(CH_2)_{10}[N(CH_2PO_3H_2)_2]_2$	P	1966, 212
$P_4C_{15}H_{36}O_{10}$	$OP[Cl_2P(O)(OC_2H_5)_2]_3$	H; P (34·0; 21·0)	1968, 298; 1969, 333
$P_4C_{15}H_{38}N_2O_{12}$	$(CH_2)_{11}[N(CH_2PO_3H_2)_2]_2$	P	1966, 212
$P_4C_{16}H_{36}$	$(Pt\text{-}C_4H_9)_4$	P (−57·8)	1966, 293
$P_4C_{16}H_{40}N_2O_{12}$	$(CH_2)_{12}[N(CH_2PO_3H_2)_2]_2$	P	1966, 212
$P_4C_{18}H_{15}Cl_5N_4$	$N_3P_3Cl_5[N=P(C_6H_5)_3]$	P (ca. 20; 15)	1968, 132
$P_4C_{18}H_{17}Cl_4N_5$	$N_3P_3Cl_4(NH_2)[N=P(C_6H_5)_3]$	P (16; 11; −2)	1967, 147; 1968, 114 132
$P_4C_{18}H_{45}Cl_3$	$[(C_2H_5)_3P]_3PCl_3$	H	1969, 344
$P_4C_{21}H_{24}O_7$	$OP[CH_2P(O)(C_6H_5)OH]_3$	P (35·3; 29·8)	1969, 333
$P_4C_{21}H_{48}O_{10}$	$OP[CH_2P(O)(Oi\text{-}C_3H_7)_2]_3$	H; P (28·9; 18·6)	1968, 298; 1969, 333
$P_4C_{24}F_{20}$	$(PC_6F_5)_4$	P (−67·0)	1967, 392
$P_4C_{24}H_{20}$	$(PC_6H_5)_4$	H	1969, 344
$P_4C_{24}H_{20}Cl_4N_4$	$N_4P_4Cl_4(C_6H_5)_4$	H	1966, 96
	$N_3P_3Cl_4(C_6H_5)[N=P(C_6H_5)_3]_3$	P (15·8; 2·4; 14)	1968, 132
$P_4C_{24}H_{20}Li_2$	$[C_6H_5PP(C_6H_5)Li]_2$	P (8·2; −86·0)	1965, 27
$P_4C_{26}H_{56}Cl_4N_8$	$N_4P_4Cl_4(C_6H_{11}NCH_3)_4$	H{P}	1968, 134
$P_4C_{27}H_{60}O_{10}$	$OP[CH_2P(O)(OC_4H_9)_2]_2$	P (29·7; 19·6)	1968, 298; 1969, 333

286 FORMULAE INDEX II

Formula	Compound	Resonance	Reference
$P_4C_{32}H_{44}N_8$	$N_4P_4(C_6H_5)_4[N(CH_3)_2]_4$	H	1966, 96
$P_4C_{32}H_{72}N_{12}$	$N_4P_4(NHCH_3)_4(C_6H_{11}NCH_3)_4$	H{P}	1968, 134
$P_4C_{33}H_{48}O_7$	$OP[CH_2P(O)(OC_4H_9)C_6H_5]_3$	H; P (29.0; 34.7)	1968, 298; 1969, 333
$P_4C_{35}H_{41}N_9O$	$N_4P_4(C_6H_5)_4(NHCH_3)_3[N(CH_3)CONHC_6H_5]$	H	1966, 96
$P_4C_{36}H_{30}$	$P[P(C_6H_5)_2]_3$	P (−16.9; −27.0)	1969, 14
$P_4C_{36}H_{50}F_{10}N_6$	$\{CH_3N(C_6H_5)P\oplus[N(CH_3)CH_2]_2\}_2(C_6H_5PF_5^\ominus)_2$	H; P (49.5; −136); F (−60.7; −57.0)	1968, 352
$P_4C_{39}H_{36}O_4$	$OP[CH_2P(O)(C_6H_5)_2]_3$	P (31.0; 24.4)	1968, 298; 1969, 333
$P_4C_{39}H_{77}N_{13}O$	$N_4P_4(NHCH_3)_3(CH_3NC_6H_{11})_4(CH_3NCONHC_6H_5)$	H{P}	1968, 134
$P_4C_{42}H_{42}O_2$	$[(C_6H_5CH_2)_2P(O)P(C_6H_5CH_2)]_2$	H	1966, 289
$P_4C_{48}H_{40}ClN$	$[(C_6H_5)_2PP(C_6H_5)_2]_2^\oplus N, Cl^\ominus$	P (28.3; 21.8; −16.5; −23.0)	1965, 11; 1967, 385
$P_4C_{49}H_{40}F_6$	$[(C_6H_5)_3P]_2C^\oplus P(C_6H_5)_2, PF_6^\ominus$	H; P (26.5; −1.4; −144.2)	1966, 78
$P_4C_{51}H_{108}O_{10}$	$OP[CH_2P(O)(OC_8H_{17})_2]_3$	H; P (29.3; 21.0)	1968, 298; 1969, 333

P_5 and Higher

Formula	Compound	Resonance	Reference
$P_5C_5F_{15}$	$(PCF_3)_5$	P (8.2); F(T)	1967, 159
$P_5C_5H_{15}$	$(PCH_3)_5$	H	1969, 328
$P_5C_6H_{14}N_5$	$N_5P_5F_9(C_6F_5)$	F	1968, 127
$P_5C_{18}H_{15}Cl_7N_5$	$N_3P_3Cl_4(N=PCl_3)[N=P(C_6H_5)_3]$	P (19; 18.5; −11.8; −20.3)	1967, 147; 1968, 114

$P_5C_{30}F_{25}$	$P(C_6F_5)_5$	$F\ (-126\cdot 4;\ -160\cdot 1;\ -149\cdot 5)$	1966, 294
$P_5C_{30}H_{25}$	$P(C_6H_5)_5$	P	1966, 295
$P_5C_{36}H_{30}Cl_4N_5$	$N_3P_3Cl_4[N{=}P(C_6H_5)_3]_2$	$P\ (13\cdot 4;\ 5\cdot 5;\ -11\cdot 6)$	1968, 132
$P_5C_{40}H_{45}N_8$	$N_4P_4(C_6H_5)_4(NHCH_3)_3[N(CH_3)P(C_6H_5)_2]$	H	1966, 96
$P_5C_{50}H_{43}F_{12}$	$[(C_6H_5)_3P]_2C^{\oplus}P(C_6H_5)_2CH_3,\ 2PF_6^{\ominus}$	$H;\ P\ (25\cdot 8;\ 22\cdot 7;\ -143\cdot 9)$	1966, 78
$P_5C_{72}H_{60}Cl_3N_3S$	$\{[(C_6H_5)_3PN]_3S\}Cl_3\cdot 2P(C_6H_5)_3$	$H;\ P\ (25;\ 11;\ 7\cdot 3)$	1968, 254
$P_6C_2Cl_{10}N_6O_4$	$[Cl_3P{=}NN{=}C(OPOCl_2)NPCl_3]_2$	$P\ (15\cdot 2;\ 6\cdot 6;\ -78\cdot 8)$	1965, 32
$P_6C_4H_{12}F_{10}N_8$	$N_6P_6F_{10}[N(CH_3)_2]$	$H;\ F\ (-59)$	1969, 151
$P_6C_6F_{16}N_6$	$N_6P_6F_{11}(C_6F_5)$	F	1968, 127
$P_6C_6H_{18}F_9N_9$	$N_6P_6F_9[N(CH_3)_2]_3$	$H;\ F$	1969, 151
$P_6C_{12}H_{10}Cl_8N_6$	$[C_6H_5P(NPCl_2)_2N]_2$	P	1968, 128
$P_6C_{72}H_{60}$–$ClGeN_3O_6$	$[(C_6H_5)_2P(O)NP(C_6H_5)_2O]_3GeCl$	$P\ (31\cdot 3)$	1968, 11
$P_6C_{72}H_{60}$–ClN_3O_6Si	$[(C_6H_5)_2P(O)NP(C_6H_5)_2O]_3SiCl$	$P\ (29\cdot 4)$	1968, 11
$P_6C_{72}H_{60}$–ClN_3O_6Sn	$[(C_6H_5)_2P(O)NP(C_6H_5)_2O]_3SnCl$	$P\ (35\cdot 7)$	1968, 11
$P_7C_6F_{18}N_7$	$N_7P_7F_{13}(C_6F_5)$	F	1968, 127
$P_8C_6F_{20}N_8$	$N_8P_8F_{15}(C_6F_5)$	F	1968, 127
$(PC_2H_6NO_2)_n$	$[NP(OCH_3)_2]_n$	$H;\ P\ (-2)$	1966, 6

Formula	Compound	Resonance	Reference
$(PC_4H_4F_6NO_2)_n$	$[NP(OCH_2CF_3)_2]_n$	H; P (−6); F (−76·6)	1966, 6
$(PC_4H_{10}NO_2)_n$	$[NP(OC_2H_5)_2]_n$	H; P (−5·8)	1966, 6
$(PC_4H_{12}N_3)_n$	$[NP(NHC_2H_5)_2]_n$	H; P (−25·1)	1966, 91
$(PC_6H_5)_n$	$(PC_6H_5)_n$	P (−9)	1965, 11
		P (−4·4)	1966, 295; 1967, 126
$(PC_6H_5O_2)_n$	$(C_6H_5PO_2)_n$	P (1)	1967, 126
$(PC_6H_7O)_n$	$(CH_3OC_6H_4P)_n$	P (−11·8)	1967, 126
$(PC_7H_9O)_n$	$(C_2H_5OC_6H_4P)_n$	P (−12·0)	1967, 126
$(PC_{10}H_6)_n$	$(\beta\text{-}C_{10}H_6P)_n$	P (3·3)	1967, 126
$(PC_{10}H_{20}N_3)_n$	$[NP(NC_5H_{10})_2]_n$	H; P (−6·2)	1966, 91
$(PC_{12}H_{10}NO_2)_n$	$[NP(OC_6H_5)_2]_n$	H; P (−17·1)	1966, 6
$(PC_{12}H_{12}N_3)_n$	$[NP(NHC_6H_5)_2]_n$	H; P (−12·5)	1966, 91
$(PC_{24}H_{53}O_4Sn_2)_n$	$\{[(C_4H_9)_2SnO]_2P(O)(C_8H_{17})O\}_n$	P{H}	1969, 140
$(P_2C_2H_2Cl_6N_6)_n$	$(Cl_3PN\!=\!C\!=\!NNHPCl_3NHN\!=\!C\!=\!N)_n$	P (−61·7; −85·0)	1965, 32
$-(PC_{10}H_{21}O_2S)_n(C_{14}H_{30}S)-$	$(C_4H_9)_2P(S)(CHC_5H_{11})[CH_2CH(COOCH_3)P(S)(C_4H_9)_2]_n$	P (58)	1966, 216

FORMULAE

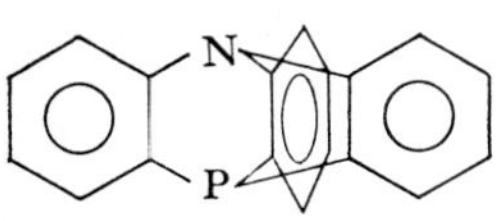

1

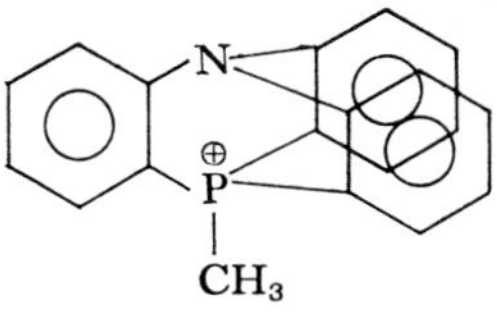

2

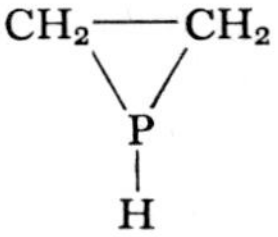

3

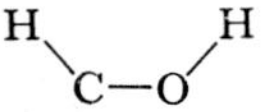

4

5
$J(^{15}N-{}^{1}H)$ +ve
$J(^{15}N-{}^{1}H')$ −ve

6
$J(^{15}N-CH)$ +ve
$J(^{15}N-CH')$ +ve

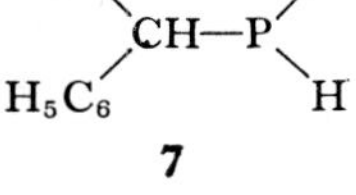

7

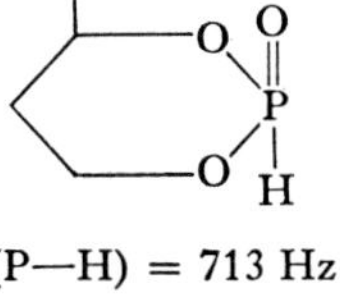

8
$^{1}J(P-H) = 713$ Hz

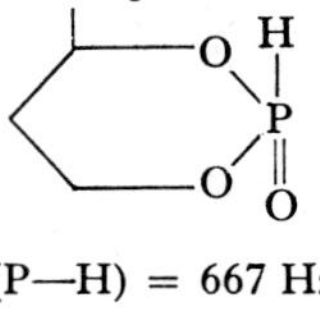

9
$^{1}J(P-H) = 667$ Hz

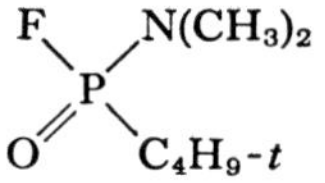

10

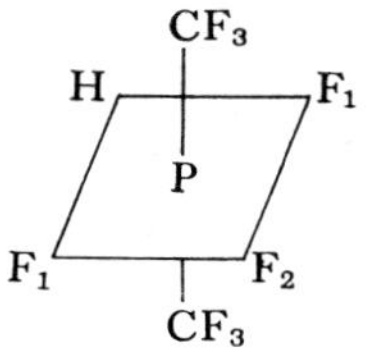

11

$(C_6H_5)_3\overset{\oplus}{P}-\overset{\ominus}{C}(CH_3)SO_2R.$

12

$(C_6H_5)_3\overset{\oplus}{P}-\overset{\ominus}{C}HSO_2R$

13

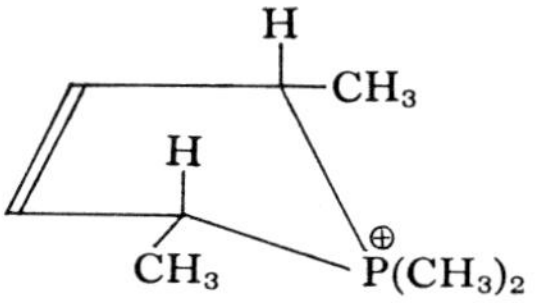

14

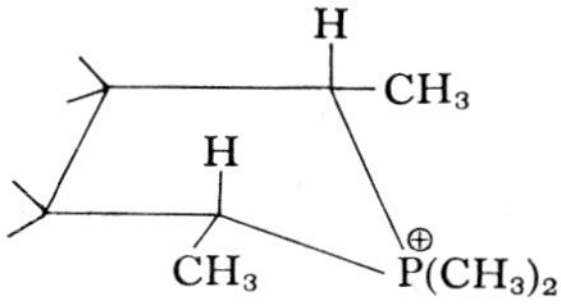

15

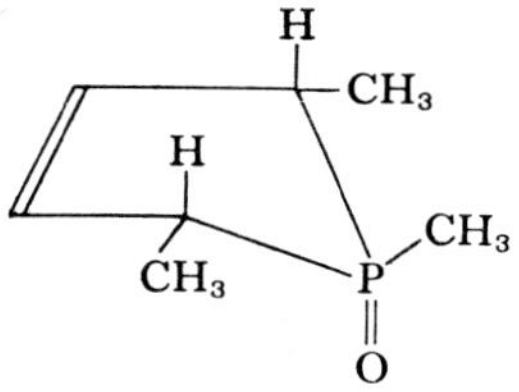

^2J(P—C—H) = 6·5 Hz

16

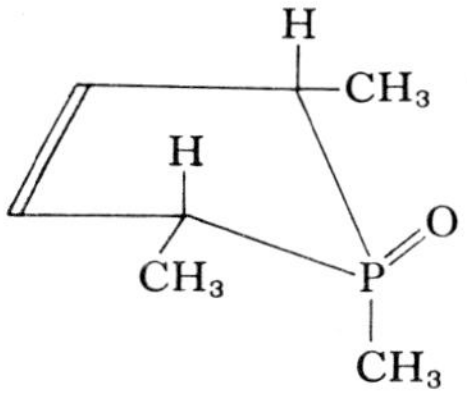

^2J(P—C—H) = 13·5 Hz

17

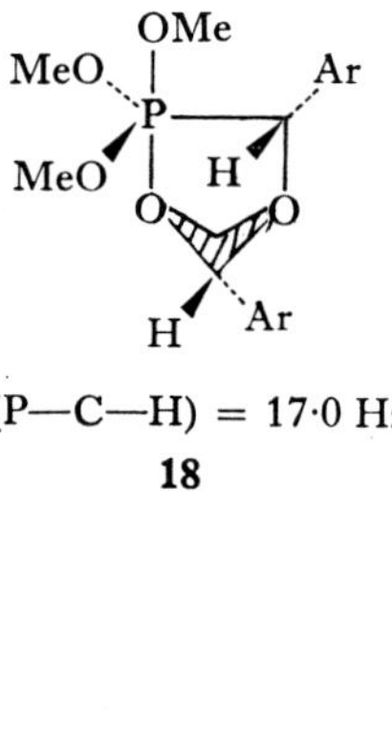

^2J(P—C—H) = 17·0 Hz

18

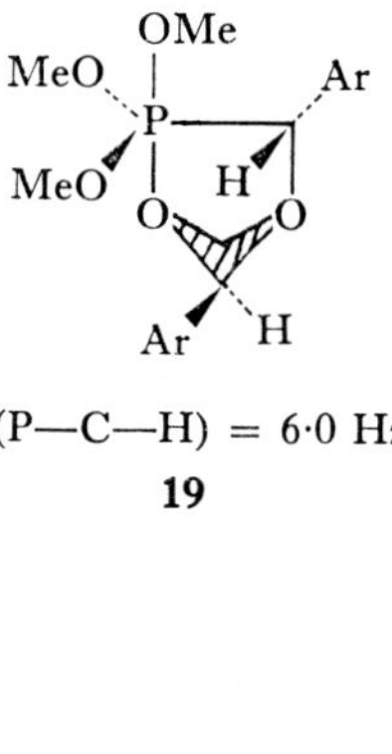

^2J(P—C—H) = 6·0 Hz

19

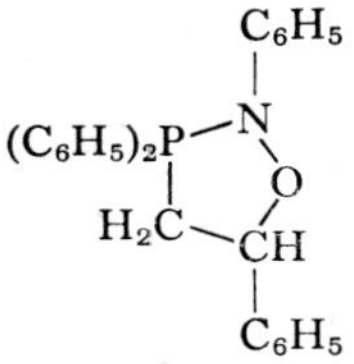

STRUCTURAL FORMULAE

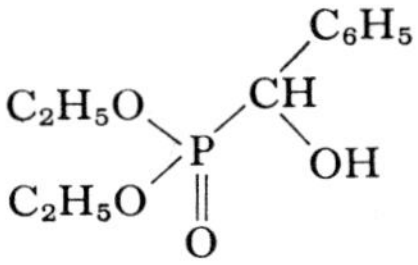

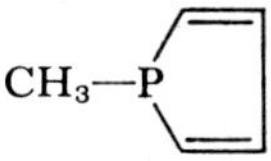

20: $(C_6H_5)_2P$–$N(C_6H_5)$–O–$CH(C_6H_5)$–CH_2 (ring)

21: $(C_2H_5O)_2P(O)$–$CH(C_6H_5)(OH)$

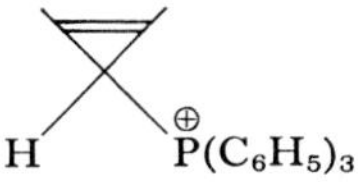

22: CH_3–P (phospholene ring)

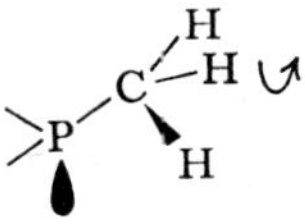

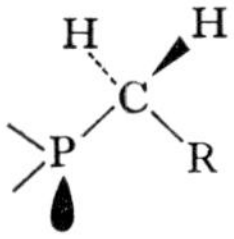

23: cyclopropene with H and $P(O)(C_6H_5)_2$
$^2J(P\text{–}H) = 40\cdot8$ Hz

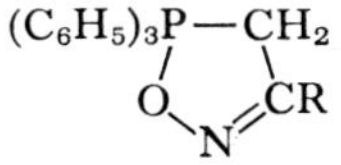

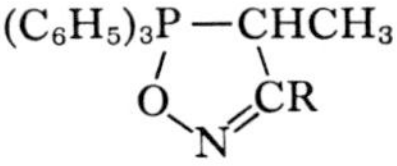

24: cyclopropene with H and $\overset{\oplus}{P}(C_6H_5)_3$
$^2J(P\text{–}H) = 51\cdot2$ Hz

25: $>P$–$C(H)(H)$–H

26: $>P$–$C(H)(H)$–R

27: $(C_6H_5)_3P$–CH_2–O–$N=CR$ (ring)
$^2J(P\text{–}H) = 11\cdot9$ Hz

28: $(C_6H_5)_3P$–$CHCH_3$–O–$N=CR$ (ring)
$^2J(P\text{–}H) = 3\cdot6$ Hz

29: $(C_6H_5)_3\overset{\oplus}{P}CH_2C(=NOH)C_6H_5$
$^2J(P\text{–}H) = 16\cdot8$ Hz

30: $(C_6H_5)_3\overset{\oplus}{P}CH(CH_3)C(=NOH)C_6H_5$
$^2J(P\text{–}H) = 13\cdot5$

31: $(CH_3O)_2P(O)$–C–$C(H)(C_6H_5)$ with O bridge
$^3J(P\text{–}C\text{–}CH) = 4\cdot5$ Hz

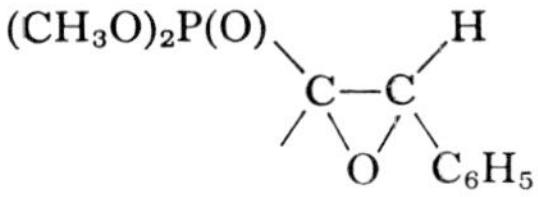

32: $(CH_3O)_2P(O)$–C–$C(C_6H_5)(H)$ with O bridge
$^3J(P\text{–}C\text{–}CH) = 5\cdot3$ Hz

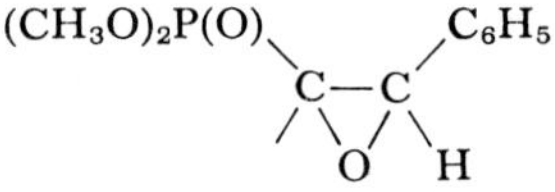

33
$^3J(\text{P—C—CH}_3) = 19\cdot5$ and $6\cdot5$ Hz

34
$^3J(\text{P—C—CH}) = 3\cdot5$ and $0\cdot0$ Hz

35
$^3J(\text{P—C—CH}_3) = 16\cdot5$ and $16\cdot2$ Hz

36
$^3J(\text{P—C—CH}_3) = 14\cdot0$ Hz

37
$^3J(\text{P—C—CH}_3) = 15\cdot0$ Hz

38
$^3J(\text{P—C—CH}_3) = ca.\ 15$ and 13 Hz

39

40
$^3J(\text{P—C—CH}) = 18$ and 21 Hz

41
$^4J(\text{P—CH}_3) = 10\cdot2$ *trans*
$7\cdot7$ *cis*

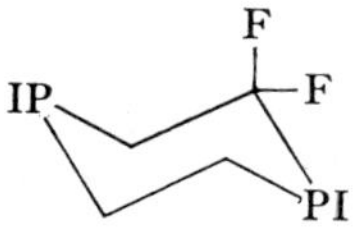

STRUCTURAL FORMULAE

42, 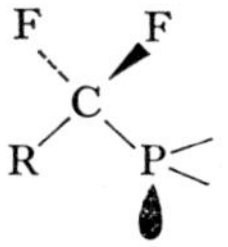 **43**, 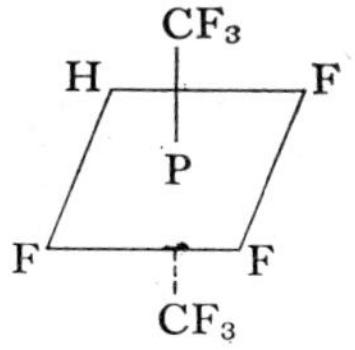 **44**

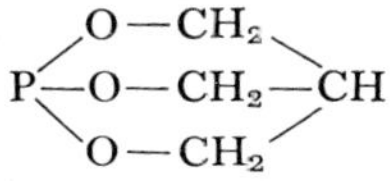

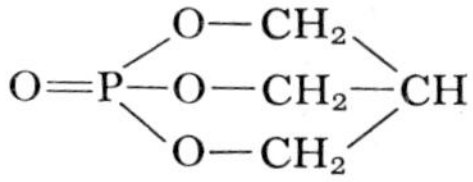

45 $^4J(P-O-C-CH) = 7.2$ Hz

46 $^4J(P-O-C-CH) = 6.2$ Hz

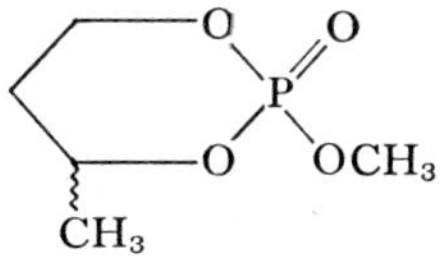

47 $^4J(P-O-C-CH) = 2$ Hz

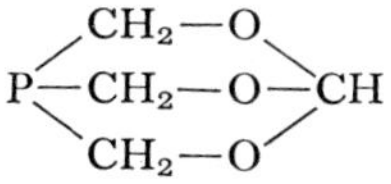

48 $^4J(P-C-O-CH) = 1.0$ Hz

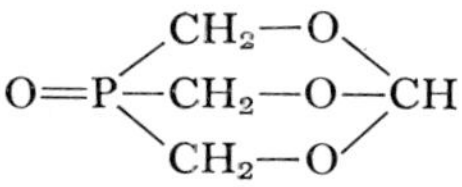

49 $^4J(P-C-O-CH) = 3.0$ Hz

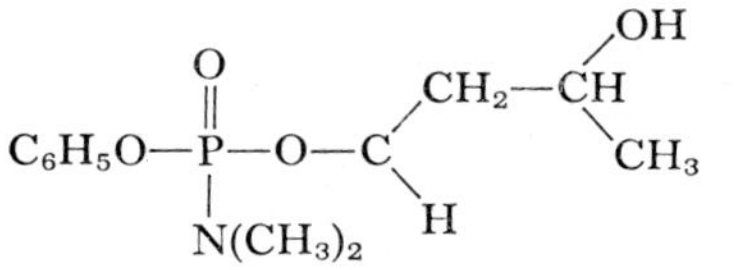

50 $^6J(P-CH_3) = 2.0$ Hz

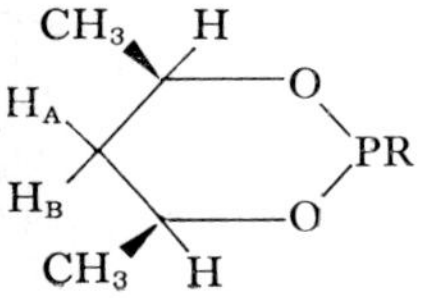

51 R = Cl or OCH$_3$

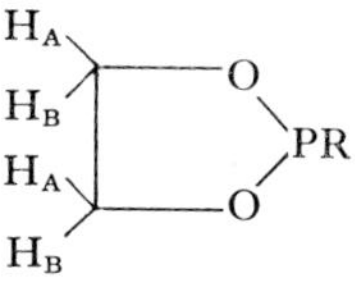

52, 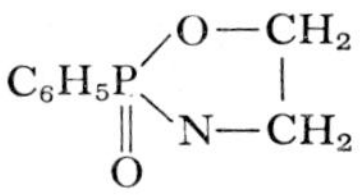 **53**

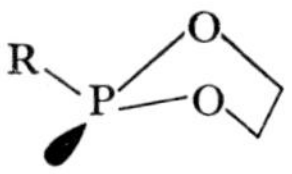

STRUCTURAL FORMULAE

54a 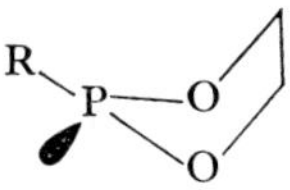 **54b** 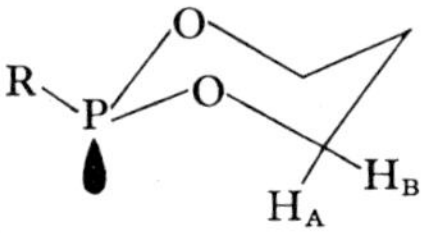 **55a** 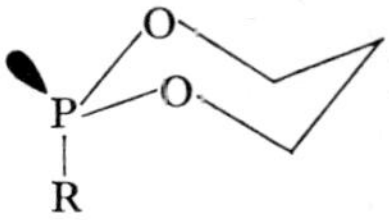 **55b**

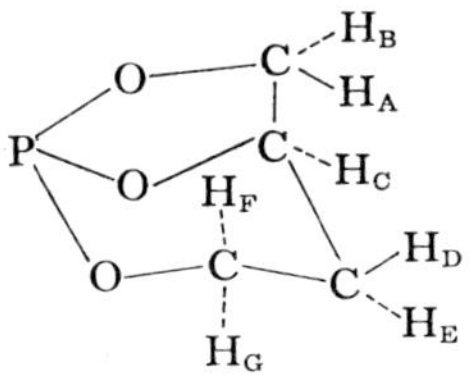

56

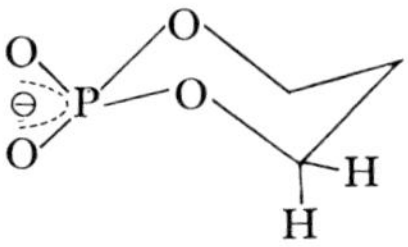

$J(trans)$ ca. 22 to 28 Hz
$J(cis)$ ca. 2 Hz
57

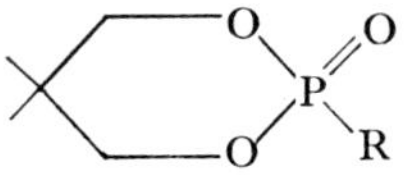

58

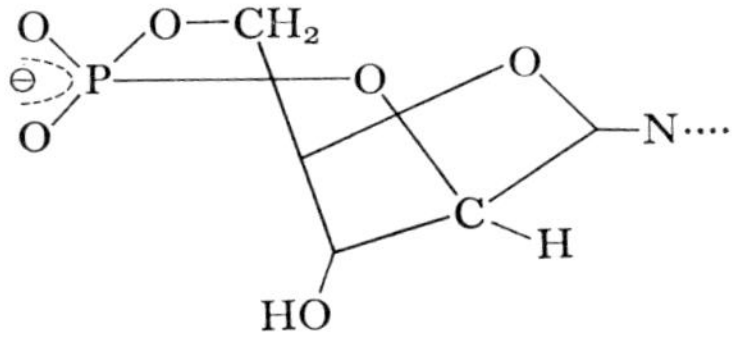

$^3J(P-O-CH) = 18$ Hz
59

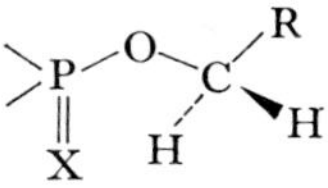

60

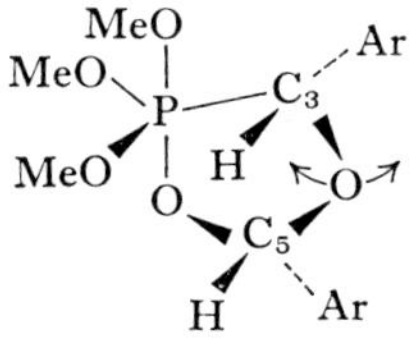

cis
61a

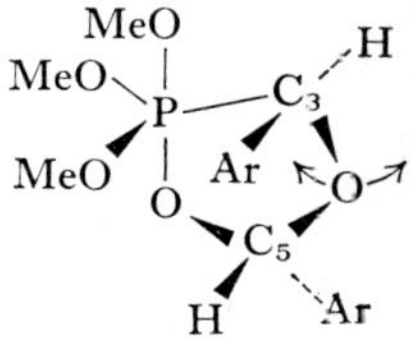

trans
61b

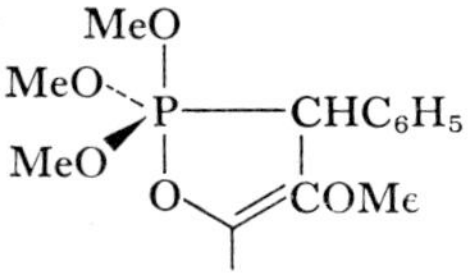

62

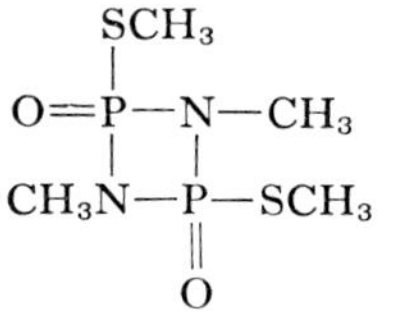

$^5J(PNPSCH) = 0.2$ Hz
63

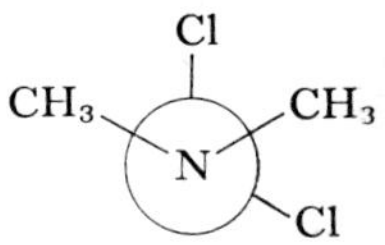

64

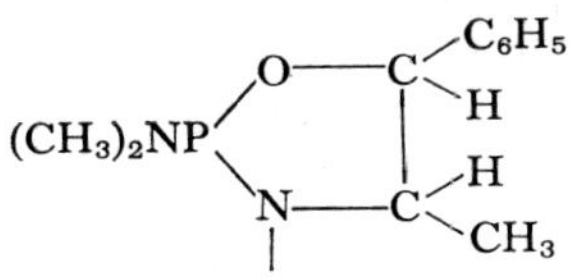

65

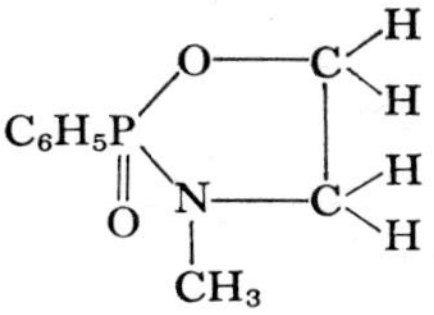

66

$(C_2H_5O)_2P(S)N = CHN(CH_3)_2$

$^4J(P\cdots CH) = 8{\cdot}5$ Hz

67

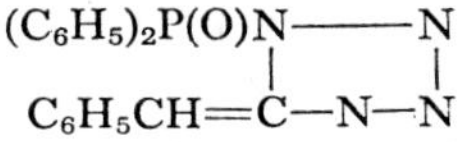

$^4J(P\cdots CH) = 10$ Hz

68

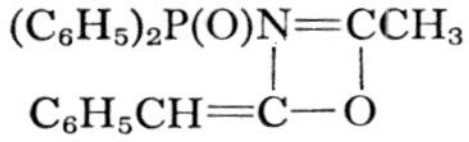

$^4J(P\cdots CH) = 12$ Hz

69

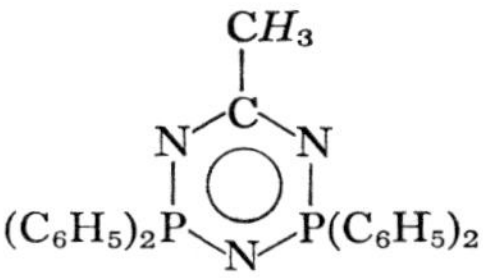

$^4J(P—CH_3) = 2{\cdot}7$ Hz

70

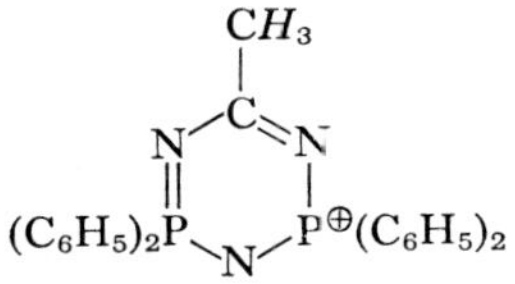

$^4J(P—CH_3) = 1{\cdot}8$ Hz

71

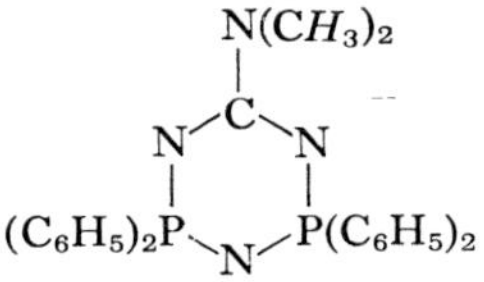

$^5J(P—CH_3) = 0{\cdot}45$ Hz

72

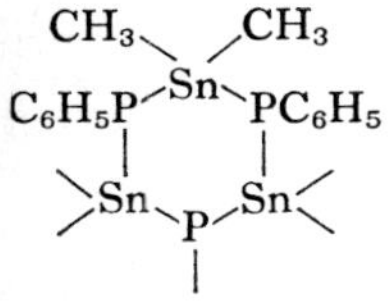

73

74 in Table XIX

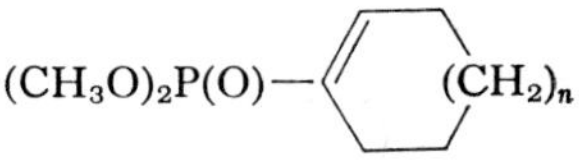

75

$(RO)_2P(O)CH=C\cdots$
$^2J(P-C-H) = ca\ 22$ Hz
76

$(RO)_2P(O)C(CH_3)=C\cdots$
$^3J(P-C-CH) = ca\ 15$ Hz
77

$Cl_2PCH=CH_2$
cis $^3J(P-C-C-H) = 13.5$ Hz
trans $^3J(P-C-C-H) = 33.0$ Hz
78

$Cl_2PCH=C(CH_3)_2$
$^4J(P-C-C-CH) = 1.6$
79

$(RO)_2PCH = C(CH_3)O$-metal
$^4J(P-C-C-CH) = 4$ Hz
80

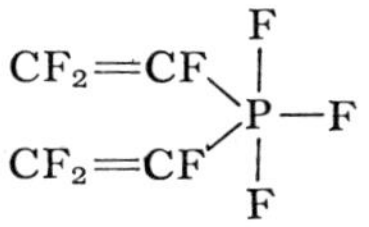

$CF_2=CF$
$CF_2=CF$ $P-F$
   F
   F

$^2J(P-CF) = 91$ Hz
cis $^3J(P-C-CF) = 15$ Hz
trans $^3J(P-C-CF) = 30$ Hz
81

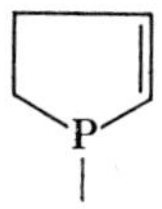

82

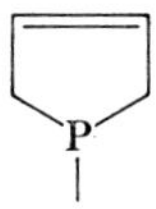

83

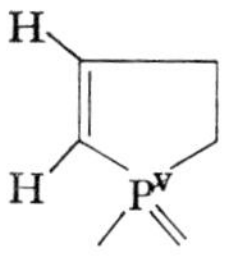

$^2J(P-CH) = 15$ to 30 Hz
$^3J(P-C-CH) = 30$ to 50 Hz
84

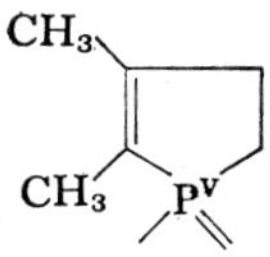

$^3J(PCCH_3)\ 10$ Hz
$^4J(PCCCH_3)\ 0$ Hz
85

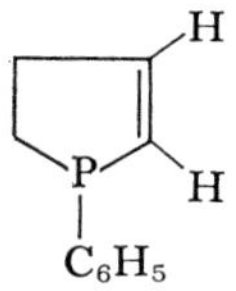

C_6H_5
$^2J(P-H) = 42$ Hz
86

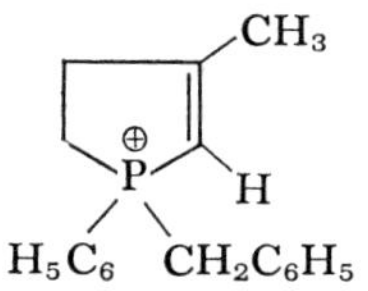

$H_5C_6\ \ CH_2C_6H_5$
$^2J(P-H) = 28.7$ Hz
87

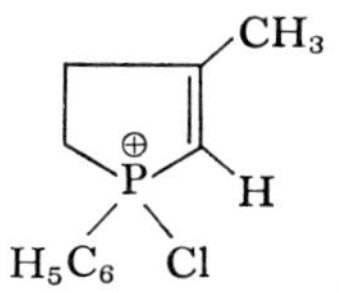

$H_5C_6\ \ Cl$
$^2J(P-H) = 32.2$ Hz
88

89
^2J(PCH) −10 to 20 Hz
^3J(PCCH) +20 to 45 Hz

90
^2J(PCH) −8 to 10 Hz
^3J(PCCH) +25 to 38 Hz

91
^2J(PCH) 5 to 8 Hz
^3J(PCCH) 5 to 8 Hz

92
^3J(P—H$_1$) 11·9 (21·2) Hz
^2J(P—H$_2$) 18·3 (ca. 14·7) Hz
^3J(P—H$_3$) ⩾ 5 (±20·7) Hz
^4J(P—H$_4$) ⩽ 5 (∓3·0) Hz

93
^3J(P—H$_1$) 3·4 (25·2) Hz
^2J(P—H$_2$) 2·2 (ca 10·3) Hz
^3J(P—H$_3$) ±13·1 (⩽5) Hz
^4J(P—H$_4$) ∓3·0 (⩾15) Hz

94
^2J(P—CH) 10·6 Hz

95
^2J(P—CH) 8·7 Hz
^3J(PCCH$_3$) 17·5 Hz

96
^2J(P—CH) 6·5 Hz
^3J(PCCH$_3$) 14·0 Hz

97
^2J(P—CH) 13·5 Hz
^3J(PCCH$_3$) 15·0 Hz

98
^2J(P—CH) 38·5 Hz
^3J(PCCH) 13·8 Hz

99
^3J(PCCH) 12·5 Hz
^3J(PCCH$_3$) 10 Hz

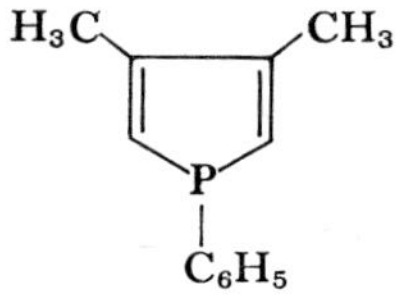

100
^2J(P—CH) 37·5 Hz
^4J(PCCCH$_3$) 3·5 Hz

101
^3J(PCCH) 40 Hz
^3J(PCCH$_3$) 13 Hz

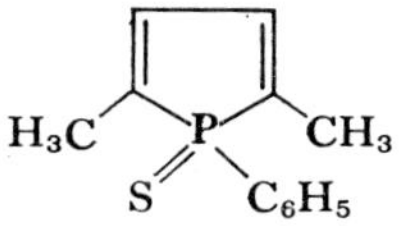

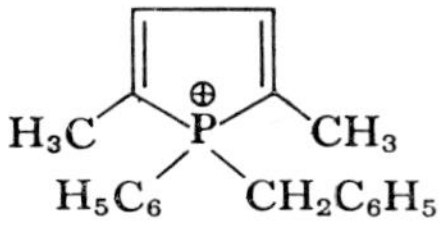

102
^3J(PCCH$_3$) 13 Hz

103
^2J(P—CH) 30·5 Hz
^4J(PCCCH$_3$) 2 Hz

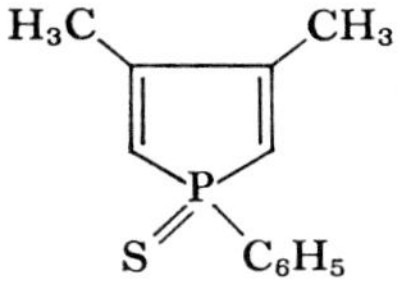

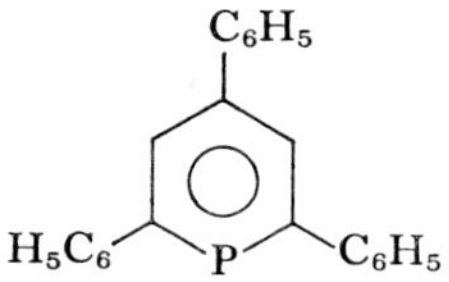

104
^3J(PCCH) 6 Hz

105
^3J(P—C—CH$_3$) 15 Hz
^3J(PCCH) 6 Hz

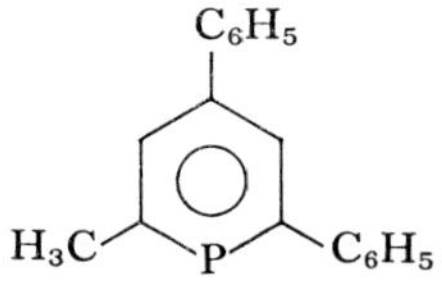

106
^3J(PCCH) 6 Hz
^5J(PCCCCH$_3$) 3·5 Hz

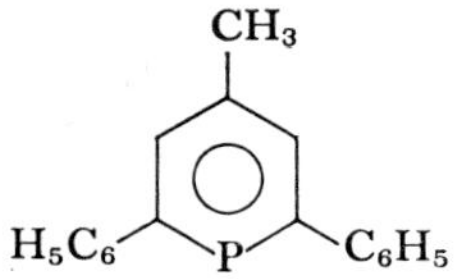

107
^3J(PCCH) 6 Hz
^4J(PCCCH$_3$) 2 Hz

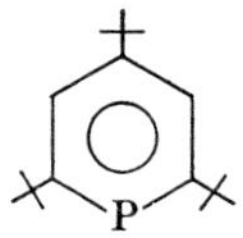

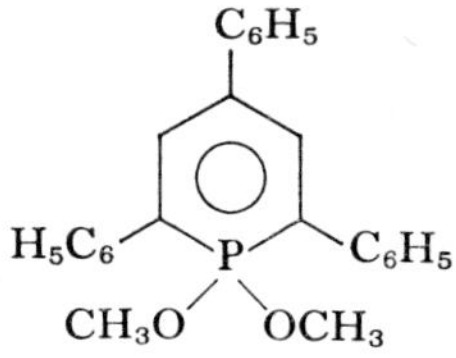

108
^3J(PCCH) 36·5 Hz

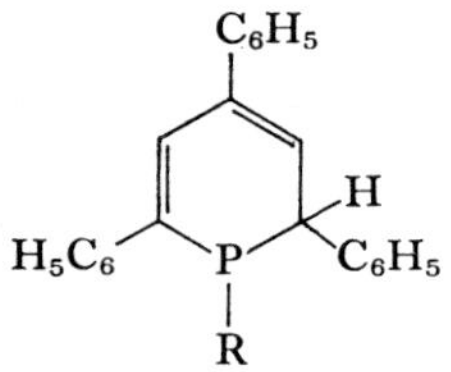

109

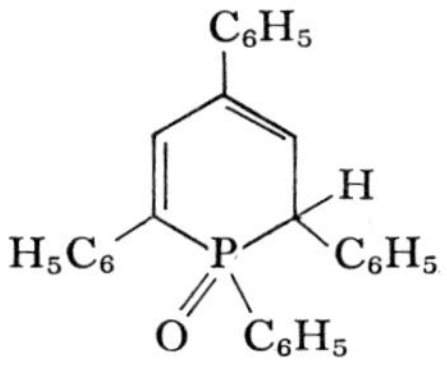

110
^2J(P—CH) 16·5 Hz
^3J(PCCH) 16·5 Hz

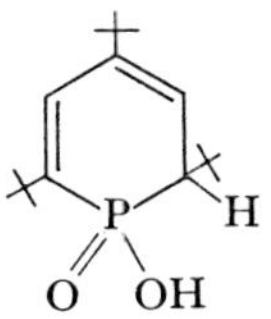

111

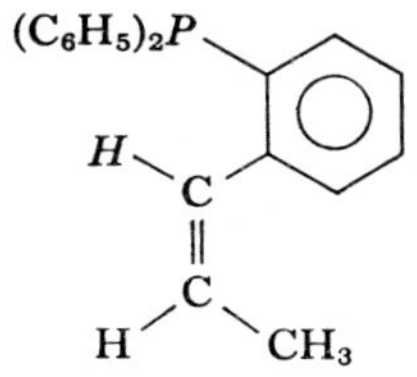

112

(C$_6$H$_5$)$_2$P—C$_6$H$_4$—C(H)=C(H)(CH$_3$)

^4J(P—H) = 4.3 Hz

113

(C$_6$H$_5$)$_2$P—C$_6$H$_4$—C(H)=C(H$_1$)(H$_2$)

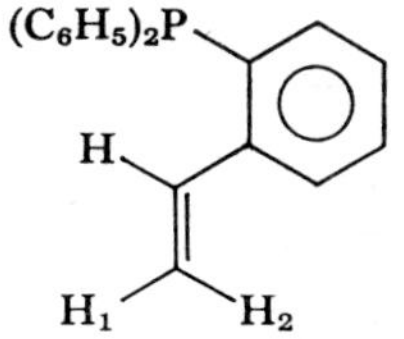

^6J(P—H$_1$) = 1.1 Hz
^6J(P—H$_2$) = 0.1 Hz

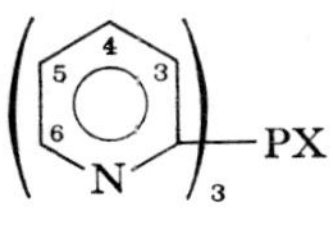

114

Pyridine-3-PX (positions 3,4,5,6)

	P:	P=O	P=S
PX =			
^3J(P—H$_3$)	1.7	5.7	6.5
^4J(P—H$_4$)	2.0	4.3	4.5
^5J(P—H$_5$)	1.2	2.4	3.0
^4J(P—H$_6$)	0.5	1.2	0.6

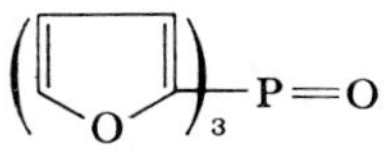

115

Furan-3-P=O

^3J(P—H$_3$) 2.0 Hz
^4J(P—H$_4$) 1.7 Hz
^4J(P—H$_5$) 2.7 Hz

116

Thiophene-3-P=O

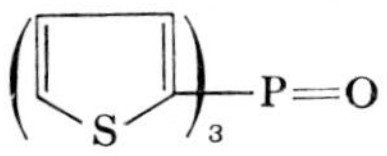

^3J(P—H$_3$) 8.0 Hz
^4J(P—H$_4$) 2.1 Hz
^4J(P—H$_5$) 4.6 Hz

117

N-Methylpyrrole-3-P=O

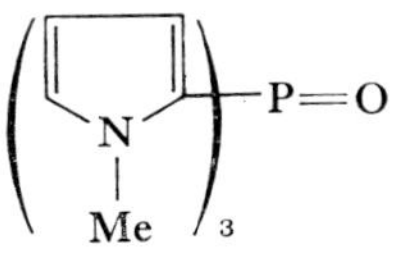

^3J(P—H$_3$) = 4.1 Hz
^4J(P—H$_4$) = 3.3 Hz
^4J(P—H$_5$) = 3.7 Hz

118

Thiophene-3-P

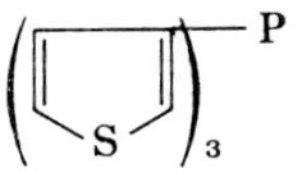

^3J(P—H$_2$) 4.2 Hz
^3J(P—H$_4$) 2.6 Hz
^4J(P—H$_5$) 1.0 Hz

119

Thiophene-3-P=S

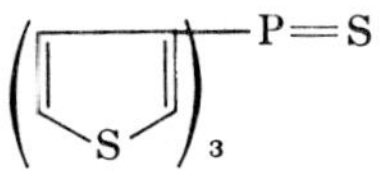

^3J(P—H$_2$) 8.8 Hz
^3J(P—H$_4$) 4.3 Hz
^4J(P—H$_5$) 2.1 Hz

120

5-Methylfuran-2-P=S

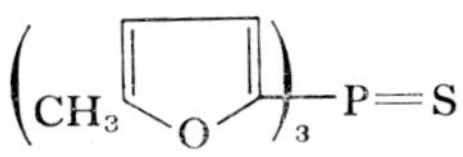

^5J(P—CH$_3$) 0.45 Hz

121

Thiophene-2-P

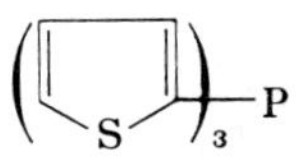

^3J(P—H$_3$) 6.1 Hz
^4J(P—H$_4$) 1.3 Hz
^4J(P—H$_5$) 0.3 Hz

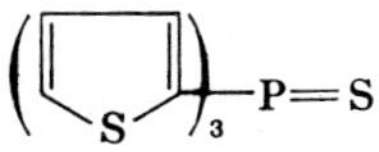

STRUCTURAL FORMULAE

122

$(\text{thiophene})_3\text{–P=S}$

$^3J(\text{P–H}_3)$ 8·9 Hz
$^4J(\text{P–H}_4)$ 1·9 Hz
$^4J(\text{P–H}_5)$ 4·4 Hz

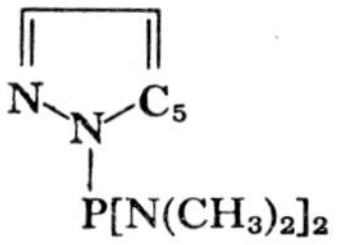

123

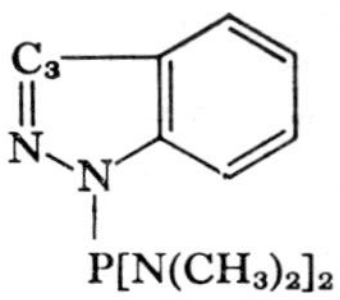

pyrazole-N–P[N(CH$_3$)$_2$]$_2$

124

benzotriazole-N–P[N(CH$_3$)$_2$]$_2$

125

$(\text{CH}_3)_3\text{P=CH–P(CH}_3)_2$
$^2J(\text{P=CH}) = 13\cdot5$ Hz
$^2J(\text{P–CH}) = 4\cdot3$ Hz

126

$(\text{CH}_3)_3\text{P=CH}_2$

127

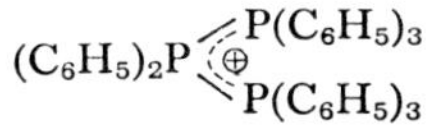

$(\text{C}_6\text{H}_5)_3\text{P=CHCHO}$

128

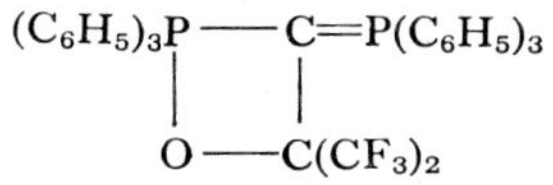

$(\text{C}_6\text{H}_5)_2\text{P}^{\oplus}\begin{smallmatrix}\text{P(C}_6\text{H}_5)_3\\ \text{P(C}_6\text{H}_5)_3\end{smallmatrix}$

129

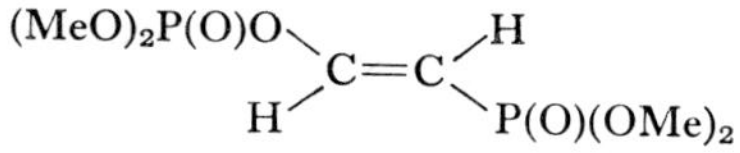

$\begin{array}{cc}(\text{C}_6\text{H}_5)_3\text{P} - \text{C=P(C}_6\text{H}_5)_3 \\ | \quad\quad | \\ \text{O} - \text{C(CF}_3)_2\end{array}$

130

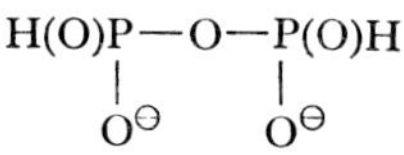

$(\text{MeO})_2\text{P(O)O}\diagdown_{}\text{C=C}\diagup^{\text{H}}_{\text{P(O)(OMe)}_2}$, H

131

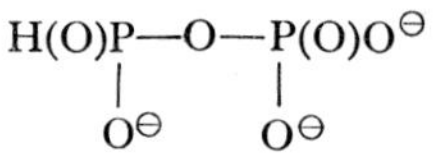

$\begin{array}{cc}\text{H(O)P–O–P(O)H} \\ | \quad\quad | \\ \text{O}^{\ominus} \quad\quad \text{O}^{\ominus}\end{array}$

132

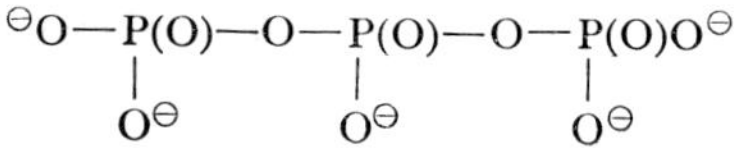

$\begin{array}{cc}\text{H(O)P–O–P(O)O}^{\ominus} \\ | \quad\quad | \\ \text{O}^{\ominus} \quad\quad \text{O}^{\ominus}\end{array}$

133

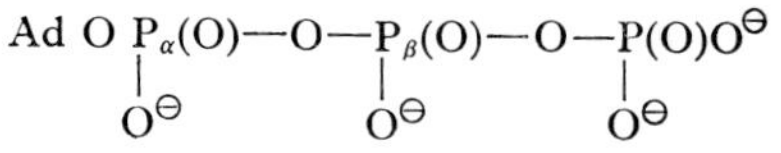

$\begin{array}{ccc}{}^{\ominus}\text{O–P(O)–O–P(O)–O–P(O)O}^{\ominus} \\ | \quad\quad | \quad\quad | \\ \text{O}^{\ominus} \quad \text{O}^{\ominus} \quad \text{O}^{\ominus}\end{array}$

134

$\begin{array}{ccc}\text{Ad O P}_\alpha(\text{O})\text{–O–P}_\beta(\text{O})\text{–O–P(O)O}^{\ominus} \\ | \quad\quad | \quad\quad | \\ \text{O}^{\ominus} \quad \text{O}^{\ominus} \quad \text{O}^{\ominus}\end{array}$

STRUCTURAL FORMULAE

$F(O)P-O-P(O)-O-P(O)O^{\ominus}$
$\quad\;\; |\qquad\quad\;\; |\qquad\quad\;\; |$
$\quad\;\;O^{\ominus}\qquad\;\;O^{\ominus}\qquad\;\;O^{\ominus}$

135

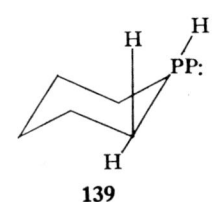

$X = S,\; ^2J(PSP) = 71$ Hz
$X = Se,\; ^2J(PSeP) = 72$ Hz

136

137

138

139

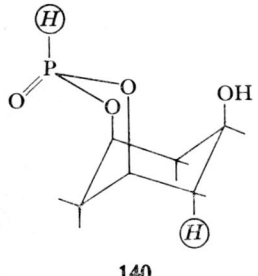

140

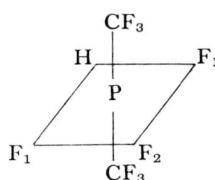

$^2J(H\text{—}F_1) = 70$ Hz
$^2J(H\text{—}F_2) = 18$ Hz
$^3J(H\text{—}CF_3) = 9\cdot 5$ Hz

141

$^2J(F_1\text{—}F_2) = 25$ Hz
$^3J(F_1\text{—}CF_3) = 12\cdot 5$ Hz
$^3J(F_2\text{—}CF_3) = 15\cdot 5$ Hz

142

$^2J(F_{ax}\text{—}F_{eq}) = 59\cdot 4$ Hz

143

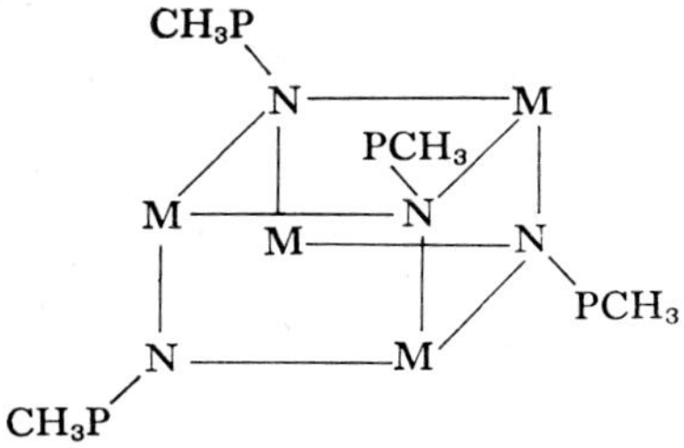

STRUCTURAL FORMULAE

144 M = Zn or Cl

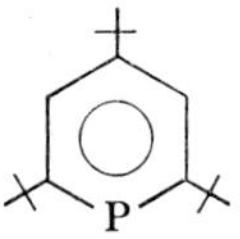

145

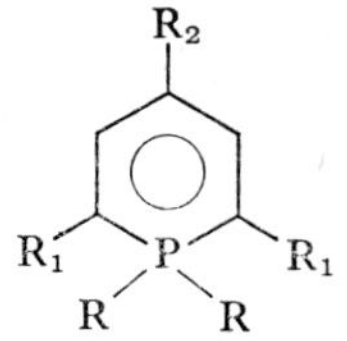

146

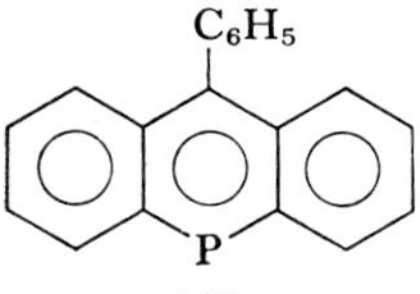

147

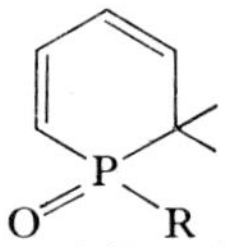

148

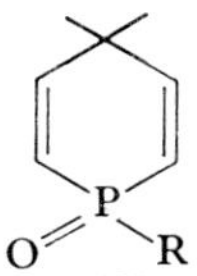

149

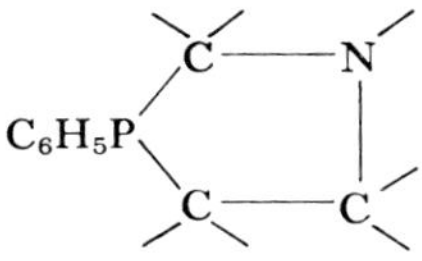

150

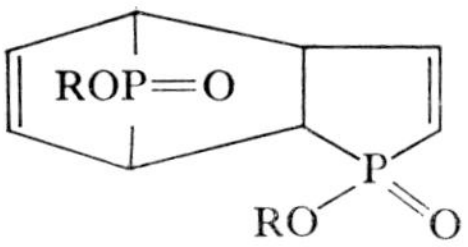

151

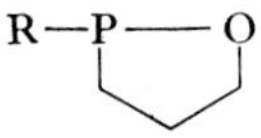

152

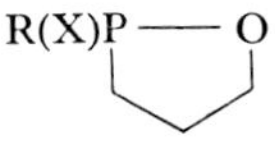

153

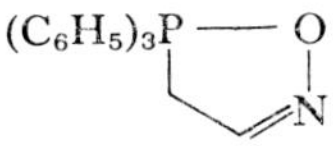

154

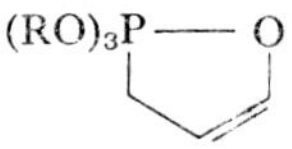

155

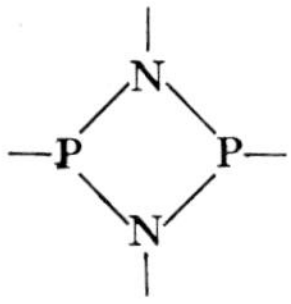

156

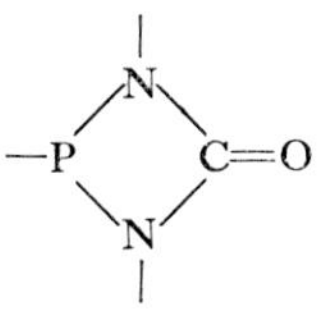

157

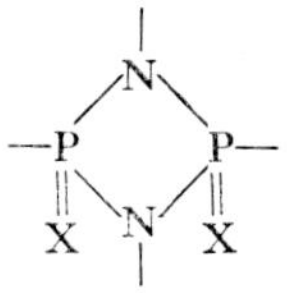

158

STRUCTURAL FORMULAE

303

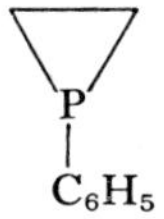

STRUCTURAL FORMULAE

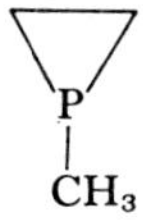

171: P–C₆H₅ phosphirane — PC_6H_5 phosphirane

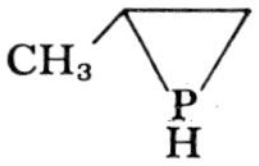

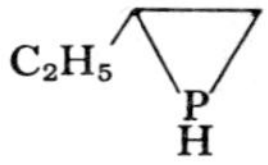

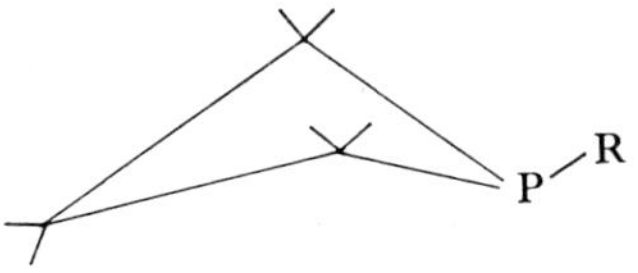

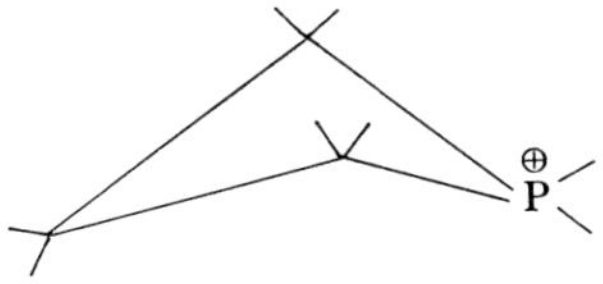

172: PCH_3 phosphirane

173: CH_3-substituted phosphirane with P–H

174: C_2H_5-substituted phosphirane with P–H

175: P–R cage structure

176: $\overset{\oplus}{P}$ cage structure

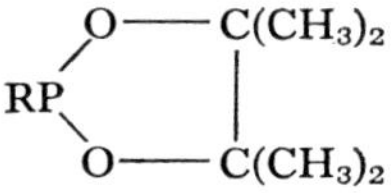

177:
$$RP\begin{matrix} O-C(CH_3)_2 \\ O-C(CH_3)_2 \end{matrix}$$

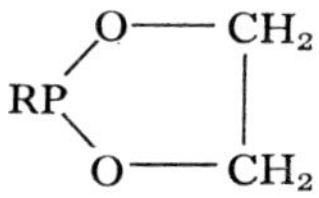

178:
$$RP\begin{matrix} O-CH_2 \\ O-CH_2 \end{matrix}$$

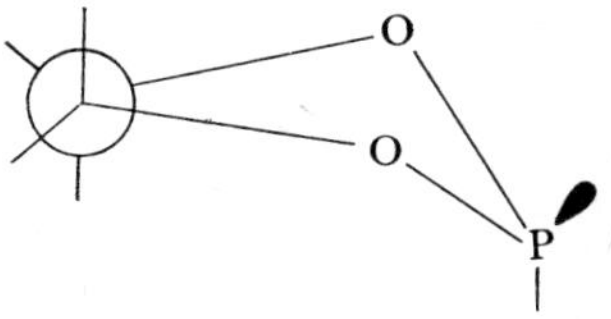

179

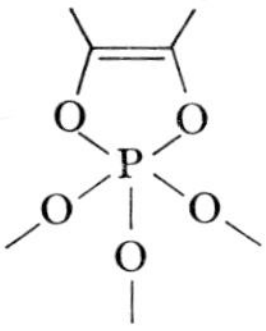

180

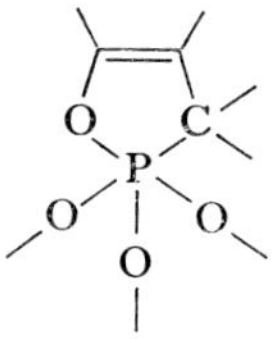

181

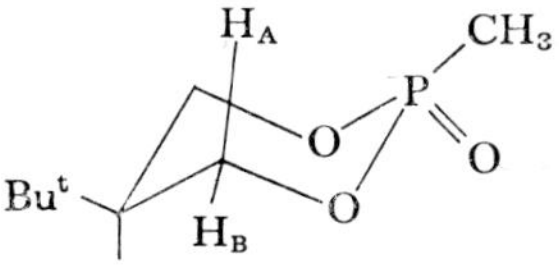

182a: $^3J(P-H_A)$ 6·4 Hz; $^3J(P-H_B)$ 22·7 Hz

182b: $^3J(P-H_A)$ 4·1 Hz; $^3J(P-H_B)$ 20·2 Hz

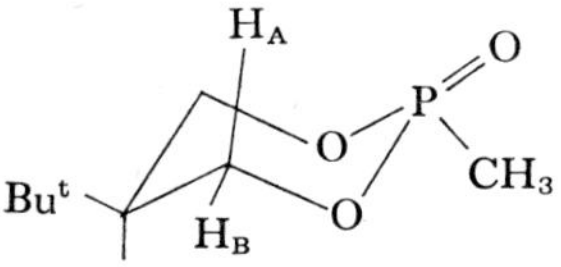

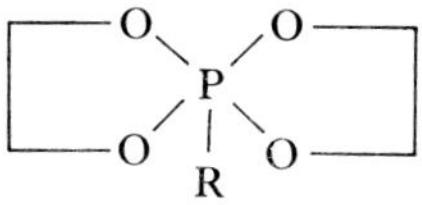

183

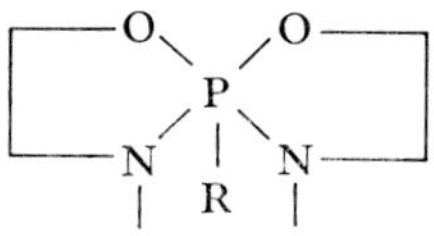

184

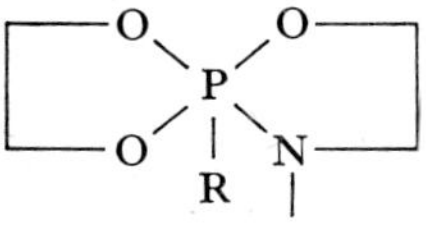

STRUCTURAL FORMULAE

185

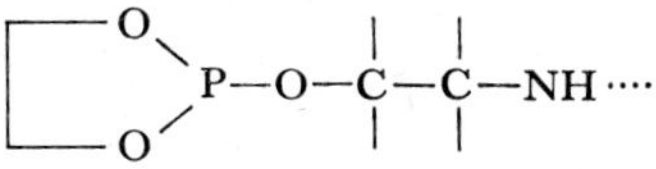

186

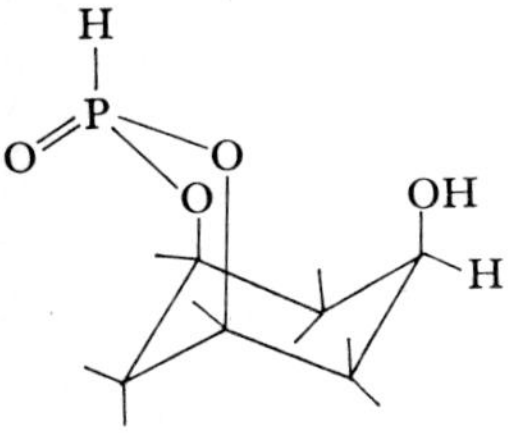

187

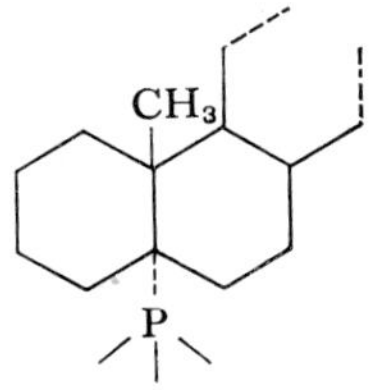

188

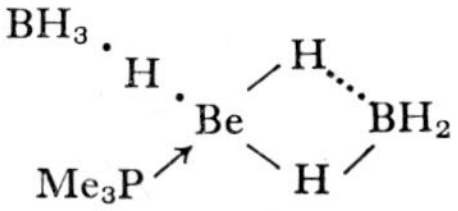

189

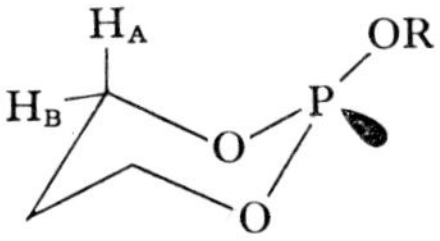

190

REFERENCES

1956
1. J. R. Wazer, C. F. Callis, J. N. Shoolery and R. C. Jones, *J. Amer. Chem. Soc.*, 1956, **78**, 5715.

1957
1. P. C. Lauterbur, *J. Chem. Phys.*, 1957, **26**, 217.
2. C. F. Callis, J. R. Van Wazer, J. N. Shoolery and W. A. Anderson, *J. Amer. Chem. Soc.*, 1957, **79**, 2719.

1958
1. D. P. Craig and N. L. Paddock, *Nature*, 1958, **181**, 1052.

1959
1. J. M. Winter, *Compt. Rend.*, 1959, **249**, 1346.

1960
1. R. S. Berry, *J. Chem. Phys.*, 1960, **32**, 933.
2. M. J. S. Dewar, E. A. C. Lucken and M. A. Whitehead, *J. Chem. Soc.*, 1960, 2423.

1961
1. H. Spiesecke and W. G. Schneider, *J. Chem. Phys.*, 1961, **35**, 722.
2. R. M. Lynden-Bell, *Trans. Faraday Soc.*, 1961, **57**, 888.
3. P. J. Bray, J. O. Edwards, J. G. O'Keefe, V. F. Ross and I. Tatsuzaki, *J. Chem. Phys.*, 1961, **35**, 435.

1962
1. G. Mavel, *J. Chim. Phys.*, 1962, **59**, 683.
2. C. G. Krespan, C. M. Langkammerer, *J. Org. Chem.*, 1962, **27**, 3584.
3. D. P. Craig and N. L. Paddock, *J. Chem. Soc.*, 1962, 4118.
4. G. Ewart, D. S. Payne, A. L. Porte and A. P. Lane, *J. Chem. Soc.*, 1962, 3984.
5. A. Davison, W. McFarlane, L. Pratt and G. Wilkinson, *J. Chem. Soc.*, 1962, 3653.
6. R. K. Harris, *J. Phys. Chem.*, 1962, **66**, 768.
7. J. V. Hatton and R. E. Richards, *Mol. Phys.*, 1962, **5**, 153.

1963
1. M. A. Fleming, *Thesis, Univ. of Michigan*; *Univ. Microfilms*, Order No. 63–6895; *Dissertation Abstra.*, 1963, **24**, 1385.
2. J. E. Drake and W. L. Jolly, *J. Chem. Phys.*, 1963, **38**, 1033.
3. R. Schmutzler, *Chem. Ber.*, 1963, **96**, 2435.
4. F. A. Cotton and R. A. Schunn, *J. Amer. Chem. Soc.*, 1963, **85**, 2394.
5. G. Weill, M. Klein and M. Calvin, *Nature (London)*, 1963, **200**, 1005.
6. R. K. Harris, in J. W. Emsley, J. Feeney and L. H. Sutcliffe, *High Resolution nuclear magnetic resonance spectroscopy*, 1966, Vol. 2, p. 959.

7. R. K. Harris, *J. Mol. Spectry.*, 1963, **10**, 309.
8. R. F. Hudson and P. A. Chopard, *Helv. Chim Acta*, 1963, **46**, 2178.

1964
1. C. J. Jameson and H. S. Gutowsky, *J. Chem. Phys.*, 1964, **40**, 1714.
2. D. Herbison-Evans and R. E. Richards, *Mol. Phys.*, 1964, **8**, 18.
3. S. Koide and E. Duval, *J. Chem. Phys.*, 1964, **41**, 315.
4. J. A. Pople and D. P. Santry, *Mol. Phys.*, 1964, **8**, 1.
5. J. G. Verkade, R. W. King and C. W. Heitsch, *Inorg. Chem.*, 1964, **3**, 884.
6. M. Beroza and A. B. Bořkovec, *J. Med. Chem.*, 1964, **7**, 44.
7. S. Koide and E. Duval, *J. Chem. Phys.*, 1964, **41**, 315.
8. B. A. Arbuzov, Yu. Yu. Samitov, A. O. Vizel' and T. V. Zykova, *Dokl. Akad. Nauk S.S.S.R.*, 1964, **159**, 1062; *Dokl. Chem.*, 1964, **159**, 1303.
9. V. Mark, *Tetrahedron Lett.*, 1964, 3139.
10. C. Osuch, J. E. Franz and Eb. Zienty, *J. Org. Chem.*, 1964, **29**, 3721.
11. W. Mahler, *J. Amer. Chem. Soc.*, 1964, **86**, 2306.
12. R. K. Harris and R. G. Hayter, *Canad. J. Chem.*, 1964, **42**, 2282.
13. T. Kruck and A. Prasch, *Z. Naturforch. B*, 1964, **19**, 669.
14. Yu. Yu. Samitov and T. V. Zykova, *Tr. Kazansk. Khim. tekhnol, Inst.* 1964, **33**, 9.
15. N. A. Razumova and I. M. Treskunova, *Zh. Obsch. Khim.*, 1964, **34**, 2949; *J. Gen. Chem. U.S.S.R.*, 1964, **34**, 2983.
16. B. A. Arbuzov, O. A. Vizel', Yu. Yu. Samitov and K. M. Ivanovskaya, *Dokl. Akad. Nauk S.S.S.R.*, 1964, **159**, 582; *Dokl. Chem.*, 1964, **159**, 1205.

1965
1. D. Purdela, *Rev. Roum. Chim.*, 1965, **10**, 949.
2. H. S. Gutowsky and J. Larmann, *J. Amer. Chem. Soc.*, 1965, **87**, 3815.
3. S. O. Grim and W. McFarlane, *Nature*, 1965, **208**, 995.
4. E. Fluck, H. Binder and F. Goldman, *Z. anorg. allgem. Chem.*, 1965, **338**, 58.
5. R. W. Taft, J. W. Rakshys, *J. Amer. Chem. Soc.*, 1965, **87**, 4387.
6. H. Ulrich and A. A. R. Sayigh, *J. Org. Chem.*, 1965, **30**, 2779
7. H. L. Retcofsky, *M.S. Thesis. Univ. Pittsburgh*, 1965.
8. R. Schmutzler and G. S. Reddy, *Z. Naturforsch. B*, 1965, **20**, 832.
9. J. A. Pople and D. P. Santry, *Mol. Phys.*, 1965, **5**, 311.
10. G. Bergerhoff and F. Knoll, *Angew. Chem.*, 1965, **77**, 1016; *Angew. Chem. Int. Ed.*, 1965, **4**, 968.
11a. E. Fluck and K. Issleib, *Chem. Ber*, 1965, **98**, 2674.
11b. H. Goldwhite, R. N. Haszeldine and D. G. Rowsell, *J. Chem. Soc.*, 1965, 6815.
12. R. Schmutzler, *J. Chem. Soc.*, 1965, 5630.
13. H. G. Horn, A. Müller and O. Glemser, *Z. Naturforsch. A*, 1965, **20**, 746.
14. R. Schmutzler, *Angew. Chem.*, 1965, **77**, 530; *Angew. Chem. Int. Ed.*, 1965, **4**, 496.
15. L. S. Bartell, K. W. Hansen, *Inorg. Chem.*, 1965, **4**, 1777.
16. K. W. Hansen and L. S. Bartell, *Inorg. Chem.*, 1965, **4**, 1775.
17. K. J. Coskran and J. G. Verkade, *Inorg. Chem.*, 1965, **4**, 1655.
18. A. H. Cowley and R. P. Pinnell, *J. Amer. Chem. Soc.*, 1965, **87**, 4454.
19. C. E. Griffin and M. Gordon, *J. Organometal. Chem.*, 1965, **3**, 414.
20. A. E. Lippman, *J. Org. Chem.*, 1965, **30**, 3217.

REFERENCES

1965 cont.
21. F. Ramirez, A. V. Patwardhan and C. P. Smith, *J. Org. Chem.*, 1965, **30**, 2575.
22. J. G. Verkade, T. J. Huttemann, M. K. Fung and R. W. King, *Inorg. Chem.*, 1965, **4**, 83.
23. R. Keat and R. A. Shaw, *J. Chem. Soc.*, 1965, 4802.
24. R. Keat, S. K. Ray and R. A. Shaw, *J. Chem. Soc.*, 1965, 7193.
25. G. S. Vasil'ev, E. N. Prilezhaeva, V. F. Bistrov and M. F. Shostakovskii, *Zh. Obshch. Khim.*, 1965, **35**, 1350; *J. Gen. Chem. U.S.S.R.*, 1965, **35**, 1356.
26. A. H. Cowley, *Chem. Revs*, 1965, **65**, 617.
27. E. Fluck and K. Issleib, *Z. anorg. allg. Chem.*, 1965, **339**, 274.
28. A. H. Cowley and H. Steinfink, *Inorg. Chem.*, 1965, **4**, 1827.
29. K. Issleib and K. Krech, *Chem. Ber.*, 1965, **98**, 2545.
30. M. M. Crutchfield and R. R. Irani, *J. Amer. Chem. Soc.*, 1965, **87**, 2815.
31. R. W. King, T. J. Huttemann and J. G. Verkade, *Chem. Comm.*, 1965, 561.
32. M. Becke-Goehring and W. Weber, *Z. anorg. allg. Chem.*, 1965, **339**, 281.
33. S. K. Das, R. Keat, R. A. Shaw and B. C. Smith, *J. Chem. Soc.*, 1965, 5032.
34. M. Becke-Goehring and W. Haubold, *Z. anorg. allg. Chem.*, 1965, **338**, 305.
35. W. G. Bentrude, *J. Amer. Chem. Soc.*, 1965, **87**, 4026.
36. R. D. Burpitt and V. W. Goodlett, *J. Org. Chem.*, 1965, **30**, 4307.
37. D. Hellwinkel, *Ber.*, 1965, **98**, 576.
38. J. G. Riess, *Bull. Soc. Chim. France*, 1965, 3552.
39. R. J. Chuck, A. G. Massey, E. W. Randall and D. Shaw, *Nuclear magnetic resonance in Chemistry* (B. Pesce, ed.), Academic Press, p. 189.
40. C. W. Heitsch, *Inorg. Chem.*, 1965, **4**, 1019.
41. J. B. De Roos and J. P. Oliver, *Inorg. Chem.*, 1965, **4**, 1741.
42. T. J. Bardos, Z. F. Chmielewicz and C. K. Navada, *J. Pharm. Sc.*, 1965, **54**, 399.
43. D. H. Brown, G. W. Fraser, A. McAuley and D. W. A. Sharp, *Chem. Ind. (London)*, 1965, 2098.
44. E. Grunwald and M. Cocivera, *Disc. Farad. Soc.*, 1965, **39**, 105.
45. A. G. Pinkus, P. G. Waldrep, S. Y. Ma, *J. Heterocyclic Chem.*, 1965, **2**, 357.
46. V. D. Lobkov, A. L. Klebanskii and E. V. Kogan, *Vysokomol. Soyed.*, 1965, **7**, 290; *Polym. Sc. U.S.S.R.*, 1965, **7**, 318.
47. F. J. Welch and H. J. Paxton, *J. Polym. Sc. A.*, 1965, **3**, 3439.
48. M. J. Gallagher and I. D. Jenkins, *Chem. Comm.*, 1965, 587.
49. E. Fluck, *Angew. Chem.*, 1965, **77**, 969.
50. E. Fluck, F. L. Goldmann and K. D. Rümpler, *Z. anorg. allg. Chem.*, 1965, **338**, 52.
51. W. Hieber and H. Duchatsch, *Chem. Ber.*, 1965, **98**, 2923.
52. T. Kruck and W. Lang, *Angew. Chem.*, 1965, **77**, 860; *Angew. Chem. Int. Ed.*, 1965, **4**, 870.
53. T. Kruck and W. Lang, *Chem. Ber*, 1965, **98**, 3060.
54. G. N. La Mar, *J. Chem. Phys.*, 1965, **43**, 235.
55. G. N. La Mar, *J. Chem. Phys.*, 1965, **43**, 1085.
56. G. N. La Mar, *J. Phys. Chem.*, 1965, **69**, 3212.
57. J. B. Wilford, A. Forster and F. G. A. Stone, *J. Chem. Soc.*, 1965, 6519.
58. P. M. Treichel and G. Werber, *Inorg. Chem.*, 1965, **4**, 1098.
59. A. D. Buckingham and P. J. Stephens, *Nuclear magnetic resonance in Chemistry* (B. Pesce, ed.), Academic Press, 1965, p. 35.
60. A. Araneo, *Gazz. Chim. Ital.*, 1965, **95**, 1431.
61. J. Chatt, R. S. Coffey and B. L. Shaw, *J. Chem. Soc.*, 1965, 7391.

1965 cont.
62. L. Malatesta, G. Gaglio and M. Angoletta, *J. Chem. Soc.*, 1965, 6974.
63. E. W. Randall and D. Shaw, *Mol. Phys.*, 1965, **10**, 41.
64. D. Shaw and E. W. Randall, *Chem. Comm.*, 1965, 82.
65. G. S. Kraihanzel and P. K. Maples, *J. Amer. Chem. Soc.*, 1965, **87**, 5267.
66. R. S. Vinal, *Thesis, Cornell Univ., Univ. Microfilms*, Order No. 65–6987; *Dissertation Abstr.*, 1965, **26**, 686.
67. J. G. Riess and J. R. Van Wazer, *J. Amer. Chem. Soc.*, 1965, **87**, 5506.
68. H. P. Fritz, I. R. Gordon, K. E. Schwarzhans and L. M. Venanzi, *J. Chem. Soc.*, 1965, 5210.
69. R. V. Lindsey, G. W. Parshall and V. G. Stolberg, *J. Amer. Chem. Soc.*, 1965, **87**, 658.
70. T. Kruck, W. Lang and N. Derner, *Z. Naturforsch. B*, 1965, **270**, 705.
71. J. R. Parker and C. V. Banks, *J. Inorg. Nucl. Chem.*, 1965, **27**, 583.
72. R. G. Shulman, H. Sternlicht and B. J. Wyluda, *J. Chem. Phys.*, 1965, **43**, 3116.
73. H. Sternlicht, R. G. Shulman and E. W. Anderson, *J. Chem. Phys.*, 1965, **43**, 3123.
74. H. Sternlicht, R. G. Shulman and E. W. Anderson, *J. Chem. Phys.*, 1965, **43**, 3133.
75. D. M. Brown and A. P. Read, *J. Chem. Soc.*, 1965, 5072.
76. W. E. Wehrli and J. G. Moffatt, *J. Amer. Chem. Soc.*, 1965, **87**, 3760.
77. W. Mahler and E. L. Muetterties, *Inorg. Chem.*, 1965, **4**, 1520.
78. H. Bock and G. Rudolph, *Chem. Ber.*, 1965, **98**, 2273.
79. K. D. Berlin and M. Nagabhushanam, *Proc. Okla. Acad. Sc.*, 1965, **45**, 111.
80. M. Green, R. N. Haszeldine, B. R. Iles and D. G. Rowsell, *J. Chem. Soc.*, 1965, 6879.
81. B. A. Arbuzov and E. N. Dianova, *Izv. Akad. Nauk S.S.S.R., Ser. Khim.*, 1965, 1584; *Bull. Acad. Sc. U.S.S.R., Div. Chem. Sc.*, 1965, 1549.
82. N. N. Mel'Nikov, A. F. Grapov, L. V. Razvodovskaya and S. L. Portnova, *Zh. Obshch. Khim.*, 1965, **35**, 1771; *J. Gen. Chem. U.S.S.R.*, 1965, **35**, 1769.
83. F. Schindler, H. Schmidbaur and G. Jonas, *Chem. Ber.*, 1965, **98**, 3345.
84. F. Ramirez, A. V. Patwardhan and C. P. Smith, *J. Amer. Chem. Soc.*, 1965, **87**, 4973.
85. F. Ramirez, D. Rhum and C. P. Smith, *Tetrahedron*, 1965, **21**, 1941.
86. L. N. Mashlyakovskii and B. I. Ionin, *Zh. Obshch. Khim.*, 1965, **35**, 1577; *J. Gen. Chem. U.S.S.R.*, 1965, **35**, 1582.
87. D. J. Martin and G. E. Griffin, *J. Org. Chem.*, 1965, **30**, 4034.
88. D. M. Brown and D. A. Usher, *J. Chem. Soc.*, 1965, 6558.
89. G. M. Bogolyubov and A. A. Petrov, *Zh. Obshch. Khim.*, 1965, **35**, 988; *J. Gen. Chem. U.S.S.R.*, 1965, **35**, 994.
90. B. I. Ionin and A. A. Petrov, *Zh. Obshch. Khim.*, 1965, **35**, 1917; *J. Gen. Chem. U.S.S.R.*, 1965, **35**, 1910.
91. H. Nöth and H. J. Vetter, *Chem. Ber.*, 1965, **98**, 1981.
92. K. D. Berlin and L. A. Wilson, *Chem. Ind. (London)*, 1965, 1522.
93. E. I. Kataev, V. V. Plemenkov and V. V. Markin, *Dokl. Akad. Nauk. S.S.S.R.*, 1965, **165**, 1313; *Dokl. Chem.*, 1965, **165**, 1208.
94. R. R. Hindersinn and R. S. Ludington, *J. Org. Chem.*, 1965, **30**, 4020.
95. J. E. Thompson, *J. Org. Chem.*, 1965, **30**, 4276.
96. G. M. Bogolyubov and A. A. Petrov, *Zh. Obshch. Khim.*, 1965, **35**, 704; *J. Gen. Chem. U.S.S.R.*, 1965, **35**, 705.

1965 cont.
97. S. A. Fuqua, W. G. Ducan and R. M. Silverstein, *J. Org. Chem.*, 1965, **30**, 2543.
98. V. Gutmann, C. Kemenater and K. Utvary, *Monatshefte*, 1965, **96**, 836.
99. D. M. Brown and D. A. Usher, *J. Chem. Soc.*, 1965, 6547.
100. B. I. Ionin and A. A. Petrov, *Zh. Obshch. Khim.*, 1965, **35**, 2255; *J. Gen. Chem. U.S.S.R.*, 1965, **35**, 2247.
101. K. Issleib and R. Rieschel, *Chem. Ber.*, 1965, **98**, 2086.
102. R. J. Sundberg, *J. Org. Chem.*, 1965, **30**, 3604.
103. E. Fluck and K. Issleib, *Z. Naturforsch. B*, 1965, **20**, 1123.
104. K. H. Büchel, H. Röchling and F. Korte, *J. Liebigs Ann. Chem.*, 1965, **685**, 10.
105. T. Ya. Medved', Yu. M. Polikarpov, K. S. Yudina and M. I. Kabachnik, *Izv. Akad. Nauk S.S.S.R., Ser. Khim.*, 1965, 1707; *Bull. Acad. Sc. U.S.S.R., Div. Chem. Sc.*, 1965, 1676.
106. K. A. Andrianov, T. V. Vasil'eva and L. K. Kozlova, *Izv. Akad. Nauk S.S.S.R., Ser. Khim.*, 1965, 381; *Bull. Acad. Sc. U.S.S.R., Div. Chem. Sc.*, 1965, 366.
107. W. R. Cullen, D. S. Dawson and G. E. Styan, *Canad. J. Chem.*, 1965, **43**, 3392.
108. D. Seyferth and G. Singh, *J. Amer. Chem. Soc.*, 1965, **87**, 4156.
109. R. D. Partos and A. J. Speziale, *J. Amer. Chem. Soc.*, 1965, **87**, 5068.
110. G. P. Schiemenz, *Angew. Chem.*, 1965, **77**, 1089; *Angew. Chem., Int. Ed.*, 1965, **4**, 1093.
111. A. M. Aguiar and D. Daigle, *J. Org. Chem.*, 1965, **30**, 3527.
112. A. M. Aguiar and D. Daigle, *J. Org. Chem.*, 1965, **30**, 2826.
113. S. Trippett, B. J. Walker and H. Hoffmann, *J. Chem. Soc.*, 1965, 7140.
114. J. P. Schaefer and D. S. Weinberg, *J. Org. Chem.*, 1965, **30**, 2635.
115. E. Zbiral, *Monatshefte*, 1965, **96**, 1967.
116. J. B. Hendrickson, C. Hall, R. Rees and J. F. Templeton, *J. Org. Chem.*, 1965, **30**, 3312.
117. H. Schumann, P. Jutzi and M. Schmidt, *Angew. Chem.*, 1965, **77**, 912; *Angew. Chem. Int. Ed.*, 1965, **4**, 869.
118. L. Maier, *Angew. Chem.*, 1965, **77**, 1032; *Angew. Chem. Int. Ed.*, 1965, **4**, 984.
119. H. Schmidbaur, D. Kuhr and U. Krüger, *Angew. Chem.*, 1965, **77**, 866; *Angew. Chem. Int. Ed.*, 1966, **4**, 877.
120. H. Koopman, F. J. Spruit, F. Van Deursen and J. Bakker, *Rec. Trav. Chim.*, 1965, **84**, 341.

1966
1. G. Mavel, *Progress in NMR Spectroscopy*, Vol. 1, p. 251, Editors, J. W. Emsley, J. Feeney and L. H. Sutcliffe, Pergamon Press, Oxford, 1966.
2. J. H. Letcher and J. R. Van Wazer, *J. Chem. Phys.*, 1966, **44**, 815.
3. J. H. Letcher and J. R. Van Wazer, *J. Chem. Phys.*, 1966, **45**, 2916.
4. L. Maier, *Helv. Chim. Acta*, 1966, **49**, 1718.
5. S. O. Grim, W. McFarlane, E. F. Davidoff and T. J. Marks, *J. Phys. Chem.*, 1966, **70**, 581.
6. H. R. Allcock, R. L. Kugel and K. J. Valan, *Inorg. Chem.*, 1966, **5**, 1709.
7. M. Fild, O. Glemser and I. Hollenberg, *Z. Naturf.*, 1966, **21B**, 920.
8. H. G. Horn and A. Müller, *Z. anorg. allg. Chem.*, 1966, **346**, 266.
9. G. Nagarajan and A. Müller, *Z. Naturforsch.*, 1966, **21B**, 505.
10. M. G. Hogben, R. S. Gay and W. A. G. Graham, *J. Amer. Chem. Soc.*, 1966, **88**, 3457.

1966 cont.
11. T. Birchall and W. L. Jolly, *Inorg. Chem.*, 1966, **5**, 2177.
12. F. Knoll and G. Bergerhoff, *Monatshefte*, 1966, **97**, 808.
13. S. L. Manatt, G. L. Juvinall, R. I. Wagner and D. D. Elleman, *J. Amer. Chem. Soc.*, 1966, **88**, 2689.
14. H. R. Hays, *J. Org. Chem.*, 1966, **31**, 3817.
15. H. R. Hays and T. J. Logan, *J. Org. Chem.*, 1966, **31**, 3391.
16. L. Maier, *Helv. Chim. Acta*, 1966, **49**, 842.
17. H. G. Aguiar, *Thesis, Tulasne Univ.*, *Univ. Microfilms*, Order No. 66–10, 749; *Diss. Abstr. B*, 1966, **27**, 1407.
18. M. J. Gallagher and I. D. Jenkins, *J. Chem. Soc. C*, 1966, 2176.
19. M. G. Barlow, M. Green, R. N. Haszeldine and H. G. Higson, *J. Chem. Soc. B*, 1966, 1025.
20. R. W. Rudolph, R. W. Parry and C. F. Farran, *Inorg. Chem.*, 1966, **5**, 723.
21. J. W. Akitt, R. H. Cragg and N. N. Greenwood, *Chem. Comm.*, 1966, 134.
22. H. Dreeskamp, H. Elser and C. Schumann, *Ber. Bunsenges.*, 1966, **70**, 751.
23. L. Maier, *Helv. Chim. Acta*, 1966, **49**, 1000.
24. R. R. Holmes and R. N. Storey, *Inorg. Chem.*, 1966, **5**, 2146.
25. D. Hellwinkel, *Angew. Chem.*, 1966, **78**, 985; *Angew. Chem. Int. Ed.*, 1966, **5**, 968.
26. R. Fields, H. Goldwhite, R. N. Haszeldine and J. Kirman, *J. Chem. Soc. C*, 1966, 2075.
27. K. Issleib and W. Gründler, *Theoret. Chim. Acta*, 1966, **6**, 64.
28. R. W. Rudolph, R. C. Taylor and R. W. Parry, *J. Amer. Chem. Soc.*, 1966, **88**, 3729.
29. D. D. Des Marteau and G. H. Cady, *Inorg. Chem.*, 1966, **5**, 1829.
30. H. G. Horn, *Z. Naturforsch. B*, 1966, **21**, 617.
31. H. J. Emeleus and T. Onak, *J. Chem. Soc. A*, 1966, 1291.
32. T. L. Charlton and R. G. Cavell, *Chem. Comm.*, 1966, 763.
33. N. M. D. Brown and P. Bladon, *Chem. Comm.*, 1966, 304.
34. M. Van Doorne, *Thesis, Univ. Michigan*, 1965; *Univ. Microfilms*, Order No. 66–6722; *Diss. Abstr. B*, 1966, **27**, 397–398.
35. R. J. Gillespie, *Inorg. Chem.*, 1966, **5**, 1634.
36. J. F. Nixon and R. Schmutzler, *Spectrochim. Acta*, 1966, **22**, 565.
37. E. J. Boros, K. J. Cockran, R. W. King and J. G. Verkade, *J. Amer. Chem. Soc.*, 1966, **88**, 1140.
38. D. F. Clemens, H. H. Sisler and W. S. Brey, *Inorg. Chem.*, 1966, **5**, 527.
39. L. Maier, *Helv. Chim. Acta*, 1966, **49**, 2458.
40. L. Maier, *Helv. Chim. Acta*, 1966, **49**, 1249.
41. A. R. Cullingworth, A. Pidcock and J. D. Smith, *Chem. Comm.*, 1966, 89.
42. J. F. Normant, *Bull. Soc. Chim. France*, 1966, 3601.
43. C. Benezra and G. Ourisson, *Bull. Soc. Chim. France*, 1966, 1825.
44. D. J. Peterson, *J. Org. Chem.*, 1966, **31**, 950.
45. L. Banford and G. E. Coates, *J. Chem. Soc. A*, 1966, 274.
46. D. Z. Denney and D. B. Denney, *J. Amer. Chem. Soc.*, 1966, **88**, 1830.
47. E. Duval and E. A. C. Lucken, *Mol. Phys.*, 1966, **10**, 499.
48. S. C. Goodman and J. G. Verkade, *Inorg. Chem.*, 1966, **5**, 498.
49. F. Heatley and S. M. Todd, *J. Chem. Soc. A*, 1966, 1152.
50. D. Gagnaire, J. B. Robert, J. Verrier and R. Wolf, *Bull. Soc. Chim. France*, 1966, 3719.
51. B. Fontal and H. Goldwhite, *Tetrahedron*, 1966, **22**, 3275.

REFERENCES

1966 cont.
52. W. H. Mueller and A. A. Oswald, *J. Org. Chem.*, 1966, **31**, 1894.
53. M. Green, R. N. Haszeldine and G. S. A. Hopkins, *J. Chem. Soc. A*, 1966, 1766.
54. J. A. Ferretti and L. Paolillo, *Ricerca Sci.* (C.N.R.), 1966, **36**, 1008.
55. K. D. Berlin and M. A. R. Khayat, *Tetrahedron*, 1966, **22**, 975.
56. D. W. Osborne, H. O. Senkbell and J. L. Wasco, *J. Org. Chem.*, 1966, **31**, 192.
57. I. Schumann and H. Blass, *Z. Naturforsch. B*, 1966, **21**, 1105.
58. C. Benezra and G. Ourisson, *Bull. Soc. Chim. France*, 1966, 2270.
59. R. W. Rudolph, J. G. Morse and R. W. Parry, *Inorg. Chem.*, 1966, **5**, 1464.
60. J. R. Little and P. F. Hartman, *J. Amer. Chem. Soc.*, 1966, **88**, 96.
61. M. P. Simonnin and B. Borecka, *Bull. Soc. Chim. France*, 1966, 3842.
62. K. D. Berlin, D. M. Hellwedge, M. Nagabhushanam and E. T. Gaudy, *Tetrahedron*, 1966, **22**, 2191.
63. M. Gordon and C. E. Griffin, *J. Org. Chem.*, 1966, **31**, 333.
64. M. Gordon, *Thesis, Univ. Pittsb.*, 1965; *Univ. Microfilms*, Order No. 66-8133; *Diss. Abstr. B*, 1966, **27**, 752.
65. G. L. Kenyon and F. H. Westheimer, *J. Amer. Chem. Soc.*, 1966, **88**, 3561.
66. F. Mathey and G. Mavel, *Compt. Rend., Ser. C*, 1966, **263**, 855.
67. T. J. Katz, C. R. Nicholson and C. A. Reilly, *J. Amer. Chem. Soc.* 1966, **88**, 3832.
68. G. Märkl, *Angew. Chem.*, 1966, **78**, 907; *Angew. Chem. Int. Ed.*, 1966, **5**, 846.
69. G. Shaw, J. K. Becconsall, R. M. Canadine and R. Murray, *Chem. Comm.*, 1966, 425.
70. C. E. Griffin, R. B. Davison and M. Gordon, *Tetrahedron*, 1966, **22**, 561.
71. J. Herweh, *J. Org. Chem.*, 1966, **31**, 4308.
72. L. V. Interrante, M. A. Bennett and R. S. Nyholm, *Inorg. Chem.*, 1966, **5**, 2212.
73. H. J. Emeleus and J. M. Miller, *J. Inorg. Nucl. Chem.*, 1966, **28**, 662.
74. W. H. Mueller, R. M. Rubin and P. E. Butler, *J. Org. Chem.*, 1966, **31**, 3537.
75. J. Herweh, *J. Org. Chem.*, 1966, **31**, 2422.
76. F. Ramirez and C. P. Smith, *Tetrahedron Lett.*, 1966, 3651.
77. E. Fluck and H. Binder, *Angew. Chem.*, 1966, **78**, 677; *Angew. Chem. Int. Ed.*, 1966, **5**, 666.
78. G. H. Birum and C. N. Mathews, *J. Amer. Chem. Soc.*, 1966, **88**, 4198.
79. R. K. Harris and C. M. Woodman, *Mol. Phys.*, 1966, **10**, 437.
80. H. Lederle, G. Ottmann and E. Kober, *Inorg. Chem.*, 1966, **5**, 1818.
81. E. F. Mooney and B. S. Thornhill, *J. Inorg. Nucl. Chem.*, 1966, **28**, 2225.
82. K. L. Freeman and M. J. Gallagher, *Austral. J. Chem.*, 1966, **19**, 2025.
83. M. Becke-Goehring, L. Leichner and B. Scharf, *Z. anorg. allg. Chem.*, 1966, **343**, 154.
84. A. P. Carroll and R. A. Shaw, *J. Chem. Soc. B*, 1966, 914.
85. S. K. Das, R. Keat, R. Shaw and B. C. Smith, *J. Chem. Soc. A*, 1966, 1677.
86. D. Dell, B. Fitzimmons, R. Keat and R. A. Shaw, *J. Chem. Soc. A*, 1966, 1680.
87. C. Hewlett and R. A. Shaw, *J. Chem. Soc. A*, 1966, 56.
88. R. Keat and R. A. Shaw, *J. Chem. Soc. A*, 1966, 908.
89. G. R. Feistel, *Thesis, Univ. of Illinois*; *Univ. Microfilms*, Order No. 66-4174; *Dissertation Abstr.*, 1966, **26**, 7021.
90. H. R. Allcock and R. L. Kugel, *Inorg. Chem.*, 1966, **5**, 1016.

1966 cont.
91. H. R. Allcock and R. L. Kugel, *Inorg. Chem.*, 1966, **5**, 1716.
92. G. Bogeat and G. Cauquis, *Bull. Soc. Chim. France*, 1966, 2735.
93. G. Engelhardt, E. Steger and R. Stahlberg, *Z. Naturforsch. B*, 1966, **21**, 586.
94. G. Engelhardt, E. Steger and R. Stahlberg, *Z. Naturforsch. B*, 1966, **21**, 1231.
95. C. T. Ford, J. M. Barr, F. E. Dickson and I. I. Bezman, *Inorg. Chem.*, 1966, **5**, 351.
96. B. Grushkin, A. J. Berlin, J. L. McClanahan and R. G. Rice, *Inorg. Chem.*, 1966, **5**, 172.
97. E. T. McBee, K. Okuhara and C. J. Morton, *Inorg. Chem.*, 1966, **5**, 456.
98. M. Becke-Goehring, K. Bayer and T. Mann, *Z. anorg. allg. Chem.*, 1966, **346**, 143.
99. J. B. Lambert and D. C. Mueller, *J. Amer. Chem. Soc.*, 1966, **88**, 3669.
100. P. C. Van Der Voorn and R. S. Drago, *J. Amer. Chem. Soc.*, 1966, **88**, 3255.
101. D. Hellwinkel, *Angew. Chem.*, 1966, **78**, 749; *Angew. Chem. Int. Ed.*, 1966, **5**, 725.
102. J. M. Jenkins, T. J. Huttemann and J. G. Verkade, *Adv. Chem. Ser.*, 1966, **42**, 604.
103. C. Benezra. *Thèse, Univ. Strasbourg*, 18 Juin 1966.
104. S. Hirai, R. G. Harvey and E. V. Jensen, *Tetrahedron*, 1966, **22**, 1625.
105. G. Pukanic, N. C. Li, W. S. Brey and G. B. Savitsky, *J. Phys. Chem.*, 1966, **70**, 2899.
106. M. Webster, *Chem. Rev.*, 1966, **66**, 87.
107. L. Kolditz and W. Rehak, *Z. anorg. allg. Chem.*, 1966, **342**, 32.
108. H. Nöth and H. Vahrenkamp, *Chem. Ber.*, 1966, **99**, 1049.
109. A. D. Norman and R. Schaeffer, *J. Amer. Chem. Soc.*, 1966, **88**, 1143.
110. S. I. Shupack and B. Wagner, *Chem. Comm.*, 1966, 547.
111. L. Maier, *Z. anorg. allg. Chem.*, 1966, **345**, 29.
112. A. R. Katritsky, F. J. Swinbourne and B. Ternai, *J. Chem. Soc. B*, 1966, 235.
113. A. M. Aguiar, H. G. Aguiar and T. G. Archibald, *Tetrahedron Lett.*, 1966, 3187.
114. J. W. Baker, R. E. Stenseth and L. C. D. Groenweghe, *J. Amer. Chem. Soc.*, 1966, **88**, 3041.
115. I. J. Borowitz, K. C. Kirby and R. Virkhaus, *J. Org. Chem.*, 1966, **31**, 4031.
116. P. A. Chopard and R. F. Hudson, *J. Chem. Soc. B*, 1966, 1089.
117. F. B. Clarke and J. W. Lyons, *J. Amer. Chem. Soc.*, 1966, **88**, 4401.
118. M. C. Demarcq, M. Guyot De La Hadrouyère, G. G. Perot and B. Rey-Coquais, *Bull. Soc. Chim. France*, 1966, 216.
119. D. B. Denney, N. Gershman and J. Giacin, *J. Org. Chem.*, 1966, **31**, 2833.
120. E. A. Dennis and F. H. Westheimer, *J. Amer. Chem. Soc.*, 1966, **88**, 3431
121. R. S. Edmundson and E. W. Mitchell, *Chem. Comm.*, 1966, 482.
122. C. T. Eyles and S. Trippett, *J. Chem. Soc. C*, 1966, 67.
123. K. L. Freeman and M. J. Gallagher, *Austral. J. Chem.*, 1966, **19**, 2159.
124. H. Goldwhite and D. G. Rowsell, *J. Amer. Chem. Soc.*, 1966, **88**, 3572.
125. N. N. Greenwood, E. J. F. Ross and A. Storr, *J. Chem. Soc. A*, 1966, 706.
126. R. F. Hudson and G. Salvadori, *Helv. Chim. Acta*, 1966, **49**, 96.
127. V. M. Ignatiev, B. I. Ionin and A. A. Petrov, *Zh. Obshch. Khim.*, 1966, **35**, 1505; *J. Gen. Chem. U.S.S.R.*, 1966, **35**, 1510.
128. A. W. Johnson and V. L. Kyllingstad, *J. Org. Chem.*, 1966, **31**, 334.
129. K. R. Martin, *Thesis, Univ. Pittsburgh*, 1965; *Univ. Microfilms*, Order No. 66–8154; *Diss. Abstr., B*, 1966, **27**, 761.

1966 cont.
130. K. R. Martin and C. E. Griffin, *J. Heterocyclic Chem.*, 1966, **3**, 92.
131. R. E. Mesmer and R. L. Caroll, *J. Amer. Chem. Soc.*, 1966, **88**, 1381.
132. D. J. Peterson and J. H. Collins, *J. Org. Chem.*, 1966, **31**, 2373.
133. K. Pilgram, *Tetrahedron*, 1966, **22**, 1241.
134. K. Pilgram, *Tetrahedron Lett.*, 1966, 3831.
135. D. Seyferth and J. Fogel, *J. Organometal. Chem.*, 1966, **6**, 205.
136. A. J. Speziale and L. J. Taylor, *J. Org. Chem.*, 1966, **31**, 2450.
137. B. G. Van Den Bos, A. Schipperheyn and F. W. Van Deursen, *Rec. Trav. Chim.*, 1966, **85**, 429.
138. J. R. Van Wazer and S. Norval, *J. Amer. Chem. Soc.*, 1966, **88**, 4415.
139. R. A. Goodrich and P. H. Treichel, *J. Amer. Chem. Soc.*, 1966, **88**, 3509.
140. H. J. Bestmann, G. Joachim, I. Lengyel, J. F. M. Oth, R. Merenyi and H. Weitkamp, *Tetrahedron Lett.*, 1966, 3355.
141. R. A. Dwek and R. E. Richards, *Chem. Comm.*, 1966, 581.
142. E. D. Jones and J. E. Hesse, *Bull. Amer. Phys. Soc.*, 1966, **11**, 172.
143. B. A. Scott, *Thesis, Penn. State Univ.*; *Univ. Microfilms*, Order No. 66-8756; *Diss. Abstr. B*, 1966, **27**, 786.
144. B. F. Stein, *Thesis, Univ. of Pennsylvania*; *Univ. Microfilms*, Order No. 66-299; *Dissertation Abstr.*, 1966, **26**, 5510.
145. B. F. Stein and R. H. Walsley, *Phys. Rev.*, 1966, **148**, 933.
146. V. I. Spitsyn and S. A. Bakhchisaraitseva, *Zh. Phys. Khim.*, 1966, **40**, 1960; *Russian J. Phys. Chem.*, 1966, **40**, 1053.
147. P. Beekenkamp and G. E. G. Hardeman, *Verres et Réfr.*, 1966, **20**, 414.
148. A. G. Lundin, E. A. Ukraintseva and V. A. Kovalchuk, *Zh. Strukt. Khim.*, 1966, **7**, 446; *J. Struct. Chem.*, 1966, **7**, 420.
149. E. R. Andrews, S. Clough, L. F. Farnell, T. D. Gledhill and I. Roberts, *Physics Lett.*, 1966, **21**, 505.
150. W. Wieker and A. R. Grimmer, *Z. Naturforsch., B*, 1966, **21**, 1103.
151. C. Kowala and J. M. Swan, *Austral. J. Chem.*, 1966, **19**, 999.
152. W. De W. Horrocks and L. H. Pignolet, *J. Amer. Chem. Soc.*, 1966, **88**, 5929.
153. R. B. King, *Inorg. Chem.*, 1966, **5**, 82.
154. T. Kruck and W. Lang, *Z. anorg. allg. Chem.*, 1966, **343**, 181.
155. H. G. Schuster-Woldan and F. Basolo, *J. Amer. Chem. Soc.*, 1966, **88**, 1657.
156. M. E. Vol'pin and I. S. Kolomnikov, *Izv. Akad. Nauk S.S.S.R.*, 1966, 2041; *Bull. Acad. Sc. U.S.S.R., Div. Chem. Sc.*, 1966, 1980.
157. B. B. Wayland and R. S. Drago, *J. Amer. Chem. Soc.*, 1966, **88**, 4597.
158. D. G. Hendricker, R. E. McCarley, R. W. King and J. G. Verkade, *Inorg. Chem.*, 1966, **5**, 639.
159. J. P. Bibler and A. Wojcicki, *Inorg. Chem.*, 1966, **5**, 889.
160. E. O. Fischer and E. Moser, *J. Organometal. Chem.*, 1966, **5**, 63.
161. T. Kruck and W. Lang, *Chem. Ber.*, 1966, **99**, 3794.
162. J. M. Jenkins and B. L. Shaw, *J. Chem. Soc. A.*, 1966, 1407.
163. E. W. Randall and D. Shaw, *Mol. Phys.*, 1966, **11**, 395.
164. R. C. Taylor, J. F. Young and G. Wilkinson, *Inorg. Chem.*, 1966, **5**, 20.
165. B. L. Booth and R. N. Haszeldine, *J. Chem. Soc. A*, 1966, 157.
166. F. A. Hartman and A. Wojcicki, *Inorg. Nucl. Chem. Lett.*, 1966, **2**, 303.
167. H. Nöth and G. Schmid, *Z. anorg. allg. Chem.*, 1966, **345**, 69.
168. C. G. Barlow and J. F. Nixon, *Inorg. Nucl. Chem. Lett.*, 1966, **2**, 323.
169. M. Lenzi and R. Poilblanc, *Compt. Rend. Ser. C*, 1966, **263**, 674.
170. J. R. Moss and B. L. Shaw, *J. Chem. Soc. A*, 1966, 1793.

1966 cont.
171. G. N. Rao and N. C. Li, *J. Inorg. Nucl. Chem.*, 1966, **28**, 2931.
172. G. R. Van Hecke and W. De W. Horrocks, *Inorg. Chem.*, 1966, **5**, 1968.
173. D. Walter and G. Wilke, *Angew. Chem.*, 1966, **78**, 941; *Angew. Chem. Int. Ed.*, 1966, **5**, 897.
174. A. B. Burg and G. B. Street, *Inorg. Chem.*, 1966, **5**, 1532.
175. J. F. Nixon, *Chem. Comm.*, 1966, 34.
176. J. G. Riess and J. R. Van Wazer, *Bull. Soc. Chim. France*, 1966, 1846.
177. J. G. Riess and J. R. Van Wazer, *J. Amer. Chem. Soc.*, 1966, **88**, 2166.
178. J. G. Riess and J. R. Van Wazer, *J. Amer. Chem. Soc.*, 1966, **88**, 2339.
179. J. M. Jenkins and B. L. Shaw, *J. Chem. Soc. A*, 1966, 770.
180. H. P. Fritz and K. E. Schwarzhans, *J. Organometal. Chem.*, 1966, **5**, 103.
181. F. J. Hopton, A. J. Rest, D. T. Rosevear and F. G. A. Stone, *J. Chem. Soc. A*, 1966, 1326.
182. H. C. Volger and K. Vrieze, *J. Organometal. Chem.*, 1966, **6**, 297.
183. M. Green, R. B. L. Osborn, A. J. Rest and F. G. A. Stone, *Chem. Comm.*, 1966, 502.
184. G. W. Parshall, *J. Amer. Chem. Soc.*, 1966, **88**, 704.
185. A. Pidcock, R. E. Richards and L. M. Venanzi, *J. Chem. Soc. A*, 1966, 1707.
186. T. Kruck and A. Engelmann, *Angew. Chem.*, 1966, **78**, 820; *Angew. Chem., Int. Ed.*, 1966, **5**, 836.
187. M. C. Baird, D. N. Lawson, J. T. Mague, J. A. Osborn and G. Wilkinson, *Chem. Comm.*, 1966, 129.
188. M. A. Bennett, R. Bramley and P. A. Longstaff, *Chem. Comm.*, 1966, 806.
189. J. Chatt and R. S. Coffey, *Chem. Comm.*, 1966, 545.
190. D. N. Lawson, J. A. Osborn and G. Wilkinson, *J. Chem. Soc. A*, 1966, 1733.
191. J. A. Osborn, F. H. Jardine, J. F. Young and G. Wilkinson, *J. Chem. Soc. A*, 1966, 1711.
192. S. O. Grim and R. A. Ference, *Inorg. Nucl. Chem. Lett.*, 1966, **2**, 205.
193. J. M. Jenkins, M. S. Lupin and B. L. Shaw, *J. Chem. Soc. A*, 1966, 1787.
194. M. Becke-Goehring and A. Slawick, *Z. anorg. allg. Chem.*, 1966, **346**, 295.
195. S. O. Grim, W. McFarlane and D. A. Wheatland, *Inorg. Nucl. Chem. Lett.*, 1966, **2**, 49.
196. J. A. Happe and M. Morales, *J. Amer. Chem. Soc.*, 1966, **88**, 2077.
197. B. T. Allen, D. Chapman and N. J. Salisbury, *Nature (London)*, 1966, **212**, 282.
198. H. Sigel, *Experientia*, 1966, **22**, 497.
199. L. Yengoyan and D. H. Rammler, *Biochemistry*, 1966, **5**, 3629.
200. D. Chapman, P. Byrne and G. G. Shipley, *Proc. Royal Soc. A*, 1966, **290**, 115.
201. D. Chapman and A. Morrison, *J. Biol. Chem.*, 1966, **241**, 5044.
202. C. Ho and R. J. Kurland, *J. Biol. Chem.*, 1966, **241**, 3002.
203. K. Onodera and S. Hirano, *Biochem. Biophys. Res. Comm.*, 1966, **25**, 239.
204. S. Åkerfeldt, *Acta Chem. Scand.*, 1966, **20**, 1783.
205. J. Ullrich and A. Mannschreck, *Bioch. Biophys. Acta*, 1966, **115**, 46.
206. F. Ramirez, O. P. Madan and C. P. Smith, *Tetrahedron*, 1966, **22**, 567.
207. Yu. A. Buslaev and V. A. Shcherbakov, *Zh. Strukt. Khim.*, 1966, **7**, 345; *J. Struct. Chem.*, 1966, **7**, 332.
208. J. G. Riess and J. R. Van Wazer, *Inorg. Chem.*, 1966, **5**, 178.
209. M. Lustig, J. K. Ruff and C. B. Colburn, *J. Amer. Chem. Soc.*, 1966, **88**, 3875.
210. F. Seel, K. Rudolph and R. Budenz, *Z. anorg. allg. Chem.*, 1966, **341**, 196.
211. B. Fontal, H. Goldwhite and D. G. Rowsell, *J. Org. Chem.*, 1966, **31**, 2424.

1966 cont.
212. K. Moedritzer and R. R. Irani, *J. Org. Chem.*, 1966, **31**, 1603.
213. A. M. Schoffstall and H. Tieckelman, *Tetrahedron*, 1966, **22**, 399.
214. S. G. Warren, *J. Chem. Soc. C.* 1966, 1349.
215. J. Dyer and J. Lee, *Trans. Farad. Soc.*, 1966, **62**, 257.
216. K. Takahashi, T. Yamasaki and G. Miyazima, *Bull. Chem. Soc. Japan*, 1966, **39**, 2787.
217. D. H. Brown, G. W. Fraser and D. W. A. Sharp, *J. Chem. Soc. A*, 1966, 171.
218. A. E. Lippman, *J. Org. Chem.*, 1966, **31**, 471.
219. K. I. Novitskii, N. A. Razumova and A. A. Petrov, *Zh. Obshch. Khim.*, 1966, **36**, 1649; *J. Gen. Chem. U.S.S.R.*, 1966, **36**, 1649.
220. N. A. Razumova, A. A. Petrov, A. Kh. Voznesenskaya and K. I. Novitskii, *Zh. Obshch. Khim.*, 1966, **36**, 244; *J. Gen. Chem. U.S.S.R.*, 1966, **36**, 255.
221. D. B. Denney and S. Varga, *Tetrahedron Lett.*, 1966, 4935.
222. B. Fontal and H. Goldwhite, *J. Org. Chem.*, 1966, **31**, 3804.
223. T. D. Smith, *J. Chem. Soc. A*, 1966, 841.
224. F. Ramirez, A. V. Patwardhan, H. J. Kugler and C. P. Smith, *Tetrahedron Lett.*, 1966, 3053.
225. F. Ramirez, C. P. Smith, A. S. Gulati and A. V. Patwardhan, *Tetrahedron Lett.*, 1966, 2151.
226. F. Herail, *Compt. Rend. Ser. C*, 1966, **262**, 1624.
227. L. Kugel and M. Halmann, *J. Amer. Chem. Soc.*, 1966, **88**, 3566.
228. G. S. Reddy and R. Schmutzler, *Inorg. Chem.*, 1966, **5**, 164.
229. V. S. Tsivunin, S. V. Fridland, T. V. Zykova and G. Kh. Kamai, *Zh. Obshch. Khim.*, 1966, **36**, 1424; *J. Gen. Chem. U.S.S.R.*, 1965, **36**, 1431.
230. C. G. Barlow and J. F. Nixon, *J. Chem. Soc. A*, 1966, 228.
231. R. J. Hartle, *J. Org. Chem.*, 1966, **31**, 4288.
232. A. Schmidpeter and H. Groeger, *Z. anorg. allg. Chem.*, 1966, **345**, 106.
233. H. P. Latscha, *Z. anorg. allg. Chem.*, 1966, **346**, 166.
234. A. N. Pudovik, T. Kh. Gazizov, Yu. Yu. Samitov and T. V. Zykova, *Dokl. Akad. Nauk, S.S.S.R.*, 1966, **166**, 615; *Dokl. Chem.*, 1966, **166**, 124.
235. K. D. Berlin and M. A. R. Khayat, *Tetrahedron*, 1966, **22**, 987.
236. J. J. Dunand, A. Rousseau and P. Servoz-Gavin, *Compt. Rend. Ser. B*, 1966, **262**, 515.
237. D. D. Magnelli, G. Tesi, J. U. Lowe and W. E. McQuistion, *Inorg. Chem.*, 1966, **5**, 457.
238. E. Fluck, H. Gross, H. Binder and J. Gloede, *Z. Naturforsch. B*, 1966, **21**, 1125.
239. A. G. Pinkus and P. G. Waldrep, *J. Org. Chem.*, 1966, **31**, 575.
240. H. L. Retcofsky and C. E. Griffin, *Tetrahedron Lett.*, 1966, 1975.
241. G. L. Kenyon and F. H. Westheimer, *J. Amer. Chem. Soc.*, 1966, **88**, 3557.
242. F. Ramirez, S. B. Bathia and C. P. Smith, *J. Org. Chem.*, 1966, **31**, 4105.
243. C. Charrier, W. Chodkiewicz and P. Cadiot, *Bull. Soc. Chim. France*, 1966, 1002.
244. W. M. Daniewski, M. Gordon and C. E. Griffin, *J. Org. Chem.*, 1966, **31**, 2083.
245. R. H. Churi and C. E. Griffin, *J. Amer. Chem. Soc.*, 1966, **88**, 1824.
246. P. E. Sonnet and A. B. Bŏrkovec, *J. Org. Chem.*, 1966, **31**, 2962.
247. M. P. Simonnin, *J. Organometal Chem.*, 1966, **5**, 155.
248. F. Ramirez, A. V. Patwardhan and C. P. Smith, *J. Org. Chem.*, 1966, **31**, 3159.

1966 cont.
249. R. G. Harvey, *Tetrahedron*, 1966, **22**, 2561.
250. I. A. Nuretdinov, R. R. Shagidullin, Yu. Ya. Shamonin and N. P. Grechkin, *Izv. Akad. Nauk S.S.S.R., Ser. Khim.*, 1966, 839; *Bull. Acad. Sc., U.S.S.R., Div. Chem. Sc.*, 1966, 803.
251. T. Nishiwaki, *Tetrahedron*, 1966, **22**, 711.
252. D. Seyferth, J. T. P. Mui and G. Singh, *J. Organometal. Chem.*, 1966, **5**, 185.
253. R. B. Davison, *Thesis, Univ. Pittsburgh*, 1965; *Univ. Microfilms*, Order No. 66–10,056; *Diss. Abstr. B*, 1966, **27**, 1807.
254. B. G. Liober and A. I. Razumov, *Zh. Obshch. Khim.*, 1966, **36**, 314; *J. Gen. Chem. U.S.S.R.*, 1966, **36**, 323.
255. C. H. Yoder and J. J. Zuckerman, *J. Amer. Chem. Soc.*, 1966, **88**, 2170.
256. W. M. Daniewski and C. E. Griffin, *J. Org. Chem.*, 1966, **31**, 3236.
257. G. Sosnovsky, D. J. Rawlinson and E. H. Zaret, *Chem. Comm.*, 1966, 453.
258. F. Ramirez, A. V. Patwardhan and C. P. Smith, *J. Org. Chem.*, 1966, **31**, 474.
259. E. Cherbuliez, A. Buchs, S. Jaccard, D. Janjic and J. Rabinowitz, *Helv. Chem. Acta*, 1966, **49**, 2395.
260. R. W. Moen and W. H. Mueller, *J. Org. Chem.*, 1966, **31**, 1971.
261. A. G. Pinkus and L. Y. C. Meng, *J. Org. Chem.*, 1966, **31**, 1038.
262. I. B. Johns, H. R. Di Pietro, R. H. Nealey and J. V. Pustinger, *J. Phys. Chem.*, 1966, **70**, 924.
263. A. Mustafa, M. M. Sidky and F. M. Soliman, *Tetrahedron*, 1966, **22**, 393.
264. D. W. Osborne, *J. Org. Chem.*, 1966, **31**, 197.
265. M. P. Simonnin, *Bull. Soc. Chim. France*, 1966, 1774.
266. R. S. Davidson, R. A. Sheldon and S. Trippett, *J. Chem. Soc. C*, 1966, 722.
267. M. P. Savage and S. Trippett, *J. Chem. Soc. C*, 1966, 1843.
268. J. W. Rave and H. R. Hays, *J. Org. Chem.*, 1966, **31**, 2894.
269. A. M. Aguiar and H. Aguiar, *J. Amer. Chem. Soc.*, 1966, **88**, 4090.
270. C. N. Mathews and G. H. Birum, *Tetrahedron Lett.*, 1966, 5707.
271. A. R. Katritzky and B. Ternai, *J. Chem. Soc. B*, 1966, 631.
272. K. Dimroth and P. Hoffmann, *Chem. Ber.* 1966, **99**, 1325.
273. S. Trippett and B. J. Walker, *J. Chem. Soc. C*, 1966, 887.
274. G. Smolinsky and B. I. Feuer, *J. Org. Chem.*, 1966, **31**, 3882.
275. A. M. Aguiar and T. G. Archibald, *Tetrahedron Letters*, 1966, 5471.
276. K. Khaleeluddin and J. M. W. Scott, *Chem. Comm.*, 1966, 511.
277. E. E. Schweizer and J. G. Thompson, *Chem. Comm.*, 1966, 666.
278. R. F. Stockel, *Tetrahedron Lett.*, 1966, 2833.
279. B. D. Faithful, R. D. Gillard, D. G. Tuck and R. Ugo, *J. Chem. Soc. A*, 1966, 1185.
280. R. D. Partos and K. W. Ratts, *J. Amer. Chem. Soc.*, 1966, **88**, 4996.
281. M. Schlosser, T. Kadibelban and G. Steinhoff, *Angew. Chem.*, 1966, **78**, 1018; *Angew. Chem. Int. Ed.*, 1966, **5**, 968.
282. A. M. Aguiar and T. G. Archibald, *Tetrahedron Letters*, 1966, 5541.
283. C. C. Price, T. Parasaran and T. V. Lakshmirayan, *J. Amer. Chem. Soc.*, 1966, **88**, 1034.
284. K. L. Paciorek and R. H. Kratzer, *J. Org. Chem.*, 1966, **31**, 2426.
285. H. Hoffmann and P. Schellenbeck, *Chem. Ber.*, 1966, **99**, 1134.
286. L. Kaplan, *J. Org. Chem.*, 1966, **31**, 3454.
287. G. Märkl and H. Olbrich, *Angew. Chem.*, 1966, **78**, 598; *Angew. Chem. Int. Ed.*, 1966, **5**, 589.
288. V. Gutmann, K. Utvary and M. Bermann, *Monatshefte*, 1966, **97**, 1745.

1966 cont.
289. L. D. Quin and H. G. Anderson, *J. Org. Chem.*, 1966, **31**, 1206.
290. C. N. Mathews, J. S. Driscoll and G. H. Birum, *Chem. Comm.*, 1966, 736.
291. P. A. Chopard, *J. Org. Chem.*, 1966, **31**, 107.
292. K. Issleib and E. Fluck, *Angew. Chem.*, 1966, **78**, 597; *Angew. Chem. Int. Ed.*, 1966, **5**, 587.
293. K. Issleib and M. Hoffmann, *Chem. Ber.*, 1966, **99**, 1320.
294. A. H. Cowley and R. P. Pinnell, *J. Amer. Chem. Soc.*, 1966, **88**, 4533.
295. E. Fluck and K. Issleib, *Z. Naturforsch. B*, 1966, **21**, 736.

1967
1. M. M. Crutchfield, C. H. Dungan, J. H. Letcher, V. Mark and J. R. Van Wazer, *Topics in Phosphorus Chemistry*, Vol. 5, Editors, M. Grayson and E. J. Griffith, Interscience Publishing, 1967.
2. V. E. Bel'skii, T. A. Zyablikova, A. R. Panteleeva and I. M. Shermergorn, *Doklady Akad. Nauk S.S.S.R.*, 1967, **177**, 340; *Dokl. Chem.*, 1967, **177**, 1014.
3. A. N. Pudovik, T. Kh. Gazizov, Yu. Yu. Samitov and T. V. Zykova, *Zhur. Obshchei. Khim.*, 1967, **37**, 706; *J. Gen. Chem. (U.S.S.R.)*, 1967, **37**, 662.
4. A. N. Pudovik, I. V. Gur'yanova, S. P. Perevezentseva and T. V. Zykova, *Zhur. Obshchei. Khim.*, 1967, **37**, 1317; *J. Gen. Chem. (U.S.S.R.)*, 1967, **37**, 1246.
5. M. P. Klein and D. A. Phelps, *Appl. Spectroscopy Rev.*, 1967, **1**, 131, Editors, R. E. Lundin, R. H. Elksen, R. A. Flath and K. Ternanishi.
6. K. Siegbahn, C. Nordling, A. Fahlman, R. Nordberg, K. Hamrin, J. Hedman, G. Johansson, T. Bergmark, S. E. Karlson, I. Lindgren and B. Lindberg, *ESCA, Atomic, Molecular and Solid State Structure Studied by means of electron spectroscopy*, Almqvist and Wiksells, Uppsala, 1967.
7. J. G. Riess, J. R. Van Wazer and J. H. Letcher, *J. Phys. Chem.*, 1967, **71**, 1925.
8. S. O. Grim, W. McFarlane and E. F. Davidoff, *J. Org. Chem.*, 1967, **32**, 781.
9. A. Schmidpeter and H. Brecht, *Angew. Chem.*, 1967, **79**, 946; *Angew. Chem. Int. Ed.*, 1967, **6**, 945.
10. R. I. Wagner, L. D. Freeman, H. Goldwhite and D. G. Rowsell, *J. Amer. Chem. Soc.*, 1967, **89**, 1102.
11. M. Fild, I. Hollenberg and O. Glemser, *Z. Naturforsch.*, 1967, **22B**, 253.
12. R. J. Grabenstetter, O. T. Quimby and T. J. Flautt, *J. Phys. Chem.*, 1967, **71**, 4194.
13. A. A. Neimysheva and I. L. Knunyants, *Dokl. Akad. Nauk S.S.S.R.*, 1967, **177**, 856; *Dokl. Chem.*, 1967, **177**, 1123.
14. V. V. Sheluchenko, M. A. Landau, S. S. Dubov, A. A. Neimysheva and I. L. Knunyants, *Dokl. Akad. Nauk*, 1967, **177**, 376; *Dokl. Chemistry*, 1967, **177**, 1050.
15. E. N. Tsvetkov, G. K. Semin, T. A. Babushkina, D. F. Lobanov and M. I. Kabachnik, *Izv. Akad. Nauk S.S.S.R., Ser. Khim.*, 1967, 2375; *Bull. Acad. Sc. U.S.S.R., Div. Chem. Sc.*, 1967, 2267.
16. W. A. Sheppard, *Trans. N.Y. Acad. Sci.*, 1967, **29**, 700.
17. H. Schindlbauer, *Chem. Ber.*, 1967, **100**, 3432.
18. R. G. Kostyanovskii, I. I. Chervin, V. V. Yakshin and A. U. Stepanyants, *Izv. Akad. Nauk S.S.S.R., Ser. Khim.*, 1967, 1629; *Bull. Acad. Sc. U.S.S.R., Div. Chem. Sc.*, 1967, 1577.
19. P. Haake, R. D. Cook and G. H. Hurst, *J. Amer. Chem. Soc.*, 1967, **89**, 2650.

1967 cont.
20. W. Drenth and D. Rosenberg, *Rec. Trav. Chim.*, 1967, **86**, 26.
21. G. Mavel and R. Favelier, *J. Chim. Phys.*, 1967, **64**, 627.
22. G. Mavel, R. Mańkowski-Favelier and G. Sturtz, *J. Chim. Phys.*, 1967, **64**, 1686.
23. W. H. N. Vriezen and F. Jellinek, *Chem. Phys. Letters*, 1967, **1**, 284.
24. T. Yonezawa, *Bull. Chem. Soc. Japan*, 1967, **40**, 487.
25. J. L. Aubagnac, P. Bouchet, J. Elguero, R. Jacquier and C. Marzin, *J. Chim. Phys.*, 1967, **64**, 1649.
26. T. H. Siddall and C. A. Prohaska, *Appl. Spectroscopy*, 1967, **21**, 9.
27. M. Green, R. J. Mawby and G. Swinden, *Chem. Comm.*, 1967, 127.
28. A. H. Cowley, W. D. White and S. L. Manatt, *J. Amer. Chem. Soc.*, 1967, **89**, 6433.
29. E. Fluck, H. Bürger and U. Goetze, *Z. Naturforsch. B*, 1967, **22**, 912.
30. E. Fluck and H. Binder, *Z. Naturforsch. B*, 1967, **22**, 1001.
31. R. B. Johannesen, *J. Chem. Phys.*, 1967, **47**, 3088.
32. R. W. Rudolph and R. W. Parry, *J. Amer. Chem. Soc.*, 1967, **89**, 1621.
33. T. L. Charlton and R. G. Cavell, *Inorg. Chem.*, 1967, **6**, 2204.
34. M. G. Voronkov, N. F. Orlov, B. L. Kaufman and V. A. Pestunovich, *Zh. Obshch. Khim.*, 1967, **37**, 2065; *J. Gen. Chem. U.S.S.R.*, 1967, **37**, 1958.
35. J. D. Murray, G. Nicklees and F. H. Pollard, *J. Chem. Soc. A*, 1967, 1726.
36. P. M. Treichel, R. A. Goodrich and S. B. Pierce, *J. Amer. Chem. Soc.*, 1967, **87**, 2017.
37. G. I. Drozd, S. Z. Ivin and V. V. Sheluchenko, *Zhurn. Vses. Khim. Obsch. Mendel.*, 1967, **12**, 474.
38. G. I. Drozd, S. Z. Ivin, V. V. Sheluchenko and B. I. Tetel'Baum, *Zh. Obshch. Khim.*, 1967, **37**, 958; *J. Gen. Chem. U.S.S.R.*, 1967, **37**, 906.
39. R. Burgada, D. Houalla and R. Wolf, *Compt. Rend. Ser. C*, 1967, **264**, 356.
40. S. L. Manatt, D. D. Elleman, A. H. Cowley and A. B. Burg, *J. Amer. Chem. Soc.*, 1967, **89**, 4544.
41. J. E. Drake and J. Simpson, *Inorg. Nucl. Chem. Lett.*, 1967, **3**, 87.
42. E. Fluck and H. Binder, *Z. Naturforsch., B*, 1967, **22**, 805.
43. E. A. V. Ebsworth and G. M. Sheldrick, *Trans. Farad. Soc.*, 1967, **63**, 1071.
44. Yu. L. Kleiman, N. V. Morkovin and B. I. Ionin, *Zh. Obshch. Khim.*, 1967, **37**, 2791; *J. Gen. Chem. U.S.S.R.*, 1967, **37**, 2661.
45. K. Moedritzer, *Inorg. Chem.*, 1967, **6**, 936.
46. G. M. Sheldrick, *Trans. Farad. Soc.*, 1967, **63**, 1065.
47. G. M. Sheldrick, *Trans. Farad. Soc.*, 1967, **63**, 1077.
48. J. F. Nixon, *J. Chem. Soc. A*, 1967, 1136.
49. F. A. Johnson and R. W. Rudolph, *J. Chem. Phys.*, 1967, **47**, 5449.
50. A. Müller, E. Niecke and O. Glemser, *Z. anorg. allg. Chem.*, 1967, **350**, 256.
51. B. I. Tetel'Baum, V. V. Sheluchenko, S. S. Dubov, G. I. Drozd and S. Z. Ivin, *Zh. Vses. Khim. Obshchchest.*, 1967, **12**, 351.
52. G. I. Drozd, S. Z. Ivin, V. V. Sheluchenko, B. I. Tetel'Baum, G. M. Luganskii and A. D. Varshavskii, *Zh. Obshch. Khim.*, 1967, **37**, 1343; *J. Gen. Chem. U.S.S.R.*, 1967, **37**, 1269.
53. H. W. Roesky, *Chem. Ber.*, 1967, **100**, 2142.
54. F. Seel, K. Rudolph and W. Gombler, *Angew. Chem.*, 1967, **79**, 686; *Angew. Chem. Int. Ed.*, 1967, **6**, 708.
55. A. V. Cunliffe, E. G. Finer, R. K. Harris and W. McFarlane, *Mol. Phys.*, 1967, **12**, 497.

1967 cont.
56. A. Müller, E. Niecke and O. Glemser, *Z. anorg. allg. Chem.*, 1967, **350**, 256.
57. R. R. Dean and W. McFarlane, *Chem. Comm.*, 1967, 840.
58. H. W. Roesky, *Chem. Ber.*, 1967, **100**, 1447.
59. G. C. Demitras and A. G. MacDiarmid, *Inorg. Chem.*, 1967, **6**, 1903.
60. A. J. Downes and R. Schmutzler, *Spectrochim. Acta*, 1967, **23A**, 681.
61. R. A. Mitsch, *J. Amer. Chem. Soc.*, 1967, **89**, 6297.
62. J. E. Griffiths, *Inorg. Chim. Acta*, 1967, **1**, 127.
63. O. Glemser, E. Niecke and J. Stenzel, *Angew. Chem.*, 1967, **79**, 723; *Angew. Chem. Int. Ed.* 1967, **6**, 709.
64. M. A. Landau, V. V. Sheluchenko, G. I. Drozd, S. S. Dubov and S. Z. Ivin, *Zh. Struktur. Khim.*, 1967, **8**, 1097; *J. Structur. Chem.*, 1967, **8**, 971.
65. S. R. Jain, W. S. Brey and H. H. Sisler, *Inorg. Chem.* 1967, **6**, 515.
66. W. McFarlane, *Chem. Comm.*, 1967, 58.
67. S. R. Jain, L. K. Krannich, R. E. Highsmith and H. H. Sisler, *Inorg. Chem.*, 1967, **6**, 1058.
68. H. Hoffmann and P. Schellenbeck, *Chem. Ber.*, 1967, **100**, 692.
69. J. S. Cohen, *J. Amer. Chem. Soc.*, 1967, **79**, 2543.
70. G. Märkl and R. Potthast, *Angew. Chem.*, 1967, **79**, 58; *Angew. Chem. Int. Ed.*, 1967, **6**, 86.
71. D. Gagnaire, J. B. Robert and J. Verrier, *Chem. Comm.*, 1967, 819.
72. H. Christol, M. Levy and C. Marty, *Compt. Rend. Ser. C*, 1967, **265**, 1511.
73. H. Schmidbaur and R. Jonas, *Chem. Ber.*, 1967, **100**, 1120.
74. Y. Ito, M. Okano and R. Oda, *Tetrahedron*, 1967, **23**, 2137.
75. H. E. Shook and L. D. Quin, *J. Amer. Chem. Soc.*, 1967, **89**, 1841.
76. J. Wulff and R. Huisgen, *Angew. Chem.*, 1967, **79**, 472; *Angew. Chem. Int. Ed.*, 1967, **6**, 457.
77. D. R. Rowsell, *J. Mol. Spectry.*, 1967, **23**, 32.
78. L. D. Quin and J. G. Bryson, *J. Amer. Chem. Soc.*, 1967, **89**, 5984.
79. S. O. Grim, D. A. Wheatland and W. McFarlane, *J. Amer. Chem. Soc.*, 1967, **89**, 5573.
80. M. P. Simonnin, J. J. Basselier and C. Charrier, *Bull. Soc. Chim. France*, 1967, 3544.
81. R. H. Churi, *Thesis, Univ. Pittsburgh*, 1966; *Univ. Microfilms*, Order No. 67-9710; *Diss. Abstr. B*, 1967, **28**, 574.
82. S. E. Cremer and R. J. Chorvat, *J. Org. Chem.*, 1967, **32**, 4066.
83. A. H. Cowley, *J. Amer. Chem. Soc.*, 1967, **89**, 5990.
84. G. Miyazima, K. Takahashi and T. Yamasaki, *Bull. Chem. Soc. Japan*, 1967, **40**, 1540.
85. E. Duval and G. J. Bene, *Helv. Phys. Acta*, 1967, **40**, 501.
86. D. Swank, C. N. Caughlan, F. Ramirez, O. P. Madan and C. P. Smith, *J. Amer. Chem. Soc.*, 1967, **89**, 6503.
87. R. S. Davidson, *J. Chem. Soc. C*, 1967, 2131.
88. D. Gagnaire and J. B. Robert, *Bull. Soc. Chim. France*, 1967, 2240.
89. J. E. Anderson and J. M. Lehn, *J. Amer. Chem. Soc.*, 1967, **89**, 1649.
90. M. Tsuboi, F. Kuriyagawa, K. Matsuo and Y. Kyogoku, *Bull. Chem. Soc. Japan*, 1967, **40**, 1813.
91. K. D. Bartle, R. S. Edmundson and D. W. Jones, *Tetrahedron*, 1967, **23**, 1701 and 3226.
92. A. Schmidpeter and H. Groeger, *Chem. Ber.*, 1967, **100**, 3979.
93. S. Kongpricha and W. C. Preusse, *Inorg. Chem.*, 1967, **6**, 1915.

1967 cont.
94. J. M. Kanamueller and H. H. Sisler, *Inorg. Chem.*, 1967, **6**, 1765.
95. A. P. Lane, D. A. Morton-Blake and D. S. Payne, *J. Chem. Soc. A*, 1967, 1492.
96. J. F. Nixon, *Chem. Comm.*, 1967, 699.
97. A. Schmidpeter and J. Ebeling, *Angew. Chem.*, 1967, **79**, 100; *Angew. Chem. Int. Ed.*, 1967, **6**, 87.
98. S. Cradock, E. A. V. Ebsworth, G. Davidson and L. A. Woodward, *J. Chem. Soc. A*, 1967, 1229.
99. J. E. Lancaster, *Spectrochim. Acta A*, 1967, **23**, 1449.
100. C. Benezra, S. Nśeič and G. Ourisson, *Bull. Soc. Chim. France*, 1967, 1140.
101. G. Mavel and R. Mańkowski-Favelier, *J. Chim. Phys.*, 1967, **64**, 1808.
102. G. H. Birum and C. N. Mathews, *J. Org. Chem.*, 1967, **321**, 3554.
103. W. R. Cullen and D. S. Dawson, *Canad. J. Chem.*, 1967, **45**, 2887.
104. C. Charrier and M. P. Simonnin, *Compt. Rend. Ser. C*, 1967, **265**, 1347.
105. V. B. Lebedev and B. I. Ionin, *Izv. Vyssh. Ucheb. Zaved., Priborostr.,* (*Leningrad*), 1967, **10**, 29; *Trad.* AD- 679, 952.
106. C. Charrier and M. P. Simonnin, *Compt. Rend. Ser. C*, 1967, **264**, 995.
107. D. J. Martin, M. Gordon and C. E. Griffin, *Tetrahedron*, 1967, **23**, 1831.
108. G. Mavel, R. Mańkowski-Favelier, G. Lavielle and G. Sturtz, *J. Chim. Phys.*, 1967, **64**, 1698.
109. B. A. Arbuzov, O. A. Vizel', Yu. Yu. Samitov and Ya. F. Tarenko, *Izv. Akad. Nauk. S.S.S.R., Ser. Khim.*, 1967, 672; *Bull. Acad. Sc. U.S.S.R., Div. Chem. Sc.*, 1967, 648.
110. Zh. L. Evtikov, N. A. Razumova and A. A. Petrov, *Dokl. Akad. Nauk S.S.S.R.*, 1967, **177**, 108; *Dokl. Chem.*, 1967, **177**, 972.
111. G. Märkl, F. Lieb and A. Merz, *Angew. Chem.*, 1967, **79**, 947; *Angew. Chem. Int. Ed.*, 1967, **6**, 944.
112. G. Märkl, F. Lieb and A. Merz, *Angew. Chem.*, 1967, **79**, 59; *Angew. Chem. Int. Ed.*, 1967, **6**, 87.
113. G. Mavel, R. Mańkowski-Favelier and T. N. Tanh, *J. Chim. Phys.*, 1967, **64**, 1691.
114. L. Maier and J. J. Daly, *Helv. Chim. Acta*, 1967, **50**, 1747.
115. C. E. Griffin, J. J. Burke, F. E. Dickson, M. Gordon, H. H. Hsieh, R. Obrycki and M. P. Williamson, *J. Phys. Chem.*, 1967, **71**, 4558.
116. F. Kaplan and C. O. Schulz, *Chem. Comm.*, 1967, 376.
117. C. E. Griffin and M. Gordon, *J. Amer. Chem. Soc.*, 1967, **89**, 4427.
118. L. B. Senyavina, V. I. Sheichenko, Yu. N. Sheinker, A. V. Dombrovskii, M. I. Shevchuk, L. I. Barsukov and L. D. Bergel'son, *Zh. Obshch. Khim.*, 1967, **37**, 499; *J. Gen. Chem. U.S.S.R.*, 1967, **37**, 469.
119. J. Elguero and R. Wolf, *Compt. Rend. Ser. C*, 1967, **265**, 1507.
120. R. C. De Selms and Tan-Wan Lin, *J. Org. Chim.*, 1967, **32**, 2023.
121. I. J. Borowitz, M. Anschel and S. Firstenberg, *J. Org. Chem.*, 1967, **32**, 1723.
122. H. J. Bestmann, G. Graf and H. Hartung, *J. Liebigs Ann. Chem.*, 1967, **706**, 68.
123. G. H. Birum and C. N. Mathews, *Chem. Comm.*, 1967, 137.
124. E. G. Finer and R. K. Harris, *Mol. Phys.*, 1967, **12**, 457.
125. E. G. Finer and R. K. Harris, *Mol. Phys.*, 1967, **13**, 65.
126. E. Fluck and H. Binder, *Z. anorg. allg. Chem.*, 1967, **354**, 113.
127. R. K. Harris, A. R. Katritzky, S. Musierowicz and B. Ternai, *J. Chem. Soc. A*, 1967, 37.

1967 cont.
128. B. I. Ionin, V. B. Lebedev and A. A. Petrov, *Zh. Obshch. Khim.*, 1967, **37**, 1174; *J. Gen. Chem. U.S.S.R.*, 1967, **37**, 1117.
129. R. L. Carroll and R. E. Mesmer, *Inorg. Chem.*, 1967, **6**, 1137.
130. E. G. Finer, *J. Mol. Spectry.*, 1967, **23**, 104.
131. D. D. Elleman and S. L. Manatt, Communication given at 8th E.N.C. Pittsburgh, 1967.
132. W. McFarlane, *J. Chem. Soc. A*, 1967, 1148.
133. J. W. Gilje, K. W. Morse and R. W. Parry, *Inorg. Chem.*, 1967, **6**, 1761.
134. K. W. Morse and R. W. Parry, *J. Amer. Chem. Soc.*, 1967, **89**, 172.
135. K. Issleib and R. Kümmel, *Chem. Ber.*, 1967, **100**, 3331.
136. O. Glemser, H. W. Roesky and P. R. Heinze, *Angew. Chem.*, 1967, **79**, 723; *Angew. Chem. Int. Ed.*, 1967, **6**, 710.
137. K. Issleib and H. Oehme, *Tetrahedron Lett.*, 1967, 1489.
138. M. Grayson and C. E. Farley, *Chem. Comm.*, 1967, 830.
139. R. Huisgen and J. Wulff, *Tetrahedron Lett.*, 1967, 917.
140. W. G. Bentrude and K. R. Darnall, *Tetrahedron Letters*, 1967, 2511.
141. I. J. Borowitz and M. Anschel, *Tetrahedron Lett.*, 1967, 1517.
142. E. Fluck and H. Binder, *Angew. Chem.*, 1967, **79**, 243; *Angew. Chem. Int. Ed.*, 1967, **6**, 260.
143. E. T. Kaiser and K. Kudo, *J. Amer. Chem. Soc.*, 1967, **89**, 6725.
144. C. W. Allen, I. C. Paul and T. Moeller, *J. Amer. Chem. Soc.*, 1967, **89**, 6361.
145. G. R. Feistel and T. Moeller, *J. Inorg. Nucl. Chem.*, 1967, **29**, 2731.
146. C. T. Ford, F. E. Dickson and I. I. Bezman, *Inorg. Chem.*, 1967, **6**, 1594.
147. W. Lehr, *Z. anorg. allg. Chem.*, 1967, **350**, 18.
148. W. Lehr, *Z. anorg. allg. Chem.*, 1967, **352**, 27.
149. R. Appel and D. Hänssgen, *Angew. Chem.*, 1967, **79**, 577; *Angew. Chem. Int. Ed.*, 1967, **6**, 560.
150. A. Schmidpeter, J. Ebeling and N. Schindler, *Angew. Chem.*, 1967, **79**, 1016; *Angew. Chem. Int. Ed.*, 1967, **6**, 996.
151. R. A. Lewis, O. Korpium and K. Mislow, *J. Amer. Chem. Soc.*, 1967, **89**, 4786.
152. K. D. Berlin and R. U. Pagilagan, *J. Org. Chem.*, 1967, **32**, 129.
153. J. P. Battioni, W. Chodkiewicz and P. Cadiot, *Compt. Rend. Ser. C*, 1967, **264**, 991.
154. D. Dron, M. L. Capmau and W. Chodkiewicz, *Compt. Rend. Ser. C*, 1967, **264**, 1883.
155. D. Dron, M. L. Capmau and W. Chodkiewicz, *Compt. Rend. Ser. C*, 1967, **265**, 673.
156. A. Sevin and W. Chodkiewicz, *Tetrahedron Letters*, 1967, 2975.
157. S. B. Pierce, *Thesis, Univ. Wisconsin*; *Univ. Microfilms*, Order No. 67–6819; *Diss. Abstr. B*, 1967, **28**, 868.
158. G. M. Whitesides and W. M. Bunting, *J. Amer. Chem. Soc.*, 1967, **89**, 6801.
159. E. J. Wells, H. P. K. Lee and L. K. Peterson, *Chem. Comm.*, 1967, 894.
160. D. G. Gorenstein and F. H. Westheimer, *J. Amer. Chem. Soc.*, 1967, **89**, 2762.
161. A. V. Bogat'skii, A. A. Kolesnik, Yu. Yu. Samitov and T. D. Butova, *Zh. Obshch. Khim.*, 1967, **37**, 1105; *J. Gen. Chem. U.S.S.R.*, 1967, **37**, 1048.
162. A. V. Bogat'skii, Yu. Yu. Samitov and A. A. Kolesnik, *Zh. Obshch. Khim.*, 1967, **37**, 960; *J. Gen. Chem. U.S.S.R.*, 1967, **37**, 909.
163. F. Ramirez, S. B. Bathia and C. P. Smith, *Tetrahedron*, 1967, **23**, 2067.

1967 cont.
164. R. Burgada, M. Bon and F. Mathis, *Compt. Rend. Ser. C*, 1967, **265**, 1499.
165. M. Sanchez, J. F. Brazier, D. Houalla and R. Wolf, *Bull. Soc. Chim. France*, 1967, 3930.
166. D. M. Hellwege, *Thesis, Oklah. State Univ.*, 1966; *Univ. Microfilms*, Order No. 67–7230; *Diss. Abstr.*, 1967, **27**, 4307.
167. C. Benezra and G. Ourisson, *Bull. Soc. Chim. France*, 1967, 624.
168. J. G. Riess, *Bull. Soc. Chim. France*, 1967, 3161.
169. E. W. Randall and D. Shaw, *Spectrochim. Acta*, 1967, **23A**, 1235.
170. Yu. I. Kol'tsov, V. V. Yastrebov and S. S. Korovin, *Zh. Neorg. Khim.*, 1967, **12**, 725; *Russian J. Inorg. Chem.*, 1967, **12**, 378.
171. N. C. Li, *Amer. Chem. Soc., Div. Fuel Chem., Preprints*, 1967, **11**, 241.
172. R. Wolf, D. Houalla and F. Mathis, *Spectrochim. Acta A*, 1967, **23A**, 1641.
173. J. E. Drake and J. Simpson, *Chem. Comm.*, 1967, 249.
174. J. G. Riess and J. R. Van Wazer, *J. Amer. Chem. Soc.*, 1967, **89**, 851.
175. F. Schindler and H. Schmidbaur, *Chem. Ber.*, 1967, **100**, 3655.
176. H. Schmidbaur, W. Wolfsberger and H. Kröner, *Chem. Ber.*, 1967, **100**, 1023.
177. F. Bonati, S. Cenini and R. Ugo, *J. Organometal. Chem.*, 1967, **9**, 395.
178. P. A. Chopard, *Helv. Chim. Acta*, 1967, **50**, 1021.
179. C. Dorémieux-Morin and M. J. M. Verdier, *Bull. Soc. Chim. France*, 1967, 1628.
180. J. E. Drake and J. Simpson, *Inorg. Chem.*, 1967, **6**, 1984.
181. D. S. Frank and D. A. Usher, *J. Amer. Chem. Soc.*, 1967, **87**, 6360.
182. G. Fritz and G. Becker, *Angew. Chem.*, 1967, **79**, 1068; *Angew. Chem. Int. Ed.*, 1967, **6**, 1078.
183. V. W. Gash, *J. Org. Chem.*, 1967, **32**, 2007.
184. D. G. Gorenstein and F. H. Westheimer, *Proc. Natl. Acad. Sc.*, 1967, **58**, 1747.
185. R. Greenhalgh and M. A. Weinberger, *Canad. J. Chem.*, 1967, **45**, 495.
186. D. Grant, J. R. Van Wazer and C. H. Duncan, *J. Polym. Sci.* A-1, 1967, **5**, 57.
187. C. U. Pittman, *J. Polym. Sci.* A-1, 1967, **5**, 2927.
188. G. Schramm and H. Berger, *Z. Naturforsch. B*, 1967, **22**, 587.
189. P. Bivel, F. Hossenlopp and J. P. Ebel, *Bull. Soc. Chim. France*, 1967, 1224.
190. R. L. Carroll and R. P. Carter, *Inorg. Chem.*, 1967, **6**, 401.
191. G. A. Wiley and W. R. Stine, *Tetrahedron Letters*, 1967, 2321.
192. H. J. Bestmann and J. P. Snyder, *J. Amer. Chem. Soc.*, 1967, **89**, 3936.
193. S. Brownstein, *Canad. J. Chem.*, 1967, **45**, 1711.
194. L. C. Hoskins and R. C. Lord, *J. Chem. Phys.*, 1967, **46**, 2402.
195. J. F. Soest, *Thesis, Univ. of Washington*; *Univ. Microfilms*, Order No. 68–9331; *Diss. Abstr. B*, 1968, **29**, 337.
196. E. A. Uehling, and J. F. Soest, *Bull. Amer. Phys. Soc.*, 1967, **12**, 290.
197. T. C. Farrar and J. J. Rush, *Nat. Bur. Stand., Spec. Publ.*, 1967, No. 301, 245.
198. P. W. Atkins, R. A. Dwek, J. B. Reid and R. E. Richards, *Mol. Phys.*, 1967, **13**, 175.
199. R. E. Bailey and J. F. Duncan, *Inorg. Chem.*, 1967, **6**, 1444.
200. K. R. K. Easwaran, V. U. S. Rao, R. Vijayaraghavan and U. R. K. Rao, *Phys. Lett.*, 1967, **A25**, 683.
201. F. Friedman, J. Grunzweig and M. Kuznietz, *Phys. Lett.*, 1967, **A25**, 690.
202. E. D. Jones, *Phys. Rev.*, 1967, **158**, 295.

1967 cont.
203. E. D. Jones, *Physics Lett. A*, 1967, **25**, 111.
204. H. Kessemeier and R. E. Norberg, *Phys. Rev.*, 1967, **155**, 321.
205. B. A. Scott, K. A. Gingerich and R. A. Bernheim, *Phys. Rev.*, 1967, **159**, 387.
206. M. Bhattacharya, A. Chowdhury and M. Bose, *Proc. Int. Conf. Spectrosc*, 1st, *Bombay*, 1967, **2**, 484.
207. M. Bhattacharya, A. Chowdhury and M. Bose, *Proc. Nucl. Phys. Solid State Phys. Symp.*, 11th, *Kampur, India*, 1967, *Solid State Phys.*, 59.
208. J. L. Bjorkstam, *Phys. Rev.*, 1967, **153**, 599 (*Bull. Amer. Phys. Soc. Ser.* II, 1965, **10**, 709).
209. M. Kunitomo, T. Terao, Y. Tsutsumi and T. Hashi, *J. Phys. Soc. Japan*, 1967, **22**, 945.
210. J. F. Soest and E. A. Uehling, *Bull. Amer. Phys. Soc.*, 1967, **12**, 290.
211. M. Van Den Akker and F. Jellinek, *J. Organometal. Chem.*, 1967, **10**, P37.
212. A. Kawamori, *J. Chem. Phys.*, 1967, **47**, 3091.
213. W. Wieker and A. R. Grimmer, *Z. Naturforsch. B*, 1967, **22**, 257.
214. W. Wieker and A. R. Grimmer, *Z. Naturforsch. B*, 1967, **22**, 983.
215. W. Wieker and A. R. Grimmer, *Z. Naturforsch. B*, 1967, **22**, 1220.
216. W. Wieker, A. R. Grimmer and L. Kolditz, *Z. Chem.*, 1967, **7**, 434.
217. K. J. Coskran, *Thesis, Iowa State Univ.*; *Univ. Microfilms*, Order No. 67–12, 948; *Diss. Abstr. B*, 1967, **28**, 1403.
218. K. J. Coskran, R. D. Bertrand and J. G. Verkade, *J. Amer. Chem. Soc.*, 1967, **89**, 4535.
219. R. Hüttel, U. Raffay and H. Reinheimer, *Angew. Chem.*, 1967, **79**, 859; *Angew. Chem. Int. Ed.*, 1967, **6**, 862.
220. E. O. Greaves, R. Bruce and P. M. Maitlis, *Chem. Comm.*, 1967, 860.
221. H. Groeger and A. Schmidpeter, *Chem. Ber.*, 1967, **100**, 3216.
222. W. De W. Horrocks, G. R. Van Hecke and D. De W. Hall, *Inorg. Chem.*, 1967, **6**, 694.
223. A. Misono, Y. Uchida, T. Saito and K. M. Song, *Chem. Comm.*, 1967, 419.
224. L. H. Pignolet and W. De W. Horrocks, *J. Amer. Chem. Soc.*, 1967, **90**, 922.
225. T. Kruck and W. Lang, *Angew. Chem.*, 1967, **79**, 474; *Angew. Chem. Int. Ed.*, 1967, **6**, 454.
226. M. A. Bennett, R. S. Nyholm and J. D. Saxby, *J. Organometal. Chem.*, 1967, **10**, 301.
227. J. G. Smith and D. T. Thompson, *J. Chem. Soc. A*, 1967, 1694.
228. S. O. Grim, Rpt. AD 647, 265-USCFSTI (1967).
229. G. M. Whitesides and J. S. Fleming, *J. Amer. Chem. Soc.*, 1967, **89**, 2855.
230. B. L. Shaw and E. Singleton, *J. Chem. Soc. A*, 1967, 1683.
231. B. L. Shaw and A. C. Smithies, *J. Chem. Soc. A*, 1967, 1047.
232. R. Ugo and F. Bonati, *J. Organometall. Chem.*, 1967, **8**, 189.
233. A. R. Manning, *J. Chem. Soc. A*, 1967, 1984.
234. A. Pidcock, J. D. Smith and B. W. Taylor, *J. Chem. Soc. A*, 1967, 872.
235. G. S. Reddy and R. Schmutzler, *Inorg. Chem.*, 1967, **6**, 823.
236. P. Cassoux, L. Fournes and J. P. Laurent, *Bull. Soc. Chim. France*, 1967, 4028.
237. F. Guerrieri and G. P. Chiusoli, *Chem. Comm.*, 1967, 781.
238. R. Mathieu and R. Poilblanc, *Compt. Rend. Ser. C*, 1967, **265**, 388.
239. P. A. McArdle and A. R. Manning, *Chem. Comm.*, 1967, 417.
240. J. G. Riess and J. R. Van Wazer, *J. Organometall. Chem.*, 1967, **8**, 347.

1967 cont.
241. H. Schumann and O. Stelzer, *Angew. Chem.*, 1967, **79**, 692; *Angew. Chem. Int. Ed.*, 1967, **6**, 701.
242. M. Van Den Akker and F. Jellinek, *Rec. Trav. Chim.*, 1967, **86**, 897.
243. P. Cassoux, J. F. Labarre and J. P. Laurent, *J. Chim. Phys.*, 1967, **64**, 813.
244. E. Fluck and J. Lorenz, *Z. Naturforsch. B*, 1967, **22**, 1095.
245. T. Kruck and M. Höfler, *Angew. Chem.*, 1967, **79**, 582; *Angew. Chem. Int. Ed.*, 1967, **6**, 563.
246. J. Powell and B. L. Shaw, *J. Chem. Soc. A*, 1967, 1839.
247. W. Beck, K. H. Stetter, S. Tadros and K. E. Schwarhans, *Chem. Ber.*, 1967, **100**, 3944.
248. R. D. W. Kemmitt, D. I. Nichols and R. D. Peacock, *Chem. Comm.*, 1967, 599.
249. W. McFarlane, *Chem. Comm.*, 1967, 772.
250. W. McFarlane, *J. Chem. Soc. A*, 1967, 1922.
251. D. Morelli, A. Segre, R. Ugo, G. La Monica, S. Cenini, F. Gonti and F. Bonati, *Chem. Comm.*, 1967, 524.
252. C. J. Nyman, C. E. Wymore and G. Wilkinson, *Chem. Comm.*, 1967, 407.
253. G. F. Svatos, *Advan. Chem. Ser.*, 1967, **62**, 388.
254. K. Vrieze and H. C. Volger, *J. Organometal. Chem.*, 1967, **9**, 537.
255. S. O. Grim, R. L. Keiter and W. McFarlane, *Inorg. Chem.*, 1967, **6**, 1133.
256. M. C. Baird, J. T. Mague, J. A. Osborn and G. Wilkinson, *J. Chem. Soc. A*, 1967, 1347.
257. P. R. Brookes and B. L. Shaw, *J. Chem. Soc.*, 1967, 1079.
258. W. Keim, *J. Organometal. Chem.*, 1965, **8**, P25.
259. C. A. Reilly and H. Thyret, *J. Amer. Chem. Soc.*, 1967, **89**, 5144.
260. P. S. Hallman, D. Evans, J. A. Osborn and G. Wilkinson, *Chem. Comm.*, 1967, 305.
261. V. M. Vdovenko, A. I. Skoblo, D. N. Suglobov, L. L. Shcherbakova and V. A. Shcherbakov, *Zh. Neorg. Khim.*, 1967, **12**, 2863; *Russian J. Inorg. Chem.*, 1967, **12**, 1513.
262. S. J. Lippard, *Trans. N.Y. Acad. Sci.*, 1967, **29**, 917.
263. G. G. Hammers and D. L. Miller, *J. Chem. Phys.*, 1967, **46**, 1533.
264. H. Sigel, K. Becker and D. B. McCormick, *Biochim. Biophys. Acta*, 1967, **148**, 655.
265. L. J. Durham, A. Larsson and P. Reichard, *Europ. J. Biochem.*, 1967, **1**, 92.
266. R. L. Ward and J. A. Happe, *Bioch. Biophys. Res. Comm.*, 1967, **28**, 785.
267. W. J. Wechler, *J. Med. Chem.*, 1967, **10**, 762.
268. D. Chapman, R. M. Williams and B. D. Ladbrooke, *Chem. Phys. Lipids*, 1967, **1**, 445.
269. P. O. Hagen and H. Goldfine, *J. Biol. Chem.*, 1967, **242**, 5700.
270. C. Benezra and G. Ourisson, *Bull. Soc. Chim. France*, 1967, 632.
271. N. Kurihara, H. Shibata, H. Saeki and M. Nakajima, *J. Liebigs. Ann. Chem.*, 1967, **701**, 225.
272. S. Okuda, N. Suzuki and S. Suzuki, *J. Biol. Chem.*, 1967, **242**, 958.
273. J. Ullrich and A. Mannschreck, *Europ. J. Biochem.*, 1967, **1**, 110.
274. C. Iwata and D. E. Metzler, *J. Heterocyclic Chem.*, 1967, **4**, 319.
275. J. S. Kittredge, A. F. Isbell and R. H. Hughes, *Biochemistry*, 1967, **6**, 289.
276. A. Schmidpeter and H. Brecht, *Angew. Chem.*, 1967, **79**, 535; *Angew. Chem. Int. Ed.*, 1967, **6**, 564.
277. E. Fluck and H. Binder, *Z. anorg. allg. Chemie*, 1967, **354**, 139.

1967 cont.
278. M. Lenzi, G. Sturtz and G. Lavielle, *Compt. Rend. Ser. C*, 1967, **264**, 1425.
279. J. K. Ruff, *Inorg. Chem.*, 1967, **6**, 2108.
280. H. Falius, *Chem. Ber.*, 1967, **100**, 1179.
281. H. W. Roesky, *Chem. Ber.*, 1967, **100**, 950.
282. H. W. Roesky, *Chem. Ber.*, 1967, **100**, 2147.
283. M. Lustig, *Angew. Chem.*, 1967, **79**, 980; *Angew. Chem. Int. Ed.*, 1967, **6**, 959.
284. D. C. Wingleth and A. D. Norman, *Chem. Comm.*, 1967, 1218.
285. M. Becke-Goehring and B. Scharf, *Z. anorg. allg. Chem.*, 1967, **353**, 320.
286. W. Haubold and M. Becke-Goehring, *Z. anorg. allg. Chem.*, 1967, **352**, 113.
287. H. W. Roesky, *Angew. Chem.*, 1967, **79**, 61; *Angew. Chem. Int. Ed.*, 1967, **6**, 90.
288. O. Glemser, U. Biermann and M. Fild, *Chem. Ber.*, 1967, **100**, 1082.
289. H. W. Roesky, *Angew. Chem.*, 1967, **79**, 61; *Angew. Chem. Int. Ed.*, 1967, **6**, 90.
290. G. I. Drozd, S. Z. Ivin and V. V. Sheluchenko, *Zh. Vses. Khim., Obshch. Mendel.*, 1967, **12**, 472.
291. L. Maier, *Helv. Chim. Acta*, 1967, **50**, 1723.
292. J. L. Little, J. T. Moran and L. J. Todd, *J. Amer. Chem. Soc.*, 1967, **89**, 5495.
293. V. Mark and J. R. Van Wazer, *J. Org. Chem.*, 1967, **32**, 1187.
294. M. Lustig and W. E. Hill, *Inorg. Chem.*, 1967, **6**, 1448.
295. J. E. Bissey, H. Goldwhite and D. G. Rowsell, *J. Org. Chem.*, 1967, **32**, 1542.
296. B. I. Ionin, V. M. Ignat'ev and V. B. Lebedev, *Zh. Obshch. Khim.*, 1967, **37**, 1863; *J. Gen. Chem. U.S.S.R.*, 1967, **37**, 1774.
297. L. H. Chance, D. J. Daigle and G. L. Drake, *J. Chem. Eng. Data*, 1967, **12**, 282.
298. D. D. Elleman, S. L. Manatt, A. J. R. Bourn and A. H. Cowley, *J. Amer. Chem. Soc.*, 1967, **89**, 4542.
299. J. Lorbert, H. Krapf and H. Nöth, *Chem. Ber.*, 1967, **100**, 3511.
300. A. T. Blomquist and V. J. Hruby, *J. Amer. Chem. Soc.*, 1967, **89**, 4996.
301. L. Maier, *Helv. Chim. Acta*, 1967, **50**, 1742.
302. H. Teichmann, *J. Liebigs Ann. Chem.*, 1967, **703**, 31.
303. A. A. Petrov, N. A. Razumova and Zh. L. Evtikov, *Zh. Obshch. Khim.*, 1967, **37**, 1410; *J. Gen. Chem. U.S.S.R.*, 1967, **37**, 1340.
304. L. D. Quin and H. E. Shook, *J. Org. Chem.*, 1967, **32**, 1604.
305. O. J. Scherer and J. Wokulat, *Z. Naturforsch. B*, 1967, **221**, 474.
306. H. Schmidbaur and W. Tronich, *Angew. Chem.*, 1967, **79**, 412; *Angew. Chem. Internat. Ed.*, 1967, **6**, 448.
307. W. H. Mueller and A. A. Oswald, *J. Org. Chem.*, 1967, **32**, 1730.
308. H. Teichmann, M. Jatkowski and G. Hilgetag, *Angew. Chem.* 1967, **79**, 379; *Angew. Chem. Int. Ed.*, 1967, **6**, 372.
309. R. G. Cavell, *Canad. J. Chem.*, 1967, **45**, 1307.
310. L. N. Mashlyakovskii. B. I. Ionin, I. S. Akrimenko and A. A. Petrov, *Zh. Obshch. Khim.*, 1967, **37**, 1307; *J. Gen. Chem. U.S.S.R.*, 1967, **37**, 1237.
311. K. I. Novitskii, N. A. Razumova and A. A. Petrov, *Khim. Org. Soedin. Fosfora, Akad. Nauk S.S.S.R., Otd. Obshch. Tekh. Khim.*, 1967, 248–255 (C.A. **69**, 43981n)

1967 cont.
312. A. N. Pudovik, V. K. Khairullin, Yu. Yu. Samitov and R. R. Shagidullin, *Zh. Obshch. Khim.*, 1967, **37**, 865; *J. Gen. Chem. U.S.S.R.*, 1967, **37**, 814.
313. N. A. Razumova, K. I. Novitskii and M. V. Kivi, *Zh. Obshch. Khim.*, 1967, **37**, 1136; *J. Gen. Chem. U.S.S.R.*, 1967, **37**, 1078.
314. H. Ulrich, B. Turcker and A. A. R. Sayigh, *J. Org. Chem.*, 1967, **32**, 1360.
315. R. P. Carter, R. L. Carroll and R. R. Irani, *Inorg. Chem.*, 1967, **6**, 939.
316. H. Schmidbaur and W. Wolfsberger, *Angew. Chem.*, 1967, **79**, 411; *Angew. Chem. Intern. Ed.*, 1967, **6**, 448.
317. H. G. Horn and O. Glemser, *Chem. Ber.*, 1967, **100**, 2258.
318. E. H. M. Ibrahim and R. A. Shaw, *Chem. Comm.*, 1967, 244.
319. V. S. Zavgorodnii, B. I. Ionin and A. A. Petrov, *Zh. Obshch. Khim.*, 1967, **37**, 949; *J. Gen. Chem. U.S.S.R.*, 1967, **37**, 898.
320. E. Gryszkiewicz-Trochimowski, *Bull. Soc. Chim. France*, 1967, 4289.
321. R. Wolf, R. Burgada, M. Sanchez and M. Bon, *Bull. Soc. Chim. France*, 1967, 1483.
322. H. Schmidbaur and W. Wolfsberger, *Chem. Ber.*, 1967, **100**, 1000.
323. M. Lenzi, G. Sturtz and G. Lavielle, *Compt. Rend. Ser. C*, 1967, **264**, 1329.
324. G. Sturtz, *Bull. Soc. Chim. France*, 1967, 1345.
325. V. A. Kukhtin, Yu. Yu. Samitov and K. M. Kirillova, *Izv. Akad. Nauk. Ser. Khim.*, 1967, 356; *Bull. Acad. U.S.S.R., Div. Chem. Sc.*, 1967, 337.
326. G. Lavielle, G. Sturtz and H. Normant, *Bull. Soc. Chim. France*, 1967, 4186.
327. D. J. Martin, M. Gordon and C. E. Griffin, *Tetrahedron*, 1967, **23**, 1831.
328. H. Schmidbaur and W. Tronich, *Chem. Ber.*, 1967, **100**, 1032.
329. D. J. Peterson, *J. Organometal. Chem.*, 1967, **8**, 199.
330. M. Wieber and J. Otto, *Chem. Ber.*, 1967, **100**, 974.
331. W. R. Cullen, D. S. Dawson and P. S. Dhaliwal, *Canad. J. Chem.*, 1967, **45**, 683.
332. S. Z. Ivin, V. K. Promonenkov and B. I. Tetel'Baum, *Zh. Obshch. Khim.*, 1967, **37**, 486; *J. Gen. Chem. U.S.S.R.*, 1967, **37**, 454.
333. T. Nishiwaki, *J. Chem. Soc. C*, 1967, 2680.
334. B. L. Laube, R. D. Bertrand, G. A. Casedy, R. D. Compton and J. G. Verkade, *Inorg. Chem.*, 1967, **6**, 173.
335. F. Ramirez and C. P. Smith, *Chem. Comm.*, 1967, 662.
336. J. I. G. Cadogan, R. K. Mackie and J. A. Maynard, *J. Chem. Soc. C*, 1967, 1356.
337. G. Engelhardt, P. Reich and H. Schumann, *Z. Naturforsch. B*, 1967, **22**, 352.
338. K. D. Berlin, D. H. Burpo, R. U. Pagilagan and D. Bude, *Chem. Comm.*, 1967, 1060.
339. A. J. Fry, *J. Org. Chem.*, 1967, **32**, 2025.
340. V. Abramov, Z. S. Druzhina and T. V. Zykova, *Zh. Obshch. Khim.*, 1967, **37**, 1332; *J. Gen. Chem. U.S.S.R.*, 1967, **37**, 1260.
341. S. Kh. Nuretdinov, V. S. Tsivunin, T. V. Zykova and G. Kh. Kamai, *Zh. Obshch. Khim.*, 1967, **37**, 692; *J. Gen. Chem. U.S.S.R.*, 1967, **37**, 648.
342. J. Satge and C. Couret, *Compt. Rend. Ser. C.*, 1967, **264**, 2169.
343. K. Issleib and B. Walther, *Angew. Chem.*, 1967, **79**, 59; *Angew. Chem. Int. Ed.*, 1967, **6**, 88.
344. A. Schmidpeter, H. Brecht and H. Groeger, *Chem. Ber.*, 1967, **100**, 3063.
345. W. R. Stine, *Thesis, Syracuse Univ.*, 1967, *Univ. Microfilm*, Order No. 67–12, 083; *Diss. Abstr. B*, 1967, **28**, 1382.
346. R. G. Kostyanovskii, V. V. Yakshin and S. L. Zimont, *Izv. Akad. Nauk S.S.S.R., Ser. Khim.*, 1967, 1398; *Bull. Acad. Ser., U.S.S.R., Div. Chem. Sc.*, 1967, 1362.

1967 cont.
347. D. L. Venezky and C. F. Poranski, *J. Org. Chem.*, 1967, **32**, 838.
348. D. W. Allen, I. T. Millar and F. G. Mann, *J. Chem. Soc. C*, 1967, 1869.
349. A. M. Aguiar, K. C. Hansen and J. I. Mague, *J. Org. Chem.*, 1967, **32**, 2383.
350. F. Ramirez, S. B. Bhatia, A. V. Patwardhan and C. P. Smith, *J. Org. Chem.*, 1967, **32**, 2194.
351. F. Ramirez, S. B. Bhatia, A. V. Patwardhan and C. P. Smith, *J. Org. Chem.*, 1967, **32**, 3547.
352. A. Takamizawa, K. Hirai and Y. Hamashima, *Tetrahedron Letters*, 1967, 5081.
353. A. Schmidpeter and H. Groeger, *Chem. Ber.*, 1967, **100**, 3052.
354. A. M. Hamid and S. Trippett, *J. Chem. Soc. C*, 1967, 2625.
355. A. Schmidpeter and J. Ebeling, *Angew. Chem.*, 1967, **79**, 534; *Angew. Chem. Int. Ed.*, 1967, **6**, 565.
356. S. E. Fishwick, J. Flint, W. Hawes and S. Trippett, *Chem. Comm.*, 1967, 1113.
357. T. W. Rave and T. J. Logan, *J. Org. Chem.*, 1967, **32**, 1629.
358. E. M. Richards and J. C. Tebby, *Chem. Comm.*, 1967, 957.
359. M. Schlosser and K. F. Christmann, *J. Liebigs Ann. Chem.*, 1967, **708**, 1.
360. J. Satge, C. Couret and M. Lesbre, *Bull. Soc. Chim. France*, 1967, 744.
361. E. J. Kupchik and V. A. Perciaccante, *J. Organometal. Chem.*, 1967, **10**, 181.
362. D. W. Allen and I. T. Millar, *Chem. Ind. (London)*, 1967, 2178.
363. S. O. Grim, W. McFarlane and T. J. Marks, *Chem. Comm.*, 1967, 1191.
364. A. M. Aguiar, K. C. Hansen and G. S. Reddy, *J. Amer. Chem. Soc.*, 1967, **89**, 3067.
365. M. L. Blanchard, H. Strzelecka, G. J. Martin and M. Simalty, *Bull. Soc. Chim. France*, 1967, 2677.
366. A. R. Hands and A. J. H. Mercer, *J. Chem. Soc. C*, 1967, 1099.
367. J. Gosselck, H. Schenk and H. Ahlbrecht, *Angew. Chem.*, 1967, **79**, 242; *Angew. Chem. Int. Ed.*, 1967, **6**, 249.
368. R. S. Davidson, R. A. Sheldon and S. Trippett, *J. Chem. Soc. C*, 1967, 1547.
369. P. A. Chopard, *Helv. Chim. Acta*, 1967, **50**, 1016.
370. R. F. Hudson and R. J. G. Searle, *Chem. Comm.*, 1967, 1249.
371. J. J. Brophy and M. J. Gallagher, *Austral. J. Chem.*, 1967, **20**, 503.
372. M. A. Shaw, J. C. Tebby, J. Ronayne and D. H. Williams, *J. Chem. Soc. C*, 1967, 944.
373. B. I. Ivanov, A. B. Ageeva and Yu. Yu. Samitov, *Dokl. Akad. Nauk S.S.S.R.*, 1967, **174**, 846; *Dokl. Chem.*, 1967, **174**, 513.
374. M. Rabinovitz, I. Agranat and E. D. Bergmann, *J. Chem. Soc. B*, 1967, 1281.
375. A. Umani-Ronchi, M. Acampora, G. Gaudiano and A. Selva, *Chim. Ind. (Milano)*, 1967, **49**, 388.
376. D. W. Allen and J. C. Tebby, *Tetrahedron*, 1967, **23**, 2795.
377. C. W. Rigby, E. Lord and C. D. Hall, *Chem. Comm.*, 1967, 714.
378. M. A. Shaw, J. C. Tebby, R. S. Ward and D. H. Williams, *J. Chem. Soc. C*, 1967, 2442.
379. H. Schmidbaur and W. Wolfsberger, *Chem. Ber.*, 1967, **100**, 1016.
380. O. T. Quimby, J. B. Prentice and D. A. Nicholson, *J. Org. Chem.*, 1967, **31**, 4111.
381. D. J. Mowthorpe and A. C. Chapman, *Spectrochim. Acta*, 1967, **23A**, 451.

1967 cont.
382. T. Kesavadas and D. S. Payne, *J. Chem. Soc. A*, 1967, 1001.
383. P. Tavs, *Chem. Ber.*, 1967, **100**, 1571.
384. J. Songstad, *Acta Chem. Scand.*, 1967, **21**, 1681.
385. H. Nöth and L. Meinel, *Z. anorg. allg. Chem.*, 1967, **349**, 225.
386. E. Fluck and H. Binder, *Angew. Chem.*, 1967, **79**, 803; *Angew. Chem. Int. Ed.*, 1967, **6**, 883.
387. A. J. Carty and R. K. Harris, *Chem. Comm.*, 1967, 234.
388. R. A. Shaw and E. H. M. Ibrahim, *Angew. Chem.*, 1967, **79**, 575; *Angew. Chem. Int. Ed.*, 1967, **6**, 556.
389. J. J. Brophy and M. J. Gallagher, *Chem. Comm.*, 1967, 344.
390. N. Wiberg and K. H. Schmid, *Angew. Chem.*, 1967, **79**, 938; *Angew. Chem. Int. Ed.*, 1967, **6**, 953.
391. R. P. Carter, M. M. Crutchfield and R. R. Irani, *Inorg. Chem.*, 1967, **6**, 942.
392. M. Fild, I. Hollenberg and O. Glemser, *Naturwiss.*, 1967, **54**, 89.
393. E. E. Nifant'ev and M. P. Korotsev, *Zh. Obshch. Khim.*, 1967, **37**, 1366; *J. Gen. Chem. U.S.S.R.*, 1967, **37**, 1293.

1968
1. W. McFarlane, *Ann. Rev. NMR Spectroscopy*, Vol. 1, p. 135, Academic Press, London, 1968.
2. R. Kosfeld, G. Hägele and W. Kuchen, *Angew. Chem.*, 1968, **80**, 794; *Angew. Chem. Intern. Ed.*, 1968, **7**, 814.
3. W. McFarlane, *Proc. Roy. Soc. A*, 1968, **306**, 185.
4. D. Purdela, *Rev. Roum. Chim.*, 1968, **13**, 1415.
5. B. I. Ionin, *Zhur. Obshch. Khim.*, 1968, **38**, 1659; *J. Gen. Chem. U.S.S.R.*, 1968, **38**, 1618.
6. E. F. Davidoff, *Thesis, Univ. Maryland*, 1967; *Univ. Microfilms* Order No. 68–3351; *Diss. Abstr. B*, 1968, **28**, 3639.
7. S. O. Grim and W. McFarlane, *Canad. J. Chem.*, 1968, **46**, 2071.
8. A. Schmidpeter and H. Brecht, *Z. Naturforsch.*, 1968, **23B**, 1529.
9. M. Fild, *Z. Naturforsch. B*, 1968, **23**, 604.
10. A. Schmidpeter and H. Brecht, *Inorg. Nucl. Chem. Lett.*, 1968, **4**, 563.
11. A. Schmidpeter and K. Stoll, *Angew. Chem.*, 1968, **80**, 558; *Angew. Chem. Int. Ed.*, 1968, **7**, 549.
12. V. V. Sheluchenko, S. S. Dubov, G. I. Drozd and S. Z. Ivin, *Zh. Strukt. Khim.*, 1968, **9**, 909; *J. Struct. Chem. U.S.S.R.*, 1968, **9**, 805.
13. A. W. Johnson and H. L. Jones, *J. Amer. Chem. Soc.*, 1968, **90**, 5232.
14. J. W. Rakshys, R. W. Taft and W. A. Sheppard, *J. Amer. Chem. Soc.*, 1968, **90**, 5236.
15. G. P. Schiemenz, *Angew. Chem.*, 1968, **80**, 559; *Angew. Chem. Intern. Ed.*, 1968, **7**, 544.
16. G. P. Schiemenz, *Angew. Chem.*, 1968, **80**, 559; *Angew. Chem. Intern. Ed.*, 1968, **7**, 545.
17. P. Bucci, *J. Amer. Chem. Soc.*, 1968, **90**, 252.
18. M. I. Kabachnik, T. Ya. Medved', P. V. Petrovskii, E. I. Fedin and K. S. Yudina, *Izv. Akad. Nauk S.S.S.R., Ser. Khim.*, 1968, 419; *Bull. Acad. Sc. U.S.S.R., Div. Chem. Sc.*, 1968, 411.
19. A. Müller, E. Niecke and B. Krebs, *Mol. Phys.*, 1968, **14**, 591.
20. G. Mavel and M. J. Green, *Chem. Comm.*, 1968, 742.

1968 cont.
21. D. Rosenberg, J. W. de Haan and W. Drenth, *Rec. Trav. Chim.*, 1968, **87**, 1387.
22. F. J. Weigert and J. D. Roberts, *Priv. comm.*, Sept. 68.
23. A. B. Burg and H. Heinen, *Inorg. Chem.*, 1968, **7**, 1021.
24. N. N. Greenwood and B. H. Robinson, *J. Chem. Soc. A*, 1968, 226.
25. M. P. Williamson and C. E. Griffin, *J. Phys. Chem.*, 1968, **72**, 4043.
26. R. K. Holmes and R. M. Deiters, *Inorg. Chem.*, 1968, **7**, 2229.
27. E. F. Kiefer, W. Gericke and S. T. Amimoto, *J. Amer. Chem. Soc.*, 1968, **90**, 6246.
28. J. P. Albrand, D. Gagnaire and J. B. Robert, *J. Mol. Spectr.*, 1968, **27**, 428.
29. T. H. Siddall and W. E. Stewart, *Spectroch. Acta*, 1968, **24A**, 81.
30. G. Mavel, *J. Chim. Phys.*, 1968, **65**, 1692.
31. R. W. Rudolph and H. W. Schiller, *J. Amer. Chem. Soc.*, 1968, **90**, 3581.
32. J. E. Drake and C. Riddle, *J. Chem. Soc. A*, 1968, 2709.
33. J. P. Albrand, D. Gagnaire and J. B. Robert, *Bull. Soc. Chim. France*, 1968, **11**, 479.
34. L. F. Centofanti and R. W. Parry, *Inorg. Chem.*, 1968, **7**, 1005.
35. T. L. Emmick and R. L. Letsinger, *J. Amer. Chem. Soc.*, 1968, **90**, 3459.
36. L. Maier, *Helv. Chim. Acta*, 1968, **51**, 1608.
37. W. McFarlane, *J. Chem. Soc. A*, 1968, 1715.
38. R. A. Goodrich, *Thesis, Univ. Wisconsin*; *Univ. Microfilms*, Order No. 67–12, 122; *Diss. Abstr. B*, 1968, **28**, 3204.
39. R. A. Goodrich and P. M. Treichel, *Inorg. Chem.*, 1968, **7**, 694.
40. A. Schmidpeter and J. Ebeling, *Angew. Chem.*, 1968, **80**, 197; *Angew. Chem. Int. Ed.*, **7**, 209.
41. J. F. Nixon and J. R. Swain, *Chem. Comm.*, 1968, 997.
42. D. Solan and P. L. Timms, *Chem. Comm.*, 1968, 1540.
43. K. Cohn and R. W. Parry, *Inorg. Chem.*, 1968, **7**, 46.
44. J. E. Griffiths, *Spectrochim. Acta*, 1968, **24A**, 115.
45. T. L. Charlton and R. G. Cavell, *Inorg. Chem.*, 1968, **7**, 2195.
46. R. G. Cavell, *Canad. J. Chem.*, 1968, **46**, 613.
47. R. C. Dobbie, L. F. Doty and R. G. Cavell, *J. Amer. Chem. Soc.*, 1968, **90**, 2015.
48. S. S. Chan and C. J. Willis, *Canad. J. Chem.*, 1968, **46**, 1237.
49. R. Rogowski and K. Cohn, *Inorg. Chem.*, 1968, **7**, 2193.
50. C. W. Allen and T. Moeller, *Inorg. Chem.*, 1968, **7**, 2177.
51. W. T. Raynes, T. A. Sutherley, H. J. Buttery and C. M. Fenton, *Mol. Phys.*, 1968, **14**, 599.
52. W. J. Vullo, *J. Org. Chem.*, 1968, **33**, 3665.
53. E. J. Boros, R. D. Compton and J. G. Verkade, *Inorg. Chem.*, 1968, **7**, 165.
54. H. H. Sisler and S. R. Jain, *Inorg. Chem.*, 1968, **7**, 104.
55. R. G. Cavell and R. C. Dobbie, *J. Chem. Soc. A*, 1968, 1406.
56. M. J. Gallagher, *Austral. J. Chem.*, 1968, **21**, 1197.
57. R. K. Harris and E. G. Finer, *Bull. Soc. Chim. France*, 1968, 2805.
58. H. Goldwhite and D. G. Rowsell, *J. Phys. Chem.*, 1968, **72**, 2666.
59. H. Goldwhite and D. G. Rowsell, *Chem. Comm.*, 1968, 1665.
60. D. W. Allen, I. T. Millar and J. C. Tebby, *Tetrahedron Letters*, 1968, 745.
61. A. Bond, M. Green and S. C. Pearson, *J. Chem. Soc. B*, 1968, 929.
62. J. P. Albrand, D. Gagnaire and J. B. Robert, *Chem. Comm.*, 1968, 1469.
63. D. T. Longone and E. S. Alexander, *Tetrahedron Letters*, 1968, 5815.

1968 cont.
64. O. J. Scherer and G. Schieder, *Angew. Chem.*, 1968, **80**, 83; *Angew. Chem. Int. Ed.*, 1968, **7**, 75.
65. P. Haake and P. S. Ossip, *Tetrahedron*, 1968, **24**, 565.
66. W. McFarlane, *Chem. Comm.*, 1968, 229.
67. S. E. Cremer and R. J. Chorvat, *Tetrahedron Letters*, 1968, 413.
68. S. E. Fishwick and J. A. Flint, *Chem. Comm.*, 1968, 182.
69. H. G. Ang and H. J. Emeleus, *J. Chem. Soc. A*, 1968, 1334.
70. J. E. Griffiths, *Spectrochim. Acta*, 1968, **24A**, 303.
71. H. Goldwhite and D. G. Rowsell, *J. Mol. Spectr.*, 1968, **27**, 364.
72. K. E. Daugherty, W. A. Eychaner and J. I. Stevens, *Appl. Spectrosc.*, 1968, **22**, 95.
73. F. Ramirez, K. Tasaka, N. B. Desai and C. P. Smith, *J. Amer. Chem. Soc.*, 1968, **90**, 751.
74. R. S. Edmundson and E. W. Mitchell, *J. Chem. Soc. C*, 1968, 2091.
75. K. Dimroth and W. Städe, *Angew. Chem.*, 1968, **80**, 966; *Angew. Chem. Int. Ed.*, 1968, **8**, 881.
76. J. Devillers, F. Mathis and J. Navech, *Compt. Rend. Ser. C*, 1968, **267**, 849.
77. P. Haake, J. P. McNeal and E. J. Goldsmith, *J. Amer. Chem. Soc.*, 1968, **90**, 715.
78. J. H. Hargis and W. G. Bentrude, *Tetrahedron Letters*, 1968, 5365.
79. D. W. White, G. K. McEwen and J. G. Verkade, *Tetrahedron Letters*, 1968, 5369.
80. D. Gagnaire, J. B. Robert and J. Verrier, *Bull. Soc. Chim. France*, 1968, 2392.
81. M. Tsuboi, M. Kainosho and A. Nakamura, in *"Recent developments of magnetic resonance in biological systems"* (S. Fujiwara, L. H. Piette, ed., Hirokawa Publ. Co., Tokyo) 1968, p. 43.
82. A. Nudelman and D. J. Cram, *J. Amer. Chem. Soc.*, 1968, **90**, 3869.
83. L. D. Hall and R. B. Malcolm, *Chem. Ind. (London)*, 1968, 92.
84. F. Ramirez, *Accounts Chem. Res.*, 1968, **1**, 168.
85. A. C. Vandenbroucke, E. J. Boros and J. G. Verkade, *Inorg. Chem.*, 1968, **7**, 1469.
86. A. Schmidpeter and C. Weingand, *Angew. Chem.*, 1968, **80**, 234; *Angew. Chem. Int. Ed.*, 1968, **7**, 210.
87. J. Gelan and M. Anteunis, *Bull. Soc. Chim. Belges.*, 1968, **77**, 447.
88. K. D. Berlin, R. Ranganathan and H. Haberlein, *J. Heter. Chem.*, 1968, **5**, 813.
89. J. F. Nixon, *J. Chem. Soc. A*, 1968, 2689.
90. D. B. Denney, D. Z. Denney and L. A. Wilson, *Tetrahedron Letters*, 1968, 85.
91. R. Keat and R. A. Shaw, *J. Chem. Soc. A*, 1968, 703.
92. D. Imbery and H. Friebolin, *Z. Naturforsch. B*, 1968, **23**, 759.
93. H. Gross and B. Costisella, *Angew. Chem.*, 1968, **80**, 364; *Angew. Chem. Int. Ed.*, 1968, **7**, 391.
94. A. Schmidpeter and J. Ebeling, *Chem. Ber.*, 1968, **101**, 3883.
95. H. Schumann and H. Benda, *Angew. Chem.*, 1968, **80**, 845; *Angew. Chem. Int. Ed.*, 1968, **7**, 812.
96. H. Siebert, J. Eints and E. Fluck, *Z. Naturforschung. B*, 1968, **23**, 1006.
97. M. P. Williamson, S. Castellano and C. E. Griffin, *J. Phys. Chem.*, 1968, **72**, 175.
98. H. Goldwhite, D. G. Rowsell and C. Valdez, *J. Organometal. Chem.*, 1968, **12**, 133.

1968 cont.
99. M. P. Simonnin and C. Charrier, *Compt. Rend. Ser. C*, 1968, **267**, 550.
100. L. D. Quin, J. P. Gratz and T. P. Barket, *J. Org. Chem.*, 1968, **33**, 1034.
101. K. Dimroth and W. Mach, *Angew. Chem.*, 1968, **80**, 489; *Angew. Chem. Int. Ed.*, 1968, **7**, 460.
102. K. Dimroth, K. Vogel, W. Mach and U. Schoeller, *Angew. Chem.*, 1968, **80**, 359; *Angew. Chem. Int. Ed.*, 1968, **7**, 371.
103. R. Keat, *Chem. Ind. (London)*, 1968, 1362.
104. A. G. Moritz, J. D. Saxby and S. Sternhell, *Austral. J. Chem.*, 1968, **21**, 2565.
105. H. Schmidbaur and W. Tronich, *Chem. Ber.*, 1968, **101**, 595.
106. H. Schmidbaur and W. Tronich, *Chem. Ber.*, 1968, **101**, 3356.
107. W. Tronich, *Inaugural-Dissertation, Julius-Maximilians-Universität, Würzburg*.
108. J. C. J. Bart, *Angew. Chem.*, 1968, **80**, 697; *Angew. Chem. Int. Ed.*, 1968, **7**, 730.
109. P. M. Borodin, E. N. Sventitskii and V. I. Chizhik, *Yad. Magn. Rezonans, Leningrad. Gos. Univ.*, 1968, (2) 69.
110. G. H. Birum and C. N. Mathews, *J. Amer. Chem. Soc.*, 1968, **90**, 3842.
111. A. Schmidpeter, H. Brecht and J. Ebeling, *Chem. Ber.*, 1968, **101**, 3902.
112. H. Falius, *Angew. Chem.*, 1968, **80**, 616; *Angew. Chem. Int. Ed.*, 1968, **7**, 622.
113. E. G. Finer and R. K. Harris, *Chem. Comm.*, 1968, 110.
114. H. P. Latscha, *Z. anorg. allg. Chem.*, 1968, **362**, 7.
115. M. Rhodes, D. W. Aksnes and J. H. Strange, *Mol. Phys.*, 1968, **15**, 541.
116. R. L. Keiter and S. O. Grim, *Chem. Comm.*, 1968, 521.
117. D. M. Nimrod, D. R. Fitzwater and J. G. Verkade, *Inorg. Chim. Acta*, 1968, **2**, 149.
118. A. R. Hands and A. J. H. Mercer, *J. Chem. Soc. C*, 1968, 2448.
119. C. G. Barlow, R. Jefferson and J. F. Nixon, *J. Chem. Soc. A*, 1968, 2692.
120. P. DeKoe, R. Van Veen and F. Bickelhaupt, *Angew. Chem.*, 1968, **80**, 486; *Angew. Chem. Int. Ed.*, 1968, **7**, 465.
121. S. E. Cremer, R. J. Chorvat, C. H. Chang and D. W. Davis, *Tetrahedron Lett.*, 1968, 5799.
122. W. Hawes and S. Trippett, *Chem. Comm.*, 1968, 295.
123. F. Ramirez, J. F. Pilot, O. P. Madan and C. P. Smith, *J. Amer. Chem. Soc.*, 1968, **90**, 1275 (correction, ibid., p. 3299).
124. F. Ramirez, *Accounts Chem. Res.*, 1968, **1**, 168.
125. M. Becke-Goehring, H. J. Wald and H. Weber, *Naturwiss.*, 1968, **55**, 491.
126. J. Devillers, M. Willson and R. Burgada, *Bull. Soc. Chim. France*, 1968, 4670.
127. T. Chivers and N. L. Paddock, *Chem. Comm.*, 1968, 704.
128. M. Biddlestone and R. A. Shaw, *Chem. Comm.*, 1968, 407.
129. R. Keat and R. A. Shaw, *Angew. Chem.*, 1968, **80**, 192; *Angew. Chem. Ind. Ed.*, 1968, **7**, 212.
130. C. W. Allen, F. Y. Tsang and T. Moeller, *Inorg. Chem.*, 1968, **7**, 2183.
131. M. K. Feldt, *Thesis, Univ. of Illinois*, 1967; *Univ. Microfilms*, Order No. 68–8067; *Diss. Abstr. B*, 1968, **28**, 4916.
132. M. K. Feldt and T. Moeller, *J. Inorg. Nucl. Chem.*, 1968, **30**, 2351.
133. B. S. Manhas, S. K. Chu and T. Moeller, *J. Inorg. Nucl. Chem.*, 1968, **30**, 322.
134. A. J. Berlin, B. Grushkin and L. R. Moffett, *Inorg. Chem.*, 1968, **7**, 589.
135. J. N. Rapko and G. R. Feistel, *Chem. Comm.*, 1968, 474.

1968 cont.
136. M. Becke-Goehring and H. J. Müller, *Z. anorg. allg. Chem.*, 1968, **362**, 51.
137. A. Schmidpeter and N. Schindler, *Angew. Chem.*, 1968, **80**, 1030; *Angew. Chem. Intern. Ed.*, 1968, **7**, 943.
138. M. J. Gallagher and I. D. Jenkins, *Topics. Stereoch.*, 1968, **3**, 1.
139. F. H. Westheimer, *Accounts Chem. Res.*, 1968, **1**, 70.
140. R. A. Lewis, O. Korpium and K. Mislow, *J. Amer. Chem. Soc.*, 1968, **90**, 4847.
141. J. B. Lambert, G. J. Jackson and D. C. Mueller, *J. Amer. Chem. Soc.*, 1968, **90**, 6401.
142. K. D. Berlin, R. T. Claunch and E. T. Gaudy, *J. Org. Chem.*, 1968, **33**, 3090.
143. M. P. Williamson and C. E. Griffin, *J. Phys. Chem.*, 1968, **72**, 2678.
144. N. N. Greenwood, B. H. Robinson and B. P. Straughan, *J. Chem. Soc. A*, 1968, 230.
145. R. Keat, W. Sim and D. S. Payne, *Chem. Comm.*, 1968, 191.
146. W. Chodkiewicz, M. L. Capmau and Boisard-Gerde, *Compt. Rend. Ser. C*, 1968, **267**, 911.
147. P. C. Lauterbur and F. Ramirez, *J. Amer. Chem. Soc.*, 1968, **90**, 6722.
148. E. A. Cohen and C. D. Cornwell, *Inorg. Chem.*, 1968, **7**, 398.
149. R. S. Edmundson and E. W. Mitchell, *J. Chem. Soc. C*, 1968, 3033.
150. J. P. Majoral, A. Munoz and J. Navech, *Compt. Rend. Ser. C.*, 1968, **266**, 235.
151. D. Hellwinkel, *Chimia (Aarau)*, 1968, **22**, 488.
152. F. Ramirez, *Trans. N.Y. Acad. Sci.*, 1968, **30**, 410.
153. F. Ramirez, A. V. Patwardhan, H. J. Kugler and C. P. Smith, *Tetrahedron*, 1968, **24**, 2275.
154. R. Burgada and H. Germa, *Compt. Rend. Ser. C*, 1968, **267**, 270.
155. A. Francina, A. Lamotte and J. C. Merlin, *Compt. Rend. Ser. C*, 1968, **267**, 763.
156. E. N. Sventitskii and G. P. Savoskina, *Yad. Magn. Rezonans, Leningrad. Gos. Univ.*, 1968, 2183.
157. S. Nishimura, C. H. Ke and N. C. Li, *J. Amer. Chem. Soc.*, 1968, **90**, 234.
158. J. E. Drake and C. Riddle, *J. Chem. Soc. A*, 1968, 1675.
159. J. E. Drake and J. Simpson, *J. Chem. Soc. A*, 1968, 974.
160. J. G. Riess and J. R. Van Wazer, *Bull. Soc. Chim. France*, 1968, 3087.
161. H. Demarne and P. Cadiot, *Bull. Soc. Chim. France*, 1968, 211.
162. H. D. Gillman and T. E. Haas, *Chem. Comm.*, 1968, 777.
163. V. K. Das and W. Kitching, *J. Organometal. Chem.*, 1968, **13**, 523.
164. S. R. Jain and H. H. Sisler, *Inorg. Chem.*, 1968, **7**, 2204.
165. A. Francina, A. Lamotte and J. C. Merlin, *Compt. Rend. Ser. C*, 1968, **266**, 1050.
166. E. F. Moran, *J. Inorg. Nucl. Chem.*, 1968, **30**, 1405.
167. J. A. K. du Plessis and S. Norval, *J. South. Afric. Chem. Inst.*, 1968, **21**, 1.
168. D. B. Denney, D. Z. Denney and B. C. Chang, *J. Amer. Chem. Soc.*, 1968, **90**, 6332.
169. H. J. Bestmann, H. G. Liberda and J. P. Snyder, *J. Amer. Chem. Soc.*, 1968, **90**, 2963.
170. P. Crews, *J. Amer. Chem. Soc.*, 1968, **90**, 2961.
171. D. M. Crouse, A. T. Wehman and E. E. Schweizer, *Chem. Comm.*, 1968, 866.
172. F. E. Hruska and S. S. Danyluk, *J. Amer. Chem. Soc.*, 1968, **90**, 3266.
173. S. Brownstein and J. Bornais, *Canad. J. Chem.*, 1968, **46**, 225.

1968 cont.
174. D. W. Aksnes, M. Rhodes and J. G. Powles, *Mol. Phys.*, 1968, **14**, 333.
175. D. J. Mowthorpe and A. C. Chapman, *Mol. Phys.* 1968, **15**, 429.
176. Yu. I. Mitchenko and V. V. Frolov, *Yad. Magn. Rezonans, Leningrad. Gos. Univ.*, 1968, (2), 184.
177. S. S. Zumdahl and R. S. Drago, *Inorg. Chem.*, 1968, **7**, 2162.
178. R. Blinc and S. Zumer, *Phys. Rev. Letters.*, 1968, **21**, 1004.
179. P. Pyykko, *Chem. Phys. Lett.*, 1968, **2**, 559.
180. T. Tsang, T. C. Farrar and J. J. Rush, *J. Chem. Phys.*, 1968, **49**, 4403.
181. E. D. Jones, *J. Appl. Phys.*, 1968, **39**, 1090.
182. E. D. Jones, *Phys. Lett. A*, 1968, **27**, 204.
183. M. Kuznietz, *J. Chem. Phys.*, 1968, **49**, 3731.
184. M. Kuznietz, G. A. Matzkanin and Y. Baskin, *Phys. Lett. A*, 1968, **28**, 122.
185. B. A. Scott, G. R. Eulenberger and R. A. Bernheim, *J. Chem. Phys.*, 1968, **48**, 263.
186. M. Bhattacharya, A. Chowdhuri and M. Bose, *J. Phys. Soc. Jap.*, 1968, **25**, 1731.
187. R. Blinc, V. Dimic, D. Kolar, G. Lahajnar, J. Stepisnik, S. Zumer, N. Vene and D. Hadzi, *J. Chem. Phys.*, 1968, **49**, 4996.
188. D. J. Genin, D. E. O'Reilly and T. Tsang, *Phys. Rev.*, 1968, **167**, 445.
189. L. C. Gupta and V. U. S. Rao, *Phys. Lett. A*, 1968, **28**, 187.
190. I. N. Penkov, I. A. Safin, Yu. V. Yablokov and A. N. Didenko, *Geokhimya*, 1968, 618.
191. M. Pollak-Stachurowa, W. Borownicki and M. Szustakowski, *Acta Phys. Pol.*, 1968, **34**, 141.
192. R. C. Gupta, *Indian J. Phys.*, 1968, **42**, 389.
193. B. D. Mosel, W. Müller-Warmuth and G. W. Schulz, *Z. Naturforsch. A*, 1968, **23**, 1224.
194. M. A. Bennett, W. R. Kreen and R. S. Nyholm, *Inorg. Chem.*, 1968, **7**, 552.
195. N. Davies, P. H. Bird and M. G. H. Wallbridge, *J. Chem. Soc. A*, 1968, 2269.
196. J. W. Dawson and L. M. Venanzi, *J. Amer. Chem. Soc.*, 1968, **90**, 7229.
197. S. Attali and R. Poilblanc, *Compt. Rend. Ser. C*, 1968, **267**, 718.
198. R. E. Ball and T. F. Endicott, *Chem. Comm.*, 1968, 51.
199. R. H. Fischer and W. De W. Horrocks, *Inorg. Chem.*, 1968, **7**, 2659.
200. H. A. O. Hill, K. G. Morallee, G. Pellizer, G. Mestroni and G. Costa, *J. Organometal. Chem.*, 1968, **11**, 167.
201. R. B. King, *Inorg. Chim. Acta.*, 1968, **2**, 454.
202. R. B. King and K. H. Pannell, *Inorg. Chem.*, 1968, **7**, 273.
203. W. Kruse and R. H. Atalla, *Chem. Comm.*, 1968, 921.
204. L. H. Pignolet, D. Forster and W. De W. Horrocks, *Inorg. Chem.*, 1968, **7**, 828.
205. L. H. Pignolet and W. De W. Horrocks, *Chem. Comm.*, 1968, 1012.
206. H. Weingarten and M. G. Miles, *J. Inorg. Nucl. Chem.*, 1968, **30**, 668.
207. E. O. Fischer, E. Louis, W. Bathelt, E. Moser and J. Müller, *J. Organometal. Chem.*, 1968, **14**, P9.
208. E. Moser and E. O. Fischer, *J. Organometal. Chem.*, 1968, **15**, 157.
209. W. E. Stanclift and D. G. Hendricker, *Inorg. Chem.*, 1968, **7**, 1242.
210. S. O. Grim, D. A. Wheatland and P. R. McAllister, *Inorg. Chem.*, 1968, **7**, 161.
211. R. Mathieu, M. Lenzi and R. Poilblanc, *Compt. Rend. Ser. C*, 1968, **266**, 806.

1968 cont.
212. E. O. Fischer, E. Louis and R. J. J. Schneider, *Angew. Chem.*, 1968, **80**, 122; *Angew. Chem. Int. Ed.*, 1968, **7**, 136.
213. R. J. Haines, A. L. Du Preez and G. T. W. Wittmann, *Chem. Comm.*, 1968, 611.
214. G. Hata, H. Kondo and A. Miyake, *J. Amer. Chem. Soc.*, 1968, **90**, 2278.
215. R. B. King and K. H. Pannell, *Inorg. Chem.*, 1968, **7**, 1510.
216. G. R. Davies, R. H. B. Mais, P. G. Owston and D. T. Thompson, *J. Chem. Soc. A*, 1968, 1251.
217. J. Chatt, G. J. Leigh, D. M. P. Mingos, E. W. Randall and D. Shaw, *Chem. Comm.*, 1968, 419.
218. M. J. Church and M. J. Mays, *Chem. Comm.*, 1968, 435.
219. M. J. Church and M. J. Mays, *J. Chem. Soc. A*, 1968, 3074.
220. J. P. Collman and C. T. Sears, *Inorg. Chem.*, 1968, **7**, 27.
221. J. Powell and B. L. Shaw, *J. Chem. Soc. A*, 1968, 617.
222. B. L. Shaw and A. C. Smithies, *J. Chem. Soc. A*, 1968, 2784.
223. H. C. Volger, K. Vrieze and A. P. Praat, *J. Organometal. Chem.*, 1968, **14**, 429.
224. K. Vrieze and H. C. Volger, *J. Organometal. Chem.*, 1968, **11**, P17.
225. T. M. Chen, *Thesis, Univ. Texas*, 1967; *Univ. Microfilms*, Order No. 68–4265; *Diss. Abstr. B*, 1968, **28**, 4085.
226. C. S. Kraihanzel and P. K. Maples, *Inorg. Chem.*, 1968, **7**, 1806.
227. P. K. Maples and C. S. Kraihanzel, *Chem. Comm.*, 1968, 922.
228. P. K. Maples and C. S. Kraihanzel, *J. Amer. Chem. Soc.*, 1968, **90**, 6645.
229. R. B. King and K. H. Pannell, *Inorg. Chem.*, 1968, **7**, 2356.
230. F. Klamberg and E. L. Muetterties, *J. Amer. Chem. Soc.*, 1968, **90**, 3296.
231. M. J. Mays and S. M. Pearson, *J. Chem. Soc. A*, 1968, 2291.
232. W. S. Tsang, D. W. Meek and A. Wojcicki, *Inorg. Chem.*, 1968, **7**, 1263.
233. P. S. Shetty, P. Jose and Q. Fernando, *Chem. Comm.*, 1968, 788.
234. R. M. Lynden-Bell, *Mol. Phys.*, 1968, **15**, 523.
235. D. Forster, K. Moedritzer and J. R. Van Wazer, *Inorg. Chem.*, 1968, **7**, 1138.
236. F. L'Eplattenier and F. Calderazzo, *Inorg. Chem.*, 1968, **7**, 1290.
237. M. A. Bennett, W. R. Kreen and R. S. Nyholm, *Inorg. Chem.*, 1968, **7**, 556.
238. P. Fitton, M. P. Johnson and J. E. McKeon, *Chem. Comm.*, 1968, 6.
239. P. Fitton and J. E. McKeon, *Chem. Comm.*, 1968, 4.
240. A. Pidcock, *Chem. Comm.*, 1968, 92.
241. R. C. Taylor, G. R. Dubson and R. A. Kolodny, *Inorg. Chem.*, 1968, **7**, 1886.
242. K. Vrieze, P. Cossee, A. P. Praat and C. W. Hilbers, *J. Organometal. Chem.*, 1968, **11**, 353.
243. K. Vrieze, A. P. Praat and P. Cossee, *J. Organometal. Chem.*, 1968, **12**, 533.
244. R. J. Goodfellow, *Chem. Comm.*, 1968, 114.
245. A. J. Rest, *J. Chem. Soc. A*, 1968, 2212.
246. D. M. Roundhill and G. Wilkinson, *J. Chem. Soc. A*, 1968, 506.
247. F. H. Allen and A. Pidcock, *J. Chem. Soc. A*, 1968, 2700.
248. P. W. Atkins, J. C. Green and M. L. H. Green, *J. Chem. Soc. A*, 1968, 2275.
249. J. Chatt and B. T. Heaton, *J. Chem. Soc. A*, 1968, 2745.
250. J. Chatt and A. D. Westland, *J. Chem. Soc. A*, 1968, 88.
251. R. R. Dean and J. C. Green, *J. Chem. Soc. A*, 1968, 3047.
252. F. Glockling and K. A. Hooton, *J. Chem. Soc. A*, 1968, 826.
253. D. A. Harbourne and F. G. A. Stone, *J. Chem. Soc. A*, 1968, 1765.
254. S. Otsuka, A. Nakamura and K. Tani, *J. Organometal. Chem.*, 1968, **14**, P30.

1968 cont.
255. M. Green, R. B. L. Osborn, A. J. Rest and F. G. A. Stone, *J. Chem. Soc. A*, 1968, 2525.
256. B. T. Healon and A. Pidcock, *J. Organometal. Chem.*, 1968, **14**, 235.
257. W. McFarlane, *Chem. Comm.*, 1968, 393.
258. J. F. Biellman, *Bull. Soc. Chim. France*, 1968, 3055.
259. K. C. Dewhirst, W. Keim and C. A. Reilly, *Inorg. Chem.*, 1968, **7**, 546.
260. D. R. Eaton and S. R. Stuart, *J. Amer. Chem. Soc.*, 1968, **90**, 4170.
261. K. C. Ramey, D. C. Line and W. B. Wise, *J. Amer. Chem. Soc.*, 1968, **90**, 4275.
262. K. Vrieze, H. C. Volger and A. P. Praat, *J. Organometal. Chem.*, 1968, **15**, 195.
263. R. D. W. Kemmitt, D. I. Nichols and R. D. Peacock, *J. Chem. Soc. A*, 1968, 1898.
264. M. I. Bruce and F. G. A. Stone, *Angew. Chem.*, 1968, **80**, 460; *Angew. Chem. Int. Ed.*, 1968, **7**, 427.
265. M. S. Lupin and B. L. Shaw, *J. Chem. Soc. A*, 1968, 741.
266. P. Kalck and R. Poilblanc, *Compt. Rend. Ser. C*, 1968, **267**, 536.
267. J. R. Moss and B. L. Shaw, *Chem. Comm.*, 1968, 632.
268. F. N. Tebbe and E. L. Muetterties, *Inorg. Chem.*, 1968, **7**, 172.
269. S. O. Grim and D. A. Wheatland, *Inorg. Nucl. Chem. Lett.*, 1968, **4**, 187.
270. R. P. Agarwal and I. Feldman, *J. Amer. Chem. Soc.*, 1968, **90**, 6635; *ibid*, 1969, **91**, 2411.
271. H. Sternlicht, D. E. Jones and K. Kustin, *J. Amer. Chem. Soc.*, 1968, **90**, 7110.
272. I. Feldman and R. P. Agarwal, *J. Amer. Chem. Soc.*, 1968, **90**, 7329.
273. B. E. Griffin, M. Jarman and C. B. Reese, *Tetrahedron*, 1968, **24**, 639.
274. R. H. Sarma, V. Ross and N. O. Kaplan, *Biochemistry*, 1968, **7**, 3052.
275. M. P. Schweizer and A. D. Broom, *J. Amer. Chem. Soc.*, 1968, **90**, 1042.
276. P. O. P. Ts'o and M. Schwizer, *Biochemistry*, 1968, **7**, 2963.
277. S. A. Penkett, A. G. Flook and D. Chapman, *Chem. Phys. Lipids*, 1968, **2**, 273.
278. O. Renkonen, *Biochem. Biophys. Acta*, 1968, **152**, 114.
279. N. J. Salisbury and D. Chapman, *Biochem. Biophys. Acta*, 1968, **163**, 314.
280. M. Smith and C. D. Jardetzky, *J. Mol. Spectr.*, 1968, **28**, 70.
281. R. Miller, A. Mildvan, H. Chang, R. Esterday, H. Maruyama and M. D. Lane, *J. Biol. Chem.*, 1968, **243**, 6030.
282. H. Babad, T. N. Taylor and M. C. Goldberg, *Anal. Chim. Acta*, 1968, **40**, 387.
283. M. C. Goldberg, H. Babad, D. Groothius and H. R. Christianson, *U.S. Geol. survey Prof.*, 1968, Paper 600–D, D20, D23.
284. H. Babad, W. Herbert and M. C. Goldberg, *Anal. Chim. Acta*, 1968, **41**, 259.
285. L. H. Keith, A. W. Garrison and A. L. Alford, *J. Assoc. Official Anal. Chem.*, 1968, **51**, 1063.
286. M. A. Slutskin, *Zap. Leningrad. Gorn. Inst.*, 1968, **56**, 39.
287. A. S. Tompa and R. D. Barefoot, *Anal. Chem.*, 1968, **40**, 650.
288. H. P. Latscha, *Z. Naturforsch, B*, 1968, **23**, 139.
289. M. Murray and R. Schmutzler, *Z. Chemie*, 1968, **8**, 241.
290. H. Vahrenkamp and H. Nöth, *J. Organometal. Chem.*, 1968, **12**, 281.
291. S. C. Peake and R. Schmutzler, *Chem. Comm.*, 1968, 1662 (Correction *ibid*, 1969, 1080).

1968 cont.
292. T. T. Conway, A. Shoeb and L. Bauer, *J. Pharm. Sc.*, 1968, **57**, 455.
293. I. Love, *Mol. Phys.*, 1968, **15**, 93.
294. K. Gosling and A. B. Burg, *J. Amer. Chem. Soc.*, 1968, **90**, 2011.
295. A. A. Neimysheva and I. L. Knunyants, *Reakst. Sposonost. Org. Soedin.*, Tartu Gos. Univ., 1968, **5**, 127.
296. K. D. Berlin, N. K. Roy, R. T. Claunch and D. Bude, *J. Amer. Chem. Soc.*, 1968, **90**, 4494.
297. L. I. Petrovskaya, M. V. Proskurnina, Z. S. Novikova and I. F. Lutsenko, *Izv. Akad. Nauk S.S.S.R., Ser. Khim.*, 1968, 1277; *Bull. Acad. Sc. U.S.S.R., Div. Chem. Sc.*, 1968, 1206.
298. L. Maier, *Angew. Chem.*, 1968, **80**, 401; *Angew. Chem. Int. Ed.*, 1968, **7**, 385.
299. L. Maier, *Angew. Chem.*, 1968, **80**, 401; *Angew. Chem. Int. Ed.*, 1968, **7**, 385.
300. L. Maier, *Angew. Chem.*, 1968, **80**, 400; *Angew. Chem Int. Ed.*, 1968, **7**, 384.
301. F. Ramirez, M. Nagabushanam and C. P. Smith, *Tetrahedron*, 1968, **24**, 1785.
302. F. Ramirez, C. P. Smith and J. F. Pilot, *J. Amer. Chem. Soc.*, 1968, **90**, 6726.
303. F. Ramirez, C. P. Smith, J. F. Pilot and A. S. Gulati, *J. Org. Chem.*, 1968, **33**, 3787.
304. L. R. Provost and R. V. Jardine, *J. Chem. Educ.*, 1968, **45**, 675.
305. H. Schmidbaur and W. Tronich, *Chem. Ber.*, 1968, **101**, 604.
306. J. A. Potenza, P. J. Caplan and E. H. Poindexter, *J. Chem. Phys.*, 1968, **49**, 2461.
307. R. G. Kostyanovsky, V. V. Yakshin and S. L. Zimont, *Tetrahedron*, 1968, **24**, 2995.
308. H. W. Roesky, *Chem. Ber.*, 1968, **101**, 636.
309. F. Ramirez, K. Tasaka, N. B. Desai and C. P. Smith, *J. Org. Chem.*, 1968, **33**, 25.
310. F. Ramirez, A. S. Gulati and C. P. Smith, *J. Org. Chem.*, 1968, **33**, 13.
311. M. Verny and R. Vessiere, *Bull. Soc. Chim. France*, 1968, 3004.
312. F. Ramirez, H. J. Kugler, A. V. Patwardhan and C. P. Smith, *J. Org. Chem.*, 1968, **33**, 1185.
313. F. Ramirez, S. B. Bhatia, A. J. Bigler and C. P. Smith, *J. Org. Chem.*, 1968, **33**, 1192.
314. A. A. Petrov, B. I. Ionin and V. M. Ignatiev, *Tetrahedron Lett.*, 1968, 15.
315. F. Ramirez, S. B. Bhatia, A. V. Patwardhan, E. H. Chen and C. P. Smith, *J. Org. Chem.*, 1968, **33**, 20.
316. S. C. Peake and R. Schmutzler, *Chem. Comm.*, 1968, 665.
317. R. Obrycki and C. E. Griffin, *J. Org. Chem.*, 1968, **33**, 632.
318. F. Ramirez, H. J. Kugler and C. P. Smith, *Tetrahedron*, 1968, **24**, 1931.
319. D. Danion and R. Carrie, *Compt. Rend. Ser. C.*, 1968, **267**, 735.
320. K. Issleib, B. Walther and E. Fluck, *Z. Chem.*, 1968, **8**, 67.
321. Y. Charbonnel, R. Burgada and J. Barrans, *Compt. Rend. Ser. C*, 1968, **266**, 1241.
322. H. Schmidbaur and W. Tronich, *Angew. Chem.*, 1968, **80**, 239; *Angew. Chem. Int. Ed.*, 1968, **7**, 220.
323. D. Danion and R. Carrie, *Tetrahedron Lett.*, 1968, 4537.
324. R. Burgada, *Compt. Rend. Ser. C*, 1968, **267**, 1854.
325. G. Sturtz, *Angew. Chem.*, 1968, **80**, 635; *Angew. Chem. Int. Ed.*, 1968, **7**, 647.

1968 cont.
326. R. E. Goldsberry, D. E. Lewis and K. Cohn, *J. Organometal. Chem.*, 1968, **15**, 491.
327. H. Bürger and U. Goetze, *J. Organometal. Chem.*, 1968, **12**, 451.
328. L. Merritt, *M.S. Thesis, Oklahoma State U.*, May, 1968.
329. T. Saegusa, Y. Ito and S. Kobayashi, *Tetrahedron Lett.*, 1968, 935.
330. K. C. Nainan and G. E. Ryschkewitsch, *Inorg. Chem.*, 1968, **7**, 1316.
331. J. Satge and C. Couret, *Compt. Rend. Ser. C*, 1968, **266**, 173.
332. D. B. Denney, D. Z. Denney and L. A. Wilson, *Tetrahedron Lett.*, 1968, 85.
333. E. E. Schweizer, C. J. Berninger and J. G. Thompson, *J. Org. Chem.*, 1966, **33**, 336.
334. C. N. Lieske, J. W. Hovanec, G. M. Steinberg and P. Blumbergs, *Chem. Comm.*, 1968, 13.
335. Yu. L. Kruglyak, G. A. Leibovskaya, I. I. Sretenskaya, V. V. Sheluchenko and I. V. Martynov, *Zhurn. Obshch. Khim.*, 1968, **38**, 943; *J. Gen. Chem., U.S.S.R.*, 1968, **38**, 908.
336. L. Maier, *Helv. Chim. Acta*, 1968, **51**, 405.
337. S. O. Grim and J. H. Ambrus, *J. Org. Chem.*, 1968, **33**, 2992.
338. G. H. Birum and C. N. Mathews, *Chem. Ind. (London)*, 1968, 653.
339. G. Märkl and F. Lieb, *Angew. Chem.*, 1968, **80**, 702; *Angew. Chem. Int. Ed.*, 1968, **7**, 733.
340. P. De Koe and F. Bickelhaupt, *Angew. Chem.*, 1968, **80**, 912; *Angew. Chem. Int. Ed.*, 1968, **7**, 889.
341. J. B. Levy, L. D. Freedman and G. O. Doack, *J. Org. Chem.*, 1968, **33**, 474.
342. Y. E. Fenton and J. J. Zuckerman, *Inorg. Chem.*, 1968, **7**, 1323.
343. G. Gaudiano, R. Mondelli, P. P. Ponti, G. Ticozzi and A. Umani-Ronchi, *J. Org. Chem.*, 1968, **33**, 4431.
344. E. Hedaya and S. Theodoropulos, *Tetrahedron*, 1968, **24**, 2241.
345. E. Zbiral and L. Berner-Fenz, *Tetrahedron*, 1968, **24**, 1363.
346. G. Pattenden and B. C. L. Weedon, *J. Chem. Soc. C*, 1968, 1984.
347. R. Mondelli, *Nato Summer School on Nuclear Magnetic Resonance Spectroscopy (September* 1968, *Coimbra, Portugal)*, Abstracts, p. 82.
348. J. L. Work and J. E. Herweh, *J. Polym. Sci.*, A-1, 1968, **6**, 2022.
349. O. T. Quimby, J. D. Curry, D. A. Nicholson, J. B. Prentice and C. H. Roy, *J. Organometal. Chem.*, 1968, **13**, 199.
350. H. W. Roesky, *Angew. Chem.*, 1968, **80**, 844; *Angew. Chem. Int. Ed.*, 1968, **7**, 815.
351. D. R. Mathieson and N. E. Miller, *Inorg. Chem.*, 1968, **7**, 709.
352. R. Schmutzler, *Inorg. Chem.*, 1968, 1327.
353. H. Gross and B. Costisella, *Angew. Chem.*, 1968, **80**, 445; *Angew. Chem. Int. Ed.*, 1968, **7**, 463.
354. H. Prakasch and H. H. Sisler, *Inorg. Chem.*, 1968, **7**, 2200.
355. H. W. Roesky and D. Bormann, *Chem. Ber.*, 1968, **101**, 630.
356. F. Seel, W. Gombler and K. H. Rudolph, *Zeitr. Naturf.*, 1968, **23B**, 387.

1969
1. J. F. Nixon and A. Pidcock, *Ann. Rev. NMR Spectroscopy*, Vol. 2., p. 345 Academic Press, London, 1969.
2. H. Elser and H. Dreeskamp, *Ber. Bunseng.*, 1969, **73**, 619.
3. W. McFarlane and J. A. Nash, *Chem. Comm.*, 1969, 127.
4. W. McFarlane and J. A. Nash, *Chem. Comm.*, 1969, 913.

1969 cont.
5a. W. Derbyshire, J. P. Stuart and D. Warner, *Mol. Phys.* 1969, **17**, 449.
5b. R. De Ketelaere, E. Muylle, W. Vanermen, E. Claeys and G. P. Van der Kelen, *Bull. Soc. chim. belges.*, 1969, **78**, 219.
6. D. L. Vanderhart, *Thesis*, *Univ. Ill.*, Univ. Microfilm, Order No. 69–1472; *Diss. Abstr.*, *B*, 1969, **29**, 2837.
7. D. L. Vanderhart, H. S. Gutowsky and T. C. Farrar, *J. Chem. Phys.*, 1969, **50**, 1058 (Rept. AD 672807).
8. E. A. C. Lucken and D. F. Williams, *Mol. Phys.*, 1969, **16**, 17.
9. C. Deverell, *Mol. Phys.*, 1969, **17**, 551.
10. S. P. Ionov and G. V. Ionova, *Zhur. Fiz. Khim.*, 1969, **43**, 825; *Russian J. Phys. Chem.*, 1969, **43**, 458.
11. K. B. Dillon and T. C. Waddington, *Chem. Comm.*, 1969, 1317.
12. G. A. Olah and W. McFarlane, *J. Org. Chem.*, 1969, **34**, 1832.
13. D. Hellwinkel and W. Schenk, *Angew. Chem.*, 1969, **81**, 1049; *Angew. Chem. Int. Ed.*, 1969, **8**, 987.
14. H. Schumann, *Angew. Chem.*, 1969, **81**, 970; *Angew. Chem. Int. Ed.*, 1969, **8**, 937.
15. A. Schmidpeter and H. Rossknecht, *Angew. Chem.*, 1969, **81**, 572; *Angew. Chem. Int. Ed.*, 1969, **8**, 614.
16. M. G. Hogben, R. S. Gay, A. J. Oliver, J. A. S. Thompson and W. A. G. Graham, *J. Amer. Chem. Soc.*, 1969, **91**, 291.
17. M. G. Hogben and W. A. G. Graham, *J. Amer. Chem. Soc.*, 1969, **91**, 283.
18. W. Prikoszovich and H. Schindlbauer, *Chem. Ber.*, 1969, **102**, 2922.
19. H. Schindlbauer and W. Prikoszovich, *Chem. Ber.*, 1969, **102**, 2914.
20. G. Mavel and R. Mańkowski-Favelier, *Unpubl. data* 1969.
21. F. Weigert and J. D. Roberts, *J. Amer. Chem. Soc.*, 1969, **21**, 4940
22. E. F. Mooney and P. H. Winson, *Ann. Rev. NMR Spectroscopy*, 1969, **2**, 153.
23. W. Drenth, *Colloque International C.N.R.S. Sur la Chimie Organique du Phosphore*, Paris, Mai 1969.
24. W. G. Henderson and E. F. Mooney, *Ann. Rev. NMR Spectroscopy*, 1969, **2**, 219.
25. L. J. Todd, J. L. Little and H. T. Silverstein, *Inorg. Chem.*, 1968, **8**, 1698.
26. G. L. Smith and H. C. Kelly, *Inorg. Chem.*, 1969, **8**, 2000.
27. A. B. Burg and J. S. Basi, *J. Amer. Chem. Soc.*, 1969, **91**, 1937.
28. R. G. Kidd and D. R. Truax, *J. Chem. Soc. D*, (*Chem. Comm.*,). 1969, 160.
29. T. Yonezawa, *Bull. Chem. Soc. Japan*, 1969, **42**, 1248.
30. C. P. Rader, *J. Amer. Chem. Soc.*, 1969, **91**, 3248.
31. R. W. Rudolph and R. A. Newmark, *J. Amer. Chem. Soc.*, 1970, **92**, 1195.
32. M. Barfield and M. Karplus, *J. Amer. Chem. Soc.*, 1969, **91**, 1.
33. G. R. Miller, A. W. Yankowsky and S. O. Grim, *J. Chem. Phys.*, 1969, **51**, 3185.
34. J. P. Albrand, Rpt. CEA-R-3733 (1969).
35. D. Crepaux, J. M. Lehn and R. R. Dean, *Mol. Phys.*, 1969, **16**, 225.
36. A. H. Cowley and W. D. White, *J. Amer. Chem. Soc.*, 1969, **91**, 1913.
37. A. H. Cowley and W. D. White, *J. Amer. Chem. Soc.*, 1969, **91**, 1917.
38. A. H. Cowley, W. D. White and M. C. Damasco, *J. Amer. Chem. Soc.*, 1969, **91**, 1922.
39. C. J. Jameson and H. S. Gutowsky, *J. Chem. Phys.*, 1969, **51**, 2790; Rpt-AD 684904.
40. C. J. Jameson, *J. Amer. Chem. Soc.*, 1969, **91**, 6232.

1969 cont.
41. S. L. Manatt, E. A. Cohen and A. H. Cowley, *J. Amer. Chem. Soc.*, 1969, **91**, 5919.
42. E. Fluck and V. Novobilsky, *Forschr. chem. Forsch.*, 1969, **13**, 125.
43. R. Fields, M. Green and A. Jones, *J. Chem. Soc. A*, 1969, 2740.
44. P. Tavs, *Angew. Chem.*, 1969, **81**, 742; *Angew. Chem. Int. Ed.*, 1969, **8**, 751.
45. M. Sanchez, L. Beslier and R. Wolf, *Bull. Soc. Chim. France*, 1969, 2778.
46. H. P. Latscha, P. B. Hormuth and H. Vollmer, *Z. Naturforsch. B*, 1969, **24**, 1237.
47. W. McFarlane, *Quart. Rev.*, 1969, **23**, 187.
48. H. Spiesecke, *Private Comm.*,
49. L. I. Vinogradov and S. N. Nikolaev, *Izv. Vyssh. Ucheb. Zaved., Fiz.*, 1969, **12**, 144.
50. W. McFarlane and R. F. M. White, *J. Chem. Soc. D (Chem. Comm.)*, 1969, 744.
51. D. Gagnaire and M. St. Jaques, *J. Phys. Chem.*, 1969, **73**, 1678.
52. M. Mikołajczyk, *J. Chem. Soc. D (Chem. Comm.)*, 1969, 1221.
53. L. I. Vinogradov, Yu. Yu. Samitov, E. G. Yarkova and A. A. Muratova, *Opt. Spektrosk.*, 1969, **26**, 959.
54. A. H. Cowley and M. W. Taylor, *J. Amer. Chem. Soc.*, 1969, **91**, 1026.
55. A. H. Cowley and M. W. Taylor, *J. Amer. Chem. Soc.*, 1969, **91**, 1929.
56. G. G. Flaskerud, K. E. Pullen and J. M. Shreeve, *Inorg. Chem.*, 1969, **8**, 728.
57. R. G. Cavell, T. L. Charlton and A. A. Pinkerton, *J. Chem. Soc. D (Chem. Comm.)*, 1969, 424.
58. D. D. Des Marteau, *J. Amer. Chem. Soc.*, 1969, **91**, 6211.
59. T. L. Charlton and R. G. Cavell, *Inorg. Chem.*, 1969, **8**, 2436.
60. T. R. Johnson and J. F. Nixon, *J. Chem. Soc. A*, 1969, 2518.
61. J. F. Nixon, *J. Chem. Soc. A*, 1969, 1087.
62. W. E. Hill, D. W. A. Sharp and C. B. Colburn, *J. Chem. Phys.*, 1969, **50**, 612.
63. L. Maier and R. Schmutzler, *J. Chem. Soc. D (Chem. Comm.)*, 1969, 961.
64. G. M. Whitesides and H. L. Mitchell, *J. Amer. Chem. Soc.*, 1969, **91**, 5384.
65. M. Becke-Goehring and H. Weber, *Z. anorg. allg. Chem.*, 1969, **365**, 185.
66. J. Nixon, *J. Inorg. Nucl. Chem.*, 1969, **31**, 1615.
67. J. Jander, D. Börner and U. Engelhardt, *J. Liebigs Ann. Chem.*, 1969, **726**, 19.
68. C. W. Haigh, *Priv. Comm.*, 1969.
69. E. Fluck and G. H. Heckman, *Z. Naturforsch. B*, 1969, **24**, 953.
70. Y. Morino, K. Kuchitsu and T. Moritani, *Inorg. Chem.*, 1969, **8**, 867.
71. R. L. Keiter and J. G. Verkade, *Inorg. Chem.*, 1969, **8**, 2115.
72. J. P. Laurent, G. Jugie and G. Commenges, *J. Inorg. Nucl. Chem.*, 1969, **31**, 1353.
73. K. D. Crosbie, C. Glidewell and G. M. Sheldrick, *J. Chem. Soc. A*, 1969, 1861.
74. R. V. Jardine, L. R. Provost and A. S. Hansen, *Private Comm.*, Feb. 1969,
75. F. Ramirez, J. F. Pilot, C. P. Smith, S. B. Bhatia and S. A. Gulati, *J. Org. Chem.*, 1969, **34**, 3385.
76. J. P. Albrand, D. Gagnaire, J. Martin and J. B. Robert. *Bull. Soc. Chim. France*, 1969, 40.

1969 cont.
77. L. D. Quin, J. G. Bryson and C. B. Moreland, *J. Amer. Chem. Soc.*, 1969, **91**, 3308.
78. H. Schumann, O. Stelzer and U. Niederreuther, *J. Organometal. Chem.*, 1969, **16**, P64.
79. A. A. Bothner-By and R. H. Cox, *J. Phys. Chem.*, 1969, **73**, 1830.
80. C. Benezra, *Tetrahedron Lett.*, 1969, 4471.
81. R. H. Cox and R. B. Adelman, *Tetrahedron Lett.*, 1969, 4017.
82. C. E. Griffin and S. K. Kundu, *J. Org. Chem.*, 1969, **34**, 1532.
83. J. A. Ross and M. D. Martz, *J. Org. Chem.*, 1969, **34**, 399.
84. A. B. Burg and I. B. Mishra, *Inorg. Chem.*, 1969, **8**, 1199.
85. H. G. Ang, *J. Inorg. Nucl. Chem.*, 1969, **31**, 3311.
86. J. H. Finley, D. Z. Denney and D. B. Denney, *J. Amer. Chem. Soc.*, 1969, **91**, 5826.
87. A. Schmidpeter and H. Brecht, *Z. Naturforsch. B*, 1969, **24**, 179.
88. J. P. Albrand, D. Gagnaire, J. B. Robert and M. Haemers, *Bull. Soc. Chim. France*, 1969, 3496.
89. J. R. Durig and J. S. Diyorio, *J. Mol. Structure*, 1969, **3**, 179.
90. M. Kainosho and A. Nakamura, *Tetrahedron*, 1969, **25**, 4071.
91. J. B. Robert, *C.E.A. Rapp.*, 1969, *C.E.A. −R*−3671.
92. C. Bodkin and P. Simpson, *J. Chem. Soc. D (Chem. Comm.)*, 1969, 829.
93. R. H. Cox, B. S. Campbell and M. G. Newton, *Private Comm.*, 1969,
94. D. Z. Denney, G. Y. Chen and D. B. Denney, *J. Chem. Soc.*, 1969, **91**, 6838.
95. M. Kainosho, A. Nakamura and M. Tsuboi, *Bull. Chem. Soc. Japan*, 1969, **42**, 1713.
96. M. Tsuboi, S. Takahashi, Y. Kyogoku, H. Hagatsu, T. Ukita and M. Kainosho, *Science*, 1969, **166**, 1504.
97. M. Kainosho and T. Shimozawa, *Tetrahedron Lett.*, 1969, 865.
98. W. J. Wechter, *J. Org. Chem.*, 1969, **34**, 244.
99. T. L. Charlton and R. G. Cavell, *Inorg. Chem.*, 1969, **8**, 281.
100. A. Schmidpeter and C. Weingand, *Z. Naturforsch. B*, 1969, **24**, 177.
101. L. S. Frankel, J. Cargioli, H. Klapper and R. Danielson, *Canad. J. Chem.*, 1969, **47**, 3167.
102. L. Elegant, M. Azzaro, R. Mańkowski-Favelier and G. Mavel, *Org. Magn. Resonance*, 1969, **1**, 471.
103. A. E. Goya, M. D. Rosario and J. W. Gilje, *Inorg. Chem.*, 1969, **8**, 725.
104. R. B. King, L. W. Houk and R. N. Kapoor, *Inorg. Chem.*, 1969, **8**, 1792.
105. W. E. Slinkard and D. W. Meek, *Inorg. Chem.*, 1969, **8**, 1811.
106. A. Schmidpeter and N. Schindler, *Chem. Ber.*, 1969, **102**, 856.
107. H. Goldwhite and D. G. Rowsell, *J. Chem. Soc. D (Chem. Comm.)*, 1969, 713.
108. E. D. Morris and C. E. Nordman, *Inorg. Chem.*, 1969, **8**, 1673.
109. R. Burgada, *Colloque International du C.N.R.S. sur la Chimie Organique du Phosphore, Paris, May* 1969.
110. A. W. Garrison, L. H. Keith and A. L. Alford, *Spectrochim. Acta*, 1969, **25A**, 77.
111. A. H. Cowley, M. J. S. Dewar, W. B. Jennings and W. R. Jackson, *J. Chem. Soc. D. (Chem. Comm.)*, 1969, 482.
112. O. J. Scherer and P. Klusmann, *Angew. Chem.*, 1969, **81**, 743; *Angew. Chem. Int. Ed.*, 1969, **8**, 752.
113. J. E. Drake and N. Goddard, *J. Chem. Soc. A*, 1969, 662.
114. A. H. Cowley and M. W. Taylor, *J. Amer. Chem. Soc.*, 1969, **91**, 1934.

1969 cont.
115. M. P. Simonnin and C. Charrier, *Org. Magn. Resonance*, 1969, **1**, 27.
116. K. Bergesen, *Acta Chem. Scand.*, 1969, **23**, 2556.
117. U. Thewalt, *Angew. Chem.*, 1969, **81**, 783; *Angew. Chem. Int. Ed.*, 1969, **8**, 769.
118. W. McFarlane, *Org. Magn. Resonance*, 1969, **1**, 3.
119. J. I. G. Cadogan, D. J. Sears and D. M. Smith, *J. Chem. Soc. C*, 1969, 1314.
120. R. DeKetelaere, E. Muylle, W. Vanermen, E. Claeys and G. P. Van Der Kelen, *Bull. Soc. Chim. Belges*, 1969, **78**, 219.
121. D. I. Nichols, *J. Chem. Soc. A*, 1969, 1471.
122. W. A. Thomas, *Comm. NMRDG meeting, Bristol*, April 1969.
123. R. H. Kemp, W. A. Thomas, M. Gordon and C. E. Griffin, *J. Chem. Soc. B*, 1969, 527.
124. H. J. Jakobsen and H. Lund, *private comm.*, 1969.
125. H. J. Jakobsen and J. A. Nielsen, *Acta Chem. Scand.*, 1969, **23**, 1070.
126. H. J. Jakobsen and J. A. Nielsen, *J. Mol. Spectry.*, 1969, **31**, 230.
127. G. Peiffer, A. Guillemonat and G. Buono, *Bull. Soc. Chim. France*, 1969, 946.
128. K. L. Lundberg, R. J. Rowatt and N. E. Miller, *Inorg. Chem.*, 1969, **8**, 1336.
129. O. Glemser, U. Biermann and S. V. von Halasz, *Inorg. Nucl. Chem. Letters*, 1969, **5**, 501.
130. R. A. Dwek, R. E. Richards, D. Taylor, G. J. Penney and G. M. Sheldrick, *J. Chem. Soc. A*, 1969, 935.
131. C. G. Barlow and G. C. Holywell, *J. Organometal. Chem.*, 1969, **16**, 439.
132. R. D. Bertrand, F. B. Ogilvie and J. G. Verkade, *J. Chem. Soc. D (Chem. Comm.)*, 1969, 756.
133. F. B. Ogilvie, R. J. Clark and J. G. Verkade, *Inorg. Chem.*, 1969, **8**, 1904.
134. F. B. Ogilvie, R. L. Keiter, G. Wulfsberg and J. G. Verkade, *Inorg. Chem.*, 1969, **8**, 2346.
135. H. Binder and E. Fluck, *Z. anorg. allg. Chem.*, 1969, **365**, 166.
136. R. A. Newmark, A. D. Norman and R. W. Rudolph, *J. Chem. Soc. D (Chem. Comm.)* 1969, 893.
137. A. Schmidpeter and C. Weingand, *Angew. Chem.*, 1969, **81**, 573; *Angew. Chem. Int. Ed.*, 1969, **8**, 615.
138. J. G. Verkade, *Colloque International du C.N.R.S. sur la Chimie Organique du Phosphore, Paris, May* 1969.
139. J. H. Strange and R. E. Morgan, *Proc. XVth Colloque Ampère*, 1969, 260 (*North Holland Publ. by.*; P. Averbuch, ed.).
140. R. E. Ridenour and E. E. Flagg, *J. Organometal. Chem.*, 1969, **16**, 393.
141. L. Centofanti, G. Kodama and R. W. Parry, *Inorg. Chem.*, 1969, **8**, 2072.
142. J. B. Lambert, W. L. Oliver and G. F. Jackson, *Tetrahedron Lett.*, 1969, 2027.
143. J. F. Nixon and J. R. Swain, *Inorg. Nucl. Chem. Lett.*, 1969, **5**, 295.
144. K. D. Berlin and D. M. Hellwege, *Topics in Phosphorus Chemistry*, 1969, **6**, 1.
145. K. Dimroth, A. Hettche, W. Städe and F. W. Steuber, *Angew. Chem.*, 1969, **81**, 784; *Angew. Chem. Int. Ed.*, 1969, **8**, 770.
146. R. L. Kluger and F. H. Westheimer, *J. Amer. Chem. Soc.*, 1969, **91**, 4143.
147. W. G. Bentrude and K. R. Darnall, *J. Chem. Soc. D (Chem. Comm.)*, 1969, 862.
148. J. Brelivet, P. Appriou and J. Teste, *Compt. Rend. Ser. C*, 1969, **268**, 2231.

1969 cont.
149. J. Ebeling and A. Schmidpeter, *Angew. Chem.*, 1969, **81**, 707; *Angew. Chem. Int. Ed.*, 1969, **8**, 674.
150. V. B. Desai, R. A. Shaw and B. C. Smith, *J. Chem. Soc. A*, 1969, 1977.
151. T. Chivers and N. L. Paddock, *J. Chem. Soc. D (Chem. Comm.)*, 1969, 337.
152. O. Glemser, E. Niecke and H. W. Roesky, *J. Chem. Soc. D (Chem. Comm.)*, 1969, 282.
153. B. Green and D. B. Sowerby, *J. Chem. Soc. D (Chem. Comm.)*, 1969, 628.
154. R. A. Lewis, K. Naumann, K. E. De Bruin and K. Mislow, *J. Chem. Soc. D (Chem. Comm.)*, 1969, 1010.
155. D. W. Allen, I. T. Millar, F. G. Mann, R. M. Canadine and J. Walker, *J. Chem. Soc. A*, 1969, 1097.
156. F. Caesar and W. D. Balzer, *Chem. Ber.*, 1969, **102**, 1665.
157. W. McFarlane and J. A. Nash, *J. Chem. Soc. D (Chem. Comm.)*, 1969, 524.
158. G. P. Schiemenz and H. Rast, *Tetrahedron Letters*, 1969, 2165.
159. L. S. Frankel, H. Klapper and J. Cargioli, *J. Phys. Chem.*, 1969, **73**, 91.
160. R. V. Jardine, A. H. Gray and J. B. Reesor, *Canad. J. Chem.*, 1969, **47**, 35.
161. K. D. Berlin and L. Merritt, *J. Chem. Eng. Data*, 1969, **14**, 130.
162. R. Marty, D. Houalla, R. Wolf and J. Riess, *Org. Magn. Resonance*, 1970, **2**, 141.
163. T. N. Timofeeva, B. I. Ionin and A. A. Petrov, *Zh. Obshch. Khim.*, 1969, **39**, 354.
164. J. P. Battioni and W. Chodkiewicz, *Compt. Rend. Ser. C*, 1969, **269**, 1556.
165. A. Sevin and W. Chodkiewicz, *Bull. Soc. Chim. France*, 1969, 4016.
166. A. Sevin and W. Chodkiewicz, *Bull. Soc. Chim. France*, 1969, 4023.
167. J. D. Macomber, *J. Magn. Resonance*, 1969, **1**, 676.
168. M. Gielen and J. Nasielski, *Bull. Soc. Chim. Belges*, 1969, **78**, 339.
769. S. Chan, H. Goldwhite, H. Keyzer, D. G. Rowsell and R. Tang, *Tetrahedron*, 1969, **25**, 1097.
110. S. E. Cremer, R. J. Chorvat and B. C. Trivedi, *J. Chem. Soc. D. (Chem. Comm.)*, 1969, 769.
171. R. Greenhalgh, J. E. Newberry, R. Woodcock and R. F. Hudson, *J. Chem. Soc. D (Chem. Comm.)*, 1969, 22.
172. W. G. Bentrude, J. H. Hargis and P. E. Rusek, *J. Chem. Soc. D (Chem. Comm.)* 1969, 296.
173. W. G. Bentrude and J. H. Hargis, *J. Chem. Soc. D (Chem. Comm.)*, 1969, 1113.
174. R. S. Edmundson, *Tetrahedron Letters*, 1969, 1905.
175. J. P. Majoral and J. Navech, *Compt. Rend., Ser. C*, 1969, **268**, 2117.
176. M. Mikołajczyk, *Angew. Chem.*, 1969, **81**, 495; *Angew. Chem. Int. Ed.*, 1969, **8**, 511.
177. M. Mikołajczyk and H. M. Schiebel, *Angew. Chem.*, 1969, **81**, 494; *Angew. Chem. Int. Ed.*, 1969, **8**, 511.
178. M. Sanchez, L. Beslier, J. Roussel and R. Wolf, *Bull. Soc. Chim. France*, 1969, 3053.
179. D. Houalla, R. Wolf, D. Gagnaire and J. B. Robert, *J. Chem. Soc. D (Chem. Comm.)*, 1969, 443.
180. G. Heckmann and E. Fluck, *Z. Naturforsch. B*, 1969, **24**, 1092.
181. J. F. Normant and H. Deshayes, *Bull. Soc. Chim. France*, 1969, 1001.
182. E. N. Sventitskii and G. P. Savoskina, *Yad. Magn. Rezonans*, 1969, 69.
183. V. M. Vdovenko, L. S. Bulyanitsa, G. P. Savoskina and E. N. Sventitskii, *Radiokhimya*, 1969, **11**, 609; *Radiochimie*, 1969, **11**, 580.

1969 cont.
184. G. B. Sergeev, S. V. Zenin, V. A. Batyuk, L. P. Karunina and T. D. Nekipelova, *Zh. Fiz. Khim.*, 1969, **43**, 985; *Russian J. Phys. Chem.*, 1969, **43**, 544.
185. I. A. Shevchuk, T. N. Simonova and L. M. Kaplan, *Zh. Neorg. Khim*, 1969, **14**, 2479.
186. L. Lunazzi and S. Brownstein, *J. Magn. Reson.*, 1969, **1**, 119.
187. A. H. Cowley and J. L. Mills, *J. Amer. Chem. Soc.*, 1969, **91**, 2911.
188. L. Elegant, *Thèse, Fac. Sciences, Nice*, 1969,
189. L. Elegant, J. F. Gal and M. Azzaro, *Bull. Soc. Chim. France*, 1969, 4273.
190. L. Elegant, R. Wolf and M. Azzaro, *Bull. Soc. Chim. France*, 1969, 4269.
191. G. Jugie, J. P. Pouyanne and J. P. Laurent, *Compt. Rend. Ser. C*, 1969, **268**, 1377.
192. J. P. Laurent, G. Jugie and R. Wolf, *J. Chim. Phys.*, 1969, **66**, 409.
193. J. P. Laussac, G. Jugie and J. P. Laurent, *Compt. Rend. Ser. C*, 1969, **269**, 698.
194. E. Mayer and R. E. Hester, *Spectrochim. Acta*, 1969, **25A**, 237.
195. J. R. Moss and W. A. Graham, *J. Chem. Soc. D (Chem. Comm.)*, 1969, 800.
196. J. G. Riess, *Rev. Chim. Minérale*, 1969, **6**, 643.
197. B. M. Cohen, A. L. Cullingworth and J. D. Smith, *J. Chem. Soc. A*, 1969, 2193.
198. K. L. Henold, J. B. De Roos and J. P. Oliver, *Inorg. Chem.*, 1969, **8**, 2035.
199. S. F. Spangenberg and H. H. Sisler, *Inorg. Chem.*, 1969, **8**, 1004.
200. Y. Iwashita, F. Tamura and A. Nakamura, *Inorg. Chem.*, 1969, **8**, 1179.
201. B. Castro, R. Burgada, G. Lavielle and J. Villieras, *Bull. Soc. Chim. France*, 1969, 2770.
202. B. Castro, R. Burgada, G. Lavielle and J. Villieras, *Compt. Rend. Ser. C*, 1969, **268**, 1067.
203. J. H. Van Dalen and S. Norval, *J. South. Afric. Chem. Inst.*, 1969, **22**, 147.
204. J. A. K. du Plessis and S. Norval, *J. South. Afric. Chem. Inst.*, 1969, **22**, 98 and 104.
205. E. L. Uhlenhopp, J. A. Glasel and A. I. Krasna, *J. Org. Chem.*, 1969, **34**, 2237.
206. H. J. Bestmann, R. Saalfrank and J. P. Snyder, *Angew. Chem.*, 1969, **81**, 227; *Angew. Chem. Int. Ed.*, 1969, **8**, 216.
207. H. I. Zeliger, J. P. Snyder and H. J. Bestmann, *Tetrahedron Letters*, 1969, 2199.
208. D. W. Aksnes, *Acta. Chem. Scand.*, 1969, **23**, 1078.
209. Yu. I. Mitchenko and V. V. Frolov, *Teor. Eksp. Khim.*, 1969, **5**, 286.
210. P. Pyykko, *Proc.* XVth *Colloque Ampère*, 1969, 424 (*North Holland Publ. by*; P. Averbuch, ed.).
211. F. Y. Fradin, *Solid State Comm.*, 1969, **7**, 759.
212. M. Kuznietz and G. A. Matzkanin, *Phys. Rev.*, 1969, **178**, 580.
213. R. A. Dwek, R. E. Richards and D. Taylor, *Ann. Rev. NMR Spectroscopy*, 1969, **2**, 293.
214. R. A. Dwek, P. J. Caplan, E. H. Poindexter and J. A. Potenza, *Chem. Phys. Lett.*, 1969, **3**, 283.
215. J. A. Potanza, E. H. Poindexter, P. J. Caplan and R. A. Dwek, *J. Amer. Chem. Soc.*, 1969, **91**, 4356.
216. R. A. Dwek, N. L. Paddock, J. A. Potenza and E. H. Poindexter, *J. Amer. Chem. Soc.*, 1969, **91**, 5436.
217. W. Derbyshire and J. P. Stuart, *Proc.* XVth *Colloque Ampère*, 1969, 403.

1969 cont.
218. S. L. Carr, C. Long, W. G. Moulton and M. Kuznietz, *Phys. Rev. Lett.*, 1969, **23**, 786.
219. S. K. Malik and R. Vijayaraghavan, *Phys. Lett. A*, 1969, **28**, 648.
220. R. J. Atkinson and C. V. Stager, *Canad. J. Phys.*, 1969, **47**, 1557.
221. R. G. Knubovets, M. L. Afans'ev and S. P. Gabuda, *Spectrosc. Lett.*, 1969, **2**, 121.
222. Y. Tsutsumi, M. Kunimoto, T. Terao and T. Hashi, *J. Phys. Soc. Jap.*, 1969, **26**, 16.
223. W. Van Der Lugt, R. A. Young and D. I. M. Knottnerus, *Proc. Kon. Ned. Akad. Wetensch., Ser. B*, 1969, **62**, 246.
224. R. A. Young, W. Van Der Lugt and J. E. Elliott, *Nature (London)*, 1969, **223**, 729.
225. L. A. Baidakov and V. A. Shcherbakov, *Izv. Akad. Nauk. S.S.S.R., Neorg. Mater.*, 1969, **5**, 1881.
226. A. G. Brekhunets, V. V. Mank, L. N. Shchegrov and V. V. Pechkovskii, *Dokl. Akad. Nauk Beloruss. S.S.R.*, 1969, **13**, 1005.
227. I. Todo and I. Tatsuzaki, *Phys. Status Solidi*, 1969, **32**, 263.
228. A. Hartland, *Proc. Phys. Soc. (J. Phys. C.) Ser.*, 2, 1969, **2**, 264.
229. J. F. Nixon and A. Pidcock, *Ann. Rev. NMR Spectroscopy* (E. F. Mooney, ed.; Academic Press), 1969, **2**, 346.
230. G. Beguin, J. J. Delpuech and A. Peguy, *Mol. Phys.*, 1969, **17**, 317.
231. B. M. Cohen and J. D. Smith, *J. Chem. Soc. A*, 1969, 2087.
232. W. Wolfsberger and H. Schmidbaur, *J. Organometal. Chem.*, 1969, **17**, 41.
233. D. A. Duddell, J. G. Evans, P. L. Goggin, R. J. Goodfellow, A. J. Rest and J. G. Smith, *J. Chem. Soc. A*, 1969, 2134.
234. N. S. Angermann and R. B. Jordan, *Inorg. Chem.*, 1969, **8**, 2579.
235. J. M. Campbell and F. G. A. Stone, *Angew. Chem.*, 1969, **81**, 120; *Angew. Chem. Intern. Ed.*, 1969, **8**, 140.
236. J. P. Collman, J. N. Cawse and J. W. Kang, *Inorg. Chem.*, 1969, **8**, 2574.
237. L. S. Frankel, *Inorg. Chem.*, 1969, **8**, 1784.
238. L. S. Frankel, *J. Chem. Phys.*, 1969, **50**, 943.
239. R. B. King, L. W. Houk and K. H. Pannell, *Inorg. Chem.*, 1969, **8**, 1042.
240. R. B. King and R. N. Kapoor, *J. Inorg. Nucl. Chem.*, 1969, **31**, 2169.
241. R. B. King, R. N. Kapoor and L. W. Houk, *J. Inorg. Nucl. Chem.*, 1969, **31**, 2179.
242. J. Lorberth, H. Nöth and P. V. Rinze, *J. Organometal. Chem.*, 1969, **16**, P1.
243. I. H. Sabherwal and A. B. Burg, *J. Chem. Soc. D (Chem. Comm.)*, 1969, 853.
244. H. Stierand and J. Grobe, *Z. anorg. allg. Chem.*, 1969, **371**, 99.
245. C. A. Udovich and R. J. Clark, *J. Amer. Chem. Soc.*, 1969, **91**, 526.
246. R. L. Martin and A. H. White, *Nature (London)*, 1969, **223**, 394.
247. H. Brunner, *J. Organometal. Chem.*, 1969, **16**, 119.
248. E. O. Fischer, E. Louis and C. G. Kreiter, *Angew. Chem.*, 1969, **81**, 397; *Angew. Chem. Int. Ed.*, 1969, **8**, 377.
249. S. O. Grim and D. A. Wheatland, *Inorg. Chem.*, 1969, **8**, 1716.
250. R. Jefferson, H. F. Klein and J. F. Nixon, *J. Chem. Soc. D (Chem. Comm.)*, 1969, 536.
251. P. R. McAllister, *Thesis, Maryland Univ.*, 1968; *Univ. Microfilms*, Order No. 69–721; *Diss. Abstr. B*, 1969, **29**, 2339.
252. M. Brini, M. M. Geistel and A. Pousse, *Compt. Rend. Ser. C*, 1969, **268**, 2040.
253. F. L. Lafferty, R. C. Jensen and J. C. Sheppard, *Inorg. Chem.*, 1969, **8**, 1875.

1969 cont.
254. E. S. Bolton, G. R. Knox and C. G. Robertson, *J. Chem. Soc. (Chem. Comm.)*, 1969, 664.
255. H. Brunner and E. Schmidt, *Angew. Chem.*, 1969, **81**, 570; *Angew. Chem. Int. Ed.*, 1969, **8**, 616.
256. A. J. Carty, T. W. Ng, W. Carter, G. J. Palenik and T. Birchall, *J. Chem. Soc. D (Chem. Comm.)*, 1969, 1101.
257. J. D. Cotton, D. Doddrell, R. L. Heazlewood and W. Kitching, *Austral. J. Chem.*, 1969, **22**, 1785.
258. W. R. Cullen, D. A. Harbourne, B. V. Liengme and J. R. Sams, *Inorg. Chem.*, 1969, **8**, 1464.
259. R. J. Haines and A. L. Du Preez, *Inorg. Chem.*, 1969, **8**, 1459.
260. K. H. Pannell, *J. Chem. Soc. D (Chem. Comm.)*, 1969, 1346.
261. M. Angoletta and G. Caglio, *Gazz. Chim. Ital.*, 1969, **99**, 46.
262. J. Ashley-Smith, M. Green, N. Mayne and F. G. A. Stone, *J. Chem. Soc. D (Chem. Comm.)*, 1969, 409.
263. J. Chatt, G. J. Leigh and D. M. P. Mingos, *J. Chem. Soc. A*, 1969, 1674.
264. A. J. Deeming and B. L. Shaw, *J. Chem. Soc. A*, 1969, 1128.
265. A. J. Deeming and B. L. Shaw, *J. Chem. Soc. A*, 1969, 1562.
266. R. E. De Simone and R. S. Drago, *Inorg. Chem.*, 1969, **8**, 2517.
267. F. Glockling and M. D. Wilby, *J. Chem. Soc. D (Chem. Comm.)*, 1969, 286.
268. G. Yagupsky and G. Wilkinson, *J. Chem. Soc. A*, 1969, 725.
269. R. B. King and A. Efraty, *Inorg. Chem.*, 1969, **8**, 2374.
270. E. P. Ross and G. R. Dobson, *J. Chem. Soc. D (Chem. Comm.)*, 1969, 1229.
271. G. M. Whitesides and G. Maglio, *J. Amer. Chem. Soc.*, 1969, **91**, 4980.
272. W. J. Miles, B. B. Garrett and R. J. Clark, *Inorg. Chem.*, 1969, **8**, 2817.
273. K. W. Barnett, *Inorg. Chem.*, 1969, **8**, 2009.
274. J. W. Faller, A. S. Anderson and C. C. Chen, *J. Organometal. Chem.*, 1969, **17**, P7.
275. M. Graziani, J. P. Bibler, R. M. Montesano and A. Wojcicki, *J. Organometal. Chem.*, 1969, **16**, 507.
276. C. G. Barlow, J. F. Nixon and J. R. Swain, *J. Chem. Soc. A*, 1969, 1082.
277. J. D. Duncan, J. C. Green, M. L. H. Green and K. A. McLauchlan, *Discuss. Faraday Soc.*, 1969, No. 47, 178.
278. K. Jonas and G. Wilke, *Angew. Chem.*, 1969, **81**, 534; *Angew. Chem. Int.* 1969, **8**, 519.
279. G. N. La Mar and E. O. Sherman, *J. Chem. Soc. D (Chem. Comm.)*, 1969, 809.
280. M. Meier, F. Basolo and R. G. Pearson, *Inorg. Chem.*, 1969, **8**, 795.
281. M. D. Rausch, Y. F. Chang and H. B. Gordon, *Inorg. Chem.*, 1969, **8**, 1335.
282. G. I. Skobnevskaya, L. Radich, Yu. N. Molin, E. E. Zaev and N. D. Chuvylkin, *Theor. Eksp. Khim.*, 1969, **5**, 378.
283. J. F. Nixon and M. D. Sexton, *J. Chem. Soc. A*, 1969, 1089.
284. J. M. Savariault, P. Cassoux and J. F. Labarre, *Compt. Rend. Ser. C*, 1969, **269**, 496.
285. J. P. Fackler, J. A. Fetchin and W. C. Seidel, *J. Amer. Chem. Soc.*, 1969, **91**, 1217.
286. I. H. Sabherwal and A. B. Burg, *Inorg. Nucl. Chem. Lett.*, 1969, **5**, 259.
287. G. J. Leigh, J. J. Levison and S. D. Robinson, *J. Chem. Soc. D (Chem. Comm.)*, 1969, 705.
288. F. L'Eplattenier, *Inorg. Chem.*, 1969, **8**, 965.
289. E. W. Randall and D. Shaw, *J. Chem. Soc. A*, 1969, 2867.

1969 cont.
290. D. W. Allen, I. T. Millar and F. G. Mann, *J. Chem. Soc.*, 1969, 1101.
291. J. P. Fackler, J. A. Fetchin, J. Mayhew, W. C. Seidel, T. J. Swift and M. Weeks, *J. Amer. Chem. Soc.*, 1969, **91**, 1941.
292. J. P. Fackler and W. C. Seidel, *Inorg. Chem.*, 1969, **8**, 1631.
293. M. Sakakibara, Y. Takahashi, S. Sakai and Y. Ishii, *Inorg. Nucl. Chem. Letters*, 1969, **5**, 427.
294. A. F. Clemmit and F. Glockling, *J. Chem. Soc. A*, 1969, 2163.
295. J. H. Nelson, H. B. Jonassen and D. M. Roundhill, *Inorg. Chem.*, 1969, **8**, 2591.
296. D. M. Roundhill, *J. Chem. Soc. D (Chem. Comm.)*, 1969, 567.
297. K. Vrieze, H. C. Volger and A. P. Praat, *J. Organometal. Chem.*, 1969, **16**, 19.
298. D. M. Barlex, R. D. W. Kemmitt and G. W. Littlecott, *J. Chem. Soc. D (Chem. Comm.)*, 1969, 613.
299. J. F. Nixon and M. D. Sexton, *J. Chem. Soc. D (Chem. Comm.)*, 1969, 827.
300. J. Chatt, J. R. Dilworth and G. J. Leigh, *J. Chem. Soc. D (Chem. Comm.)*, 1969, 687.
301. J. Chatt and R. S. Coffey, *J. Chem. Soc. A*, 1969, 1963.
302. A. J. Deeming and B. L. Shaw, *J. Chem. Soc. A*, 1969, 597.
303. J. T. Mague and J. P. Mitchener, *Inorg. Chem.*, 1969, **8**, 119.
304. J. T. Moelwyn-Hughes and A. W. B. Garner, *J. Chem. Soc. D (Chem. Comm.)*, 1969, 1309.
305. K. C. Nainan and G. E. Ryschewitsch, *Inorg. Chem.*, 1969, **8**, 2671.
306. T. H. Brown and J. P. Green, *J. Amer. Chem. Soc.*, 1969, **91**, 3378.
307. M. Cooke and M. Green, *J. Chem. Soc. A*, 1969, 651.
308. S. O. Grim, P. R. McAllister and R. M. Singer, *J. Chem. Soc. D (Chem. Comm.)*, 1969, 38.
309. M. Tsutsui, M. Hancock, J. Ariyoshi and M. N. Levy, *Angew. Chem.*, 1969, **81**, 453; *Angew. Chem. Int. Ed.*, 1969, **8**, 411.
310. L. S. Frankel, *J. Mol. Spectry.*, 1969, **29**, 273.
311. R. G. Bryant, *J. Amer. Soc.*, 1969, **91**, 1870.
312. P. T. Thyrum, R. J. Luchi and E. M. Thyrum, *Nature (London)*, 1969, **223**, 747.
313. A. M. Bobst, F. Rottman and P. A. Cerutti, *J. Amer. Chem. Soc.*, 1969, **91**, 4603.
314. A. S. Mildvan and H. Weiner, *J. Biol. Chem.*, 1969, **244**, 2465.
315. S. I. Chan and J. H. Nelson, *J. Amer. Chem. Soc.*, 1969, **91**, 168.
316. D. P. Hollis, *Org. Magn. Resonance*, 1969, **1**, 305.
317. J. H. Prestegard and S. I. Chan, *J. Amer. Chem. Soc.*, 1969, **91**, 2843.
318. H. Rüterjans, *Angew. Chem.*, 1969, **81**, 402; *Angew. Chem. Int. Ed.*, 1969, **8**, 392.
319. P. O. P. Ts'o, N. S. Kondo, M. P. Schweizer and D. P. Hollis, *Biochemistry*, 1969, **8**, 997.
320. D. Chapman, R. B. Leslie, R. Hirz and A. Scanu, *Biochim. Biophys. Acta*, 1969, **176**, 524.
321. M. S. Feather and M. J. Lybyer, *Biochim. Biophys. Res. Comm.*, 1969, **35**, 538.
322. L. H. Keith and A. L. Alford, *Anal. Chim. Acta*, 1969, **44**, 447.
323. R. V. Ammon, *Angew. Chem.*, 1969, **81**, 908; *Angew. Chem. Int. Ed.*, 1969, **8**, 889.
324. T. H. Siddall and W. E. Stewart, *J. Inorg. Nucl. Chem.*, 1969, **31**, 3557.

1969 cont.
325. H. W. Roesky and L. F. Grimm, *Inorg. Nucl. Chem. Letters*, 1969, **5**, 13.
326. H. W. Roesky and W. Grosse Böwing, *Inorg. Nucl. Chem. Letters*, 1969, **5**, 597.
327. F. Knoll and J. R. Van Wazer, *J. Inorg. Nucl. Chem.*, 1969, **31**, 2620.
328. M. Lustig, *Inorg. Chem.*, 1969, **8**, 443.
329. P. Mansfield and K. H. B. Richards, *Chem. Phys. Lett.*, 1969, **3**, 169.
330. K. D. Crosbie and G. M. Sheldrick, *J. Inorg. Nucl. Chem.*, 1969, **31**, 3684.
331. L. Maier, *Helv. Chim. Acta*, 1969, **52**, 1337.
332. L. Maier, *Helv. Chim. Acta*, 1969, **52**, 827.
333. L. Maier, *Helv. Chim. Acta*, 1969, **52**, 858.
334. L. Maier, *Helv. Chim. Acta*, 1969, **52**, 845.
335. L. N. Shitov and B. M. Gladshtein, *Zhurn. Obshch. Khim.*, 1969, **39**, 1251.
336. M. Durand and J. P. Laurent, *Bull. Soc. Chim. France*, 1969, 48.
337. F. Ramirez, S. L. Glaser, A. J. Bigler and J. F. Pilot, *J. Amer. Chem. Soc.*, 1969, **91**, 496.
338. E. Guibé-Jampel and M. Wakselman, *J. Chem. Soc. D (Chem. Comm.)*, 1969, 720.
339. F. Mathey and G. Mavel, *Compt. Rend. Ser. C*, 1969, **268**, 1902.
340. R. Burgada and J. Roussel, *Bull. Soc. Chim. France*, 1970, 192.
341. J. Satge and C. Couret, *Bull. Soc. Chim. France*, 1969, 333.
342. R. M. Lequan and M. P. Simonnin, *Compt. Rend., Ser. C*, 1969, **268**, 1400.
343. M. Revel and J. Navech, *Compt. Rend., Ser. C*, 1969, **268**, 121.
344. S. F. Spangenberg and H. H. Sisler, *Inorg. Chem.*, 1969, **8**, 1006.
345. M. F. Chasle, M. Soenen and A. Foucaud, *Compt. Rend., Ser. C*, 1969, **269**, 499.
346. H. Gross, B. Costisella and W. Bürger, *J. Prakt. Chem.*, 1969, **311**, 563.
347. W. E. Slinkard and D. W. Meek, *J. Chem. Soc. D (Chem. Comm.)*, 1969, 361.
348. K. Utvary and M. Bermann, *Inorg. Chem.*, 1969, **8**, 1038.
349. G. Pattenden and B. J. Walker, *J. Chem. Soc. A*, 1969, 531.
350. D. J. H. Smith and S. Trippett, *J. Chem. Soc. D (Chem. Comm.)*, 1969, 855.
351. Yu. L. Krugelyak, M. A. Laundau, G. A. Leibovskaya, I. V. Martynov, L. I. Slatykova and M. A. Sokal'ski, *Zhurn. Obshch. Khim.*, 1969, **39**, 215.
352. P. Savignac and P. Chabrier, *Compt. Rend., Ser. C*, 1969, **268**, 861.
353. F. Ramirez and G. V. Loewengart, *J. Amer. Chem. Soc.*, 1969, **91**, 2293.
354. A. H. Cowley and J. L. Mills, *J. Amer. Chem. Soc.*, 1969, **91**, 2915.
355. G. Peiffer, A. Guillemonat, J. C. Traynard and E. Gaydou, *Bull. Soc. Chim. France*, 1969, 1304.
356. L. D. Quin, J. H. Somers and R. H. Prince, *J. Org. Chem.*, 1969, **34**, 3700.
357. Y. Charbonnel, J. Barrans and R. Burgada, *Private Comm.*, 1969.
358. F. Mathey, *Compt. Rend., Ser. C*, 1969, **269**, 1066.
359. M. Revel, M. T. Boisdon and J. Navech, *Compt. Rend., Ser. C*, 1969, **269**, 182.
360. E. Brunn and R. Huisgen, *Angew. Chem.*, 1969, **81**, 534; *Angew. Chem. Int. Ed.*, 1969, **8**, 512.
361. R. F. Hudson and A. Mancuso, *J. Chem. Soc. D (Chem. Comm.)*, 1969, 522.
362. R. Burgada, *Bull. Soc. Chim. France*, 1971, 136.
363. L. Legrand and N. Lozac'h, *Bull. Soc. Chim. France*, 1969, 1173.
364. W. Siebert, W. E. Davidsohn and M. C. Henry, *J. Organometal. Chem.*, 1969, **17**, 65.
365. J. P. Battioni and W. Chodkiewicz, *Bull. Soc. Chim. France*, 1969, 981.

1969 cont.
366. K. D. Bartle, D. W. Jones and R. S. Matthews, *Tetrahedron*, 1969, **25**, 2701.
367. R. E. Ireland and G. Pfister, *Tetrahedron Letters*, 1969, 2145.
368. H. Binder and E. Fluck, *Z. anorg. allg. Chem.*, 1969, **365**, 170.
369. E. M. Richards and J. C. Tebby, *J. Chem. Soc. D (Chem. Comm.)*, 1969, 494.
370. R. Jefferson, J. F. Nixon and T. M. Painter, *J. Chem. Soc. D (Chem. Comm.)*, 1969, 622.
371. V. V. Sheluchenko, I. M. Filatova, E. L. Zaitseva and A. Ya. Yakubovich, *Zh. Obshch. Khim.*, 1969, **39**, 194.
372. K. P. Lannert and M. D. Joesten, *Inorg. Chem.*, 1969, **8**, 1775.
373. J. Grobe and U. Moller, *J. Organometal. Chem.*, 1969, **17**, 263.
374. S. R. Jain and H. H. Sisler, *Inorg. Chem.*, 1969, **8**, 1243.
375. T. A. Blazer, R. Schmutzler and I. K. Gregor, *Z. Naturforsch. B*, 1969, **24**, 1081.
376. M. Bermann and K. Utvary, *J. Inorg. Nucl. Chem.*, 1969, **31**, 271.
377. H. Schumann and H. Benda, *Angew. Chem.*, 1969, **81**, 1049; *Angew. Chem. Int. Ed.*, 1969, **8**, 989.
378. E. F. Moran and D. P. Reider, *Inorg. Chem.*, 1969, **8**, 1550.
379. B. Green and D. B. Sowerby, *Inorg. Nucl. Chem. Letters*, 1969, **5**, 989.
380. H. P. Latscha and W. Klein, *Angew. Chem.*, 1969, **81**, 291; *Angew. Chem. Int. Ed.*, 1969, **8**, 278.
381. D. Voigt, P. Swysen and M. C. Labarre, *Bull. Soc. Chim. France*, 1969, 3383.

1970
1. J. P. Fayet, *Compt. rend.*, 1970, **270C**, 9.
2. M. Ellenberger, L. Brehamet, M. Villemin and F. Toma, *FEBS Lett.*, 1970, **8**, 195.

1971
1. R. Haque and D. R. Buhler, *Ann. Rep. NMR Spectroscopy*, 1971, **4**, 237.

INDEXES

AUTHOR INDEX

Numbers in parentheses are reference numbers and are included to assist in locating references when authors' names are not mentioned in the text. Numbers in *italics* refer to the page on which the reference is listed.

A

Abramov, V., 176 (1967, 340), 192 (1967, 340), 193 (1967, 340), 207 (1967, 340), *328*

Acampora, M., 246 (1967, 375), 247 (1967, 375), 255 (1967, 375), *329*

Adelman, R. B., 120 (1969, 81), 126 (1969, 81), 129 (1969, 81), 137 (1969, 81), 138 (1969, 81), *342*

Afans'ev, M. L., 88 (1969, 221), *346*

Agarwal, R. P., 89 (1968, 270), 94 (1968, 272), *337*

Ageeva, A. B., 244 (1967, 373), *329*

Agranat, I., 245 (1967, 374), 252 (1967, 374)) 253 1967, 374), *329*

Aguiar, A. M., 86 (1966, 113), 203 (1967, 349), 210 (1967, 349), 219 (1966, 269), 230 (1965, 111; 1966, 275; 1967, 364), 231 (1965), 112), 232 (1966, 275), 235 (1967, 349), 246 (1966, 282), 271 (1967, 349), 272 (1967, 349), 273 (1967, 349, 364), 274 (1967, 349), 275 (1966, 113, 269; 1967, 349, 364), *311, 314, 318, 329*

Aguiar, H. G., 18 (1966, 17), 86 (1966, 113), 219 (1966, 269), 275 (1966, 113, 269), *312, 314, 318*

Ahlbrecht, H., 235 (1967, 367), 236 (1967, 367), 241 (1967, 367), *329*

Åkerfeldt, S., 94 (1966, 204), 113 (1966, 204), 187 (1966, 204), 273 (1966, 204), 274 (1966, 204), *316*

Akitt, J. W., 18 (1966, 21), 55 (1966, 21), 131 (1966, 21), 132 (1966, 21), *312*

Akrimento, I. S., 123 (1967, 310), 145 (1967, 310), 149 (1967, 310), *327*

Aksnes, D. W., 73 (1968, 115), 87 (1968, 115, 174; 1969, 208), 97 (1968, 115), 98 (1968, 174) 99 (1968, 174), *333, 335, 345*

Albrand, J. P., 13 (1968, 28; 1969, 34), 17 (1968, 28), 18 (1968, 28, 33; 1969, 34), 19 (1969, 34), 28 (1968, 28, 33; 1969, 34), 29 (1968, 62; 1969, 76), 30 (1968, 28; 1969, 34), 38 (1969, 34), 40 (1969, 34, 88), 74 (1968, 28; 1969, 34), 80 (1969, 34), 92 (1969, 34), 105 (1968, 33), 109 (1968, 33), 118 (1968, 28), 125 (1969, 88), 137 (1969, 88), 143 (1968, 33; 1969, 34), 153 (1968, 28, 33; 1969, 34), 154 (1968, 33), 174 (1969, 34), 180 (1969, 76), 187 (1969, 76), 188 (1969, 76), 196 (1969, 34), 198 (1969, 76), 276 (1968, 33), *331, 340, 341, 342*

Alexander, E. S., 29 (1968, 63), 247 (1968, 63), 248 (1968, 63), 250 (1968, 63), 253 (1968, 63), *331*

Alford, A. L., 47 (1969, 110), 81 (1969, 110), 94 (1968, 285; 1969, 322), 117 (1968, 285; 1969, 322), 118 (1968, 285; 1969, 322), 128 (1968, 285), 140 (1968, 285; 1969, 322), 141 (1969, 110), 145 (1968, 285; 1969, 322), 149 (1968, 285; 1969, 110, 322), 151 (1968, 285; 1969, 322), 153 (1968, 285; 1969, 322), 157 (1968, 285; 1969, 322), 160 (1968, 285; 1969, 322), 161 (1969, 110), 163 (1968, 285; 1969, 110), 171 (1968, 285), 172 (1968, 285; 1969, 110, 322), 173 (1968, 285), 174 (1968, 285; 1969, 322), 175 (1968, 285), 176 (1968, 285; 1969, 322), 179 (1968, 285; 1969, 322), 180 (1968, 285), 181 (1968, 285; 1969, 110, 322), 187 (1968, 285), 191

(1968, 285; 1969, 322), 192 (1968, 285; 1969, 322), 194 (1968, 285), 202 (1968, 285), 203 (1968, 285; 1969, 322), 204 (1968, 285), 242 (1968, 285), 263 (1968, 285; 1969, 322), 264 (1969, 322), 265 (1968, 285), 266 (1968, 285), *337, 342, 348*

Allcock, H. R., 7 (1966, 6), 79 (1966, 6, 90, 91), 101 (1966, 6), 278 (1966, 6, 90), 279 (1966, 6, 90), 281 (1966, 90), 283 (1966, 6, 90), 287 (1966, 6), 288 (1966, 6, 91), *311, 313*

Allen, B. T., 89 (1966, 197), *316*

Allen, C. W., 23 (1968, 50), 79 (1967, 144; 1968, 50, 130), 279 (1967, 144; 1968, 50, 130), 280 (1968, 130), 281 (1968, 130), *323, 331, 333*

Allen, D. W., 28, 80 (1969, 155), 92 (1969, 155, 290), 112 (1968, 60), 201 (1968, 60), 202 (1968, 60), 203 (1967, 348), 209 (1968, 60), 227 (1967, 362), 228 (1967, 348; 1968, 60), 231 (1968, 60), 246 (1967, 376), 247 (1967, 376), 250 (1967, 376), 253 (1967, 376), 269 (1969, 155), *329, 331, 344, 348*

Allen, F. H., 92 (1968, 247), *336*

Ambrus, J. H., 223 (1968, 337), 227 (1968, 337), 231 (1968, 337), 235 (1968, 337), 241 (1968, 337), 243 (1968, 337), *339*

Amimoto, S. T., 13 (1968, 27), *331*

Ammon, R. V., 94 (1969, 323), *348*

Anderson, A. S., 91 (1969, 274), *347*

Anderson, E. W., 89 (1965, 73, 74), *310*

Anderson, H. G., 273 (1966, 289), 286 (1966, 289), *319*

Anderson, J. E., 40 (1967, 89), *321*

Anderson, W. A., 65 (1957, 2), *307*

Andrews, E. R., 88 (1966, 149), *315*

Andrianov, K. A., 210 (1965, 106), *311*

Ang, H. G., 34 (1968, 69), 35 (1969, 85), 77 (1968, 69), 82 (1969, 85), 116 (1968, 69; 1969, 85), 123 (1968, 69), 130 (1968, 69), *332, 342*

Angermann, N. S., 90 (1969, 234), 92 (1969, 234), *346*

Angoletta, M., 91 (1965, 62; 1969, 261), *310, 347*

Anschel, M., 58 (1967, 121), 78 (1967, 121, 141), 86 (1967, 121, 141), 172 (1967, 121), 183 (1967, 121), 189 (1967, 121), 190 (1967, 121), 198 (1967, 121), 199 (1967, 121), 205 (1967, 121), 225 (1967, 121), 252 (1967, 141), *322, 323*

Antennis, M., 44 (1968, 87), *332*

Appel, R., 79 (1967, 149), 272 (1967, 149), *323*

Appriou, P., 78 (1969, 148), 151 (1969, 148), 163 (1969, 148), *343*

Araneo, A., 91 (1965, 60), *309*

Arbuzov, B. A., 53 (1964, 8; 1967, 109), 109 (1965, 81), 116 (1964, 8; 1967, 109), 121 (1965, 81), 123 (1964, 8), 124 (1964, 8; 1967, 109), 133 (1964, 16), 134 (1964, 8, 16), 135 (1964, 8, 16), 144 (1964, 8, 16), 145 (1964, 16), 251 (1967, 109), *308, 310, 322*

Archibald, T. G., 86 (1966, 113), 230 (1966, 275), 232 (1966, 275), 246 (1966, 282), 275 (1966, 113), *314, 318*

Ariyoshi, J., 89 (1969, 309), *348*

Ashley-Smith, J., 91 (1969, 262), *347*

Atalla, R. H., 90 (1968, 203), 91 (1968, 203), *335*

Atkins, P. W., 87 (1967, 198), 92 (1968, 248), 109 (1967, 198), 114 (1967, 198), 119 (1967, 198), 120 (1967, 198), 131 (1967, 198), 140 (1967, 198), 142 (1967, 198), 186 (1967, 198), 223 (1967, 198), 224 (1967, 198), 227 (1967, 198), 252 (1967, 198), *324, 336*

Atkinson, R. J., 88 (1969, 220), *346*

Attali, S., 90 (1968, 197), *335*

Aubagnac, J. L., 13 (1967, 25), *320*

Azzaro, M., 44 (1969, 102), 84 (1969, 102, 189, 190), 85 (1969, 102), 86 (1969, 102), 143 (1969, 190), 216 (1969, 190), *342, 345*

B

Babad, H., 94 (1968, 282, 283, 284), 117 (1968, 284), 118 (1968, 283, 284), 128 (1968, 283, 284), 140 (1968, 284), 145 (1968, 284), 146 (1968, 284), 149 (1968, 284), 151 (1968,

284), 153 (1968, 284), 154 (1968, 284), 157 (1968, 283, 284), 160 (1968, 282, 284), 163 (1968, 284), 171 (1968, 284), 172 (1968, 284), 173 (1968, 284), 174 (1968, 283, 284), 176 (1968, 284), 179 (1968, 284), 181 (1968, 284), 182 (1968, 283), 188 (1968, 284), 189 (1968, 284), 191 (1968, 284), 192 (1968, 284), 202 (1968, 284), 203 (1968, 284), 204 (1968, 284), 206 (1968, 284), 262 (1968, 284), 263 (1968, 282, 284), *337*

Babushkina, T. A., 7 (1967, 15), *319*

Baidakov, L. A., 88 (1969, 225), *346*

Brailey, R. E., 88 (1967, 199), *324*

Baird, M. C., 92 (1966, 187; 1967, 256), 131 (1967, 256), *316, 326*

Baker, J. W., 86 (1966, 114), 104 (1966, 114), 107 (1966, 114), 111 (1966, 114), 112 (1966, 114), 118 (1966, 114), 131 (1966, 114), 134 (1966, 114), 143 (1966, 114), 171 (1966, 114), *314*

Bakhchisaraitseva, S. A., 88 (1966, 146), *315*

Bakker, J., 99 (1965, 120), *311*

Ball, R. E., 90 (1968, 198), *335*

Balzer, W. D., 80 (1969, 156), 220 (1969, 156), 226 (1969, 156), 229 (1969, 156), 232 (1969, 156), 235 (1969, 156), 239 (1969, 156), 241 (1969, 156), 243 (1969, 156), 247 (1969, 156), 253 (1969, 156), *344*

Banford, L., 32 (1966, 45), 115 (1966, 45), *312*

Banks, C. V., 93 (1965, 71), 271 (1965, 71), *310*

Bardos, T. J., 86 (1965, 42), 146 (1965, 42), 167 (1965, 42), 183 (1965, 42), 200 (1965, 42), 213 (1965, 42), *309*

Barefoot, R. D., 94 (1968, 287), 141 (1968, 287), 142 (1968, 287), 143 (1968, 287), *337*

Barfield, M., 13, *340*

Barket, T. P., 53 (1968, 100), 124 (1968, 100), 130 (1968, 100), 131 (1968, 100), 134 (1968, 100), 135 (1968, 100), 144 (1968, 100), 179 (1968, 100), 180 (1968, 100), 189 (1968, 100), 198 (1968, 100), 205 (1968, 100), 224 (1968, 100), *333*

Barlex, D. M., 92 (1969, 298), *348*

Barlow, C. G., 64 (1969, 131), 76 (1968, 119), 91 (1968, 168; 1968, 119; 1969, 131, 276), 92 (1966, 168), 105 (1968, 119), 118 (1966, 230), 120 (1966, 230), 125 (1966, 230), 221 (1969, 276), *315, 317, 333, 343, 347*

Barlow, M. G., 18 (1966, 19), 21 (1966, 19), 27 (1966, 19), 45 (1966, 19), 57 (1966, 19), 76 (1966, 19), 129 (1966, 19), 130 (1966, 19), 150 (1966, 19), 171 (1966, 19), 185 (1966, 19), 201 (1966, 19), 205 (1966, 19), 214 (1966, 19), 221 (1966, 19), 227 (1966, 19), *312*

Barnett, K. W., 91 (1969, 273), *347*

Barr, J. M., 79 (1966, 95), 281 (1966, 95), 282 (1966, 95), *314*

Barrans, J., 161 (1968, 321), 168 (1969, 357), 169 (1968, 321), 170 (1968, 321; 1969, 257), 178 (1969, 321), 190 (1969, 357), 191 (1969, 357), 199 (1968, 321), 206 (1968, 321), *338, 349*

Barsukov, L. I., 56 (1967, 118), 234 (1967, 118), 235 (1967, 118), 244 (1967, 118), 246 (1967, 118), 247 (1967, 118), *322*

Bart, J. C. J., 59 (1968, 108), 227 (1968, 108), *333*

Bartell, L. S., 25 (1965, 15, 16), *308*

Bartle, K. D., 41 (1967, 91), 66 (1967, 91), 125 (1967, 91), 126 (1967, 91), 134 (1967, 91), 137 (1967, 91), 148 (1967, 91), 158 (1967, 91), 167 (1967, 91), 168 (1967, 91), 177 (1967, 91), 181 (1967, 91), 184 (1967, 91), 189 (1967, 91), 190 (1967, 91), 198 (1967, 91), 219 (1969, 366), 241 (1969, 366), 243 (1967, 91), 264 (1967, 91), *321, 350*

Basi, J. S., 10 (1969, 27), 88 (1969, 27), 116 (1969, 27), 259 (1969, 27), 278 (1969, 27), *340*

Basolo, F., 90 (1966, 155), 92 (1966, 155; 1969, 280), *315, 347*

Basselier, J. J., 31 (1967, 80), 47 (1967, 80), 81 (1967, 80), 122 (1967, 80),

153 (1967, 80), 173 (1967, 80), 174 (1967, 80), 180 (1967, 80), 197 (1967, 80), *321*

Bathelt, W., 90 (1968, 207), 91 (1968, 207), 93 (1968, 207), *335*

Bathia, S. B., 39 (1967, 163), 83 (1967, 163), 133 (1966, 242), 144 (1966, 242), 164 (1966, 242), 165 (1966, 242), 215 (1967, 163), 219 (1966, 242; 1967, 163), 224 (1966, 242), 225 (1967, 163), 229 (1966, 242), 236 (1966, 242), *317, 323*

Battioni, J. P., 81 (1967, 153; 1969, 164), 217 (1969, 365), 232 (1969, 365), 235 (1967, 153), 236 (1969, 164), 239 (1967, 153; 1969, 164), 243 (1969, 164), 246 (1969, 164), 248 (1969, 164), *323. 344, 349*

Batyuk, V. A., 84 (1969, 184), *345*

Bauer, L., 105 (1968, 292), 110 (1968, 292), 149 (1968, 292), 154 (1968, 292), *338*

Bayer, K., 79 (1966, 98), 256 (1966, 98), *314*

Becconsall, J. K., 55 (1966, 69), 223 (1966, 69), *313*

Beck, W., 92 (1967, 247), *326*

Becke-Goehring, M., 22 (1969, 65), 78 (1965, 32; 1966, 83; 1968, 125; 1969, 65), 79 (1965, 34; 1966, 98; 1968, 136), 93 (1966, 194), 98 (1968, 136), 100 (1967, 285, 286), 101 (1968, 136), 108 (1968, 136), 109 (1968, 125; 1969, 65), 122 (1968, 125, 136; 1969, 65), 127 (1968, 125; 1969, 65), 132 (1967, 286), 187 (1965, 34), 256 (1966, 98), 257 (1966, 83; 1967, 285; 1968, 136; 1969, 65), 259 (1966, 83), 260 (1966, 83), 261 (1966, 83), 265 (1967, 286), 267 (1966, 83), 278 (1968, 136), 279 (1968, 136), 283 (1965, 34), 284 (1965, 32; 1966, 83); 287 (1965, 32), 288 (1965, 32), *309, 313, 314, 316, 327, 333, 334, 341*

Beckenkamp, P., 88 (1966, 147), 257 (1966, 147), *315*

Becker, G., 86 (1967, 182), 263 (1967, 182), 264 (1967, 182), 279 (1967, 182), *324*

Becker, K., 89 (1967, 264), *326*

Beguin, G., 90 (1969, 230), *346*

Bel'skii, V. E., 2 (1967, 2), 279 (1967, 2), 280 (1967, 2), 281 (1967, 2), *319*

Benda, H., 48 (1968, 95), 281 (1968, 95), 282 (1969, 377), *332, 350*

Bene, C. J., 37 (1967, 85), 140 (1967, 85), 169 (1967, 85), *321*

Benezra, C., 31 (1966, 43), 32 (1969, 80), 33, 40, 47 (1966, 43), 50 (1966, 43, 58; 1967, 100), 83 (1966, 103; 1967, 100, 167), 93 (1966, 103), 94 (1967, 167, 270), 137 (1967, 100), 144 (1966, 43), 145 (1967, 100), 146 (1966, 43), 147 (1967, 100), 149 (1967, 100), 155 (1966, 43; 1967, 100), 165 (1969, 80), 166 (1969, 80), 177 (1966, 43), 183 (1969, 80), 207 (1967, 100), 213 (1967, 100), 221 (1967, 100), 230 (1976, 100), 237 (1967, 167), 241 (1967, 167), 244 (1967, 270), 248 (1967, 270), 250 (1966, 43; 1967, 167, 270), 251 (1967, 167), 252 (1966, 43; 1967, 167), 255 (1967, 167), 272 (1967, 100), *312, 313, 314, 322, 324, 326, 342*

Bennett, M. A., 56 (1966, 72), 90 (1966, 72; 1967, 226; 1968, 194), 91 (1966, 72; 1967, 226), 92 (1966, 188; 1968, 237), 93 (1966, 72; 1967, 226), 234 (1966, 72; 1968, 194), 238 (1968, 194), *313, 316, 325, 335, 336*

Bentrude, W. G., 40 (1968, 78), 78 (1967, 140; 1969, 147), 79 (1965, 35), 80 (1965, 35), 82 (1967, 140; 1969, 172), 83 (1968, 78; 1969, 172, 173), 133 (1969, 172), 137 (1968, 78), 144 (1965, 35), 146 (1968, 78), 148 (1969, 147), 157 (1968, 78; 1969, 172, 173), 177 (1969, 147), 197 (1967, 140), 203 (1967, 140), 216 (1967, 140), *309, 323, 332, 343, 344*

Bergel'son, L. D., 56 (1967, 118), 234 (1967, 118), 235 (1967, 118), 244 (1967, 118), 246 (1967, 118), 247 (1967, 118), *322*

Berger, H., 86 (1967, 188), 88 (1967, 188), 187 (1967, 188), *324*
Bergerhoff, G., 18 (1965, 10; 1966, 12), 99 (1965, 10; 1966, 12), *308, 312*
Bergesen, K., 53 (1969, 116), 78 (1969, 116), 86 (1969, 116), *343*
Bergmann, E. D., 245 (1967, 374), 252 (1967, 374), 253 (1967, 374), *329*
Bergmark, T., 4 (1967, 6), *319*
Berlin, A. J., 79 (1966, 96; 1968, 134), 281 (1966, 96), 285 (1966, 96; 1968 134), 286 (1966, 96; 1968, 134), 287 (1966, 96), *314, 333*
Berlin, K. D., 44 (1968, 88), 74 (1966, 55), 48 (1968, 88), 52 (1966, 62), 78, 80 (1967, 152; 1968, 142; 1969, 161), 82 (1969, 144), 86 (1966, 62), 103 (1966, 62), 105 (1965, 79), 109 (1965, 79), 110 (1968, 296), 122 (1968, 296), 126 (1966, 235), 128 (1966, 235), 129 (1968, 296), 132 (1965, 79), 144 (1968, 142), 149 (1965, 92), 154 (1968, 142, 296), 159 (1966, 55), 161 (1968, 296), 165 (1968, 142), 168 (1966, 55, 235), 170 (1965, 92; 1966, 55, 235; 1968, 296), 171 (1967, 338), 179 (1968, 142), 181 (1967, 338), 182 (1965, 92; 1968, 142, 296), 186 (1965, 79), 187 (1965, 79; 1968, 142), 188 (1967, 338), 190 (1969, 161), 191 (1968), 142, 296), 192 (1968, 296), 196 (1966, 235), 197 (1965, 79; 1968, 142), 200 (1966, 55; 1968, 142), 204 (1969, 161), 206 (1969, 161), 210 (1968, 142), 211 (1968, 142), 212 (1968, 142, 296), 214 (1965, 79), 215 (1966, 55), 216 (1967, 152), 217 (1969, 161), 220 (1967, 152), 227 (1966, 62), 230 (1968, 88), 231 (1968, 88; 1969, 161), 238 (1968, 88), 240 (1966, 62), 244 (1969, 161), 248 (1968, 88), 251 (1968, 88), *310, 313, 317, 323, 328, 332, 334, 338, 343, 344*
Bermann, M., 118 (1966, 288), 153 (1969, 348), 156 (1969, 348), 173 (1969, 348), 176 (1969, 348), 202 (1969, 348), 205 (1969, 348), 206 (1969, 348), 207 (1969, 348), 257 (1966, 288), 258 (1966, 288), 259 (1966, 288), 260 (1966, 288), 261 (1966, 288), 262 (1966, 288), 263 (1966, 288), 264 (1966, 288), 265 (1966, 288), 267 (1966, 288), 268 (1966, 288), 269 (1966, 288), 282 (1969, 376), *318, 349, 350*
Berner-Fenz, L., 240 (1968, 345), 241 (1968, 345), 242 (1968, 345), 243 (1968, 345), 245 (1968, 345), *339*
Bernheim, R. A., 88 (1967, 205; 1968, 185), *325, 335*
Berninger, C. J., 208 (1968, 333), 234 (1968, 333), *339*
Berry, R. S., 25, *307*
Bertrand, R. D., 64 (1969, 132), 89 (1969, 132), 90 (1967, 218; 1969, 132), 92 (1967, 218), 114 (1967, 218), 133 (1967, 218), 135 (1967, 218), 145 (1967, 218), 159 (1967, 334), 161 (1967, 334), 220 (1967, 334), *325, 328, 343*
Beslier, L., 19 (1969, 45), 83 (1969, 178), 107 (1969, 45), 119 (1969, 45), 120 (1969, 45), 121 (1969, 45), 127 (1969, 45), 130 (1969, 45), 139 (1969, 45, 178), 143 (1969, 45), 152 (1969, 45), 153 (1969, 45), 162 (1969, 45), 163 (1969, 45, 178), 172 (1969, 45, 178), 173 (1969, 178), 175 (1969, 45), 181 (1969, 45, 178), 190 (1969, 178), 199 (1969, 178), 202 (1969, 45, 178), 209 (1969, 45, 178), 216 (1969, 45), 233 (1969, 45), *341, 344*
Bestmann, H. J., 59 (1967, 122), 86 (1966, 140; 1967, 192; 1968, 169; 1969, 206, 207), 227 (1967, 192; 1968, 169), 228 (1968, 169), 230 (1967, 192), 231 (1967, 192; 1968, 169), 234 (1967, 192), 238 (1966, 140; 1967, 192), 241 (1969, 206), 242 (1967, 122), 243 (1967, 192; 1969, 206), 249 (1967, 192), *315, 322, 324, 334, 345*
Bezman, I. I., 79 (1966, 95; 1967, 146), 278 (1967, 146), 279 (1967, 146), 280 (1967, 146), 281 (1966, 95), 282 (1966, 95), *314, 323*
Bhatia, S. B., 29 (1969, 75), 144 (1968,

313), 150 (1968, 315), 151 (1968, 313), 156 (1968, 313), 161 (1968, 315), 166 (1968, 313), 196 (1968, 315), 197 (1968, 313, 315), 198 (1968, 313), 204 (1968, 313), 205 (1967, 350, 351; 1968, 313), 211 (1967, 350, 351), 218 (1969, 75), 219 (1967, 350; 1968, 315; 1969, 75), 224 (1968, 315), 228 (1968, 315), 269 (1967, 350, 351; 1968, 315), 270 (1967, 350), 283 (1967, 350), *329*, *338*, *341*

Bhattacharya, M., 88 (1967, 206, 207; 1968, 186), 189 (1968, 186), *325*, *335*

Bibler, J. P., 91 (1966, 159; 1969, 275), *315*, *347*

Bickelhaupt, F., 78 (1968, 120), 195 (1968, 120), 196 (1968, 120), 227 (1968, 340), *333*, *339*

Biddlestone, M., 79 (1968, 128), 287 (1968, 128), *333*

Biellman, J. F., 92 (1968, 258), *337*

Biermann, U., 60 (1969, 129), 67 (1969, 129), 98 (1969, 129), 100 (1969, 129), 103 (1967, 288), 115 (1969, 129), *327*, *343*

Bigler, A. J., 119 (1969, 337), 123 (1969, 337), 127 (1969, 337), 132 (1969, 337), 135 (1969, 337), 144 (1968, 313), 150 (1968, 315), 151 (1968, 313), 156 (1968, 313), 166 (1968, 313), 180 (1969, 337), 197 (1968, 313), 198 (1968, 313), 204 (1968, 313), 205 (1968, 313), *338*, *349*

Binder, H., 7 (1965, 4), 13 (1967, 30), 18 (1967, 30), 20 (1967, 30, 42), 63 (1966, 77; 1967, 126), 65 (1969, 135), 78 (1967, 142), 97 (1965, 4; 1967, 277), 98 (1965, 4; 1967, 277), 99 (1967, 42), 103 (1965, 4), 105 (1965, 4), 106 (1965, 4), 109 (1967, 42, 277), 110 (1967, 277), 111 (1967, 277), 114 (1967, 277), 115 (1965, 4), 118 (1967, 277), 121 (1967, 42), 123 (1969, 135), 130 (1966, 238; 1967, 277), 131 (1966, 238; 1967, 277), 132 (1967, 30, 277), 136 (1967, 277), 140 (1967, 277), 141 (1967, 277), 142 (1967, 277), 143 (1967, 126), 151 (1966, 77, 238), 155 (1967, 126), 160 (1967, 277), 169 (1967, 277), 172 (1967, 126), 175 (1967, 126), 178 (1967, 277), 179 (1966, 77; 1967, 126), 185 (1966, 238), 186 (1967, 277), 187 (1967, 30, 42, 277), 193 (1967, 277), 194 (1967, 277), 203 (1967, 277), 204 (1967, 277), 206 (1967, 277), 212 (1967, 277), 213 (1967, 277), 214 (1967, 277), 218 (1967, 277), 223 (1967, 277), 224 (1967, 277), 236 (1969, 368), 244 (1967, 277), 252 (1967, 277), 256 (1967, 277), 261 (1969, 368), 262 (1967, 126), 263 (1967, 126), 265 (1967, 277, 386; 1969, 135, 368), 267 1967, 126), 268 (1967, 126, 142; 1969, 135, 368), 269 (1966, 77; 1967, 126), 270 (1967, 277), 288 (1967, 126), *308*, *313*, *317*, *320*, *323*. *326*, *330*, *343*, *350*

Birchall, T., 18 (1966, 11), 91 (1969, 256), 99 (1966, 11), *312*, *347*

Bird, P. H., 90 (1968, 195), *335*

Birum, G. H., 51 (1967, 102), 59 (1967, 102, 123; 1968, 110), 64 (1966, 78; 1967, 123), 90 (1967, 102), 223 (1966, 78, 270; 1968, 110), 224 (1968, 338), 230 (1966, 270; 1968, 110), 234 (1966, 270), 237 (1967, 102; 1968, 110), 238 (1966, 270), 242 (1967, 102), 244 (1967, 102; 1968, 110), 245 (1967, 102; 1968, 110, 338), 247 (1967, 102; 1968, 110), 249 (1968, 110), 251 (1968, 110), 254 (1968, 110), 256 (1967, 123), 269 (1967, 102), 274 (1966, 78), 275 (1966, 290; 1967, 102, 123; 1968, 110), 276 (1967, 102; 1968, 110, 338), 283 (1966, 78; 1967, 102), 284 (1966, 78), 286 (1966, 78), 287 (1966, 78), *313*, *318*, *319*, *322*, *333*, *339*

Bissey, J. E., 110 (1967, 295), 116 (1967, 295), 123 (1967, 295), 161 (1967, 295), 207 (1967, 295), 208 (1967, 295), *327*

Bistrov, V. F., 51 (1965, 25), 123 (1965, 25), *309*

AUTHOR INDEX 359

Bivel, P., 86 (1967, 189), 94 (1967, 189), 100 (1967, 189), 101 (1967, 189), *324*
Bjorkstam, J. L., 88 (1967, 208), *325*
Bladon, P., 23 (1966, 33), 39 (1966, 33), 77 (1966, 39), 123 (1966, 33), *312*
Blanchard, M. L., 234 (1967, 365), 249 (1967, 365), 253 (1967, 365), 254 (1967, 365), 255 (1967, 365), *329*
Blass, H., 49 (1966, 57), 170 (1966, 57), *313*
Blazer, T. A., 271 (1969, 375), *350*
Blinc, R., 87 (1968, 178), 88 (1968, 178, 187), *335*
Blomquist, A. T., 114 (1967, 300), 246 (1967, 300), 276 (1967, 300), *327*
Blumbergs, P., 208 (1968, 334), *339*
Bobst, A. M., 89 (1969, 313), *348*
Bock, H., 100 (1965, 78), 101 (1965, 78), 114 (1965, 78), 142 (1965, 78), 223 (1965, 78), 259 (1965, 78), 265 (1965, 78), *310*
Bodkin, C., 40 (1969, 92), 41 (1969, 92), 83 (1969, 92), 118 (1969, 92), 137 (1969, 92), *342*
Bogat'skii, A. V., 83 (1967, 161, 162), 146 (1967, 161, 162), 158 (1967, 161), 168 (1967, 161, 162), *323*
Bogeat, G., 79 (1966, 92), 278 (1966, 92), 283 (1966, 92), *314*
Bogolyubov, G. M., 125 (1965, 89), 133 (1965, 89), 142 (1965, 89), 162 (1965, 96), 171 (1965, 89), 201 (1965, 96), 218 (1965, 89), 237 (1965, 96), 242 (1965, 96), *310*
Boisard-Gerde, 245 (1968, 146), 252 (1968, 146), *334*
Boisdon, M. T., 191 (1969, 359), 197 (1969, 359), 199 (1969, 359), 202 (1969, 359), 204 (1969, 359), 210 (1969, 359), 216 (1969, 359), 232 (1969, 359), 236 (1969, 359), *349*
Bolton, E. S., 91 (1969, 254), *347*
Bon, M., 83 (1967, 164), 127 (1967, 164), 139 (1967, 164, 321), 168 (1967, 164), 205 (1967, 164), 216 (1967, 164), *324, 328*
Bonati, F., 86 (1967, 177), 91 (1967, 232), 92 (1967, 251), 262 (1967, 177), 263 (1967, 177), 272 (1967, 177), *324, 325, 326*
Bond, A., 29 (1968, 61), 32 (1968, 61), 53 (1968, 61), 135 (1968, 61), 144 (1968, 61), 150 (1968, 61), 156 (1968, 61), *331*
Booth, B. L., 91 (1966, 165), *315*
Borecka, B., 51 (1966, 61), 208 (1966, 61), 214 (1966, 61), 219 (1966, 61), 234 (1966, 61), 247 (1966, 61), *313*
Bŏrkovec, A. B., 141 (1966, 246), 149 (1966, 246), 150 (1966, 246), 161 (1966, 246), 170 (1966, 246), 178 (1966, 246), 179 (1966, 246), 184 (1966, 246), 192 (1966, 246), 195 (1966, 246), 199 (1966, 246), 200 (1966, 246), 206 (1966, 246), *317*
Bormann, D., 278 (1968, 355), *339*
Bornais, J., 86 (1968, 173), *334*
Börner, D., 23 (1969, 67), 35 (1969, 67), 244 (1969, 67), 245 (1969, 67), *341*
Borodin, P. M., 59 (1968, 109), 87 (1968, 109), 93 (1968, 109), 184 (1968, 109), 194 (1968, 109), *333*
Boros, E. J., 27 (1966, 37; 1968, 53), 28 (1968, 53), 29 (1968, 53), 37 (1966, 37), 38 (1966, 37), 42 (1966, 37), 43 (1968, 85), 91 (1968, 53), 92 (1968, 53), 117 (1966, 37; 1968, 53), 125 (1966, 37; 1968, 53, 85), 133 (1966, 37), 136 (1968, 85), 166 (1968, 53), 257 (1966, 37), 258 (1966, 37), *312, 331, 332*
Borowitz, I. J., 58 (1967, 121), 78 (1967, 121, 141), 86 (1966, 115), 86 (1967, 121, 141), 172 (1967, 121), 183 (1967, 121), 189 (1967, 121), 190 (1967, 121), 198 (1967, 121), 199 (1967, 121), 205 (1967, 121), 225 (1967, 121), 234 (1966, 115), 252 (1967, 141), *314, 322, 323*
Borownicki, W., 88 (1968, 191), *335*
Bose, M., 88 (1967, 206, 207; 1968, 186), 189 (1968, 186), *325, 335*
Bothner-By, A. A., 32, 107 (1969, 79), 173 (1969, 79), *342*
Bouchet, P., 13 (1967, 25), *320*
Bourn, A. J. R., 114 (1967, 298), *327*
Bramley, R., 92 (1966, 188), *316*

Bray, P. J., 88 (1961, 3), 97 (1961, 3), *307*
Brazier, J. F., 83 (1967, 165), 86 (1967, 165), 128 (1967, 165), 136 (1967, 165), 153 (1967, 165), 187 (1967, 165), *324*
Brecht, H., 7 (1968, 10), 38 (1969, 87), 39 (1969, 87), 45 (1969, 87), 46 (1969, 87), 59 (1968, 111), 87 (1968, 8), 97 (1967, 276), 98 (1967, 276), 109 (1967, 9), 122 (1969, 87), 128 (1967, 9), 132 (1967, 276; 1969, 87), 141 (1967, 276), 150 (1969, 87), 154 (1969, 87), 164 (1969, 87), 175 (1969, 87), 182 (1969, 87), 185 (1968, 8), 186 (1967, 276; 1968, 8), 187 (1967, 9, 344; 1968, 8), 197 (1967, 9; 1968, 8; 1969, 87), 203 (1967, 9; 1968, 8; 1969, 87), 204 (1967, 344; 1968, 8; 1969, 87), 210 (1967, 344; 1968, 8; 1969, 87), 223 (1967, 276), 228 (1969, 87), 259 (1968, 111), 260 (1968, 111), 261 (1968, 111), 267 (1968, 111), 268 (1968, 111), 270 (1967, 9, 344; 1968, 111), 271 (1967, 344; 1968, 111), 272 (1967, 9, 344; 1968, 111), 273 (1967, 9), *319, 326, 328, 330, 333, 342*
Brehamet, L., 65 (1970, 2), *350*
Brekhunets, A. G., 88 (1969, 226), *346*
Brelivet, J., 78 (1969, 148), 151 (1969, 148), 163 (1969, 148), *343*
Brey, W. S., 27 (1966, 38; 1967, 65), 45 (1967, 65), 46 (1967, 65), 84 (1966, 105), 97 (1966, 105), 110 (1967, 65), 115 (1967, 65), 122 (1966, 38), 123 (1967, 65), 129 (1967, 65), 134 (1967, 65), 140 (1966, 105), 143 (1967, 65), 175 (1967, 65), 188 (1967, 65), 194 (1966, 105), 197 (1967, 65), 204 (1967, 65), 209 (1966, 105), 224 (1967, 65), 244 (1966, 105), *312, 314*
Brini, M., 90 (1969, 252), *346*
Brookes, P. R., 92 (1967, 257), 94 (1967, 257), 241 (1967, 257), *326*
Broom, A. D., 94 (1968, 275), *337*
Brophy, J. J., 257 (1967, 371), 259 (1967, 371), 260 (1967, 371), 261 (1967, 371), 263 (1967, 371), 264 (1967, (371), 267 (1967, 371), 271 (1967, 371), 272 (1967, 371), 273 (1967, 371, 389), 274 (1967, 371), 275 (1967, 371), 276 (1967, 371), *329, 330*
Brown, D. H., 86 (1965, 43), 109 (1966, 217), 120 (1966, 217), 121 (1966, 217), 122 (1966, 217), 138 (1966, 217), 160 (1966, 217), 259 (1965, 43), *309, 317*
Brown, D. M., 94 (1965, 75), 119 (1965, 88), 127 (1965, 88), 138 (1965, 88), 162 (1965, 88), 163 (1965, 88), 167 (1965, 88), 176 (1965, 99), *310, 311*
Brown, N. M. D., 23 (1966, 33), 39 (1966, 33), 77 (1966, 33), 123 (1966, 33), *312*
Brown, T. H., 92 (1969, 306), *348*
Brownstein, S., 84 (1969, 186), 86 (1967, 193; 1968, 173), *324, 334, 345*
Bruce, M. I., 93 (1968, 264), *337*
Bruce, R., 90 (1967, 220), 92 (1967, 220), *325*
Brunn, E., 196 (1969, 360), 251 (1969, 360), *349*
Brunner, H., 90 (1969, 247), 91 (1969, 247, 255), 93 (1969, 247), 266 (1969, 247), *346, 347*
Bryant, R. G., 89 (1969, 311), *348*
Bryson, J. G., 29 (1967, 78; 1969, 77), 54 (1967, 78; 1969, 77), 123 (1967, 78), 124 (1967, 78; 1969, 77), 134 (1969, 77), 237 (1969, 77), *321, 342*
Bucci, P., 7 (1968, 17), 8, 113 (1968, 17), 139 (1968, 17), *330*
Büchel, K. H., 191 (1965, 104), *311*
Buckingham, A. D., 91 (1965, 59), 92 (1965, 59), 93 (1965, 59), *309*
Buchs, A., 189 (1966, 259), *318*
Bude, D., 110 (1968, 296), 122 (1968, 296), 129 (1968, 296), 154 (1968, 296), 161 (1968, 296), 170 (1968, 296), 171 (1967, 338), 181 (1967, 338), 182 (1968, 296), 188 (1967, 338), 191 (1968, 296), 192 (1968, 296), 212 (1968, 296), *328, 338*
Budenz, R., 104 (1966, 210), *316*
Buhler, D. R., 94 (1971, 1), *350*

Bulyanitsa, L. S., 84 (1969, 183), *344*
Bunting, W. M., 81 (1967, 158), 255 (1967, 158), *323*
Buono, G., 58 (1969, 127), 114 (1969, 127), 119 (1969, 127), 125 (1969, 127), 126 (1969, 127), 156 (1969, 127), 158 (1969, 127), *343*
Burg, A. B., 10 (1968, 23; 1969, 27), 17 (1967, 40), 20 (1967, 40), 33 (1967, 40), 37 (1967, 40), 38 (1969, 84), 60 (1967, 40), 61 (1967, 40), 63 (1967, 40), 76 (1969, 84), 88 (1969, 27), 90 (1969, 243), 92 (1966, 174; 1969, 286), 104 (1967, 40), 106 (1967, 40; 1968, 294), 110 (1967, 40; 1968, 23), 111 (1969, 84), 116 (1969, 27), 172 (1969, 243), 257 (1967, 40), 258 (1967, 40), 259 (1969, 27), 278 (1969, 27), *316, 320, 331, 338, 340, 342, 346, 347*
Burgada, R., 19 (1967, 39), 39 (1967, 39), 47 (1969, 109), 78 (1968, 126; 1969, 109), 83 (1967, 39, 164; 1968, 154; 1969, 109), 86 (1968, 126; 1969, 201, 202), 119 (1967, 39; 1968, 154), 122 (1968, 126), 127 (1967, 164), 128 (1969, 109), 133 (1969, 109), 134 (1968, 154), 136 (1969, 340), 139 (1967, 164, 321), 141 (1968, 126), 145 (1969, 109), 149 (1969, 109), 151 (1969, 109), 154 (1969, 109), 161 (1968, 321; 1969, 340), 167 (1968, 324), 168 (1967, 164; 1968, 324; 1969, 357), 169 (1968, 321; 1969, 109), 170 (1968, 321; 1969, 357), 176 (1968, 324; 1969, 340), 178 (1968, 321; 1969, 340), 181 (1968, 154; 1969, 109), 184 (1968, 324; 1969, 340), 190 (1968, 324; 1969, 340, 357), 191 (1969, 109, 357), 192 (1968, 154), 193 (1968, 324; 1969, 340), 194 (1969, 340), 199 (1968, 321), 205 (1967, 164), 206 (1968, 321; 1969, 340), 207 (1969, 362), 215 (1969, 362), 216 (1967, 164; 1969, 109), 217 (1969, 109), 218 (1969, 362), 220 (1969, 340), 227 (1969, 340), 229 (1968, 154), 260 (1968, 154), 262 (1968, 126), *320, 324,* *328, 333, 334, 338, 342, 345, 349*

Bürger, H., 13 (1967, 29), 18 (1967, 29), 49 (1967, 29), 115 (1967, 29), 143 (1967, 29), 171 (1967, 29; 1968, 327), 204 (1968, 327), *320, 339*
Bürger, W., 144 (1969, 346), 157 (1969, 346), 169 (1969, 346), 177 (1969, 346), 178 (1969, 346), 264 (1969, 346), 266 (1969, 346), 267 (1969, 346), *349*
Burke, J. J., 56 (1967, 115), 82 (1967, 115), 152 (1967, 115), 222 (1967, 115), *322*
Burpitt, R. D., 80 (1965, 36), 161 (1965, 36), 170 (1965, 36), 178 (1965, 36), *309*
Burpo, D. H., 171 (1967, 338), 181 (1967, 338), 188 (1967, 338), *328*
Buslaev, Yu, A., 99 (1966, 207), *316*
Butler, P. E., 58 (1966, 74), 120 (1966, 74), 125 (1966, 74), 127 (1966, 74), 130 (1966, 74), 131 (1966, 74), 136 (1966, 74), 137 (1966, 74), 146 (1966, 74), 148 (1966, 74), *313*
Butova, T. D., 83 (1967, 161), 146 (1967, 161), 158 (1967, 161), 168 (1967, 161), *323*
Buttery, H. J., 24 (1968, 51), 83 (1968, 51), 99 (1968, 51), *331*
Bryne, P., 94 (1966, 200), *316*

C

Cadiot, P., 81 (1967, 153), 85 (1968, 161), 134 (1966, 243), 144 (1966, 243), 201 (1966, 243), 208 (1966, 243), 218 (1966, 243), 227 (1966, 243), 235 (1967, 153), 239 (1967, 153), *317, 323, 334*
Cadogan, J. I. G., 55 (1969, 119), 132 (1969, 119), 151 (1969, 119), 153 (1969, 119), 164 (1969, 119), 165 (1969, 119), 169 (1967, 336), 171 (1969, 119), 173 (1969, 119), 189 (1969, 119), 190 (1969, 119), 203 (1969, 119), 223 (1969, 119), 224 (1969, 119), *328. 343*
Cady, G. H., 21 (1966, 29), 99 (1966, 29), *312*

Caesar, F., 80 (1969, 156), 220 (1969, 156), 226 (1969, 156), 229 (1969, 156), 232 (1969, 156), 235 (1969, 156), 239 (1969, 156), 241 (1969, 156), 243 (1969, 156), 247 (1969, 156), 253 (1969, 156), *344*
Caglio, G., 91 (1969, 261), *347*
Calderazzo, F., 92 (1968, 236), 93 (1968, 236), *336*
Callis, C. F., 5 (1956, 1), 65 (1957, 2), *307*
Calvin, M., 86 (1963, 5), 284 (1963, 5), *307*
Campbell, B. S., 40 (1969, 93), 82 (1969, 93), 89 (1969, 93), 126 (1969, 93), 147 (1969, 93), 148 (1969, 93), 157 (1969, 93), 163 (1969, 93), 171 (1969, 93), *342*
Campbell, J. M., 90 (1969, 235), *346*
Canadine, R. M., 55 (1966, 69), 80 (1969, 155), 92 (1969, 155), 223 (1966, 69), 269 (1969, 155), *313, 344*
Caplan, P. J., 87 (1969, 214, 215), 121 (1968, 306), 140 (1968, 306), *345*
Capmau, M. L., 81 (1967, 154, 155), 225 (1967, 154), 229 (1967, 154, 155), 232 (1967, 155), 236 (1967, 155), 239 (1967, 155), 242 (1967, 155), 245 (1967, 155; 1968, 146), 248 (1967, 155), 252 (1968, 146), *323, 334*
Cargioli, J., 42 (1969, 101), 80 (1969, 101, 159), 92 (1969, 159), 140 (1969, 101), 159 (1969, 101), 169 (1969, 101), 178 (1969, 101), 191 (1969, 101), 208 (1969, 159), 215 (1969, 159), *342, 344*
Caroll, R. L., 86 (1966, 131), 99 (1966, 131), 101 (1966, 131), *315*
Carr, S. L., 88 (1969, 218), *346*
Carrie, R., 154 (1968, 319), 165 (1968, 323), 175 (1968, 323), 179 (1968, 319), 197 (1968, 323), 204 (1968, 323), 211 (1968, 323), 218 (1968, 319), *338*
Carroll, A. P., 79 (1966, 84), 278 (1966, 84), *313*
Carroll, R. L., 86 (1967, 129, 190), 100 (1967, 190), 101 (1967, 129, 190), 129 (1967, 315), 259 (1967, 315), 277 (1967, 315), *323, 324, 328*
Carter, R. P., 86 (1967, 190), 100 (1967, 190), 101 (1967, 190), 129 (1967, 315), 259 (1967, 315), 277 (1967, 315, 391), *324, 328, 330*
Carter, W., 91 (1969, 256), *347*
Carty, A. J., 91 (1969, 256), 271 (1967, 387), 272 (1967, 387), 273 (1967, 387), *330, 347*
Casedy, G. A., 159 (1967, 334), 161 (1967, 334), 220 (1967, 334), *328*
Cassoux, P., 92 (1967, 236, 243; 1969, 284), 104 (1967, 236), 107 (1967, 236), 112 (1967, 236), 118 (1967, 236), *325, 326, 347*
Castellano, S., 49 (1968, 97), 135 (1968, 97), 137 (1968, 97), *332*
Castro, B., 86 (1969, 201, 202), *345*
Caughlan, C. N., 38 (1967, 86), 39 (1967, 86), 114 (1967, 86), 125 (1967, 86), 138 (1967, 86), 148 (1967, 86), *321*
Cauquis, G., 79 (1966, 92), 278 (1966, 92), 283 (1966, 92), *314*
Cavell, R. G., 13 (1967, 33), 18 (1967, 33), 19 (1967, 33), 21 (1967, 33; 1968, 45; 1969, 57, 59), 22 (1966, 32; 1967, 33; 1968, 45, 46, 47), 23 (1968, 46), 27 (1968, 55), 34 (1968, 47, 55; 1969, 57), 39 (1968, 45), 42 (1969, 99), 43 (1968, 45), 60 (1968, 55), 61 (1968, 55; 1969, 57, 59), 63 (1968, 55), 75 (1966, 32; 1967, 33), 76 (1968, 45, 46, 47, 55; 1969, 99), 77 (1968, 46), 81 (1968, 46), 84 (1968, 45), 86 (1968, 45), 98 (1966, 32; 1967, 33; 1968, 45; 1969, 99), 99 (1968, 46), 100 (1969, 57), 101 (1969, 57, 59), 104 (1968, 45), 105 (1968, 47), 106 (1968, 47), 107 (1968, 47), 115 (1968, 55), 116 (1968, 55), 117 (1968, 47), 122 (1967, 309; 1968, 46), 131 (1968, 45), 132 (1968, 45), 142 (1969, 59), 244 (1968, 46), 258 (1968, 55; 1969, 57), *312, 320, 327, 331, 341, 342*
Cawse, J. N., 90 (1969, 236), 91 (1969, 236), 92 (1969, 236), *346*

Cenini, S., 86 (1967, 177), 92 (1967, 251), 262 (1967, 177), 263 (1967, 177), 272 (1967, 177), *324, 326*

Centofanti, L., 18 (1968, 34), 20 (1968, 34), 21 (1968, 34), 22 (1968, 34), 65 (1968, 34), 74 (1968, 34; 1969, 141), 84 (1969, 141), 92 (1969, 141), 97 (1969, 141), 98 (1968, 34), 100 (1968, 34), 110 (1969, 141), *327, 331, 343*

Cerutti, P. A., 89 (1969, 313), *348*

Chabrier, P., 159 (1969, 352), 202 (1969, 352), *349*

Chan, S., 82 (1969, 169), 94 (1969, 315, 317), 107 (1969, 169), 112 (1969, 169), 118 (1969, 169), 151 (1969, 169), *344, 348*

Chan, S. S., 22 (1968, 48), 103 (1968, 48), 106 (1968, 48), 110 (1968, 48), *331*

Chance, L. H., 39 (1967, 297), 112 (1967, 297), *327*

Chang, B. C., 86 (1968, 168), 186 (1968, 168), 218 (1968, 168), 223 (1968, 168), *334*

Chang, C. H., 78 (1968, 121), 82 (1968, 121), 166 (1968, 121), 192 (1968, 121), 204 (1968, 121), *333*

Chang, H., 94 (1968, 281), 110 (1968, 281), *337*

Chang, Y. F., 92 (1969, 281), *347*

Chapman, A. C., 87 (1968, 175), 259 (1967, 381), 262 (1967, 381), 263 (1967, 381), *329, 335*

Chapman, D., 89 (1966, 197), 93 (1966, 199, 200), 94 (1967, 268; 1968, 277, 279; 1969, 320), 152 (1968, 277), *316, 326, 337, 348*

Charbonnel, Y., 161 (1968, 321), 168 (1969, 357), 169 (1968, 321), 170 (1968, 321; 1969, 357), 178 (1968, 321), 190 (1969, 357), 191 (1969, 357), 199 (1968, 321), 206 (1968, 321), *338, 349*

Charlton, T. L., 13 (1967, 33), 18 (1967, 33), 19 (1967, 33), 21 (1967, 33; 1968, 45; 1969, 57, 59), 22 (1966, 32; 1967, 33; 1968, 45), 34 (1969, 57), 39 (1968, 45), 42 (1969, 99), 43 (1968, 45), 61 (1969, 57, 59), 75 (1966, 32; 1967, 33), 76 (1968, 45; 1969, 99), 84 (1968, 45), 86 (1968, 45), 98 (1966, 32; 1967, 33; 1968, 45; 1969, 99), 100 (1969, 57), 101 (1969, 57, 59), 104 (1968, 45), 131 (1968, 45), 132 (1968, 45), 142 (1969, 59), 258 (1969, 57), *312, 320, 331, 341, 342*

Charrier, C., 31 (1967, 80), 47 (1967, 80), 51 (1967, 104; 1968, 99; 1969, 115), 53 (1967, 106), 81 (1967, 80), 111 (1967, 104; 1968, 99; 1969, 115), 116 (1968, 99; 1969, 115), 122 (1967, 80), 123 (1967, 104), 133 (1967, 106), 134 (1966, 243), 136 (1967, 106), 144 (1966, 243; 1967, 104; 1968, 99; 1969, 115), 145 (1968, 99; 1969, 115), 146 (1967, 104, 106; 1968, 99; 1969, 115), 153 (1967, 80), 154 (1967, 106), 157 (1967, 104; 1968, 99; 1969, 115), 158 (1967, 106), 162 (1967, 104; 1969, 115), 166 (1967, 104; 1969, 115), 167 (1969, 115), 173 (1967, 80), 174 (1967, 80), 179 (1967, 106), 180 (1967, 80, 106; 1968, 99; 1969, 115), 197 (1967, 80, 106), 201 (1966, 243), 208 (1966, 243; 1967, 104; 1968, 99; 1969, 115), 214 (1969, 115), 218 (1966, 243), 227 (1966, 243), *317, 321, 322, 333, 343*

Chasle, M. F., 142 (1969, 345), 204 (1969, 345), 229 (1969, 345), 242 (1969, 345), *349*

Chatt, J., 89 (1968, 217), 91 (1965, 61; 1968, 217; 1969, 263), 92 (1966, 189; 1968, 217, 249, 250; 1969, 263, 300, 301), 93 (1968, 217; 1969, 263), *309, 316, 336, 347, 348*

Chen, C. C., 91 (1969, 274), *347*

Chen, E. H., 150 (1968, 315), 161 (1968, 315), 196 (1968, 315), 197 (1968, 315), 219 (1968, 315), 224 (1968, 315), 228 (1968, 315), 269 (1968, 315), *338*

Chen, G. Y., 40 (1969, 94), 41 (1969, 94), 42 (1969, 94), 84 (1969, 94), 85 (1969, 94), 112 (1969, 94), 119 (1969, 94), 126 (1969, 94), 136 (1969, 94), 137 (1969, 94), 138

(1969, 94), 145 (1969, 94), 147 (1969, 94), 148 (1969, 94), *242*
Chen, T. M., 91 (1968, 225), *336*
Cherbuliez, E., 189 (1966, 259), *318*
Chervin, I. I., 7 (1967, 18), *319*
Chiusoli, G. P., 92 (1967, 237), 171 (1967, 237), *325*
Chivers, T., 79 (1968, 127; 1969, 151), 101 (1969, 151), 102 (1969, 151), 278 (1968, 127; 1969, 151), 284 (1968, 127), 286 (1968, 127), 287 (1968, 127; 1969, 151), *333, 344*
Chizhik, V. I., 59 (1968, 109), 87 (1968, 109), 93 (1968, 109), 184 (1968, 109), 194 (1968, 109), *333*
Chmielewicz, Z. F., 86 (1965, 42), 146 (1965, 42), 167 (1965, 42), 183 (1965, 42), 200 (1965, 42), 213 (1965, 42), *309*
Chodkiewicz, W., 81 (1967, 153, 154, 155, 156; 1969, 164, 165, 166), 134 (1966, 243), 144 (1966, 243), 201 (1966, 243), 207 (1969, 166), 208 (1966, 243), 217 (1969, 365), 218 (1966, 243), 220 (1967, 156), 225 (1967, 154), 227 (1966, 243), 229 (1967, 154, 155; 1969, 165), 232 (1967, 155, 156; 1969, 165, 365), 233 (1967, 156), 235 (1967, 153), 236 (1967, 155; 1969, 164, 165), 237 (1969, 165), 239 (1967, 153, 155; 1969, 164, 165), 240 (1969, 165), 241 (1969, 165, 166), 242 (1967, 155), 243 (1969, 164, 165), 245 (1967, 155, 156; 1968, 146; 1969, 165), 246 (1969, 164, 165), 248 (1967, 155; 1969, 164), 252 (1968, 146; 1969, 165), *317, 323, 334, 344, 349*
Chopard, P. A., 86 (1966, 116; 1967, 178), 237 (1967, 369), 238 (1963, 8), 243 (1966, 116), 261 (1967, 178), 275 (1966, 291), 276 (1967, 369), *308, 314, 319, 324, 329*
Chorvat, R. J., 32 (1967, 82; 1968, 67), 67 (1968, 67), 78 (1967, 82; 1968, 67, 121), 82 (1968, 121; 1969, 170), 166 (1968, 121), 167 (1967, 82), 177 (1967, 82), 189 (1967, 82), 192 (1968, 121), 198 (1967, 82), 199 (1967, 82), 204 (1968, 121), 205 (1967, 82), 211 (1968, 67), 212 (1967, 82; 1968, 67), 217 (1967, 82; 1969, 170), 236 (1968, 67; 1969, 170), 237 (1968, 67), 240 (1968, 67), *32, 332, 333, 344*
Chowdhuri, A., 88 (1968, 186), 189 (1968, 186), *335*
Chowdhury, A., 88 (1967, 206, 207), *325*
Christianson, H. R., 94 (1968, 283), 118 (1968, 283), 128 (1968, 283), 157 (1968, 283), 174 (1968, 283), 182 (1968, 283), *337*
Christmann, K. F., 152 (1967, 359), 223 (1967, 359), 235 (1967, 359), 242 (1967, 359), 247 (1967, 359), *329*
Christol, H., 28 (1967, 72), 109 (1967, 72), 113 (1967, 72), 114 (1967, 72), 121 (1967, 72), 129 (1967, 72), *321*
Chu, S. K., 79 (1968, 133), 280 (1968, 133), *333*
Chuck, R. J., 83 (1965, 39), 139 (1965, 39), 160 (1965, 39), *309*
Church, M. J., 91 (1968, 218, 219), 92 (1968, 218, 219), *336*
Churi, R. H., 32, 137 (1966, 245; 1967, 81), 166 (1967, 81), 172 (1967, 81), 181 (1967, 81), 183 (1966, 245; 1967, 81); 189 (1967, 81), 190 (1966, 245; 1967, 81), 191 (1967, 81), 198 (1966, 245; 1967, 81), 205 (1967, 81), 215 (1966, 245), 225 (1966, 245; 1967, 81), *317, 321*
Chuvylkin, N. D., 92 (1969, 282), 169 (1969, 282), *347*
Claeys, E., 4 (1969, 56), 88 (1969, 56), 101 (1969, 56), 131 (1969, 120), 185 (1969, 120), 222 (1969, 120), 223 (1969, 120), *340, 343*
Clark, R. J., 64 (1969, 133), 89 (1969, 133), 90 (1969, 133, 245), 91 (1969, 133, 272), 92 (1969, 133), 93 (1969, 133, 272), 220 (1969, 272), *343, 346, 347*
Clarke, F. B., 86 (1966, 117), 100 (1966, 117), 121 (1966, 117), *314*
Claunch, R. T., 80 (1968, 142), 110 (1968, 296), 122 (1968, 296), 129 (1968, 296), 144 (1968, 142), 154

(1968, 142, 196), 161 (1968, 296), 165 (1968, 142), 170 (1968, 296), 179 (1968, 142), 182 (1968, 142, 296), 187 (1968, 142), 191 (1968, 142, 296), 192 (1968, 296), 197 (1968, 142), 200 (1968, 142), 209 (1968, 142), 211 (1968, 142), 212 (1968, 142, 296), *334, 338*
Clemens, D. F., 27 (1966, 38), 122 (1966, 38), *312*
Clemmit, A. F., 92 (1969, 294), *348*
Clough, S., 88 (1966, 149), *315*
Coates, G. E., 32 (1966, 45), 115 (1966, 45), *312*
Cocivera, M., 86 (1965, 44), *309*
Cockran, K. J., 27 (1966, 37), 37 (1966, 37), 38 (1966, 37), 42 (1966, 37), 117 (1966, 37), 125 (1966, 37), 137 (1966, 37), 257 (1966, 37), 258 (1966, 37), *312*
Coffey, R. S., 91 (1965, 61), 92 (1966, 189; 1969, 301), *309, 316, 348*
Cohen, B. M., 85 (1969, 197), 90 (1969, 231), *345, 346*
Cohen, E. A., 16 (1969, 41), 70 (1969, 41), 74 (1969, 41), 82 (1969, 148), 99 (1969, 41), 100 (1969, 41), 105 (1969, 41), 110 (1969, 41), 132 (1969, 41); *334, 341*
Cohen, J. S., 27 (1967, 69), 38 (1967, 69), 108 (1967, 69), 112 (1967, 69), 113 (1967, 69), 114 (1967, 69), 119 (1967, 69), 120 (1967, 69), 122 (1967, 69), 127 (1967, 69), 140 (1967, 69), 149 (1967, 69), 228 (1967, 69), 263 (1967, 69), *321*
Cohn, K., 21 (1968, 43), 22 (1968, 49), 46 (1968, 43), 76 (1968, 43), 90 (1968, 43), 97 (1968, 49), 98 (1968, 49), 108 (1968, 43), 170 (1968, 326), *331, 339*
Colburn, C. B., 21 (1969, 62), 22 (1969, 62), 25 (1969, 62), 43 (1969, 62), 61 (1969, 62), 66 (1969, 62), 75 (1969, 62), 76 (1969, 62), 77 (1969, 62), 100 (1966, 209; 1969, 62), 101 (1969, 62), 104 (1969, 62), 108 (1969, 62), 109 (1969, 62), 113 (1969, 62), 114 (1969, 62), 132 (1969, 62), 152 (1969, 62), 153 (1969, 62), 186 (1969, 62), 196 (1969, 62), 202 (1969, 62), 222 (1969, 62), 223 (1969, 62), *316, 341*
Collins, J. H., 86 (1966, 132), 152 (1966, 132), 153 (1966, 132), 182 (1966, 132), 207 (1966, 132), 221 (1966, 132), *315*
Collman, J. P., 90 (1969, 236), 91 (1968, 220; 1969, 236), 92 (1969, 236), 135 (1968, 220), *336, 346*
Commenges, G., 27 (1969, 72), 45 (1969, 72), 84 (1969, 72), 85 (1969, 72), 122 (1969, 72), 142 (1969, 72), *341*
Compton, R. D., 27 (1968, 53), 28 (1968, 53), 29 (1968, 53), 91 (1968, 53), 92 (1968, 53), 117 (1968, 53), 125 (1968, 53), 159 (1967, 334), 161 (1967, 334), 166 (1968, 53), 220 (1967, 334), *328, 331*
Conway, T. T., 105 (1968, 292), 110 (1968, 292), 149 (1968, 292), 154 (1968, 292), *338*
Cook, R. D., 7 (1967, 19), 109 (1967, 19), 110 (1967, 19), 114 (1967, 19), 153 (1967, 19), 154 (1967, 19), 196 (1967, 19), 197 (1967, 19), *319*
Cooke, M., 93 (1969, 307), *348*
Cornwell, C. D., 82 (1968, 148), *334*
Coskran, K. J., 25 (1965, 17), 90 (1967, 217, 218), 92 (1967, 217, 218), 114 (1967, 218), 133 (1967, 218), 135 (1967, 218), 145 (1967, 218), *308, 325*
Cosee, P., 89 (1968, 243), 92 (1968, 242, 243), *336*
Costa, G., 90 (1968, 200), *335*
Costisella, B., 47 (1968, 93), 144 (1969, 346), 157 (1969, 346), 169 (1969, 346), 177 (1969, 346), 178 (1969, 346), 190 (1968, 93), 205 (1968, 93), 264 (1968, 353; 1969, 346), 265 (1968, 93, 353), 266 (1968, 353; 1969, 346), 267 (1969, 346), *332, 339, 349*
Cotton, F. A., 50 (1963, 4), *307*
Cotton, J. D., 91 (1969, 257), *347*
Couret, C., 136 (1969, 341), 184 (1967, 342), 193 (1969, 341), 207 (1968, 331), 217 (1967, 342), 220 (1968,

331), 225 (1969, 341), 226 (1967, 342, 360), 230 (1968, 331), 243 (1969, 341), *328, 329, 339, 349*

Cowley, A. H., 14 (1967, 28; 1969, 36, 37, 38), 15, 16 (1969, 41), 17 (1967, 40), 20 (1967, 28, 40), 21 (1969, 54, 55), 27 (1965, 18), 33 (1967, 40), 34 (1967, 83), 37 (1967, 40), 45 (1965, 18), 46 (1966, 18), 47 (1969, 111), 50 (1969, 54, 55), 51 (1969, 114), 60 (1965, 26; 1967, 40), 61 (1967, 40, 83), 62, 63 (1965, 28; 1967, 40; 1969, 36, 37), 65 (1965, 28), 70 (1969, 41), 74 (1969, 41), 76 (1969, 55), 77 (1969, 55, 114), 84 (1969, 187), 99 (1967, 28; 1969, 41), 100 (1967, 28; 1969, 41), 101 (1967, 28), 104 (1967, 40, 83), 105 (1969, 41), 106 (1967, 40; 1969, 54, 55, 114), 108 (1965, 18; 1969, 55), 110 (1967, 40; 1969, 41), 114 (1967, 298), 115 (1969, 55, 114), 117 (1969, 55), 119 (1965, 18), 120 (1965, 18), 122 (1965, 18), 129 (1965, 18; 1969, 55), 132 (1969, 41, 55), 136 (1969, 55), 142 (1965, 18), 150 (1965, 18), 160 (1965, 18), 165 (1969, 354), 169 (1969, 354), 191 (1969, 354), 192 (1969, 111), 193 (1969, 354), 195 (1965, 18), 208 (1969, 354), 209 (1969, 354), 219 (1967, 28), 223 (1969, 354), 257 (1967, 40), 258 (1967, 28, 40), 259 (1965, 28; 1967, 28, 83), 278 (1967, 83), 287 (1966, 294), *308, 309, 319, 320, 321, 327, 340, 341, 342, 345, 349*

Cox, R. H., 32, 40 (1969, 93), 82, 89 (1969, 93), 107 (1969, 79), 120 (1969, 81), 126 (1969, 81, 93), 129 (1969, 81), 137 (1969, 81), 138 (1969, 81), 147 (1969, 93), 148 (1969, 93), 157 (1969, 93), 163 (1969, 93), 171 (1969, 93), 173 (1969, 79), *342*

Cradock, S., 48 (1967, 98), 99 (1967, 98), 101 (1967, 98), *322*

Cragg, R. H., 18 (1966, 21), 55 (1966, 21), 131 (1966, 31), 132 (1966, 21), *312*

Craig, D. P., 79 (1958, 1; 1962, 3), *307*

Cremer, S. E., 32 (1967, 82; 1968, 67), 67 (1968, 67), 78 (1967, 82; 1968, 67, 121), 82 (1968, 121; 1969, 170), 166 (1968, 121), 167 (1967, 82), 177 (1967, 82), 189 (1967, 82), 192 (1968, 121), 198 (1967, 82), 199 (1967, 82), 204 (1968, 121), 205 (1967, 82), 211 (1968, 67), 212 (1967, 82; 1968, 67), 217 (1967, 82; 1969, 170), 236 (1968, 67; 1969, 170), 237 (1968, 67), 240 (1968, 67), *321, 332, 333, 344*

Crepaux, D., 13 (1969, 35), *340*

Crews, P., 86 (1968, 170), 228 (1968, 170), 234 (1968, 170), 238 (1968, 170), 244 (1968, 170), *334*

Crosbie, K. D., 27 (1969, 73), 28 (1969, 73), 48 (1969, 73), 74 (1969, 73), 75 (1969, 73), 105 (1969, 73, 330), 109 (1969, 330), 110 (1969, 73), *341, 349*

Crouse, D. M., 86 (1968, 171), 238 (1968, 171), 241 (1968, 171), 249 (1968, 171), *334*

Crutchfield, M. M., 1 (1967, 1), 2 (1967, 1), 4 (1967, 1), 65 (1965, 30; 1967, 1), 90 (1965, 30), 91 (1965, 30), 101 (1965, 30), 120 (1965, 30), 270 (1967, 391), *309, 319, 330*

Cullen, W. R., 51 (1967, 103), 91 (1969, 258), 104 (1967, 103), 152 (1967, 331), 153 (1967, 103), 186 (1967, 103, 331), 214 (1965, 107), 261 (1967, 103, 331), 273 (1967, 103, 331), *311, 322, 328, 347*

Cullingworth, A. R., 29 (1966, 41), 85 (1966, 41; 1969, 197), 90 (1966, 41), 113 (1966, 41), 153 (1966, 41), 196 (1966, 41), *312, 345*

Cunliffe, A. V., 20 (1967, 55), 21 (1967, 55), 33 (1967, 55), 34 (1967, 55), 42 (1967, 55), 77 (1967, 55), 106 (1967, 55), 110 (1967, 55), *320*

Curry, J. D., 256 (1968, 349), 257 (1968, 349), 258 (1968, 349), 259 (1968, 349), 260 (1968, 349), 261 (1968, 349), 263 (1968, 349), 266 (1968, 349), *339*

D

Daigle, D., 231 (1965, 112), *311*
Daigle, D. J., 39 (1967, 297), 112 (1967, 297), *327*
Daly, J. J., 55 (1967, 114), 131 (1967, 114), *322*
Damasco, M. C., 14 (1969, 38), *340*
Danielson, R., 42 (1969, 101), 80 (1969, 101), 140 (1969, 101), 159 (1969, 101), 169 (1969, 101), 178 (1969, 101), 191 (1969, 101), *342*
Daniewski, W. M., 135 (1966, 244), 136 (1966, 244), 175 (1966, 256), 182 (1966, 256), 183 (1966, 256), 192 (1966, 256), *317, 318*
Danion, D., 154 (1968, 319), 165 (1968, 323), 175 (1968, 323), 179 (1968, 319), 197 (1968, 323), 204 (1968, 323), 211 (1968, 323), 218 (1968, 319), *338*
Danyluk, S. S., 86 (1968, 172), 94 (1968, 172), *334*
Darnall, K. R., 78 (1967, 140; 1969, 147), 82 (1967, 140), 148 (1969, 147), 177 (1969, 147), 197 (1967, 140), 203 (1967, 140), 216 (1967, 140), *323, 343*
Das, S. K., 79 (1965, 33; 1966, 85), 279 (1966, 85), 280 (1965, 33; 1966, 85), *309, 313*
Das, V. K., 85 (1968, 163), *334*
Daugherty, K. E., 37 (1968, 72), 105 (1968, 72), 114 (1968, 72), 129 (1968, 72), *332*
Davidoff, E. F., 5 (1967, 8; 1968, 6), 6 (1966, 5), 112 (1968, 6), 122 (1966, 5), 129 (1966, 5; 1968, 6), 153 (1968, 6), 158 (1968, 6), 159 (1968, 6), 164 (1968, 6), 169 (1968, 6), 172 (1968, 6), 173 (1968, 6), 174 (1968, 6), 178 (1966, 5; 1968, 6), 191 (1967, 8; 1968, 6), 193 (1967, 8; 1968, 6), 194 (1968, 6), 199 (1968, 6), 200 (1968, 6), 201 (1967, 8; 1968, 6), 203 (1965, 5), 206 (1966, 5; 1967, 8; 1968, 6), 207 (1966, 5), 208 (1967, 8), 209 (1967, 8; 1968, 6), 212 (1966, 5; 1967, 8; 1968, 6), 216 (1966, 5; 1967, 8; 1968, 6), 217 (1966, 5; 1967, 8; 1968, 6), 218 (1966, 5), 219 (1967, 8; 1968, 6); 220 (1966, 5), 224 (1967, 8; 1968, 6), 225 (1966, 5; 1968, 6), 226 (1966, 5; 1967, 8; 1968, 6), 228 (1966, 5; 1967, 8; 1968, 6), 229 (1966, 5; 1968, 6), 230 (1968, 6), 231 (1966, 5; 1968, 6), 233 (1966, 5; 1968, 6), 233 (1966, 5; 1968, 6), 234 (1966, 5), 235 (1966, 5; 1967, 8; 1968, 6), 236 (1968, 6), 238 (1966, 5), 239 (1966, 5; 1968, 6), 241 (1966, 5; 1968, 6), 242 (1966, 5), 243 (1966, 5; 1968, 6), 244 (1966, 5), 245 (1966, 5), 246 (1968, 6), 247 (1966, 5), 252 (1966, 5), 275 (1966, 5), 276 (1966, 5), *311, 319, 330*
Davidsohn, W. E., 212 (1969, 364), 217 (1969, 364), *349*
Davidson, G., 48 (1967, 98), 99 (1967, 98), 101 (1967, 98), *322*
Davidson, R. S., 39 (1967, 87), 154 (1967, 87), 197 (1967, 87), 203 (1966, 266), 209 (1966, 266), 236 (1967, 368), 246 (1966, 266), 274 (1967, 368), *318, 321, 329*
Davies, G. R., 91 (1968, 216), *336*
Davies, N., 90 (1968, 195), *335*
Davis, D. W., 78 (1968, 121), 82 (1968, 121), 166 (1968, 121), 192 (1968, 121), 204 (1968, 121), *333*
Davison, A., 91 (1962, 5), 93 (1962, 5), *307*
Davison, R. B., 55 (1966, 70), 56 (1966, 70), 163 (1966, 253), 164 (1966, 70, 253), 181 (1966, 253), 228 (1966, 70, 253), 234 (1966, 253), 268 (1966, 253), *313, 318*
Dawson, D. S., 51 (1967, 103), 104 (1967, 103), 152 (1967, 331), 153 (1967, 103), 186 (1967, 103, 331), 214 (1965, 107), 261 (1967, 103, 331), 273 (1967, 103, 331), *311, 322, 328*
Dawson, J. W., 90 (1968, 196), 93 (1968, 196), *335*
Dean, R. R., 13 (1969, 35), 20 (1967, 57), 22 (1967, 57), 33 (1967, 57), 66 (1967, 57), 92 (1968, 251), *321, 336, 340*

De Bruin, K. E., 80 (1969, 154), 182 (1969, 154), *344*
Deeming, A. J., 91 (1969, 264, 265), 92 (1969, 302), *347, 348*
de Haan, J. W., 8 (1968, 21), 175 (1968, 21), *331*
De Ketelaere, 4 (1969, 56), 88 (1969, 56), 101 (1969, 56), 131 (1969, 120), 185 (1969, 120), 186 (1969, 120), 222 (1969, 120), 223 (1969, 120), *340, 343*
De Koe, P., 78 (1968, 120), 195 (1968, 120), 196 (1968, 120), 227 (1968, 340), *333, 339*
Dell, D., 79 (1966, 86), 107 (1966, 86), 280 (1966, 86), 281 (1966, 86), 282 (1966, 86), 283 (1966, 86), *313*
Delpuech, J. J., 90 (1969, 230), *346*
Demarne, M., 85 (1968, 161), *334*
Demarcq, M. C., 86 (1966, 118), *314*
Demitras, G. C., 22 (1967, 59), 46 (1967, 59), 76 (1967, 59), 109 (1967, 59), 257 (1967, 59), *321*
Denney, D. B., 37 (1966, 46; 1969, 86), 38 (1969, 86), 40 (1969, 94), 41 (1969, 94), 42 (1969, 94), 45 (1968, 90), 82 (1966, 46), 83 (1966, 46), 84 (1969, 86, 94), 85 (1969, 94), 86 (1966, 119; 1968, 168), 90 (1966, 221), 111 (1966, 221), 112 (1969, 94), 114 (1969, 86), 118 (1966, 46), 119 (1966, 46; 1969, 94), 122 (1969, 86), 126 (1966, 46, 119; 1969, 94), 136 (1969, 94), 137 (1966, 46; 1969, 94), 138 (1969, 94), 140 (1969, 86), 145 (1969, 94), 147 (1969, 94), 148 (1969, 86, 94), 157 (1966, 119), 158 (1966, 119; 1969, 86), 160 (1969, 86), 168 (1969, 86), 171 (1966, 119), 177 (1969, 86), 186 (1968, 168), 195 (1968, 907, 207 (1968, 332), 218 (1968, 168, 332), 223 (1968, 168, 332), 227 (1966, 119), 231 (1968, 332), 239 (1968, 332), 246 (1969, 119), *312, 314, 317, 332, 334, 339, 342*
Denney, D. Z., 37 (1966, 46; 1969, 86), 38 (1969, 86), 40 (1969, 94), 41 (1969, 94), 42 (1969, 94), 45 (1968, 90), 82 (1966, 46), 83 (1966, 46), 84 (1969, 86, 94), 85 (1969, 94), 86 (1968, 168), 112 (1969, 94), 114 (1969, 86), 118 (1966, 46), 119 (1966, 46; 1969, 94), 122 (1969, 86), 126 (1966, 46; 1969, 94), 136 (1969, 94), 137 (1966, 46; 1969, 94), 138 (1969, 94), 140 (1969, 86), 145 (1969, 94), 147 (1969, 94), 148 (1969, 86, 94), 158 (1969, 86), 160 (1969, 86), 168 (1969, 86), 177 (1969, 86), 186 (1968, 168), 195 (1968, 90), 207 (1968, 332), 218 (1968, 168, 332), 223 (1968, 168, 332), 231 (1968, 332), 239 (1968, 332), *312, 332, 334, 339, 342*
Dennis, E. A., 86 (1966, 120), 136 (1966, 120), 140 (1966, 120), 147 (1966, 120), *314*
Derbyshire, W., 4 (1969, 5a), 88 (1969, 5a, 217), 101 (1969, 5a), *340, 345*
Derner, N., 92 (1965, 70), *310*
De Roos, J. B., 85 (1965, 41; 1969, 198), 142 (1969, 198), *309, 345*
Desai, N. B., 38 (1968, 73), 114 (1968, 73), 125 (1968, 73), 135 (1968, 309), 138 (1968, 73), 148 (1968, 73), 199 (1968, 73), 215 (1968, 309), 219 (1968, 309), 229 (1968, 73), 233 (1968, 73), 235 (1968, 73), 236 (1968, 309), 245 (1968, 73), 253 (1968, 309), 262 (1968, 309), 268 (1968, 309), 269 (1968, 309), 274 (1968, 309), *332, 338*
Desai, V. B., 79 (1969, 150), 279 (1969, 150), 280 (1969, 150), 281 (1969, 150), 282 (1969, 150), 283 (1969, 150), *344*
De Selms, R. C., 58 (1967, 120), 114 (1967, 120), 119 (1967, 120), 137 (1967, 120), *322*
Deshayes, H., 84 (1969, 181), 143 (1969, 181), *344*
De Simone, R. E., 91 (1969, 266), *347*
Des Marteau, D. D., 21 (1966, 29; 1969, 58), 22 (1969, 58), 99 (1966, 29), 100 (1969, 58), *312, 341*
Deverell, C., 4 (1969, 9), 87 (1969, 9), *340*
Devillers, J., 40 (1968, 76), 47 (1968,

76), 78 (1968, 76, 126), 82 (1968, 76), 86 (1968, 126), 122 (1968, 126), 141 (1968, 126), 164 (1968, 76), 215 (1968, 76), 216 (1968, 76), 262 (1968, 126), *332, 333*

Dewar, M. J. S., 47 (1969, 111), 79 (1960, 2), 192 (1969, 111), *307, 342*

Dewhirst, K. C., 92 (1968, 259), 93 (1968, 259), 196 (1968, 259), *337*

Dhaliwal, P. S., 152 (1967, 331), 186 (1967, 331), 261 (1967, 331), 273 (1967, 331), *328*

Diagle, D., 230 (1965, 111), *311*

Dianova, E. N., 109 (1965, 81), 121 (1965, 81), *310*

Dickson, F. E., 56 (1967, 115), 79 (1966, 95; 1967, 146), 82 (1967, 115), 152 (1967, 115), 222 (1967, 115), 278 (1967, 146), 279 (1967, 146), 280 (1967, 146), 281 (1966, 95), 282 (1966, 95), *314, 322, 323*

Didenko, A. N., 88 (1968, 190), *335*

Dillon, K. B., 5 (1969, 11), 194 (1969, 11), 218 (1969, 11), *340*

Dilworth, J. R., 92 (1969, 300), *348*

Dimic, V., 88 (1968, 187), *335*

Dimroth, K., 39 (1968, 75), 53 (1968, 101), 54 (1968, 75, 101, 102), 78 (1968, 75, 101, 102; 1969, 145), 119 (1969, 145), 171 (1968, 101), 188 (1968, 101), 206 (1968, 101), 219 (1968, 101), 224 (1966, 272), 226 (1968, 102), 229 (1968, 101), 232 (1966, 272; 1968, 75), 238 (1966, 272), 240 (1968, 102), 242 (1968, 75), 245 (1968, 75), 248 (1968, 75), 250 (1968, 75), 256 (1969, 145), 269 (1968, 101), 276 (1968, 102), *318, 332, 333, 343*

Di Pietro, H. R., 195 (1966, 262), 196 (1966, 262), 223 (1966, 262), 227 (1966, 262), *318*

Diyorio, J. S., 40 (1969, 89), *342*

Doack, G. O., 227 (1968, 341), 231 (1968, 341), *339*

Dobbie, R. C., 22 (1968, 47), 27 (1968, 55), 34 (1968, 47, 55), 60 (1968, 55), 61 (1968, 55), 63 (1968, 55), 76 (1968, 47, 55), 105 (1968, 47), 106 (1968, 47), 107 (1968, 47), 112 (1968, 47), 115 (1968, 55), 116 (1968, 55), 117 (1968, 47), 258 (1968, 55), *331*

Dobson, G. R., 91 (1969, 270), 173 (1969, 270), *347*

Doddrell, D., 91 (1969, 257), *347*

Dombrovskii, A. V., 56 (1967, 118), 234 (1967, 118), 235 (1967, 118), 244 (1967, 118), 246 (1967, 118), 247 (1967, 118), *322*

Dorémieux-Morin, C., 86 (1967, 179), *324*

Doty, L. F., 22 (1968, 47), 34 (1968, 47), 76 (1968, 47), 105 (1968, 47), 106 (1968, 47), 107 (1968, 47), 112 (1968, 47), 117 (1968, 47), *331*

Downes, A. J., 22 (1967, 60), 28 (1967, 60), 75 (1967, 60), 113 (1967, 60), *321*

Drago, R. S., 82 (1966, 100), 87 (1968, 177), 90 (1966, 157), 91 (1969, 266), *314, 315, 335, 347*

Drake, G. L., 39 (1967, 297), 112 (1967, 297), *327*

Drake, J. E., 17 (1963, 2), 18 (1968, 32), 20 (1967, 41), 48 (1968, 32; 1969, 113), 84 (1967, 41, 173; 1968, 158, 159; 1969, 113), 86 (1967, 180), 97 (1967, 41, 173; 1968, 158; 1969, 113), 99 (1968, 32, 158), 100 (1968, 158; 1969, 113), *307, 320, 324, 331, 334, 342*

Dreeskamp, H., 2 (1969, 2), 8 (1969, 2), 15 (1969, 2), 18 (1966, 22), 27 (1966, 22; 1969, 2), 28 (1969, 2), 30 (1969, 2), 31 (1966, 22; 1969, 2), 68 (1969, 2), 69 (1969, 2), 72 (1969, 2), 73, 74 (1969, 2), 75 (1969, 2), 100 (1966, 22), 105 (1966, 22), 110 (1966, 22), 113 (1969, 2), 114 (1966, 22), 121 (1966, 22), 122 (1969, 2), 141 (1966, 22), 160 (1966, 22), 210 (1969, 2), *312, 339*

Drenth, W., 8 (1967, 20; 1968, 21; 1969, 23), 134 (1967, 20), 139 (1967, 30), 140 (1967, 20), 175 (1967, 20; 1968, 21), 177 (1967, 20), 178 (1967, 20), *320, 331, 340*

Driscoll, J. S., 275 (1966, 290), *319*

Dron, D., 81 (1967, 154, 155), 225 (1967, 154), 229 (1967, 154, 155), 232 (1967, 155), 236 (1967, 155), 239 (1967, 155), 242 (1967, 155), 245 (1967, 155), 248 (1967, 155), *323*

Drozd, G. I., 7 (1968, 12), 19 (1967, 37, 38), 20 (1968, 12), 21 (1967, 37, 38, 51, 52; 1968, 12), 25 (1967, 64), 99 (1967, 51), 103 (1967, 51), 104 (1967, 38, 51, 290; 1968, 12), 107 (1967, 38, 51, 290; 1968, 12), 108 (1968, 12), 109 (1967, 64), 113 (1968, 12), 114 (1967, 38), 119 (1967, 64; 1968, 12), 125 (1967, 37), 127 (1968, 12), 128 (1967, 37, 38, 64; 1968, 12), 130 (1967, 51, 52, 290), 131 (1967, 37), 132 (1967, 51, 52, 290), 136 (1967, 64; 1968, 12), 138 (1967, 51), 139 (1967, 64), 143 (1967, 52, 64, 290), 146 (1967, 64), 152 (1968, 12), 153 (1967, 64), 157 (1967, 64), 169 (1967, 64; 1968, 12), 173 (1967, 64), 180 (1967, 64), 181 (1967, 64), 209 (1967, 52), *320, 321, 327, 330*

Druzhina, Z. S., 176 (1967, 340), 192 (1967, 340), 193 (1967, 340), 207 (1967, 340), *328*

Dubov, S. S., 7 (1967, 14; 1968, 12), 20 (1968, 12), 21 (1967, 51; 1968, 12), 25 (1967, 64), 97 (1967, 14), 98 (1967, 14), 99 (1967, 14, 51), 103 (1967, 14, 51), 104 (1967, 14, 51; 1968, 12), 105 (1967, 14), 107 (1967, 14, 51; 1968, 12), 108 (1967, 14; 1968, 12), 109 (1967, 64), 112 (1967, 14), 113 (1967, 14; 1968, 12), 118 (1967, 14), 119 (1967, 64), 120 (1967, 14; 1968, 12), 122 (1967, 14), 127 (1967, 14; 1968, 12), 128 (1967, 64; 1968, 12), 130 (1967, 51), 132 (1967, 14, 51), 136 (1967, 14, 64; 1968, 12), 138 (1967, 14, 51), 139 (1967, 64), 143 (1967, 64), 146 (1967, 64), 152 (1967, 14; 1968, 12), 153 (1967, 64), 157 (1967, 64), 158 (1967, 14), 169 (1967, 64; 1968, 12), 172 (1967, 14), 173 (1967, 64), 180 (1967, 64), 181 (1967, 64), 186 (1967, 14), *319, 320, 321, 330*

Dubson, G. R., 92 (1968, 241), *336*

Ducan, W. G., 167 (1965, 97), *311*

Duchstach, H., 90 (1965, 51), 91 (1965, 51), *309*

Duddell, D. A., 90 (1969, 233), 92 (1969, 233), *346*

Dunand, J. J., 127 (1966, 236), *317*

Duncan, C. H., 86 (1967, 186), 104 (1967, 186), 108 (1967, 186), 114 (1967, 186), 131 (1967, 186), 132 (1967, 186), 258 (1967, 186), 259 (1967, 186), 278 (1967, 186), *324*

Duncan, J. D., 92 (1969, 277), *347*

Duncan, J. F., 88 (1967, 199), *324*

Dungan, C. H., 1 (1967, 1), 2 (1967, 1), 4 (1967, 1), 65 (1967, 1), *319*

du Plessis, J. A. K., 86 (1968, 167; 1969, 204), *334, 345*

Du Preez, A. L., 91 (1968, 213; 1969, 259), 148 (1968, 213), *336, 347*

Durand, M., 114 (1969, 336), 140 (1969, 336), *349*

Durham, L. J., 94 (1967, 265), *326*

Durig, J. R., 40 (1969, 89), *342*

Duval, E., 13 (1964, 3), 37 (1966, 47; 1967, 85), 52 (1964, 7), 140 (1966, 47; 1967, 85), 169 (1967, 85), *308, 312*

Dwek, R. A., 63 (1969, 130), 66 (1969, 130), 87 (1966, 141; 1967, 198; 1969, 130, 213, 214, 215), 88 (1969, 216), 101 (1969, 216), 102 (1969, 216), 109 (1967, 198), 114 (1966, 141; 1967, 198), 119 (1967, 198), 120 (1967, 198), 131 (1967, 198), 140 (1966, 141; 1967, 198), 142 (1967, 198), 186 (1966, 141; 1967, 198); 223 (1966, 141; 1967, 198), 224 (1966, 141; 1967, 198), 227 (1967, 198), 252 (1967, 198), *315, 324, 343, 345*

Dyer, J., 107 (1966, 215), 117 (1966, 215), *317*

E

Easwaran, K. R. K., 88 (1967, 200), *324*

Eaton, D. R., 92 (1968, 260), 235 (1968, 260), *337*
Ebel, J. P., 86 (1967, 189), 94 (1967, 189), 100 (1967, 189), 101 (1967, 189), *324*
Ebeling, J., 19 (1968, 40), 48 (1967, 97; 1968, 94), 59 (1968, 111), 78 (1969, 149), 79 (1967, 150), 209 (1967, 355), 258 (1969, 149), 259 (1967, 355; 1968, 94, 111), 260 (1968, 94, 111), 261 (1968, 94, 111), 262 (1969, 149), 265 (1968, 94; 1969, 149), 267 (1968, 94, 111), 268 (1968, 111), 270 (1968, 111), 271 (1967, 97; 1968, 94, 111), 272 (1968, 111), 273 (1968, 94), 274 (1967, 97; 1968, 94), 282 (1968, 40), *322, 323, 329, 331, 332, 333, 344*
Ebsworth, E. A. V., 20 (1967, 43), 48 (1967, 98), 83 (1967, 43), 99 (1967, 43, 98), 101 (1967, 98), *320, 322*
Edmundson, R. S., 39 (1968, 74), 41 (1967, 91), 66 (1967, 91), 83 (1968, 149; 1969, 174), 86 (1966, 121), 117 (1966, 121), 124 (1968, 149), 125 (1967, 91; 1968, 74), 126 (1967, 91), 134 (1967, 91; 1968, 74), 137 (1967, 91; 1968, 74), 148 (1967, 91), 158 (1967, 91), 167 (1967, 91), 168 (1967, 91; 1968, 74), 175 (1968, 149), 176 (1968, 149), 177 (1967, 91), 180 (1966, 121), 181 (1967, 91; 1968, 74), 184 (1967, 91), 189 (1967, 91), 190 (1967, 91; 1968, 74), 198 (1967, 91; 1968, 74), 211 (1968, 74), 212 (1968, 74), 243 (1967, 91; 1968, 74), 264 (1967, 91), *314, 321, 332, 334, 344*
Edwards, J. O., 88 (1961, 3), 97 (1961, 3), *307*
Efraty, A., 91 (1969, 269), *347*
Eints, J., 48 (1968, 96), 73 (1968, 96), 99 (1968, 96), 100 (1968, 96), 102 (1968, 96), *332*
Elegant, L., 44 (1969, 102), 84 (1969, 102, 188, 189, 190), 85 (1969, 102, 188), 86 (1969, 102), 143 (1969, 190), 216 (1969, 190), *342, 345*
Elguero, J., 13 (1967, 25), 58, 147 (1967, 119), 157 (1967, 119), 167 (1967, 119), 182 (1967, 119), *320, 322*
Elleman, D. D., 17 (1966, 13; 1967, 40), 18 (1966, 13), 20 (1967, 40), 25 (1966, 13), 28 (1966, 13), 33 (1967, 40), 37 (1967, 40), 44 (1966, 13), 60 (1967, 40), 61 (1967, 40), 63 (1967, 40), 68 (1967, 131), 99 (1966, 13), 104 (1967, 40), 105 (1966, 13), 106 (1967, 40), 109 (1966, 13), 110 (1967, 40), 113 (1966, 13), 114 (1967, 298), 257 (1967, 40), 258 (1967, 40), *312, 320, 323, 327*
Ellerberger, M., 65, *350*
Elliott, J. E., 88 (1969, 224), *346*
Elser, H., 2 (1969, 2), 8 (1969, 2), 15 (1969, 2), 18 (1966, 22), 27 (1966, 22; 1969, 2), 28 (1969, 2), 30 (1969, 2), 31 (1966, 22; 1969, 2), 68 (1969, 2), 69 (1969, 2), 72 (1969, 2), 73 (1969, 2), 74 (1969, 2), 75 (1969, 2), 100 (1966, 22), 105 (1966, 22), 110 (1966, 22), 113 (1969, 2), 115 (1966, 22), 121 (1966, 22), 122 (1969, 2), 141 (1966, 22), 160 (1966, 22), 210 (1969, 2), *312, 339*
Emeleus, H. J., 22 (1966, 31), 34 (1968, 69), 44 (1966, 31), 57 (1966, 73), 77 (1966, 31; 1968, 69), 106 (1966, 31) 116 (1968, 69), 123 (1968, 69), 130 (1968, 69), 215 (1966, 73), 221 (1966, 73), *312, 313, 332*
Emmick, T. L., 19 (1968, 35), 74 (1968, 35), 173 (1968, 35), 181 (1968, 35), 182 (1968, 35), 187 (1968, 35), 196 (1968, 35), 203 (1968, 35), 223 (1968, 35), 241 (1968, 35), *331*
Endicott, T. F., 90 (1968, 198), *335*
Engelhardt, G., 79 (1966, 93, 94), 101 (1966, 93, 94), 117 (1966, 94), 170 (1967, 337), 251 (1967, 337), 256 (1967, 337), 277 (1966, 93), 278 (1966, 93), *314, 328*
Engelhardt, U., 23 (1969, 67), 35 (1969, 67), 244 (1969, 67), 245 (1969, 67), *341*
Englemann, A., 92 (1966, 186), *316*

Esterday, R., 94 (1968, 281), 110 (1968, 281), *337*
Eulenberger, G. R., 88 (1968, 185), *335*
Evans, D., 93 (1967, 260), *326*
Evans, J. G., 90 (1969, 233), 92 (1969, 233), *346*
Evtikov, Zh.L., 53 (1967, 110), 75 (1967, 110), 116 (1967, 110), 117 (1967, 110, 303), 123 (1967, 110), 124 (1967, 110, 303), 134 (1967, 110), *322, 327*
Ewart, G., 85 (1962, 4), 212 (1962, 4), *307*
Eychaner, W. A., 37 (1968, 72), 105 (1968, 72), 114 (1968, 72), 129 (1968, 72), *332*
Eyles, C. T., 86 (1966, 122), 236 (1966, 122), *314*

F

Fackler, J. P., 89, 92 (1969, 291, 292), *348*
Fahlman, A., 4 (1967, 6), *319*
Faithful, B. D., 242 (1966, 279), *318*
Falius, H., 61 (1968, 112), 66 (1968, 112), 98 (1967, 280), 100 (1968, 112), *326, 333*
Faller, J. W., 91 (1969, 274), *347*
Farley, C. E., 78 (1967, 138), 162 (1967, 138), 166 (1967, 138), *323*
Farnell, L. F., 88 (1966, 149), *315*
Farran, C. F., 18 (1966, 20), 84 (1966, 20), *312*
Farrar, T. C., 4 (1969, 7), 20 (1969, 7), 87 (1967, 197; 1968, 180), 88 (1969, 7), 97 (1969, 7), *324, 335, 340*
Favelier, R., 8 (1967, 21), 52 (1967, 21), 124 (1967, 21), 125 (1967, 21), 126 (1967, 21), 135 (1967, 21), 144 (1967, 21), 145 (1967, 21), 147 (1967, 21), 149 (1967, 21), 155 (1967, 21), 165 (1967, 21), 166 (1967, 21), 175 (1967, 21), 177 (1967, 21), 182 (1967, 21), 183 (1967, 21), 188 (1967, 21), 192 (1967, 21), 207 (1967, 21), *320*
Fayet, J. P., 11 (1970, 1), *350*
Feather, M. S., 94 (1969, 321), 101 (1969, 321), 121 (1969, 321), 138 (1969, 321), *348*
Fedin, E. I., 7 (1968, 18), 271 (1968, 18), 273 (1968, 18), *330*
Feistel, G. R., 79 (1966, 89; 1967, 145; 1968, 135), 101 (1967, 145), 102 (1967, 145), 279 (1968, 135), *313, 323, 333*
Feldman, I., 89 (1968, 270), 94 (1968, 272), *337*
Feldt, M. K., 79 (1968, 131, 132), 285 (1968, 132), 287 (1968, 132), *333*
Fenton, C. M., 24 (1968, 51), 83 (1968, 51), 99 (1968, 51), *331*
Fenton, Y. E., 227 (1968, 342), *339*
Ference, R. A., 92 (1966, 192), *316*
Fernando, Q., 92 (1968, 233), *336*
Fetchin, J. A., 89 (1969, 291), 92 (1969, 291), *348*
Feuer, B. I., 229 (1966, 274), *318*
Fields, R., 18 (1969, 43), 34 (1969, 43), 35 (1969, 43), 75 (1969, 43), 106 (1966, 26; 1969, 43), 185 (1966, 26), *312, 341*
Filatova, I. M., 261 (1969, 371), 264 (1969, 371), 266 (1969, 371), 267 (1969, 371), 268 (1969, 371), 269 (1969, 371), *350*
Fild, M., 7 (1966, 7; 1967, 11; 1968, 9), 103 (1967, 288), 105 (1967, 11), 109 (1967, 11), 129 (1966, 7), 130 (1966, 7; 1967, 11), 132 (1967, 11), 150 (1967, 11; 1968, 9); 171 (1967, 11), 184 (1966, 7), 185 (1966, 7; 1967, 11), 187 (1967, 11), 195 (1967, 11; 1968, 9), 201 (1967, 11), 205 (1966, 7), 214 (1966, 7), 221 (1966, 6; 1967, 11; 1968, 9), 222 (1967, 11), 285 (1967, 392), *311, 319, 327, 330*
Finer, E. G., 20 (1967, 55), 21 (1967, 55), 28 (1968, 57), 33 (1967, 55), 34 (1967, 55), 42 (1967, 55), 60 (1967, 124, 125), 62 (1967, 124, 125; 1968, 57, 113), 63 (1967, 124, 125; 1968, 113), 64 (1968, 57), 66 (1967, 130; 1968, 57), 68 (1967, 125), 70, 72 (1968, 57), 77 (1967, 55), 79 (1967, 130), 90 (1968, 57), 100 (1968, 113), 101 (1967, 130; 1968, 113), 106 (1967, 55), 110

(1967, 55), 257 (1968, 57), 259 (1967, 124, 125; 1968, 57, 113), 261 (1968, 57), 262 (1968, 113), 263 (1968, 113), 268 (1968, 113), 270 (1968, 113), 277 (1967, 130), 278 (1967, 130), 279 (1967, 130; 1968, 57), 281 (1967, 130; 1968, 57), 282 (1967, 130; 1968, 57), *320, 322, 323, 331, 333*
Finley, J. H., 37 (1969, 86), 38 (1969, 86), 84 (1969, 86), 114 (1969, 86), 122 (1969, 86), 140 (1969, 86), 148 (1969, 86), 158 (1969, 86), 160 (1969, 86), 168 (1969, 86), 177 (1969, 86), *342*
Firstenberg, S., 58 (1967, 121), 78 (1967, 121), 86 (1967, 121), 172 (1967, 121), 183 (1967, 121), 189 (1967, 121), 190 (1967, 121), 198 (1967, 121), 199 (1967, 121), 205 (1967, 121), 225 (1967, 121), *322*
Fischer, E. O., 90 (1968, 207, 208; 1969, 248), 91 (1966, 160; 1968, 207, 208, 212), 93 (1966, 160; 1968, 207, 208, 212), *315, 335, 336, 346*
Fischer, R. H., 90 (1968, 199), 92 (1968, 199), *335*
Fishwick, S. E., 32 (1968, 68), 34 (1968, 68), 80 (1968, 80), 199 (1968, 68), 211 (1967, 356), 212 (1967, 356), *329, 332*
Fitton, P., 92 (1968, 238, 239), *336*
Fitzimmons, B., 79 (1966, 86), 107 (1966, 86), 280 (1966, 86), 281 (1966, 86), 282 (1966, 86), 283 (1966, 86), *313*
Fitzwater, D. R., 75 (1968, 117), 83 (1968, 117), 89 (1968, 117), 135 (1968, 117), *333*
Flagg, E. E., 73 (1969, 140), 86 (1969, 140), 129 (1969, 140), 245 (1969, 140), 274 (1969, 140), 288 (1969, 140), *343*
Flaskerud, G. G., 21 (1969, 56), 76 (1969, 56), 103 (1969, 56), 106 (1969, 56), 107 (1969, 56), 110 (1969, 56), 116 (1969, 56), *341*
Flautt, T. J., 7 (1967, 12), 256 (1967, 12), 257 (1967, 12), 258 (1967, 12), *319*

Fleming, J. S., 90 (1967, 229), *325*
Fleming, M. A., 11, 12, 73, 98 (1963, 1), 100 (1963, 1), 109 (1963, 1), 114 (1963, 1), 140 (1963, 1), 160 (1963, 1), *307*
Flint, J., 32 (1968, 68), 34 (1968, 68), 80 (1968, 68), 199 (1968, 68), 211 (1967, 356), 212 (1967, 356), *329, 332*
Flock, A. G., 94 (1968, 277), 152 (1968, 277), *337*
Fluck, E., 7 (1965, 4), 13 (1967, 29, 30), 18 (1965, 11a; 1967, 29, 30; 1969, 42), 20 (1965, 11a; 1967, 30, 42), 24 (1969, 69), 44 (1969, 69), 48 (1968, 96), 49 (1967, 29), 60, 61 (1969, 69), 63 (1965, 11a, 27; 1966, 77; 1967, 126), 65 (1969, 135), 67 (1969, 69), 73 (1968, 96), 77 (1967, 142), *290, 291, 293, 295, 299, 301, 302, 304, 305, 308, 312, 314, 320, 323, 325, 326, 332*, 83 (1969, 180), 84 (1969, 42), 86 (1965, 49, 50), 92 (1967, 244), 97 (1965, 4; 1967, 277), 98 (1965, 4; 1967, 277), 99 (1967, 42; 1968, 96), 100 (1968, 96; 1969, 69), 101 (1969, 69), 102 (1968, 96), 103 (1965, 4), 105 (1965, 4), 106 (1965, 4), 109 (1967, 42, 277), 110 (1967, 277), 111 (1965, 11a; 1967, 277), 113 (1967, 244), 114 (1967, 277), 115 (1965, 4, 50; 1967, 29), 118 (1967, 277), 121 (1967, 42), 122 (1967, 244), 123 (1969, 135), 130 (1966, 238; 1967, 277), 131 (1966, 238; 1967, 277), 132 (1967, 30, 277), 136 (1967, 277), 139 (1967, 244), 140 (1967, 277), 141 (1967, 277), 142 (1967, 277), 143 (1967, 29, 126), 149 (1967, 244), 151 (1966, 77, 238), 155 (1967, 126), 158 (1968, 320), 159 (1968, 320), 160 (1967, 244, 277), 164 (1967, 244), 165 (1967, 244), 169 (1967, 244, 277), 171 (1967, 29), 172 (1967, 126), 173 (1965, 11a), 175 (1967, 126, 244), 178 (1967, 244, 277), 179 (1966, 77; 1967, 126), 185 (1966, 238), 186 (1965, 103; 1967, 277; 1968, 320), 187 (1965,

103; 1967, 30, 42, 277; 1968, 320), 191 (1967, 244), 193 (1967, 244, 277), 194 (1967, 277), 196 (1967, 244), 199 (1967, 244), 201 (1967, 244), 203 (1967, 277), 204 (1967, 277), 206 (1967, 244, 277), 209 (1967, 244), 212 (1967, 244, 277), 213 (1967, 277), 214 (1967, 244, 277), 217 (1967, 244), 218 (1967, 277), 220 (1967, 244), 223 (1965, 103; 1967, 244, 277), 224 (1967, 277), 226 (1967, 244), 228 (167, 244), 229 (1967, 244), 230 (1967 244), 233 (1967, 244), 235 (1967, 244), 236 (1969, 368), 237 (1967, 244), 239 (1967, 244), 244 (1967, 244, 277), 245 (1967, 244), 252 (1967, 277), 256 (1967, 277), 258 (1965, 27; 1969, 69), 259 (1965, 11a), 260 (1965, 27; 1969, 69), 261 (1969, 368), 262 (1965, 11a; 1967, 126), 263 (1967, 126), 265 (1967, 277, 386; 1969, 69, 135, 368), 266 (1965, 11a), 267 (1965, 11a; 1967, 126), 268 (1965, 11a; 1967, 126, 142; 1969, 135, 368), 269 (1966, 77; 1967, 126), 270 (1965, 11a; 1967, 277), 278 (1965, 27), 280 (1966, 292), 284 (1965, 27), 285 (1965, 27), 286 (1965, 11a), 287 (1966, 295), 288 (1965, 11a; 1966, 295; 1967, 126), *308, 309, 311, 313, 317, 319, 320, 322, 323, 326, 330, 332, 338, 341, 343, 344, 350*
- Fogel, J., 86 (1966, 135), 138 (1966, 135), 231 (1966, 135), 234 (1966, 135), 235 (1966, 135), 238 (1966, 135), 239 (1966, 135), *315*
- Fontal, B., 40 (1966, 51), 82 (1966, 51), 105 (1966, 211), 111 (1966, 222), 133 (1966, 222), 134 (1966, 222), 135 (1966, 51), 140 (1966, 51), 146 (1966, 222), 147 (1966, 51), 156 (1966, 222), 158 (1966, 51), 189 (1966, 51), 190 (1966, 51), 218 (1966, 51), *312, 316, 317*
- Ford, C. T., 79 (1966, 95; 1967, 146), 278 (1967, 146), 279 (1967, 146), 280 (1967, 146), 281 (1966, 95), 282 (1966, 95), *314, 323*

Forster, A., 90 (1965, 57), *309*
Forster, D., 89 (1968, 235), 90 (1968, 204), 91 (1968, 204), 92 (1968, 204, 235), *335, 336*
Foucaud, A., 142 (1969, 345), 204 (1969, 345), 229 (1969, 345), 242 (1969, 345), *349*
Fournes, L., 92 (1967, 236), 104 (1967, 236), 107 (1967, 236), 112 (1967, 236), 118 (1967, 236), *325*
Fradin, F. Y., 87 (1969, 211), *345*
Francina, A., 83 (1968, 155), 86 (1968, 165), 100 (1968, 155), 121 (1968, 165), *334*
Frank, D. S., 86 (1967, 181), 125 (1967, 181), 137 (1967, 181), *324*
Frankel, L. S., 42 (1969, 101), 80 (1969, 101, 159), 89 (1969, 237, 238, 310), 90 (1969, 237, 238), 91 (1969, 237), 92 (1969, 159), 140 (1969, 101), 159 (1969, 101), 169 (1969, 101), 178 (1969, 101), 191 (1969, 101), 208 (1969, 159), 215 (1969, 159), *342, 344, 346, 348*
Franz, J. E., 59 (1964, 10), 237 (1964, 10), 240 (1964, 10), 243 (1964, 10), *308*
Fraser, G. W., 86 (1965, 43), 109 (1966, 217), 120 (1966, 217), 121 (1966, 217), 122 (1966, 217), 138 (1966, 217), 160 (1966, 217), 259 (1965, 43), *309, 317*
Freedman, L. D., 227 (1968, 341), 231 (1968, 341), *339*
Freeman, K. L., 77 (1966, 82), 86 (1966, 123), 122 (1966, 82), 227 (1966, 123), 228 (1966, 123), 231 (1966, 82), 244 (1966, 82), 272 (1966, 82), *313, 314*
Freeman, L. D., 7 (1967, 10), 18 (1967, 10), 20 (1967, 10), *319*
Fridland, S. V., 116 (1966, 229), 155 (1966, 229), *317*
Friebolin, H., 46 (1968, 92), 47 (1968, 92), 80 (1968, 92), 153 (1968, 92), 173 (1968, 92), 191 (1968, 92), 196 (1968, 92), 202 (1968, 92), 206 (1968, 92), 216 (1968, 92), *332*
Friedman, F., 88 (1967, 201), *324*

AUTHOR INDEX

Fritz, G., 86 (1967, 182), 263 (1967, 182), 264 (1967, 182), 279 (1967, 182), *324*
Fritz, H. P., 92 (1965, 68; 1966, 180), *310, 316*
Frolov, V. V., 87 (1968, 176; 1969, 209), *335, 345*
Fry, A. J., 172 (1967, 339), *328*
Fung, M. K., 40 (1965, 22), *309*
Fuqua, S. A., 167 (1965, 97), *311*

G

Gabuda, S. P., 88 (1969, 221), *346*
Gaglio, G., 91 (1965, 62), *310*
Gagnaire, D., 13 (1968, 28), 17 (1968, 28), 18 (1968, 28, 33), 20 (1969, 51), 27 (1967, 71), 28 (1968, 28, 33), 29 (1968, 62; 1969, 76), 30 (1968, 28), 40 (1966, 50; 1967, 88; 1968, 80; 1969, 88), 74 (1968, 28), 82 (1966, 50), 83 (1969, 179), 105 (1968, 33), 107 (1966, 50), 109 (1968, 33), 112 (1966, 50), 118 (1966, 50; 1968, 28), 125 (1968, 80; 1969, 88), 137 (1967, 88; 1968, 80; 1969, 88), 139 (1968, 80), 143 (1968, 33), 148 (1968, 80), 151 (1966, 50), 153 (1968, 28, 33, 80; 1969, 51), 154 (1968, 33), 158 (1969, 179), 162 (1966, 50), 180 (1967, 88; 1968, 80; 1969, 76), 187 (1969, 76), 188 (1969, 76), 191 (1969, 51), 193 (1969, 179), 198 (1969, 76), 276 (1968, 33), *312, 321, 331, 332, 341, 342, 344*
Gal, J. F., 84 (1969, 189), *345*
Gallagher, M. J., 18 (1966, 18), 19 (1966, 18), 26, 77 (1966, 82), 80, 86 (1965, 48; 1966, 123), 122 (1966, 82), 127 (1968, 56), 132 (1966, 18), 142 (1966, 18), 143 (1968, 56), 174 (1966, 18), 182 (1966, 18), 196 (1968, 56), 197 (1968, 56), 208 (1968, 56), 217 (1966, 18), 219 (1968, 56), 227 (1966, 123), 228 (1966, 123; 1968, 56), 231 (1966, 82; 1968, 56), 234 (1968, 56), 244 (1966, 82; 1968, 56), 257 (1967, 371; 1968, 56), 259 (1967, 371), 260 (1967, 371), 261 (1967, 371), 263 (1967, 371; 1968, 56), 264 (1967, 371), 265 (1965, 48; 1966, 18), 267 (1967, 371), 271 (1967, 371), 272 (1966, 82; 1967, 371), 273 (1967, 371, 389), 274 (1967, 371), 275 (1967, 371), 276 (1967, 371), *309, 312, 313, 314, 329, 330, 331, 334*
Gambler, W., 113 (1968, 356), *339*
Garner, A. W. B., 92 (1969, 304), *348*
Garrett, B. B., 91 (1969, 272), 220 (1969, 272), *347*
Garrison, A. W., 47 (1969, 110), 81 (1969, 110), 94 (1968, 285), 117 (1968, 285), 118 (1968, 285), 128 (1968, 285), 140 (1968, 285), 141 (1969, 110), 145 (1968, 285), 149 (1968, 285; 1969, 110), 151 (1968, 285), 153 (1968, 285), 157 (1968, 285), 160 (1968, 285), 161 (1969, 110), 163 (1968, 285; 1969, 110), 171 (1968, 285), 172 (1968, 285; 1969, 110), 173 (1968, 285), 174 (1968, 285), 175 (1968, 285), 176 (1968, 285), 179 (1968, 285), 180 (1968, 285), 181 (1968, 285; 1969, 110), 187 (1968, 285), 191 (1968, 285), 192 (1968, 285), 194 (1968, 285), 202 (1968, 285), 203 (1968, 285), 204 (1968, 285), 242 (1968, 285), 263 (1968, 285), 265 (1968, 265), 266 (1968, 285), *337, 342*
Gash, V. W., 86 (1967, 183), *324*
Gaudiano, G., 233 (1968, 343), 235 (1968, 343), 246 (1967, 375; 1968, 343), 247 (1967, 375; 1968, 343), 250 (1968, 343), 251 (1968, 343), 255 (1967, 375), *329, 339*
Gaudy, E. T., 52 (1966, 62), 80 (1968, 142), 86 (1966, 62), 103 (1966, 62), 144 (1968, 142), 154 (1968, 142), 165 (1968, 142), 179 (1968, 142), 182 (1968, 142), 187 (1968, 142), 191 (1968, 142), 197 (1968, 142), 200 (1968, 142), 209 (1968, 142), 211 (1968, 142), 212 (1968, 142), 227 (1966, 62), 240 (1966, 62), *313, 334*
Gay, R. S., 7 (1966, 10; 1969, 16), 150

(1966, 10), 221 (1966, 10), 222 (1966, 10), *311, 340*
Gaydou, E., 166 (1969, 355), 176 (1969, 355), 183 (1969, 355), *349*
Gazizov, T. Kh., 2 (1967, 3), 7 (1967, 3), 125 (1966, 234), 127 (1966, 234), 134 (1967, 3), 147 (1966, 234), 155 (1966, 234; 1967, 3), 156 (1967, 3), 175 (1967, 3), 176 (1967, 3), 192 (1967, 3), 214 (1966, 234), 254 (1967, 3), *317, 319*
Geistel, M. M., 90 (1969, 252), *346*
Gelan, J., 44 (1968, 87), *332*
Genin, D. J., 88 (1968, 188), *335*
Gericke, W., 13 (1968, 27), *331*
Germa, H., 83 (1968, 154), 119 (1968, 154), 134 (1968, 154), 181 (1968, 154), 192 (1968, 154), 229 (1968, 154), 260 (1968, 154), *334*
Gershman, N., 86 (1966, 119), 126 (1966, 119), 157 (1966, 119), 158 (1966, 119), 171 (1966, 119), 227 (1966, 119), 246 (1966, 119), *314*
Giacin, J., 86 (1966, 119), 126 (1966, 119), 157 (1966, 119), 158 (1966, 119), 171 (1966, 119), 227 (1966, 119), 246 (1966, 119), *314*
Gielen, M., 81 (1969, 168), *344*
Gilje, J. W., 45 (1969, 103), 74 (1967, 133), 76 (1969, 103), 84 (1967, 133), 108 (1969, 103), 109 (1969, 103), 113 (1969, 103), 122 (1969, 103), 142 (1969, 103), *323, 342*
Gillard, R. D., 242 (1966, 279), *318*
Gillespie, R. J., 25 (1966, 35), *312*
Gillman, H. D., 85 (1968, 162), 90 (1968, 162), 97 (1968, 162), 99 (1968, 162), *334*
Gingerich, K. A., 88 (1967, 205), *325*
Gladshtein, B. M., 113 (1969, 335), 127 (1969, 335), 138 (1969, 335), 152 (1969, 335), *349*
Glasel, J. A., 86 (1969, 205), *345*
Glaser, S. L., 119 (1969, 337), 123 (1969, 337), 127 (1969, 337), 132 (1969, 337), 135 (1969, 337), 180 (1969, 337), *349*
Gledhill, T. D., 88 (1966, 149), *315*
Glemser, O., 7 (1966, 7; 1967, 11), 21 (1967, 50, 56), 22 (1965, 13; 1967, 56), 23 (1967, 63), 60, 66 (1967, 50), 67 (1969, 129), 77 (1967, 136), 79 (1969, 152), 97 (1967, 50, 56), 98 (1967, 50; 1969, 129), 99 (1965, 13; 1967, 136), 100 (1969, 129), 101 (1969, 152), 103 (1967, 288), 105 (1967, 11), 109 (1967, 11), 115 (1969, 129), 129 (1966, 7), 130 (1966, 7; 1967, 11), 132 (1967, 11, 317), 143 (1967, 63), 150 (1967, 11), 153 (1967, 50), 171 (1967, 11), 184 (1966, 7), 185 (1966, 7; 1967, 11), 187 (1967, 11), 195 (1967, 11, 63), 201 (1967, 11), 205 (1966, 7), 214 (1966, 7), 221 (1966, 7; 1967, 11), 222 (1967, 11), 277 (1969, 152), 285 (1967, 392), *308, 311, 319, 320, 321, 323, 327, 328, 330, 343, 344*
Glidewell, C., 27 (1969, 73), 28 (1969, 73), 48 (1969, 73), 74 (1969, 73), 75 (1969, 73), 105 (1969, 73), 110 (1969, 73), *341*
Glockling, F., 91 (1969, 267), 92 (1968, 252; 1969, 294), *336, 347, 348*
Gloede, J., 130 (1966, 238), 131 (1966, 238), 151 (1966, 238), 185 (1966, 238), *317*
Goddard, N., 48 (1969, 113), 84 (1969, 113), 97 (1969, 113), 100 (1969, 113), *342*
Goetze, U., 13 (1967, 29), 18 (1967, 29), 49 (1967, 29), 115 (1967, 29), 143 (1967, 29), 171 (1967, 29; 1968, 327), 204 (1968, 327), *320, 339*
Goggin, P. L., 90 (1969, 233), 92 (1969, 233), *346*
Goldberg, M. C., 94 (1968, 282, 283, 284), 117 (1968, 284), 118 (1968, 283, 284), 128 (1968, 283, 284), 140 (1968, 284), 145 (1968, 284), 146 (1968, 284), 149 (1968, 284), 151 (1968, 284), 153 (1968, 284), 154 (1968, 284), 157 (1968, 283, 284), 160 (1968, 282, 284), 163 (1968, 284), 171 (1968, 284), 172 (1968, 284), 173 (1968, 284), 174 (1968, 283, 284), 176 (1968, 284), 179 (1968, 284), 181 (1968, 284), 182 (1968, 283), 188 (1968, 284), 189 (1968, 284), 191 (1968, 284), 192

(1968, 284), 202 (1968, 284), 203 (1968, 284), 204 (1968, 284), 206 (1968, 284), 262 (1968, 284), 263 (1968, 282, 284), *337*

Goldfine, H., 94 (1967, 269), *326*

Goldman, F., 7 (1965, 4), 97 (1965, 4), 98 (1965, 4), 103 (1965, 4), 105 (1965, 4), 106 (1965, 4), 115 (1965, 4), *308*

Goldmann, F. L., 86 (1965, 50), 115 (1965, 50), *309*

Goldsberry, R. E., 170 (1968, 326), *339*

Goldsmith, E. J., 40 (1968, 77), 82 (1968, 77), 107 (1968, 77), 151 (1968, 77), *332*

Goldwhite, H., 18 (1965, 116), 20 (1965, 116), 28 (1968, 58, 59), 29 (1968, 58, 59), 35 (1968, 71), 40 (1966, 51), 44 (1969, 107), 51 (1968, 98), 63 (1965, 116), 76 (1968, 98), 80 (1968, 58, 59), 82 (1966, 51; 1969, 169), 86 (1966, 124), 91 (1968, 98), 105 (1966, 211; 1968, 58), 106 (1966, 26), 107 (1969, 169), 108 (1966, 124), 110 (1967, 295), 111 (1965, 116; 1966, 222), 112 (1969, 169), 113 (1968, 59; 1969, 107), 115 (1966, 124), 116 (1967, 295), 117 (1968, 71; 1969, 107), 118 (1969, 169), 122 (1966, 124), 123 (1967, 295; 1968, 98), 127 (1968, 59), 128 (1969, 107), 133 (1966, 222), 134 (1966, 222), 135 (1966, 51), 140 (1966, 51), 146 (1966, 222), 147 (1966, 51), 151 (1969, 169), 154 (1966, 124), 156 (1966, 222), 158 (1966, 51, 124), 173 (1965, 116), 185 (1966, 26), 189 (1966, 51), 190 (1966, 51), 207 (1967, 295; 1968, 98), 208 (1967, 295), 218 (1966, 51), 259 (1965, 116), 262 (1965, 116), 266 (1965, 116), 267 (1965, 116), 268 (1965, 116), 270 (1965, 116), 286 (1965, 116), 288 (1965, 116), *308, 312, 314, 316, 317, 327, 331, 332, 342, 344*

Gombler, W., 21 (1967, 54), 108 (1967, 54), 109 (1967, 54), 259 (1967, 54), *320*

Gonti, F., 92 (1967, 251), *326*

Goodfellow, R. J., 90 (1969, 233), 92 (1968, 244; 1969, 233), *336, 346*

Goodlett, V. W., 80 (1965, 36), 161 (1965, 36), 170 (1965, 36), 178 (1965, 36), *309*

Goodman, S. C., 38 (1966, 48), 166 (1966, 48), *312*

Goodrich, R. A., 19 (1967, 36; 1968, 38, 39), 22 (1967, 36; 1968, 38, 39), 28 (1968, 38, 39), 31 (1968, 39), 75 (1967, 36; 1968, 38, 39), 77 (1968, 38, 39), 86 (1966, 139), 99 (1967, 36), 105 (1968, 38), 109 (1966, 139; 1968, 38, 39), *315, 320, 331*

Gordon, H. B., 92 (1969, 281), *347*

Gordon, I. R., 92 (1965, 68), *310*

Gordon, M., 26 (1965, 19), 53 (1966, 63, 64; 1967, 107), 55 (1966, 70; 1967, 117), 56 (1966, 70; 1967, 115, 117), 57 (1969, 123), 58 (1967, 117), 82 (1967, 115), 124 (1966, 63), 133 (1969, 123), 135 (1966, 244), 136 (1966, 244), 145 (1966, 63), 147 (1967, 327), 152 (1967, 115), 157 (1967, 327), 163 (1967, 117), 164 (1966, 70; 1967, 117), 174 (1967, 117), 176 (1967, 327), 185 (1969, 123), 192 (1967, 327), 200 (1967, 327), 204 (1969, 123), 210 (1969, 123), 211 (1967, 327), 221 (1966, 63), 222 (1967, 115), 228 (1966, 70; 1967, 117), 232 (1967, 327), 233 (1967, 117), 235 (1967, 117), 238 (1967, 117), 244 (1967, 117), 246 (1967, 117), 247 (1967, 117), *308, 313, 317, 322, 328, 343*

Gorenstein, D. G., 82 (1967, 160), 86 (1967, 184), 148 (1967, 160), 211 (1967, 160), *323, 324*

Gosling, K., 106 (1968, 294), *338*

Gosselck, J., 235 (1967, 367), 236 (1967, 367), 241 (1967, 367), *329*

Goya, A. E., 45 (1969, 103), 76 (1969, 103), 108 (1969, 103), 109 (1969, 103), 113 (1969, 103), 122 (1969, 103), 142 (1969, 103), *342*

Grabenstetter, R. J., 7 (1967, 12), 256 (1967, 12), 257 (1967, 12), 258 (1967, 12), *319*

Graf, G., 59 (1967, 122), 242 (1967, 122), *322*

Graham, W. A. G., 7 (1966, 10; 1969, 16, 17), 22 (1969, 17), 84 (1969, 195), 92 (1969, 195), 129 (1969, 17), 150 (1966, 10; 1969, 17), 185 (1969, 17), 221 (1966, 10; 1969, 17), 222 (1966, 10; 1969, 17), *311, 340, 345*

Grant, D., 86 (1967, 186), 104 (1967, 186), 108 (1967, 186), 114 (1967, 186), 131 (1967, 186), 132 (1967, 186), 258 (1967, 186), 259 (1967, 186), 278 (1967, 186), *324*

Grapov, A. F., 113 (1965, 82), 165 (1965, 82), 197 (1965, 82), *310*

Gratz, J. P., 53 (1968, 100), 124 (1968, 100), 130 (1968, 100), 131 (1968, 100), 134 (1968, 100), 135 (1968, 100), 144 (1968, 100), 179 (1968, 100), 180 (1968, 100), 189 (1968, 100), 198 (1968, 100), 205 (1968, 100), 224 (1968, 100), *333*

Gray, A. H., 80 (1969, 160), 148 (1969, 160), 173 (1969, 160), 198 (1969, 160), 199 (1969, 160), *344*

Grayson, M., 78 (1967, 138), 162 (1967, 138), 166 (1967, 138), *323*

Graziani, M., 91 (1969, 275), *347*

Greaves, E. O., 90 (1967, 220), 92 (1967, 220), *325*

Grechkin, N. P., 151 (1966, 250), 171 (1966, 250), 180 (1966, 250), 189 (1966, 250), 190 (1966, 250), 198 (1966, 250), *318*

Green, B., 79 (1969, 153), 92 (1969, 153), 277 (1969, 153), 284 (1969, 379), *344, 350*

Green, J. C., 92 (1968, 248, 251; 1969, 277), *336, 347*

Green, J. P., 92 (1969, 306), *348*

Green, M., 13 (1967, 27), 18 (1966, 19; 1969, 43), 21 (1966, 19), 27 (1966, 19), 29 (1968, 61), 32 (1968, 61), 34 (1969, 43), 35 (1969, 43), 44 (1966, 53), 45 (1966, 19), 46 (1966, 53), 53 (1968, 61), 57 (1966, 19), 67 (1966, 53), 75 (1969, 43), 76 (1966, 19), 90 (1967, 27), 91 (1969, 262), 92 (1966, 183; 1968, 255), 93 (1969, 307), 106 (1969, 43), 107 (1965, 80), 119 (1965, 80), 129 (1966, 19), 130 (1966, 19), 135 (1968, 61), 139 (1965, 80), 144 (1968, 61), 145 (1968, 20), 150 (1966, 19; 1968, 61), 156 (1968, 61), 171 (1966, 19), 185 (1966, 19), 201 (1966, 19), 205 (1966, 19), 214 (1966, 19), 221 (1966, 19), 227 (1966, 19), 257 (1966, 53), 258 (1966, 53), 259 (1966, 53), 260 (1966, 53), 261 (1966, 53), 267 (1966, 53), *310, 312, 313, 316, 320, 331, 337, 341, 347, 348*

Green, M. J., 8 (1968, 20), 9 (1968, 20), 68 (1968, 20), 69 (1968, 20), 72 (1968, 20), 104 (1968, 20), 147 (1968, 20), *330*

Green, M. L. H., 92 (1968, 248; 1969, 277), *336, 347*

Greenhalgh, R., 82 (1969, 171), 86 (1967, 185), 121 (1969, 171), 138 (1967, 185), 141 (1967, 185), 161 (1967, 185), 163 (1969, 171), 180 (1969, 171), 215 (1969, 171), 263 (1967, 185), *324, 344*

Greenwood, N. N., 10 (1968, 24), 18 (1966, 21), 55 (1966, 21), 80 (1968, 144), 86 (1966, 125), 101 (1968, 144), 110 (1966, 125; 1968, 24, 144), 117 (1968, 144), 131 (1966, 21), 132 (1966, 21), 133 (1968, 24), 134 (1968, 144), 150 (1968, 144), 260 (1968, 24), *312, 314, 331, 334*

Gregor, I. K., 271 (1969, 375), *350*

Griffin, B. E., 94 (1968, 273), *337*

Griffin, C. E., 11 (1968, 25), 26 (1965, 19), 32 (1969, 82), 37 (1969, 82), 49 (1968, 97), 53 (1966, 63; 1967, 107), 55 (1966, 70; 1967, 117), 56 (1966, 70; 1967, 115, 117), 57 (1969, 123), 58 (1967, 117), 80 (1968, 25, 143), 82 (1967, 115), 86 (1966, 130), 107 (1968, 25), 119 (1968, 25; 1969, 82), 120 (1968, 25), 121 (1968, 25), 124 (1966, 63), 131 (1966, 240), 133 (1966, 130; 1969, 123), 135 (1966, 244; 1968, 25, 97), 136 (1966, 244), 137 (1966, 245; 1968, 25, 97; 1969, 82), 139

(1969, 82), 140 (1968, 25), 141 (1968, 25; 1969, 82), 145 (1966, 63), 147 (1967, 327), 148 (1969, 82), 151 (1966, 130), 152 (1967, 115), 154 (1966, 240; 1968, 317), 157 (1967, 327), 160 (1969, 82), 163 (1967, 117; 1968, 317), 164 (1966, 70; 1967, 117; 1968, 317), 165 (1969, 82), 169 (1969, 82), 173 (1969, 82), 174 (1966, 240; 1967, 117; 1968, 25, 317), 175 (1966, 256; 1969, 82), 176 (1967, 327; 1969, 82), 178 (1969, 82), 182 (1966, 256; 1968, 317), 183 (1966, 245, 256; 1969, 82), 185 (1969, 123), 186 (1966, 240), 190 (1966, 245), 191 (1968, 317; 1969, 82), 192 (1966, 256; 1967, 327), 198 (1966, 245), 199 (1965, 87; 1969, 82), 200 (1967, 327), 203 (1966, 240; 1968, 25), 204 (1969, 123), 210 (1969, 123), 211 (1967, 327), 215 (1966, 245), 221 (1966, 63), 222 (1967, 115), 223 (1966, 240), 225 (1966, 245), 228 (1966, 70; 1967, 117), 232 (1967, 327), 233 (1967, 117), 235 (1967, 117), 238 (1967, 117), 244 (1967, 117), 246 (1967, 247 (1967, 117), 262 (1968, 25), 264 (1968, 143), *308, 313, 315, 317, 318, 322, 328, 331, 332, 334, 338, 342, 343*

Griffin, G. E., 119 (1965, 87), 138 (1965, 87), 177 (1965, 87), 202 (1965, 87), 208 (1965, 87), 218 (1965, 87), 220 (1965, 87), 231 (1965, 87), 247 (1965, 87), *310*

Griffiths, J. E., 21 (1968, 44), 23 (1967, 62), 34 (1968, 44, 70), 35 (1967, 62), 77 (1967, 62), 103 (1967, 62; 1968, 44, 70), 104 (1967, 62), *321, 331, 332*

Grim, S. O., 5 (1965, 3; 1967, 8), 6 (1966, 5; 1968, 7), 13 (1969, 33), 18 (1968, 7), 27 (1968, 7), 30 (1967, 79), 31 (1967, 79; 1968, 7), 57 (1969, 33), 74 (1967, 79; 1968, 116), 87 (1968, 7), 90 (1967, 79, 228; 1968, 210; 1969, 249), 91 (1967, 79, 228; 1968, 210; 1969, 249), 92 (1966, 192; 1967, 228, 255), 93 (1966, 195; 1967, 79, 228; 1968, 210, 269; 1969, 249, 307), 109 (1965, 3), 113 (1965, 3; 1968, 7), 115 (1968, 7), 121 (1965, 3), 122 (1966, 5), 127 (1965, 3), 129 (1966, 5), 139 (1965, 3; 1968, 7), 141 (1968, 7), 148 (1969, 249), 153 (1965, 3; 1968, 7), 154 (1968, 7), 159 (1965, 3), 169 (1965, 3; 1968, 7), 173 (1965, 3; 1968, 7), 174 (1968, 7), 178 (1966, 5), 187 (1965, 3), 191 (1965, 3; 1967, 8; 1968, 7), 193 (1965, 3; 1967, 8; 1968, 7; 1969, 249), 194 (1968, 7), 196 (1965, 3; 1967, 79; 1968, 7), 201 (1967, 8), 202 (1965, 3; 1967, 79; 1968, 7), 203 (1966, 5; 1968, 7), 206 (1965, 3; 1966, 5; 1967, 8; 1968, 7; 1969, 249), 207 (1966, 5), 208 (1967, 8), 209 (1966, 5; 1967, 8; 1968, 7), 212 (1966, 5; 1967, 8), 216 (1965, 3; 1966, 5; 1967, 8, 79; 1968, 7; 1969, 249), 217 (1966, 5; 1967, 8), 218 (1966, 5), 219 (1967, 8), 220 (1966, 5), 223 (1965, 3; 1968, 337), 224 (1965, 3, 1967, 8; 1968, 7), 225 (1966, 5; 1968, 7), 226 (1965, 3; 1966, 5; 1967, 8; 1968, 7), 227 (1967, 363; 1968, 337; 1969, 33), 228 (1966, 5; 1967, 8, 363), 229 (1966, 5), 230 (1969, 33), 231 (1966, 5; 1967, 8, 363; 1968, 337), 233 (1966, 5), 234 (1966, 5; 1969, 33), 235 (1966, 5; 1967, 8, 363; 1968, 337), 238 (1966, 5), 239 (1966, 5; 1967, 363), 241 (1966, 5; 1968, 337), 242 (1966, 5), 243 (1966, 5; 1968, 337), 244 (1966, 5), 245 (1966, 5), 247 (1966, 5), 252 (1966, 5), 275 (1966, 5), 276 (1966, 5), *308, 311, 316, 319, 321, 325, 326, 329, 330, 333, 335, 337, 339, 340, 346, 348*

Grimm, L. F., 94 (1969, 325), 97 (1969, 325), 100 (1969, 325), 101 (1969, 325), 109 (1969, 325), 120 (1969, 325), 121 (1969, 325), *349*

Grimmer, A. R., 88 (1966, 150; 1967, 213, 214, 215, 216), 202 (1967, 216), *315, 325*

Grobe, J., 90 (1969, 244), 264 (1969, 373), *346, 350*

Groeger, H., 43 (1967, 92), 90 (1967, 221), 91 (1967, 221), 92 (1967, 221), 121 (1966, 232), 187 (1967, 344), 203 (1967, 92), 204 (1967, 92, 344, 353), 209 (1967, 221), 210 (1967, 221, 344), 216 (1967, 353), 220 (1967, 221, 353), 223 (1967, 92), 228 (1967, 92), 270 (1966, 232; 1967, 92, 344), 271 (1966, 232; 1967, 92, 344), 272 (1966, 232; 1967, 344), 281 (1967, 92), *317, 321, 325, 329*

Groenweghe, L. C. D., 86 (1966, 114), 104 (1966, 114), 107 (1966, 114), 111 (1966, 114), 112 (1966, 114), 118 (1966, 114), 131 (1966, 114), 134 (1966, 114), 143 (1966, 114), 171 (1966, 114), *314*

Groothius, D., 94 (1968, 283), 118 (1968, 283), 128 (1968, 283), 157 (1968, 283), 174 (1968, 283), 182 (1968, 283), *337*

Gross, H., 47 (1968, 93), 130 (1966, 238), 131 (1966, 238), 144 (1969, 346), 151 (1966, 238), 157 (1969, 346), 169 (1969, 346), 177 (1969, 346), 178 (1969, 346), 185 (1966, 238), 190 (1968, 93), 205 (1968, 93), 264 (1968, 353; 1969, 346), 265 (1968, 93, 353), 266 (1968, 353; 1969, 346), 267 (1969, 346), *317, 332, 339, 349*

Grosse Böwing, W., 98 (1969, 326), *349*

Gründler, W., 20 (1966, 27), *312*

Grunwald, E., 86 (1965, 44), *309*

Grunzweig, J., 88 (1967, 201), *324*

Grushkin, B., 79 (1966, 96; 1968, 134), 281 (1966, 96), 285 (1966, 96; 1968, 134), 286 (1966, 96; 1968, 134), 287 (1966, 96), *314, 333*

Gryszkiewicz-Trochimowski, E., 134 (1967, 320), *328*

Guerrieri, F., 92 (1967, 237), 171 (1967, 237), *325*

Guibé-Jampel, E., 129 (1969, 338), 221 (1969, 338), *349*

Guillemonat, A., 58 (1969, 127), 114 (1969, 127), 119 (1969, 127), 125 (1969, 127), 126 (1969, 127), 156 (1969, 127), 158 (1969, 127), 166 (1969, 355), 176 (1969, 355), 183 (1969, 355), *343, 349*

Gulati, A. S., 22 (1968, 303), 113 (1968, 303), 114 (1966, 225), 139 (1968, 303), 140 (1966, 225; 1968, 303), 141 (1968, 303, 310), 148 (1966, 225), 153 (1968, 303), 159 (1968, 303, 310), 162 (1966, 225; 1968, 310), 172 (1968, 310), 173 (1966, 225; 1968, 303), 177 (1966, 225), 188 (1966, 225; 1968, 303), 193 (1968, 303), 201 (1968, 303), 202 (1968, 303), 203 (1966, 225), 204 (1968, 303), 206 (1966, 225), 219 (1966, 225; 1968, 303), 223 (1966, 225; 1968, 303), 224 (1966, 225), 225 (1966, 225), 226 (1966, 225; 1968, 303), 230 (1966, 225; 1968, 303), 231 (1968, 303), 232 (1966, 225; 1968, 303), 233 (1966, 225), 238 (1966, 225; 1968, 303), 240 (1968, 303, 310), 242 (1966, 225; 1968, 303), 246 (1968, 303), 247 (1966, 225), 248 (1968, 303), 249 (1966, 225; 1968, 303), 252 (1966, 225; 1968, 303), 253 (1966, 225), *317, 338*

Gulati, S. A., 29 (1969, 75), 218 (1969, 75), 219 (1969, 75), *341*

Gupta, L. C., 88 (1968, 189), *335*

Gupta, R. C., 88 (1968, 192), *335*

Gur'yanova, I. V., 2 (1967, 4), 156 (1967, 4), 182 (1967, 4), 200 (1967, 4), 213 (1967, 4), 221 (1967, 4), *319*

Gutmann, V., 118 (1966, 288), 173 (1965, 98), 206 (1965, 98), 226 (1965, 98), 257 (1965, 98; 1966, 288), 258 (1965, 98; 1966, 288), 259 (1966, 288), 260 (1965, 98; 1966, 288), 261 (1966, 288), 262 (1965, 98; 1966, 288), 263 (1966, 288), 264 (1966, 288), 265 (1966, 288), 267 (1966, 288), 268 (1966, 288), 269 (1966, 288), *318*

Gutowsky, H. S., 4 (1969, 7), 5 (1964, 1; 1965, 2), 14, 20 (1969, 7), 88 (1969, 7), 97 (1965, 2; 1969, 7), 98 (1965,

AUTHOR INDEX

2), 99 (1965, 2), 100 (1965, 2), 113 (1965, 2), *308, 340*
Guyot De La Hadrouyere, M., 86 (1966, 118), *314*

H

Haake, P., 7 (1967, 19), 31 (1968, 65), 40, 82, 107 (1968, 77), 109 (1967, 19; 1968, 65), 110 (1967, 19), 114 (1967, 19; 1968, 65), 121 (1968, 65), 122 (1968, 65), 128 (1968, 65), 140 (1968, 65), 149 (1968, 65), 151 (1968, 77), 153 (1967, 19), 154 (1967, 19), 159 (1968, 65), 169 (1968, 65), 179 (1968, 65), 184 (1968, 65), 195 (1968, 65), 196 (1967, 19), 197 (1967, 19), *319, 332*
Haas, T. E., 85 (1968, 162), 90 (1968, 162), 97 (1968, 162), 99 (1968, 162), *334*
Haberlein, H., 44 (1968, 88), 48 (1968, 88), 230 (1968, 88), 231 (1968, 88), 238 (1968, 88), 248 (1968, 88), 251 (1968, 88), *332*
Hadzi, D., 88 (1968, 187), *335*
Haemers, M., 40 (1969, 88), 125 (1969, 88), 137 (1969, 88), *342*
Hagatsu, H., 41 (1969, 96), 83 (1969, 96), 94 (1969, 96), *342*
Hägele, G., 2 (1968, 2), 28 (1968, 2), 31 (1968, 2), 127 (1968, 2), 128 (1968, 2), 129 (1968, 2), *330*
Hagen, P. O., 94 (1967, 269), *326*
Haigh, C. W., 40 (1969, 68), 44, 66 (1969, 68), *341*
Haines, R. J., 91 (1968, 213; 1969, 259), 148 (1968, 213), *336, 347*
Hall, C., 251 (1965, 116), 253 (1965, 116), *311*
Hall, C. D., 248 (1967, 377), 253 (1967, 377), 254 (1967, 377), *329*
Hall, D. De W., 90 (1967, 222), *325*
Hall, L. D., 42 (1968, 83), 83 (1968, 83), 163 (1968, 83), 172 (1968, 83), 181 (1968, 83), 190 (1968, 83), 211 (1968, 83), *332*
Hallman, P. S., 93 (1967, 260), *326*
Halman, M., 115 (1966, 227), *317*

Hamashima, Y., 206 (1967, 352), 217 (1967, 352), 241 (1967, 352), *329*
Hamid, A. M., 209 (1967, 354), *329*
Hammers, G. G., 89 (1967, 263), *326*
Hamrin, K., 4 (1967, 6), *319*
Hancock, M., 89 (1969, 309), *348*
Hands, A. R., 75 (1968, 118), 209 (1968, 118), 235 (1967, 366), *329, 333*
Hansen, A. S., 27 (1969, 74), 38 (1969, 74), 75 (1969, 74), 93 (1969, 74), 94 (1969, 74), 104 (1969, 74), 109 (1969, 74), 112 (1969, 74), 113 (1969, 74), 120 (1969, 74), 133 (1969, 74), 136 (1969, 74), 138 (1969, 74), 140 (1969, 74), 148 (1969, 74), 149 (1969, 74), 154 (1969, 74), 155 (1969, 74), 174 (1969, 74), 194 (1969, 74), 202 (1969, 74), 220 (1969, 74), 235 (1969, 74), *341*
Hansen, K. C., 203 (1967, 349), 210 (1967, 349), 230 (1967, 364), 235 (1967, 349), 271 (1967, 349), 272 (1967, 349), 273 (1967, 349, 364), 274 (1967, 349), 275 (1967, 249, 364), *329*
Hansen, K. W., 25 (1965, 15, 16), *308*
Hänssgen, D., 79 (1967, 149), 272 (1967, 149), *323*
Happe, J. A., 89 (1966, 196), 94 (1967, 266), *316, 326*
Haque, R., 94 (1971, 1), *350*
Harbourne, D. A., 91 (1969, 258), 92 (1968, 253), *336, 347*
Hardeman, G. E. G., 88 (1966, 147), 257 (1966, 147), *315*
Hargis, J. H., 40 (1968, 78), 82 (1969, 172), 83 (1968, 78; 1969, 172, 173), 133 (1969, 172), 137 (1968, 78), 146 (1968, 78), 157 (1968, 78; 1969, 172, 173), *332, 344*
Harris, R. K., 20 (1967, 55), 21 (1967, 55), 28 (1968, 57), 33 (1967, 55), 34 (1967, 55), 42 (1967, 55), 60 (1967, 124, 125), 62 (1967, 124, 125; 1968, 57, 113), 63 (1964, 12; 1967, 124, 125, 127; 1968, 113), 64 (1966, 79; 1968, 57), 65 (1966, 79), 66 (1968, 57), 68 (1967, 125), 70, 72 (1968, 57), 77 (1967, 55), 90 (1968,

57), 98 (1966, 79), 100 (1968, 113), 101 (1966, 79; 1968, 113), 106 (1962, 6; 1963, 6, 7; 1967, 55), 109 (1967, 55), 257 (1966, 79; 1968, 57), 259 (1967, 124, 125; 1968, 57, 113), 261 (1968, 57), 262 (1967, 127; 1968, 113), 263 (1967, 127; 1968, 113), 268 (1968, 113), 270 (1968, 113), 271 (1967, 387), 272 (1967, 387), 273 (1967, 387), 279 (1968, 57), 281 (1968, 57), 282 (1968, 57), *307, 308, 313, 320, 322, 330, 331, 333*

Hartland, A., 88 (1969, 228), *346*

Hartle, R. J., 121 (1966, 231), 126 (1966, 231), *317*

Hartman, F. A., 91 (1966, 166), *315*

Hartman, P. F., 50 (1966, 60), 86 (1966, 60), *313*

Hartung, H., 59 (1967, 122), 242 (1967, 122), *322*

Harvey, R. G., 83 (1966, 104), 93 (1966, 104), 148 (1966, 249), 158 (1966, 249), 167 (1966, 249), 184 (1966, 249), 205 (1966, 249), 211 (1966, 249), 252 (1966, 104), *314, 318*

Hashi, T., 88 (1967, 209; 1969, 222), 135 (1967, 209), *325, 346*

Haszeldine, R. N., 18 (1965, 116; 1966, 19), 20 (1965, 116), 21 (1966, 19), 27 (1966, 19), 44 (1966, 53), 45 (1966, 19), 46 (1966, 53), 57 (1966, 19), 63 (1965, 116), 67 (1966, 53), 76 (1966, 19), 91 (1966, 165), 106 (1966, 26), 107 (1965, 80), 111 (1965, 116), 119 (1965, 80), 129 (1966, 19), 130 (1966, 19), 139 (1965, 80), 150 (1966, 19), 171 (1966, 19), 173 (1965, 116), 185 (1966, 19, 26), 201 (1966, 19), 205 (1966, 19), 214 (1966, 19), 221 (1966, 19), 227 (1966, 19), 257 (1966, 53), 258 (1966, 53), 259 (1965, 116; 1966, 53), 260 (1966, 53), 261 (1966, 53), 262 (1965, 116), 266 (1965, 116), 267 (1965, 116; 1966, 53), 268 (1965, 116), 270 (1965, 116), 286 (1965, 116), 288 (1965, 116), *308, 310, 312, 313, 315*

Hata, G., 91 (1968, 214), *336*

Hatton, J. V., 108 (1962, 7), *307*

Haubold, W., 79 (1965, 34), 100 (1967, 286), 132 (1967, 286), 187 (1965, 34), 265 (1967, 286), 283 (1965, 34), *309, 327*

Hawes, W., 78 (1968, 122), 205 (1968, 122), 211 (1967, 356), 212 (1967, 356), *329, 333*

Hays, H. R., 18 (1966, 14, 15), 74 (1966, 14), 177 (1966, 14), 193 (1966, 14), 201 (1966, 14), 207 (1966, 14, 268), 214 (1966, 14, 268), 218 (1966, 14), 221 (1966, 268), 227 (1966, 14), 233 (1966, 14), 244 (1966, 14), 245 (1966, 14), 247 (1966, 14), 254 (1966, 14), 255 (1966, 14), 256 (1966, 15), 257 (1966, 15), 258 (1966, 15), 259 (1966, 15), 260 (1966, 15), 263 (1966, 15), 264 (1966, 15), 265 (1966, 15), 266 (1966, 15), 267 (1966, 15), 268 (1966, 15), 269 (1966, 15), 271 (1966, 15), 273 (1966, 15, 268), 274 (1966, 15), *312, 318*

Hayter, R. G., 63 (1964, 12), *308*

Healon, B. T., 92 (1968, 256), *337*

Heatley, F., 39 (1966, 49), 79 (1966, 49), 101 (1966, 49), 277 (1966, 49), *312*

Heaton, B. T., 92 (1968, 249), *336*

Heazlewood, R. L., 91 (1969, 257), *347*

Heckman, G. H., 24 (1969, 69), 60, 61 (1969, 69), 67 (1969, 69), 100 (1969, 69), 101 (1969, 69), 258 (1969, 69), 260 (1969, 69), 265 (1969, 69), *341*

Heckmann, G., 83 (1969, 180), *344*

Hedaya, E., 237 (1968, 344), 238 (1968, 344), 240 (1968, 344), 246 (1968, 344), 248 (1968, 344), 250 (1968, 344), *339*

Hedman, J., 4 (1967, 6), *319*

Heinen, H., 10 (1968, 23), 110 (1968, 23), *331*

Heinze, P. R., 77 (1967, 136), 99 (1967, 136), *323*

Heitsch, C. W., 40 (1964, 5), 84 (1965, 40), *308, 309*

Helbrege, D. M., 78, 82 (1969, 144), *343*

Hellwedge, D. M., 52 (1966, 62), 83 (1967, 166), 86 (1966, 62), 103

(1966, 62), 137 (1967, 166), 148 (1967, 166), 155 (1967, 166), 158 (1967, 166), 166 (1967, 166), 168 (1967, 166), 176 (1967, 166), 183 (1967, 166), 192 (1967, 166), 200 (1967, 166), 204 (1967, 166), 227 (1966, 62), 240 (1966, 62), *313, 324*
Hellwinkel, D., 6 (1969, 13), 19 (1966, 25), 83 (1965, 37; 1966, 25, 101; 1968, 151), 222 (1969, 13), 227 (1969, 13), 228 (1969, 13), 242 (1966, 25), 252 (1968, 151), 255 (1966, 101; 1968, 151), *309, 312, 314, 334, 340*
Henderson, W. G., 104 (1969, 24), *340*
Hendricker, D. G., 90 (1966, 158; 1968, 209), 91 (1966, 158; 1968, 209), 93 (1966, 158; 1968, 209), 125 (1968, 209), 145 (1968, 209), *315, 335*
Hendrickson, J. B., 251 (1965, 116), 253 (1965, 116), *311*
Henold, K. L., 85 (1969, 198), 142 (1969, 198), *345*
Henry, M. C., 212 (1969, 364), 217 (1969, 364), *349*
Herail, F., 114 (1966, 226), *317*
Herbert, W., 94 (1968, 284), 117 (1968, 284), 118 (1968, 284), 128 (1968, 284), 140 (1968, 284), 145 (1968, 284), 146 (1968, 284), 149 (1968, 284), 151 (1968, 284), 153 (1968, 284), 154 (1968, 284), 157 (1968, 284), 160 (1968, 284), 163 (1968, 284), 171 (1968, 284), 172 (1968, 284), 173 (1968, 284), 174 (1968, 284), 176 (1968, 284), 179 (1968, 284), 181 (1968, 284), 188 (1968, 284), 189 (1968, 284), 191 (1968, 284), 192 (1968, 284), 202 (1968, 284), 203 (1968, 284), 204 (1968, 284), 206 (1968, 284), 262 (1968, 284), 263 (1968, 284), *337*
Herbison-Evans, D., 11 (1964, 2), 142 (1964, 2), *308*
Herweh, J., 56 (1966, 71), 58 (1966, 75), 86 (1966, 71), 165 (1966, 71), 182 (1966, 71), 196 (1966, 75), 197 (1966, 71), 222 (1966, 75), 225 (1966, 71), 249 (1968, 248), *313, 339*

Hesse, J. E., 88 (1966, 142), *315*
Hester, R. E., 84 (1969, 194), 97 (1969, 194), *345*
Hettche, A., 78 (1969, 145), 119 (1969, 145), 256 (1969, 145), *343*
Hewlett, C., 79 (1966, 87), 279 (1966, 87), 281 (1966, 87), 282 (1966, 87), *313*
Hieber, W., 90 (1965, 51), 91 (1965, 51), *309*
Highsmith, R. E., 27 (1967, 67), 142 (1967, 67), 150 (1967, 67), *321*
Higson, H. G., 18 (1966, 19), 21 (1966, 19), 27 (1966, 19), 45 (1966, 19), 57 (1966, 19), 76 (1966, 19), 129 (1966, 19), 130 (1966, 19), 150 (1966, 19), 171 (1966, 19), 185 (1966, 19), 201 (1966, 19), 205 (1966, 19), 214 (1966, 19), 221 (1966, 19), 227 (1966, 19), *312*
Hilbers, C. W., 92 (1968, 242), *336*
Hilgetag, G., 122 (1967, 308), 129 (1967, 308), 141 (1967, 308), 149 (1967, 308), 160 (1967, 308), 178 (1967, 308), *327*
Hill, H. A. O., 90 (1968, 200), *335*
Hill, W. E., 21 (1969, 62), 22 (1969, 62), 25 (1969, 62), 43 (1969, 62), 61 (1969, 62), 66 (1969, 62), 75 (1969, 62), 76 (1969, 62), 77 (1969, 62), 100 (1969, 62), 101 (1969, 62), 104 (1969, 62), 108 (1969, 62), 109 (1969, 62), 110 (1967, 294), 111 (1967, 294), 113 (1969, 62), 114 (1969, 62), 132 (1969, 62), 152 (1969, 62), 153 (1969, 62), 186 (1969, 62), 196 (1969, 62), 202 (1969, 62), 222 (1969, 62), 223 (1969, 62), *327, 341*
Hindersinn, R. R., 156 (1965, 94), *310*
Hirai, K., 206 (1967, 352), 217 (1967, 352), 241 (1967, 352), *329*
Hirai, S., 83 (1966, 104), 93 (1966, 104), 252 (1966, 104), *314*
Hirano, S., 94 (1966, 203), *316*
Hirz, R., 94 (1969, 320), *348*
Ho, C., 94 (1966, 202), *316*
Hoffmann, H., 27 (1967, 68), 31 (1967, 68), 194 (1967, 68), 201 (1967, 68),

231 (1965, 113), 252 (1966, 285), 255 (1966, 285), *311, 318, 321*
Hoffmann, M., 121 (1966, 293), 285 (1966, 293), *319*
Hoffmann, P., 224 (1966, 272), 232 (1966, 272), 238 (1966, 272), *318*
Höfler, M., 92 (1967, 245), *326*
Hogben, M. G., 7 (1966, 10; 1969, 16, 17), 22 (1969, 17), 129 (1969, 17), 150 (1966, 10; 1969, 17), 185 (1969, 17), 221 (1966, 10; 1969, 17), 222 (1966, 10; 1969, 17), *311, 340*
Hollenberg, I., 7 (1966, 7; 1967, 11), 105 (1967, 11), 109 (1967, 11), 129 (1966, 7), 130 (1966, 7; 1967, 11), 132 (1967, 11), 150 (1967, 11), 171 (1967, 11), 184 (1966, 7), 185 (1966, 7; 1967, 11), 187 (1967, 11), 195 (1967, 11), 201 (1967, 11), 205 (1966, 7), 214 (1966, 7), 221 (1966, 7; 1967, 11), 222 (1967, 11), 285 (1967, 392), *311, 319, 330*
Hollis, D. P., 94 (1969, 316, 319), *348*
Holmes, R. R., 19 (1966, 24), 22 (1966, 24), 75 (1966, 24), 99 (1966, 24), *312*
Holywell, G. C., 64 (1969, 131), 91 (1969, 131), *343*
Hooton, K. A., 92 (1968, 252), *336*
Hopkins, G. S. A., 44 (1966, 53), 46 (1966, 53), 67 (1966, 53), 257 (1966, 53), 258 (1966, 53), 259 (1966, 53), 260 (1966, 53), 261 (1966, 53), 267 (1966, 53), *313*
Hopton, F. J., 92 (1966, 181), *316*
Hormuth, P. B., 19 (1969, 46), *341*
Horn, H. G., 7 (1966, 8), 21 (1966, 30), 22 (1965, 13; 1966, 8, 30), 31 (1966, 30), 76 (1966, 30), 97 (1966, 8), 98 (1966, 8), 99 (1965, 13; 1966, 8), 109 (1966, 30), 118 (1966, 30), 120 (1966, 30), 132 (1967, 317), 136 (1966, 30), 138 (1966, 30), *308, 311, 312, 328*
Horrocks, W. De W., 90 (1966, 152; 1967, 222, 224; 1968, 199, 204, 205), 91 (1968, 204), 92 (1966, 152, 172; 1967, 224; 1968, 199, 204, 205), *315, 316, 325, 335*

Hoskins, L. C., 86 (1967, 194), *324*
Hossenlopp, F., 86 (1967, 189), 94 (1967, 189), 100 (1967, 189), 101 (1967, 189), *324*
Houalla, D., 19 (1967, 39), 39 (1967, 39), 80 (1969, 162), 83 (1967, 39, 165; 1969, 179), 84 (1967, 172), 86 (1967, 165), 109 (1967, 172), 114 (1967, 172; 1969, 162), 119 (1967, 39), 122 (1967, 172), 128 (1967, 165; 1969, 162), 136 (1967, 165), 143 (1967, 172), 153 (1967, 165; 1969, 162), 154 (1969, 162), 157 (1967, 172), 158 (1969, 179), 161 (1967, 172), 164 (1969, 162), 170 (1967, 172; 1969, 162), 187 (1967, 165), 191 (1969, 162), 192 (1967, 172), 193 (1969, 179), 203 (1967, 172), 206 (1969, 162), 217 (1969, 162), 221 (1969, 162), *329, 324, 344*
Houk, L. W., 45 (1969, 104), 90 (1969, 239, 241), 91 (1969, 104, 239, 241), *342, 346*
Hovanec, J. W., 208 (1968, 334), *339*
Hruby, V. J., 114 (1967, 300), 246 (1967, 300), 276 (1967, 300), *327*
Hruska, F. E., 86 (1968, 172), 94 (1968, 172), *334*
Hsich, H. H., 56 (1967, 115), 82 (1967, 115), 152 (1967, 115), 222 (1967, 115), *322*
Hudson, R. F., 82 (1969, 171), 86 (1966, 116, 126), 114 (1966, 126), 121 (1966, 126; 1969, 171), 163 (1969, 171), 180 (1969, 171), 199 (1969, 361), 215 (1969, 171), 238 (1963, 8), 243 (1966, 116), *308, 314, 344, 349*
Hughes, R. H., 94 (1967, 275), *326*
Huisgen, R., 29 (1967, 76), 32 (1967, 76), 78 (1967, 76, 139), 86 (1967, 139), 196 (1969, 360), 246 (1967, 139), 250 (1967, 139), 251 (1969, 360), 253 (1967, 76, 139), 255 (1967, 76, 139), 256 (1967, 139), *321, 323, 349*
Hurst, G. H., 7 (1967, 19), 109 (1967, 19), 110 (1967, 19), 114 (1967, 19),

AUTHOR INDEX

153 (1967, 19), 154 (1967, 19), 196 (1967, 19), 197 (1967, 19), *319*
Hüttel, R., 90 (1967, 219), *325*
Huttemann, T. J., 40 (1965, 22), 74 (1965, 31), 83 (1966, 102), 90 (1965, 31), 135 (1966, 102), *309*, *314*

I

Ibrahim, E. H. M., 133 (1967, 318), 155 (1967, 318), 175 (1967, 318), 192 (1967, 318), 206 (1967, 318), 224 (1967, 318), 272 (1967, 388), *328*, *330*
Ignatiev, V. M., 86 (1966, 127), 111 (1966, 127; 1967, 296), 123 (1966, 127), 124 (1967, 296), 144 (1967, 296), 145 (1967, 296), 146 (1967, 296), 157 (1968, 314), 167 (1968, 314), 177 (1968, 314), 222 (1968, 314), *314*, *327*, *338*
Iles, B. R., 107 (1965, 80), 119 (1965, 80), 139 (1965, 80), *310*
Imbery, D., 46 (1968, 92), 47 (1968, 92), 80 (1968, 92), 153 (1968, 92), 173 (1968, 92), 191 (1968, 92), 196 (1968, 92), 202 (1968, 92), 206 (1968, 92), 216 (1968, 92), *332*
Interrante, L. V., 56 (1966, 72), 90 (1966, 72), 91 (1966, 72), 93 (1966, 72), 234 (1966, 72), *313*
Ionin, B. I., 5 (1968, 5), 20 (1967, 44), 42 (1967, 44), 52, 72 (1967, 128), 81 (1969, 163), 86 (1966, 127), 109 (1967, 44), 111 (1966, 127; 1967, 105, 296), 116 (1965, 86), 123 (1965, 86; 1966, 127; 1967, 310), 124 (1967, 296), 134 (1965, 86; 1967, 319), 134 (1965, 90), 135 (1967, 319), 136 (1965, 90), 144 (1967, 296), 145 (1967, 296, 310), 146 (1967, 296), 149 (1967, 310), 150 (1965, 86), 155 (1965, 90), 157 (1968, 314; 1969, 163), 166 (1969, 163), 167 (1968, 314), 176 (1965, 100), 177 (1965, 100; 1968, 314), 93 (1967, 319, 222 (1968, 314), 236 (1969, 163), 260 (1967, 128), 264 (1965, 90), *310*, *311*, *314*, 320, 322, 323, 327, 328, 330, 338, 344
Ionov, S. P., 4 (1969, 10), *340*
Ionova, G. V., 4 (1969, 10), *340*
Irani, R. R., 65 (1965, 30), 90 (1965, 30), 91 (1965, 30), 101 (1965, 30), 105 (1966, 212), 115 (1966, 212), 120 (1965, 30), 129 (1967, 315), 258 (1966, 212), 259 (1967, 315), 277 (1966, 212; 1967, 315, 391), 284 (1966, 212), 285 (1966, 212), *309*, *317*, *328*, *330*
Ireland, R. E., 226 (1969, 367), 254 (1969, 367), *350*
Isbell, A. F., 94 (1967, 275), *326*
Ishii, Y., 92 (1969, 293), *348*
Issleib, K., 18 (1965, 11a), 20 (1965, 11a, 1966, 27), 63 (1965, 11a, 27, 29), 74 (1967, 135), 78 (1967, 137), 108 (1967, 135), 111 (1965, 11a), 112 (1967, 135), 119 (1967, 135), 121 (1966, 293), 152 (1967, 135), 158 (1968, 320), 159 (1968, 320), 163 (1967, 137), 172 (1967, 137), 173 (1965, 11a), 178 (1965, 101), 181 (1967, 137), 184 (1967, 343), 186 (1965, 103; 1968, 320), 187 (1965, 103; 1968, 320), 188 (1967, 137), 194 (1965, 101), 198 (1967, 137), 199 (1967, 137), 205 (1967, 137), 209 (1967, 137), 210 (1967, 343), 211 (1967, 137), 215 (1967, 137), 219 (1967, 137), 223 (1965, 103), 233 (1967, 343), 247 (1967, 343), 249 (1967, 137), 251 (1967, 343), 258 (1965, 27), 259 (1965, 11a), 260 (1965, 27), 262 (1965, 21a), 265 (1965, 29), 266 (1965, 11a, 29), 267 (1965, 11a), 268 (1965, 11a, 29), 270 (1965, 11a, 29), 278 (1965, 27, 29), 280 (1966, 292), 282 (1967, 343), 283 (1967, 343), 284 (1965, 27, 29), 285 (1965, 27; 1966, 293), 286 (1965, 11a), 287 (1966, 295), 288 (1965, 11a; 1966, 295), *308*, *311*, *312*, *319*, *323*, *328*, *338*
Ito, Y., 29 (1967, 74), 183 (1968, 329), 249 (1967, 74), 250 (1967, 74), *321*, *339*
Ivanov, B. I., 244 (1967, 373), *329*

Ivanovskaya, K. M., 133 (1964, 16), 134 (1964, 16), 135 (1964, 16), 144 (1964, 16), 145 (1964, 16), *308*
Ivin, S. F., 21 (1967, 51), 99 (1957, 51), 103 (1967, 51), 104 (1967, 51), 107 (1967, 51), 130 (1967, 51), 132 (1967, 51), 138 (1967, 51), *320*
Ivin, S. Z., 7 (1968, 12), 19 (1967, 37, 38), 20 (1968, 12), 21 (1967, 37, 38, 52; 1968, 12), 25 (1967, 64), 104 (1967, 38, 290; 1968, 12), 107 (1967, 38, 290; 1968, 12), 108 (1968, 12), 109 (1967, 64), 113 (1968, 12), 114 (1967, 38), 119 (1967, 64), 120 (1968, 12), 125 (1967, 37), 127 (1968, 12), 128 (1967, 37, 38, 64; 1968, 12), 130 (1967, 52, 290), 131 (1967, 37), 132 (1967, 52, 290), 136 (1967, 64; 1968, 12), 139 (1967, 64), 143 (1967, 52, 64, 290), 146 (1967, 64), 152 (1968, 12), 153 (1967, 64), 154 (1967), 332), 157 (1967, 64, 332), 169 (1967, 64; 1968, 12), 173 (1967, 64), 180 (1967, 64), 181 (1967, 64), 209 (1967, 52), *320, 321, 327, 328, 330*
Iwashita, Y., 85 (1969, 200), *345*
Iwata, C., 94 (1967, 274), *326*

J

Jaccard, S., 189 (1966, 259), *318*
Jackson, G. F., 74 (1969, 142), 126 (1969, 142), 167 (1969, 142), *343*
Jackson, G. J., 80 (1968, 141), 267 (1968, 141), *334*
Jackson, W. R., 47 (1969, 111), 192 (1969, 111), *342*
Jacquier, R., 13 (1967, 25), *320*
Jain, S. R., 27 (1967, 65, 67; 1968, 54), 28 (1968, 54), 45 (1967, 65; 1968, 54), 46 (1967, 65), 85 (1968, 164), 110 (1967, 65), 113 (1968, 54), 115 (1967, 65; 1968, 54), 122 (1968, 54), 123 (1967, 65), 129 (1967, 65; 1968, 54), 134 (1967, 65), 142 (1967, 67), 143 (1967, 65), 150 (1967, 67), 153 (1968, 54), 154 (1968, 54), 174 (1968, 54), 175 (1967, 65), 188 (1967, 65), 193 (1968, 164), 196 (1968, 54), 197 (1967, 65; 1968, 54), 204 (1967, 65), 210 (1968, 54), 224 (1967, 65), 271 (1969, 374), 272 (1969, 374), 274 (1969, 374), *321, 331, 334, 350*
Jakobsen, H. J., 58 (1969, 124, 125, 126), 69 (1969, 124), 185 (1969, 124, 125, 126), 210 (1969, 124), *343*
Jameson, C. J., 5 (1964, 1), 14, 16, *308, 340*
Jander, J., 23 (1969, 67), 35 (1969, 67), 244 (1969, 67), 245 (1969, 67), *341*
Janjic, D., 189 (1966, 259), *318*
Jardetzky, C. D., 94 (1968, 280), *337*
Jardine, F. H., 92 (1966, 191), *316*
Jardine, R. V., 27 (1969, 74), 38 (1969, 74), 75 (1969, 74), 80 (1969, 160), 93 (1969, 74), 94 (1969, 74), 104 (1969, 74), 109 (1969, 74), 112 (1969, 74), 113 (1969, 74), 114 (1968, 304), 119 (1968, 304), 120 (1969, 74), 121 (1968, 304), 133 (1969, 74), 136 (1969, 74), 138 (1969, 74), 139 (1968, 304), 140 (1969, 74), 148 (1969, 74, 160), 149 (1969, 74), 154 (1969, 74), 155 (1968, 304; 1969, 74), 159 (1968, 304), 164 (1968, 304), 173 (1969, 160), 174 (1968, 304; 1969, 74), 193 (1968, 304), 194 (1969, 74), 198 (1969, 160), 199 (1969, 160), 202 (1969, 74), 213 (1968, 304), 220 (1969, 74), 235 (1969, 74), 239 (1968, 304), *338, 341, 344*
Jarman, M., 94 (1968, 273), *337*
Jatkowski, M., 122 (1967, 308), 129 (1967, 308), 141 (1967, 308), 149 (1967, 308), 160 (1967, 308), 178 (1967, 308), *327*
Jefferson, R., 76 (1968, 119), 90 (1969, 250), 91 (1968, 119; 1969, 250), 93 (1969, 250), 105 (1968, 119), 256 (1969, 370), *333, 346, 350*
Jellinek, F., 11 (1967, 23), 74 (1967, 23), 88 (1967, 211), 90 (1967, 23), 92 (1967, 211, 242), 93 (1967, 211), 113 (1967, 23), 139 (1967, 23), 223 (1967, 23), *320, 325, 326*

Jenkins, I. D., 18 (1966, 18), 19 (1966, 18), 80, 86 (1965, 48), 132 (1966, 18), 142 (1966, 18), 174 (1966, 18), 182 (1966, 18), 217 (1966, 18), 265 (1965, 48; 1966, 18), *309, 312, 334*
Jenkins, J. M., 83 (1966, 102), 91 (1966, 162), 92 (1966, 179), 93 (1966, 193), 135 (1966, 102), *314, 315, 316*
Jennings, W. B., 47 (1969, 111), 192 (1969, 111), *342*
Jensen, E. V., 83 (1966, 104), 93 (1969, 104), 252 (1966, 104), *314*
Jensen, R. C., 90 (1969, 253), *346*
Joachim, G., 86 (1966, 140), 238 (1966, 140), *315*
Joesten, M. D., 263 (1969, 372), *350*
Johannesen, R. B., 13 (1967, 31), 17 (1967, 31), 18 (1967, 31), 20 (1967, 31), 21 (1967, 31), 75 (1967, 31), *320*
Johansson, G., 4 (1967, 6), *319*
Johns, I. B., 195 (1966, 262), 196 (1966, 262), 223 (1966, 262), 227 (1966, 262), *318*
Johnson, A. W., 7 (1968, 13), 86 (1966, 128), 128 (1968, 13), 164 (1968, 13), 222 (1968, 13), 223 (1968, 13), 227 (1968, 13), 228 (1968, 13), *314, 330*
Johnson, F. A., 21 (1967, 49), 60 (1967, 49), 61 (1967, 49), 100 (1967, 49), *320*
Johnson, M. P., 92 (1968, 238), *336*
Johnson, T. R., 21 (1969, 60), 60 (1969, 60), 61 (1969, 60), 64 (1969, 60), 90 (1969, 60), 91 (1969, 60), 93 (1969, 60), 257 (1969, 60), *341*
Jolly, W. L., 17 (1963, 2), 18 (1966, 11), 99 (1966, 11), *307, 312*
Jonas, K., 92 (1969, 278), *347*
Jonassen, H. B., 92 (1969, 295), *348*
Jones, A., 18 (1969, 43), 34 (1969, 43), 35 (1969, 43), 75 (1969, 43), 106 (1969, 43), *341*
Jones, D. E., 89 (1968, 271), *337*
Jones, D. W., 41 (1967, 91), 66 (1967, 91), 125 (1967, 91), 126 (1967, 91), 134 (1967, 91), 137 (1967, 91), 148 (1967, 91), 158 (1967, 91), 167 (1967, 91), 168 (1967, 91), 177 (1967, 91), 181 (1967, 91), 184 (1967, 91), 189 (1967, 91), 190 (1967, 91), 198 (1967, 91), 219 (1969, 366), 241 (1969, 366), 243 (1967, 91), 264 (1967, 91), *321, 350*
Jones, E. D., 88 (1966, 142; 1967, 202, 203; 1968, 181, 182), *315, 324, 325, 335*
Jones, G., 114 (1965, 83), 140 (1965, 83), 142 (1965, 83), 161 (1965, 83), 170 (1965, 83), 195 (1965, 83), *310*
Jones, H. L., 7 (1968, 13), 128 (1968, 13), 164 (1968, 13), 222 (1968, 13), 223 (1968, 13), 227 (1968, 13), 228 (1968, 13), *330*
Jones, R., 28 (1967, 73), 142 (1967, 73), 156 (1967, 73), 170 (1967, 73), 236 (1967, 73), *321*
Jones, R. C., 5 (1956, 1), *307*
Jordan, R. B., 90 (1969, 234), 92 (1969, 234), *346*
Jose, P., 92 (1968, 233), *336*
Jugie, G., 27 (1969, 72), 45 (1969, 72), 84 (1969, 72, 191, 192, 193), 85 (1969, 72, 192), 122 (1969, 72), 142 (1969, 72), 149 (1969, 192), 160 (1969, 192), 170 (1969, 192), *341, 345*
Jutzi, P., 254 (1965, 117), 255 (1965, 117), *311*
Juvinall, G. L., 17 (1966, 13), 18 (1966, 13), 25 (1966, 13), 27 (1966, 13), 44 (1966, 13), 99 (1966, 13), 105 (1966, 13), 109 (1966, 13), 113 (1966, 13), *312*

K

Kabachnik, M. I., 7 (1967, 15; 1968, 18), 271 (1968, 18), 273 (1968, 18), *319, 330*
Kadibelban, T., 244 (1966, 281), 249 (1966, 281), 251 (1966, 281), 252 (1966, 281), *318*
Kainosho, M., 40, 41 (1968, 81; 1969, 90, 95, 96, 97), 83 (1969, 90, 95, 96, 97), 94 (1969, 96), 104 (1968, 81), 108 (1969, 95), 117 (1969, 90), 125 (1969, 95), 126 (1969, 95), 158 (1969, 97), 162 (1969, 95), 259 (1969, 95), *332, 342*

Kaiser, E. T., 78 (1967, 143), 210 (1967, 143), *323*
Kalck, P., 93 (1968, 266), *337*
Kamai, G. Kh., 116 (1966, 229), 155 (1966, 229), 176 (1967, 341), *317, 328*
Kanamueller, J. M., 44 (1967, 94), 45 (1967, 94), 165 (1967, 94), 171 (1967, 94), 192 (1967, 94), 272 (1967, 94), 283 (1967, 94), *322*
Kang, J. W., 90 (1969, 236), 91 (1969, 236), 92 (1969, 236), *346*
Kanzansr, Tr., 109 (1964, 14), 119 (1964, 14), 126 (1964, 14), 147 (1964, 14), *308*
Kaplan, F., 55 (1967, 116), 130 (1967, 116), 151 (1967, 116), 152 (1967, 116), 162 (1967, 116), 171 (1967, 116), 180 (1967, 116), 181 (1967, 116), *322*
Kaplan, L. M., 84 (1969, 185), *345*
Kaplan, N. O., 94 (1968, 274), 158 (1968, 274), 253 (1966, 286), *318, 337*
Kapoor, R. N., 45 (1969, 104), 90 (1969, 240, 241), 91 (1969, 104, 240, 241), *342, 346*
Karlson, S. E., 4 (1967, 6), *319*
Karplus, M., 13, *340*
Karunina, L. P., 84 (1969, 184), *345*
Kataev, E. I., 150 (1965, 93), 156 (1965, 93), 167 (1965, 93), 176 (1965, 93), *310*
Katritzky, A. R., 63 (1967, 127), 85 (1966, 112), 224 (1966, 271), 242 (1966, 271), 262 (1967, 127), 263 (1967, 127), *314, 318, 322*
Katz, T. J., 53, 201 (1966, 67), 202 (1966, 67), *313*
Kaufman, B. L., 19 (1967, 34), 89 (1967, 34), 140 (1967, 34), 159 (1967, 34), 161 (1967, 34), 170 (1967, 34), 179 (1967, 34), 192 (1967, 34), 195 (1967, 34), 207 (1967, 34), 227 (1967, 34), *320*
Kawamori, A., 88 (1967, 212), *325*
Ke, C. H., 84 (1968, 157), 194 (1968, 157), *334*
Keat, R., 45 (1965, 23; 1968, 91), 46 (1965, 23; 1968, 91), 55 (1968, 103), 67 (1965, 24), 79 (1965, 24, 33; 1966, 85, 86, 88; 1968, 91, 129), 81 (1968, 145), 86 (1968, 145), 92 (1968, 145), 107 (1966, 86), 108 (1965, 23; 1968, 91), 117 (1968, 145), 122 (1965, 23; 1968, 91), 123 (1965, 23; 1968, 91), 131 (1968, 103), 142 (1965, 23; 1968, 91), 161 (1968, 91), 174 (1965, 23; 1968, 91), 175 (1965, 23; 1968, 91), 179 (1968, 91), 186 (1968, 103), 192 (1968, 145), 203 (1965, 23; 1968, 91), 204 (1968, 91), 210 (1968, 145), 223 (1968, 103), 235 (1968, 103), 277 (1965, 24; 1968, 91), 278 (1965, 24; 1966, 88; 1968, 91, 129), 279 (1965, 24; 1966, 85, 88; 1968, 91, 129), 280 (1965, 33; 1966, 85, 86, 88), 281 (1966, 86, 88), 282 (1966, 86, 88), 283 (1966, 86), *309, 313, 332, 333, 334*
Keim, W., 92 (1967, 258; 1968, 259), 93 (1968, 259), 196 (1968, 259), *326, 337*
Keiter, R. L., 27 (1969, 71), 43 (1969, 71), 64 (1969, 134), 74 (1968, 116), 91 (1969, 134), 92 (1967, 255), 93 (1969, 71), 147 (1969, 71), 157 (1969, 71), 166 (1969, 71), 177 (1969, 71), 193 (1969, 71), *326, 333, 341, 343*
Keith, L. H., 47 (1969, 110), 81 (1969, 110), 94 (1968, 285; 1969, 322), 117 (1968, 285; 1969, 322), 118 (1968, 285; 1969, 322), 128 (1968, 285), 140 (1968, 285; 1969, 322), 141 (1969, 110), 145 (1968, 285; 1969, 322), 149 (1968, 285; 1969, 110, 322), 151 (1968, 285; 1969, 322), 153 (1968, 285; 1969, 322), 157 (1968, 285; 1969, 322), 160 (1968, 285; 1969, 322), 161 (1969, 110), 163, (1968, 285; 1969, 110), 171 (1968, 285), 172 (1968, 285; 1969, 110, 322), 173 (1968, 285), 174 (1968, 285; 1969, 322), 175 (1968, 285), 176 (1968, 285; 1969, 322), 179 (1968, 285; 1969, 322), 180 (1968, 285), 181 (1968, 285; 1969, 110, 322), 187 (1968, 285), 191

(1968, 285; 1969, 322), 192 (1968, 285; 1969, 322), 194 (1968, 285), 202 (1968, 285), 203 (1968, 285; 1969, 322), 204 (1968, 285), 242 (1968, 285), 263 (1968, 285; 1969, 322), 264 (1969, 322), 265 (1968, 285), 266 (1968, 285), *337, 342, 348*

Kelly, H. C., 10 (1969, 26), 11 (1969, 26), 74 (1969, 26), 92 (1969, 26), 142 (1969, 26), 236 (1969, 26), 261 (1969, 26), 269 (1969, 26), *340*

Kemenater, C., 173 (1965, 98), 206 (1965, 98), 226 (1965, 98), 257 (1965, 98), 258 (1965, 98), 260 (1965, 98), 262 (1965, 98), *311*

Kemmitt, R. D. W., 92 (1967, 248; 1968, 263; 1969, 298), 221 (1968, 263), 222 (1968, 263), *326, 337, 348*

Kemp, R. H., 57 (1969, 123), 133 (1969, 123), 185 (1969, 123), 204 (1969, 123), 210 (1969, 123), *343*

Kenyon, G. L., 53 (1966, 65), 133 (1966, 241), 135 (1966, 241), 151 (1966, 241), 163 (1966, 241), 202 (1966, 241), 280 (1966, 65), *313, 317*

Kesavadas, T., 260 (1967, 382), *330*

Kessemeier, H., 88 (1967, 204), 169 (1967, 204), *325*

Keyzer, H., 82 (1969, 169), 107 (1969, 169), 112 (1969, 169), 118 (1969, 169), 151 (1969, 169), *344*

Khairullin, V. K., 123 (1967, 312), *328*

Khaleeluddin, K., 238 (1966, 276), 249 (1966, 276), *318*

Khayat, M. A. R., 47 (1966, 55), 126 (1966, 235), 128 (1966, 235), 159 (1966, 55), 168 (1966, 55, 235), 170 (1966, 55, 235), 196 (1966, 235), 200 (1966, 55), 215 (1966, 55), *313, 317*

Kidd, R. G., 11 (1969, 28), 12 (1969, 28), 84 (1969, 28), 85 (1969, 28), 86 (1969, 29), *340*

Kiefer, E. F., 13 (1968, 27), *331*

King, R. B., 45 (1969, 104), 90 (1966, 153; 1968, 201, 202; 1969, 239, 240, 241), 91 (1968, 201, 202, 215, 229; 1969, 104, 239, 240, 241, 269), *315, 335, 336, 342, 346, 347*

King, R. W., 27 (1966, 37), 37 (1966, 37), 38 (1966, 37), 40 (1964, 5; 1965, 22), 42 (1966, 37), 74 (1965, 31), 90 (1965, 31; 1966, 158), 91 (1966, 158), 93 (1966, 158), 117 (1966, 37), 125 (1966, 37), 133 (1966, 37), 257 (1966, 37), 258 (1966, 37), *308, 309, 312, 315*

Kirby, K. C., 86 (1966, 115), 234 (1966, 115), *314*

Kirillova, K. M., 145 (1967, 325), 155 (1967, 325), *328*

Kirman, J., 106 (1966, 26), 185 (1966, 26), *312*

Kitching, W., 85 (1968, 163), 91 (1969, 257), *334, 347*

Kittredge, J. S., 94 (1967, 275), *326*

Kivi, M. V., 117 (1967, 313), 124 (1967, 313), 125 (1967, 313), 138 (1967, 313), 146 (1967, 313), 167 (1967, 313), 168 (1967, 313), *328*

Klamberg, F., 91 (1968, 230), 93 (1968, 230), *336*

Klapper, H., 42 (1969, 101), 80 (1969, 101, 159), 92 (1969, 159), 140 (1969, 101), 159 (1969, 101), 169 (1969, 101), 178 (1969, 101), 191 (1969, 101), 208 (1969, 159), 215 (1969, *342, 344*

Klebanskii, A. L., 86 (1965, 46), *309*

Kleiman, Yu. L., 20 (1967, 44), 42 (1967, 44), 109 (1967, 44), *320*

Klein, H. F., 90 (1969, 250), 91 (1969, 250), 93 (1969, 250), *346*

Klein, M., 2, 86 (1963, 5), 284 (1963, 5), *307, 319*

Klein, W., 117 (1969, 380), *350*

Kluger, R. L., 78 (1969, 146), 83 (1969, 146), 86 (1969, 146), 135 (1969, 146), 261 (1969, 146), 263 (1969, 146), 265 (1969, 146), *343*

Klusmann, P., 47 (1969, 112), 120 (1969, 112), 128 (1969, 112), 159 (1969, 112), 262 (1969, 112), *342*

Knoll, F., 18 (1965, 10; 1966, 12), 98 (1969, 327), 99 (1965, 10; 1966, 12; 1969, 327), *308, 312, 349*

Knottnerus, D. I. M., 88 (1969, 223), *346*
Knox, G. R., 91 (1969, 254), *347*
Knubovets, R. G., 88 (1969, 221), *346*
Knuck, T., 91 (1964, 13), *308*
Knunyants, I. L., 7 (1967, 13, 14), 97 (1967, 14), 98 (1967, 14), 99 (1967, 14), 103 (1967, 14), 104 (1967, 14), 105 (1967, 14), 107 (1967, 14), 108 (1967, 13, 14; 1968, 295), 112 (1967, 14), 113 (1967, 13, 14; 1968, 295), 118 (1967, 14), 120 (1967, 13, 14; 1968, 295), 122 (1967, 14), 127 (1967, 13, 14; 1968, 295), 128 (1968, 295), 132 (1967, 14), 136 (1967, 14), 138 (1967, 13, 14; 1968, 295), 152 (1967, 13, 14; 1968, 295), 158 (1967, 13, 14; 1968, 295), 163 (1967, 13; 1968, 295), 172 (1967, 14), 186 (1967, 14), 252 (1967, 13), *319*, *338*
Kobayashi, S., 183 (1968, 329), *339*
Kober, E., 67 (1966, 80), 79 (1966, 80), 279 (1966, 80), 280 (1966, 80), 281 (1966, 80), *313*
Kodama, G., 74 (1969, 141), 84 (1969, 141), 92 (1969, 141), 97 (1969, 141), 110 (1969, 141), *343*
Kogan, E. V., 86 (1965, 46), *309*
Koide, S., 13 (1964, 3), 52 (1964, 7), *308*
Kolar, D., 88 (1968, 187), *335*
Kolditz, L., 84 (1966, 107), 88 (1967, 216), 202 (1967, 216), *314*, *325*
Kolesnik, A. A., 83 (1967, 161, 162), 146 (1967, 161, 162), 158 (1967, 161), 168 (1967, 161, 162), *323*
Kolodny, R. A., 92 (1968, 241), *336*
Kolomnikov, I. S., 90 (1966, 156), *315*
Kol'tsov, Yu. I., 84 (1967, 170), 252 (1967, 170), *324*
Kondo, H., 91 (1968, 214), *336*
Kondo, N. S., 94 (1969, 319), *348*
Kongpricha, S., 44 (1967, 93), 98 (1967, 93), *321*
Koopman, H., 99 (1965, 120), *311*
Korotsev, M. P., 118 (1967, 393), *330*
Korovin, S. S., 84 (1967, 170), 252 (1967, 170), *324*
Korpium, O., 80 (1967, 151; 1968, 140), 205 (1967, 157; 1968, 140), 217 (1967, 151), 220 (1967, 151; 1968, 140), 221 (1967, 151; 1968, 140), 230 (1967, 151; 1968, 140), 240 (1967, 151; 1968, 140), 247 (1968, 140), 250 (1967, 151), *323*, *334*
Korte, F., 191 (1965, 104), *311*
Kosfeld, R., 2 (1968, 2), 28 (1968, 2), 31 (1968, 2), 127 (1968, 2), 128 (1968, 2), 129 (1968, 2), *330*
Kostyanovskii, R. G., 7 (1967, 18), 119 (1967, 346; 1968, 307), 128 (1968, 307), 202 (1967, 346; 1968, 307), *319*, *328*, *338*
Kovalchuk, V. A., 88 (1966, 148), *315*
Kowala, C., 90 (1966, 151), *315*
Kozlova, L. K., 210 (1965, 106), *311*
Kraihanzel, C. S., 91 (1968, 226, 227, 228), *336*
Kraihanzel, G. S., 91 (1965, 65), 92 (1965, 65), *310*
Krannich, L. K., 27 (1967, 67), 142 (1967, 67), 150 (1967, 67), *321*
Krapf, H., 114 (1967, 299), 142 (1967, 299), 170 (1967, 299), 195 (1967, 299), 243 (1967, 299), *327*
Krasna, A. I., 86 (1969, 205), *345*
Kratzer, R. H., 251 (1966, 284), 256 (1966, 284), *318*
Krebbs, B., 7 (1968, 19), 97 (1968, 19), 98 (1968, 19), 99 (1968, 19), *330*
Krech, K., 63 (1965, 29), 265 (1965, 29), 266 (1965, 29), 268 (1965, 29), 270 (1965, 29), 278 (1965, 29), 284 (1965, 29), *309*
Kreen, W. R., 90 (1968, 194), 92 (1968, 237), 234 (1968, 194), 238 (1968, 194), *335*, *336*
Kreiter, C. G., 90 (1969, 248), *346*
Krespan, C. G., 35 (1962, 2), 115 (1962, 2), 116 (1962, 2), 258 (1962, 2), *307*
Kröner, H., 85 (1967, 176), 141 (1967, 176), 142 (1967, 176), 179 (1967, 176), 261 (1967, 176), 264 (1967, 176), *324*
Kruck, T., 90 (1965, 52, 53; 1966, 154; 1967, 225), 91 (1965, 52; 1966, 161), 92 (1965, 70; 1966, 186; 1967, 245), *309*, *310*, *315*, *316*, *325*, *326*
Krüger, U., 275 (1965, 119), *311*

Kruglyak, Yu. L., 159 (1969, 351), 213 (1968, 335), *339, 349*
Kruse, W., 90 (1968, 203), 91 (1968, 203), *335*
Kuchen, W., 2 (1968, 2), 28 (1968, 2), 31 (1968, 2), 127 (1968, 2), 128 (1968, 2), 129 (1968, 2), *330*
Kuchitsu, K., 25 (1969, 70), *341*
Kudo, K., 78 (1967, 143), 210 (1967, 143), *323*
Kugel, L., 115 (1966, 227), *317*
Kugel, R. L., 7 (1966, 6), 79 (1966, 6, 90, 91), 101 (1966, 6), 278 (1966, 6, 90), 279 (1966, 6, 90), 281 (1966, 90), 283 (1966, 6, 90), 287 (1966, 6), 288 (1966, 6, 91), *311, 313*
Kugler, H. J., 83 (1968, 153), 112 (1966, 224), 114 (1966, 224), 119 (1966, 224), 128 (1966, 224; 1968, 153), 135 (1968, 312), 141 (1966, 224; 1968, 153), 142 (1968, 153), 144 (1968, 312), 145 (1968, 312), 154 (1968, 318), 155 (1968, 312), 156 (1968, 312), 159 (1968, 224; 1968, 153), 164 (1968, 318), 169 (1968, 318), 176 (1968, 312), 177 (1968, 312), 179 (1966, 224), 199 (1968, 318), 200 (1968, 318), 211 (1968, 312), 220 (1968, 318), 229 (1968, 153), 233 (1966, 224; 1968, 153), 240 (1968, 153), 241 (1968, 153), 247 (1968, 153), 252 (1968, 318), *317, 334, 338*
Kuhr, D., 275 (1965, 119), *311*
Kukhtin, V. A., 145 (1967, 325), 155 (1967, 325), *328*
Kümmel, R., 74 (1967, 135), 108 (1967, 135), 112 (1967, 135), 119 (1967, 135), 152 (1967, 135), *323*
Kundu, S. K., 32 (1969, 82), 37 (1969, 82), 119 (1969, 82), 137 (1969, 82), 139 (1969, 82), 141 (1969, 82), 148 (1969, 82), 160 (1969, 82), 165 (1969, 82), 169 (1969, 82), 173 (1969, 82), 175 (1969, 82), 176 (1969, 82), 178 (1969, 82), 183 (1969, 82), 191 (1969, 82), 199 (1969, 82), *342*
Kunitomo, M., 88 (1967, 209; 1969, 222), 135 (1967, 209), *325, 346*

Kupchik, E. J., 227 (1967, 361), 234 (1967, 361), *329*
Kurihara, N., 94 (1967, 271), *326*
Kuriyagawa, F., 41 (1967, 90), 83 (1967, 90), 104 (1967, 90), 108 (1967, 90), 165 (1967, 90), 201 (1967, 90), 257 (1967, 90), 262 (1967, 90), 265 (1967, 90), 268 (1967, 90), *321*
Kurland, R. J., 94 (1966, 202), *316*
Kustin, K., 89 (1968, 271), *337*
Kuznietz, M., 87 (1969, 212), 88 (1967, 201; 1968, 183, 184; 1969, 212, 218), *324, 335, 345, 346*
Kyllingstad, V. L., 86 (1966, 128), *314*
Kyogoku, Y., 41 (1967, 90; 1969, 96), 83 (1967, 90; 1969, 96), 84 (1969, 96), 104 (1967, 90), 108 (1969, 96), 165 (1967, 90), 201 (1967, 90), 257 (1967, 90), 262 (1967, 90), 265 (1967, 90), 268 (1967, 90), *321, 342*

L

Labarre, J. F., 92 (1967, 243; 1969, 284), *326, 347*
Labarre, M. C., 112 (1969, 381), *350*
Ladbrooke, B. D., 94 (1967, 268), *326*
Lafferty, F. L., 90 (1969, 253), *346*
Lahajnar, G., 88 (1968, 187), *335*
Lakshmirayan, T. V., 250 (1966, 283), *318*
La Mar, G. N., 45 (1965, 54), 89 (1965, 54, 55, 56; 1969, 279), 90 (1965, 54, 55, 56), 92 (1965, 54, 55, 56; 1969, 279), *309, 347*
Lambert, J. B., 74 (1969, 142), 80 (1966, 99; 1968, 141), 126 (1969, 142), 167 (1969, 142), 267 (1966, 99; 1968, 141), *314, 334, 343*
La Monica, G., 92 (1967, 251), *326*
Lamotte, A., 83 (1968, 155), 86 (1968, 165), 100 (1968, 155), 121 (1968, 165), *334*
Lancaster, J. E., 49 133 (1967, 99), 137 (1967, 99), 144 (1967, 99), 158 (1967, 99), 172 (1967, 99), 231 (1967, 99), *322*
Landau, M. A., 7 (1967, 14), 25 (1967, 64), 97 (1967, 14), 98 (1967, 14), 99

(1967, 14), 103 (1967, 14), 104 (1967, 14), 105 (1967, 14), 107 (1967, 14), 108 (1967, 14), 109 (1967, 64), 112 (1967, 14), 113 (1967, 14), 118 (1967, 14), 119 (1967, 64), 120 (1967, 14), 122 (1967, 14), 127 (1967, 14), 128 (1967, 64), 132 (1967, 14), 136 (1967, 14, 64), 138 (1967, 14), 139 (1967, 64), 143 (1967, 64), 146 (1967, 64), 152 (1967, 14), 153 (1967, 64), 157 (1967, 64), 158 (1967, 14), 169 (1967, 64), 172 (1967, 14), 173 (1967, 64), 180 (1967, 64), 181 (1967, 64), 186 (1967, 14), *319, 321*

Lane, A. P., 44 (1967, 95), 85 (1962, 4), 174 (1967, 95), 175 (1967, 95), 182 (1967, 95), 191 (1967, 95), 192 (1967, 95), 200 (1967, 95), 206 (1967, 95), 212 (1962, 4; 1967, 95), *322*

Lane, M. D., 94 (1968, 281), 110 (1968, 281), *337*

Lang, W., 90 (1965, 52, 53; 1966, 154; 1967, 225), 91 (1965, 52; 1966, 161), 92 (1965, 70), *309, 310, 315, 325*

Langkammerer, C. M., 35 (1962, 2), 115 (1962, 2), 116 (1962, 2), 258 (1962, 2), *307*

Lannert, K. P., 263 (1969, 372), *350*

Larmann, J., 5 (1965, 2), 97 (1965, 2), 98 (1965, 2), 99 (1965, 2), 100 (1965, 2), 113 (1965, 2), *308*

Larsson, A., 94 (1967, 265), *326*

Latscha, H. P., 19 (1969, 46), 67 (1968, 114), 79 (1968, 114), 98 (1968, 288), 99 (1968, 288), 101 (1968, 114), 117 (1969, 380), 124 (1966, 233), 126 (1966, 233), 131 (1968, 288), 186 (1968, 288), 223 (1968, 288), 251 (1968, 288), 277 (1968, 114), 278 (1968, 114), 281 (1968, 114), 282 (1968, 114), 283 (1968, 114), 285 (1968, 114), 286 (1968, 114), *317, 333, 337, 341, 350*

Laube, B. L., 159 (1967, 334), 161 (1967, 334), 220 (1967, 334), *328*

Laundau, M. A., 159 (1969, 351), *349*

Laurent, J. P., 27 (1969, 72), 45 (1969, 72), 84 (1969, 72, 191, 192, 193), 85 (1969, 72, 192), 92 (1967, 236, 243), 104 (1967, 236), 107 (1967, 236), 112 (1967, 236), 114 (1969, 336), 118 (1967, 236), 122 (1969, 72), 140 (1969, 336), 142 (1969, 72), 149 (1969, 192), 160 (1969, 192), 170 (1969, 192), *325, 326, 341, 345, 349*

Laussac, J. P., 84 (1969, 193), *345*

Lauterbur, P. C., 7 (1957, 1), 81 (1968, 147), *307, 334*

Lavielle, G., 53 (1967, 108), 86 (1969, 201, 202), 125 (1967, 108), 126 (1967, 108), 144 (1967, 108, 323), 145 (1967, 108, 326), 146 (1967, 108, 323, 326), 147 (1967, 278), 156 (1967, 108, 323, 326), 158 (1967, 278), 166 (1967, 323), 168 (1967, 108, 278), 177 (1967, 108, 278, 326), 180 (1967, 108), 184 (1967, 108, 278), 198 (1967, 108), 200 (1967, 108), 205 (1967, 108), 225 (1967, 108), 264 (1967, 278), 266 (1967, 278), *322, 327, 328, 345*

Lawson, D. N., 92 (1966, 187, 190), *316*

Lebedev, V. B., 52 (1967, 105), 72 (1967, 128), 111 (1967, 105, 296), 124 (1967, 296), 144 (1967, 296), 145 (1967, 296), 146 (1967, 296), 260 (1967, 128), *322, 323, 327*

Lederle, H., 67 (1966, 80), 79 (1966, 80), 279 (1966, 80), 280 (1966, 80), 281 (1966, 80), *313*

Lee, H. P. K., 82 (1967, 159), 284 (1967, 159), 286 (1967, 159), *323*

Lee, J., 107 (1966, 215), 117 (1966, 215), *317*

Legrand, L., 208 (1969, 363), 215 (1969, 363), *349*

Lehn, J. M., 13 (1969, 35), 40 (1967, 89), *321, 340*

Lehr, W., 79 (1967, 147, 148), 101 (1967, 147), 277 (1967, 148), 285 (1967, 147), 286 (1967, 147), *323*

Leibovskaya, G. A., 159 (1969, 351), 213 (1968, 335), *339, 349*

Leichner, L., 78 (1966, 83), 257 (1966, 83), 259 (1966, 83), 260 (1966, 83),

261 (1966, 83), 267 (1966, 83), 284 (1966, 83), *313*
Leigh, G. J., 89 (1968, 217), 91 (1968, 217; 1969, 263), 92 (1968, 217; 1969, 263, 287, 300), 93 (1968, 217; 1969, 263), *336, 347, 348*
Lengyel, I., 86 (1966, 140), 238 (1966, 140), *315*
Lenzi, M., 90 (1968, 211), 91 (1968, 211), 92 (1966, 169; 1968, 211), 93 (1968, 211), 108 (1968, 211), 113 (1968, 211), 114 (1966, 169; 1968, 211), 144 (1967, 323), 146 (1967, 323), 147 (1967, 278), 156 (1967, 323), 158 (1967, 278), 166 (1967, 323), 168 (1967, 278), 177 (1967, 278), 184 (1967, 278), 193 (1968, 211), 264 (1967, 278), 266 (1967, 278), *315, 327, 328, 335*
L'Eplattenier, F., 92 (1968, 236; 1969, 288), 93 (1968, 236), 117 (1969, 288), *336, 347*
Lequan, R. M., 137 (1969, 342), 149 (1969, 342), 157 (1969, 342), 159 (1969, 342), *349*
Lesbre, M., 226 (1967, 360), *329*
Leslie, R. B., 94 (1969, 320), *348*
Letcher, J. H., 1 (1967, 1), 2 (1967, 1), 4 (1967, 1, 7), 65 (1967, 1), 103 (1967, 7), 104 (1967, 7), 107 (1967, 7), 108 (1967, 7), 111 (1967, 7), 180 (1967, 7), 131 (1967, 7), 132 (1967, 7), 133 (1967, 7), 135 (1967, 7), 149 (1967, 7), 151 (1967, 7), 160 (1967, 7), 169 (1967, 7), 173 (1967, 7), 196 (1967, 7), 201 (1967, 7), 207 (1967, 7), 213 (1967, 7), 214 (1967, 7), 218 (1967, 7), 228 (1967, 7), 231 (1967, 7), 234 (1967, 7), 235 (1967, 7), 239 (1967, 7), 257 (1967, 7), 258 (1967, 7), *311, 319*
Letsinger, R. L., 19 (1968, 35), 74 (1968, 35), 173 (1968, 35), 181 (1968, 35), 182 (1968, 35), 187 (1968, 35), 196 (1968, 35), 203 (1968, 35), 223 (1968, 35), 241 (1968, 35), *331*
Levison, J. J., 92 (1969, 287), *347*
Levy, J. B., 227 (1968, 341), 231 (1968, 341), *339*
Levy, M., 28 (1967, 72), 109 (1967, 72), 113 (1967, 72), 114 (1967, 72), 121 (1967, 72), 129 (1967, 72), *321*
Levy, M. N., 89 (1969, 309), *348*
Lewis, D. E., 170 (1968, 326), *339*
Lewis, R. A., 80 (1967, 151; 1968, 140; 1969, 154), 182 (1969, 154), 205 (1967, 151; 1968, 140), 217 (1967, 151), 220 (1967, 151; 1968, 140), 221 (1967, 151; 1968, 140), 230 (1967, 151; 1968, 140), 240 (1967, 151; 1968, 140), 247 (1968, 140), 250 (1967, 151), *323, 334, 344*
Li, N. C., 84 (1966, 105; 1967, 171; 1968, 157), 90 (1966, 171), 92 (1966, 171), 97 (1966, 105), 140 (1966, 105), 194 (1966, 105; 1968, 157), 209 (1966, 105), 244 (1966, 105), *314, 316, 324*
Liberda, H. G., 86 (1968, 169), 227 (1968, 169), 228 (1968, 169), 231 (1968, 169), *334*
Lieb, F., 54 (1967, 111, 112), 225 (1968, 339), 247 (1968, 339), 248 (1967, 112), 249 (1967, 112), *322, 339*
Liengme, B. V., 91 (1969, 258), *347*
Lieske, C. N., 208 (1968, 334), *339*
Lingberg, B., 4 (1967, 6), *319*
Lindgren, I., 4 (1967, 6), *319*
Lindsey, R. V., 92 (1965, 69), *310*
Line, D. C., 92 (1968, 261), *337*
Liober, B. G., 121 (1966, 254), 165 (1966, 254), 166 (1966, 254), *318*
Lippard, S. J., 89 (1967, 262), *326*
Lippman, A. E., 38 (1965, 20), 43 (1965, 20), 108 (1965, 20), 114 (1965, 20; 1966, 218), 119 (1965, 20), 120 (1965, 20), 121 (1966, 218), 140 (1965, 20; 1966, 218), 153 (1965, 20), 173 (1965, 20), 259 (1966, 218), 262 (1966, 218), *308, 317*
Little, J. L., 10 (1969, 25), 105 (1967, 292; 1969, 25), 110 (1969, 25), 141 (1967, 292), *327, 340*
Little, J. R., 50 (1966, 60), 86 (1966, 60), *313*
Littlecott, G. W., 92 (1969, 298), *348*
Lobanov, D. F., 7 (1967, 15), *319*
Lobkov, V. D., 86 (1965, 46), *309*

Loewengart, G. V., 164 (1969, 353), 166 (1969, 353), 174 (1969, 353), 175 (1969, 353), 203 (1969, 353), 209 (1969, 353), 210 (1969, 353), 231 (1969, 353), 232 (1969, 353), *349*

Logan, T. J., 18 (1966, 15), 218 (1967, 357), 255 (1967, 357), 256 (1966, 15), 257 (1966, 15), 258 (1966, 15), 259 (1966, 15), 260 (1966, 15), 263 (1966, 15), 264 (1966, 15), 265 (1966, 15), 266 (1966, 15), 267 (1966, 15), 268 (1966, 15), 269 (1966, 15), 271 (1966, 15), 273 (1966, 15), 274 (1966, 15), *312, 329*

Long, C., 88 (1969, 218), *346*

Longone, D. T., 29 (1968, 63), 247 (1968, 63), 248 (1968, 63), 250 (1968, 63), 253 (1968, 63), *331*

Longstaff, P. A., 92 (1966, 188), *316*

Lorbert, J., 114 (1967, 299), 142 (1967, 299), 170 (1967, 299), 195 (1967, 299), 243 (1967, 299), *327*

Lorberth, J., 90 (1969, 242), *346*

Lord, E., 248 (1967, 377), 253 (1967, 377), 254 (1967, 377), *329*

Lord, R. C., 86 (1967, 194), *324*

Lorenz, J., 92 (1967, 244), 113 (1967, 244), 122 (1967, 244), 139 (1967, 244), 149 (1967, 244), 160 (1967, 244), 164 (1967, 244), 165 (1967, 244), 169 (1967, 244), 175 (1967, 244), 178 (1967, 244), 191 (1967, 244), 193 (1967, 244), 196 (1967, 244), 199 (1967, 244), 201 (1967, 244), 206 (1967, 244), 209 (1967, 244), 212 (1967, 244), 214 (1967, 244), 217 (1967, 244), 220 (1967, 244), 223 (2967, 244), 226 (1967, 244), 228 (1967, 244), 229 (1967, 244), 230 (1967, 244), 233 (1967, 244), 235 (1967, 244), 237 (1967, 244), 239 (1967, 244), 244 (1967, 244), 245 (1967, 244), *326*

Louis, E., 90 (1968, 207; 1969, 248), 91 (1968, 207, 212), 93 (1968, 207, 212), *335, 336, 346*

Love, I., 106 (1968, 293), 110 (1968, 293), *338*

Lowe, J. U., 130 (1966, 237), 221 (1966, 237), *317*

Lozac'h, N., 208 (1969, 363), 215 (1969, 363), *349*

Luchi, R. J., 89 (1969, 312), 94 (1969, 312), *348*

Lucken, E. A. C., 4, 37 (1966, 47), 79 (1960, 2), 115 (1969, 8), 140 (1966, 47), *307, 312, 340*

Ludington, R. S., 156 (1965, 94), *310*

Luganskii, G. M., 21 (1967, 52), 130 (1967, 52), 132 (1967, 52), 143 (1967, 52), 209 (1967, 52), *320*

Lunazzi, L., 84 (1969, 186), *345*

Lund, H., 58 (1969, 124), 69 (1969, 124), 185 (1969, 124), 210 (1969, 124), *343*

Lundberg, K. L., 60 (1969, 128), 121 (1969, 128), 129 (1969, 128), 141 (1969, 128), 260 (1969, 128), 278 (1969, 128), *343*

Lundin, A. G., 88 (1966, 148), *315*

Lupin, M. S., 93 (1966, 193; 1968, 265), *316, 337*

Lustig, M., 98 (1969, 328), 99 (1967, 283; 1969, 328), 100 (1966, 209), 101 (1969, 328), 110 (1967, 294), 111 (1967, 294), 286 (1969, 328), *316, 327, 349*

Lutsenko, I. F., 111 (1968, 297), 126 (1968, 297), 134 (1968, 297), 137 (1968, 297), 145 (1968, 297), 146 (1968, 297), 147 (1968, 297), 148 (1968, 297), 149 (1968, 297), 155 (1968, 297), 156 (1968, 297), 162 (1968, 297), 165 (1968, 297), 167 (1968, 297), 176 (1968, 297), 183 (1968, 297), 188 (1968, 297), 192 (1968, 297), 208 (1968, 297), 209 (1968, 297), 271 (1968, 297), *338*

Lybyer, M. J., 94 (1969, 321), 101 (1969, 321), 121 (1969, 321), 138 (1969, 321), *348*

Lynden-Bell, R. M., 63 (1961, 2), 92 (1968, 234), *307, 336*

Lyons, J. W., 86 (1966, 117), 100 (1966, 117), 121 (1966, 117), *314*

M

Ma, S. Y., 86 (1965, 45), 131 (1965, 45), 148 (1965, 45), 195 (1965, 45), 219

(1965, 45), 224 (1965, 45), 244 (1965, 45), *309*
MacDiarmid, A. G., 22 (1967, 59), 46 (1967, 59), 76 (1967, 59), 109 (1967, 59), 257 (1967 59), *321*
Mach, W., 53 (1968, 101), 54 (1968, 101, 102), 78 (1968, 101, 102), 171 (1968, 101), 188 (1968, 101), 206 (1968, 101), 219 (1968, 101), 226 (1968, 102), 229 (1968, 101), 240 (1968, 102), 269 (1968, 101), 276 (1968, 102), *333*
Mackie, R. K., 169 (1967, 336), *328*
Macomber, J. D., 81 (1969, 167), 91 (1969, 167), 99 (1969, 167), *344*
Madan, O. P., 38 (1967, 86), 39 (1967, 86), 78 (1968, 123), 82 (1968, 123), 97 (1966, 206), 114 (1966, 206; 1967, 86), 125 (1967, 86), 138 (1967, 86), 142 (1966, 206), 148 (1967, 86), 173 (1966, 206), 174 (1966, 206), 193 (1966, 206), *316, 321, 333*
Maglio, G., 91 (1969, 271), *347*
Magnelli, D. D., 130 (1966, 237), 221 (1966, 237), *317*
Mague, J. I., 203 (1967, 349), 210 (1967, 349), 235 (1967, 349), 271 (1967, 349), 272 (1967, 349), 273 (1967, 349), 274 (1967, 349), 275 (1967, 349), *329*
Mague, J. T., 92 (1966, 187; 1967, 256; 1969, 303), 131 (1967, 256), 277 (1969, 303), *316, 326, 348*
Mahler, W., 63 (1964, 11), 97 (1965, 77), 98 (1965, 77), *308, 310*
Maier, L., 5, 18 (1966, 16), 19 (1966, 23; 1969, 36), 22 (1969, 63), 27 (1966, 39), 28 (1966, 40), 55 (1967, 114), 85 (1966, 111), 97 (1966, 111), 99 (1969, 63), 105 (1966, 4; 1967, 291 (107 (1966, 111; 1969, 331, 332), 109 (1966, 4, 40; 1967, 291), 111 (1966, 4), 112 (1966, 4; 1967, 301; 1968, 298, 299, 300; 1969, 332, 333, 334), 113 (1967, 291), 115 (1966, 4; 1967, 291, 301), 118 (1968, 299, 300; 1969, 63, 332, 334), 119 (1966, 111), 121 (1966,' 40), 122 (1966, 39), 126 (1969, 332), 127 (1967, 301), 129 (1967, 291, 301), 131 (1966, 111; 1967, 114), 132 (1966, 4), 133 (1966, 4), 136 (1967, 291; 1968, 300; 1969, 332), 138 (1967, 301), 141 (1967, 291), 151 (1968, 299; 1969, 334), 153 (1966, 40), 159 (1966, 4), 160 (1966, 23, 40), 164 (1966, 4), 170 (1967, 291), 176 (1968, 299; 1969, 334), 179 (1966, 4; 1967, 291), 183 (1966, 40), 186 (1966, 111), 187 (1966, 23, 40; 1968, 36), 188 (1967, 291), 191 (1969, 333, 334), 193 (1966, 4; 1967, 301), 194 (1966, 40), 201 (1966, 40), 202 (1966, 40), 206 (1966, 4; 1969, 333), 207 (1968, 299; 1969, 334), 209 (1966, 40), 213 (1966, 4; 1967, 291), 214 (1966, 4; 1967, 291), 216 (1968, 336; 1969, 333), 217 (1966, 4), 218 (1966, 40), 219 (1966, 40), 220 (1966, 4; 40; 1968, 36), 224 (1966, 111), 225 (1966, 40), 227 (1967, 291; 1968, 36), 230 (1968, 336), 233 (1966 40), 237 (1968, 36), 240 (1968, 336), 244 (1966, 4, 40), 248 (1967, 291), 251 (1966, 40), 255 (1967, 291), (256 (1965, 118; 1966, 16), 257 (1965, 118; 1969, 332), 258 (1965, 118; 1966, 16), 259 (1966, 16), 260 (1969, 332), 261 (1967, 301), 263 (1969, 332), 264 (1965, 118; 1969, 332), 266 (1965, 118; 1968, 336; 1969, 332), 267 (1965, 118), 269 (1966, 16; 1968, 336), 270 (1966, 16; 1968, 336), 271 (1966, 16), 272 (1968, 336), 273 (1968, 336), 274 (1968, 336), 277 (1968, 300; 1969, 334), 278 (1969, 332, 334), 279 (1968, 299, 300; 1969, 332, 334), 280 (1968, 300; 1969, 332, 334), 281 (1968, 299, 300; 1969, 332, 334), 282 (1969, 332, 334), 283 (1968, 299; 1969, 334), 284 (1968, 298; 1969, 333), 285 (1968, 298; 1969, 333), 286 (1968, 298; 1969, 333), *311, 312, 314, 322, 327, 331, 338, 339, 341, 349*

Mais, R. H. B., 91 (1968, 216), *336*
Maitlis, P. M., 90 (1967, 220), 92 (1967, 220), *325*
Majoral, J. P., 83 (1968, 150; 1969, 175), 172 (1968, 150; 1969, 175), 181 (1968, 150; 1969, 175), 190 (1968, 150; 1969, 175), 199 (1969, 175), 205 (1968, 150), 234 (1969, 175), *334, 344*
Malatesta, L., 91 (1965, 62), *310*
Malcolm, R. B., 42 (1968, 83), 83 (1968, 83), 163 (1968, 83), 172 (1968, 83), 181 (1968, 83), 190 (1968, 83), 211 (1968, 83), *332*
Malik, S. K., 88 (1969, 219), *346*
Manatt, S. L., 14 (1967, 28), 16, 17 (1966, 13; 1967, 40), 18 (1966, 13), 20 (1967, 28, 40), 25 (1966, 13), 28 (1966, 13), 33 (1967, 40), 37, 44, 60 (1967, 40), 61 (1967, 40), 62, 63 (1967, 40), 68 (1967, 131), 70 (1969, 41), 74, 99 (1966, 13; 1967, 28; 1969, 41), 100 (1967, 28; 1969, 41), 101 (1967, 28), 104 (1967, 40), 105 (1966, 13; 1969, 41), 106 (1967, 40), 109 (1966, 13), 110 (1967, 40; 1969, 41), 113 (1966, 13), 114 (1967, 298), 132 (1969, 41), 219 (1967, 28), 257 (1967, 40), 258 (1967, 28, 40), 259 (1967, 28), *312, 320, 323, 327, 341*
Mancuso, A., 199 (1969, 361), *349*
Mańhas, B. S., 79 (1968, 133), 280 (1968, 133), *333*
Mank, V. V., 88 (1969, 226), *346*
Mańkowski-Favelier, R., 8 (1967, 22; 1969, 20), 9 (1969, 20), 27 (1969, 20), 28 (1969, 20), 44 (1969, 20; 102), 45 (1969, 20), 50 (1967, 101), 53 (1967, 101, 108; 1969, 20), 54 (1969, 20), 55 (1967, 113), 59 (1969, 20), 60 (1969, 20), 72 (1969, 20), 75 (1969, 20), 84 (1969, 102), 85 (1969, 102), 86 (1969, 102), 125 (1967, 108), 126 (1967, 108), 139 (1967, 22), 143 (1967, 113), 144 (1967, 101, 108), 145 (1967, 108), 146 (1967, 101, 108), 147 (1967, 22), 149 (1967, 22), 153 (1967, 113), 156 (1967, 101, 108), 158 (1967, 22), 168 (1967, 22, 108), 177 (1967, 108), 178 (1967, 22), 179 (1967, 108), 180 (1967, 101, 108; 1969, 20), 182 (1967, 113), 184 (1967, 22, 108), 187 (1969, 20), 188 (1969, 189 (1967, 113; 1969, 20), 197 (1967, 113), 198 (1967, 101, 108), 200 (1967, 101, 108), 201 (1967, 113), 205 (1967, 108), 206 (1967, 113), 212 (1967, 113), 225 (1967, 108), *320, 322, 340, 342*
Mann, F. G., 80 (1969, 155), 92 (1969, 155, 290), 203 (1967, 348), 228 (1967, 348), 269 (1969, 155), *329, 344, 348*
Mann, T., 79 (1966, 98), 256 (1966, 98), *314*
Manning, A. R., 91 (1967, 233), 92 (1967, 239), 196 (1967, 239), *325*
Mannschreck, A., 94 (1966, 205; 1967, 273), *316*
Mansfield, P., 101 (1969, 329), *349*
Maples, P. K., 91 (1965, 65; 1968, 226, 227, 228), 92 (1965, 65), *310, 336*
Mark, V., 1 (1967, 1), 2 (1967, 1), 4 (1967, 1), 58 (1964, 9), 65 (1967, 1), 109 (1967, 293), 114 (1967, 293), 129 (1967, 293), 136 (1967, 293,) 140 (1964, 9; 1967, 293), 141 (1964, 9), 142 (1964, 9), 147 (1967, 293), 149 (1964, 9; 1967, 293), 154 (1967, 293), 157 (1967, 293), 160 (1967, 293), 167 (1967, 293), 169 (1964, 9; 1967, 293), 175 (1967, 293), 178 (1967, 293), 190 (1967, 293), 192 (1967, 293), 194 (1967, 293), 212 (1967, 293), 213 (1967, 293), 214 (1967, 293), *308, 319, 327*
Markin, V. V., 150 (1965, 93), 156 (1965, 93), 167 (1965, 93), 176 (1965, 93), *310*
Märkl, G., 27 (1967, 70), 54 (1966, 68; 1967, 70, 111, 112), 187 (1967, 70), 225 (1968, 339), 240 (1966, 68), 242 (1967, 70), 246 (1967, 70), 247 (1968, 339), 248 (1967, 112), 249 (1967, 112), 254 (1966, 287), *313, 318, 321, 322, 339*
Marks, T. J., 6 (1966, 5), 122, (1966, 5), 129 (1966, 5), 178 (1966, 5), 203

AUTHOR INDEX

(1966, 5), 206 (1966, 5), 207 (1966, 5), 212 (1966, 5), 216 (1966, 5), 217 (1966, 5), 218 (1966, 5), 220 (1966, 5), 225 (1966, 5), 226 (1966, 5), 227 (1967, 363), 228 (1966, 5; 1967, 363), 229 (1966, 5), 231 (1966, 5; 1967, 363), 233 (1966, 5), 234 (1966, 5), 235 (1966, 5; 1967, 363), 238 (1966, 5), 239 (1966, 5; 1967, 363), 241 (1966, 5; 1967, 363), 242 (1966, 5), 243 (1966, 5; 1967, 363), 244 (1966, 5), 245 (1966, 5), 247 (1966, 5), 252 (1966, 5), 275 (1966, 5), 276 (1966, 5), *311, 329*

Martin, D. J., 53 (1967, 107), 119 (1965, 87), 138 (1965, 87), 147 (1967, 327), 157 (1967, 327), 176 (1967, 327), 177 (1965, 87), 192 (1967, 327), 199 (1965, 87), 200 (1967, 327), 202 (1965, 87), 208 (1965, 87), 211 (1967, 327), 218 (1965, 87), 220 (1965, 87), 231 (1965, 87), 232 (1967, 327), 247 (1965, 87), *310, 322, 328*

Martin, G. J., 234 (1967, 365), 249 (1967, 365), 253 (1967, 365), 254 (1967, 365), 255 (1967, 365), *329*

Martin, J., 29 (1969, 76), 180 (1969, 76), 187 (1969, 76), 188 (1969, 76), 198 (1969, 76), *341*

Martin, K. R., 86 (1966, 129, 130), 134 (1966, 130), 151 (1966, 130), 280 (1966, 129), *314, 315*

Martin, R. L., 90 (1969, 246), *346*

Marty, C., 28 (1967, 72), 109 (1967, 72), 113 (1967, 72), 114 (1967, 72), 121 (1967, 72), 129 (1967, 72), *321*

Marty, R., 80 (1969, 162), 114 (1969, 162), 128 (1969, 162), 153 (1969, 162), 154 (1969, 162), 164 (1969, 162), 170 (1969, 162), 191 (1969, 162), 206 (1969, 162), 217 (1969, 162), 221 (1969, 162), *344*

Martynov, I. V., 159 (1969, 351), 213 (1968, 335), *349*

Martz, M. D., 32 (1969, 83), 33 (1969, 83), 83 (1969, 83), *342*

Maruyama, H., 94 (1968, 281), 110 (1968, 281), *337*

Marzin, C., 13 (1967, 25), *320*

Mashlyakovskii, L. N., 116 (1965, 86), 123 (1965, 86; 1967, 310), 133 (1965, 86), 145 (1967, 310), 149 (1967, 310), 150 (1965, 86), *310, 327*

Massey, A. G., 83 (1965, 39), 139 (1965, 39), 160 (1965, 39), *309*

Mathews, C. N., 51 (1967, 102), 59 (1967, 102, 123; 1968, 110), 64 (1966, 78, 79; 1967, 123), 90 (1967, 102), 223 (1966, 78, 270; 1968, 110), 224 (1968, 338), 230 (1966, 270; 1968, 110), 234 (1966, 270), 237 (1967, 102; 1968, 110), 238 (1966, 270), 242 (1967, 102), 244 (1967, 102; 1968, 110), 245 (1967, 102; 1968, 110, 338), 247 (1967, 102; 1968, 110), 249 (1968, 110), 251 (1968, 110), 254 (1968, 110), 256 (1967, 123), 269 (1967, 102), 274 (1966, 78), 275 (1966, 290; 1967, 102, 123; 1968, 110), 276 (1967, 102; 1968, 110, 338), 283 (1966, 78; 1967, 102), 284 (1966, 78), 286 (1966, 78), 287 (1966, 78), *313, 318, 319, 322, 333, 339*

Mathey, F., 53 (1966, 66), 134 (1966, 66; 1969, 339), 135 (1966, 66; 1969, 339), 145 (1966, 66), 187 (1969, 358), 188 (1969, 358), *313, 349*

Mathieson, D. R., 261 (1968, 351), *339*

Mathieu, R., 90 (1968, 211), 91 (1968, 211), 92 (1967, 238; 1968, 211), 93 (1968, 211), 108 (1968, 211), 113 (1968, 211), 114 (1968, 211), 193 (1968, 211), *325, 335*

Mathis, F., 40 (1968, 76), 47 (1968, 76), 78 (1968, 76), 82 (1968, 76), 83 (1967, 164), 84 (1967, 172), 109 (1967, 172), 114 (1967, 172), 122 (1967, 172), 127 (1967, 164), 139 (1967, 164), 143 (1967, 172), 157 (1967, 172), 161 (1967, 172), 164 (1968, 76), 168 (1967, 164), 170 (1967, 172), 192 (1967, 172), 203 (1967, 172), 205 (1967, 164), 215 (1968, 76), 216 (1967, 164; 1968, 76), *324, 332*

Matsuo, K., 41 (1967, 90), 83 (1967, 90), 104 (1967, 90), 108 (1967, 90), 165 (1967, 90), 201 (1967, 90), 257 (1967, 90), 262 (1967, 90), 265 (1967, 90), 268 (1967, 90), *321*

Matthews, R. S., 219 (1969, 366), 241 (1969, 366), *350*

Matzkanin, G. A., 87 (1969, 212), 88 (1969, 212), *345*

Mavel, G., 1 (1966, 1), 8 (1967, 21, 22; 1968, 20; 1969, 20), 9 (1968, 20; 1969, 20), 16 (1962, 1; 1966, 1), 17 (1968, 30), 24 (1968, 30), 25 (1966, 1; 1968, 30), 26 (1966, 1), 27 (1969, 20), 28 (1968, 30; 1969, 20), 29 (1968, 30), 31 (1966, 1), 34 (1968, 30), 37 (1966, 1), 42 (1966, 1), 43 (1966, 1), 44 (1969, 20, 102), 45 (1969, 20), 50 (1967, 101), 52 (1967, 21), 53 (1966, 66; 1967, 101, 108; 1969, 20), 54 (1969, 20), 55 (1967, 113), 59 (1969, 20), 60 (1969, 20), 66 (1966, 1), 68 (1968, 20), 69 (1968, 20), 72 (1968, 20; 1969, 20), 75 (1969, 20), 81 (1966, 1), 84 (1969, 102), 85 (1966, 1; 1969, 102), 86 (1969, 102), 88 (1966, 1), 90 (1966, 1), 94 (1966, 1), 104 (1968, 20), 124 (1967, 21), 125 (1967, 21, 108), 126 (1967, 21, 108), 134 (1966, 66; 1969, 339), 135 (1966, 66; 1967, 21; 1969, 339), 139 (1967, 22), 143 (1967, 113), 144 (1967, 21, 101, 108), 145 (1966, 66; 1967, 21, 108; 1968, 20), 146 (1967, 101, 108), 147 (1967, 21, 22; 1968, 20), 149 (1967, 21, 22), 154 (1967, 113), 155 (1967, 21), 156 (1967, 101, 108), 158 (1967, 22), 165 (1967, 21), 166 (1967, 21), 168 (1967, 22, 108), 175 (1967, 21), 177 (1967, 21, 108), 178 (1967, 22), 179 (1967, 108), 180 (1967, 101, 108; 1969, 20), 182 (1967, 21, 113), 183 (1967, 21), 184 (1967, 22, 108), 187 (1969, 20), 188 (1967, 21; 1969, 20), 189 (1967, 113; 1969, 20), 192 (1967, 21), 197 (1967, 113), 198 (1967, 101, 108), 200 (1967, 101, 108), 201 (1967, 113), 205 (1967, 108), 206 (1967, 113), 207 (1967, 21), 225 (1967, 108), *307, 311, 313, 320, 322, 331, 340, 342, 349*

Mawby, R. J., 13 (1967, 27), 90 (1967, 27), *320*

Mayer, E., 84 (1969, 194), 97 (1969, 194), *345*

Mayhew, J., 89 (1969, 291), 92 (1969, 291), *348*

Maynard, J. A., 169 (1967, 336), *328*

Mayne, N., 91 (1969, 262), *347*

Mays, M. J., 91 (1968, 218, 219, 231), 92 (1968, 218, 219), 124 (1968, 231), *336*

McAllister, P. R., 90 (1968, 210; 1969, 251), 91 (1968, 210; 1969, 251), 93 (1968, 210; 1969, 251, 307), *335, 346, 348*

McArdle, P. A., 92 (1967, 239), 196 (1967, 239), *325*

McAuley, A., 86 (1965, 43), 259 (1965, 43), *309*

McBee, E. T., 79 (1966, 97), 281 (1966, 97), *314*

McCarley, R. E., 90 (1966, 158), 91 (1966, 158), 93 (1966, 158), *315*

McClanahan, J. L., 79 (1966, 96), 281 (1966, 96), 285 (1966, 96), 286 (1966, 96), 287 (1966, 96), *314*

McCormick, D. B., 89 (1967, 264), *326*

McEwen, G. K., 40 (1968, 79), 83 (1968, 79), 84 (1968, 79), *332*

McFarlane, W., 2 (1968, 1, 3; 1969, 3, 4), 5 (1965, 3; 1967, 8), 6 (1966, 5; 1968, 7; 1969, 12), 8 (1968, 3; 1969, 3, 4), 9 (1968, 3), 15 (1968, 3), 17 (1969, 47), 18 (1968, 7), 19 (1968, 37), 20 (1967, 55, 57; 1969, 50), 21 (1967, 55), 22 (1967, 57), 26 (1969, 50), 27 (1967, 66; 1968, 7; 1969, 3, 4), 28 (1969, 3, 4), 30 (1967, 79), 31 (1967, 79; 1968, 7), 32 (1968, 66), 33 (1967, 55, 57), 34 (1967, 55), 42 (1967, 55), 56 (1969, 118), 60 (1968, 37), 62 (1969, 47), 47), 64, 66 (1967, 57; 1968, 37), 68 (1967, 66; 1969, 3, 47), 69 (1967, 66; 1968, 3; 1969, 3, 4), 70 (1968, 3), 72 (1967, 132; 1968, 3; 1969, 3, 4), 73 (1969, 4), 74 1967, 79), 77

(1967, 55), 80 (1969, 157), 87 (1968, 7; 1969, 12), 90 (1967, 79), 91 (1962, 5; 1967, 79), 92 (1967, 249, 250, 255; 1968, 257), 93 (1962, 5; 1966, 195; 1967, 79), 101 (1968, 37), 104 (1968, 3), 106 (1967, 55), 107 (1968, 3), 109 (1965, 3; 1968, 3), 110 (1967, 55), 113 (1965, 3; 1968, 3, 7), 114 (1968, 3), 115 (1968, 3, 7; 1969, 4, 50), 120 (1968, 3), 121 (1965, 3; 122 (1966, 5; 1968, 3), 128 (1965, 3), 129 (1966, 5), 131 (1969, 50), 132 (1969, 50), 139 (1965, 3; 1968, 3, 7), 140 (1968, 3; 1969, 50), 141 (1968, 7; 1969, 50), 151 (1969, 50), 153 (1965, 3; 1967, 66; 1968, 7; 1969, 50(, 154 (1967, 66; 1968, 3, 7; 1969, 50), 155 (1969, 50), 159 (1965, 3), 169 (1965, 3; 1968, 7; 1969, 50), 170 (1969, 50), 173 (1965, 3; 1968, 7), 174 (1968, 7), 178 (1966, 5), 186 (1969, 50), 187 (1965, 3; 1969, 50), 191 (1965, 3; 1967, 8; 1968, 7, 66), 193 (1965, 3; 1967, 8; 1968, 7; 1969, 50), 194 (1968, 7; 1969, 12), 195 (1969, 50), 196 (1965, 3; 1967, 79; 1968, 7; 1969, 50), 197 (1969, 50), 201 (1967, 8), 202 (1965, 3; 1967, 79; 1968, 7), 203 (1966, 5; 1968, 7), 206 (1965, 3; 1966, 5; 1967, 8; 1968, 7), 207 (1966, 5), 208 (1967, 8), 209 (1965, 3; 1967, 8; 1968, 7), 212 (1966, 5; 1967, 8), 216 (1965, 3; 1966, 5; 1967, 8, 79; 1968, 7; 1969, 157), 217 (1966, 5; 1967, 8), 218 (1966, 5), 219 (1967, 8), 220 (1966, 5), 222 (1969, 118), 223 (1965, 3; 1969, 50), 224 (1965, 3; 1967, 8; 1968, 7; 1969, 50), 225 (1966, 5; 1968, 7), 226 (1965, 3; 1966, 5; 1967, 8; 1968, 7), 227 (1967, 363; 1969, 50), 228 (1966, 5; 1967, 8, 363), 229 (1966, 5), 231 (1966, 5; 1967, 8, 363), 233 (1966, 5), 234 (1966, 5), 235 (1966, 5; 1967, 8, 363; 1969, 118), 238 (1966, 5), 239 (1966, 5; 1967, 363; 1969, 118), 241 (1966, 5; 1967, 363), 242 (1966, 5), 243 (1966, 5; 1967, 363), 244 (1966, 5), 245 (1966, 5), 246 (1969, 118), 247 (1966, 5), 252 (1966, 5), 257 (1968, 3; 1969, 3), 275 (1966, 5), 276 (1966, 5), 284 (1966, 5), *307, 308, 311, 316, 319, 320, 321, 323, 326, 329, 330, 331, 332, 337, 339, 340, 341, 343, 344*

McKeon, J. E., 92 (1968, 238, 239), *336*
McLauchlan, K. A., 92 (1969, 277), *347*
McNeal, J. P., 40 (1968, 77), 82 (1968, 77), 107 (1968, 77), 151 (1968, 77), *332*
McQuistion, W. E., 130 (1966, 237), 221 (1966, 237), *317*
Medved', T. Ya., 7 (1968, 18), 271 (1968, 18), 273 (1968, 18), *330*
Meek, D. W., 44 (1969, 105), 45 (1969, 105), 91 (1968, 232), 142 (1969, 105), 150 (1969, 105, 347), 203 (1969, 105), 204 (1969, 105 (228 (1968, 232), 231 (1968, 232), 235 (1968, 232), 266 (1969, 105), 271 (1969, 347), *336, 342, 349*
Meier, M., 92 (1969, 280), *347*
Meinel, L., 265 (1967, 385), 282 (1967, 385), 286 (1967, 385), *330*
Mel'Nikov, N. N., 113 (1965, 82), 165 (1965, 82), 197 (1965, 82), *310*
Meng, L. Y. C., 195 (1966, 261), 244 (1966, 261), *318*
Mercer, A. J. H., 75 (1968, 118), 209 (1968, 118), 235 (1967, 366), *329, 333*
Merenyi, R., 86 (1966, 140), 238 (1966, 140), *315*
Merlin, J. C., 83, (1968, 155), 86 (1968, 165), 100 (1968, 155), 121 (1968, 165), *334*
Merritt, L., 80 (1969, 161), 173 (1968, 328), 190 (1968, 328; 1969, 161), 204 (1968, 328; 1969, 161), 206 (1968, 328; 1969, 161), 217 (1968, 328; 1969, 161), 231 (1968, 328; 1969, 161), 244 (1968, 328; 1969, 161), *339, 344*
Merz, A., 54 (1967, 111, 112), 248 (1967, 112), 249 (1967, 112), *322*

Mesmer, R. E., 86 (1966, 131; 1967, 129), 99 (1966, 131), 101 (1966, 131; 1967, 129), *315, 323*
Mestroni, G., 90 (1968, 200), *335*
Metzler, D. E., 94 (1967, 274), *326*
Mikolajczyk, M., 20, 42 (1969, 52), 83 (1969, 176, 177), 118 (1969, 177), 119 (1969, 52), 124 (1969, 177), 126 (1969, 176), 134 (1969, 177), 161 (1969, 176), 165 (1969, 177), *341, 344*
Mildvan, A. S., 89 (1969, 314), 94 (1968, 281), 110 (1968, 281), *337, 348*
Miles, W. J., 91 (1969, 272), 220 (1969, 272), *347*
Millar, I. T., 28 (1968, 60), 80 (1969, 155), 92 (1969, 155, 290), 112 (1968, 60), 201 (1968, 60), 202 (1968, 60), 203 (1967, 348), 209 (1968, 60), 227 (1967, 362), 228 (1967, 348; 1968, 60), 231 (1968, 60), 269 (1969, 155), *329, 331, 344, 348*
Miller, D. L., 89 (1967, 263), *326*
Miller, G. R., 13 (1969, 33), 57 (1969, 33), 227 (1969, 33), 230 (1969, 33), 234 (1969, 33), *340*
Miller, J. M., 57 (1966, 73), 215 (1966, 73), 221 (1966, 73), *313*
Miller, N. E., 60 (1969, 128), 121 (1969, 128), 129 (1969, 128), 141 (1969, 128), 260 (1969, 128), 261 (1968, 351), 278 (1969, 128), *339, 343*
Miller, R., 94 (1968, 281), 110 (1968, 281), *337*
Mills, J. L., 84 (1969, 187), 165 (1969, 354), 169 (1969, 354), 191 (1969, 354), 193 (1969, 354), 208 (1969, 354), 209 (1969, 354), 223 (1969, 354), *345, 349*
Mingos, D. M. P., 89 (1968, 217), 91 (1968, 217; 1969, 263), 92 (1968, 217; 1969, 263), 93 (1968, 217; 1969, 263), *336, 347*
Mishra, I. B., 38 (1969, 84), 76 (1969, 84), 111 (1969, 84), *342*
Mislow, K., 80 (1967, 151; 1968, 140; 1969, 154), 182 (1969, 154), 205 (1967, 151; 1968, 140), 217 (1967, 151), 220 (1967, 151; 1968, 140), 221 (1967, 151; 1968, 140), 230 (1967, 151; 1968, 140), 240 (1967, 151; 1968, 140), 247 (1968, 140), 250 (1967, 151), *323, 334, 344*
Misono, A., 90 (1967, 223), *325*
Mitchell, E. W., 39 (1968, 74), 83 (1968, 149), 86 (1966, 121), 117 (1966, 121), 124 (1968, 141), 125 (1968, 74), 134 (1968, 74), 137 (1968, 74), 168 (1968, 74), 175 (1968, 149), 176 (1968, 149), 180 (1966, 121), 181 (1968, 74), 190 (1968, 74), 198 (1968, 74), 211 (1968, 74), 212 (1968, 74), 243 (1968, 74), *314, 332, 334*
Mitchell, H. L., 22 (1969, 64), 25 (1969, 64), 109 (1969, 64), *341*
Mitchener, J. P., 92 (1969, 303), 277 (1969, 303), *348*
Mitchenko, Yu. I., 87 (1968, 176; 1969, 209), *335, 345*
Mitsch, R. A., 22 (1967, 61), 39 (1967, 61), 46 (1967, 61), 114 (1967, 61), 129 (1967, 61), 142 (1967, 61), 193 (1967, 61), 214 (1967, 61), 223 (1967, 61), *321*
Miyake, A., 91 (1968, 214), *336*
Miyazima, G., 34 (1967, 84), 37 (1967, 84), 107 (1966, 216; 1967, 84), 119 (1966, 216), 120 (1966, 216; 1967, 84), 121 (1967, 84), 140 (1966, 216; 1967, 84), 288 (1966, 216), *317, 321*
Moedritzer, K., 20 (1967, 45), 89 (1968, 235), 92 (1968, 235), 100 (1967, 45), 105 (1966, 212), 115 (1966, 212), 143 (1967, 45), 258 (1966, 212), 277 (1966, 212), 284 (1966, 212), 285 (1966, 212), *317, 320, 336*
Moeller, T., 23 (1968, 50), 79 (1967, 144, 145; 1968, 50, 130, 132, 133), 101 (1967, 145), 102 (1967, 145), 279 (1967, 144; 1968, 50, 130), 280 (1968, 130, 133), 281 (1968, 130), 285 (1968, 132), 287 (1968, 132), *323, 331, 333*
Moelwyn-Hughes, J. T., 92 (1969, 304), *348*

Moen, R. W., 191 (1966, 260), 198 (1966, 260), *318*
Moffatt, J. G., 94 (1965, 76), *310*
Moffett, L. R., 79 (1968, 134), 285 (1968, 134), 286 (1968, 134), *333*
Molin, Yu. N., 92 (1969, 282), 169 (1969, 282), *347*
Moller, U., 264 (1969, 373), *350*
Mondelli, R., 174 (1968, 347), 233 (1968, 343), 235 (1968, 343), 246 (1968, 343, 347), 247 (1968, 343, 347), 250 (1968, 343), 251 (1968, 343), *339*
Montesano, R. M., 91 (1969, 275), *347*
Mooney, E. F., 74 (1966, 81), 84 (1966, 81), 104 (1969, 24), *313, 340*
Morales, M., 89 (1966, 196), *316*
Morallee, K. G., 90 (1968, 200), *335*
Moran, E. F., 86 (1968, 166), 102 (1968, 166), 131 (1968, 166), 283 (1969, 378), *334, 350*
Moran, J. T., 105 (1967, 292), 141 (1967, 292), *327*
Moreland, C. B., 29 (1969, 77), 54 (1969, 77), 124 (1969, 77), 134 (1969, 77), 237 (1969, 77), *342*
Morelli, D., 92 (1967, 251), *326*
Morgan, R. E., 73 (1969, 139), 87 (1969, 139), *343*
Morino, Y., 25 (1969, 70), *341*
Moritani, T., 25 (1969, 70), *341*
Moritz, A. G., 56 (1968, 104), 230 (1968, 104), 234 (1968, 104), *333*
Morkovin, N. V., 20 (1967, 44), 42 (1967, 44), 109 (1967, 44), *320*
Morris, E. D., 44 (1969, 108), *342*
Morrison, A., 93 (1966, 201), *316*
Morse, J. G., 50 (1966, 59), 98 (1966, 59), *313*
Morse, K. W., 74 (1967, 133, 134), 84 (1967, 133, 134), *323*
Morton, C. J., 79 (1966, 97), 281 (1966, 97), *314*
Morton-Blake, D. A., 44 (1967, 95), 174 (1967, 95), 175 (1967, 95), 182 (1967, 95), 191 (1967, 95), 192 (1967, 95), 200 (1967, 95), 206 (1967, 95), 212 (1967, 95), *322*
Mosel, B. D., 88 (1968, 193), 97 (1968, 193), *335*
Moser, E., 90 (1968, 207, 208), 91 (1966, 160; 1968, 207, 208), 93 (1966, 160; 1968, 207, 208), *315, 335*
Moss, J. R., 84 (1969, 195), 92 (1966, 170; 1969, 195), 93 (1968, 267), *315, 337, 345*
Moulton, W. G., 88 (1969, 218), *346*
Mowthorpe, D. J., 87 (1968, 175), 259 (1967, 381), 262 (1967, 381), 263 (1967, 381), *329, 335*
Mueller, D. C., 80 (1966, 99; 1968, 141), 267 (1966, 99; 1968, 141), *314, 334*
Mueller, W. H., 43 (1966, 52), 58 (1966, 74), 120 (1966, 74), 121 (1967, 307), 125 (1966, 74), 127 (1966, 74), 130 (1966, 74), 131 (1966, 74), 136 (1966, 74), 137 (1966, 74), 140 (1966, 52; 1967, 307), 146 (1966, 74), 147 (1967, 307), 148 (1966, 74), 157 (1967, 307), 160 (1966, 52), 167 (1966, 52), 168 (1966, 52; 1967, 307), 169 (1967, 307), 177 (1967, 307), 183 (1966, 52), 184 (1967, 307), 191 (1966, 52, 260), 198 (1966, 52, 260), *313, 318, 327*
Muetterties, E. L., 91 (1968, 230), 93 (1968, 230, 268), 97 (1965, 77), 98 (1965, 77), 186 (1968, 268), *310, 336, 337*
Mui, J. T. P., 162 (1966, 252), 163 (1966, 252), 181 (1966, 252), *318*
Müller, A., 7 (1966, 8, 9; 1968, 19), 21 (1967, 50, 56), 22 (1965, 13; 1966, 8; 1967, 56), 66 (1967, 50), 97 (1966, 8, 9; 1967, 50, 56; 1968, 19), 98 (1966, 8, 9; 1967, 50; 1968, 19), 99 (1965, 13; 1966, 8, 9; 1968, 19), 153 (1967, 50), *308, 311, 320, 321, 330*
Müller, H. J., 79 (1968, 136), 98 (1968, 136), 101 (1968, 136), 108 (1968, 136), 122 (1968, 136), 257 (1968, 136), 278 (1968, 136), 279 (1968, 136), *334*
Müller, J., 90 (1968, 207), 91 (1968, 207), 93 (1968, 207), *335*
Müller-Warmuth, W., 88 (1968, 193), 97 (1968, 193), *335*

Munoz, A., 83 (1968, 150), 172 (1968, 150), 181 (1968, 150), 190 (1968, 150), 205 (1968, 150), *334*

Muratova, A. A., 20 (1969, 53), 83 (1969, 53), 93 (1969, 53), *341*

Murray, J. D., 19 (1967, 35), 100 (1967, 35), *320*

Murray, M., 106 (1968, 289), *337*

Murray, R., 55 (1966, 69), 223 (1966, 69), *313*

Musierowicz, S., 63 (1967, 127), 262 (1967, 127), 263 (1967, 127), *322*

Mustafa, A., 198 (1966, 263), 239 (1966, 263), *318*

Muylle, E., 4 (1969, 5b), 88 (1969, 5b), 101 (1969, 5b), 131 (1969, 120), 185 (1969, 120), 186 (1969, 120), 222 (1969, 120), 223 (1969, 120), *340, 343*

N

Nagabhushanam, M., 52 (1966, 62), 86 (1966, 62), 103 (1966, 62), 105 (1965, 79), 109 (1965, 79), 120 (1968, 301), 127 (1968, 301), 132 (1965, 79), 145 (1968, 301), 148 (1968, 301), 151 (1968, 301), 157 (1968, 301), 164 (1968, 301), 180 (1968, 301), 186 (1965, 79), 187 (1965, 79), 188 (1968, 301), 189 (1968, 301), 190 (1968, 301), 197 (1965, 79), 198 (1968, 301), 205 (1968, 301), 211 (1968, 301), 214 (1965, 79), 218 (1968, 301), 219 (1968, 301), 224 (1968, 301), 227 (1966, 62), 229 (1968, 301), 233 (1968, 301), 238 (1968, 301), 240 (1966, 62), 245 (1968, 301), 247 (1968, 301), 253 (1968, 301), *310, 313, 338*

Nagarajan, G., 7 (1966, 9), 97 (1966, 9), 98 (1966, 9), 99 (1966, 9), *311*

Nainan, K. C., 92 (1969, 305), 189 (1968, 330), 204 (1968, 330), *339 348*

Nakajima, M., 94 (1967, 271), *326*

Nakamura, A., 40, 41 (1968, 81; 1969, 90, 95), 83 (1969, 90, 95), 85 (1969, 200), 92 (1968, 254), 104 (1968, 81), 108 (1969, 95), 117 (1969, 90), 125 (1969, 95), 126 (1969, 95), 162 (1969, 95), 259 (1969, 95), 274 (1968, 254), 284 (1968, 254), 287 (1968, 254), *332, 336, 342, 345*

Nash, J. A., 2 (1969, 3, 4), 8 (1969, 3, 4), 27 (1969, 3, 4), 28 (1969, 3, 4), 68 (1969, 3), 69 (1969, 3, 4), 72 (1969, 3, 4), 73 (1969, 4), 80 (1969, 157), 115 (1969, 4), 216 (1969, 157), 257 (1969, 3), *339, 344*

Nasielski, J., 81 (1969, 168), *344*

Naumann, K., 80 (1969, 154), 182 (1969, 154), *344*

Navada, C. K., 86 (1965, 42), 146 (1965, 42), 167 (1965, 42), 183 (1965, 42), 200 (1965, 42), 213 (1965, 42), *309*

Navech, J., 37 (1969, 343), 40 (1968, 76), 47 (1968, 76), 78 (1968, 76), 82 (1968, 76), 83 (1968, 150; 1969, 175), 139 (1969, 343), 151 (1969, 343), 159 (1969, 343), 163 (1969, 343), 164 (1968, 76), 172 (1968, 150; 1969, 175), 174 (1969, 343), 181 (1968, 150; 1969, 175), 182 (1969, 343), 190 (1968, 150; 1969, 175, 343), 191 (1969, 359), 197 (1969, 359), 199 (1969, 175, 359), 202 (1969, 359), 203 (1969, 359), 204 (1969, 359), 205 (1968, 150), 206 (1969, 343), 210 (1969, 359), 215 (1968, 76), 216 (1968, 76; 1969, 359), 232 (1969, 359), 234 (1969, 175), 236 (1969, 359), *332, 334, 344, 349*

Nealey, R. H., 195 (1966, 262), 196 (1966, 262), 223 (1966, 262), 227 (1966, 262), *318*

Neimysheva, A. A., 7 (1967, 13, 14), 97 (1967, 14), 98 (1967, 14), 99 (1967, 14), 103 (1967, 14), 104 (1967, 14), 105 (1967, 14), 107 (1967, 14), 108 1967, 13, 14; 1968, 295), 112 (1967, 14), 113 (1967, 13, 14; 1968, 295), 118 (1967, 14), 120 (1967, 13, 14; 1968, 295), 122 (1967, 14), 127 (1967, 13, 14; 1968, 295), 128 (1968, 295), 132 (1967, 14), 136 (1967, 14), 138 (1967, 13, 14; 1968, 295), 152 (1967, 13, 14; 1968, 295), (1967, 13, 14; 1968, 295), 163

(1967, 13; 1968, 295), 172 (1967, 14), 186 (1967, 14), 252 (1967, 13), *319, 338*

Nekipelova, T. D., 84 (1969, 184), *345*

Nelson, J. H., 92 (1969, 295), 94 (1969, 315), *348*

Newberry, J. E., 82 (1969, 171), 121 (1969, 171), 163 (1969, 171), 180 (1969, 171), 215 (1969, 171), *344*

Newmark, R. A., 14 (1969, 31), 15 (1969, 31), 20 (1969, 31), 21 (1969, 31), 24 (1969, 31), 60, 61 (1969, 31), 62, 63 (1969, 31), 64 (1969, 31), 65 (1969, 31, 136), 66 (1969, 31), 67 (1969, 31), 76 (1969, 31), 80, 100 (1969, 31), 101 (1969, 31, 136), 256 (1969, 31), 257 (1969, 136), *340, 343*

Newton, M. G., 40 (1969, 93), 82 (1969, 93), 89 (1969, 93), 126 (1969, 93), 147 (1969, 93), 148 (1969, 93), 163 (1969, 93), 171 (1969, 93), *342*

Ng, T. W., 91 (1969, 256), *347*

Nichols, D. I., 57, 92 (1967, 248; 1968, 263; 1969, 121), 221 (1968, 263; 1969, 121), 222 (1968, 263; 1969, 121), *326, 337, 343*

Nicholson, C. R., 53 (1966, 67), 201 (1966, 67), 202 (1966, 67), *313*

Nicholson, D. A., 256 (1967, 380; 1968, 349), 257 (1967, 380; 1968, 349), 258 (1968, 349), 259 (1968, 349), 260 (1968, 349), 261 (1968, 349), 263 (1968, 349), 266 (1967, 380; 1968, 349), *329, 339*

Nicklees, G., 19 (1967, 35), 100 (1967, 35), *320*

Niecke, E., 7 (1968, 19), 21 (1967, 50, 56), 22 (1967, 56), 23 (1967, 63), 66 (1967, 50), 79 (1969, 152), 97 (1967, 50, 56; 1968, 19), 98 (1967, 50; 1968, 19), 99 (1968, 19), 101 (1969, 152), 143 (1967, 63), 153 (1967, 50), 277 (1969, 152), *320, 321, 330, 344*

Niederreuther, U., 31 (1969, 78), 90 (1969, 78), 91 (1969, 78), 92 (1969, 78), 93 (1969, 78), 193 (1969, 78), *342*

Nielsen, J. A., 58 (1969, 125, 126), 185 (1969, 125, 126), *343*

Nifant'ev, E. E., 118 (1967, 393), *330*

Nikolaev, S. N., 17, *341*

Nimrod, D. M., 75 (1968, 117), 83 (1968, 117), 89 (1968, 117), 135 (1968, 117), *333*

Nishimura, S., 84 (1968, 157), 194 (1968, 157), *334*

Nishiwaki, T., 158 (1967, 333), 160 (1966, 251), 178 (1966, 251), 194 (1966, 251), 207 (1966, 251), 218 (1966, 251), 227 (1966, 251), *318, 328*

Nixon, J. F., 2 (1969, 1), 19 (1968, 41), 21 (1967, 48; 1969, 60, 61), 22 (1969, 66), 23 (1968, 41), 25 (1966, 36; 1968, 41), 26 (1966, 36), 27 (1966, 36), 34 (1967, 48), 35 (1968, 41; 1969, 66), 37 (1968, 41), 45 (1967, 96; 1968, 89), 60 (1969, 60, 61), 61 (1969, 60, 61), 64 (1969, 60), 68 (1969, 61), 75 (1966, 36; 1968, 41), 75 (1969, 143), 76 (1968, 119; 1969, 61), 77 (1968, 41), 84 (1966, 36), 86 (1966, 36), 88 (1969, 61), 89, 90 (1969, 60, 250), 91 (1966, 168; 1968, 119; 1969, 60, 250, 276), 92 (1966, 168, 175; 1967, 48; 1969, 283, 299), 93 (1969, 60, 250), 99 (1967, 48; 1969, 143), 103 (1966, 36; 1967, 48; 1969, 66), 104 (1966, 36), 105 (1966, 36; 1968, 110), 106 (1967, 48; 1968, 41), 108 (1966, 36), 109 (1966, 36), 110 (1969, 66), 113 (1966, 36), 114 (1966, 36), 115 (1966, 36), 118 (1966, 36, 230), 119 (1969, 61), 120 (1966, 230; 1967, 48; 1969, 61), 122 (1966, 36), 125 (1966, 230; 1967, 48), 126 (1969, 61), 127 (1966, 36), 130 (1969, 66), 132 (1966, 36), 138 (1969, 61), 142 (1966, 36), 143 (1966, 36), 148 (1966, 36), 153 (1966, 36), 154 (1966, 36), 158 (1969, 61), 174 (1966, 36), 175 (1966, 36), 197 (1968, 89), 203 (1966, 36), 221 (1969, 276), 243 (1969, 299), 256 (1967, 96; 1968, 89; 1969, 61,

370), 257 (1967, 96; 1968, 89; 1969, 60, 61), 258 (1966, 36), 260 (1966, 36; 1969, 61), 267 (1966, 36), 277 (1966, 36), *312, 315, 316, 317, 320, 322, 331, 332, 333, 339, 341, 343, 346, 347, 348, 350*
Norberg, R. E., 88 (1967, 204), 169 (1967, 204), *325*
Nordberg, R., 4 (1967, 6), *319*
Nordberg, C., 4 (1967, 6), *319*
Nordman, C. E., 44 (1969, 108), *342*
Norman, A. D., 65 (1969, 136), 84 (1966, 109), 86 (1966, 109), 97 (1966, 109), 99 (1967, 284), 101 (1969, 136), 257 (1969, 136), *314, 327, 343*
Normant, H., 145 (1967, 326), 146 (1967, 326), 156 (1967, 326), 177 (1967, 326), *328*
Normant, J. F., 29 (1966, 42), 84 (1969, 181), 143 (1969, 181), 174 (1966, 42), 182 (1966, 42), 220 (1966, 42), *312, 344*
Norval, S., 86 (1966, 138; 1968, 167; 1969, 203, 204), 99 (1969, 203), 109 (1966, 138), 140 (1966, 138), 258 (1966, 138), 278 (1966, 138), *315, 334, 345*
Nöth, H., 84 (1966, 108), 90 (1969, 242), 91 (1966, 167), 104 (1968, 290), 108 (1968, 290), 113 (1968, 290), 114 (1967, 299; 1968, 290), 122 (1968, 290), 129 (1968, 290), 142 (1965, 91; 1967, 299), 150 (1965, 91), 170 (1967, 299), 195 (1967, 299), 243 (1967, 299), 265 (1967, 385), 282 (1967, 385), 286 (1967, 385), *310, 314, 315, 327, 330, 337, 346*
Novikova, Z. S., 111 (1968, 297), 126 (1968, 297), 134 (1968, 297), 137 (1968, 297), 145 (1968, 297), 147 (1968, 297), 148 (1968, 297), 149 (1968, 297), 155 (1968, 297), 156 (1968, 297), 162 (1968, 297), 165 (1968, 297), 167 (1968, 297), 176 (1968, 297), 183 (1968, 297), 188 (1968, 297), 192 (1968, 297), 208 (1968, 297), 209 (1968, 297), 271 (1968, 297), *338*

Novitskii, K. I., 107 (1966, 220), 111 (1966, 219), 116 (1966, 220), 117 (1966, 219; 1967, 313), 123 (1967, 311), 124 (1967, 311, 313), 125 (1967, 313), 134 (1967, 311), 135 (1967, 311), 137 (1967, 311), 138 (1966, 220; 1967, 313), 145 (1967, 311), 146 (1966, 220; 1967, 313), 148 (1967, 311), 156 (1966, 220; 1967, 311), 158 (1967, 311), 167 (1966, 220; 1967, 311, 313), 168 (1967, 313), 176 (1966, 220; 1967, 311), 188 (1966, 220), 190 (1966, 220), 197 (1966, 220), 204 (1966, 220), *317, 327, 318*
Novobilsky, V., 18 (1969, 42), 84 (1969, 42), *341*
Nśeič, S., 50 (1967, 100), 83 (1967, 100), 137 (1967, 100), 145 (1967, 100), 147 (1967, 100), 149 (1967, 100), 155 (1967, 100), 207 (1967, 100), 213 (1967, 100), 221 (1967, 100), 230 (1967, 100), 272 (1967, 100), *322*
Nuretdinov, I. A., 151 (1966, 250), 171 1966, 250), 180 (1966, 250), 189 (1966, 250), 190 (1966, 250), 198 (1966, 250), *318*
Nuretdinov, S. Kh., 176 (1967, 341), *328*
Nyholm, R. S., 56 (1966, 72), 90 (1966, 72; 1967, 226; 1968, 194), 91 (1966, 72; 1967, 226), 92 (1968, 237), 93 (1966, 72; 1967, 226), 234 (1966, 72; 1968, 194), 238 (1968, 194), *313, 325, 335, 336*
Nyman, C. J., 92 (1967, 252), *326*

O

Obrycki, R., 56 (1967, 115), 82 (1967, 115), 152 (1967, 115), 154 (1968, 317), 163 (1968, 317), 164 (1968, 317), 174 (1968, 317), 182 (1968, 317), 191 (1968, 317), 222 (1967, 115), *322, 338*
Oda, R., 29 (1967, 74), 249 (1967, 74), 250 (1967, 74), *321*
Oehme, H., 78 (1967, 137), 163 (1967, 137), 172 (1967, 137), 181 (1967,

137), 188 (1967, 137), 198 (1967, 137), 199 (1967, 137), 205 (1967, 137), 209 (1967, 137), 211 (1967, 137), 215 (1967, 137), 219 (1967, 137), 249 (1967, 137), *323*
Ogilvie, F. B., 64 (1969, 132, 133, 134), 89 (1969, 132, 133), 90 (1969, 132, 133), 91 (1969, 133, 134), 92 (1969, 133), 93 (1969, 133), *343*
Okano, M., 29 (1967, 74), 249 (1967, 74), 250 (1967, 74), *321*
O'Keefe, J. G., 88 (1961, 3), 97 (1961, 3), *307*
Okuda, S., 94 (1967, 272), *326*
Okuhara, K., 79 (1966, 97), 281 (1966, 97), *314*
Olah, G. A., 6 (1969, 12), 87 (1969, 12), 194 (1969, 12), *340*
Olbrich, H., 254 (1966, 287), *318*
Oliver, A. J., 7 (1969, 16), *340*
Oliver, J. P., 85 (1965, 41; 1969, 198), 142 (1969, 198), *309, 345*
Oliver, W. L., 74 (1969, 142), 126 (1969, 142), 167 (1969, 142), *343*
Onak, T., 22 (1966, 31), 44 (1966, 31), 77 (1966, 31), 106 (1966, 31), *312*
Onodera, K., 94 (1966, 203), *316*
O'Reilly, D. E., 88 (1968, 188), *335*
Orlov, N. F., 19 (1967, 34), 89 (1967, 34), 140 (1967, 34), 159 (1967, 34), 161 (1967, 34), 170 (1967, 34), 179 (1967, 34), 192 (1967, 34), 195 (1967, 34), 207 (1967, 34), 227 (1967, 34), *320*
Osborn, J. A., 92 (1966, 187, 190, 191; 1967, 256), 93 (1967, 260), 131 (1967, 256), *316, 326*
Osborn, R. B. L., 92 (1966, 183; 1968, 255), *316, 337*
Osborne, D. W., 48 (1966, 56), 149 (1966, 56), 188 (1966, 56), 198 (1966, 264), 208 (1966, 56), 211 (1966, 264), 225 (1966, 264), 268 (1966, 264), *313, 318*
Ossip, P. S., 31 (1968, 65), 109 (1968, 65), 114 (1968, 65), 121 (1968, 65), 122 (1968, 65), 128 (1968, 65), 140 (1968, 65), 149 (1968, 65), 159 (1968, 65), 169 (1968, 65), 179

(1968, 65), 184 (1968, 65), 195 (1968, 65), *332*
Osuch, C., 59 (1964, 10), 237 (1964, 10), 240 (1964, 10), 243 (1964, 10), *308*
Oswald, A. A., 43 (1966, 52), 121 (1967, 307), 140 (1966, 52; 1967, 307), 147 (1967, 307), 157 (1967, 307), 160 (1966, 52), 167 (1966, 52), 168 (1966, 52; 1967, 307), 169 (1967, 307), 177 (1967, 307), 183 (1966, 52), 184 (1967, 307), 191 (1966, 52), 198 (1966, 52), *313, 327*
Oth, J. F. M., 86 (1966, 140), 238 (1966, 140), *315*
Otsuka, S., 92 (1968, 254), 274 (1968, 254), 284 (1968, 254), 287 (1968, 254), *336*
Ottmann, G., 67 (1966, 80), 79 (1966, 80), 279 (1966, 80), 280 (1966, 80), 281 (1966, 80), *313*
Otto, J., 152 (1967, 330), *328*
Ourisson, G., 31 (1966, 43), 40 (1966, 43), 47 (1966, 43), 50 (1966, 43, 58; 1967, 100), 83 (1967, 100, 167), 94 (1967, 167, 270), 137 (1967, 100), 144 (1966, 43), 145 (1967, 100), 146 (1966, 43), 147 (1967, 100), 149 (1967, 100), 155 (1966, 43; 1967, 100), 177 (1966, 43), 207 (1967, 100), 213 (1967, 100), 221 (1967, 100), 230 (1967, 100), 237 (1967, 167), 241 (1967, 167), 244 (1967, 270), 248 (1967, 270), 250 (1966, 43; 1967, 167, 270), 251 (1967, 167), 252 (1966, 43; 1967, 167), 155 (1967, 167), 272 (1967, 100), *312, 313, 322, 324, 326*
Owston, P. G., 91 (1968, 216), *336*

P

Paciorek, K. L., 251 (1966, 284), 256 (1966, 284), *318*
Paddock, N. L., 79 (1958, 1; 1962, 3; 1968, 127; 1969, 151), 88 (1969, 216), 101 (1969, 151, 216), 102 (1969, 151, 216), 278 (1968, 127; 1969, 151), 284 (1968, 127), 286

(1968, 127); 287 (1968, 127; 1969, 151), *307, 333, 344, 345*
Pagilagan, R. U., 80 (1967, 152), 171 (1967, 338), 181 (1967, 338), 188 (1967, 338), 216 (1967, 152), 220 (1967, 152), *323, 328*
Painter, T. M., 256 (1969, 370), *350*
Palenik, G. J., 91 (1969, 256), *347*
Pannell, K. H., 90 (1968, 202; 1969, 239), 91 (1968, 202, 215, 229; 1969, 239, 260), *335, 336, 346, 347*
Panteleeva, A. R., 2 (1967, 2), 279 (1967, 2), 280 (1967, 2), 281 (1967, 2), *319*
Parasaran, T., 250 (1966, 283), *318*
Parker, J. R., 93 (1965, 71), 271 (1965, 71), *310*
Parry, R. W., 18 (1966, 20; 1967, 32; 1968, 34), 21 (1966, 28; 1967, 32; 1968, 34, 43), 22 (1968, 34), 46 (1968, 43), 50 (1966, 59), 65 (1968, 34), 66 (1966, 28), 74 (1967, 32, 133, 134; 1968, 34; 1969, 141), 76 (1968, 43), 84 (1966, 20; 1967, 32, 133, 134; 1969, 141), 90 (1968, 43), 92 (1969, 141), 97 (1969, 141), 98 (1966, 59; 1968, 34), 99 (1966, 28), 100 (1966, 28; 1968, 34), 103 (1966, 28), 108 (1968, 43), 110 (1969, 141), *312, 313, 320, 323, 331, 343*
Parshall, G. W., 92 (1965, 69; 1966, 184), *310, 316*
Partos, R. D., 224 (1965, 109), 239 (1965, 109), 242 (1966, 280), 253 (1965, 109; 1966, 280), 255 (1965, 109; 1966, 280), *311, 318*
Pattenden, G., 156 (1969, 349), 213 (1969, 349), 217 (1969, 349), 218 (1969, 349), 234 (1969, 349), 238 (1969, 349), 240 (1968, 346), 249 (1968, 346), *339, 349*
Patwardhan, A. V., 39 (1965, 21), 83 (1968, 153), 112 (1966, 224), 114 (1965, 21, 84; 1966, 224, 225), 119 (1966, 224), 128 (1966, 224; 1968, 153), 135 (1968, 312), 140 (1966, 225), 141 (1966, 224; 1968, 153), 142 (1965, 84; 1968, 153), 144 (1968, 312), 145 (1968, 312), 148 (1965, 21; 1966, 225, 248), 150 (1968, 315), 155 (1968, 312), 156 (1968, 312), 159 (1966, 244; 1968, 153), 161 (1968, 315), 162 (1966, 225), 173 (1966, 225, 248), 176 (1968, 312), 177 (1966, 225; 1968, 312), 179 (1966, 224), 181 (1965, 21), 183 (1966, 248, 258), 184 (1965, 21), 188 (1966, 225), 190 (1966, 258), 196 (1968, 315), 197 (1968, 315), 200 (1965, 84; 1966, 248), 203 (1966, 225), 205 (1967, 350, 351), 206 (1966, 225), 211 (1965, 21; 1968, 248, 258; 1967, 350, 351; 1968, 312), 212 (1966, 248), 217 (1966, 258), 219 (1965, 21, 84; 1966, 225; 1967, 350; 1968, 315), 223 (1966, 225), 224 (1965, 21, 84; 1966, 225; 1968, 315), 225 (1966, 225), 226 (1966, 225), 228 (1968, 315), 229 (1965, 21; 1966, 258; 1968, 153), 230 (1966, 225), 232 (1960, 225, 258), 233 (1965, 84; 1966, 224, 225; 1968, 153), 236 (1965, 21, 84; 1966, 258), 238 (1966, 225), 240 (1965, 84; 1968, 153), 241 (1968, 153), 242 (1966, 225), 243 (1965, 21), 247 (1966, 225; 1968, 153), 249 (1966, 225), 252 (1966, 225), 253 (1965, 21; 1966, 225), 269 (1967, 350, 351; 1968, 315), 270 (1967, 350), 283 (1967, 350), *309, 310, 317, 318, 329, 334, 338*
Paul, I. C., 79 (1967, 144), 279 (1967, 144), *323*
Paxton, H. J., 86 (1965, 47), 208 (1965, 47), *309*
Payne, D. S., 44 (1967, 95), 81 (1968, 145), 85 (1962, 4), 86 (1968, 145), 92 (1968, 145), 117 (1968, 145), 174 (1967, 95), 175 (1967, 95), 182 (1967, 95), 191 (1967, 95), 192 (1967, 95; 1968, 145), 200 (1967, 95), 206 (1967, 95), 210 (1968, 145), 212 (1962, 4; 1967, 95), 260 (1967, 382), *307, 322, 330, 334*
Peacock, R. D., 92 (1967, 248; 1968, 263), 221 (1968, 263), 222 (1968, 263), *326, 337*

Peake, S. C., 104 (1968, 291), 108 (1968, 291), 132 (1968, 291), 152 (1968, 152 (1968, 316), 186 (1968, 316), 196 (1968, 316), 202 (1968, 316), 223 (1968, 316), *337*, *338*

Pearson, R. G., 92 (1969, 280), *347*

Pearson, S. C., 29 (1968, 61), 32 (1968, 61), 53 (1968, 61), 135 (1968, 61), 144 (1968, 61), 150 (1968, 61), 156 (1968, 61), *331*

Pearson, S. M., 91 (1968, 231), 124 (1968, 231), *336*

Pechkovskii, V. V., 88 (1969, 226), *346*

Peguy, A., 90 (1969, 230), *346*

Peiffer, G., 58 (1969, 127), 114 (1969, 127), 119 (1969, 127), 125 (1969, 127), 126 (1969, 127), 156 (1969, 127), 158 (1969, 127), 166 (1969, 355), 176 (1969, 355), 183 (1969, 355), *343*, *349*

Pellizer, G., 90 (1968, 200), *335*

Penkett, S. A., 94 (1968, 277), 152 (1968, 277), *337*

Penkov, I. N., 88 (1968, 190), *335*

Penney, G. J., 63 (1969, 130), 66 (1969, 130), 87 (1969, 130), *343*

Perciaccante, V. A., 227 (1967, 361), 234 (1967, 361), *329*

Perevezentseva, S. P., 2 (1967, 4), 156 (1967, 4), 182 (1967, 4), 200 (1967, 4), 213 (1967, 4), 221 (1967, 4), *319*

Perot, G. G., 86 (1966, 118), *314*

Pestunvich, V. A., 19 (1967, 34), 89 (1967, 34), 140 (1967, 34), 159 (1967, 34), 161 (1967, 34), 170 (1967, 34), 179 (1967, 34), 192 1967, 34), 195 (1967, 34), 207 (1967, 34), 227 (1967, 34), *320*

Peterson, D. J., 32 (1966, 44), 86 (1966, 132), 151 (1967, 329), 152 (1966, 132; 1967, 329), 153 (1966, 132; 1967, 329), 177 (1966, 44), 182 (1966, 132), 196 (1967, 329), 205 (1967, 329), 207 (1966, 132; 1967, 329), 313 (1967, 329), 221 (1966, 132), 223 (1967, 329), 225 (1966, 44), 229 (1966, 44), 246 (1967, 32), 248 (1967, 329), 271 (1967, 329), *312*, *315*, *328*

Peterson, L. K., 82 (1967, 159), 284 (1967, 159), 286 (1967, 159), *323*

Petrov, A. A., 53 (1967, 110), 72 (1967, 128), 75 (1967, 110), 81 (1969, 163), 86 (1966, 127), 107 (1966, 220), 111 (1966, 127, 219, 220), 116 (1967, 110), 117 (1966, 219; 1967, 110, 303), 123 (1966, 127; 1967, 110, 310, 311), 124 (1967, 110, 303, 311), 125 (1965, 89), 133 (1965, 89; 1967, 319), 134 (1965, 90; 1967, 110, 311), 135 (1967, 311, 319), 136 (1965, 90), 137 (1967, 311), 138 (1966, 220), 143 (1965, 89), 145 (1967, 310, 311), 146 (1966, 220), 148 (1967, 311), 149 (1967, 310), 155 (1965, 90), 156 (1966, 220; 1967, 311), 157 (1968, 314; 1969, 163), 158 (1967, 311), 162 (1965, 96), 166 (1969, 163), 167 (1966, 220; 1967, 311; 1968, 314), 172 (1965, 89), 176 (1965, 100; 1966, 220; 1967, 311), 177 (1965, 100; 1968, 314), 188 (1966, 220), 190 (1966, 220), 193 (1967, 319), 197 (1966, 220), 201 (1965, 96), 204 (1966, 220), 218 (1965, 89), 222 (1968, 314), 236 (1969, 163), 237 (1965, 96), 242 (1965, 96), 260 (1967, 128), 264 (1965, 90), *310*, *311*, *314*, *317*, *322*, *323*, *327*, *328*, *338*, *344*

Petrovskaya, L. I., 111 (1968, 297), 126 (1968, 297), 134 (1968, 297), 137 (1968, 297), 145 (1968, 297), 146 (1968, 297), 147 (1968, 297), 148 (1968, 297), 149 (1968, 297), 155 (1968, 297), 156 (1968, 297), 162 (1968, 297), 165 (1968, 297), 167 297), 176 (1968, 297), 183 (1968, 297), 188 (1968, 297), 192 (1968, 297), 208 (1968, 297), 209 (1968, 197), 271 (1968, 297), *338*

Petrovskii, P. V., 7 (1968, 18), 271 (1968, 18), 273 (1968, 18), *330*

Pfister, G., 226 (1969, 367), 254 (1969, 367) *350*

Phelps, D. A., 2, *319*

Pidcock, A., 2 (1969, 1), 29 (1966, 41), 85 (1966, 41), 89, 90 (1966, 41), 91

(1967, 234), 92 (1966, 185; 1968, 240, 247, 256), 113 (1966, 41), 153 (1966, 41), 196 (1966, 41), *312, 316, 325, 336, 337, 339, 346*

Pierce, S. B., 19 (1967, 36), 22 (1967, 36), 75 (1967, 36), 81 (1967, 157), 99 (1967, 36, 157), 105 (1967, 157), *320, 323*

Pignolet, L. H., 90 (1966, 152; 1967, 224; 1968, 204, 205), 91 (1968, 204), 92 (1966, 152; 1967, 224; 1968, 204, 205), *315, 325, 335*

Pilgram, K., 86 (1966, 133, 134), 114 (1966, 133), 121 (1966, 133), 124 (1966, 133), 129 (1966, 133), 135 (1966, 133), 140 (1966, 133), 154 (1966, 133), 189 (1966, 134), 190 (1966, 134), 198 (1966, 134), *315*

Pilot, J. F., 22 (1968, 303), 29 (1969, 75), 78 (1968, 123), 82 (1968, 123), 113 (1968, 302, 303), 119 (1969, 337), 123 (1968, 302; 1969, 337), 127 (1969, 337), 132 (1969, 337), 135 (1969, 337), 139 (1968, 302, 303), 140 (1968, 303), 141 (1968, 303), 144 (1968, 302), 153 (1968, 303), 159 (1968, 303), 161 (1968, 302), 162 (1968, 302), 173 (1968, 302, 303), 179 (1968, 302), 180 (1969, 337), 188 (1968, 302, 303), 193 (1968, 303), 201 (1968, 303), 202 (1968, 302, 303), 204 (1968, 303), 208 (1968, 302), 214 (1968, 302, 303), 218 (1969, 75), 219 (1968, 303; 1969, 75), 223 (1968, 303), 226 (1968, 303), 230 (1968, 302, 303), 231 (1968, 303), 232 (1968, 303), 238 (1968, 303), 240 (1968, 303), 242 (1968, 303), 246 (1968, 303), 248 (1968, 303), 249 (1968, 303), 252 (1968, 303), *333, 338, 341, 349*

Pinkerton, A. A., 21 (1969, 57), 34 (1969, 57), 61 (1969, 57), 100 (1969, 57), 101 (1969, 57), 258 (1969, 57), *341*

Pinkus, A. G., 86 (1965, 45), 131 (1965, 45; 1966, 239), 143 (1966, 239), 148 (1965, 45; 1966, 239), 186 (1965, 45), 195 (1965, 45; 1966, 261), 1965, 45; 1966, 239), 224 (1965, 45; 1966, 239), 244 (1965, 45; 1966, 239, 261), *309, 317, 318*

Pinnell, R. P., 27 (1965, 18), 45 (1965, 18), 46 (1965, 18), 108 (1965, 18), 119 (1965, 18), 120 (1965, 18), 122 (1965, 18), 129 (1965, 18), 142 (1965, 18), 150 (1965, 18), 160 (1965, 18), 195 (1965, 18), 287 (1966, 294), *308, 319*

Pittman, C. U., 86 (1967, 187), *324*

Plemenkov, V. V., 150 (1965, 93), 156 (1965, 93), 167 (1965, 93), 176 (1965, 93), *310*

Poilblanc, R., 90 (1968, 197, 211), 91 (1968, 211), 92 (1966, 169; 1967, 238; 1968, 211), 93 (1968, 211, 211, 266), 108 (1968, 211), 113 (1968, 211), 114 (1966, 169; 1968, 211), 193 (1968, 211), *315, 325, 335, 337*

Poindexter, E. H., 87 (1969, 214, 215), 88 (1969, 216), 101 (1969, 216), 102 (1969, 216), 121 (1968, 306), 140 (1968, 306), *338, 345*

Pollak-Stachurowa, M., 88 (1968, 191), *335*

Pollard, F. H., 19 (1967, 35), 100 (1967, 35), *320*

Ponti, P. P., 233 (1968, 343), 235 (1968, 343), 246 (1968, 343), 247 (1968, 343), 250 (1968, 343), 251 (1968, 343), *339*

Pople, J. A., 14 (1964, 4; 1965, 9), *308*

Poranski, C. F., 202 (1967, 347), *329*

Porte, A. L., 85 (1962, 4), 212 (1962, 4), *307*

Portnova, S. L., 113 (1965, 82), 165 (1965, 82), 197 (1965, 82), *310*

Potenza, J. A., 87 (1969, 214, 215), 88 (1969, 216), 101 (1969, 216), 102 (1969, 216), 121 (1968, 306), 140 (1968, 306), *338, 345*

Potthast, R., 27 (1967, 70), 54 (1967, 70), 187 (1967, 70), 242 (1967, 70), 246 (1967, 70), *321*

Pousse, A., 90 (1969, 252), *346*

Pouyanne, J. P., 84 (1969, 191), *345*

Powell, J., 91 (1968, 221), 92 (1967, 246), *326, 336*
Powles, J. G., 87 (1968, 174), 98 (1968, 174), 99 (1968, 174), *335*
Praat, A. P., 89 (1968, 223), 91 (1968, 223), 92 (1968, 242, 243, 262; 1969, 297), *336, 337, 348*
Prasch, A., 91 (1964, 13), *308*
Pratt, L., 91 (1962, 5), 93 (1962, 5), *307*
Prentice, J. B., 256 (1967, 380; 1968, 349), 257 (1967, 380; 1968, 349), 258 (1968, 349), 259 (1968, 349), 260 (1968, 349), 261 (1968, 349), 263 (1968, 349), 266 (1967, 380; 1968, 349), *329, 339*
Prestegard, J. H., 94 (1969, 317), *348*
Preusse, W. C., 44 (1967, 93), 98 (1967, 93), *321*
Price, C. C., 250 (1966, 283), *318*
Prikoszovich, W., 7 (1969, 18, 19), 131 (1969, 18, 19), 132 (1969, 18), 152 (1969, 18, 19), 164 (1969, 18, 19), 172 (1969, 18, 19), 185 (1969, 18, 19), 196 (1969, 18, 19), 215 (1969, 18, 19), 222 (1969, 18, 19), 223 (1969, 18, 19), 225 (1969, 18, 19), 234 (1969, 19), *340*
Prilezhaeva, E. N., 51 (1965, 25), 123 (1965, 25), *309*
Prince, R. H., 167 (1969, 356), 190 (1969, 356), 199 (1969, 356), 225 (1969, 356), *349*
Prohaska, C. A., 13 (1967, 26), 128 (1967, 26), 139 (1967, 26), 148 (1967, 26), 159 (1967, 26), 168 (1967, 26), 169 (1967, 26), 177 (1967, 26), 184 (1967, 26), 190 (1967, 26), 200 (1967, 26), 207 (1967, 26), 212 (1967, 26), 213 (1967, 26), 221 (1967, 26), 237 (196, 26), 241 (1967, 26), *320*
Promonenkov, V. K., 154 (1967, 332), 157 (1967, 332), *328*
Proskurnina, M. V., 111 (1968, 297), 126 (1968, 297), 134 (1968, 297), 137 (1968, 297), 145 (1968, 297), 146 (1968, 297), 147 (1968, 297), 148 (1968, 297), 149 (1968, 297), 155 (1968, 297), 156 (1968, 297), 162 (1968, 297), 165 (1968, 297), 167 (1968, 297), 176 (1968, 297), 183 (1968, 297), 188 (1968, 297), 192 (1968, 297), 208 (1968, 297), 209 (1968, 297), 271 (1968, 297), *338*
Provost, L. R., 27 (1969, 74), 38 (1969, 74), 75 (1969, 74), 93 (1969, 74), 94 (1969, 74), 104 (1969, 74), 109 (1969, 74), 112 (1969, 74), 113 (1969, 74), 114 (1968, 304), 119 (1968, 304), 120 (1969, 74), 121 (1968, 304), 133 (1969, 74), 136 (1969, 74), 138 (1969, 74), 139 (1968, 304), 140 (1969, 74), 148 (1969, 74), 149 (1969, 74), 154 (1969, 74), 155 (1968, 304; 1969, 74), 159 (1968, 304), 164 (1968, 304), 174 (1968, 304; 1969, 74), 193 (1968, 304), 194 (1969, 74), 202 (1969, 74), 213 (1968, 304), 220 (1969, 74), 235 (1969, 74), 239 (1968, 304), *338, 341*
Pudovik, A. N., 2 (1967, 3,4), 7 (1967, 3), 123 (1967, 312), 125 (1966, 234), 127 (1966, 234), 134 (1967, 3), 147 (1966, 234), 155 (1966, 234; 1967, 3), 156 (1967, 3, 4), 175 (1967, 3), 176 (1967, 3), 182 (1967, 4), 192 (1967, 3), 200 (1967, 4), 213 (1967, 4), 214 (1966, 234), 221 (1967, 4), 254 (1967, 3), *317, 319, 328*
Pukanic, G., 84 (1966, 105), 97 (1966, 105), 140 (1966, 105), 194 (1966, 105), 209 (1966, 105), 244 (1966, 105), *314*
Pullen, K. E., 21 (1969, 56), 76 (1969, 56), 103 (1969, 56), 106 (1969, 56), 107 (1969, 56), 110 (1969, 56), 116 (1969, 56), *341*
Purdela, D., 4 (1965, 1; 1968, 4), 97 (1965, 1), 98 (1965, 1), 99 (1965, 1), 100 (1965, 1), 102 (1965, 1), 113 (1965, 1), 229 (1965, 1), *308, 330*
Pustinger, J. V., 195 (1966, 262), 196 (1966, 262), 223 (1966, 262), 227 (1966, 262), *318*
Pyykko, P., 87 (1968, 179; 1969, 210), *335, 345*

Q

Quimby, O. T., 7 (1967, 12), 256 (1967, 12, 380; 1968, 349), 257 (1967, 12, 380; 1968, 349), 258 (1967, 12; 1968, 349), 259 (1968, 349), 260 (1968, 349), 261 (1968, 349), 263 (1968, 349), 266 (1967, 380; 1968, 349), *319, 329, 339*

Quin, L. D., 29 (1967, 75, 78; 1969, 77), 53 (1968, 100), 54 (1967, 78; 1969, 77), 123 (1967, 78), 124 (1967, 78; 1968, 100; 1969, 77), 130 (1968, 100), 131 (1968, 100), 134 (1967, 75; 1968, 100; 1969, 77), 135 (1968, 100), 136 (1967, 75), 144 (1968, 100), 157 (1967, 75), 167 (1969, 356), 176 (1967, 75), 179 (1968, 100), 180 (1968, 100), 188 (1967, 75), 189 (1967, 75; 1968, 100), 190 (1969, 356), 198 (1968, 100), 199 (1969, 256), 205 (1968, 100), 224 (1968, 100), 225 (1969, 356), 237 (1969, 77), 273 (1966, 289), 286 (1966, 289), *319, 321, 333, 342, 349*

R

Rabinowitz, J., 189 (1966, 259), *318*
Rabinovitz, M., 245 (1967, 374), 252 (1967, 374), 253 (1967, 374), *329*
Rader, C. P., 13 (1969, 30), *340*
Radich, L., 92 (1969, 282), 169 (1969, 282), *347*
Raffay, U., 90 (1967, 219), *325*
Raksjys, J. W., 7 (1965, 5; 1968, 14), 128 (1968, 14), 131 (1968, 14), 132 (1968, 14), 150 (1968, 14), 152 (1968, 14), 164 (1965, 5), 174 (1968, 14), 195 (1968, 14), 214 (1968, 14), 223 (1968, 14), 227 (1968, 14), 228 (1968, 14), 242 (1968, 14), *308, 330*
Ramey, K. C., 92 (1968, 261), *337*
Ramirez, F., 22 (1968, 303), 29 (1969, 75), 38 (1967, 86; 1968, 73), 39 (1965, 21; 1967, 86, 163), 78 (1968, 123, 124), 81 (1968, 147), 82 (1968, 123), 83 (1967, 163; 1968, 152, 153), 97 (1966, 206), 98 (1968, 124), 112 (1966, 224), 113 (1968, 302, 303), 114 (1965, 21; 84, 85; 1966, 206, 224, 225; 1967, 86; 1968, 73), 119 (1966, 224; 1969, 337), 120 (1968, 301), 123 (1968, 302; 1969, 337), 125 (1967, 86; 1968, 73), 127 (1968, 301; 1969, 337), 128 (1966, 224; 1968, 153), 132 (1969, 337), 133 (1966, 242), 135 (1968, 309, 312; 1969, 337), 138 (1967, 86; 1968, 73), 139 (1968, 302, 303), 140 (1966, 225; 1968, 303), 141 (1966, 224; 1968, 153, 303), 142 (1965, 84; 1966, 76, 206; 1968, 124, 153, 310), 144 (1966, 242; 1968, 302, 312, 313), 145 (1968, 301, 312), 148 (1965, 21; 1966, 225, 248; 1967, 86; 1968, 73, 301), 150 (1968, 315), 303), 154 (1968, 318), 155 (1968, 312), 156 (1968, 312, 313), 157 (1968, 301), 159 (1966, 224; 1968, 153, 303, 310), 161 (1968, 302, 315), 162 (1966, 225; 1967, 335; 1968, 124, 302, 310), 164 (1966, 242; 1968, 124, 301, 318; 1969, 353), 165 (1966, 242), 166 (1968, 313; 1969, 353), 169 (1968, 318), 172 (1968, 310), 173 (1966, 206, 225, 248; 1968, 302, 303), 174 (1966, 206; 1969, 353), 175 (1969, 353), 176 (1968, 312), 177 (1966, 225; 1968, 312), 179 (1966, 224; 1968, 302), 180 (1968, 301; 1969, 337), 181 (1965, 21), 183 (1966, 248, 258), 184 (1965, 21), 188 (1966, 225; 1967, 335; 1968, 301, 302, 303), 189 (1968, 124, 301), 190 (1966, 258; 1968, 301), 193 (1966, 206; 1968, 303), 194 (1965, 85), 196 (1968, 315), 197 (1968, 313, 315), 198 (1968, 301, 313), 199 (1968, 73, 318), 200 (1965, 84; 1966, 248; 1968, 318), 201 (1968, 303), 202 (1968, 302, 303), 203 (1965, 85; 1966, 225; 1969, 353), 204 (1968, 303, 313), 205 (1967, 350, 351; 1968, 301, 313), 206 (1966, 225), 208 (1968, 302), 209 (1969, 353), 210 (1969, 353), 211 (1965, 21; 1966, 248, 258; 1967,

AUTHOR INDEX

350, 351; 1968, 124, 152, 301, 312), 212 (1966, 248), 214 (1965, 85; 1968, 302, 303), 215 (1965, 85; 1967, 163; 1968, 309), 217 (1966, 258), 218 (1968, 301; 1969, 75), 219 (1965, 21, 84; 1966, 225, 242; 1967, 163, 350; 1968, 124, 152, 301, 303, 309, 315; 1969, 75), 220 (1968, 124, 318), 222 (1968, 124), 223 (1965, 85; 1966, 225; 1968, 303), 224 (1965, 21, 84, 1966, 225, 242; 1968, 301, 315), 225 (1966, 225; 1967, 163), 226 (1966, 225; 1968, 303), 228 (1968, 315), 229 (1965, 21; 1966, 242, 258; 1968, 73, 153, 301), 230 (1966, 225; 1968, 124, 302, 303), 231 (1968, 124, 303; 1969, 353), 232 (1966, 225, 258; 1968, 124, 152, 301, 303; 1969, 353), 233 (1965, 84; 1966, 224, 225; 1968, 73, 124, 153, 301), 235 (1968, 73), 236 (1965, 21, 84, 1966, 242, 258; 1968, 309), 238 (1966, 225; 1968, 301, 303), 240 (1965, 84; 1968, 153, 303, 310), 241 (1968, 153), 242 (1966, 225; 1967, 335; 1968, 124, 303), 243 (1965, 21), 245 (1968, 73, 124, 152, 301), 246 (1968, 303), 247 (1966, 225; 1968, 153, 301), 248 (1968, 124, 303), 249 (1966, 225; 1968, 303), 250 (1967, 335), 251 (1968, 124), 252 (1966, 225; 1967, 335; 1968, 303, 318), 253 (1965, 21; 1966, 225; 1967, 335; 1968, 301, 309), 262 (1968, 309), 268 (1968, 309), 269 (1967, 350, 351; 1968, 309, 315), 270 (1967, 350), 271 (1967, 335), 272 (1965, 85), 274 (1968, 309), 283 (1967, 350), *308, 309, 310, 313, 316, 317, 318, 321, 323, 328, 329, 332, 333, 334, 338, 341, 349*

Rammler, D. H., 94 (1966, 199), *316*
Randall, E. W., 83 (1965, 39; 1967, 169), 89 (1965, 63, 64; 1968, 217), 91 (1965, 63, 64; 1966, 163; 1968, 217), 92 (1965, 63, 64; 1966, 163; 1969, 289), 93 (1965, 64; 1968, 217), 139 (1965, 39), 141 (1967, 169), 149 (1967, 169), 160 (1965, 39, 1967, 169), *309, 310, 315, 324, 336, 347*
Ranganathan, R., 44 (1968, 88), 48 (1968, 88), 230 (1968, 88), 231 (1968, 88), 238 (1968, 88), 248 (1968, 88), 251 (1968, 88), *332*
Rao, G. N., 90 (1966, 171), 92 (1966, 171), *316*
Rao, U. R. K., 88 (1967, 200), *324*
Rao, V. U. S., 88 (1967, 200; 1968, 189), *324, 335*
Rapko, J. N., 79 (1968, 135), 279 (1968, 135), *333*
Rast, H., 80 (1969, 158), *344*
Ratts, K. W., 242 (1966, 280), 253 (1966, 280), 255 (1966, 280), *318*
Rausch, M. D., 92 (1969, 281), *347*
Rave, J. W., 207 (1966, 268), 214 (1966, 268), 221 (1966, 268), 273 (1966, 268), *318*
Rave, T. W., 218 (1967, 357), 253 (1967, 357), *329*
Rawlinson, D. J., 178 (1966, 257), 216 (1966, 257), *318*
Ray, S. K., 67 (1965, 24), 79 (1965, 24), 277 (1965, 24), 278 (1965, 24), 279 (1965, 24), *309*
Raynes, W. T., 24 (1968, 51), 83 (1968, 51), 99 (1968, 51), *331*
Razumov, A. I., 121 (1966, 254), 165 (1966, 254), 166 (1966, 254), *318*
Razumova, N. A., 53 (1967, 110), 75 (1967, 110), 107 (1966, 220), 111 (1966, 219, 220), 116 (1967, 110), 117 (1966, 219; 1967, 110, 303, 313), 123 (1967, 110, 311), 124 (1964, 15; 1967, 110, 303, 311, 313), 125 (1967, 313), 134 (1967, 110, 311), 135 (1964, 15; 1967, 311), 137 (1967, 311), 138 (1966, 220; 1967, 313), 145 (1964, 15; 1967, 311), 146 (1966, 220; 1967, 313), 148 (1967, 311), 155 (1964, 15), 156 (1966, 220; 1967, 311), 158 (1967, 311), 167 (1966, 220; 1967, 311, 313), 168 (1967, 313), 176 (1966, 220; 1967, 311), 188 (1966, 220), 190 (1966, 220), 197 (1966, 220), 204 (1966, 220), *308, 317, 322, 327, 328*

Razvodovskaya, L. V., 113 (1965, 82), 165 (1965, 82), 197 (1965, 82), *310*
Read, A. P., 94 (1965, 75), *310*
Reddy, G. S., 13 (1965, 8), 21 (1965, 8), 31 (1965, 8), 32 (1965, 8), 47 (1965, 8), 76 (1965, 8), 80 (1965, 8), 91 1967, 235), 92 (1967, 235), 98 (1967, 235), 99 (1967, 235), 104 (1967, 235), 112 (1967, 235), 114 (1966, 228), 118 (1965, 8), 120 (1967, 235), 121 (1965, 8), 127 (1965, 8), 130 (1967, 235), 131 (1967, 235), 132 (1967, 235), 138 (1965, 8), 158 (1965, 8), 172 (1965, 8), 230 (1967, 364), 261 (1966, 228), 268 (1966, 228), 273 (1967, 364), 275 (1967, 364), *308, 317, 325, 329*
Rees, R., 251 (1965, 116), 253 (1965, 116), *311*
Reese, C. B., 94 (1968, 273), *337*
Reesor, J. B., 80 (1969, 160), 148 (1969, 160), 173 (1969, 160), 198 (1969, 160), 199 (1969, 160), *344*
Rehak, W., 84 (1966, 107), *314*
Reich, P., 170 (1967, 337), 251 (1967, 337), 256 (1967, 337), *328*
Reichard, P., 94 (1967, 265), *326*
Reid, J. B., 87 (1967, 198), 109 (1967, 198), 114 (1967, 198), 119 (1967, 198), 120 (1967, 198), 131 (1967, 198), 140 (1967, 198), 142 (1967, 198), 186 (1967, 198), 223 (1967, 198), 224 (1967, 198), 227 (1967, 198), 252 (198), *324*
Reider, D. P., 283 (1969, 378), *350*
Reilly, C. A., 53 (1966, 67), 92 (1967, 259; 1968, 259), 93 (1968, 259), 196 (1968, 259), 201 (1966, 67), 202 (1966, 67), *313, 326, 337*
Reinheimer, H., 90 (1967, 219), *325*
Renkonen, O., 94 (1968, 278), *337*
Rest, A. J., 90 (1969, 233), 92 (1966, 181, 183; 1968, 245, 255; 1969, 233), *316, 336, 337, 346*
Retcofsky, H. L., 9 (1965, 7), 114 (1965, 7), 131 (1965, 7; 1966, 240), 154 (1965, 7; 1966, 240), 174 (1965, 7; 1966, 240), 186 (1965, 7; 1966, 240), 203 (1965, 7; 1966, 240), 223 (1965, 7; 1966, 240), *308, 317*
Revel, M., 37 (1969, 343), 139 (1969, 343), 151 (1969, 343), 159 (1969, 343), 163 (1969, 343), 174 (1969, 343), 182 (1969, 343), 190 (1969, 343), 191 (1969, 359), 197 (1969, 359), 199 (1969, 359), 202 (1969, 359), 203 (1969, 359), 204 (1969, 359), 206 (1969, 343), 210 (1969, 359), 216 (1969, 359), 232 (1969, 359), 236 (1969, 359), *349*
Rey-Coquais, B., 86 (1966, 118), *314*
Rhodes, M., 73 (1968, 115), 87 (1968, 115, 174), 97 (1968, 115), 98 (1968, 174), 99 (1968, 174), *333, 335*
Rhum, D., 114 (1965, 85), 194 (1965, 85), 203 (1965, 85), 214 (1965, 85), 215 (1965, 85), 223 (1965, 85), 272 (1965, 85), *310*
Rice, R. G., 79 (1966, 96), 281 (1966, 96), 285 (1966, 96), 286 (1966, 96), 287 (1966, 96), *314*
Richards, E. M., 219 (1967, 358), 238 (1969, 369), 246 (1969, 369), *328, 350*
Richards, K. H. B., 101 (1969, 329), *349*
Richards, R. E., 11 (1964, 2), 63 (1969, 130), 66 (1969, 130), 87 (1966, 141; 1967, 198; 1969, 130, 213), 92 (1966, 185), 107 (1962, 7), 109 (1967, 198), 114 (1966, 141; 1967, 198), 119 (1967, 198), 120 (1967 198), 131 (1967, 198), 140 (1966, 141; 1967, 198), 142 (1964, 2; 1967, 198), 186 (1966, 141; 1967, 198), 223 (1966, 141; 1967, 198), 224 (1966, 141; 1967, 198), 227 (1967, 198), 252 (1967, 198), *307, 308, 315, 316, 324, 343, 345*
Riddle, C., 18 (1968, 32), 48 (1968, 32), 84 (1968, 158), 97 (1968, 158), 99 (1968, 32, 158), 100 (1968, 158), *331, 334*
Ridenour, R. E., 73 (1969, 140), 86 (1969, 140), 129 (1969, 140), 245 (1969, 140), 274 (1969, 140), 288 (1969, 140), *343*
Rieschel, R., 178 (1965, 101), 194 (1965, 101), *311*

Riess, J., 80 (1969, 162), 114 (1969, 162), 128 (1969, 162), 153 (1969, 162), 154 (1969, 162), 164 (1969, 162), 170 (1969, 162), 191 (1969, 162), 206 (1969, 162), 217 (1969, 162), 221 (1969, 162), *334, 344*

Riess, J. G., 4 (1967, 7), 83 (1965, 38; 1967, 168), 84 (1967, 174; 1968, 160; 1969, 196), 92 (1965, 67; 1966, 176, 177, 178; 1967, 174, 240), 94 (1965, 38; 1967, 168), 100 (1966, 208), 101 (1966, 208), 103 (1967, 7), 104 (1967, 7), 107 (1967, 7), 108 (1967, 7), 111 (1967, 7), 114 (1966, 208) 115 (1966, 208), 130 (1967, 7), 131 (1967, 7), 132 (1967, 7), 133 (1967, 7), 135 (1967, 7), 149 (1967, 7), 151 (1967, 7), 160 (1967, 7), 169 (1967, 7), 173 (1967, 7), 183 (1967, 168), 192 (1967, 168), 196 (1967, 7), 201 (1967, 7), 207 (1967, 7), 213 (1967, 7), 214 (1967, 7), 218 (1967, 7), 228 (1967, 7), 231 (1967, 7), 233 (1967, 168), 234 (1967, 7), 235 (1967, 7), 237 (1967, 168), 239 (1967, 7), 241 (1967, 168), 250 (1967, 168), 252 (1967, 168), 257 (1967, 7), 258 (1967, 7), 269 (1967, 168), *309, 310, 316, 319, 324, 325, 345*

Rigby, C. W., 248 (1967, 277), 253 (1967, 277), 254 (1967, 277), *329*

Rinze, P. V., 90 (1969, 242), *346*

Robert, J. B., 13 (1968, 28), 17 (1968, 28), 18 (1968, 28, 33), 27 (1967, 71), 28 (1968, 28, 33), 29 (1968, 62; 1969, 76), 30 (1968, 28), 40 (1966, 50; 1967, 88; 1968, 80; 1969, 88, 91), 74 (1968, 28), 82 (1966, 50), 83 (1969, 179), 105 (1968, 33), 107 (1966, 50), 109 (1968, 33), 112 (1966, 50), 118 (1966, 50; 1968, 28), 125 (1968, 80; 1969, 88), 137 (1967, 88; 1968, 80; 1969, 88), 139 (1968, 80), 143 (1968, 33), 148 (1968, 80), 151 (1966, 50), 153 (1968, 28, 33, 80), 154 (1968, 33), 158 (1969, 179), 162 (1966, 60), 180 (1967, 88; 1968, 80; 1969, 76), 187 (1969, 76), 188 (1969, 76), 193 (1969, 179), 198 (1969, 76), 276 (1968, 33), *312, 321, 331, 332, 341, 342, 344*

Roberts, I., 88 (1966, 149), *315*

Roberts, J. D., 8 (1968, 22; 1969, 21), 9 (1968, 22), 15 (1968, 22), 68 (1969, 21), 69 (1968, 22; 1969, 21), 70 (1969, 21), 71 (1968, 22; 1969, 21), 72 (1968, 22; 1969, 21), 105 (1968, 22), 109 (1968, 22), 113 (1968, 22), 114 (1968, 22), 119 (1968, 22), 140 (1968, 22), 160 (1968, 22), 192 (1968, 22; 1969, 21), 194 (1968, 22), 213 (1968, 22), 218 (1968, 22; 1969, 21), 223 (1968, 22), 224 (1968, 22), 228 (1968, 22), 242 (1969, 21), 259 (1968, 22), *331, 340*

Robertson, C. G., 91 (1969, 254), *347*

Robinson, B. H., 10 (1968, 24), 80 (1968, 144), 101 (1968, 144), 111 (1968, 24, 144), 117 (1968, 144), 133 (1968, 24), 134 (1968, 144), 150 (1968, 144), 260 (1968, 24), *331, 334*

Robinson, S. D., 92 (1969, 287), *347*

Röchling, H., 191 (1965, 104), *311*

Roesky, H. W., 21 (1967, 53), 22 (1967, 53, 58), 73 (1967, 53), 77 (1967, 136), 79 (1969, 152), 94 (1969, 325), 97 (1967, 58; 1969, 325), 98 (1967, 58, 281, 282; 1969, 326), 99 (1967, 136), 100 (1969, 325), 101 (1967, 287; 1969, 152, 325), 103 (1967, 53, 287, 289), 106 (1967, 53, 287, 289), 107 (1967, 282), 109 (1969, 325), 114 (1967, 282), 115 (1967, 289), 120 (1969, 325), 121 (1969, 325), 131 (1968, 308), 132 (1968, 308), 152 (1968, 308), 153 (1968, 308), 173 (1968, 308), 186 (1968, 308), 193 (1967, 282), 194 (1967, 281), 258 (1968, 350), 269 (1967, 281), 270 (1967, 58), 277 (1969, 152), 278 (1968, 355), *320, 321, 323, 327, 339, 344, 349*

Rogowski, R., 22 (1968, 49), 97 (1968, 49), 98 (1968, 49), *331*

Ronayne, J., 243 (1967, 372), 246 (1967, 372), 276 (1967, 372), *329*

Rosario, M. D., 45 (1969, 103), 76 (1969, 103), 108 (1969, 103), 109 (1969, 103), 113 (1969, 103), 122 (1969, 103), 142 (1969, 103), *342*

Rosenberg, D., 8 (1967, 20; 1968, 21), 134 (1967, 20), 139 (1967, 20), 140 (1967, 20), 175 (1967, 20; 1968, 21), 177 (1967, 20), 178 (1967, 20), *320, 331*

Rosevear, D. T., 92 (1966, 181), *316*

Ross, E. J. F., 86 (1966, 125), 110 (1966, 125), *314*

Ross, E. P., 91 (1969, 270), 173 (1969, 270), *347*

Ross, J. A., 32 (1969, 83), 33 (1969, 83), 83 (1969, 83), *342*

Ross, V., 94 (1968, 274), 158 (1968, 274), *337*

Ross, V. F., 88 (1961, 3), 97 (1961, 3), *307*

Rossknecht, H., 7 (1969, 15), 19 (1969, 15), 67 (1969, 15), 77 (1969, 15), 175 (1969, 15), 182 (1969, 15), 268 (1969, 15), *340*

Rottman, F., 89 (1969, 313), *348*

Roundhill, D. M., 92 (1968, 246; 1969, 295, 296), *336, 348*

Rousseau, A., 127 (1966, 236), *317*

Roussel, J., 83 (1969, 178), 136 (1969, 340), 139 (1969, 178), 161 (1969, 340), 163 (1969, 178), 172 (1969, 178), 173 (1969, 178), 176 (1969, 340), 178 (1969, 340), 181 (1969, 178), 184 (1969, 340), 190 (1969, 178, 340), 193 (1969, 340), 194 (1969, 340), 199 (1969, 178), 202 (1969, 178), 206 (1969, 340), 209 (1969, 178), 220 (1969, 340), 227

Roussel, J., (1969, 340), *344, 349*

Rowatt, R. J., 60 (1969, 128), 121 (1969, 128), 129 (1969, 128), 141 (1969, 128), 260 (1969, 128), 278 (1969, 128), *343*

Rowsell, D. G., 7 (1967, 10), 18 (1965, 11b; 1967, 10), 20 (1965, 11b; 1967, 10), 28 (1968, 58, 59), 29 (1968, 58, 59), 35 (1968, 71), 44 (1969, 107), 51 (1968, 98), 63 (1965, 11b), 76 (1968, 98 80), (1968, 58, 59), 82 (1969, 169), 86 (1966, 124), 91 (1968, 98), 105 (1966, 211; 1968, 58), 107 (1965, 80; 1969, 169), 108 (1966, 124), 110 (1967, 295), 111 (1965, 11b), 112 (1969, 169), 113 (1968, 59; 1969, 107), 115 (1966, 124), 116 (1967, 295), 117 (1968, 71; 1969, 107), 118 (1969, 169), 119 (1965, 80), 122 (1966, 124), 123 (1967, 295; 1968, 98), 127 (1968, 59), 128 (1969, 107), 139 (1965, 80), 151 (1969, 169), 154 (1966, 124), 158 (1966, 124), 161 (1967, 295), 173 (1965, 11b), 207 (1967, 295; 1968, 98), 208 (1967, 295), 259 (1965, 11b), 262 (1965, 11b), 266 (1965, 11b), 267 (1965, 11b), 268 (1965, 11b), 270 (1965, 11b), 286 (1965, 11b), 288 (1965, 11b), *308, 310, 314, 316, 319, 227, 331, 332, 342, 344*

Rowsell, D. R., 29 (1967, 77), 112 (1967, 77), 113 (1967, 77), 127 (1967, 77), 128 (1967, 77), *321*

Roy, C. H., 256 (1968, 349), 257 (1968, 349), 258 (1968, 349), 259 (1968, 349), 260 (1968, 349), 261 (1968, 349), 263 (1968, 349), 266 (1968, 349), *339*

Roy, N. K., 110 (1968, 296), 122 (1968, 296), 129 (1968, 296), 154 (1968, 296), 161 (1968, 296), 170 (1968, 296), 182 (1968, 296), 191 (1968, 296), 192 (196), 212 (1968, 296), *338*

Rubin, R. M., 58 (1966, 74), 120 (1966, 74), 125 (1966, 74), 127 (1966, 74), 130 (1966, 74), 131 (1966, 74), 136 (1966, 74), 137 (1966, 74), 146 (1966, 74), 148 (1966, 74), *313*

Rudolph, G., 100 (1965, 78), 101 (1965, 78), 114 (1965, 78), 142 (1965, 78), 223 (1965, 78), 259 (1965, 78), 265 (1965, 78), *310*

Rudolph, K., 21 (1967, 54), 104 (1966, 210), 108 (1967, 54), 109 (1967, 54), 259 (1967, 54), *316, 320*

Rudolph, K. H., 113 (1968, 356), *339*

Rudolph, R. W., 14 (1969, 31), 15 (1969, 31), 18 (1966, 20; 1967, 32; 1968, 31), 20 (1969, 31), 21 (1966, 28; 1967, 32, 49; 1968, 31; 1969, 31), 24 (1969, 31), 50 (1966, 59), 60 (1967, 49; 1968, 31; 1969, 31), 61 (1967, 49; 1968, 31; 1969, 31), 62, 63 (1968, 31; 1969, 31), 64 (1969, 31), 65 (1969, 31, 136), 66 (1966, 28; 1969, 31), 67 (1969, 31), 74 (1967, 32), 76 (1969, 31), 80, 84 (1966, 20; 1967, 32), 98 (1966, 59), 99 (1966, 28), 100 (1966, 28; 1967, 1968, 31; 1969, 31), 101 (1969, 31, 136), 103 (1966, 28), 256 (1969, 31), 257 (1969, 136), *312, 313, 320, 331, 340, 343*

Ruff, J. K., 98 (1967, 279), 100 (1966, 209), *316, 327*

Rümpler, K. D., 86 (1965, 50), 115 (1965, 50), *309*

Rusek, P. E., 82 (1969, 172), 83 (1969, 172), 133 (1969, 172), 157 (1969, 172), *344*

Rush, J. J., 87 (1967, 197; 1968, 180), *324, 335*

Rüterjans, H., 94 (1969, 318), *348*

Ryschkewitsch, G. E., 92 (1969, 305), 189 (1968, 330), 204 (1968, 330), *339, 348*

S

Saalfrank, R., 86 (1969, 206), 241 (1969, 206), 243 (1969, 206), *345*

Sabherwal, I. H., 90 (1969, 243), 92 (1969, 286), 172 (1969, 243), *346, 347*

Saegusa, T., 183 (1968, 329), *339*

Saeki, H., 94 (1967, 271), *326*

Safin, I. A., 88 (1968, 190), *335*

Saito, T., 90 (1967, 223), *325*

Sakakibara, M., 92 (1969, 293), *348*

Sakai, S., 92 (1969, 293), *348*

Salisbury, N. J., 89 (1966, 197), 94 (1968, 279), *316, 337*

Salvadori, G., 86 (1966, 126), 114 (1966, 126), 121 (1966, 126), *314*

Samitov, Yu. Yu., 2 (1967, 3), 7 (1967, 3), 20 (1969, 53), 53 (1964, 8; 1967, 109), 83 (1967, 161, 162; 1969, 53), 93 (1969, 53), 109 (1964, 14), 116 (1964, 8; 1967, 109), 119 (1964, 14), 123 (1964, 8; 1967, 312) 124 (1964, 8; 1967, 109), 125 (1966, 234), 126 (1964, 14), 127 (1966, 234), 133 (1964, 16), 134 (1964, 8, 16; 1967, 3), 135 (1964, 8, 16), 136 (1964, 14), 144 (1964, 8, 16), 145 (1964, 16; 1967, 325), 146 (1967, 161, 162), 147 (1964, 14; 1966, 234), 155 (1966, 234; 1967, 3, 325), 156 (1967, 3), 158 (1967, 161), 168 (1967, 161, 162), 175 (1967, 3), 176 (1967, 3), 192 (1967, 3), 214 (1966, 234), (244 1967, 373), 251 (1967, 109), 254 (1967, 3), *308, 317, 319, 322, 323, 328, 329, 341*

Sams, J. R., 91 (1969, 258), *347*

Sanchez, M., 19 (1969, 45), 83 (1967, 165; 1969, 178), 86 (1967, 165), 107 (1969, 45), 119 (1969, 45), 120 (1969, 45), 121 (1969, 45), 127 (1969, 45), 128 (1967, 165), 130 (1969, 45), 136 (1967, 165), 139 (1967, 321; 1969, 45, 178), 143 (1969, 45), 152 (1969, 45), 153 (1967, 165; 1969, 45), 162 (1969, 45), 163 (1969, 45, 178), 172 (1969, 45, 178), 173 (1969, 178), 175 (1969, 45), 181 (1969, 45, 178), 187 (1967, 165), 190 (1969, 178), 199 (1969, 178), 202 (1969, 45, 178), 209 (1969, 45, 178), 216 (1969, 45), 233 (1969, 45), *324, 328, 341, 344*

Santry, D. P., 14 (1964, 4; 1965, 9), *308*

Sarma, R. H., 94 (1968, 274), 158 (1968, 274), *337*

Satge, J., 136 (1969, 341), 184 (1967, 342), 193 (1969, 341), 207 (1968, 331), 217 (1967, 342), 220 (1968, 331), 225 (1969, 341), 226 (1967, 342, 360), 230 (1968, 331), 243 (1969, 341), *328, 329, 339, 349*

Savage, M. P., 204 (1966, 267), 239 (1966, 267), 240 (1966, 267), 248 (1966, 267), *318*

Savariault, J. M., 92 (1969, 284), *347*

Savignac, P., 159 (1969, 352), 202 (1969, 352), *349*

Savitsky, G. B., 84 (1966, 105), 97 (1966, 105), 140 (1966, 105), 194 (1966, 105), 209 (1966, 105), 244 (1966, 105), *314*

Savoskina, G. P., 84 (1968, 156; 1969, 182, 183), 184 (1968, 156), *334, 344*

Saxby, J. D., 56 (1968, 104), 90 (1967, 226), 91 (1967, 226), 93 (1967, 226), 230 (1968, 104), 234 (1968, 104), *325, 333*

Sayigh, A. A. R., 7 (1965, 6), 111 (1965, 6), 125 (1967, 314), 127 (1967, 314), 139 (1967, 314), 141 (1967, 314), 149 (1967, 314), 167 (1965, 6), 173 (1967, 314), 180 (1965, 6), 181 (1967, 314), *308, 328*

Scanu, A., 94 (1969, 320), *348*

Schaefer, J. P., 245 (1965, 114), *311*

Schaeffer, R., 84 (1966, 109), 86 (1966, 109), 97 (1966, 109), *314*

Scharf, B., 78 (1966, 83), 100 (1967, 285), 257 (1966, 83; 1967, 285), 259 (1966, 83), 260 (1966, 83), 261 (1966, 83), 267 (1966, 83), 284 (1966, 83), *313, 327*

Schellenbeck, P., 27 (1967, 68), 31 (1967, 68), 194 (1967, 68), 201 (1967, 68), 252 (1966, 285), 255 (1966, 285), *318, 321*

Schenk, H., 235 (1967, 367), 236 (1967, 367), 241 (1967, 367), *329*

Schenk, W., 6 (1969, 13), 222 (1969, 13), 227 (1969, 13), 228 (1969, 13), *340*

Scherer, O. J., 31 (1968, 64), 47 (1968, 64; 1969, 112), 119 (1967, 305), 120 (1969, 112), 128 (1969, 112), 129 (1967, 305), 139 (1967, 305), 141 (1967, 305), 159 (1969, 112), 160 (1968, 64), 170 (1967, 305), 184 (1968, 64), 207 (1968, 64), 262 (1969, 112), 266 (1967, 305), *327, 332, 342*

Schiebel, H. M., 83 (1969, 177), 118 (1969, 177), 124 (1969, 177), 134 (1969, 177), 165 (1969, 177), *344*

Schieder, G., 31 (1968, 64), 47 (1968, 64), 160 (1968, 64), 184 (1968, 64), 207 (1968, 64), *332*

Schiemenz, G. P., 7 (1968, 15, 16), 80 (1969, 158), 143 (1968, 15), 164 (1968, 15), 209 (1968, 15), 216 (1968, 15), 227 (1965, 110), 228 (1968, 15, 16), 231 (1968, 15, 16), 232 (1968, 15, 16), 235 (1968, 15, 16), 236 (1968, 15, 16), 239 (1968, 15, 16), 241 (1968, 15), 247 (1968, 15), *311, 330, 344*

Schiller, H. W., 18 (1968, 31), 21 (1968, 31), 60 (1968, 31), 61 (1968, 31), 63 (1968, 31), 100 (1968, 31), *331*

Schindlbauer, H., 7 (1967, 17; 1969, 18, 19), 131 (1967, 17; 1969, 18, 19), 132 (1967, 17; 1969, 18), 152 (1967, 17; 1969, 18, 19), 164 (1969, 18, 19), 172 (1967, 17; 1969, 18, 19), 185 (1969, 18, 19), 196 (1969, 18, 19), 215 (1969, 18, 19), 222 (1967, 17; 1969, 18, 19), 223 (1967, 17; 1969, 18, 19), 225 (1969, 18, 19), 234 (1969, 19), *319, 340*

Schindler, F., 85 (1967, 175), 114 (1965, 83; 1967, 175), 140 (1965, 83), 142 (1965, 83), 161 (1965, 83), 170 (1965, 83), 195 (1965, 83), *310, 324*

Schindler, N., 46 (1969, 106), 67 (1969, 106), 79 (1967, 150; 1968, 137), 193 (1968, 137), 200 (1968, 137), 217 (1968, 137), 220 (1968, 137), 226 (1968, 137), 229 (1968, 137), 233 (1968, 137), 250 (1968, 137), 260 (1969, 106), 261 (1969, 106), 264 (1969, 106), 266 (1969, 106), 267 (1969, 106), 268 (1969, 106), 269 (1969, 106), *323, 334, 342*

Schipperheyn, A., 86 (1966, 137), 212 (1966, 137), *315*

Schlosser, M., 152 (1967, 359), 223 (1967, 359), 235 (1967, 359), 242 (1967, 359), 244 (1966, 281), 247 (1967, 359), 249 (1966, 281), 251 (1966, 281), 252 (1966, 281), *318, 329*

Schmid, G., 91 (1966, 167), *315*

Schmid, K. H., 274 (1967, 390), *330*

Schmidbaur, H., 28 (1967, 73), 59 (1968, 105, 106), 85 (1967, 175, 176), 90 (1969, 232), 114 (1965, 83; 1967, 175), 120 (1968, 106), 121

(1967, 306; 1968, 105, 305), 127 (1968, 106), 128 (1968, 105, 106, 305), 129 (1967, 316), 139 (1968, 106), 140 (1965, 83), 141 (1967, 176), 142 (1965, 83; 1967, 73, 176, 322), 149 (1967, 328; 1968, 105, 305), 150 (1968, 105), 156 (1967, 73), 159 (1968, 105, 305), 161 (1965, 83; 1968, 105, 322), 170 (1965, 83; 1967, 73, 322), 171 (1967, 322), 179 (1967, 176, 316; 1968, 105), 195 (1968, 83; 1967, 322), 214 (1967, 322), 236 (1967, 73, 322), 239 (1967, 328), 243 (1967, 322), 245 (1967, 328), 254 (1967, 379), 255 (1967, 379), 261 1967, 176), 263 (1967, 316), 264 (1967, 176), 265 (1967, 316), 275 (1965, 119), *310, 311, 321, 324, 327, 328, 329, 333, 338, 346*

Schmidpeter, A., 7 (1968, 10, 11; 1969, 15), 19 (1968, 40; 1969, 15), 38 (1969, 87), 39 (1969, 87), 43 (1967, 92; 1968, 86; 1969, 100), 45 (1969, 87), 46 (1969, 87, 106), 48 (1967, 97; 1968, 94), 59 (1968, 111) 67 (1969, 15, 106, 137), 77 (1969, 15) 78 (1969, 149), 79 (1967, 150; 1968, 137), 87 (1968, 8), 90 (1967, 221), 91 (1967, 221), 92 (1967, 221), 97 (1967, 276), 98 (1967, 276), 100 (1968, 86; 1969, 137), 105 (1968, 86), 109 (1967, 9), 121 (1966, 232), 122 (1969, 87), 128 (1967, 9), 132 (1967, 276; 1969, 87), 141 (1967, 276), 150 (1969, 87), 154 (1969, 87), 164 (1969, 87), 175 (1969, 15, 87), 182 (1969, 15, 87), 185 (1968, 8), 186 (1967, 276; 1968, 8), 187 (1967, 9, 344; 1968, 8), 193 (1968, 137), 197 (1967, 9; 1968, 8; 1969, 87), 200 (1968, 137), 203 (1967, 9, 92; 1968, 8; 1969, 87), 204 (1967, 92, 344; 1968, 8; 1969, 87), 209 (1967, 221, 353, 355), 210 (1967, 221, 344; 1968, 8; 1969, 87), 216 (1967, 353), 217 (1968, 137), 220 (1967, 221, 353; 1968, 137), 223 (1967, 92, 276), 226 (1968, 137), 228 (1967, 92; 1969, 87), 229 (1968, 137), 233 (1968, 137), 250 (1968, 137), 258 (1969, 149), 259 (1967, 355; 1968, 94, 111), 260 (1968, 11, 94, 111; 1969, 106), 261 (1968, 94, 111; 1969, 106), 262 (1969, 149), 264 (1969, 106), 265 (1968, 94; 1969, 149), 266 (1969, 106), 267 (1968, 94; 111; 1969, 106), 268 (1968, 111; 1969, 15, 106), 269 (1969, 106), 270 (1966, 232; 1967, 9, 92, 344; 1968, 111), 271 (1966, 232; 1967, 92, 97, 344; 1968, 94, 111), 272 (1966, 232; 1967, 9, 344; 1968, 111), 273 (1967, 9; 1968, 94), 274 (1967, 97; 1968, 94), 280 (1969, 100), 281 (1967, 92), 282 (1968, 40; 1969, 100), 283 (1969, 100), 287 (1968, 11), *317, 319, 321, 322, 323, 325, 326, 328, 329, 330, 331, 332, 333, 334, 340, 342, 343, 344*

Schmidt, E., 91 (1969, 255), *347*
Schmidt, M., 254 (1965, 117), 255 (1965, 117), *311*

Schmutzler, R., 13 (1965, 8), 21 (1963, 3; 1965, 8, 12), 22 (1965, 12; 1967, 60; 1969, 63), 25 (1965, 12, 14; 1966, 36), 26 (1966, 36), 27 (1966, 36), 28 (1967, 60), 31 (1965, 8), 32 (1965, 8), 47 (1965, 8), 75 (1966, 36; 1967, 60), 76 (1965, 8), 80 (1965, 8), 81, 84 (1966, 36), 86 (1966, 36), 91 (1967, 235), 92 (1967, 235), 98 (1967, 235), 99 (1967, 235; 1969, 63), 103 (1966, 36), 104 (1965, 12; 1966, 36; 1967, 235; 1968, 291), 105 (1966, 36), 106 (1968, 289), 107 (1963, 3), 108 (1966, 36; 1968, 291), 109 (1966, 36), 111 (1963, 3), 112 (1963, 3; 1967, 235), 113 (1965, 12; 1966, 36; 1967, 60), 114 (1965, 12; 1966, 36), 115 (1966, 36, 228), 118 (1965, 8; 1966, 36; 1969, 63), 120 (1967, 235), 121 (1965, 8), 122 (1965, 12, 14; 1966, 36), 127 (1965, 8; 1966, 36), 129 (1965, 12), 130 (1967, 235), 131 (1967, 235), 132 (1963, 3; 1965, 12; 1966, 36; 1967, 235; 1968, 291), 139 (1965, 8), 142

(1965, 12; 1966, 36), 143 (1966, 36), 148 (1966, 36), 152 (1968, 316), 153 (1965, 12; 1966, 36), 154 (1966, 36), 158 (1965, 8), 172 (1965, 8), 173 (1965, 12), 174 (1965, 12; 1966, 36), 175 (1966, 36), 186 (1968, 316), 196 (1968, 316), 202 (1968, 316), 203 (1966, 36), 223 (1968, 316), 257 (1963, 3), 258 (1966, 36), 260 (1963, 3; 1966, 36), 261 (1966, 228), 262 (1968, 352), 267 (1966, 36; 1968, 352), 268 (1966, 228), 271 (1969, 375), 277 (1966, 36), 286 (1968, 352), *307, 308, 312, 317, 321, 325, 337, 338, 339, 341, 350*

Schneider, R. J. J., 91 (1968, 212), 93 (1968, 212), *336*

Schneider, W. G., 8 (1961, 1), *307*

Schoeller, U., 54 (1968, 102), 78 (1968, 102), 226 (1968, 102), 240 (1968, 102), 276 (1968, 102), *333*

Schoffstall, A. M., 105 (1966, 213), 155 (1966, 213), 157 (1966, 213), 166 (1966, 213), 183 (1966, 213), 204 (1966, 213), 237 (1966, 213), *317*

Schramm, G., 86 (1967, 188), 88 (1967, 188), 187 (1967, 188), *324*

Schulz, C. O., 55 (1967, 116), 130 (1967, 116), 151 (1967, 116), 152 (1967, 116), 162 (1967, 116), 171 (1967, 116), 180 (1967, 116), 181 (1967, 116), *322*

Schulz, G. W., 88 (1968, 193), 97 (1968, 193), *335*

Schumann, C., 18 (1966, 22), 27 (1966, 22), 31 (1966, 22), 100 (1966, 22), 105 (1966, 22), 110 (1966, 22), 115 (1966, 22), 121 (1966, 22), 141 (1966, 22), 160 (1966, 22), *312*

Schumann, H., 7, 31 (1969, 78), 48 (1968, 95), 90 (1969, 14, 78), 91 (1969, 14, 78), 92 (1967, 241; 1969, 78), 93 (1969, 14, 78), 170 (1967, 337; 1969, 14), 171 (1969, 14), 193 (1969, 78), 195 (1969, 14), 210 (1969, 14), 251 (1967, 337; 1969, 14), 254 (1965, 117; 1969, 14), 255 (1965, 117), 256 (1967, 337; 1969, 14), 273 (1969, 14), 281 (1968, 95), 282 (1969, 14, 377), 283 (1969, 14), 286 (1969, 14), *311, 326, 328, 332, 340, 342, 350*

Schumann, I., 49 (1966, 57), 170 (1966, 57), *313*

Schunn, R. A., 50 (1963, 4), *307*

Schuster-Woldan, H. G., 90 (1966, 155), 92 (1966, 155), *315*

Schwarzhans, K. E., 92 (1965, 68; 1966, 180; 1967, 247), *310, 316, 326*

Schweizer, E. E., 86 (1968, 171), 121 (1966, 277), 132 (1966, 277), 208 (1968, 333), 234 (1968, 333), 238 (1968, 171), 241 (1966, 277; 1968, 171), 249 (1968, 171), *318, 334, 339*

Schweizer, H. P., 94 (1968, 275), *337*

Schweizer, M. P., 94 (1969, 319), *348*

Schwizer, M., 94 (1968, 276), 223 (1968, 276), *337*

Scott, B. A., 88 (1966, 143; 1967, 205; 1968, 185), *315, 325, 335*

Scott, J. M. W., 238 (1966, 276), 249 (1966, 276), *318*

Sears, C. T., 91 (1968, 220), 135 (1968, 220), *336*

Sears, D. J., 55 (1969, 119), 132 (1969, 119), 151 (1969, 119), 153 (1969, 119), 164 (1969, 119), 165 (1969, 119), 171 (1969, 119), 173 (1969, 119), 189 (1969, 119), 190 (1969, 119), 203 (1969, 119), 223 (1969, 119), 224 (1969, 119), *343*

Seel, F., 21 (1967, 54), 104 (1966, 210), 108 (1967, 54), 109 (1967, 54), 113 (1968, 356), 259 (1967, 54), *316, 320, 339*

Segre, A., 92 (1967, 251), *326*

Seidel, W. C., 89 (1969, 291), 92 (1969, 291, 292), *348*

Selva, A., 246 (1967, 375), 247 (1967, 375), 255 (1967, 375), *329*

Semin, G. K., 7 (1967, 15), *319*

Senkbell, H. O., 48 (1966, 56), 149 (1966, 56), 188 (1966, 56), 208 (1966, 56), *313*

Senyavina, L. B., 56 (1967, 118), 234 (1967, 118), 235 (1967, 118), 244 (1957, 118), 246 (1967, 118), 247 (1967, 118), *322*

Sergeev, G. B., 84 (1969, 184), *345*

Servoz-Gavin, P., 127 (1966, 236), *317*
Sevin, A., 81 (1967, 156; 1969, 165, 166), 207 (1969, 166), 220 (1967, 156), 229 (1969, 165), 232 (1967, 156; 1969, 165), 233 (1967, 156), 236 (1969, 165), 237 (1969, 165), 239 (1969, 165), 240 (1969, 165), 241 (1969, 165, 166), 243 (1969, 165), 245 (1967, 156; 1969, 165), 246 (1969, 165), 252 (1969, 165), *323, 344*
Sexton, M. D., 92 (1969, 283, 299), 243 (1969, 299), *347, 348*
Seyferth, D., 86 (1966, 135), 138 (1966, 135), 162 (1966, 252), 163 (1966, 252), 181 (1966, 252), 219 (1965, 108), 231 (1966, 135), 234 (1966, 135), 235 (1966, 135), 238 (1966, 135), 239 (1965, 108; 1966, 135), 241 (1965, 108), 243 (1965, 108), *311, 315, 318*
Shagidullin, R. R., 123 (1967, 312), 151 (1966, 250), 171 (1966, 250), 180 (1966, 250), 189 (1966, 250), 190 (1966, 250), 198 (1966, 250), *318, 328*
Shamonin, Yu. Ya., 151 (1966, 250), 171 (1966, 250), 180 (1966, 250), 189 (1966, 250), 190 (1966, 250), 198 (1966, 250), *318*
Sharp, D. W. A., 21 (1969, 62), 22 (1969, 62), 25 (1969, 62), 43 (1969, 62), 61 (1969, 62), 66 (1969, 62), 75 (1969, 62), 76 (1969, 62), 77 (1969, 62), 86 (1965, 43), 100 (1969, 62), 101 (1969, 62), 104 (1969, 62), 108 (1969, 62), 109 (1966, 217; 1969, 62), 113 (1969, 62), 114 (1969, 62), 120 (1966, 217), 121 (1966, 217), 122 (1966, 217), 132 (1969, 62), 138 (1966, 217), 152 (1969, 62), 153 (1969, 62), 160 (1966, 217), 186 (1969, 62), 196 (1969, 62), 202 (1969, 62), 223 (1969, 62), 259 (1965, 43), *309, 317, 341*
Shaw, B. L., 91 (1965, 61; 1966, 162; 1967, 230, 231; 1968, 221, 222; 1969, 264, 265), 92 (1966, 170, 179; 1967, 246, 257; 1969, 302), 93 (1966, 193; 1968, 265, 267), 94 (1967, 257), 152 (1967, 231), 241 (1967, 257), *309, 315, 316, 325, 326, 336, 337, 347, 348*
Shaw, D., 83 (1965, 39; 1967, 169), 89 (1965, 63, 64; 1968, 217), 91 (1965, 63, 64; 1966, 163; 1968, 217), 92 (1965, 63, 64; 1966, 163; 1968, 217; 1969, 289), 93 (1965, 64; 1968, 217), 139 (1965, 39), 141 (1967, 169), 149 (1967, 169), 160 (1965, 39; 1967, 169), *309, 310, 315, 324, 336, 347*
Shaw, G., 55 (1966, 69), 223 (1966, 69), *313*
Shaw, M. A., 243 (1967, 372), 246 (1967, 372), 249 (1967, 378), 274 (1967, 378), 275 (1967, 378), 276 (1967, 372), *328*
Shaw, R., 79 (1966, 85), 279 (1966, 85), 280 (1966, 85), *313*
Shaw, R. A., 45 (1965, 23; 1968, 91), 46 (1965, 23; 1968, 91), 67 (1965, 24), 79 (1965, 24, 33; 1966, 84, 86, 87, 88; 1968, 91, 128, 129; 1969, 150), 107 (1966, 86), 108 (1965, 23; 1968, 91), 122 (1965, 23; 1968, 91), 123 (1965, 23; 1968, 91), 133 (1967, 318), 142 (1965, 23; 1968, 91), 155 (1967, 318), 161 (1968, 91), 174 (1965, 23; 1968, 91), 175 (1965, 23; 1967, 318; 1968, 91), 179 (1968, 91), 192 (1967, 318), 203 (1965, 23; 1968, 91), 204 (1968, 91), 206 (1967, 318), 224 (1967, 318), 272 (1967, 388), 277 (1965, 24; 1968, 91), 278 (1965, 24; 1966, 84, 88; 1968, 91, 129), 279 (1965, 24; 1966, 87, 88; 1968, 91, 129; 1969, 150), 280 (1965, 33; 1966, 86, 88; 1969, 150), 281 (1966, 86, 87, 88; 1969, 150), 282 (1966, 86, 87, 88; 1969, 150), 283 (1966, 86; 1969, 150), 287 (1968, 128), *309, 313, 328, 330, 332, 333, 344*
Shcherbakov, V. A., 88 (1969, 225), 93 (1967, 261), 99 (1966, 207), *316, 326, 346*
Shcherbakova, L. L., 93 (1967, 261), *326*

Shchegrov, L. N., 88 (1969, 226), *346*
Sheichenko, V. I., 56 (1967, 118), 234 (1967, 118), 235 (1967, 118), 244 (1967, 118), 246 (1967, 118), 247 (1967, 118), *322*
Sheinker, Yu. N., 56 (1967, 118), 234 (1967, 118), 235 (1967, 118), 244 (1967, 118), 246 (1967, 118), 247 (1967, 118), *322*
Sheldon, R. A., 203 (1966, 266), 209 (1966, 266), 236 (1967, 368), 246 (1966, 266), 274 (1967, 368), *318, 329*
Sheldrick, G. M., 20 (1967, 43, 46, 47), 27 (1969, 73), 28 (1969, 73), 48 (1969, 73), 63 (1969, 130), 66 (1969, 130), 74 (1969, 73), 75 (1969, 73), 83 (1967, 43), 87 (1969, 130), 99 (1967, 43, 46, 47), 100 (1967, 47), 105 (1969, 73, 330), 109 (1969, 330), 110 (1969, 73), *320, 341, 343, 349*
Sheluchenko, V. V., 7 (1967, 14; 1968, 12), 19 (1967, 37, 38), 20 (1968, 12), 21 (1967, 37, 38, 51, 52; 1968, 12), 25 (1967, 64), 97 (1967, 14), 98 (1967, 14), 99 (1967, 14, 51), 103 (1967, 14, 51), 104 (1967, 14, 38, 51, 290; 1968, 12), 105 (1967, 14), 107 (1967, 14, 38, 51, 290; 1968, 12), 108 (1967, 14; 1968, 12), 109 (1967, 64), 112 (1967, 14), 113 (1967, 14; 1968, 12), 114 (1967, 38), 118 (1967, 14), 119 (1967, 64), 120 (1967, 14; 1968, 12), 122 (1967, 14), 125 (1967, 37), 127 (1967, 14; 1968, 12), 128 (1967, 37, 38, 64; 1968, 12), 130 (1967, 51, 52, 290), 131 (1967, 37), 132 (1967, 14, 51, 52, 290), 136 (1967, 14, 64; 1968, 12), 138 (1967, 14, 51), 139 (1967, 64), 143 (1967, 52, 64, 290), 146 (1967, 64), 152 (1967, 14; 1968, 12). 153 (1967, 64), 157 (1967, 64), 158 (1967, 14), 169 (1967, 64; 1968, 12), 172 (1967, 14), 173 (1967, 64), 180 (1967, 64), 181 (1967, 64), 186 (1967, 14), 209 (1967, 52), 213 (1968, 335), 261 (1969, 371), 264 (1969, 371), 266 (1969, 371), 267 (1969, 371), 286 (1969, 371), 269 (1969, 371), *319, 320, 321, 327, 330, 339, 350*
Sheppard, J. C., 90 (1969, 253), *346*
Sheppard, W. A., 7 (1968, 14), 128 (1968, 14), 131 (1968, 14), 132 (1968, 14), 150 (1968, 14), 152 (1968, 14), 174 (1968, 14), 195 (1968, 14), 201 (1967, 16), 214 (1968, 14), 223 (1968, 14), 227 (1968, 14), 228 (1968, 14), 242 (1968, 14), *319, 330*
Sherman, E. O., 89 (1969, 279), 92 (1969, 279), *347*
Shermergorn, I. M., 2 (1967, 2), 279 (1967, 2), 280 (1967, 2), 281 (1967, 2), *319*
Shetty, P. S., 92 (1968, 233), *336*
Shevchuk, I. A., 84 (1969, 185), *345*
Shevchuk, M. I., 56 (1967, 118), 234 (1967, 118), 235 (1967, 118), 244 (1967, 118), 246 (1967, 118), 247 (1967, 118), *322*
Shibata, H., 94 (1967, 271), *326*
Shimozawa, T., 41 (1969, 97), 83 (1969, 97), 158 (1969, 97), *342*
Shipley, G. G., 94 (1966, 200), *316*
Shitov, L. N., 113 (1969, 335), 127 (1969, 335), 138 (1969, 335), 152 (1969, 335), *349*
Shoeb, A., 105 (1968, 292), 110 (1968, 292), 149 (1968, 292), 154 (1968, 292), *338*
Shook, H. E., 29 (1967, 75), 134 (1967, 75), 136 (1967, 75), 157 (1967, 75), 176 (1967, 75), 188 (1967, 75), 189 (1967, 75), *321*
Shoolery, J. N., 5 (1956, 1), 65 (1957, 2), *307*
Shostakovskii, M. F., 51 (1965, 25), 123 (1965, 25), *309*
Shreeve, J. M., 21 (1969, 56), 76 (1969, 56), 103 (1969, 56), 106 (1969, 56), 107 (1969, 56), 110 (1969, 56), 116 (1969, 56), *341*
Shulman, R. G., 89 (1965, 72, 73, 74), *310*
Shupack, S. I., 84 (1966, 110), 90 (1966, 110), 91 (1966, 110), 92 (1966, 110), *314*

Siddall, T. H., 13 (1967, 26; 1968, 29), 44 (1968, 29), 47 (1968, 29), 94 (1969, 324), 128 (1967, 26), 139 (1967, 26; 1968, 29), 148 (1967, 26), 159 (1967, 26), 168 (1967, 26; 1968, 29), 169 (1967, 26), 177 (1967, 26; 1968, 29), 178 (1968, 29), 184 (1967, 26; 1968, 29), 190 (1967, 26; 1968, 29), 200 (1967, 26; 1968, 29), 207 (1967, 26), 212 (1967, 26; 1968, 29), 213 (1967, 26), 217 (1968, 29), 221 (1967, 26; 1968, 29), 226 (1968, 29), 230 (1968, 29), 233 (1968, 29), 237 (1967, 26; 1968, 29), 239 (1969, 324), 241 (1967, 26), 256 (1969, 324), *320, 331, 348*

Sidky, M. M., 198 (1966, 263), 239 (1966, 263), *318*

Siebert, H., 48 (1968, 96), 73 (1968, 96), 99 (1968, 96), 100 (1968, 96), 102 (1968, 96), *332*

Siebert, W., 212 (1969, 364), 217 (1969, 364), *349*

Siegbahn, K., 4 (1967, 6), *319*

Sigel, H., 89 (1966, 198; 1967, 264), 121 (1966, 198), *316, 326*

Silverstein, H. T., 10 (1969, 25), 105 (1969, 25), 110 (1969, 25), *340*

Silverstein, R. M., 167 (1965, 97), *311*

Sim, W., 81 (1968, 145), 86 (1968, 145), 92 (1968, 145), 117 (1968, 145), 192 (1968, 145), 210 (1968, 145), *334*

Simalty, M., 234 (1967, 365), 249 (1967, 365), 253 (1967, 365), 254 (1967, 365), 255 (1967, 365), *329*

Simonnin, M. P., 31 (1967, 80), 47 (1967, 80), 51 (1966, 61; 1967, 104; 1968, 99; 1969, 115), 53 (1967, 106), 81 (1967, 80), 111 (1967, 104; 1968, 99; 1969, 115), 116 (1968, 99; 1969, 115), 122 (1967, 80), 123 (1967, 104), 133 (1967, 106), 136 (1967, 106), 137 (1969, 342), 144 (1966, 247; 1967, 104; 1968, 99; 1969, 115), 145 (1968, 99; 1969, 115), 146 (1967, 104, 106; 1968, 99; 1969, 115), 149 (1969, 342), 153 (1967, 80), 154 (1967, 106), 157 (1967, 104; 1968, 99; 1969, 115, 342), 158 (1967, 106), 159 (1967, 106), 162 (1967, 104; 1969, 115), 166 (1967, 104; 1969, 115), 167 (1969, 115), 173 (1967, 80), 174 (1967, 80), 175 (1966, 247), 179 (1967, 106), 180 (1967, 80, 106; 1968, 99; 1969, 115), 197 (1967, 80, 106), 201 (1966, 247, 265), 208 (1966, 61, 247, 265; 1967, 104; 1968, 99; 1969, 115), 214 (1966, 61, 247; 1969, 115), 215 (1966, 247), 218 (1966, 247), 219 (1966, 61), 227 (1966, 247), 234 (1966, 61), 247 (1966, 61), 255 (1966, 247), *313, 317, 318, 321, 322, 333, 343, 349*

Simonova, T. N., 84 (1969, 185), *345*

Simpson, J., 20 (1967, 41), 84 (1967, 41, 173; 1968, 159), 86 (1967, 180), 97 (1967, 41, 173), *320, 324, 334*

Simpson, P., (1969, 92), 41 (1969, 92), 83 (1969, 92), 118 (1969, 92), 137 (1969, 92), *342*

Singer, R. M., 93 (1969, 308), *348*

Singh, G., 162 (1966, 252), 163 (1966, 252), 181 (1966, 252), 219 (1965, 108), 239 (1965, 108), 241 (1965, 108), 243 (1965, 108), *311, 318*

Singleton, E., 91 (1967, 230), *325*

Sisler, H. H., 27 (1966, 38; 1967, 65, 67; 1968, 54), 28 (1968, 54), 44 (1967, 94), 45 (1967, 65, 94; 1968, 54), 46 (1967, 65), 85 (1968, 164; 1969, 199), 90 (1969, 199), 110 (1967, 65), 113 (1968, 54), 115 (1967, 65; 1968, 54), 122 (1966, 38; 1968, 54), 123 (1966, 38), 129 (1967, 65; 1968, 54), 134 (1967, 65), 139 (1969, 344), 142 (1967, 67), 143 (1967, 65), 150 (1967, 67), 153 (1968, 54), 154 (1968, 54), 165 (1967, 94), 171 (1967, 94), 174 (1968, 54), 175 (1967, 65), 188 (1967, 65), 192 (1967, 94), 193 (1968, 164), 196 (1968, 54), 197 (1967, 65; 1968, 54), 201 (1967, 65), 210 (1968, 54), 224 (1967, 65), 244 (1969, 344), 261 (1969, 344), 263 (1969, 344), 265 (1969,

344), 266 (1969, 344), 267 (1969, 199, 344), 268 (1969, 199), 270 (1969, 344), 271 (1969, 374), 272 (1967, 94; 1969, 344, 374), 274 (1969, 374), 283 (1967, 94), 285 (1969, 344), *312, 322, 331, 334, 345, 349, 350*

Skoblo, A. I., 93 (1967, 261), *326*

Skobnevskaya, G. I., 92 (1969, 282), 169 (1969, 282), *347*

Slatykova, L. I., 159 (1969, 351), *349*

Slawick, A., 93 (1966, 194), *316*

Slinkard, W. E., 44 (1969, 105), 45 (1969, 105), 142 (1969, 105), 150 (1969, 105, 347), 203 (1969, 105), 204 (1969, 105), 266 (1969, 105), 271 (1969, 347), *342, 349*

Slutskin, M. A., 94 (1968, 286), *337*

Smith, B. C., 79 (1965, 33; 1966, 85; 1969, 150), 279 (1966, 85; 1969, 150), 280 (1966, 85; 1969, 150), 281 (1969, 150), 282 (1969, 150), 283 (1969, 150), *309, 313, 344*

Smith, C. P., 22 (1968, 303), 29 (1969, 75), 38 (1967, 86; 1968, 73), 39 (1965, 21; 1967, 86, 163), 58 (1966, 76), 78 (1968, 123), 82 (1968, 123), 83 (1967, 163; 1968, 153), 97 (1966, 206), 112 (1966, 224), 113 (1968, 302, 303), 114 (1965, 21, 84, 85; 1966, 206, 224, 225; 1967, 86; 1968, 73), 119 (1966, 224), 120 (1968, 301), 123 (1968, 302), 125 (1967, 86; 1968, 73), 127 (1968, 301), 128 (1966, 224; 1968, 153), 133 (1966, 242), 135 (1968, 309, 312), 138 (1967, 86; 1968, 73), 139 (1968, 302, 303), 140 (1966, 225; 1968, 303), 141 (1966, 224; 1968, 153, 303), 142 (1965, 84; 1966, 76, 206; 1968, 153, 310), 144 (1966, 242; 1968, 302, 312, 313), 145 (1968, 301, 312), 148 (1965, 21; 1966, 225, 248; 1967, 86; 1968, 73, 301), 150 (1968, 315), 151 (1968, 301, 313), 153 (1968, 303), 154 (1968, 318), 155 (1968, 312), 156 (1968, 312, 313), 157 (1968, 301), 159 (1966, 224; 1968, 153, 303, 310), 161 (1968, 302, 315), 162 (1966, 225; 1967, 335; 1968, 302, 310), 164 (1966, 242; 1968, 301, 318; 1969, 119), 165 (1966, 242), 166 (1968, 313), 169 (1968, 318), 172 (1968, 310), 173 (1966, 206, 225, 248; 1968, 302, 303), 174 (1966, 206), 176 (1968, 312), 177 (1966, 225; 1968, 312), 178 (1966, 224), 179 (1968, 302), 180 (1968, 301), 181 (1965, 21), 183 (1966, 248, 258), 184 (1965, 21), 188 (1966, 225; 1967, 335; 1968, 301, 302, 303), 189 (1968, 301), 190 (1966, 258; 1968, 301), 193 (1966, 206; 1968, 303), 194 (1965, 85), 196 (1968, 315), 197 (1968, 313, 315), 198 (1968, 301, 313), 199 (1968, 73, 318), 200 (1965, 84; 1966, 248; 1968, 318), 201 (1968, 303), 202 (1968, 302, 303), 203 (1965, 85; 1966, 225), 204 (1968, 303, 313), 205 (1967, 350, 351; 1968, 301, 313), 206 (1966, 225), 208 (1968, 302), 211 (1965, 21; 1966, 248, 258; 1967, 350, 351; 1968, 301, 312), 212 (1966, 248), 214 (1965, 85; 1968, 302, 303), 215 (1965, 85; 1967, 163; 1968, 309), 217 (1966, 258; 1968, 301; 1969, 75), 219 (1965, 21, 84; 1966, 225, 242; 1967, 163, 350; 1968, 301, 303, 309, 315; 1969, 75), 220 (1968, 318), 223 (1965, 85; 1966, 225; 1968, 303), 224 (1965, 21, 84; 1966, 225, 242; 1968, 301, 315), 225 (1966, 225; 1967, 163), 226 (1966, 225; 1968, 303), 228 (1968, 315), 229 (1965, 21; 1966, 242, 258; 1968, 73, 153, 301), 230 (1966, 225; 1968, 302, 303), 231 (1968, 303), 232 (1966, 225, 258; 1968, 301, 303), 233 (1965, 84; 1966, 224, 225; 1968, 73, 153, 301), 235 (1968, 73), 236 (1965, 21, 84; 1966, 242, 258; 1968, 309), 238 (1966, 225; 1968, 301, 303), 240 (1965, 84; 1968, 153, 303, 310), 241 (1968, 153), 242 (1966, 225; 1967, 335; 1968, 303), 243 (1965, 21), 245

(1968, 73, 301), 246 (1968, 303), 247 (1966, 225; 1968, 153, 301), 248 (1968, 303), 249 (1966, 225; 1968, 303), 250 (1967, 335), 252 (1966, 225; 1967, 335; 1968, 303, 318), 253 (1965, 21; 1966, 225; 1967, 335; 1968, 301, 309), 262 (1968, 309), 268 (1968, 309), 269 (1967, 350, 351; 1968, 309, 315), 270 (1967, 350), 271 (1967, 335), 272 (1965, 85), 274 (1968, 309), 283 (1967, 350), *309, 310, 313, 316, 317, 318, 321, 323, 328, 329, 332, 333, 334, 338, 341*

Smith, D. J. H., 156 (1969, 350), 212 (1969, 350), *349*

Smith, D. M., 55 (1969, 119), 132, (1969, 119), 151 (1969, 119), 153 (1969, 119), 165 (1969, 119), 171 (1969, 119), 173 (1969, 119), 189 (1969, 119), 190 (1969, 119), 203 (1969, 119), 223 (1969, 119), 224 (1969, 119), *343*

Smith, G. L., 10 (1969, 26), 11 (1969, 26), 74 (1969, 26), 92 (1969, 26), 142 (1969, 26), 236 (1969, 26), 261 (1969, 26), 269 (1969, 26), *340*

Smith, J. D., 29 (1966, 41), 85 (1966, 41; 1969, 197), 90 (1966, 41; 1969, 231), 91 (1967, 234), 113 (1966, 41), 196 (1966, 41), *312, 325, 345, 346*

Smith, J. G., 90 (1967, 227; 1969, 233), 91 (1967, 227), 92 (1969, 233), 93 (1967, 227), *325, 346*

Smith, M., 94 (1968, 280), *337*

Smith, T. D., 112 (1966, 223), 138 (1966, 223), *317*

Smithies, A. C., 91 (1967, 231; 1968, 222), 152 (1967, 231), *325, 336*

Smolinsky, G., 229 (1966, 274), *318*

Snyder, J. P., 86 (1967, 192; 1968, 169; 1969, 206, 207), 227 (1967, 192; 1968, 169), 228 (1968, 169), 230 (1967, 192), 231 (1967, 192; 1968, 169), 234 (1967, 192), 241 (1969, 206), 243 (1967, 192; 1969, 206), 249 (1967, 192), *324, 334, 345*

Soenen, M., 142 (1969, 345), 204 (1969, 345), 229 (1969, 345), 242 (1969, 345), *349*

Soest, J. F., 87 (1967, 195, 196), 88 (1967, 195, 210), 271 (1967, 196), *324. 325*

Sokal'ski, M. A., 159 (1969, 351), *349*

Solan, D., 21 (1968, 42), 22 (1968, 42), 28 (1968, 42), 60 (1968, 42), 61 (1968, 42), 63 (1968, 42), 64 (1968, 42), 74 (1968, 42), 75 (1968, 42), 77 (1968, 42), 98 (1968, 42), 99 (1968, 42), 104 (1968, 42), 105 (1968, 42), 108 (1968, 42), 109 (1968, 42), 113 (1968, 42), *331*

Soliman, F. M., 198 (1966, 263), 239 (1966, 263), *318*

Somers, J. H., 167 (1969, 356), 190 (1969, 356), 199 (1969, 356), 225 (1969, 356), *349*

Song, K. M., 90 (1967, 223), *325*

Songstad, J., 264 (1967, 384), *330*

Sonnet, P. E., 141 (1966, 246), 149 (1966, 246), 150 (1966, 246), 161 (1966, 246), 170 (1966, 246), 178 (1966, 246), 179 (1966, 246), 184 (1966, 246), 192 (1966, 246), 195 (1966, 246), 199 (1966, 246), 200 (1966, 246), 206 (1966, 246), *317*

Sosnovsky, G., 178 (1966, 257), 216 (1966, 257), *318*

Sowerby, D. B., 79 (1969, 153), 92 (1969, 153), 277 (1969, 153), 284 (1969, 379), *344, 350*

Spangenberg, S. F., 85 (1969, 199), 90 (1969, 199), 139 (1969, 344), 244 (1969, 344), 261 (1969, 344), 263 (1969, 344), 265 (1969, 344), 266 (1969, 344), 267 (1969, 199, 344), 268 (1969, 199), 270 (1969, 344), 272 (1969, 344), 285 (1969, 344), *345, 349*

Speziale, A. J., 86 (1966, 136), 224 (1965, 109), 239 (1965, 109), 253 (1965, 109), 255 (1965, 109), *311, 315*

Spiesecke, H., 8 (1961, 1), 17 (1969, 48), *307, 341*

Spitsyn, V. I., 88 (1966, 146), *315*

Spruit, F. J., 99 (1965, 120), *311*

Sretenskaya, I. I., 213 (1968, 335), *339*

Städe, W., 39 (1968, 75), 54 (1968, 75), 78 (1968, 75; 1969, 145), 119 (1969, 145), 232 (1968, 75), 242 (1968, 75), 245 (1968, 75), 248 (1968, 75), 250 (1968, 75), 256 (1969, 145), *332, 343*

Stager, C. V., 88 (1969, 220), *346*

Stahlberg, R., 79 (1966, 93, 94), 101 (1966, 93, 94), 117 (1966, 94), 277 (1966, 93), 278 (1966, 93), *314*

Stanclift, W. E., 90 (1968, 209), 91 (1968, 209), 93 (1968, 209), 125 (1968, 209), 145 (1968, 209), *335*

Steger, E., 79 (1966, 93, 94), 101 (1966, 93, 94), 117 (1966, 94), 277 (1966, 93), 278 (1966, 93), *314*

Stein, B. F., 88 (1966, 144, 145), *315*

Steinberg, G. M., 208 (1968, 334), *339*

Steinfink, H., 63 (1965, 28), 65 (1965, 28), 259 (1965, 28), *309*

Steinhoff, G., 244 (1966, 281), 249 (1966, 281), 251 (1966, 281), 252 (1966, 281), *318*

Stelzer, O., 31 (1969, 78), 90 (1969, 78), 91 (1969, 78), 92 (1967, 241; 1969, 78), 93 (1969, 78), 193 (1969, 78), *326, 342*

Stenseth, R. E., 86 (1966, 114), 104 (1966, 114), 107 (1966, 114), 111 (1966, 114), 112 (1966, 114), 118 (1966, 114), 131 (1966, 114), 134 (1966, 114), 143 (1966, 114), 171 (1966, 114), *314*

Stenzel, J., 23 (1967, 63), 143 (1967, 63), 195 (1967, 63), *321*

Stepanyants, A. U., 7 (1967, 18), *319*

Stephens, P. J., 91 (1965, 59), 92 (1965, 59), 93 (1965, 59), *309*

Stepisnik, J., 88 (1968, 187), *335*

Sternhell, S., 56 (1968, 104), 230 (1968, 104), 234 (1968, 104), *333*

Sternlicht, H., 89 (1965, 72, 73, 74; 1968, 271), *310, 337*

Stetter, K. H., 92 (1967, 247), *326*

Steuber, F. W., 78 (1969, 145), 119 (1969, 145), 256 (1969, 145), *343*

Stevens, J. I., 37 (1968, 72), 105 (1968, 72), 114 (1968, 72), 129 (1968, 72), *332*

Stewart, W. E., 13 (1968, 29), 44 (1968, 29), 47 (1968, 29), 94 (1969, 324), 139 (1968, 29), 168 (1968, 29), 177 (1968, 29), 178 (1968, 29), 184 (1968, 29), 190 (1968, 29), 200 (1968, 29), 212 (1968, 29), 217 (1968, 29), 221 (1968, 29), 226 (1968, 29), 230 (1968, 29), 233 (1968, 29), 237 (1968, 29), 238 (1969, 324), 256 (1969, 324), *331, 348*

Stierand, H., 90 (1969, 244), *346*

Stine, W. R., 86 (1967, 191), 193 (1967, 191, 345), 194 (1967, 345), 221 (1967, 345), 223 (1967, 191, 345), 241 (1967, 345), *324, 328*

St. Jacques, M., 20 (1969, 51), 153 (1969, 51), 191 (1969, 51), *341*

Stockel, R. F., 97 (1966, 278), 242 (1966, 278), *318*

Stolberg, V. G., 92 (1965, 69), *310*

Stoll, K., 7 (1968, 11), 260 (1968, 11), 287 (1968, 11), *330*

Stone, F. G. A., 90 (1965, 57; 1969, 235), 91 (1969, 262), 92 (1966, 181, 183; 1968, 253, 255), 93 (1968, 264), *309, 316, 336, 337, 346, 347*

Storey, R. N., 19 (1966, 24), 22 (1966, 24), 75 (1966, 24), 99 (1966, 24), *312*

Storr, A., 86 (1966, 125), 110 (1966, 125), *316*

Strange, J. H., 73 (1968, 115; 1969, 139), 87 (1968, 115; 1969, 139), 97 (1968, 115), *333, 343*

Straughan, B. P., 80 (1968, 144), 101 (1968, 144), 111 (1968, 144), 117 (1968, 144), 134 (1968, 144), 150 (1968, 144), *334*

Street, G. B., 92 (1966, 174), *316*

Strzelecka, H., 234 (1967, 365), 249 (1967, 365), 253 (1967, 365), 254 (1967, 365), 255 (1967, 365), *329*

Stuart, J. P., 4 (1969, 5a), 88 (1969, 5a, 217), 101 (1969, 5a), *340, 345*

Stuart, S. R., 92 (1968, 260), 235 (1968, 260), *337*

Sturtz, G., 53 (1967, 108), 125 (1967, 108), 126 (1967, 108), 139 (1967, 22), 144 (1967, 108, 323, 324), 145 (1967, 108, 324, 326), 146 (1967,

108, 323, 324, 326), 147 (1967, 22, 278), 149 (1967, 22), 155 (1967, 324), 156 (1967, 108, 323, 326), 158 (1967, 278), 166 (1967, 323), 168 (1967, 22, 108, 278, 324; 1968, 325), 177 (1967, 108, 278, 326), 178 (1967, 22), 180 (1967, 108), 183 (1968, 325), 184 (1967, 22, 108, 278; 1968, 325), 192 (1968, 325), 198 (1967, 108), 200 (1967, 108), 205 (1967, 108), 225 (1967, 108), 264 (1967, 278), 266 (1967, 278), *322, 327, 328, 338*

Styan, G. E., 214 (1965, 107), *311*
Suglobov, D. N., 93 (1967, 261), *326*
Sundberg, R. J., 181 (1965, 102), 191 (1965, 102), 206 (1965, 102), 224 (1965, 102), *311*
Sutherley, T. A., 24 (1968, 51), 83 (1968, 51), 99 (1968, 51), *331*
Suzuki, N., 94 (1967, 272), *326*
Suzuki, S., 94 (1967, 272), *326*
Svatos, G. F., 92 (1967, 253), *326*
Sventitskii, E. N., 59 (1968, 109), 84 (1968, 156; 1969, 182, 183), 87 (1968, 109), 93 (1968, 109), 184 (1968, 109, 156), 194 (1968, 109), *333, 334, 344*
Swain, J. R., 19 (1968, 41), 23 (1968, 41), 25 (1968, 41), 35 (1968, 41), 37 (1968, 41), 75 (1968, 41; 1969, 143), 77 (1968, 41), 91 (1969, 276), 99 (1969, 143), 106 (1968, 41), 221 (1969, 276), *331, 343, 347*
Swan, J. M., 90 (1966, 151), *315*
Swank, D., 38 (1967, 86), 39 (1967, 86), 114 (1967, 86), 125 (1967, 86), 138 (1967, 86), 148 (1967, 86), *321*
Swift, T. J., 89 (1969, 291), 92 (1969, 291), *348*
Swinbourne, F. J., 85 (1966, 112), *314*
Swinden, G., 13 (1967, 27), 90 (1967, 27), *320*
Swysen, P., 112 (1969, 381), *350*
Szustakowski, M., 88 (1968, 191), *335*

T

Tadros, S., 92 (1967, 247), *326*
Taft, R. W., 7 (1965, 5, 1968, 14), 128 (1968, 14), 131 (1968, 14), 132 (1968, 14), 150 (1968, 14), 152 (1968, 14), 164 (1965, 5), 174 (1968, 14), 195 (1968, 14), 214 (1968, 14), 223 (1968, 14), 227 (1968, 14), 228 (1968, 14), 242 (1968, 14), *308, 330*
Takahashi, K., 34 (1967, 84), 37 (1967, 84), 107 (1966, 216; 1967, 84), 119 (1966, 216), 120 (1966, 216; 1967, 84), 121 (1967, 84), 140 (1966, 216; 1967, 84), 288 (1966, 216), *317, 321*
Takahashi, S., 41 (1969, 96), 83 (1969, 96), 94 (1969, 96), *342*
Takahashi, Y., 92 (1969, 293), *348*
Takamizawa, A., 206 (1967, 352), 217 (1967, 352), 241 (1967, 352), *329*
Tamura, F., 85 (1969, 200), *345*
Tang, R., 82 (1969, 169), 107 (1969, 169), 112 (1969, 169), 118 (1969, 169), 151 (1969, 169), *344*
Tanh, T. N., 55 (1967, 113), 143 (1967, 113), 154 (1967, 113), 182 (1967, 113), 189 (1967, 113), 197 (1967, 113), 201 (1967, 113), 206 (1967, 113), 212 (1967, 113), *322*
Tani, K., 92 (1968, 254), 274 (1968, 254), 284 (1968, 254), 287 (1968, 254), *336*
Tan-Wan Lin, 58 (1967, 120), 114 (1967, 120), 119 (1967, 120), 137 (1967, 120), *322*
Tarenko, Ya. F., 53 (1967, 109), 116 (1967, 109), 124 (1967, 109), 251 (1967, 109), *322*
Tasaka, K., 38 (1968, 73), 114 (1968, 73), 125 (1968, 73), 135 (1968, 309), 138 (1968, 73), 148 (1968, 73), 199 (1968, 73), 215 (1968, 309), 219 (1968, 309), 229 (1968, 73), 233 (1968, 73), 235 (1968, 73), 236 (1968, 309), 245 (1968, 73), 253 (1968, 309), 262 (1968, 309), 268 (1968, 309), 269 (1968, 309), 274 (1968, 309), *332, 338*
Tatsuzaki, I., 88 (1961, 3; 1969, 227), 97 (1961, 3), *307, 346*
Tavs, P., 18 (1969, 44), 119 (1969, 44), 193 (1969, 44), 218 (1969, 44), 259 (1969, 44), 264 (1967, 383), *330, 341*
Taylor, B. W., 91 (1967, 234), *325*

Taylor, D., 63 (1969, 130), 66 (1969, 130), 87 (1969, 130, 213), *343, 345*
Taylor, L. J., 86 (1966, 136), *315*
Taylor, M. W., 21 (1969, 54, 55), 50 (1969, 54, 55), 51 (1969, 114), 76 (1969, 55), 77 (1969, 55, 114), 106 (1969, 54, 55, 114), 108 (1969, 55), 115 (1969, 55, 114), 117 (1969, 55), 129 (1969, 55), 132 (1969, 55), 136 (1969, 55), *341, 342*
Taylor, R. C., 21 (1966, 28), 66 (1966, 28), 91 (1966, 164), 92 (1968, 241), 99 (1966, 28), 100 (1966, 28), 103 (1966, 28), *312, 315, 336*
Taylor, T. N., 94 (1968, 282), 160 (1968, 282), 263 (1968, 282), *337*
Tebbe, F. N., 93 (1968, 268), 186 (1968, 268), *337*
Tebby, J. C., 28 (1968, 60), 112 (1968, 60), 201 (1968, 60), 202 (1968, 60), 209 (1968, 60), 219 (1967, 358), 228 (1968, 60), 231 (1968, 60), 238 (1969, 369), 243 (1967, 372), 246 (1967, 372, 376; 1969, 369), 247 (1967, 376), 249 (1967, 378), 250 (1967, 376), 253 (1967, 376), 274 (1967, 378), 275 (1967, 378), 276 (1967, 372), *329, 331, 350*
Teichmann, H., 115 (1967, 302), 122 (1967, 308), 129 (1967, 308), 141 (1967, 308), 149 (1967, 308), 160 (1967, 308), 173 (1967, 302), 178 (1967, 308), *327*
Templeton, J. F., 251 (1965, 116), 253 (1965, 116), *311*
Terao, T., 88 (1967, 209; 1969, 222), 135 (1967, 209), *325, 346*
Ternai, B., 63 (1967, 127), 85 (1966, 112), 224 (1966, 271), 242 (1966, 271), 262 (1967, 127), 263 (1967, 127), *314, 318, 322*
Tesi, G., 130 (1966, 237), 221 (1966, 237), *317*
Teste, J., 78 (1969, 148), 151 (1969, 148), 163 (1969, 148), *343*
Tetel'Baum, B. I., 19 (1967, 38), 21 (1967, 38, 51, 52), 99 (1967, 51), 103 (1967, 51), 104 (1967, 38, 51), 107 (1967, 38, 51), 114 (1967, 38), 128 (1967, 38), 130 (1967, 51, 52), 132 (1967, 51, 52), 138 (1967, 51), 143 (1967, 52), 154 (1967, 332), 157 (1967, 332), 209 (1967, 52), *320, 328*
Theodoropulos, S., 237 (1968, 344), 238 (1968, 344), 240 (1968, 344), 246 (1968, 344), 248 (1968, 344), 250 (1968, 344), *339*
Thewalt, U., 54 (1969, 117), *343*
Thomas, W. A., 57, 133 (1969, 123), 185 (1969, 123), 204 (1969, 123), 208 (1969, 122), 210 (1969, 123), *343*
Thompson, D. T., 90 (1967, 227), 91 (1967, 227; 1968, 216), 93 (1967, 227), *325, 336*
Thompson, J. A. S., 7 (1969, 16), *340*
Thompson, J. E., 156 (1965, 95), 183 (1965, 95), *310*
Thompson, J. G., 121 (1966, 277), 132 (1966, 277), 208 (1968, 333), 234 (1968, 333), 241 (1966, 277), *318, 339*
Thornhill, B. S., 74 (1966, 81), 84 (1966, 81), *313*
Thyret, H., 92 (1967, 259), *326*
Thyrum, E. M., 89 (1969, 312), 94 (1969, 312), *348*
Thyrum, P. T., 89 (1969, 312), 94 (1969, 312), *348*
Ticozzi, G., 233 (1968, 343), 235 (1968, 343), 246 (1968, 343), 247 (1968, 343), 250 (1968, 343), 251 (1968, 343), *339*
Tieckelman, H., 105 (1966, 213), 155 (1966, 213), 157 (1966, 213), 166 (1966, 213), 183 (1966, 213), 204 (1966, 213), 237 (1966, 213), *317*
Timms, P. L., 21 (1968, 42), 22 (1968, 42), 28 (1968, 42), 60 (1968, 42), 61 (1968, 42), 63 (1968, 42), 64 (1968, 42), 74 (1968, 42), 75 (1968, 42), 77 (1968, 42), 98 (1968, 42), 99 (1968, 42), 104 (1968, 42), 105 (1968, 42), 108 (1968, 42), 109 (1968, 42), 113 (1968, 42), *331*
Timofeeva, T. N., 81, 157 (1969, 163), 166 (1969, 163), 236 (1969, 163), *344*

Todd, L. J., 10 (1969, 25), 105 (1967, 292; 1969, 25), 110 (1969, 25), 141 (1967, 292), *327, 340*
Todd, S. M., 39 (1966, 49), 79 (1966, 49), 101 (1966, 49), 277 (1966, 49), *312*
Todo, I., 88 (1969, 227), *346*
Toma, F., 65 (1970, 2), *350*
Tompa, A. S., 94 (1968, 287), 141 (1968, 287), 142 (1968, 287), 143 (1968, 287), *337*
Traynard, J. C., 166 (1969, 355), 176 (1969, 355), 183 (1969, 355), *349*
Treichel, P. H., 86 (1966, 139), 109 (1966, 139), *315*
Treichel, P. M., 19 (1967, 36; 1968, 39), 22 (1967, 36; 1968, 39), 28 (1968, 39), 31 (1968, 39), 75 (1967, 36; 1968, 39), 77 (1968, 39), 90 (1965, 58), 99 (1967, 36), 109 (1968, 39), *309, 320, 331*
Treskunova, I. M., 124 (1964, 15), 135 (1964, 15), 145 (1964, 15), 155 (1964, 15), *308*
Trippett, S., 78 (1968, 122), 86 (1966, 122), 156 (1969, 350), 203 (1966, 266), 204 (1966, 267), 205 (1968, 122), 209 (1966, 266; 1967, 354), 211 (1967, 356), 212 (1967, 356; 1969, 350), 229 (1966, 273), 231 (1965, 113), 236 (1966, 122; 1967, 368), 239 (1966, 267), 240 (1966, 267), 243 (1966, 273), 246 (1966, 266), 248 (1966, 267), 274 (1967, 368), *311, 314, 318, 329, 333, 349*
Trivedi, B. C., 82 (1969, 170), 217 (1969, 170), 236(1969, 170), *344*
Tronich, W., 59 (1968, 105, 106, 107), 60 (1968, 107), 120 (1968, 106, 107), 121 (1967, 306; 1968, 105, 107, 305), 122 (1968, 107), 127 (1968, 106), 128 (1968, 105, 106, 107, 305), 129 (1968, 107), 139 (1968, 106, 107), 141 (1968, 107), 149 (1967, 328; 1968, 105, 107, 305), 150 (1968, 105, 107), 159 (1968, 105, 107, 305), 161 (1968, 105, 107, 322), 179 (1968, 105, 107), 245 (1967, 328), 260 (1968, 107), 263 (1968, 107), 264 (1968, 107), *327, 328, 333, 338*
Truax, D. R., 11 (1969, 28), 12 (1969, 28), 84 (1969, 28), 85 (1969, 28), 86 (1969, 28), *340*
Tsang, F. Y., 79 (1968, 130), 279 (1968, 130), 280 (1968, 130), 281 (1968, 130), *333*
Tsang, T., 87 (1968, 180), 88 (1968, 188), *335*
Tsang, W. S., 91 (1968, 232), 228 (1968, 232), 231 (1968, 232), 235 (1968, 232), *336*
Tsivunin, V. S., 116 (1966, 229), 155 (1966, 229), 176 (1967, 341), *317, 328*
Ts'o, P. O. P., 94 (1968, 276; 1969, 319), 223 (1968, 276), *337, 348*
Tsuboi, M., 41 (1967, 90; 1968, 81; 1969, 95, 96), 83 (1967, 90; 1969, 95, 96), 94 (1969, 96), 104 (1967, 90; 1968, 81), 108 (1967, 90; 1969, 95), 125 (1969, 95), 126 (1969, 95), 162 (1969, 95), 165 (1967, 90), 201 (1967, 90), 257 (1967, 90), 259 (1969, 95), 262 (1967, 90), 265 (1967, 90), 268 (1967, 90), *321, 332*
Tsutsui, M., 89 (1969, 309), *348*
Tsutsumi, Y., 88 (1967, 209; 1969, 222), 135 (1967, 209), *325, 346*
Tsvetkov, E. N., 7 (1967, 15), *319*
Tuck, D. G., 242 (1966, 279), *318*
Turcker, B., 125 (1967, 314), 127 (1967, 314), 139 (1967, 314), 141 (1967, 314), 149 (1967, 314), 173 (1967, 314), 181 (1967, 314), *328*

U

Uchida, Y., 90 (1967, 223), *325*
Udovich, C. A., 90 (1969, 245), *346*
Uehling, E. A., 87 (1967, 196, 210), 271 (1967, 196), *324, 325*
Ugo, R., 86 (1967, 177), 91 (1967, 232), 92 (1967, 251), 242 (1966, 279), 262 (1967, 177), 263 (1967, 177), 272 (1967, 177), *318, 324, 325, 326*
Uhlenhopp, E. L., 86 (1969, 205), *345*
Ukitaa, T., 41 (1969, 96), 83 (1969, 96), 94 (1969, 96), *342*

Ukraintseva, E. A., 88 (1966, 148), *315*
Ullrich, J., 94 (1966, 205; 1967, 273), *316*, *326*
Ulrich, H., 7 (1965, 6), 111 (1965, 6), 125 (1967, 314), 127 (1967, 314), 139 (1967, 314), 141 (1967, 314), 149 (1967, 314), 167 (1965, 6), 173 (1967, 314), 180 (1965, 6), 181 (1967, 314), *308*, *328*
Umani-Ronchi, A., 233 (1968, 343), 235 (1968, 343), 246 (1967, 375; 1968, 343), 247 (1967, 375); 1968, 343), 250 (1968, 343), 251 (1968, 343), 255 (1967, 375), *329*, *339*
Usher, D. A., 86 (1967, 181), 119 (1965, 88), 125 (1967, 181), 127 (1965, 88), 137 (1967, 181), 138 (1965, 88), 162 (1965, 88), 163 (1965, 88), 167 (1965, 88), 176 (1965, 99), *310*, *311*, *324*
Utvary, K., 118 (1966, 288), 153 (1969, 348), 156 (1969, 348), 173 (1965, 98; 1969, 348), 176 (1969, 348), 202 (1969, 348), 205 (1969, 348), 206 (1965, 98; 1969, 348), 207 (1969, 348), 226 (1965, 98), 257 (1965, 98; 1966, 288), 258 (1965, 98; 1966, 288), 259 (1966, 288), 260 (1965, 98; 1966, 288), 261 (1966, 288), 262 (1965, 98; 1966, 288), 263 (1966, 288), 264 (1966, 288), 265 (1966, 288), 267 (1966, 288), 268 (1966, 288), 269 (1966, 288), 282 (1969, 376), *311*, *318*, *349*, *350*

V

Vahrenkamp, H., 84 (1966, 108), 104 (1968, 290), 108 (1968, 290), 113 (1968, 290), 114 (1968, 290), 122 (1968, 290), 129 (1968, 290), *314*, *337*
Valan, K. J., 7 (1966, 6), 79 (1966, 6), 101 (1966, 6), 278 (1966, 6), 279 (1966, 6), 283 (1966, 6), 287 (1966, 6), 288 (1966, 6), *311*
Valdez, C., 51 (1968, 98), 76 (1968, 98), 91 (1968, 98), 123 (1968, 98), 207 (1968, 98), *332*

Van Dalen, J. H., 86 (1969, 203), 99 (1969, 203), *345*
Van Den Akker, M., 88 (1967, 211), 92 (1967, 211, 242), 93 (1967, 211), *325*, *326*
Van Den Bos, B. G., 86 (1966, 137), 212 (1966, 137), *315*
Vandenbroucke, A. C., 43 (1968, 85), 125 (1968, 85), 136 (1968, 85), *332*
Vanderhart, D. L., 4 (1969, 6, 7), 20 (1969, 6, 7), 88 (1969, 6, 7), 97 (1969, 7), *340*
Van der Kelen, G. P., 4 (1969, 56), 88 (1969, 56), 101 (1969, 56), 131 (1969, 120), 185 (1969, 120), 186 (1969, 120), 222 (1969, 120), 223 (1969, 120), *340*, *343*
Van Der Lugt, W., 88 (1969, 223, 224), *346*
Van Der Voorn, P. C., 82 (1966, 100), *314*
Van Deursen, F. W., 86 (1966, 137), 99 (165, 120), 212 (1966, 137), *311*, *315*
Van Doorne, M., 24 (1966, 34), 45 (1966, 34), 47 (1966, 34), 268 (1966, 34), *312*
Vanermen, W., 4 (1969, 5b), 88 (1969, 5b), 101 (1969, 5b), 131 (1969, 120) 185 (1969, 120), 186 (1969, 120), 222 (1969, 120), 223 (1969, 120), *340*, *343*
Van Hecke, G. R., 90 (1967, 222), 92 (1966, 172), *316*, *325*
Van Veen, R., 78 (1968, 120), 195 (1968, 120), 196 (1968, 120), *333*
Van Wazer, J. R., 1, 2, 4 (1967, 1, 7), 65 (1957, 2; 1967, 1), 84 (1967, 174; 1968, 160), 86 (1966, 138; 1967, 186), 89 (1968, 235), 92 (1965, 67; 1966, 176, 177, 178; 1967, 174, 240; 1968, 235), 98 (1969, 327), 99 (1969, 327), 100 (1966, 208), 101 (1966, 208), 103 (1967, 7), 104 (1967, 7, 186), 107 (1967, 7), 108 (1967, 7, 186), 109 (1966, 138; 1967, 293), 111 (1967, 7), 114 (1966, 208; 1967, 186, 293), 115 (1966, 208), 129 (1967, 293), 130 (1967, 7), 131 (1967, 7, 186), 132

(1967, 7, 186), 133 (1967, 7), 135 (1967, 7), 136 (1967, 293), 140 (1966, 138; 1967, 293), 147 (1967, 293), 149 (1967, 7, 293), 151 (1967, 7), 154 (1967, 293), 157 (1967, 293), 160 (1967, 7, 293), 167 (1967, 293), 169 (1967, 7, 293), 173 (1967, 7), 175 (1967, 293), 178 (1967, 293), 190 (1967, 293), 192 (1967, 293), 194 (1967, 293), 196 (1967, 7), 201 (1967, 7), 207 (1967, 7), 212 (1967, 293), 213 (1967, 7, 293), 214 (1967, 7, 293), 218 (1967, 7), 228 (1967, 7), 231 (1967, 7), 234 (1967, 7), 235 (1967, 7), 239 (1967, 7), 257 (1967, 7), 258 (1966, 138; 1967, 7, 186), 259 (1967, 186), 278 (1966, 138; 1967, 186), *307, 310, 311, 315, 316, 319, 324, 325, 327, 334, 336, 349*

Varga, S., 90 (1966, 221), 111 (1966, 221), *317*

Varshavskii, A. D., 21 (1967, 52), 130 (1967, 52), 132 (1967, 52), 143 (1967, 52), 209 (1967, 52), *320*

Vasil'ev, G. S., 51 (1965, 25), 123 (1965, 25), *309*

Vasil'eva, T. V., 210 (1965, 106), *311*

Vdovenko, V. M., 84 (1969, 183), 93 (1967, 261), *326, 344*

Venanzi, L. M., 90 (1968, 196), 92 (1965, 68; 1966, 185), 93 (1968, 196), *310, 316, 335*

Vene, N., 88 (1968, 187), *335*

Venezky, D. L., 202 (1967, 347), *329*

Verdier, M. J. M., 86 (1967, 179), *324*

Verkade, J. G., 25 (1965, 17), 27 (1966, 37; 1968, 53; 1969, 71), 28 (1968, 53), 29 (1968, 53), 37 (1966, 37), 38 (1966, 37, 48), 40 (1964, 5; 1965, 22; 1968, 79), 42 (1966, 37), 43 (1968, 85; 1969, 71), 64 (1969, 132, 133, 134), 68 (1969, 138), 74 (1965, 31), 75 (1968, 117), 83 (1966, 102; 1968, 79, 117), 84 (1968, 79), 89 (1968, 117; 1969, 132, 133), 90 (1965, 31; 1966, 158; 1967, 218; 1969, 132, 133), 91 (1966, 158; 1968, 53; 1969, 133, 134), 92 (1967, 218; 1968, 53;

1969, 133), 93 (1966, 158; 1969, 71, 133, 138), 114 (1967, 218), 117 (1966, 37; 1968, 53), 125 (1966, 37; 1968, 53, 85), 133 (1966, 37; 1967, 218), 135 (1966, 102; 1967, 218; 1968, 117), 136 (1968, 85), 145 (1967, 218), 147 (1969, 71), 157 (1969, 71), 159 (1967, 334), 166 (1966, 48; 1968, 53; 1969, 71), 177 (1969, 71), 193 (1969, 71), 220 (1967, 334), 257 (1966, 37), 258 (1966, 37), *308, 309, 312, 314, 315, 325, 328, 331, 332, 333, 341, 343*

Verny, M., 143 (1968, 311), 236 (1968, 311), *338*

Verrier, J., 27 (1967, 71), 40 (1966, 50; 1968, 80), 82 (1966, 50), 107 (1966, 50), 112 (1966, 50), 118 (1966, 50), 125 (1968, 80), 137 (1968, 80), 139 (1968, 80), 148 (1968, 80), 151 (1966, 50), 153 (1968, 80), 162 (1966, 50), 180 (1968, 80), *312, 321, 332*

Vessiere, R., 143 (1968, 311), 236 (1968, 311), *338*

Vetter, H. J., 142 (1965, 91), 150 (1965, 91), *310*

Vijayaraghavan, R., 88 (1967, 200; 1969, 219), *324, 346*

Villemin, M., 65 (1970, 2), *350*

Villieras, J., 86 (1969, 201, 202), *345*

Vinal, R. S., 92 (1965, 66), *311*

Vinogradov, L. I., 17, 20 (1969, 53), 83 (1969, 53), 93 (1969, 53), *341*

Virkhaus, R., 86 (1966, 115), 234 (1966, 115), *314*

Vizel', A. O., 53 (1964, 8), 116 (1964, 8), 123 (1964, 8), 124 (1964, 8), 134 (1964, 8), 135 (1964, 8), 144 (1964, 8), *308*

Vizel', O. A., 53 (1967, 109), 116 (1967, 109), 124 (1967, 109), 133 (1964, 16), 134 (1964, 16), 135 (1964, 16), 144 (1964, 16), 145 (1964, 16), 251 (1967, 109), *308, 322*

Vogel, K., 54 (1968, 102), 78 (1968, 102), 226 (1968, 102), 240 (1968, 102), 276 (1968, 102), *333*

Voigt, D., 112 (1969, 381), *350*

Volger, H. C., 91 (1968, 223, 224), 92

(1966, 182; 1967, 254; 1968, 224, 262; 1969, 297), *316, 326, 336, 337, 348*
Vollmer, H., 19 (1969, 46), *341*
Vol'pin, M. E., 90 (1966, 156), *315*
von Halasz, S. V., 60 (1969, 129), 67 (1969, 129), 98 (1969, 129), 100 (1969, 129), 115 (1969, 129), *343*
Voronkov, M. G., 19 (1967, 34), 89 (1967, 34), 140 (1967, 34), 159 (1967, 34), 161 (1967, 34), 170 (1967, 34), 179 (1967, 34), 192 (1967, 34), 195 (1967, 34), 207 (1967, 34), 227 (1967, 34), *320*
Voznesenskaya, A.Kh., 107 (1966, 220), 111 (1966, 220), 138 (1966, 220), 146 (1966, 220), 156 (1966, 220), 167 (1966, 220), 176 (1966, 220), 188 (1966, 220), 190 (1966, 220), 197 (1966, 220), 204 (1966, 220), *317*
Vrieze, K., 11 (1967, 23), 89 (1968, 243), 91 (1968, 223, 224), 92 (1966, 182; 1967, 254; 1968, 224, 242, 243, 262; 1969, 297), *316, 320, 326, 336, 337, 348*
Vriezen, W. H. N., 11 (1967, 23), 74 (1967, 23), 90 (1967, 23), 113 (1967, 23), 139 (1967, 23), 223 (1967, 23), *320*
Vullo, W. J., 27 (1968, 52), *331*

W

Waddington, T. C., 5 (1969, 11), 194 (1969, 11), 218 (1969, 11), *340*
Wagner, B., 84 (1966, 110), 90 (1966, 110), 91 (1966, 110), 92 (1966, 110), *314*
Wagner, R. I., 7 (1967, 10), 17 (1966, 13), 18 (1966, 13; 1967, 10), 20 (1967, 10), 25 (1966, 13), 28 (1966, 13), 44 (1966, 13), 99 (1966, 13), 105 (1966, 13), 109 (1966, 13), 113 (1966, 13), *312, 319*
Wakselman, M., 129 (1969, 338), 221 (1969, 338), *349*
Wald, H. J., 78 (1968, 125), 109 (1968, 125), 122 (1968, 125), 127 (1968, 125), *333*

Waldrep, P. G., 86 (1965, 45), 131 (1965, 45; 1966, 239), 143 (1966, 239), 148 (1965, 45; 1966, 239), 186 (1965, 45), 195 (1965, 45), 219 (1965, 45; 1966, 239), 224 (1965, 45; 1966, 239), 244 (1965, 45; 1966, 239), *309, 317*
Walker, B. J., 156 (1969, 349), 213 (1969, 349), 217 (1969, 349), 218 (1969, 349), 229 (1966, 273), 231 (1965, 113), 234 (1969, 349), 238 (1969, 349), 243 (1966, 273), *311, 318, 349*
Walker, J., 80 (1969, 155), 92 (1969, 155), 269 (1969, 155), *344*
Wallbridge, M. G. H., 90 (1968, 195), *335*
Walsley, R. H., 88 (1966, 145), *315*
Walter, D., 92 (1966, 173), *316*
Walther, B., 158 (1968, 320), 159 (1968, 320), 184 (1967, 343), 186 (1968, 320), 187 (1968, 320), 210 (1967, 343), 233 (1967, 343), 247 (1967, 343), 251 (1967, 343), 282 (1967, 343), 283 (1967, 343), *328, 338*
Ward, R. L., 94 (1967, 266), *326*
Ward, R. S., 249 (1967, 378), 274 (1967, 378), 275 (1967, 378), *329*
Warner, D., 4 (1969, 5a), 88 (1969, 5a), 101 (1969, 5a), *340*
Warren, S. G., 105 (1966, 214), *317*
Wasco, J. L., 48 (1966, 56), 149 (1966, 56), 188 (1966, 56), 208 (1966, 56), *313*
Wayland, B. B., 90 (1966, 157), *315*
Wazer, J. R., 5 (1956, 1), *307*
Weber, H., 22 (1969, 65), 78 (1968, 125; 1969, 65), 109 (1968, 125; 1969, 65), 122 (1968, 125; 1969, 65), 127 (1968, 125; 1969, 65), 257 (1969, 65), *333, 341*
Weber, W., 78 (1965, 32), 284 (1965, 32), 287 (1965, 32), 288 (1965, 32), *309*
Webster, M., 84, 194 (1966, 106), 203 (1966, 106), 211 (1966, 106), 223 (1966, 106), 224 (1966, 106), 229 (1966, 106), 240 (1966, 106), 243 (1966, 106), 245 (1966, 106), 246 (1966, 106), 247 (1966, 106), 249

(1966, 106), 250 (1966, 106), 251 (1966, 106), 254 (1966, 106), 255 (1966, 106), *314*
Wechler, W. J., 94 (1967, 267), *326*
Wechter, W. J., 41 (1969, 98), 94 (1969, 98), 164 (1969, 98), 215 (1969, 98), *342*
Weedon, B. C. L., 240 (1968, 346), 249 (1968, 346), *339*
Weeks, M., 89 (1969, 291), 92 (1969, 291), *348*
Wehman, A. T., 86 (1968, 171), 238 (1968, 171), 241 (1968, 171), 249 (1968, 171), *334*
Wehrli, W. E., 94 (1965, 76), *310*
Weigert, F. J., 8 (1968, 21, 22), 9 (1968, 22), 15 (1968, 22), 68 (1969, 21), 69 (1968, 22; 1969, 21), 70 (1969, 21), 71 (1968, 22), 72 (1968, 22; 1969, 21), 105 (1968, 22), 109 (1968, 22), 113 (1968, 22), 114 (1968, 22), 119 (1968, 22), 140 (1968, 22), 160 (1968, 22), 193 (1968, 22; 1969, 21), 194 (1968, 22), 213 (1968, 22), 218 (1968, 22; 1969, 21), 223 (1968, 22), 224 (1968, 22), 228 (1968, 22), 242 (1969, 21), 259 (1968, 22), *331, 340*
Weill, G., 86 (1963, 5), 284 (1963, 5), *307*
Weinberg, D. S., 245 (1965, 114), *311*
Weinberger, M. A., 86 (1967, 185), 138 (1967, 185), 141 (1967, 185), 161 (1967, 185), 263 (1967, 185), *324*
Weiner, H., 89 (1969, 314), *348*
Weingand, C., 43 (1968, 86; 1969, 100), 67 (1969, 137), 100 (1968, 86; 1969, 137), 105 (1968, 86), 280 (1969, 100), 282 (1969, 100), 283 (1969, 100), *332, 342, 343*
Weitkamp, H., 86 (1966, 140), 238 (1966, 140), *315*
Welch, F. J., 86 (1965, 47), 208 (1965, 47), *309*
Wells, E. J., 82 (1967, 159), 284 (1967, 159), 286 (1967, 159), *323*
Werber, G., 90 (1965, 158), *309*
Westheimer, F. H., 53 (1966, 65), 78 (1969, 146), 80 (1968, 139), 82 (1967, 160), 83 (1969, 146), 86 (1966, 120; 1967, 184; 1969, 146), 133 (1966, 241), 135 (1966, 241; 1969, 146), 136 (1966, 120), 140 (1966, 120), 144 (1966, 120), 148 (1967, 160), 151 (1966, 241), 163 (1966, 241), 202 (1966, 241), 211 (1967, 160), 261 (1969, 146), 263 (1969, 146), 265 (1969, 146), 280 (1966, 65), *313, 314, 317, 323, 324, 334, 343*
Westland, A. D., 92 (1968, 250), *336*
Wheatland, D. A., 30 (1967, 79), 31 (1967, 79), 74 (1967, 79), 90; 1967, 79; 1968, 210; 1969, 249), 91 (1967, 79; 1968, 210; 1969, 249), 93 (1966, 195; 1967, 79; 1968, 210, 269; 1969, 249), 148 (1969, 249), 193 (1969, 249), 196 (1967, 79), 202 (1967, 79), 206 (1969, 249), 216 (1967, 79; 1969, 249), *316, 321, 335, 337, 346*
White, A. H., 90 (1969, 246), *346*
White, D. W., 40 (1968, 79), 83 (1968, 79), 84 (1968, 79), *332*
White, R. F. M., 20 (1969, 50), 26 (1969, 50), 115 (1969, 50), 131 (1969, 50), 132 (1969, 50), 140 (1969, 50), 141 (1969, 50), 151 (1969, 50), 153 (1969, 50), 154 (1969, 50), 155 (1969, 50), 169 (1969, 50), 170 (1969, 50), 186 (1969, 50), 187 (1969, 50), 193 (1969, 50), 195 (1969, 50), 196 (1969, 50), 197 (1969, 50), 223 (1969, 50), 224 (1969, 50), 227 (1969, 50), *341*
White, W. D., 14 (1967, 28; 1969, 36, 37, 38), 15, 20 (1967, 28), 62, 63 (1969, 36, 37), 99 (1967, 28), 100 (1967, 28), 101 (1967, 28), 219 (1967, 28), 258 (1967, 28), 259 (1967, 28), *320, 340*
Whitehead, M. A., 79 (1960, 2), *307*
Whitesides, G. M., 22 (1969, 64), 25 (1969, 64), 81 (1967, 158), 90 (1967, 229), 91 (1969, 271), 109 (1969, 64), 255 (1967, 158), *323, 325, 341, 347*
Wiberg, N., 274 (1967, 390), *330*
Wieber, M., 152 (1967, 330), *328*

Wieker, W., 88, (1966, 150; 1967, 213, 214, 215, 216), 202 (1967, 216), *315, 325*
Wilby, M. D., 91 (1969, 267), *347*
Wiley, G. A., 86 (1967, 191), 193 (1967, 191), 223 (1967, 191), *324*
Wilford, J. B., 90 (1965, 57), *309*
Wilke, G., 92 (1966, 173; 1969, 278), *316, 347*
Wilkinson, G., 91 (1962, 5; 1966, 164; 1969, 268), 92 (1966, 187, 190, 191; 1967, 252, 256; 1968, 246), 93 (1962, 5; 1967, 260), 131 (1967, 256), *307, 315, 316, 326, 336, 347*
Williams, D. F., 4, 115 (1969, 8), *340*
Williams, D. H., 243 (1967, 372), 246 (1967, 372), 249 (1967, 378), 274 (1967, 378), 275 (1967, 378), 276 (1967, 372), *329*
Williams, R. M., 94 (1967, 268), *326*
Williamson, M. P., 11 (1968, 25), 49, 56 (1967, 115), 80 (1968, 25, 143), 82 (1967, 115), 107 (1968, 25), 119 (1968, 25), 120 (1968, 25), 121 (1968, 25), 135 (1968, 25, 97), 137 (1968, 25, 97), 140 (1968, 25), 141 (1968, 25), 152 (1967, 115), 174 (1968, 25), 203 (1968, 25), 222 (1967, 115), 262 (1968, 25), 264 (1968, 143), *322, 331, 332, 334*
Willis, C. J., 22 (1968, 48), 103 (1968, 48), 106 (1968, 48), 110 (1968, 48), *331*
Willson, M., 78 (1968, 126), 86 (1968, 126), 122 (1968, 126), 141 (1968, 126), 262 (1968, 126), *333*
Wilson, L. A., 45 (1968, 90), 149 (1965, 92), 170 (1965, 92), 182 (1965, 92), 195 (1968, 90), 207 (1968, 332), 218 (1968, 332), 223 (1968, 332), 231 (1968, 332), 239 (1968, 332), *310, 332, 339*
Wingleth, D. C., 99 (1967, 284), *327*
Winter, J. M., 73, 87, 97 (1959, 1), 98 (1959, 1), *307*
Wise, W. B., 92 (1968, 261), *337*
Wittmann, G. T. W., 91 (1968, 213), 148 (1968, 213), *336*
Wojcicki, A., 91 (1966, 159, 166; 1968, 232; 1969, 275), 228 (1968, 232), 231 (1968, 232), 235 (1968, 232), *315, 336, 347*
Wokulat, J., 119 (1967, 305), 129 (1967, 305), 139 (1967, 305), 141 (1967, 305), 170 (1967, 305), 266 (1967, 305), *327*
Wolf, R., 19 (1967, 39; 1969, 45), 39 (1967, 39), 40 (1966, 50), 58, 80 (1969, 162), 82 (1966, 50), 83 (1967, 39, 165; 1969, 178, 179), 84 (1967, 172; 1969, 190, 192), 85 (1969, 192), 86 (1967, 165), 107 (1966, 50; 1969, 45), 109 (1967, 172), 112 (1966, 50), 114 (1967, 172; 1969, 162), 118 (1966, 50), 119 (1967, 39; 1969, 45), 120 (1969, 45), 121 (1969, 45), 122 (1967, 172), 127 (1969, 45), 128 (1967, 165; 1969, 162), 130 (1969, 45), 136 (1967, 165), 139 (1967, 321; 1969, 45, 178), 143 (1967, 172; 1969, 45, 190), 147 (1967, 119), 149 (1969, 192), 151 (1966, 50), 152 (1969, 45), 153 (1967, 165; 1969, 45, 162), 154 (1969, 162), 157 (1967, 119, 172), 159 (1969, 179), 160 (1969, 192), 161 (1967, 172), 162 (1966, 50; 1969, 45), 163 (1969, 45, 178), 164 (1969, 162), 167 (1967, 119), 170 (1967, 172; 1969, 162, 192), 172 (1969, 45, 178), 173 (1969, 178) 175 (1969, 45), 181 (1969, 45, 178), 182 (1967, 119), 187 (1967, 165), 190 (1969, 178), 191 (1969, 162), 192 (1967, 172), 193 (1969, 179), 199 (1969, 178), 202 (1969, 45, 178), 203 (1967, 172), 206 (1969, 162), 209 (1969, 45, 178), 216 (1969, 45, 190), 217 (1969, 162), 221 (1969, 162), 233 (1969, 45), *312, 320, 322, 324, 328, 341, 344, 345*
Wolfsberger, W., 85 (1967, 176), 90 (1969, 232), 129 (1967, 316), 141 (1967, 176), 142 (1967, 176, 322), 170 (1967, 322), 171 (1967, 322), 179 (1967, 176, 316), 195 (1967, 322), 214 (1967, 322), 236 (1967, 322), 243 (1967, 322), 254 (1967, 379), 255 (1967, 379), 261 (1967,

176), 263 (1967, 316), 264 (1967, 176), 265 (1967, 316), *324, 328, 329, 346*
Woodcock, R., 82 (1969, 171), 121 (1969, 171), 163 (1969, 171), 180 (1969, 171), 215 (1969, 171), *344*
Woodman, C. M., 64 (1966, 79), 65 (1966, 79), 98 (1966, 79), 101 (1966, 79), 257 (1966, 79), *313*
Woodward, L. A., 48 (1967, 98), 99 (1967, 98), 101 (1967, 98), *322*
Work, J. L., 249 (1968, 348), *339*
Wulff, J., 29 (1967, 76), 32 (1967, 76), 78 (1967, 76, 139), 86 (1967, 139), 246 (1967, 139), 250 (1967, 139), 253 (1967, 76, 139), 255 (1967, 76, 139), 256 (1967, 139), *321, 323*
Wulfsberg, G., 64 (1969, 134), 91 (1969, 134), *343*
Wyluda, B. J., 89 (1965, 72), *310*
Wymore, C. E., 92 (1967, 252), *326*

Y

Yablokov, Yu.Y., 88 (1968, 190), *335*
Yagupsky, G., 91 (1969, 268), *347*
Yakshin, V. V., 7 (1967, 18), 119 (1967, 346; 1968, 307), 128 (1968, 307), 202 (1967, 346; 1968, 307), *319, 328, 338*
Yakubovich, A. Ya., 261 (1969, 371), 264 (1969, 371), 266 (1969, 371), 267 (1969, 371), 268 (1969, 371), 269 (1969, 371), *350*
Yamasaki, T., 34 (1967, 84), 37 (1967, 84), 107 (1966, 216; 1967, 84), 119 (1966, 216), 120 (1966, 216; 1967, 84), 121 (1967, 84), 140 (1966, 216; 1967, 84), 288 (1966, 216), *317, 321*
Yankowsky, A. W., 13 (1969, 33), 57 (1969, 33), 227 (1969, 33), 230 (1969, 33), 234 (1969, 33), *340*
Yarkova, E. G., 20 (1969, 53), 83 (1969, 53), 93 (1969, 53), *341*
Yastrebov, V. V., 84 (1967, 170), 252 (1967, 170), *324*
Yengoyan, L., 94 (1966, 199), *316*

Yonezawa, T., 12, *320, 340*
Young, J. F., 91 (1966, 164), 92 (1966, 191), *315, 316*
Young, R. A., 88 (1969, 223, 224), *346*
Yudina, K. S., 7 (1968, 18), 271 (1968, 18), 273 (1968, 18), *330*

Z

Zaev, E. E., 92 (1969, 282), 169 (1969, 282), *347*
Zaitseva, E. L., 261 (1969, 371), 264 (1969, 371), 266 (1969, 371), 267 (1969, 371), 268 (1969, 371), 269 (1969, 371), *350*
Zaret, E. H., 178 (1966, 257), 216 (1966, 257), *318*
Zavgorodnii, V. S., 133 (1967, 319), 135 (1967, 319), 193 (1967, 319), *328*
Zbiral, E., 240 (1968, 345), 241 (1968, 345), 242 (1968, 345), 243 (1968, 345), 245 (1965, 115; 1968, 345), 251 (1965, 115), 252 (1965, 115), *311, 339*
Zeliger, H. I., 86 (1969, 207), *345*
Zenin, S. V., 84 (1969, 184), *345*
Zienty, Eb., 59 (1964, 10), 237 (1964, 10), 240 (1964, 10), 243 (1964, 10), *308*
Zimont, S. L., 119 (1967, 346; 1968, 307), 128 (1968, 307), 202 (1967, 346; 1968, 307), *328, 338*
Zuckerman, J. J., 227 (1968, 342), *339*
Zumdahl, S. S., 87 (1968, 177), *335*
Zumer, S., 87 (1968, 178), 88 (1968, 178, 187), *335*
Zyablikova, T. A., 2 (1967, 2), 279 (1967, 2), 280 (1967, 2), 281 (1967, 2), *319*
Zykova, T. V., 2 (1967, 3, 4), 7 (1967, 3), 53 (1964, 8), 109 (1964, 14), 116 (1964, 8; 1966, 229), 119 (1964, 14), 123 (1964, 8), 124 (1964, 8), 125 (1966, 234), 126 (1964, 14), 127 (1966, 234), 134 (1964, 8; 1967, 3), 135 (1964, 8), 136 (1964, 14), 144 (1964, 8), 147 (1964, 14; 1966,

234), 155 (1966, 229, 234; 1967, 3), 156 (1967, 3, 4), 175 (1967, 3), 176 (1967, 3, 340, 341), 182 (1967, 4), 192 (1967, 3, 340), 193 (1967, 340), 200 (1967, 4), 207 (1967, 340), 213 (1967, 4), 214 (1966, 234), 221 (1967, 4), 254 (1967, 3), *308, 317, 319, 328*

SUBJECT INDEX

The numbers in **bold** indicate the pages on which the topic is discussed in detail.

A

Acetylenic phosphorus compounds, 52
Acid-catalysed exchange, 86
Activation parameters, of diphosphines, 80
Acyclic, penta-coordinated compounds, stereochemistry of, **81**
Additivity constants, σ^P, 5
Adduct formation, **84**
Aliphatic compounds, stereochemistry of, **80**
Allenic-acetylenic isomerism, 52
Allenic phosphines, 51
Allenic phosphine oxides, 81
Allenic phosphorus systems, 51
Allylic phosphorus compounds, 53
Aluminium-27 shifts, 11
Aluminium-phosphorus adducts, 11, 85
Aminophosphines, 64
Analytical uses, of ^{31}P NMR, 94
Anisotropic effects, 9
Anisotropy, of boron ligands, 85
and conformation, 81
Application of intermolecular studies, **83**
Application of ^{31}P NMR, 78
Aromatic character, in phosphiran, 7
Aromaticity, 79
Aromatic phosphorus compounds, 55
A.T.P., effect of pH on J(POP) in, 65
Fourier transform spectrum, 2
Azaphosphatriptycene, 6

B

Basic strengths, relative order of, 85
Biological mechanisms, use of ^{31}P in, 80
Biological problems, studied using ^{31}P NMR, **89**
Bonds, relationship to shifts, 6

Bond angles, 4
and ^2J(F–P–H), 75
Borane adducts, 10
Boron-11 shifts, 10
Boron adducts, of phosphorus, 84
Boron halide phosphorus adducts, 10
Boron-phosphorus compounds, 10

C

Carbohydrates, phosphorylated, 94
Carbon-13 satellites, 68
Carbon-13 shifts, **7**
Carbon-phosphorus coupling, **68**
Carbon-phosphorus heterocyclics, **78**
Chemical shift anisotropy, 2, 88
Chemical shifts, in phosphorus compounds, **2**
Cholestanyl phosphonate, 33
cis and *trans* isomers, distinguishing of, 79
Complex formation, effect of on J(F–P), 20
effect of in ^2J(H–C–P), 29
and removal of stereospecific J, 41
Conformation, in cyclic compounds, 74
from J(H–P), 33
Conformational equilibria, 41, 80
Conjugation effects in phosphorus compounds, 7
Coupling through phosphorus, **74**
Coupling transmission, 66
Cytidine phosphate, 41

D

$d\pi$–$p\pi$ conjugation, 80
between P and N, 44
Decoupling, by complex formation, 41

SUBJECT INDEX

Determination of ^{31}P shifts, 2
Diastereoisomerism, in spiro compounds, 83
Dihedral angles, and J(F–P), 35
  and ^2J(H–C–P), 29
  and ^3J(HC–O–P), 41
Dihedral dependence, of J values, 83
  of ^3J(HC–C–P), 32
Dihydrophosphazenes, 54
Dioxaphospholanes, conformation of, 40
Dioxaphosphorinanes, conformation of, 40
Diphosphines, rotational isomerism, 80
Double quanta transitions and sign of J, 58
Double resonance studies, 55
  of CaHPO$_4$, 88
Dynamic polarisation, 87

E

Electronegativity, and J(F–P), 35
  effects of on ^2J(P–N–P), 67
  effects of on ^2J(P–O–P), 65
  of phosphorus, 9
  in phosphorus compounds, 7
Electronic structure of phosphides, 87
Empirical correlation of ^{31}P shifts, 5
Enzymes, phosphorylated, 94
Ephedrine derivatives, 47
Epoxyalkylphosphonates, 32
ESCA spectroscopy, 4
Ethylene phosphites, conformation of, 40
Exchange studies, 81
Extraction studies, of transition metals ions, 94

F

Factors effecting ^1J(P–P), **62**
Ferroelectric charges in KH$_2$PO$_4$, 87
Fluorinated vinyl compounds, 50
Fluorine chemical shifts, 7
Fluoroaromatic phosphorus compounds, 56
Fluorophosphoranes, 25
  ^2J(F–F) in, 77
  stereochemistry of, 81

Fluorovinyl phosphines, 50
Fluorovinylphosphorus compounds, ^4J(F–F) in, 77
Formula Indices, 95
Fourier transform, 2

G

Gallium-phosphorus adducts, 85
Geometrical dependence, of J values, 64
  of J(H–C–P), 30
Germyl phosphines, J(H–P) in, 48
Group contributions to J(F–P), 20
Group electronegativities, 7

H

Halogen exchange, 84
  in complexes, 85
  followed by ^{27}Al resonance, 11
Heterocyclics, containing phosphorus, 78
Heteronuclear tickling, 68
  and sign of J, 64
Hindered rotation in phosphines, 29
Hormones, phosphorylated, 94
Hund coupling mechanism, 13
Hybridisation, effect of on ^2J(HPH), 74
  effect of on ^2J(POP), 65
  of phosphorus and J values, 12
Hydrogen bonding, 81, 84
Hydrogen phosphonates, 84
Hydroxyapatite, 88

I

Indazoles, 58
Indium phosphorus adducts, 85
INDOR technique, 2, 9, 73
Inorganic compounds, formulae index of, 97
Intermolecular effects, applications of, 83
Intramolecular rearrangements, 89
Inversion at phosphorus atom, 44
Ion-pair mechanism, 89
Isotopic effects, 20

J

J values, comparison of calculated and experimental, 14
  comparison of through oxygen and sulphur, 44
  dependence upon substituent electronegativity, 62
  geometrical dependence of, 64
$J(^{27}\text{Al}-^{31}\text{P})$, 11, 74
  in complexes, 85
$J(^{11}\text{B}-^{31}\text{P})$, 74
  in boron-phosphorus adducts, 84
$J(\text{Br}-^{31}\text{P})$, in PBr_3, 73
$^3J(^{13}\text{C}-\text{P}-\text{CH})$, 74
$^1J(^{13}\text{C}-^{31}\text{P})$, 68
$^2J(^{13}\text{C}-^{31}\text{P})$, **71**
$^2J(^{13}\text{C}-\text{O}-^{31}\text{P})$, 71
$^3J(^{13}\text{C}-\text{C}-\text{C}-^{31}\text{P})$, 71
$^3J(^{13}\text{C}-\text{C}-\text{O}-^{31}\text{P})$, 71
$^4J(^{13}\text{C}-\text{C}-\text{C}-\text{C}-^{31}\text{P})$, 71
$J(^{35}\text{Cl}-^{31}\text{P})$, in PCl_3, 73
$J(\text{Cu}-^{31}\text{P})$, 74
$J(\text{F}-\text{F})$, stereospecificity of, 13
$^3J(\text{FP}-\text{PF})$, 77
$^4J(\text{F}-\text{F})$, 77
  in fluorovinyl phosphines, 50
$^5J(\text{F}-\text{F})$, 77
$J(\text{F}-\text{H})$, stereospecificity of, 13
$^2J(\text{F}-\text{P}-\text{H})$, 75
$^3J(\text{FP}-\text{CH})$, 75
$^3J(\text{FP}-\text{NH})$, 76
$^3J(\text{FP}(\text{S})\text{NH})$, 76
$^3J(\text{FP}(\text{S})\text{SH})$, 76
$^4J(\text{FP}-\text{N}-\text{CH})$, 76
$^5J(\text{F}-\text{H})$, 76
$^6J(\text{F}-\text{H})$, 76
$J(\text{F}-\text{P})$, **20, 33**
  insensitivity of on adduct formation, 85
  relation to s character, 17
  stereospecificity of, 13
  typical values of, 34
$^1J(\text{F}-\text{P})$, in structural analysis, 79
  table of, 21
$^2J(\text{F}-\text{P})$, 61
  cis and trans values, 51
  in fluorovinyl phosphines, 50
$^2J(\text{FB}-\text{P})$, 84
$^3J(\text{F}-\text{P})$, **37**
  cis and trans values, 51
  in fluorovinyl phosphines, 50
  typical values of in pentafluorophenyl phosphines, 57
$^3J(\text{FCP}-\text{P})$, 61
$^3J(\text{FC}-\text{S}-\text{P})$, 42
$^3J(\text{FP}=\text{NP})$, 61
$^3J(\text{FP}-\text{N}=\text{P})$, 61
$^3J(\text{FP}-\text{N}-\text{P})$, 61
$^3J(\text{FP}-\text{O}-\text{P})$, 61
$^3J(\text{FP}-\text{S}-\text{P})$, 61
$^4J(\text{F}-\text{P})$, cis and trans values, 51
  in pentafluorophenylphosphorus compounds, 57
  in 2-trifluoromethylphenyl phosphine, 57
$^4J(\text{FC}-\text{P}-\text{S}-\text{P})$, 61
$^5J(\text{F}-\text{P})$, in fluoroaromatic phosphorus compounds, 56
  in pentafluorophenylphosphorus compounds, 57
$J(\text{H}-\text{H})$, stereospecificity of, 13
$^2J(\text{H}-\text{P}-\text{H})$, 74
  sign of, 17
$^3J(\text{HC}-\text{P}-\text{H})$, 74
$^3J(\text{HP}-\text{Si}-\text{H})$, 74
$^4J(\text{HC}-\text{P}-\text{CH})$, in $\text{P}(\text{CH}_3)_3$, 74
$^4J(\text{HC}-\text{P}-\text{SiH})$, 75
$^5J(\text{H}-\text{H})$, and configuration, 83
$J(\text{H}-^{31}\text{P})$, sensitivity of to complex formation, 85
  stereospecificity of, 13
  through oxygen, **37**
  through nitrogen, **44**
  through other nuclei, **48**
  through sulphur, **42**
  in vinyl phosphorus compounds, 49
$^1J(\text{H}-\text{P})$, **17**
  table of values of, 18
$^2J(\text{H}-\text{P})$, in allenic phosphines, 51
  in phospholenes, 53
$^2J(\text{H}_3\text{C}-\text{P})$, 62
$^3J(\text{H}-\text{P})$, in allenic phosphines, 51
  in aromatic phosphorus compounds, 55
  effect of conjugation on, 54
  in phenylphosphines, 56
  in 3-phospholenes, 53
  in structural analysis, 79
  in vinyl esters, 58

in ylides, 59
^4J(H–P), in allenic phosphines, 51
  and configurations, 83
  in phenylphosphines, 56
  stereospecificity of, 32
  in vinyl esters, 58
  in ylides, 59
^5J(H–P), in allenic phosphines, 51
  in ylides, 59
^2J(HB–P), 84
^2J(HC–P), **26**
  stereospecificity of, **29**
  table of, 27
  temperature dependence of, 86
^3J(H$_2$C–P), in phosphorus esters, 13
^4J(H$_3$C–P), in phosphorus esters, 13
^4J(o-H$_3$C–P), in tolyl phosphines, 55
^7J(H$_3$C–P), 56
^3J(H–C–C–P), 29, **30**
  dihedral dependence of, 32
  table of values of, 31
^3J(HC–Ge–P), 49
^3J(HC–O–P), **37**, 71
  comparison of in phosphites and phosphates, 42
  in phosphorinane, 41
  sign of, 37
  typical values of, 38
^4J(HC–C–O–P), 71
  in phosphorinanes, 42
  sign of, 37
^3J(HC–N–P), 47
  typical values of, 45
^4J(H$_3$C–N–C–P), 60
^4J(H$_3$C–P–C–P), 60
^3J(H$_3$C–P–P), 60, 62
^3J(HC–S–P), typical values of, 43
^4J(H$_3$C–Se–P), 60
^3J(HC–Si–P), 49
^3J(H$_3$C–Sn–P), 48
^2J(HN–P), 44
^3J[(HO)C–P], 37
^2J(H–P–P), 60
^3J(H–P–O–P), 60
^2J(H$_2$X–P), in silylphosphines, 49
^3J(H$_3$X–X–P), in silylphosphines, 49
J(^{199}Hg–^{31}P), 74
J(^{14}N–^{31}P), 73
^1J(^{14}N–^{31}P), in (EtO)$_2$P(O)NHEt, 73
^2J(^{14}N–^{31}P), in F$_2$PCNS, 73
J(^{17}O–^{31}P), 73
J(^{31}P–^{31}P), sign of, 14
  and stereochemistry, 89
^1J(^{31}P–^{31}P), 62
  sign of in P$_2$F$_4$, 64
  values of, 63
^2J(^{31}P–^{31}P), **64**
^2J(P–C–P), 64
^2J(P–N–P), 66
^2J(P–O–P), 64
^2J(P–P–P), in P(PF$_2$)$_3$, 64
^2J(P–S–P), 65
^3J(P–C–O–P), **68**
J(^{31}P–^{77}Se), 73
J(^{31}P–^{29}Si), 73
^1J(^{31}P–^{119}Sn), 73
^2J(^{31}P–O–^{119}Sn), 73
J(^{31}P–^{183}W), 74

K

Karplus relationships, 13
  of ^2J(H–C–P), 29
  of J(H–P), 50

L

Lewis bases, phosphorus compounds acting as, 84
Lifetime studies, of phosphonium ions, 86
Ligand exchange, 86
  mechanism of, 85, 89
  in metal complexes, 87
Line shape analysis, 88
Long-range couplings, **59**
  of J(H–P), 48
  and conformation, 83

M

Metal complexes, **88**
  of phosphorus compounds, table of, 90
Metathetical reactions, 86
Methanes, methyl substituted, ^{13}C shifts of, 10
Methylene non-equivalence in phosphines, 28

SUBJECT INDEX

Methylene phosphoranes, 58
  rearrangements of, 86
Methyl group, non-equivalence of, 13
Methyl phosphates, pH dependence of, 86
Methyl phosphine, ^{13}C shifts of, 10
1-Methyl phosphorinanes, 29
Methyl polyphosphates, 86
Microwave spectra and symmetry, 82
Molecular geometry, and $J(F-P)$, 35
Molecular rearrangements, 88

N

^{14}N decoupling, 48
N-inversion, 44
Nitrogen-14 shifts, 11
Noise decoupling, 9
Nuclear Overhauser effect, 68
Nucleoside phosphates, 94
Nucleotides, 94

O

Organic compounds, formulae index of, 103
Organometallic phosphines, 7
Oxazaphospholanes, $J(H-P)$ in, 40
Oxygen-17 shifts, 11
  table of in phosphorus compounds, 12
Oxyphosphoranes, 78

P

$p\pi-p\pi$ bonding, 55
P–C bond, σ-character of, 6
P–N bonds, restricted rotation about, 68
pH, effect of on $J(POP)$ in ATP, 65
pH dependence, of molecular species, 86
pK values, 7
Pentafluorophenyl phosphorus compounds, 57
Perfluoroaromatic phosphorus compounds, 56
Pesticides, 94
Phase transitions, 88
1-Phenyl-dimethylphospholes, 54

Phenyl phosphinidine, interconversion in, 86
Phosphabenzenes, 54, 78
Phosphetane derivatives, 32, 78
Phosphides, 88
Phosphines, correlation of shifts in, 5
Phosphiran, 6, 20
Phospholanes, $J(H-P)$ in, 40
Phospholenic compounds, 53
Phospholes, 54
Phospholipids, 94
Phosphonates, stereospecificity of $J(H-P)$ in, 31
Phosphonitrilics, **78**
  $J(F-P)$ in, 20
  $^{2}J(P-N-P)$ in, 66
Phosphonium compounds, shift correlation in, 6
Phosphonium ions, $^{2}J(H-C-P)$ in, 28
Phosphonyl group, conjugation of, 9
Phosphoproteins, 94
Phosphoric acid esters, conformation of, 41
Phosphorinanes, $J(H-P)$ in, 40
Phosphorus chemical shifts, **2**
Phosphorus couplings, effect of adduct formation on, 85
  to other nuclei, **71**
Phosphorus heterocyclics, 78
Phosphorus-metal hybridisation, 64
Phosphorus-phosphorus couplings, **60**
Phosphoryl compounds, $^{2}J(H-C-P)$ in, 28
Phosphoryl group, anisotropy of, 9
Polymer characterisation, 85
Polymerisation studies, 86
Propargylic phosphorus compounds, 53
Proton chemical shifts, **7**
Protonation, of phosphines, 87
Pseudorotation, and exchange, 81
  and fluorine exchange, 25
Pyrazoles, 58
Pyridine derivatives, 57
Pyrophosphates, study of in solid state, 2

R

Reaction studies, application of ^{31}P NMR, **85**

Reduced couplings, K, 15
Rehybridisation of phosphorus, 9
Relative signs of J(F–P), 20
  of ^3J(F–P), 60
Relaxation rate mechanisms, 89
Relaxation studies, 73, 87
Restricted rotation about N–P bonds, 68
Rotamers in acyclic compounds, 40
Rotational equilibrium, 66
Rotational isomerism, in diphosphines, 80

S

σ-Bonded adducts, 84
σ-π rearrangements, 89
Selenium shifts, indirect measurement of, 12
Selenophosphates, 84
Sign of coupling, charges in, 14
  dependence of valency upon, 62
  by double quantum transitions, 58
  by heteronuclear tickling, 64
Sign of, J(F–P), 33, 60
  ^3J(FC–S–P), 42
  ^3J(HC–P–H), 74
  J(H–P), in allenic phosphines, 51
  in vinyl phosphorus compounds, 49
  ^2J(HC–P), 28
  ^4J(H–P), 37
  ^2J(P–O–P), 64
  J(P–S–P), 65
Sign reversal in, J(P–P), 60
Silicon shifts, indirect measurement of, 12
Silyl phosphines, J(H–P) in, 48
Solid samples, studies of, 88
Solvent dependence, of chemical exchange, 89
  of conformation, 42
  of J(F–P), 35
  of ^2J(H–H), 20
Solvent effects, 80, 83
  on aromatic phosphorus compounds, 55
  on J(F–P), 24
Spectral accumulation, 9
Spin decoupling, by chemical exchange, 89

Spin-rotation constant, 4
Spin-spin coupling in phosphorus compounds, **12**
Spiro compounds, stereochemistry of, 83
Spirophosphoranes, 83
  interconversion in, 86
Spirophosphorus compounds, 83
Stereochemical problems, application of ^{31}P NMR, **79**
Stereochemistry, of one-ring compounds, **82**
  of saturated aliphatic compounds, **80**
  of spiro-compounds, **83**
  of unsaturated compounds, 81
Stereospecificity and J values, 12
  of J(HCPH), 74
  of J(H–P), 42
  of ^2J(H–C–P), 29
  of J(HC–O–P), 40
Steric hindrance, 80
Steroids, phosphorylated, 94
Structural problems, application of ^{31}P NMR, 78
Substituent constants, effect on J(F–P), 24
Substituent electronegatives, 4
  dependence of J upon, 62
Sugars, phosphorylated, 94
Syn-anti isomerism, 82

T

Table of, ^2J(^{13}C–P), 72
  of metal-phosphorus complexes, 90
Tautomerism of $P^{III} \to P^V$, 83
Temperature dependence, of confirmation, 42
  of exchange, 89
  of ^4J(F–P), 57
Temperature effects, 83
  on ^2J(H–H), 20
Thermodynamic parameters, 84
Theoretical aspects, of ^1J(P–P), **62**
  of ^{31}p shifts, **4**
Thiamine phosphate, 94
Thiono-thiolo isomerism, 84
Thiophosphates, 84
Thiophosphoryl compounds, ^2J(H–C–P) in, 28

Thiophosphoryl group, anisotropy of, 9
Through-space coupling, 13, 56
  of F–P, 51
Tickling experiments, sign of ^1J(P–P), 62
Tin shifts, indirect measurement of, 12
Transition metal ions, extraction studies of, 94
Tributyl phosphate, solvent effect on, 87
Trioxaphosphabicyclooctane, 13
  long-range J in, 40
Triple resonance and signs of coupling, 37
Tris(4-fluorophenyl) phosphine oxide, 55
Twist angle, 82
Typical values of $J(^{13}C-^{31}P)$, 69
  of ^2J(P–N–P), 67

U

Unsaturated aliphatic compounds, stereochemistry of, 81
Unsaturated systems, J(H–P) in, 49
Uranyl nitrate complex with tributyl phosphate, 87

V

Variable temperature studies, 46, 62, 84, 88
  of exchange, 81
  on J(F–P), 24
  on ^2J(P–P), 67
  of ligand-exchange, 89
  of pseudorotation, 83
  of relaxation mechanisms, 73
  of rotamers, 32
  of T_1, 87
Valence rearrangements, 86
Valency, effect of on sign of J, 62
O-Vinyl derivatives, J(H–P) in, 58
Vinyl phosphorus compounds, 49
Vitamin B6, 94

W

Walsh rule, 70

Y

Ylides, 59

Z

Zürcher rules, 83

DATE DUE

DEMCO 38-297